T0329507

Plant Tissue Culture, Development, and Biotechnology

Plant Tissue Culture, Development, and Biotechnology

Edited by
Robert N. Trigiano
University of Tennessee
Knoxville, USA

Dennis J. Gray
University of Florida
Apopka, USA

CRC Press is an imprint of the
Taylor & Francis Group, an **informa** business

CRC Press
Taylor & Francis Group
6000 Broken Sound Parkway NW, Suite 300
Boca Raton, FL 33487-2742

© 2011 by Taylor and Francis Group, LLC
CRC Press is an imprint of Taylor & Francis Group, an Informa business

No claim to original U.S. Government works

Printed in the United States of America on acid-free paper
10 9 8 7 6 5 4 3 2 1

International Standard Book Number: 978-1-4200-8326-2 (Paperback)

This book contains information obtained from authentic and highly regarded sources. Reasonable efforts have been made to publish reliable data and information, but the author and publisher cannot assume responsibility for the validity of all materials or the consequences of their use. The authors and publishers have attempted to trace the copyright holders of all material reproduced in this publication and apologize to copyright holders if permission to publish in this form has not been obtained. If any copyright material has not been acknowledged please write and let us know so we may rectify in any future reprint.

Except as permitted under U.S. Copyright Law, no part of this book may be reprinted, reproduced, transmitted, or utilized in any form by any electronic, mechanical, or other means, now known or hereafter invented, including photocopying, microfilming, and recording, or in any information storage or retrieval system, without written permission from the publishers.

For permission to photocopy or use material electronically from this work, please access www.copyright.com (http://www.copyright.com/) or contact the Copyright Clearance Center, Inc. (CCC), 222 Rosewood Drive, Danvers, MA 01923, 978-750-8400. CCC is a not-for-profit organization that provides licenses and registration for a variety of users. For organizations that have been granted a photocopy license by the CCC, a separate system of payment has been arranged.

Trademark Notice: Product or corporate names may be trademarks or registered trademarks, and are used only for identification and explanation without intent to infringe.

Library of Congress Cataloging-in-Publication Data

Plant tissue culture, development, and biotechnology / editors: Robert N. Trigiano and Dennis J. Gray.
-- 1st ed.
p. cm.
Includes bibliographical references and index.
ISBN 978-1-4200-8326-2 (alk. paper)
1. Plant tissue culture. 2. Plant tissue culture--Laboratory manuals. I. Trigiano, R. N. (Robert Nicholas), 1953- II. Gray, Dennis J. (Dennis John), 1953-

QK725.P4778 2010
571.5'38--dc22 2010016264

Visit the Taylor & Francis Web site at
http://www.taylorandfrancis.com

and the CRC Press Web site at
http://www.crcpress.com

Contents

SECTION IV Crop Improvement Techniques

SECTION V The Business of Biotechnology

Acknowledgments

Foremost, we wish to recognize the extraordinary efforts and talents of all the contributing authors—their creativity, support, advice, understanding, and patience throughout the conception and long-delayed development of this book was nothing less than phenomenal; the Institute of Agriculture at the University of Tennessee and especially Dr. Carl Jones, Head of Entomology and Plant Pathology, for providing the time and financial support necessary for RNT to complete the book. A great big thanks goes out to our videographer, Doug Edlund, and our voice person, Chuck Denny—the DVD of the three laboratory exercises would not have been possible without them. Thanks are also extended to Lisa M. Vito for her work on the figures. We thank Randy Brehm at CRC Press, whose constant encouragement and work was essential for the completion of this textbook. The high-quality work by Pat Roberson and Gail Renard at CRC Press is greatly appreciated.

Lastly, RNT sincerely thanks CAB, DJG, BHO, ANT, CGC, and especially PKR(T) for their advice, insight, friendship, love, and constant support—you've all made my life much richer.

DJG thanks Sharon D'Oria for her support and help during this project.

Editors

Robert N. Trigiano received his B.S. degree with an emphasis in biology and chemistry from Juniata College, Huntingdon, Pennsylvania, in 1975 and an M.S. in biology (mycology) from Pennsylvania State University in 1977. He was an associate research agronomist working with mushroom culture and plant pathology for the Green Giant Co., Le Sueur, Minnesota, until 1979 and then was a mushroom grower for Rol-Land Farms, Ltd., Blenheim, Ontario, Canada, during 1979 and 1980. He completed a Ph.D. in botany and plant pathology (co-majors) at North Carolina State University at Raleigh in 1983. After concluding postdoctoral work in the Plant and Soil Science Department at the University of Tennessee, he was appointed an assistant professor in the Department of Ornamental Horticulture and Landscape Design at the same university in 1987, promoted to associate professor in 1991, and to professor in 1997. He served as interim head of the department from 1999–2001. He joined the Department of Entomology and Plant Pathology at the University of Tennessee in 2002.

Dr. Trigiano is a member of the American Phytopathological Society (APS), the American Society for Horticultural Science (ASHS), and the honorary societies of Gamma Sigma Delta, Sigma Xi, and Phi Kappa Phi. He received the T.J. Whatley Distinguished Young Scientist Award (University of Tennessee Institute of Agriculture) and the Gamma Sigma Delta Research individual and team Award of Merit (University of Tennessee). He was given the ASHS publication award for the most outstanding educational paper and the Southern region ASHS L. M. Ware Distinguished Research Award. In 2006, he was elected a fellow of the American Society for Horticultural Science. Dr. Trigiano was awarded the B. Otto and Kathleen Wheeley Award of Excellence in Technology Transfer at the University of Tennessee and the Management of Creative Agricultural Technologies, LLC, in 2007. He has been an editor for *Plant Disease*; the ASHS journals *Plant Cell, Tissue and Organ Culture,* and *Plant Cell Reports*; and is currently the editor-in-chief of *Critical Reviews in Plant Sciences.* Additionally, he has co-edited eight books, including *Plant Tissue Culture Concepts and Laboratory Exercises, Plant Pathology Concepts and Laboratory Exercises,* and *Plant Development and Biotechnology.*

Dr. Trigiano has been the recipient of several research grants from the U.S. Department of Agriculture (USDA), Horticultural Research Institute, and from private industries and foundations. In addition to the patents he holds, he has published more than 200 research papers, book chapters, and popular press articles. He teaches undergraduate/graduate courses in plant tissue culture, mycology, DNA analysis, protein gel electrophoresis, technical writing, and plant microtechnique. His current research interests include molecular markers for breeding ornamental plants, diseases of ornamental plants, somatic embryogenesis and micropropagation of ornamental species, fungal physiology, population analysis, DNA profiling of fungi and plants, and gene discovery.

Dennis J. Gray is a tenured professor of developmental biology at the University of Florida/IFAS's Mid-Florida Research and Education Center (MREC), where he conducts biotechnology research designed to facilitate crop improvement. He is an authority on plant cell culture and genetic engineering with in-depth experience in the biotechnology of forage grasses, cereals, cucurbits, and, most notably, grapes. Dr. Gray received his B.A. degree in biological sciences from California State College in 1976. He received an M.S. degree in 1979 from Auburn University, where he majored in mycology with a minor in botany. He subsequently attended North Carolina State University, receiving his Ph.D. in 1982 with a major in botany and minor in plant pathology. Following a 2-year postdoctoral training period at the University of Tennessee, he became an assistant professor at the University of Florida's Institute for Food and Agricultural Sciences in 1984. He was promoted to associate professor with tenure in 1989, full professor in 1993, and University of Florida Research Foundation professor in 1998.

At the University of Florida, Dr. Gray founded and directs the Grape Biotechnology Research Laboratory. The laboratory has become the world leader in genetic engineering of grape. He has developed a number of key cellular and molecular technologies required to genetically engineer grape and other crops for disease resistance and growth enhancement. Seven U.S. patents and approximately 40 related foreign patents have been issued, with several in process. These comprise approximately 50% of the U.S. patents in grape biotechnology. Applications stemming from the technology have wide application in plants, animals, yeasts, and bacteria for plant improvement, including efficient biomass production and conversion into biofuels. Selected publications and presentations describing the research being conducted can be accessed at http://www.mrec.ifas.ufl.edu/grapes/genetics.

Dr. Gray has published over 120 articles in scientific journals. He has coedited five popular textbooks covering the topics of plant tissue culture and biotechnology. He has contributed approximately 50 chapters to various books. He is frequently invited to speak at national and international scientific meetings, which have been summarized in over 80 published abstracts.

Dr. Gray is editor in chief of *Critical Reviews in Plant Sciences*, a refereed journal currently with the 10th highest impact rating among plant-related journals. He also is a frequent reviewer for a host of other refereed journals.

In recognition of his accomplishments in intellectual property development, Dr. Gray was awarded the University of Florida's Research Foundation Professorship in 1998, and he received the 2002 USDA Secretary's Honor Awards for outstanding research accomplishments in grape biotechnology, which is the highest honor given yearly by that agency. He received UF/IFAS awards for research in 2008 and 2009.

Contributors

David W. Altman
Central City Concern
Portland, Oregon

Caula A. Beyl
College of Agricultural and Natural Resources
University of Tennessee
Knoxville, Tennessee

Sarah Boggess
Department of Entomology and Plant Pathology
University of Tennessee
Knoxville, Tennessee

J. D. Caponetti
Department of Botany
University of Tennessee
Knoxville, Tennessee

Alan C. Cassells
Department of Zoology, Ecology and Plant Science
National University of Ireland Cork
Cork, Ireland

Feng Chen
Department of Plant Sciences
University of Tennessee
Knoxville, Tennessee

Z.-M. Cheng
Department of Plant Sciences
University of Tennessee
Knoxville, Tennessee

W. K. Choo
School of Biosciences and Biotechnology
Universiti Kebangsaan
Selangor, Malaysia

Michael E. Compton
School of Agriculture
University of Wisconsin–Platteville
Platteville, Wisconsin

Bob V. Conger
Department of Plant Sciences
University of Tennessee
Knoxville, Tennessee

Rene E. DeVries
Department of Entomology and Plant Pathology
University of Tennessee
Knoxville, Tennessee

Sadanand A. Dhekney
Mid-Florida Research and Education Center
The University of Florida/IFAS
Apopka, Florida

Tom Eeckhaut
Institute for Agricultural and Fisheries Research
Melle, Belgium

Chris Eisenschenk
Saliwanchik, Lloyd and Saliwanchik
Gainesville, Florida

Jennifer A. Franklin
Department of Forestry, Wildlife and Fisheries
University of Tennessee
Knoxville, Tennessee

Robert L. Geneve
Department of Horticulture
University of Kentucky
Lexington, Kentucky

Jean H. Gould
Department of Ecosystem Science and Management
Molecular and Environmental Plant Sciences Program
Texas A&M University
College Station, Texas

E. T. Graham
Department of Plant Sciences
University of Tennessee
Knoxville, Tennessee

D. J. Gray
Mid-Florida Research and Education Center
University of Florida/Institute of Food and Agricultural Sciences (IFAS)
Apopka, Florida

Jude W. Grosser
Horticultural Science Department
Citrus Research and Education Center
University of Florida
Lake Alfred, Florida

Denita Hadziabdic
Department of Entomology and Plant Pathology
University of Tennessee
Knoxville, Tennessee

M. D. Halfhill
Biology Department
St. Ambroise University
Davenport, Iowa

L. C. Hudson
SoyMeds, Inc.
Davidson, North Carolina

Timothy Johnson
Environmental Horticulture Department
University of Florida
Gainesville, Florida

Michael Kane
Environmental Horticulture Department
University of Florida
Gainesville, Florida

Peter Kaufman
Department of Surgery and the Michigan Integrative Medicine Program
University of Michigan
Ann Arbor, Michigan

Philip Kauth
Environmental Horticulture Department
University of Florida
Gainesville, Florida

Byung-Hoon Kim
Department of Natural Sciences
Albany State University
Albany, Georgia

Ara Kirakosyan
Department of Surgery and the Michigan Integrative Medicine Program
University of Michigan
Ann Arbor, Michigan

Svetlana V. Kushnarenko
Institute of Plant Biology and Biotechnology at National Center of Biotechnology
Timiryazev Str., Almaty
Republic of Kazakhstan

Peggy G. Lemaux
Department of Plant and Microbial Biology
University of California
Berkeley, California

Zhijian T. Li
Mid-Florida Research and Education Center
The University of Florida/IFAS
Apopka, Florida

Mary Ann Lila
Plants for Human Health Institute
North Carolina State University
Kannapolis, North Carolina

K. R. Malueg
Department of Plant Sciences
University of Tennessee
Knoxville, Tennessee

Ruth C. Martin
USDA-ARS, National Forage Seed Production Research Center
Corvallis, Oregon

Ling Meng
Department of Plant and Microbial Biology
University of California
Berkeley, California

Hiroyuki Nonogaki
Department of Horticulture
Oregon State University
Corvallis, Oregon

M. N. Normah
School of Biosciences and Biotechnology
Universiti Kebangsaan
Selangor, Malaysia

Margaret A. Norton
Department of Crop Sciences
University of Illinois at Urbana–Champaign
Urbana, Illinois

Ahmad A. Omar
College of Agriculture
Biochemistry Department
Zagazig University
Zagazig, Egypt

Linas Padegimas
Department of Biological Sciences
Mississippi State University
Mississippi State, Mississippi

K. A. Pickens
Department of Plant Sciences
University of Tennessee
Knoxville, Tennessee

Barbara M. Reed
USDA-ARS National Clonal Germplasm Repository
Corvallis, Oregon

Sandra M. Reed
Floral and Nursery Plant Research Unit
U.S. National Arboretum
USDA/ARS
McMinnville, Tennessee

Nancy A. Reichert
Department of Biological Sciences
Mississippi State University
Mississippi State, Mississippi

H. A. Richards
Department of Plant Sciences
University of Tennessee
Knoxville, Tennessee

Timothy A. Rinehart
Thad Cochran Southern Horticultural Research Center
USDA/ARS
Poplarville, Mississippi

Randy B. Rogers
Department of Natural Resources and Environmental Sciences
University of Illinois
Urbana, Illinois

Naomi R. Rowland
Biology Department
Western Kentucky University
Bowling Green, Kentucky

M. C. Scott
Department of Forestry, Wildlife, and Fisheries
University of Tennessee
Knoxville, Tennessee

E. Mitchell Seymour
Department of Surgery and the Michigan Integrative Medicine Program
University of Michigan
Ann Arbor, Michigan

Robert M. Skirvin
Department of Crop Sciences
University of Illinois at Urbana–Champaign
Urbana, Illinois

Songquan Song
Institute of Botany
Chinese Academy of Sciences
Beijing, China

C. N. Stewart, Jr.
Department of Plant Sciences
University of Tennessee
Knoxville, Tennessee

G. R. L. Suttle
Microplant Nurseries, Inc.
Gervais, Oregon

R. N. Trigiano
Department of Entomology and Plant Pathology
University of Tennessee
Knoxville, Tennessee

Johan Van Huylenbroeck
Institute for Agricultural and Fisheries Research
Melle, Belgium

Katrijn Van Laere
Institute for Agricultural and Fisheries Research
Melle, Belgium

Richard E. Veilleux
Department of Horticulture
Virginia Polytechnic Institute and State University
Blacksburg, Virginia

L. M. Vito
Department of Entomology and Plant Pathology
University of Tennessee
Knoxville, Tennessee

Albrecht G. von Arnim
Department of Biochemistry and Cellular and Molecular Biology
University of Tennessee
Knoxville, Tennessee

Phillip A. Wadl
Department of Entomology and Plant Pathology
University of Tennessee
Knoxville, Tennessee

Xinwang Wang
Texas AgriLife Research and Extension Center
Texas A & M University
Dallas, Texas

David T. Webb
Department of Botany
University of Hawaii at Manoa
Honolulu, Hawaii

M. T. Windham
Department of Entomology and Plant Pathology
University of Tennessee
Knoxville, Tennessee

Xiyan Yang
National Key Laboratory of Crop Genetic Improvement
Huazhong Agricultural University
Wuhan, Hubei, China

Margaret M. Young
Elizabeth City State University
Elizabeth City, North Carolina

Shibo Zhang
Department of Plant and Microbial Biology
University of California
Berkeley, California

Xianlong Zhang
National Key Laboratory of Crop Genetic Improvement
Huazhong Agricultural University
Wuhan, Hubei, China

Section I

Introduction

1 Introduction

D. J. Gray and R. N. Trigiano

PROLOGUE—INTRODUCTION TO 2003 BOOK (ABRIDGED)

All life on earth depends on the continuous acquisition of energy. The sun provides nearly all renewable energy in the form of radiation. Radiant energy heats the earth, causing convection and contributing in great measure to weather patterns, which transfer water through the atmosphere from oceans and lakes to land and back; this creates the basic environment needed to nurture life. However, life itself depends on the availability of chemical energy, the vast majority of which also is captured from sunlight.

Plants with chlorophyll perform the function of energy capture, utilizing photons to drive the cleavage water into oxygen and hydrogen. In the reaction termed *photosynthesis*, carbon dioxide (some of it created by the respiration of plants, animals, fungi, and bacteria) is ultimately recycled by its conversion into oxygen, water, and sugar. The chemical energy of sugar then is used to drive all of the other metabolic reactions needed to perpetuate life. The carbon in sugar is utilized to produce the more complex carbohydrates, proteins, and other structural molecules that make up plant cells, tissues, and organs. In turn, the chemical energy and nutritive substances contained in plants are used as the primary energy source for animals. This fundamental ability of green plants to thrive and reproduce by capturing and utilizing energy from sunlight has led to the tremendous diversity of life on earth.

The evolutionary force of natural selection that continuously drives diversification of plants and animals is well known. However, humankind has altered the course of natural selection by recognizing some sources of food and fiber to be preferable to others, hence choosing and encouraging those deemed to be superior in one way or another. Selection of improved "varieties" probably began in an unintentional manner, but then became increasingly more organized over the millennia. Eventually, the possibility of selective breeding of plants (and animals) was recognized and became the cornerstone of modern agricultural development. In turn, agricultural development became the foundation of modern civilization, allowing humankind to switch from a lifestyle of bare subsistence to one that promoted pursuit of activities never before imagined. And, relevant to this discussion, one heretofore "unimaginable activity" included the conception, development, and pursuit of plant science.

The development of plants into such a diversity of useful products is made possible only by human endeavor. Development of improved plants by use of previously unrecognized approaches continues today. Breeding programs are being revolutionized with biotechnology, leading to plants that have properties we never anticipated. Along with such rapid change comes controversy—new discussions, fear of the unknown, optimism, and opportunities.

INTRODUCTION TO 2010 BOOK

The earth is a "water planet," so-called due to the great abundance of H_2O, creating a unique environment that is conducive to the vast array of plant life we see around us. Plants grow at or near every spot on the planet. They are able to grow and reproduce in nearly every environment—from dry to wet, hot to cold, from the tops of mountains to the depths of seas.

In adapting to environmental extremes, plants have evolved to express a dizzying array of diversity. Less noted, perhaps, are factors causing the range of adaptations of plants within their immediate environment—to each other and to other organisms. For example, consider the vast number of species coexisting in certain tropical and temperate forests. They evolved to excel in exploiting niches in their environment. The incredible diversity of plants in certain ecosystems comes in large part due to their struggle to survive in competition with others. In some instances, species cooperate and facilitate the survival of one another; in others, one-sided parasitic relationships develop. The interrelationships between species in such ecosystems are highly dynamic in themselves; add to this the influences stemming from the earth's changing environment in general and incursions of species from without. A force is thus placed on individual species to adapt and change or risk extinction. The outcome is tremendous and diverse.

Nearly all plant scientists, in one way or another, study the diversity of plants. For example, among other things, botanists study the speciation of plants, differences in their structure, and/or ecology. Agronomists and horticulturalists study the uses of plants. Geneticists create new adapted species by various means and/or seek to understand how genes function. Physiologists study the interaction of plants in their environment at the molecular, cellular, and whole plant levels.

With such a broad range of disciplines within plant science, training necessarily involves a degree of specialization. While this textbook addresses several specialized areas generically termed *biotechnology*, it is unique in its use of a broad range of species that are studied via laboratory exercises. It strives to provide detailed perspectives and hands-on training in a rather diverse field. It is based upon our previous textbooks *Plant Development and Biotechnology* and *Plant Tissue Culture Concepts and Laboratory Exercises—Second Edition.* As editors, our intention was to concentrate on plant development and its application in biotechnology, providing supporting laboratory exercises while maintaining as broad of a base as possible. Thus, we address a wide range of related topics in one unique book.

WHAT'S INSIDE

In addition to this introductory section, the textbook is divided into the following four primary areas: Supporting Methodologies, Propagation Techniques, Crop Improvement Techniques, and The Business of Biotechnology. Each section combines related facets of plant development and biotechnology. All concept chapters begin with concepts designed to alert the student to the most important ideas in the chapter. The laboratory exercises chapters detail the materials needed and protocols required to complete the experiments, but are unique in that they also provide generalized anticipated results and questions designed to challenge students to think about the exercise. Also, we have included a DVD with PowerPoint presentations, mostly in color, of the figures in the chapters. Lastly, we have included three DVD "movie" presentations highlighting three laboratory experiences in the book—more on that later.

Section II, "Supporting Methodologies and General Concepts," is very eclectic in its subject matter and contains 10 concept and laboratory exercise chapters. The section begins the process of teaching tissue culture by discussing various methods to prepare the medium (Chapter 2). This chapter provides complete and logical examples of how to make solutions and dilutions, and to accomplish sterile culture work. An experienced teacher with laboratory resources and methodologies already in place may choose to pay less attention to this chapter, whereas the instructor of a newly established course will find it indispensable. Chapter 2 was selected as one of our DVD movie experiences for students. This DVD chronicles how to make simple media and should be very helpful to those just beginning to use plant tissue culture tools. From the many comments that we have had on this chapter over the last several years, we believe that many students and researchers will applaud its inclusion as very valuable and an excellent reference. Section II continues with Chapter 3, a laboratory exercise designed to investigate nutrition of plant tissue culture media. These "tried and true" experiments have been very helpful for beginning students according to communications from those who have adopted the book. Supporting this media preparation chapter is Chapter 4, a general treatise on plant growth regulators outlining their form, function, and effects. Chapter 5 explores statistical treatment of data, a concept often ignored in courses involving tissue culture (and others). Chapters 6 and 7 emphasize common methods to visualize and document studies (microscopy/photography and histology, respectively) of plant development and other plant processes. Again, we have retained the chapter containing histological techniques because of the overwhelming response of the readers of the previous editions. A short movie depicting a simple paraffin-based histological technique is included on the accompanying DVD. The chapter on microscopes was adapted from *Plant Pathology Concepts and Laboratory Exercises,* Second Edition, and has received outstanding marks for its information content—very few books include this much-needed chapter for beginning and advanced students. Chapters 8 and 9 introduce students to the concepts of plant anatomy and morphology, and seeds and germination, respectively. We have found that students taking tissue cultures courses often lack essential basic botanical information, and these two chapters together offer a brief but thorough primer on these subjects. The section ends with two very informative chapters: Chapter 10 depicts some methods for assessing genetic diversity of plants, and Chapter 11 is an excellent treatise on molecular tools for studying plant development. These chapters were included in the present edition as a response to user's needs and requests for more molecular biology/genetic information in the context of plant biotechnology.

Section III, "Propagation and Development Concepts," contains 14 concept and laboratory chapters that encompass the essential foundations of plant tissue culture. The section is very rich in experimental protocols that should be useful in any course on the subject. In this section, the three types of commonly used culture regeneration systems are introduced and discussed from traditional viewpoints and from more modern molecular developmental aspects.

We begin Section III by discussing in "Propagation by Shoot Culture" (Chapter 12) a process that is more commonly termed *micropropagation.* This excellent chapter is supported by three easy-to-complete laboratory exercises with *Syngonium* (Chapter 13), roses (Chapter 14), and potato (Chapter 15). Micropropagation is the simplest and most commercially useful tissue culture method. The tissue culture industry uses micropropagation almost exclusively for ornamental plant production, and Chapter 16 outlines the "nuts and bolts" of industrial production. The next two chapters concern themselves with indexing for plant pathogens. Chapters 17 and 18 include both traditional methods and newer, more sophisticated molecular and serological techniques for indexing plants for pathogens and assessing cultures for bacterial and fungal contaminants of cultures (Chapter 17). The final chapter (Chapter 18) in this series explores practical culture-indexing techniques.

The second propagation system to be discussed in Section III is in "Propagation from Nonmeristematic Tissues—Organogenesis" (Chapter 19). Organogenesis is the de novo development of organs, typically shoots and/or roots, from cells and tissues that would not normally form them. The term *adventitious* has also been used to describe the plant parts formed by the process of organogenesis. Shoot organogenesis is another means of propagating plants. While not used much in commercial production, organogenesis is used extensively in genetic engineering (see later) as a means to produce plants from genetically altered cells. In Chapter 19, both the theory and developmental sequences of how cells are induced to follow such a developmental pathway are discussed. Chapter 20 covers some of the more recently discovered molecular aspects of meristem formation, development, and control, which are presented in depth and illustrates the significance of our burgeoning molecular biological knowledge to practical usage. A laboratory exercise using chrysanthemum and African violet illustrates some of the basic principles of organogenesis and plant regeneration from leaves (Chapter 21). One of the experiments illustrating producing shoots, rooting shoots, and regenerating entire chrysanthemum plants is visually demonstrated on the accompanying DVD.

The third propagation system discussed in Section III is "Propagation from Nonmeristematic Tissues—Nonzygotic Embryogenesis" (Chapter 22). *Nonzygotic embryogenesis* is a broadly defined term meant to cover all instances where embryogenesis occurs outside of the normal developmental pathway found in the seed. One type of nonzygotic embryogenesis is termed *somatic embryogenesis.*

This is a unique phenomenon exhibited by vascular plants, in which somatic (nonsexual) cells are induced to behave like zygotes. Such induced cells begin a complex, genetically programmed series of divisions and eventual differentiation to form an embryo that is more or less identical to a zygotic embryo. This type of propagation system is important since the embryos develop from single cells that can be genetically engineered and are complete individuals that are capable of germinating directly into plants. Thus, potentially, somatic embryogenesis also represents an efficient propagation system. Chapter 22 discusses the developmental processes and significance of nonzygotic embryogenesis. In addition, the process can be very efficient, and thus there has been interest in developing nonzygotic embryogenesis into a propagation system. Chapter 23 summarizes some of the newer molecular studies concerned with initiation, development, and control of the nonzygotic embryogenic process. The last two chapters in this section are laboratory exercises that are proven winners. Chapter 24 utilizes orchardgrass, a model system for monocot nonzygotic embryogenesis, and Chapter 25 involves embryogenesis from cineraria—a florists' crop.

In Section IV, "Crop Improvement Techniques," the aforementioned propagation techniques are integrated with other methodologies in order to modify and manipulate germplasm. Chapter 26 discusses the use of plant protoplasts. Protoplasts are plant cells from which the cell wall has been enzymatically removed, making them amenable to cell fusion and other methods of germplasm manipulation. Chapters 27 and 28 are laboratory exercises that illustrate some of the principles discussed in Chapter 26. Chapter 27 is a simple protocol essentially detailing how to make protoplasts and is amendable to use as a demonstration, whereas Chapter 28 provides practical experiences manipulating protoplasts. Chapter 29 explores the use of haploid culture in plant improvement. Haploid cultures usually are derived from the meiotic divisions of the microspore mother cells and result in cells, tissues, and plants with half the normal somatic cell chromosome number. Such plants are of great use in genetic studies and breeding, since all of the recessive genes are expressed and, by doubling the haploid plants back to the diploid ploidy level, dihaploid plants are produced, which are completely homozygous. True homozygous plants are time consuming and often impossible to produce by conventional breeding. Chapter 30 provides an easy to follow protocol for creating haploid tobacco and potato plants. These chapters are followed by a concept/laboratory chapter on embryo rescue, which was conspicuously lacking in one of our previous editions. Chapter 31 details the technique of embryo rescue and discusses its usage as a commonly applied methodology.

Chapters 32 and 33 discuss promoters and gene regulation and genetic engineering technologies, which are currently hot topics in agriculture. Genetic transformation, wherein genes from unrelated organisms can be integrated into plants without sexual reproduction, resulting in "transgenic plants," is the most significant application for plant tissue culture when considering its impact on humankind and is discussed in Chapter 33. Transformation and genetic engineering concept chapters are supported by the following three chapters: Chapter 34, a new concept/laboratory exercise that outlines procedures for transforming meristems; Chapter 35, detailing some simple experiments to transform chrysanthemum and tobacco; and Chapter 36, providing another, yet different, procedure to transform tobacco. These chapters are followed by a discussion on genetically modified organisms and the controversy that surrounds them, especially as it impacts food supplies, commerce, and natural ecosystems (Chapter 37). Many favorable comments convinced us to retain this chapter in the current edition.

Chapter 38 is brand new to our books and describes the use of cryopreservation of germplasm. Cryopreservation is an efficient means of safeguarding valuable plant germplasm by freezing all metabolic activity. Cryopreserved cells and tissues can be kept for extended periods of time without mutations occurring or any physiological decline. Two new laboratory chapters on this topic have been added. Chapter 39 deals with cryopreservation of shoots tips, whereas Chapter 40 involves seeds. Chapter 41 describes the production of secondary products by plant cells in culture and is new to this edition. This subject is somewhat unique in the context of previous topics, because the end product is not a regenerated plant but, rather, a chemical manufactured by plant cells. Use of plants as biofactories to produce complex pharmaceuticals is of great interest, particularly due to its potential application in health care. Chapter 42 depicts a laboratory exercise for producing anthocyanins in culture. The last chapter (Chapter 43) in this section is also new and discusses reasons that genetic and/or phenotypic variations occur as a result of plant tissue culture.

Section V, "The Business of Biotechnology," includes chapters on two topics that are exceedingly important, but often omitted from textbooks on tissue culture and biotechnology. We have had many requests to include this information in the present edition. To this end, the first chapter (Chapter 44) in this section is a look at entrepreneurship in the biotechnology field. This chapter outlines the must dos (and some of the don'ts) when considering launching a new enterprise. Chapter 45, written by a lawyer with experience in the field, explains the processes of patents, trade secrets, and other legalities of biotechnology.

Based on the many telephone calls, letters, and e-mails that we received, the first and second editions of *Plant Tissue Culture Concepts and Laboratory Exercises* successfully facilitated training of students in current principles and methodologies of our rapidly evolving field. Although the strictly tissue culture users found the previous editions useful, a significant number of readers and instructors wanted a book composed primarily of lecture topics that included some of the molecular aspects of plant development. We asked for constructive criticisms from users of the book and employed them to make a number of improvements in *Plant Development and Biotechnology*. We have now come full circle as many of the users of the previous editions wanted not only the concepts and theories but also useable and reliable laboratory exercises that illustrated and reinforce some of the principles in the concept chapters. We believe that we and our contributing authors addressed many of these needs now and, hopefully, you, the students and instructors in the field, will enjoy at least the same level of success achieved previously! As always, don't be shy—we welcome comments from colleagues and students as they put the textbook to use.

Section II

Supporting Methodologies and General Concepts

2 Getting Started with Tissue Culture—Media Preparation, Sterile Technique, and Laboratory Equipment

Caula A. Beyl

CONCEPTS

- A tissue laboratory needs adequate physical space for work and storage and equipment such as an autoclave, a distilled water source, balances, refrigerators, various laboratory instruments, culture vessels, and flow hoods, to name a few items.
- There are many kinds of growth media available, and the type of basal culture medium selected depends upon the species to be cultured. The growth of the plant in culture is also affected by the selection of plant growth regulators (PGR) and environmental (cultural) conditions.
- There are about 20 different components in tissue culture medium and these include inorganic mineral elements, various organic compounds, PGR, and support substances (e.g., agar or filter paper).
- Plant growth regulator concentrations in media are typically expressed as mg/L or μM. When comparing the effects of several PGRs on tissues in culture, prepare media using μM concentrations since an equal number of molecules of the various PGRs will be present in each of the media.

A plant tissue culture laboratory has several functional areas, whether it is designed for teaching or research and no matter what its size or how elaborate it is. It has some elements similar to a well-run kitchen and other elements that more closely resemble an operating room. There are areas devoted to preliminary handling of plant tissue destined for culture, media preparation, sterilization of media and tools, a sterile transfer hood or "clean room" for aseptic manipulations, a culture growth room, and an area devoted to washing and cleaning glassware and tools (see Chapter 16). The following chapter will serve as an introduction to what goes into setting up a tissue culture laboratory, what supplies and equipment are necessary, some basics concerning making stock solutions, calculating molar concentrations, making tissue culture media, preparing a transfer (sterile) hood, and culturing various plant cells, tissues, and organs.

EQUIPMENT AND SUPPLIES FOR A TISSUE CULTURE LABORATORY

Ideally, there should be enough bench area to allow for both preparation of media and storage space for chemicals and glassware. A tissue culture laboratory needs an assortment of glassware, which may include graduated measuring cylinders, wide-necked erlenmeyer flasks, medium bottles, test tubes with caps, petri dishes, volumetric flasks, beakers, and a range of pipettes. In general, glassware should be able to withstand repeated autoclaving. Baby food jars are inexpensive alternative tissue culture containers and well suited for teaching as well as research. Ample quantities can be obtained by preceding the recycling truck on its pickup day (provided you are not embarrassed by the practice). Some tissue culture laboratories find presterilized disposable culture containers and plastic petri dishes to be convenient, but the cost may be prohibitive for others on a tight budget. There are also reusable plastic containers available, but their longevity and resistance to wear, heat, and chemicals vary considerably.

It is also good to stock metal or wooden racks to support culture tubes both for cooling and later during their time in the culture room, metal trays (such as cafeteria trays) and carts for transport of cultures, stoppers and various closures, nonabsorbent cotton, cheesecloth, foam plugs, metal or plastic caps, aluminum foil, Parafilm™, and plastic wrap.

Some basic equipment is also necessary to operate a tissue laboratory including a pH meter, balances (one analytical to 4 decimal places and another to 2 decimal places), Bunsen burners, alcohol lamps or electric sterilizing devices, several hot plates with magnetic stirrers, a microwave oven for rapid melting of large volumes of agar medium, a compound microscope and hemocytometer for cell counting, a low g centrifuge, stereomicroscopes (ideally with fiber-optic light sources), large (10 or 25 L) plastic carboys to store high quality (purity) water, a fume hood, an autoclave (or at the very least, a pressure cooker), and a refrigerator to store media, stock solutions, plant growth regulators (PGRs), etc. A dishwasher is useful, but a large sink with drying racks, pipette and acid baths, and a forced air oven for drying glassware will also work. Also, deionized distilled water for the final rinsing of glassware is needed. Aseptic manipulations and transfers are done in multistation laminar flow hoods (one for each pair of students).

Equipment used in the sterile transfer hood usually includes a spray bottle containing 70% ethanol, spatulas (useful for transferring callus clumps), forceps (short, long, and fine-tipped), scalpel handles (#3), disposable scalpel blades (#10 and #11), a rack for holding sterile tools, a pipette bulb or pump, Bunsen burner, alcohol lamp or other sterilizing device, and a sterile surface for cutting explants (see the following text). If necessary for the experimental design, uniform-sized explants can be obtained using a sterile cork borer.

There are a number of options for providing a sterile surface for cutting explants. A stack of paper towels wrapped in aluminum foil is effective and, as each layer becomes messy, it can be peeled off and the next layer beneath it used (Figure 2.1). Others prefer reusable surfaces such as glazed ceramic tiles (local tile retailers are quite generous and will donate samples), metal commercial ashtrays, or glass petri dishes (100 × 15 mm). Sterile plastic petri dishes also can be used, but the cost may outweigh the advantages. A container is needed to hold 95% ethanol used for cooling flame-sterilized instruments. An ideal container for this purpose is a Coplin jar (slide staining) with a small wad of cheesecloth at the bottom to prevent breakage of the glass when tools are dropped in. It has the advantage that it is made of heavier glass than the typical laboratory glassware and, since the base is flared, it is not prone to tipping over. Other containers can also serve the same purpose such as test tubes in a rack or placed in a flask or beaker to prevent them from spilling. Plastic containers, which can catch fire and melt, never should be used to hold alcohol.

WATER

High-quality water is a required ingredient of plant tissue culture media. Ordinary tap water contains cations, anions, particulates of various kinds, microorganisms, and gases (chlorine) that make it unsuitable for use in tissue culture media. Various methods are used to treat water, including filtration through activated carbon to remove organics and chlorine, deionization or demineralization by passing water through exchange resins to remove dissolved ionized impurities, and distillation that eliminates most ionic and particulate impurities, volatile and nonvolatile chemicals, organic matter, and microorganisms. The process of reverse osmosis, which removes 99% of the dissolved ionized impurities, uses a semipermeable membrane through which a portion of the water is forced under pressure, and the remainder containing the concentrated impurities is rejected. The most universally reliable method of water purification for tissue culture use is a deionization treatment followed by one or two glass distillations, although simple deionization alone is sometimes successfully used.

FIGURE 2.1 A typical layout of materials in the hood showing placement of the sterile tile work surface, an alcohol lamp, spray bottle containing 80% ethanol, a cloth for wiping down the hood, and two different kinds of tool holders—a glass staining (Coplin) jar and a metal rack for holding test tubes.

In some cases, newer reverse osmosis purifying equipment (Milli-RO™, Millipore™; RO pure™, Barnstead™; Bion™, Pierce™), combined with cartridge ion exchange, adsorption, and membrane filtering equipment, has replaced the traditional glass distillation of water.

THE CULTURE ROOM

After the explants are plated on the tissue culture medium under the sterile transfer hood, they are moved to the culture room. It can be as simple as a room with shelves equipped with lights or as complex as a room with intricate climate control. Most culture rooms tend to be rather simple consisting of cool white fluorescent lights mounted to illuminate each shelf. Adjustable shelves are an asset that allows for differently sized tissue culture containers and for moving the light closer to the containers to achieve higher light intensities. Putting the lights on timers allows for photoperiod manipulation. Some cultures grow equally as well in dark or light. Temperatures of 26°C–28°C are usually optimum. Heat buildup can be a problem if the room is small, so adequate air conditioning is required. Good airflow also helps to reduce condensation occurring inside petri dishes or other culture vessels. Some laboratories purchase incubators designed for plant tissue culture. If a liquid medium is used, the culture room should be equipped with a rotary or reciprocal shaker to provide sufficient oxygenation. The optimum temperature, light, and shaker conditions vary depending upon the plant species being cultured.

CHARACTERISTICS OF SOME OF THE MORE COMMON TISSUE CULTURE MEDIA

The type of tissue culture medium selected depends upon the species to be cultured. Some species are sensitive to high salts or have different requirements for PGRs. The age of the plant also has an effect. For example, juvenile tissues generally regenerate roots more readily than adult tissues. The type of organ cultured is important; for example, roots require thiamine. Each desired cultural effect has its own unique requirements such as auxin for induction of adventitious roots and altering the cytokinin:auxin ratio for initiation and development of adventitious shoots.

Development of culture medium formulations was a result of systematic trial and experimentation. The composition of several of the most commonly used plant tissue culture media with respect to their components in mg/L and molar units is presented in Table 2.1. Murashige and Skoog (1962) medium (MS) (see Table 2.2) is suitable for many applications and the most commonly used basic tissue culture medium for plant regeneration from tissues and callus. It was developed for tobacco-based primarily upon the mineral analysis of tobacco tissue. This is a "high salt" medium due

TABLE 2.1
Composition of Five Commonly Used Tissue Culture Media in Milligrams per Liter and Molar Concentrations with Inorganic Components[a]

Compounds	Murashige and Skoog	Gamborg B-5	WPM	Nitsch and Nitsch	Schenk and Hildebrandt	White
Macronutrients in mg/L (mM)						
NH_4NO_3	1650 (20.6)	—	400 (5.0)	720 (9.0)	—	—
$NH_4H_2PO_4$	—	—	—	—	300 (2.6)	—
NH_4SO_4	—	134 (1.0)	—	—	—	—
$CaCl_2 \cdot 2H_2O$	440 (3.0)	150 (1.0)	96 (0.7)	166 (1.1)	151 (1.0)	—
$Ca(NO_3)_2 \cdot 4H_2O$	—	—	556 (2.4)	—	—	288 (1.2)
$MgSO_4 \cdot 7H_2O$	370 (1.5)	246 (1.0)	370 (1.5)	185 (0.75)	400 (1.6)	737 (3.0)
KCl	—	—	—	—	—	65 (0.9)
KNO_3	1900 (18.8)	2528 (25)	—	950 (9.4)	2500 (24.8)	80 (0.8)
K_2SO_4	—	—	990	—	—	—
KH_2PO_4	170 (1.25)	—	170 (1.3)	68 (0.5)	—	—
$NaH_2PO_4 \cdot H_2O$	—	150 (1.1)	—	—	—	19 (0.14)
Na_2SO_4	—	—	—	—	—	200 (1.4)
Micronutrients in mg/L (μM)						
H_3BO_3	6.2 (100)	3.0 (49)	6.2 (100)	10 (162)	5 (80)	1.5 (25)
$CoCl_2 \cdot 6H_2O$	0.025 (0.1)	0.025 (0.1)	—	—	0.1 (0.4)	—
$CuSO_4 \cdot 5H_2O$	0.025 (0.1)	0.025 (0.1)	0.25 (1)	0.025 (0.1)	0.2 (0.08)	0.01 (0.04)
$Na_2EDTA \cdot 2H_2O$	37.2 (100)	37.2 (100)	37.2 (100)	37.2 (100)	20.1 (54)	—
$Fe_2(SO_4)_3$	—	—	—	—	—	2.5 (6.2)
$FeSO_4 \cdot 7H_2O$	27.8 (100)	27.8 (100)	27.8 (100)	27.8 (100)	15 (54)	—
$MnSO_4 \cdot H_2O$	—	10.0 (59)	—	—	10.0 (59)	5.04 (30)
$MnSO_4 \cdot 4H_2O$	22.3 (100)	—	22.3 (100)	25.0(112)	—	—
KI	0.83 (5)	0.75 (5)	—	—	0.1 (0.6)	0.75 (5)
$NaMoO_3$	—	—	—	—	—	0.001 (0.001)
$Na_2MoO_4 \cdot 2H_2O$	0.25 (1)	0.25 (1)	0.25 (1)	0.25 (1)	0.1 (0.4)	—
$ZnSO_4 \cdot 7H_2O$	8.6 (30)	2.0 (7.0)	8.6 (30)	10 (35)	1 (3)	2.67 (9)
Organics in mg/L (μM)						
Myo-inositol	100 (550)	100 (550)	100 (550)	100 (550)	1000 (5500	—
Glycine	2.0 (26.6)	—	2.0 (26.6)	2.0 (26.6)	—	3.0 (40)
Nicotinic acid	0.5 (4.1)	1.0 (8.2)	0.5 (4.1)	5 (40.6)	5.0 (41)	0.5 (4.1)
Pyridoxine HCl	0.5 (2.4)	0.1 (0.45)	0.5 (2.4)	0.5 (2.4)	0.5 (2.4)	0.1 (0.45)
Thiamine HCl	0.1 (0.3)	10.0 (30)	1.0 (3.0)	0.5 (1.5)	5.0 (14.8)	0.1 (0.3)
Biotin	—	—	—	0.2 (0.05)	—	—

[a] Corrected per Owen, H. R., and A. R. Miller. 1992. *Plant Cell, Tissue, and Organ Culture* 28:147–150.

to its content of potassium and nitrogen salts. Linsmaier and Skoog medium (1965) is basically MS medium with respect to its inorganic portion, but only inositol and thiamine HCl are retained among the organic components. To counteract salt sensitivity of some woody species, Lloyd and McCown (1980) developed the woody plant medium (WPM).

Gamborg's B5 medium (Gamborg et al., 1968) was devised for soybean callus culture and has lesser amounts of nitrate and, particularly, ammonium salts than does the MS medium. Although B5 was originally developed for the purpose of obtaining callus or for use with suspension culture, it also works well as a basal medium for whole plant regeneration. Schenk and Hildebrandt (1972) developed SH medium for the callus culture of monocots and dicots. White's medium (1963), which was designed for the tissue culture of tomato roots, has a lower concentration of salts than MS medium. Nitsch's medium (Nitsch and Nitsch, 1969) was developed for anther culture and contains a salt concentration intermediate to that of MS and White's media.

Many companies sell packaged mixtures of the better known media recipes. These are easy to make because they merely involve dissolving the packaged mix in a specified volume of water. These can be purchased as the salts, the vitamins, or the entire mix with or without PGRs, agar, and sucrose. These are convenient, less prone to individual error, and make keeping stock solutions unnecessary. However, they are more expensive than making media from scratch.

Components of the Tissue Culture Medium

Growth and development of an explant in vitro is a product of its genetics, surrounding environment, and components of the tissue culture medium, the last of which is easiest to manipulate to our own ends. Tissue culture medium consists of 95% water, macro- and micronutrients, PGRs, vitamins, sugars (because plants in vitro are often not photosynthetically-competent), and sometimes various other simple-to-complex organic materials. All in all, about 20 different components are usually needed for tissue culture medium.

Inorganic Mineral Elements

Just as a plant growing in vivo requires many different elements from either soil or fertilizers, the plant tissue growing in vitro requires a combination of macro- and micronutrients. The choice of macro- and microsalts and their concentrations is species dependent. MS medium is very popular because most plants react to it favorably; however, it may not necessary result in the optimum growth and development for every species since the salt content is so high.

The macronutrients are required in millimolar (mM) quantities in most plant basal media. Nitrogen (N) is usually supplied in the form of ammonium (NH_4^+) and nitrate (NO_3^-) ions, although sometimes more complex organic sources such as urea, individual amino acids such as glutamine, or complex mixtures of amino acids found in casein hydrolysate are used, too. Although most plants prefer NO_3^- to NH_4^+, the right balance of the two ions for optimum in vitro growth and development for the selected species may differ considerably.

In addition to nitrogen, potassium, magnesium, calcium, phosphorus, and sulfur are provided in the medium as different components referred to as the macrosalts. $MgSO_4•7H_2O$ provides both magnesium and sulfur; $NH_4 H_2PO_4$, KH_2PO_4, or $NaH_2PO_4•H_2O$ provide phosphorus; $CaCl_2•2H_2O$ or $Ca(NO_3)_2•4H_2O$ provide calcium; and KCl, KNO_3, or KH_2PO_4 provide potassium. Chloride is provided by KCl and/or $CaCl_2•2H_2O$.

Microsalts typically include boron (H_3BO_3), cobalt ($CoCl_2•6H_2O$), iron (complex of $FeSO_4•7H_2O$ and $Na_2EDTA•2H_2O$ or rarely as $Fe_2(SO_4)_3$), manganese ($MnSO_4•H_2O$ or $MnSO_4•4H_2O$), molybdenum ($NaMoO_3•2H_2O$), copper ($CuSO_4•5H_2O$), and zinc ($ZnSO_4•7H_2O$). Microsalts are needed in much lower (micromolar; μM) concentrations than the macronutrients. Some media may contain very small amounts of iodide (KI), but sufficient quantities of many of the trace elements inadvertently may be provided because reagent-grade chemicals contain many inorganic contaminants.

Organic Compounds

Sugar (most commonly sucrose) is a very important part of any nutrient medium, and its addition is essential for in vitro growth and development of the culture. Most plant cultures are unable to photosynthesize effectively for a variety of reasons, including insufficiently organized cellular and tissue development, lack of chlorophyll, limited gas exchange and CO_2 in the tissue culture vessels, and less than optimum environmental conditions such as low light. A concentration of 20–60 g/L sucrose (a disaccharide made up of d-glucose and d-fructose) is the most often used carbon and energy source, since this sugar is also synthesized and transported naturally by the plant. Other mono- or disaccharides and sugar alcohols such as glucose, fructose, sorbitol, and maltose may be used. The sugar concentration chosen is dependent on the type and age of the explant in culture. For example, very young embryos require a relatively high sugar concentration (>3%). Fructose was better suited for proliferation of mulberry buds in vitro than sucrose, glucose, maltose, raffinose, or lactose (Coffin et al., 1976). For apple, sorbitol and sucrose supported callus initiation and growth equally as well, but sorbitol was better for peach after the fourth subculture (Oka and Ohyama, 1982).

Sugar (sucrose) that is bought from the supermarket is usually adequate, but be careful to get pure cane sugar as corn sugar is primarily fructose. Raw cane sugar is purified and according to the manufacturers analysis consists of 99.94% sucrose, 0.02% water, and 0.04% other materials (inorganic elements and also raffinose, fructose, and glucose). Nutrient salts contribute approximately 20%–50% to the osmotic potential of the medium, and sucrose is responsible for the remainder. The contribution of sucrose to the osmotic potential increases as it is hydrolyzed into glucose and fructose during autoclaving. This may be an important consideration when performing osmotic-sensitive procedures such as protoplast isolation and culture (Chapter 26).

Vitamins are organic substances that are parts of enzymes or cofactors for essential metabolic functions. Of the vitamins, only thiamine (vitamin B_1 at 0.1–5.0 mg/L) is essential in culture as it is involved in carbohydrate metabolism and the biosynthesis of some amino acids. It is usually added to tissue culture media as thiamine hydrochloride. Nicotinic acid, also known as *niacin*, *vitamin* B_3, or *vitamin PP*, forms part of a respiratory coenzyme and is used at concentrations between 0.1 to 5 mg/L. MS medium contains thiamine HCl as well as two other vitamins, nicotinic acid and pyridoxine (vitamin B_6), in the HCl form. Pyridoxine is an important coenzyme in many metabolic reactions and is used in media at concentrations of 0.1 to 1.0 mg/L. Biotin (vitamin H) is commonly added to tissue culture media at 0.01–1.0 mg/L. Other vitamins that are sometimes used are folic acid (vitamin M; 0.1–0.5 mg/L), riboflavin (vitamin B_2; 0.1–10 mg/L), ascorbic acid (vitamin C; 1–100 mg/L), pantothenic acid (vitamin B_5; 0.5–2.5 mg/L), tocopherol (vitamin E; 1–50 mg/L), and *para*-aminobenzoic acid (0.5–1.0 mg/L).

Sometimes characterized as one of the B complex vitamin group, myo-inositol is really a sugar alcohol involved in the synthesis of phospholipids, cell wall pectins, and membrane systems in cell cytoplasm. It is added to tissue culture media at a concentration of about 0.1–1.0 g/L and has been demonstrated to be necessary for some monocots, dicots, and gymnosperms.

In addition, other amino acids are sometimes used in tissue culture media. These include L-glutamine, L-asparagine, L-serine, and L-proline, which are used as sources of reduced organic nitrogen, especially for inducing and maintaining somatic or nonzygotic embryogenesis (Chapter 22). L-Glycine, the simplest amino acid, is a common additive since it is essential in purine synthesis and is a part of the porphyrin ring structure of chlorophyll.

Complex organic compounds are a group of undefined supplements such as casein hydrolysate, coconut milk (the liquid endosperm of the coconut), orange juice, tomato juice, grape juice, pineapple juice, sap from birch, banana puree, etc. These compounds are often used when no other combination of known defined components produces the desired growth or development. However, the composition of these supplements is basically unknown and may also vary from lot-to-lot causing variable responses. For example, the composition of coconut milk (used at a dilution of 50–150 mL/L), a natural source of the PGR, zeatin, not only differs between young and old coconuts, but also among coconuts of the same age.

Some complex organic compounds are used as organic sources of nitrogen such as casein hydrolysate, a mixture of about 20 different amino acids and ammonium (0.1–1.0 g/L), peptone (0.25–3.0 g/L), tryptone (0.25–2.0 g/L), and malt extract (0.5–1.0 g/L). These mixtures are very complex and contain vitamins as well as amino acids. Yeast extract (0.25–2.0 g/L) is used because of the high concentration and quality of B vitamins.

Polyamines, particularly putrescine and spermidine, are sometimes beneficial for somatic embryogenesis. Polyamines are also cofactors for adventitious root formation. Putrescine is capable of synchronizing the embryogenic process of carrot and enhances somatic embryogenesis in cotton.

Activated charcoal is useful for absorption of the brown or black pigments and oxidized and polymerized phenolic compounds (melanins). It is incorporated into the medium at concentrations of 0.2–3.0% (w/v). It is also useful for absorbing other organic compounds including PGRs, such as auxins and cytokinins, vitamins, iron, and zinc chelates (Nissen and Sutter, 1990). Explants transferred to media containing no PGRs sometimes continue to exhibit growth and development typical of the medium they were on previously that contained PGRs. These carryover effects of PGRs are minimized by adding activated charcoal when transferring explants to media without PGRs. Another feature of using activated charcoal is that it changes the light environment by darkening the medium, so it can help with root formation and growth. It may also promote nonzygotic embryogenesis and enhance growth and organogenesis of woody species.

Leached pigments and oxidized polyphenolic compounds and tannins can greatly inhibit growth and development. These are formed by some explants as a result of wounding. If charcoal does not reduce the inhibitory effects of polyphenols, addition of polyvinylpyrrolidone (PVP, 250–1000 mg/L), or antioxidants, such as citric acid, ascorbic acid, or thiourea, can be tested.

PLANT GROWTH REGULATORS (PGRs)

PGRs exert dramatic effects at low concentrations (0.001–10 μM). They control and modulate the initiation and development of shoots and roots on explants and embryos on semisolid or in liquid medium cultures. They also stimulate cell division and expansion. Sometimes, a tissue or an explant is autotrophic and can produce its own supply of PGRs. Usually PGRs must be supplied in the medium for growth and development of the tissues or cells in culture.

The most important classes of the PGRs used in tissue culture are the auxins and cytokinins. The relative effects of auxin and cytokinin ratio on morphogenesis of cultured tobacco tissues were demonstrated by Skoog and Miller (1957) and still serve as the basis for plant tissue culture manipulations today. Some of the PGRs used are hormones (naturally synthesized by higher plants), and others are synthetic (man-made) compounds. PGRs exert dramatic effects depending on the concentration used, the target tissue, and their inherent activity even though they are used in very low concentrations in the media (from 0.1 to 100 μM). The concentrations of PGRs are typically reported in mg/L or in μM units of concentration. Comparisons of PGRs based on their molar concentrations are more useful because the molar concentration is a reflection of the actual number of molecules of the PGR per unit volume (Table 2.3).

Auxins play a role in many developmental processes, including cell elongation and swelling of tissue, apical dominance, adventitious root formation, and somatic embryogenesis. Generally, when the concentration of auxin is low, root initiation is favored and, when the concentration is high, callus formation occurs. The most common synthetic auxins used are 1-naphthaleneacetic acid (NAA), 2,4-dichlorophenoxyacetic acid (2,4-D), and 4-amino-3,5,6-trichloro-2-pyridinecarboxylic acid (picloram). Naturally occurring indoleacetic acid (IAA) and indolebutyric acid (IBA) are also used. IBA was once considered synthetic, but has also been found to occur naturally in many plants including olive and tobacco (Epstein et al., 1989). Both IBA and IAA are photosensitive so that stock solutions must be stored in the dark. IAA is also easily broken down by enzymes (peroxidases and IAA oxidase). IAA is the weakest auxin and is typically used at concentrations between 0.01 to 10 mg/L. More active auxins such as IBA, NAA, 2,4-D, and picloram are used at concentrations ranging from 0.001 to 10 mg/L. Picloram and 2,4-D are examples of auxins used primarily to induce and regulate somatic embryogenesis.

Cytokinins promote cell division and shoot initiation and growth in vitro. The cytokinins most commonly used are zeatin, dihydrozeatin, kinetin, benzyladenine, thidiazuron, and 2iP. In higher concentrations (1–10 μM), they induce adventitious shoot formation but inhibit root formation. They promote axillary shoot formation by opposing apical dominance regulated by auxins. Benzyladenine has significantly stronger cytokinin activity than the naturally occurring zeatin. However, a concentration of 0.05–0.1 μM thidiazuron, a diphenyl substituted urea, is more active than 4–10 μM BA, but thidiazuron may inhibit root formation, causing difficulties in plant regeneration.

Gibberellins (GA) are less commonly used in plant tissue culture. GA_3 is the most often used, but it is very heat sensitive (after autoclaving 90% of the biological activity is lost). Typically, it is filter sterilized and added to autoclaved medium after it has cooled. Gibberellins help to stimulate elongation of internodes and have proved to be necessary for meristem growth for some species. Abscisic acid (ABA) is not normally considered an important PGR for tissue culture except for somatic embryogenesis and in the culture of some woody plants. For example, it promotes maturation and germination of somatic embryos of caraway (Ammirato, 1974), citrus (Kochba et al., 1978), and spruce (Roberts et al., 1990).

Organ and callus cultures are able to produce ethylene, a gaseous PGR. Since culture vessels are almost entirely closed, ethylene can sometimes accumulate. Many plastic containers also contribute to ethylene content in the vessel. There are contrasting reports in the literature concerning the role played by ethylene in vitro. It appears to influence embryogenesis and organ formation in some gymnosperms. Sometimes, in vitro growth can be promoted by ethylene. At other times, addition of ethylene inhibitors results in better initiation or growth. For example, the ethylene inhibitors, particularly silver nitrate, are used to enhance embryogenic culture initiation in corn. High levels of 2,4-D can induce ethylene formation.

AGAR AND ALTERNATIVE CULTURE SUPPORT SYSTEMS

Agar is used to solidify tissue culture media into a gel. It enables the explant to be placed in intimate contact with the medium (e.g., on the surface or embedded) but remain aerated. Agar is a high molecular weight polysaccharide that can bind water and is derived from seaweed. It is added to the medium in concentrations ranging from 0.5 to 1.0% (w/v). High concentrations of agar result in a harder medium. If a lower concentration of agar is used (0.4%) or if the pH is low, the medium will be too soft and will not gel properly. The consistency of the agar can also influence the growth. If it is too hard, plant growth is reduced. If it is too soft, plants that have a translucent water-soaked (hyperhydric) appearance may be the result (Singha, 1982). To gel properly, a medium with 0.6% agar must have a pH above 4.8. Sometimes, activated charcoal in the medium will interfere with gelling. Typical tissue culture agar melts easily at ~65°C and solidifies at ~45°C.

Agar also contains organic and inorganic contaminants, the amount of which varies between brands. Organic acids, phenolic compounds, and long-chain fatty acids are common contaminants. A manufacturer's analysis shows that Difco Bacto agar also contains (amounts in ppm) 0.0–0.5 cadmium, 0.0–0.1 chromium, 0.5–1.5 copper, 1.5–5.0 iron, 0.0–0.5 lead, 210.0–430.0 magnesium, 0.1–0.5 manganese, and 5.0–10.0 zinc. Generally, relatively pure, plant tissue culture-tested types of agar should be used, such as Phytagar. Poor quality agar can interfere with or inhibit the growth of cultures.

Agarose is often used when culturing protoplasts or single cells. Agarose is a purified extract of agar that leaves behind agaropectin and sulfate groups. Since its gel strength is greater, less is used to create a suitable support or suspending medium.

Gellan gums such as Gelrite and Phytagel are alternative gelling agents. They are made from a polysaccharide produced by a bacterium. Rather than being translucent (like agar), they are clear, so it is much easier to detect contamination, but they cannot be heated and gelled again, and the concentration of divalent cations such as calcium and magnesium must be within a restricted range or gelling will not occur.

Mechanical supports such as filter paper bridges or polyethylene rafts do not rely on a gelling agent. They can be used with liquid media, which then circulates better, but keeps the explant at the medium surface so that it remains oxygenated. The types of support systems that have been used are as varied as the imagination and include rock wool, cheesecloth, pieces of foam, and glass beads.

Steps in the Preparation of Tissue Culture Medium

Please view Tissue Culture Medium on the accompanying DVD before attempting to make medium. The video shows the step-by-step process of making growth medium.

The first step in making tissue culture medium is to assemble needed glassware, for example, a 1-L beaker, 1-L volumetric flask, stirring bar, balance, pipettes, and the various stock solutions (Table 2.2). The plethora of units used to measure concentration may be confusing when first encountered, so a description of the most common units and what they mean is given in the following text. Once familiar with these, you can confidently proceed to the section on making stock solutions.

Units of Concentration Clarified

Concentrations of any substance can be given in several ways and are interconvertable (see Procedures 2.1 and 2.2). The following list gives some of the methods of indicating the concentration commonly found in literature on tissue culture:

- Percentage based upon volume % (v/v): Used for coconut milk, tomato juice, and orange juice. For example, if a 100 mL of a 5% (v/v) coconut milk was desired, 5 mL of coconut milk would be diluted to 100 mL with water.
- Percentage based upon weight % (w/v): This is often used to express concentrations of agar or sugar. For example, to make a 1% (w/v) agar solution, dissolve 10 g of agar in 1 L of nutrient medium.
- Molar solution: A mole is the same number of grams as the molecular weight (Avogadro's number [6.02×10^{23}] of molecules), so a 1 molar solution represents 1 mol of the substance in 1 L of solution and 0.01 M represents 0.01 times the molecular weight in 1 L. A millimolar (mM) solution is 0.01 mol/L and a micromolar (μM) solution is 0.000001 mol/L. Substances like plant growth regulators are active at micromolar concentrations. Molar concentration is used to accurately compare relative reactivity among different compounds. For example, a 1 μM concentration of IAA would contain the same number of molecules as a 1 μM concentration of kinetin, although the same could not be said for units based upon weight.
- Milligrams per liter (mg/L): Although not an accurate means of comparing substances molecularly, this is simpler to calculate and use, since it is a direct weight. 1 mg is 10^{-3} g. Such direct measurement is commonly used for macronutrients and sometimes with PGRs. 1 mg/L means placing 1 mg of the desired substance in a final volume of 1 L of solution.
- Microgram per liter (μg/L): 1 μg = 0.001 or 10^{-3} mg = 0.000001 or 10^{-6} g. This is used with micronutrients and also sometimes with growth regulators. It means placing 1 μg of substance in 1 L of solution.
- Parts per million (ppm): Sometimes media components are expressed in ppm, which means 1 part per million or 1 mg/L.

Instructions for making media can be found in Procedure 2.3. These instructions describe MS medium preparation, but will work just as effectively for any other media that you choose. Merely follow the same steps and substitute the macro- and micronutrient stocks that you have made for the desired medium. Omit the agar to produce a liquid medium for use in suspension culture. PGRs can

TABLE 2.2
Macro- and Micronutrient 100× Stock Solutions for Murashige and Skoog (MS) Medium (1962)

Stock (100×)	Component	Amount
Nitrate	NH_4NO_3	165.0 g
	KNO_3	190.0 g
Sulfate	$MgSO_4$•$7H_2O$	37.0 g
	$MnSO_4$•$4H_2O$	2.2 g
	$ZnSO_4$•$7H_2O$	0.86 g
	$CuSO_4$•$5H_2O$	2.5 mg[a]
Halide	$CaCl_2$•$2H_2O$	44.0 g
	KI	83.0 mg
	$CoCl_2$•$6H_2O$	2.5 mg[b]
PBMo	KH_2PO_4	17.0 g
	H_3BO_3	620.0 mg
	Na_2MoO_4	25.0 mg
NaFeEDTA[c]	$FeSO_4$•$7H_2O$	2.78 g
	NaEDTA	3.72 g

Note: The number of grams or milligrams indicated in the amount column should be added to 1000 mL of deionized distilled water to make 1 L of the appropriate stock solution. For each liter of the medium made, 10 mL of each stock solution will be used. To make 100× stock solutions for any of the media listed in Table 2.1, multiply the amount of chemical listed in the table by 100 and dissolve in 1 L of deionized, distilled water.

[a] Because this amount is too small to weight conveniently, dissolve 25 mg of $CuSO_4$•$5H_2O$ in 100 mL of deionized distilled water, then add 10 mL of this solution to the sulfate stock.

[b] Because this amount is too small to weight conveniently, dissolve 25 mg of CoC_{12}•$6H_2O$ in 100 mL of deionized distilled water, then add 10 mL of this solution to the halide stock.

[c] Mix the $FeSO_4$•$7H_2O$ and NaEDTA together and heat gently until the solution becomes orange. Store in an amber bottle or protect from light.

be customized for the medium of your choice whether it is intended for initiation of callus, shoots, roots, or some other purpose.

Adjusting the pH of the medium is an essential step. Plant cells in culture prefer a slightly acidic pH, generally between 5.3 and 5.8. When pH values are lower than 4.5 or higher than 7.0, growth and development in vitro generally are greatly inhibited. This is probably due to several factors including PGRs, such as IAA and gibberellic acid becoming less stable, phosphate and ion salts precipitating, vitamin B_1 and pantothenic acid becoming less stable, reduced uptake of ammonium ions, and changes in the consistency of the agar (the agar becomes liquefied at lower pH). Adjusting the pH is the last step before adding and dissolving the agar and then distributing it into the culture vessels and autoclaving. If the pH is not what it should be, it can be adjusted using KOH to raise pH or HCl to lower it (0.1–1.0 N) depending upon if the pH is too low or too high. While NaOH can be used, it can lead to an undesirable increase in sodium ions. The pH of a culture medium generally drops by 0.3 to 0.5 units after it is autoclaved and then changes throughout the period of culture both due to oxidation and the differential uptake and secretion of substances by growing tissues.

Stock Solutions of the Mineral Salts

Mineral salts can be prepared as stock solutions 10 to 1000× the concentration specified in the medium. Mineral salts are often grouped into the following two stock solutions: one for

Procedure 2.1

Converting Molar Solutions to Milligrams/Liters and Milligrams/ Liters to Molar Solutions Using Conversion Factors

- How to determine how many mg/liter are needed for a 1.0 M concentration:

  First, look up the molecular weight of the plant growth regulator. In this example, we will use kinetin. The molecular weight is 215.2, so a 1 molar solution will consist of 215.2 g per liter (L) of solution.

  By using conversion factors and crossing out terms, you cannot go wrong!

$$1\text{ M solution} = \frac{\cancel{1\text{ mol}}}{\text{L}} \times \frac{215.2\text{ g}}{\cancel{1\text{ mol}}} = 215.2\text{ g/L}$$

  To see what a 1.0 mM solution of kinetin would consist of, multiply the grams necessary for a 1 M solution by 10^{-3} which would give you 215.2×10^{-3} g/L or 215.2 mg/L.

$$1\text{ mM solution} = \frac{\cancel{1\text{ mmol}}}{\text{L}} \times \frac{215.2\text{ mg}}{\cancel{1\text{ mmol}}} = 215.2\text{ mg/L}$$

  To see what a 1.0 μM solution of kinetin would consist of, multiply the grams necessary for a 1 molar solution by 10^{-6} which would give you 215.2×10^{-6} g/L or 215.2 μg/L.

$$1\ \mu\text{M solution} = \frac{\cancel{1\ \mu\text{mol}}}{\text{L}} \times \frac{215.2\ \mu\text{g}}{\cancel{1\ \mu\text{mol}}} = 215.2\ \mu\text{g/L or } 0.215\text{ mg/L}$$

- How to determine the molar concentration of a solution that is in mg/L:

  Let us assume you are given a 10 mg/L solution of indolebutyric acid (IBA) and you wish to know its molarity. First, look up the molecular weight of the plant growth regulator. In this example, using IBA, the molecular weight of IBA is 203.2.

  Again, using conversion factors and crossing out terms:

$$\frac{10\ \cancel{\text{mg}}\text{ IBA}}{\text{L}} \times \frac{\cancel{1\text{ g}}}{1000\ \cancel{\text{mg}}} \times \frac{1\text{ mol}}{203.2\ \cancel{\text{g}}} = \frac{0.0000492\text{ mol}}{\text{L}} = 49.2\ \mu\text{M IBA}$$

Note: Remember 1 mol = 10^3 mmol = 10^6 μmol.

macroelements and one for microelements, but unless these are kept relatively dilute (10×), precipitation can occur. In order to produce more concentrated solutions, the preferred method is to group the compounds by the ions they contain such as nitrate, sulfate, halide, P, B, and Mo (phosphorus, boron, molybdenum, respectively) and iron, and make them up as 100× stocks. Table 2.2 lists the stock solutions for MS medium made at 100× final concentration, which means that 10 mL of each stock is used to make 1 L of medium. Some of the stock solutions require extra steps to get the components into solution (e.g., NaFeEDTA stock) or may require making a serial dilution to obtain the amount of a trace component for the stock (sulfate and halide stocks).

Sometimes, the amount of a particular component needed for a tissue culture medium is extremely small, so that it is difficult to weigh out the amount even for the 100× stock solution. Because such small quantities of a substance cannot be weighed accurately, a serial dilution technique is used to achieve the proper molarity or weight of the compound. The following example illustrates how a serial dilutions can be used to obtain the correct amount of a component (in this case $CuSO_4{\bullet}5H_2O$) of the medium for its appropriate stock solution.

The stock solution calls for 2.5 mg of $CuSO_4{\bullet}5H_2O$ as a part of the sulfate stock of MS medium. Make an initial stock solution by placing 25 mg of $CuSO_4{\bullet}5H_2O$ in 100 mL of deionized/distilled water. After mixing thoroughly, use 10 mL of this solution, which will contain the desired 2.5 mg, and place it into the sulfate stock. This procedure required only one serial dilution, but any component can be subjected to one or more dilutions to obtain the desired amount.

TABLE 2.3
Plant Growth Regulators, Their Molecular Weights, Conversions of mg/L Concentrations into μM Equivalents, and Conversion μM Concentrations into mg/L Equivalents

			mg/L Equivalents for These Plant Growth Regulator mg/L Concentrations				μM Equivalents for These μM Concentrations			
	Abbreviation	**M.W.**	**0.1**	**1.0**	**10.0**	**100.0**	**0.1**	**0.5**	**1.0**	**10.0**
Abscisic acid	ABA	264.3	0.0264	0.264	2.64	26.4	0.38	1.89	2.78	37.8
Benzyladenine	BA	225.2	0.0225	0.225	2.25	22.5	0.44	2.22	4.44	44.4
Dihydrozeatin	2hZ	220.3	0.0220	0.220	2.20	22.0	0.45	2.27	4.53	45.3
Gibberellic acid	GA3	346.4	0.0346	0.346	3.46	34.6	0.29	1.44	2.89	28.9
Indoleacetic acid	IAA	175.2	0.0175	0.175	1.75	17.5	0.57	2.85	5.71	57.1
Indolebutyric acid	IBA	203.2	0.0203	0.203	2.03	20.3	0.49	2.46	4.90	49.0
Potassium salt of IBA	K-IBA	241.3	0.0241	0.241	2.41	24.1	0.41	2.07	4.14	41.4
Kinetin	Kin	215.2	0.0215	0.215	2.15	21.5	0.46	2.32	4.65	46.7
Naphthaleneacetic acid	NAA	186.2	0.0186	0.186	1.86	18.6	0.54	2.69	5.37	53.7
Picloram	Pic	241.5	0.0242	0.242	2.42	24.2	0.41	2.07	4.14	41.4
Thidiazuron	TDZ	220.3	0.0220	0.220	2.20	22.0	0.45	2.27	4.54	45.4
Zeatin	Zea	219.2	0.0219	0.219	2.19	21.9	0.46	2.28	4.56	45.6
2-Isopentenyl adenine	2-ip	203.3	0.0203	0.203	2.03	20.3	0.49	2.46	4.92	49.2
2,4-Dichlorophenoxy-acetic acid	2,4-D	221.04	0.0221	0.221	2.21	22.1	0.45	2.26	4.52	45.2

Once a stock solution is made, it should be labeled as in the example below to avoid error and inadvertently keeping a stock solution too long.

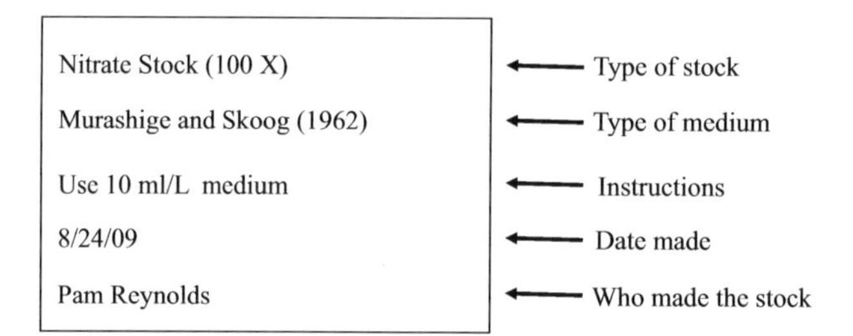

Making Stock Solutions of the PGRs and Vitamins

Vitamins and PGR stock solutions can be made up in concentrations 100–1000 X of that required in the medium. Determine the desired amount for 1 L of medium, the volume of stock solution needed to deliver that dosage of vitamin or PGR, and the volume of stock you wish to make. Procedure 2.2 gives examples of how to make vitamin and PGR stock solutions. Many of the PGRs require special handling to get them into solution.

Sterilizing Equipment and Media

Tissue culture media, in addition to providing an ideal medium for growth of plant cells, also are an ideal substrate for growth of bacteria and fungi. So, it is necessary to sterilize the media, culture vessels, tools, instruments, and to surface disinfest the explants as well. The most commonly used

Procedure 2.2
Making Stock Solutions for Vitamins and Plant Growth Regulators

Step	Instructions and Comments
1	To make a stock solution for nicotinic acid, look at how much is required for 1 L of medium. In this example, we are going to assume we are making MS medium, so we will need 0.5 mg.
2	It would be convenient to be able to add the 0.5 mg of nicotinic acid by adding a volume of the nicotinic acid stock that corresponded to 1 mL. If 0.5 mg of nicotinic acid must be in 1 mL of the stock solution, then (multiplying by 100) use 50 mg for the 100 mL of stock solution. 1 mL may be dispensed into 1.5 mL eppendorf tubes and frozen (–20°C).
3	Now prepare a PGR stock—IBA for a rooting medium. If 1.0 mg/L of IBA is required for rooting medium, then our solution must contain 1 mg in each milliliter of the IBA stock. We must weigh 100 mg of IBA for the 100 mL of IBA stock solution. IBA is not very soluble in water, so first dissolve it in a small amount (~1 mL) of a solvent such as 95% ethanol, 100% propanol, or 1N KOH. Swirl it to dissolve, and then slowly add the remainder of the water to a final volume of 100 mL. Label the stock solution and add 1 mL for every liter of medium to be made. Store in a brown bottle in the refrigerator (4°C). Note: Many growth regulators need special handling to dissolve. Indoleacetic acid, indolebutyric acid, naphthaleneacetic acid, 2,4-D, benzyladenine, 2-iP, and zeatin can be dissolved in either 95% ethanol, 100% propanol, or 1 N KOH. Kinetin and ABA are best dissolved in 1 N KOH, but thidiazuron will not dissolve in either alcohol or base, so a small amount of dimethylsulfoxide must be used. By using only 1 mL of each stock, only very small amounts of the solvent are added to the medium and will minimize toxic effects.
4	Now let us make a stock solution of IBA, but this time we want to have a 5 μM solution of IBA in the medium. Look up the molecular weight of IBA in Table 2.3 and find that it is 203.2. Using the same rationale in the preceding two examples, we wish to have 5 μmol delivered to the medium by using 1 mL of IBA stock solution. So to make 100 mL of stock solution, you must have 500 μmol or 0.5 mmol (mM) in the 100 mL of stock solution. If 1 mol is 203.2 g, 1 mmol is 203.2 mg, so 0.5 mmol IBA × 203.2 mg/1 mmol = 101.6 mg of IBA needed for 100 mL of IBA stock solution. To prepare the solution, weigh out the IBA, dissolve it in 1 N KOH as described earlier, and label the stock solution and store in the refrigerator (4°C) in a brown bottle. Add 1 mL of the IBA stock solution to deliver 5 μmol for every liter of the medium.

means of sterilizing equipment and media is by autoclaving at 121°C with a pressure of 15 psi for 15 min or longer for large volumes. Glassware and instruments are usually wrapped in heavy-duty aluminum foil or put in autoclave bags. Media (even those that contain agar) are in a liquid form in the autoclave, requiring a slow exhaust cycle to prevent the media from boiling over (superheating) when pressure is reduced. Larger volumes of media require longer autoclave times. Media should be sterilized in tissue culture vessels with some kind of closure such as caps or plugs made of nonabsorbant cotton covered by aluminum foil. This way they do not become contaminated when they are removed from the autoclave for cooling. Use of racks that tilt the tissue culture tubes during cooling can give a slanted surface to the agar medium. These can be purchased or made from scratch using a little ingenuity.

Some components of the medium may be heat labile or altered by the heat so that they become inactive. These are usually added to the medium after it has been autoclaved but before the medium has solidified. It is filtered through a bacteria-proof membrane (0.22 μm) filter and added to the sterilized medium after it has cooled enough not to harm the heat-labile compound (less than 60°C) and then thoroughly mixed before distributing it into the culture vessels. A rule of thumb is to add the filtered material at a point when the culture flask is just cool enough to be handled without burning one's hands. Some filters are available presterilized and fit on the end of a syringe for volumes ranging from 1 to 200 mL. More elaborate disposable assemblies are also available.

Preparing the Sterile Transfer Hood

Successfully transferring explants to the sterile tissue culture medium is done in a laminar flow transfer hood. A transfer hood is equipped with positive-pressure ventilation and a bacteria-proof high-efficiency particulate air (HEPA) filter. Laminar flow hoods come in two basic types. Unlike Class I hoods that protect the operator and the environment by drawing in air through the access opening and venting it through a HEPA filter, Class II hoods used for tissue culture operations provide protection for the work area, and tissue culture materials as well. Generally, these hoods have air forced into the cabinet through a dust filter and a HEPA filter, and then it is directed either downward (vertical flow unit) or outward (horizontal flow unit) over the working surface at a uniform rate. The constant flow of bacteria and fungal spore-free filtered air prevents nonfiltered air and particles from settling on the working area that must be kept clean and disinfected. The simplest transfer cabinet is an enclosed plastic box or shield with a UV light and no airflow. A glove box can also be used, but both of these low-cost, low-technology options are not convenient for large numbers of transfers.

In the transfer hood, you should have ready some standard tools for use such as a scalpel (with a sharp blade), long-handled forceps, and sometimes a spatula. Occasionally fine-pointed forceps, scissors, razor blades, or cork borers are needed for preparing explants.

Many people prefer doing sterile manipulations on the surface of a presterilized, disposable petri dish. Other alternatives that have worked well are stacks of standard laboratory-grade paper towels. These can be wrapped in aluminum foil and autoclaved. When the top sheet of the stack is used, it can be peeled off and discarded, leaving the clean one beneath it exposed to act as the next working surface. Another alternative is to use ceramic tiles. These can also be wrapped in aluminum foil and autoclaved, but tiles with very slick or very rough surfaces should be avoided.

In Procedure 2.4 is a suggested protocol to follow for preparing the sterile transfer hood. It also contains tips for keeping your work surface clean, eliminating contamination, and avoiding burns when flaming your instruments!

Storage of Culture Media

Once culture medium has been made, distributed into tissue culture vessels, and sterilized by autoclaving, it can be stored for up to 1 month provided it is kept sealed to prevent excessive evaporation

Procedure 2.3
Step-by-step Media Making

Step	Instructions and Comments
1	If you are making 1 L of media, put about half the volume of deionized, distilled water into a beaker and add a stirring bar. Place the beaker on a stirrer so that when you add different components, they will be thoroughly mixed.
2	Add 10 mL of the following stock solutions (details on how to make them is provided in Table 2.2): nitrate, sulfate, halide, PBMo, and NaFeEDTA.
3	Add the appropriate amount of the vitamin stocks. How to make the vitamin and PGR stocks is found in Procedure 2.2. At this time, also add the appropriate amounts of myo-inositol (usually 100 mg) and sucrose (usually 20 to 30 g/L).
4	Add the appropriate volume of the PGR stocks that you plan to use (details on how to make them is also found in Procedure 2.2).
5	Adjust the pH using 0.1 to 1.0 N NaOH or HCl, depending upon how high or low it is. For most media, the pH should be 5.4 to 5.8, and adjust the final volume (1 L) with deionized, distilled water.
6	If you are going to use liquid medium, you can distribute the media into the tissue culture vessels you plan to use and autoclave. If you plan to use solid medium, weigh out and add the agar or other gelling agent to the liquid. If you are going to put it into tissue culture vessels (tubes or boxes), melt the agar by putting it on a heat plate while stirring or use a microwave to melt the agar. You may have to experiment with settings, since microwaves differ in output. Once the media is distributed into the tissue culture vessels, these are capped and placed in the autoclave for sterilization. Autoclave for 15 min at 121°C. If a large quantity is being autoclaved, longer times may be necessary. If you plan to distribute the medium into tissue culture vessels after autoclaving (such as into sterile, disposable petri dishes), cap the vessel with a cotton or foam plug and cover the plug with aluminum foil. When the medium is cool enough to handle (about 60°C), it can be moved to the sterile transfer hood, uncapped, and poured into dishes.

Procedure 2.4
Getting Under the Hood

Step	Instructions and Comments
1	Turn on the transfer hood so that positive air pressure is maintained. This ensures that all of the air passing over the work surface is sterile. You should feel air flowing against your face at the opening. Make sure there are no drafts such as open windows or air conditioning vents that may interfere with the air flow coming out of the hood.
2	Use a spray bottle filled with 70% (v/v) ethanol and spray down the interior of the hood. Do not spray the HEPA filter! This is more effective than absolute alcohol for sterilizing surfaces perhaps because 70% (v/v) ethanol denatures DNA. You can also use a piece of cheesecloth saturated with 70% (v/v) ethanol to help to distribute the ethanol more uniformly. Allow it to dry. To maintain the cleanliness of the interior, anything that is now placed inside the hood must be sprayed with 70% (v/v). This includes the alcohol lamp, the slide staining jar filled with 80 to 95% (v/v) ethanol for flaming the instruments, a rack for the tools, racks of the tissue culture vessels containing medium, and all of the presterilized wrapped bundles containing your working surface (tiles, paper towels, ashtrays, etc.) and your tools (forceps, spatula, scalpel, etc.).
3	When you are ready to begin, remove jewelry and wash your hands thoroughly with soap and water and, just before placing them in the hood, spray them down with 70% ethanol.
4	You may now open your work surface by peeling back the heavy duty aluminum foil exposing the surface of the tile (or other alternative). Never block the air flow across the surface coming from the filter unit. Also do not pass your hands across the surface of the tile. Talking while you are in the unit also compromises sterility. If you must talk, turn your head to one side. If you have long hair, fasten it so that it does not dangle onto your work surface.
5	Keep any open sterile containers as far back in the hood as you can. When you open containers that have been sterilized, keep your fingers away from the opening. If you open a glass container or vessel, as general rule, pass the opening through the flame. This creates a warm updraft from the vessel helping to prevent contamination from entering it.
6	Flaming instruments can be hazardous if you forget that ethanol is flammable! When you are flame sterilizing an instrument, for example, forceps, dip them into the jar containing the 80% ethanol and, when you lift them out, keep the tip of the instrument at an angle downward (away from your fingers) so that any excess alcohol does not run onto your hand (Figure 2.1). Then pass the tip through the flame of the alcohol burner and hold the instrument parallel to the work surface. When the flame has consumed the alcohol, let the tool cool on the rack until you are ready to use it. Never place a hot tool back into the jar containing 80% ethanol! I know of an experienced scientist who momentarily forgot this simple rule and the resulting fire singed his hand and the hair off his forearm!

of water from the medium. It should also be placed in a dark, cool place to minimize degradation of light labile components such as IAA. Storage at 4°C prolongs the time that media can be stored, but condensate may form in the container and encourage contamination. By making media 5–7 days in advance, you allow time to check for any unwanted microbial contamination before explants are transferred onto the medium. However, media for certain sensitive species or operations, or that contain particularly unstable ingredients, must be used fresh and cannot be stored.

Surface Disinfesting Plant Tissues

Just as the media, instruments and tools must be sterilized, so must the plant tissue be surface disinfested before it is placed on culture medium. Many different materials have been used to surface disinfest explants, but the most commonly used are 0.5–1% (v/v) sodium hypochlorite (commercial bleach contains 5% sodium hypochlorite), 70% alcohol, or 10% hydrogen peroxide. Others include using a 7% saturated solution of calcium hypochlorite, 1% solution of bromine water, 0.2% mercuric chloride solution, and 1% silver nitrate solution. If these more rigorous techniques are used, precautions should be taken to minimize health and safety risks. Touching or breathing bromine can cause blisters on skin and mucous membranes, and even death. Mercuric chloride, in addition to being problematic as a heavy metal for disposal, is highly poisonous and can be fatal if swallowed. Silver nitrate stains clothing, utensils, and skin, reacting with proteins and nucleic acids.

The type of disinfectant used, the concentration, and the amount of exposure time (1–30 min) vary depending on the sensitivity of the tissue and how difficult it is to disinfest. Woody or field-grown plants are sometimes very difficult to disinfest and may benefit from being placed in a beaker with cheesecloth over the top and placed under running water overnight. In some cases, employing a two step protocol (70% ethanol followed by bleach) or adding a wetting agent such as Tween 20 or detergent helps to increase the effectiveness. In any case, the final step before trimming the explant and placing onto sterile medium is to rinse it several times in sterile, distilled water to eliminate the residue of the disinfesting agent.

FINAL WORD

Tissue culture is much like good cooking. There are simple recipes and then there is "haute cuisine." Cooking is also a very rewarding activity particularly when the end result is delicious. By following the procedures outlined in this chapter and in the succeeding chapters of the book, you should find that with care and attention to detail, you will be a "chef extraordinaire" and your tissue culture ventures will be successful!

REFERENCES

Ammirato, P. V. 1974. The effects of abscisic acid on the development of somatic embryos from cells of caraway (*Carum carvi* L.). *Bot. Gaz.* 135:328–337.

Coffin, R., C. D. Taper, and C. Chong. 1976. Sorbitol and sucrose as carbon source for callus culture of some species of the Rosaceae. *Can. J. Bot.* 54:547–551.

Epstein, E., K.H. Chen, and J.D. Cohen. 1989. Identification of indole-3-butyric acid as an endogenous constituent of maize kernels and leaves. *Plant Growth Regul.* 8:215–253.

Gamborg, O. L., R. A. Miller, and K. Ojima. 1968. Nutrient requirements of suspension cultures of soybean root cells. *Exp. Cell Res.* 50:151–158.

Kochba, J., P. Spiegel-Roy, H. Neumann, and S. Saad. 1978. Stimulation of embryogenesis in citrus ovular callus by ABA, Ethephon, CCC and Alar and its supression by GA_2. *Z. Pflanzenphysiol.* 89:427–432.

Linsmaier, E.M. and F. Skoog. 1965. Organic growth factor requirements of tobacco tissue cultures. *Physiol. Plant.* 18:100–127.

Lloyd, G. and B. McCown. 1980. Commercially feasible micropropagation of mountain laurel, *Kalmia latifolia*, by use of shoot tip culture. *Intl. Plant Prop. Soc. Proc.* 30:421–427.

Murashige, T. and F. Skoog. 1962. A revised medium for rapid growth and bio-assays with tobacco tissue cultures. *Physiol. Plant.* 15:473–497.

Nissen, S. J. and E. G. Sutter. 1990. Stability of IAA and IBA in nutrient medium to several tissue culture procedures. *HortScience* 25:800–802.

Nitsch, J. P. and C. Nitsch. 1969. Haploid plants from pollen grains. *Science* 163:85–87.

Oka, S. and K. Ohyama.1982. Sugar utilization in mulberry (*Morus alba* L.) bud culture. In Plant Tissue Culture. (A. Fujiwara Ed.) *Proc. 5th Int. Cong. Plant Tiss. Cell Cult., Jap. Assoc. Plant Tissue Culture*, Tokyo, Japan, pp. 67–68.

Owen, H. R. and A. R. Miller. 1992. An examination and correction of plant tissue culture basal medium formulations. *Plant Cell, Tissue, and Organ Culture* 28:147–150.

Roberts, D.R., B.S. Flinn, D.T. Webb, F.B. Webster, and B.C.S. Sutton. 1990. Abscisic acid and indole-3-butyric acid regulation of maturation and accumulation of storage proteins in somatic embryos of interior spruce. *Physiol. Plant.* 78:355–360.

Schenk, R. V. and A. C. Hildebrandt. 1972. Medium and techniques for induction and growth of monocotyledonous plant cell cultures. *Can. J. Bot.* 50:199–204.

Singha, S. 1982. Influence of agar concentration on in vitro shoot proliferation of *Malus* sp. 'Almey' and *Pyrus communis* 'Seckel'. *J. Amer. Soc. Hort. Sci.* 107:657–660.

Skoog, F. and C. O. Miller. 1957. Chemical regulation of growth and organ formation in plant tissues cultured in vitro. *Symp. Soc. Exp. Biol.* 11:118–131.

White, P. R. 1963. *The Cultivation of Plant and Animal Cells*. 2nd Ed. Ronald Press Co., New York.

3 Nutrition of Cell and Organ Cultures

J. D. Caponetti and R. N. Trigiano

Callus is a relatively undifferentiated tissue consisting primarily of parenchymatous cells. Callus tissue can serve as an experimental system to investigate and solve a broad range of basic research problems in plant cytology, physiology, morphology, anatomy, biochemistry, pathology, and genetics. It can also be used to resolve applied research problems in organogenesis and embryogenesis related to the propagation of horticultural and agronomic plants.

Tissues from various organs from many species of plants can be induced to form callus. However, many seemingly unrelated factors may determine the ability of a specific tissue to form callus. Among these chemical factors are mineral nutrition and plant growth regulators (PGRs); environmental factors, such as light, temperature, and humidity; and the genetic constitution or the genotype of the plant. A medium that induces good callus growth in one genotype or species may fail to do so in another closely related plant. The same holds true for a particular combination and concentration of PGRs added to the medium. Also, tissues from different plant species and genotypes can respond differently from one another under various conditions of light, temperature, and humidity. Therefore, extensive empirical research is required to determine which combination and concentration of medium nutritional ingredients, PGRs, and environmental factors are best for maximizing callus production for each species or genotype. An example of such studies are Murashige and Skoog (1962) and Linsmaier and Skoog (1965).

The following laboratory experiments illustrate the technique of obtaining callus from mature tissues of several angiosperm stems and roots, and compare callus growth on a standard medium formulation. Also, callus growth of one species will be contrasted on several media to which ingredients are added in a serial sequence. Specifically, the laboratory experiences will familiarize students with the following concepts: (1) that one medium may be sufficient to support callus growth from some species, but is not able to support growth from others; and (2) that some ingredients in a medium are more vital for callus tissue growth when provided in combination rather than separately. Each of the aforementioned experiments will be detailed in the following text, including procedures and anticipated results. In all cases, growth will be measured by fresh and dry weight of developed callus.

GENERAL CONSIDERATIONS

Source of Plant Material

The tobacco plants (*Nicotiana tabacum* L.) needed for these experiments can be cultivated in a greenhouse from seed obtained from a farmers' co-op store or from a biological supply company. Alternatively, tobacco plants of the appropriate size may be available from other laboratories/greenhouses on campus. Plants that are about 1.5 meters tall would be the appropriate size. The tobacco varieties that produce good results include WI 38, Xanthi, Burley 14, Burley 21, KY 9, KY 17, VA 509, SpG-28, and Hicks. The other plant material needed for these experiments can be purchased from a grocery store produce counter. Depending on what is available at the time of year that these experiments are conducted, good results can be obtained with Irish potato tuber (*Solanum tuberosum* L.), turnip tuber (*Brassica rapa* L.), Jerusalem artichoke (*Helianthus tuberosus* L.) rhizome, sweet potato root (*Ipomoea batatas* L.), carrot root (*Daucus carota* L.), and parsnip root (*Pastinaca sativa* L.).

Experiment 1. Growth Potential of Callus from Several Species of Plants on a Single Complete Medium

An important basic concept in plant tissue culture is that tissue from different species will not grow equally on one specific medium. This experiment was designed to illustrate this concept by comparing the growth of tobacco stem pith on a medium designed for it with the growth of stem and root tissues of several other plants on the same medium. The experiment can best be performed by teams consisting of two students each. The experiment requires 4 to 8 weeks to complete.

Materials

The following laboratory items are needed for each team of students:

- Several tobacco stem sections
- One or more of the plant tissues listed earlier
- Coconut milk
- Several 150 × 25 mm culture tubes with plastic or metallic caps
- Supply of wire and wooden racks
- Number 3 cork borer and glass rod
- 10% solution of commercial bleach to which several drops of dishwashing liquid have been added
- Drying ovens set for 60°C
- Aluminum foil weighing cups

The culture medium is modified from Murashige and Skoog (1962) and Linsmaier and Skoog (1965) and was designed specifically for the culture of tobacco stem pith tissue. Each liter of medium is composed of Murashige and Skoog (1962) major and minor salts (see Chapter 2 for composition), 100 mg of myo-inositol, 0.4 mg of thiamine HCl, 2.0 mg (10.7 μM) of naphthaleneacetic acid (NAA), 0.5 mg (2.3 μM) of kinetin, 30 g (88 mM) of sucrose, and 10 g of agar. Before adding the agar, adjust the pH of the medium to 5.7. Heat to melt the agar and distribute about 20 mL of medium to each of 50, 150 × 25 mm culture tubes. Cover each tube with a plastic or metallic cap, place the tubes in a metal rack, and autoclave for 20 min. For each tissue type used in the experiment, 20–30 tubes should be prepared by each team. If facilities are limited, each team could process one tissue type.

Follow the protocol outlined in Procedure 3.1 to complete this exercise.

Anticipated Results

Since the culture medium used was designed for tobacco, the tobacco explant disks will produce the largest and heaviest calluses when the experiment is terminated. The other plant tissues will produce callus but not to the extent that tobacco disks do.

Questions

- What concept of tissue growth does this experiment demonstrate?
- Why does the experiment work best with tobacco pith tissue?
- Would the experiment work with other explant tissues? Explain.
- Would raising the temperature of the growth chamber to 35°C improve the growth of the tobacco tissues? If so, why? If not, why not?

Procedure 3.1
Growth Potential of Callus from Several Species of Plants on a Single Complete Medium

Step	Instructions and Comments
1	Surface disinfest stems and roots with 10% (v/v) of liquid chlorine bleach solution for 10 min each in large beakers and rinse three times with sterile distilled water.
2	Cut out cylinders of tissue with sterile number 3 cork borer.
3	Express each tissue core from the cork borer into sterile plastic petri dishes with the sterilized metal punch.
4	Cut 5 mm thick sections with a sterile scalpel and place one section in each of about 30 culture tubes containing modified MS medium composed of MS medium as described.
5	Incubate tubes in a randomized complete block design at 25°C with 25 μmol m^{-2} s–[1] of white light.
6	Weigh 30 explant disks of each tissue type in tared aluminum cups and record the fresh weights. Divide the net weight by 30 to obtain the fresh weight of one disk. Place the cups in an oven set at 60°C for several days to obtain dry weights. Divide the net weight by 30 to obtain the dry weight of one disk.
7	When the tobacco explants have produced callus that almost fills the diameter of the tube (about 4–8 weeks), pool uncontaminated calluses of each tissue separately into tared cups. Record the fresh weight and divide by the number of calluses to obtain the fresh weight of one callus. Place the cups in the drying oven for several days to obtain dry weights and then calculate the dry weight of one callus. Don't forget to subtract fresh and dry weights of the explant disks in order to obtain accurate weight gains of calluses.
8	Student teams should share and analyze the data by one of the procedures described in Chapter 5.

Experiment 2. Initiation and Growth of Tobacco Callus on Serially Composed Media

Another important basic concept in plant tissue culture is that most tissues require a combination of specific nutrients and PGRs to produce the appropriate growth response of the tissues. This experiment was designed to illustrate this concept by starting with a medium without nutrients, and then preparing eight other media by adding one category of nutrients at a time to each medium, ending with a medium that contains all the essential nutrients. Two additional media test the effects of coconut milk, which supposedly contains complete nutrition and PGRs for growth of callus from explant tissues of tobacco.

All 11 media are different, but are based on the MS medium described previously and should be prepared as directed in Experiment 1. The 11 media all include distilled water, 0.8% phytagar, and adjusted to pH 5.7. The media are listed as follows:.

1. Water only.
2. MS salts.
3. MS salts and 3% sucrose.
4. MS salts, 3% sucrose, and 0.4 mg thiamine HCl.
5. MS salts, 3% sucrose, and 100 mg myo-inositol.
6. MS salts 3% sucrose, 0.4 mg thiamine HCl, 100 mg myo-inositol.
7. MS salts, 3% sucrose, 0.4 mg thiamine HCl, 100 mg myo-inositol, 0.5 mg (2.3 μM) kinetin.
8. MS salts, 3% sucrose, 0.4 thiamine HCl, 100 mg myo-inositol, 2.0 mg (10.7 μM) NAA.
9. MS salts, 3% sucrose, 0.4 mg thiamine HCl, 100 mg myo-inositol, 0.5 mg kinetin, and 2.0 mg NAA.
10. 20% coconut milk (200 mL of coconut milk and 800 mL of water). However, the amount of nutrients in medium 10 may not be sufficient to produce the full potential of callus growth. Therefore, it is necessary to test the performance of tobacco pith on a medium 11 that consists of all the ingredients in medium #9 and 20% coconut milk. Coconut milk can be obtained from fresh nuts, as a canned concentrate from a grocery store or Sigma.

Follow Procedure 3.2 as a guideline for this experiment.

Procedure 3.2
Initiation and Growth of Tobacco Callus on Serially Composed Media

Step	Instructions and Comments
1	Surface disinfest 6 cm stem segments of tobacco with 10% (v/v) of liquid chlorine bleach solution for 10 min in large beakers, and rinse three times with sterile distilled water.
2	Cut out cylinders of tobacco pith tissue with a sterile number 3 cork borer.
3	Express the cylinders of pith tissue from the cork borer into sterile plastic petri dishes with a sterile metal punch.
4	Cut 5 mm thick sections with a sterile scalpel, and place one section in each of about 30 culture tubes each containing one of 11 serially composed media (see list of 11 media).
5	Incubate tubes in a randomized complete block design at 25°C with 25 µmol m^{-2} s^{-1} of white light.
6	When the explants on medium 11 have produced callus that almost fills the diameter of the tube (about 4–8 weeks), pool uncontaminated calluses from each medium separately into tared cups. Record the fresh weights and divide by the number of calluses to obtain the fresh weight of one callus. Place the cups in the drying oven for several days to obtain dry weights and then calculate dry weight of one callus. Each team already has fresh and dry weights of tobacco explant tissues from Experiment 1 (Procedure 3.1, step 7). Don't forget to subtract these fresh and dry weights in order to obtain accurate weight gains of calluses from the separate 11 media.
7	Student teams should share and analyze the data by one of the procedures described in Chapter 5.

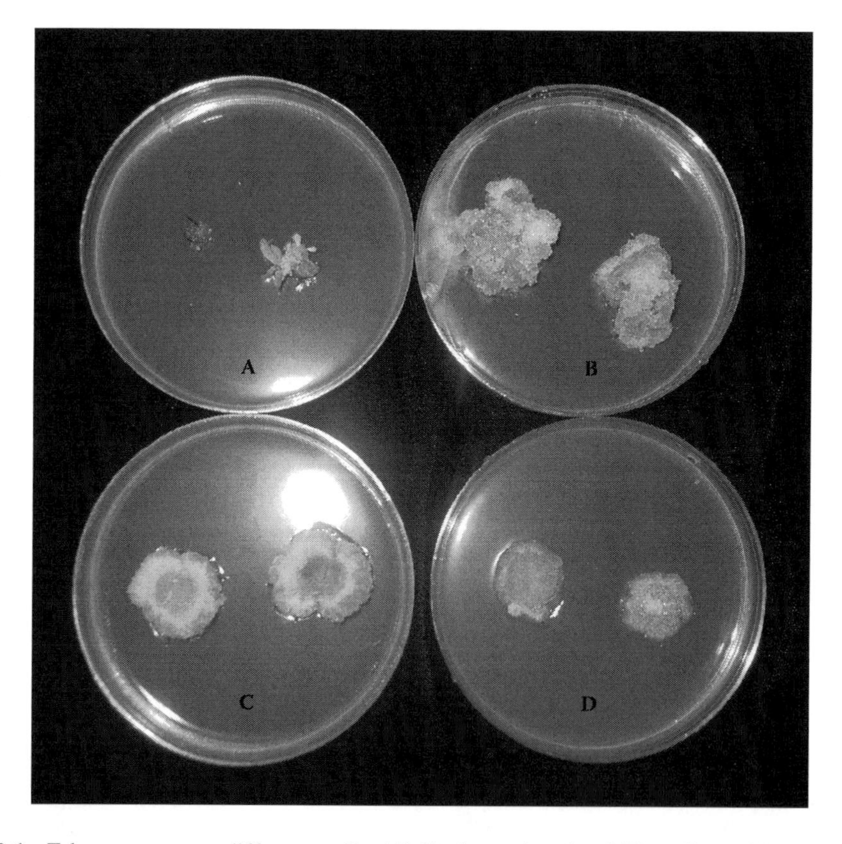

FIGURE 3.1 Tobacco grown on different media. (A). Explants placed on MS medium without growth regulators. Very little callus was formed, but several shoots formed after 3 weeks. (B). Explants placed on MS medium containing both kinetin and NAA. Abundant watery callus was formed after 3 weeks. (C). Explants placed on MS medium containing only NAA. Callus was hard and compact and for the most part lacked green coloration after 3 weeks. (D). Explants placed on MS medium containing only kinetin. Callus was hard and compact, but mostly green after 3 weeks.

Anticipated Results

Growth of callus from the disks on media 1 through 8 will be relatively poor. Growth from disks on medium 9 will be good because all essential nutrients are present. Callus growth on medium 11 will be best because it contains all the nutrients of medium 9 plus the nutrients from coconut milk. Growth of callus on medium 10 will occur but will be less than that on media 9 or 11. Plants placed on media with or without growth regulators may also differentiate shoots and roots (Figure 3.1).

Questions

- What is the function of each ingredient of the medium in the growth of callus?
- Why are fresh and dry weights recorded for calluses of each plant tissue?
- Why is it important that all explant disks be of uniform size?

REFERENCES

Linsmaier, E. M. and F. Skoog. 1965. Organic growth factor requirements of tobacco tissue cultures. *Physiol. Plant.* 18:100–127.

Murashige, T. and F. Skoog. 1962. A revised medium for rapid growth and bioassays with tobacco tissue cultures. *Physiol. Plant.* 15:473–497.

4 PGRs and Their Use in Micropropagation

Caula A. Beyl

CONCEPTS

- Plant Growth Regulators (PGRs) are organic molecules that profoundly affect different plant processes such as growth and morphogenesis in small concentrations. Those that are naturally occurring are called phytohormones or hormones, and those that are synthetic are often more resistant to enzymatic breakdown and thus exhibit heightened effectiveness.
- The classical five PGRs are auxins, cytokinins, gibberellins, abscisic acid, and ethylene, although the first two—auxins and cytokinins—are by far the most important to micropropagators.
- Additional classes of PGRs have been discovered recently, including polyamines, jasmonates, salicyclic acid, and brassinosteroids.
- The relative amounts of auxins to cytokinins are important in determining whether cultures develop shoots (high cytokinin-to-auxin ratio), roots (low cytokinin-to-auxin ratio), or callus (relatively high levels of both).
- Auxins, usually 1-naphthalene acetic acid (NAA) and indole-3-butyric acid (IBA), are used to induce rhizogenesis on proliferated shoots or microcuttings.
- Synthetic herbicidal auxins, such as 2,4-dichlorophenoxyacetic acid (2,4-D) and dicamba (DIC), are commonly used to induce somatic embryogenesis, and ABA is often used to enhance embryo maturation.

Growth regulators are those substances that are capable of exerting large effects in plants in terms of gene expression, growth, and development even though they occur in relatively low concentrations (~0.01 mg/L). Endogenous PGRs (PGRs), those that the plant biosynthesizes for itself, are called hormones. The word hormone is derived from the Greek *hormon*, which means that which sets in motion. These plant hormones or phytohormones control processes such as cell division and enlargement, root and bud initiation, dormancy, flowering, and ripening. Horticulturists make use of the unique features of synthetic growth regulators and apply them exogenously (originating outside the plant) to generate desired responses in the plant. When a plant is developing from a germinating seed, endogenous growth regulators or hormones are responsible for directing the development of shoots and roots via cell division and enlargement. In tissue culture, we make use of the ability to direct growth using growth regulators to stimulate the development of nonzygotic embryos, growth and development of callus, proliferation of axillary shoots, and development of adventitious roots. This has enabled application of tissue culture techniques to fundamental studies of how cells function, growth and development of various tissues of the plant, commercial clonal plant propagation, germplasm storage, development of pathogen-free plants, use of bioreactors to produce secondary plant products, genetic transformation, and even the production of artificial seed by means of somatic embryogenesis (Figure 4.1).

Traditionally, the following five categories of hormones have been classified: auxins, gibberellins, cytokinins, abscisic acid, and ethylene. Each has unique effects, and yet they are capable of working together synergistically or antagonistically to exert other responses within the plant. Some of the newer PGRs that have been discovered and classified are polyamines, jasmonates,

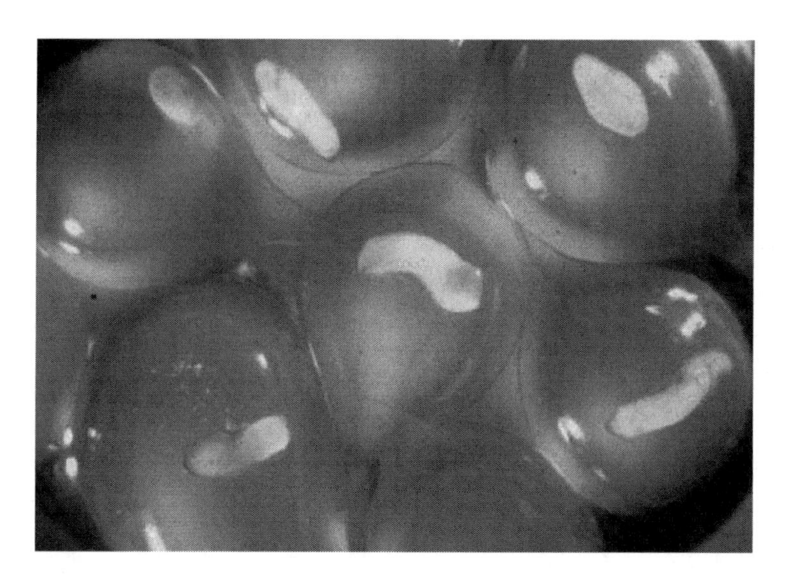

FIGURE 4.1 Chrysanthemum somatic embryos encapsulated in calcium alginate to produce "synthetic seeds" (Courtesy of R.N. Trigiano, University of Tennessee).

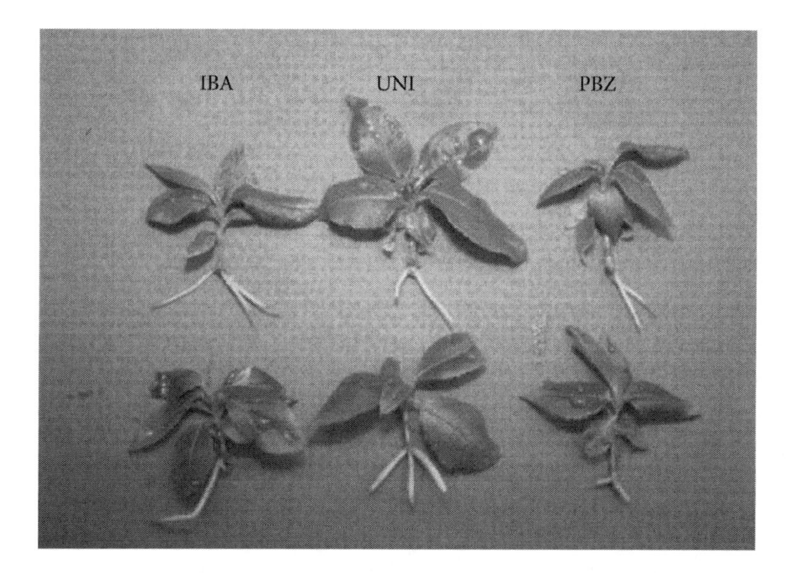

FIGURE 4.2 Appearance of plantlets of *Prunus serotina* (Eliasson and Beyl, unpublished) during acclimatization following culture with IBA, uniconazole (UNI), and paclobutrazol (PBZ). Note the larger leaves and stockier roots of the uniconazole treatment. Both triazoles enhanced plantlet survival.

brassinosteroids, salicylic acid, oligosaccharins, and triazoles. Each of these is discussed further in the following text with respect to the history of its discovery, general effects, and importance in tissue culture of plants.

Cultures of plant tissues may require a certain plant growth regulator or combination in one stage of growth, and then a completely different one or the absence of all PGRs in another. For example, in micropropagation stage I, the initiation phase, a combination of auxins and cytokinins may be needed. In stage II, the multiplication stage, most typically cytokinins are used to stimulate shoot proliferation. In stage III, the root formation stage, most often auxins are used to induce roots. In stage IV, acclimatization, the herbicidal triazoles, paclobutrazol (PBZ), and uniconazole (UNI), have been shown to enhance survival in the case of *Prunus serotina* (Figure 4.2).

Much of the research literature on in vitro culture of plant tissues reports on experimentation to determine what types and concentrations of PGRs will induce specific responses, such as adventitious shoot proliferation or root formation, or embryogenesis. Some species are quite amenable to clonal propagation via tissue culture and require only low concentrations of phytohormones to elicit the desired effects, and others, such as some woody species, require either higher concentrations or inherently stronger phytohormones to obtain results. Sometimes a culture that has previously required a specific growth regulator for growth no longer responds the same way in its presence, and this is called *habituation*. This means that the culture either no longer needs it or cannot use it if it is in the medium. Many different facets of the in vitro environment and the explants affect success in vitro, but the type, concentration, and duration of exposure to PGRs typically have the most profound effect. The choice of PGR to use is predicated on what outcome is desired, the sensitivity of the target tissue, the rates of synthesis and degradation within the explant, interactions with endogenous growth substances, and influence of the culture environment. What is interpreted as a change in sensitivity of the tissue in turn may mean that there is a change in the hormonal receptors in terms of number or receptivity, altered degradation of the phytohormone or its conversion to conjugated inactive forms, or even a change in its localization near the active site due to transport away from the receptors (Gaspar et al., 1996).

AUXINS

The traditional plant growth regulator classes encompass auxins, cytokinins, gibberellins, abscisic acid, and ethylene, with the first two being by far the most important for use in plant tissue culture media. The role that auxins play is consistent with the Greek origin of the word *auxein*, meaning to enlarge or to grow. In general, auxins promote cell enlargement and root initiation. Hints about the existence of a growth-promoting substance were given by Darwin (1880) working with canary grass. He discovered that unilateral light would cause the tip to bend in the direction of the light and that this would not occur if the tip were cut off or were covered with an opaque cap. Went (1926) later confirmed the role of a growth-regulating substance when he allowed excised tips of oat seedlings to diffuse into agar blocks, and these blocks, when placed on one side of the decapitated oat seedlings, would restore the ability of the seedling to bend in the direction away from the agar block. In the whole plant, auxins play a role in phototropism and geotropism, apical dominance, root induction, and wounding responses. Commercially, auxins are used to induce parthenocarpy to prevent fruit abscission, stimulate root formation on cuttings, and, in the case of stronger synthetic auxins, to serve as herbicides.

Auxins are characterized as having an aromatic ring separated from a carboxyl group. Indoleacetic acid (IAA) is the most commonly found naturally occurring auxin, but its use in tissue culture is limited by its sensitivity to light and its tendency to become oxidized in the media, metabolized by the tissues of the plant, or broken down by microorganisms. Other auxins shown to be naturally occurring are phenylacetic acid (PAA) and indole-3-butyric acid (IBA), most often used to induce root formation. Synthetic auxins such as naphthalene acetic acid (NAA) and strong auxins such as 2,4-dichlorophenoxyacetic acid (2,4-D), dicamba (DIC), and picloram (PIC), used horticulturally as herbicides against dicotyledonous weeds, are commonly used in tissue culture to induce callus growth and are not as labile as IAA. The tendency of IAA to be labile in culture has been used successfully for apple microcuttings, which are induced to form roots at high IAA concentrations, but growth of roots is promoted by the lower IAA concentrations that result after time (Guan et al., 1997).

The structures of some of the most common auxins used in tissue culture are shown in Figure 4.3. These have in common an aromatic or indole ring a certain distance from a carboxyl group. Within the plant, the concentration of endogenous IAA is regulated by the rate of biosynthesis, the rate of oxidation by the enzyme IAA oxidase, and the formation of conjugates with sugars and amino acids. When in the conjugated form, IAA cannot be broken down by IAA oxidase, and is stored until released again to the free active form. What concentration of auxins to use in tissue culture is

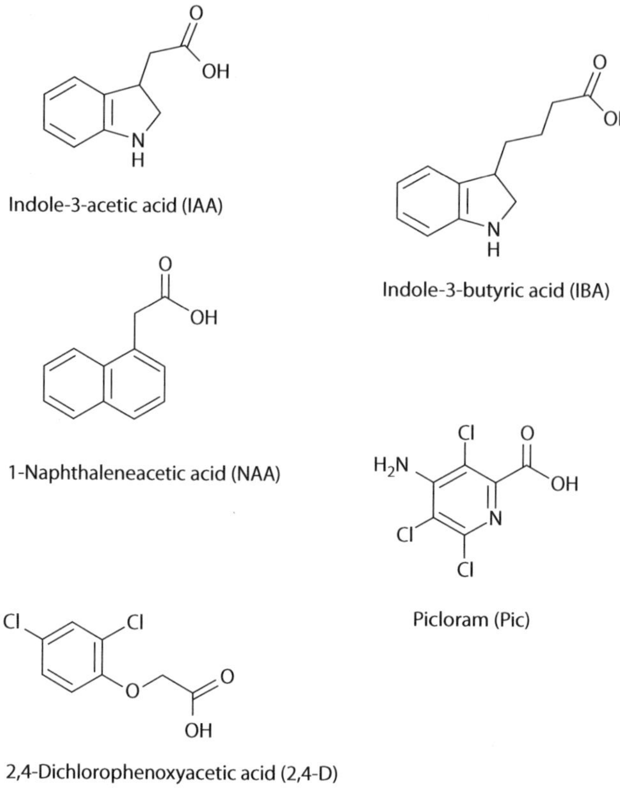

Indole-3-acetic acid (IAA)

Indole-3-butyric acid (IBA)

1-Naphthaleneacetic acid (NAA)

Picloram (Pic)

2,4-Dichlorophenoxyacetic acid (2,4-D)

FIGURE 4.3 Natural and synthetic auxins most commonly used in tissue culture, including indole-3-acetic acid (IAA), indole-3-butyric acid (IBA), 1-naphthaleneacetic acid (NAA), 2,4-dichlorophenoxyacetic acid (2,4-D), and 4-amino-3,5,6-trichloropicolinic acid or picloram (PIC).

highly dependent upon the desired response and the relative strength of the auxin. PAA and IAA are relatively weak auxins, with 2,4-D and NAA being stronger. Even stronger still are DIC and PIC.

In micropropagation, auxins are used to induce callus and, combined with cytokinins, promote growth of callus, cell suspensions, and, in high concentrations relative to cytokinins, promote roots. For woody plants, callus induction occurs with 2,4-D at concentrations between 5–15 μM often with the addition of a cytokinin. Monocots require higher concentrations of 2,4-D generally in the range of 10–50 μM (Machakova et al., 2008).

Phenolic compounds, those with one or more hydroxyl groups on an aromatic ring, are often reported as enhancing callus growth, adventitious shoot formation, shoot proliferation, and even rooting, but these effects are usually synergistic with auxins and, in particular, IAA (Machakova et al., 2008). These effects have been attributed to a possible protection of auxin by oxidative breakdown by the phenols acting as alternative substrates, or actual induction of morphogenesis by the breakdown products of the phenols.

CYTOKININS

Unlike auxins whose discovery occurred through observation on tropistic effects, cytokinins were discovered for their ability to promote cell growth in vitro. An excellent narration of the events and persons involved in the discovery of cytokinins can be found in Amasino (2005). In Folke Skoog's laboratory, research was under way looking at chemical control of new shoot formation on excised tobacco stem pieces; he called the newly formed shoots "buddies" (Amasino, 2005). Variable results

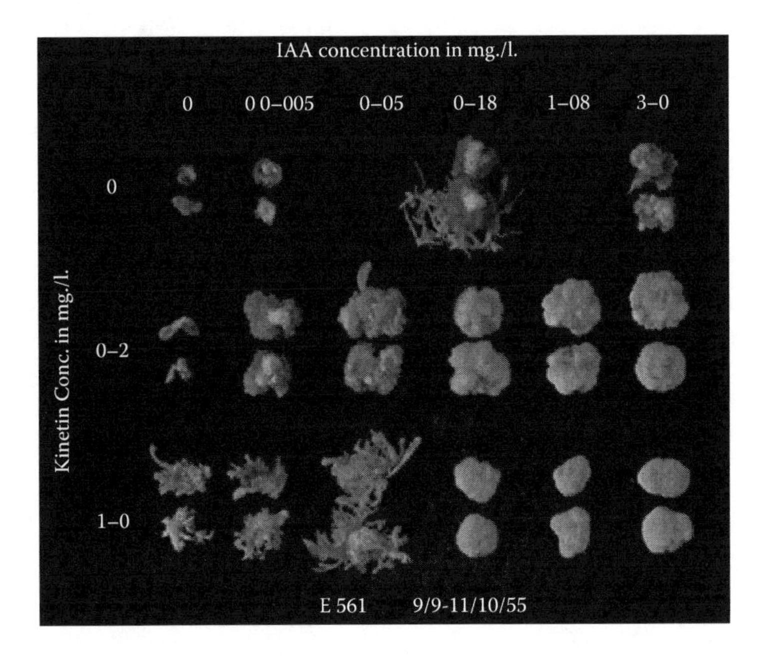

FIGURE 4.4 Classic paper of Skoog and Miller (1957) showing the interaction between various concentrations of IAA and Kin in promoting shoot formation, root formation, or callus. (From Skoog F. and C. O. Miller. 1957. *Symp Soc Exp Biol* XI:118–131. With permission.)

with tobacco stem pieces from the greenhouse led him to seek an alternative system using tobacco pith parenchyma cells. He found that they could be induced to enlarge with auxin but did not undergo cell division. Carlos Miller joined Skoog's lab as a postdoc and found that, when he supplied yeast extract, he was able to get cell division. These materials contained purines, and when he tested purine-containing herring sperm DNA, he found that aged or autoclaved herring sperm DNA stimulated cell division, whereas fresh materials did not. The active component, isolated and purified, was named "kinetin" (Miller et al., 1955). The relationship between auxins and cytokinins in regulating the formation of either shoots or roots from callus was reported in a classic publication by Skoog and Miller (1957). Interaction between various concentrations of IAA and Kinetin (KIN) are shown in Figure 4.4. If the ratio of auxin to cytokinin is high, root formation is favored, and if the ratio of cytokinin to auxin is high, shoot formation occurs. High concentrations of both promote callus formation.

In the whole plant, cytokinins play a role in a variety of processes including cell division and differentiation, delay of senescence, development of chloroplasts, resource uptake and allocation, nodulation in leguminous species, vascular development, as well as initiation and development of shoots. In tissue culture systems, they are used to alleviate apical dormancy, allowing shoot proliferation, and to stimulate cell division often in concert with auxins. Cytokinins can be divided into two general groups—adenine-type cytokinins and phenylurea cytokinins. Adenine types include KIN, zeatin (ZEA), and benzyladenine (BA). These can be further subdivided by the nature of the side chain (aromatic or isoprenoid) on the N6 terminal of the adenine moiety. The isoprenoid class includes ZEA and isopentenyladenine (IPA). Aromatic types include BA and meta-topolin (mT). Phenylurea types include cytokinins such as diphenylurea and thidiazuron (TDZ). Cytokinin receptor studies have suggested that different molecular structures affect which receptors will have affinity. In some cases, cytokinin-binding proteins, which are proposed receptors, have better affinity for the aromatic BA than for isoprenoid-type ZEA (Keim et al., 1981), and there may be a clear differentiation in binding, activity, and effects between aromatic and isoprenoid cytokinins. Structures of some of the more common cytokinins used in tissue culture are depicted in Figure 4.5.

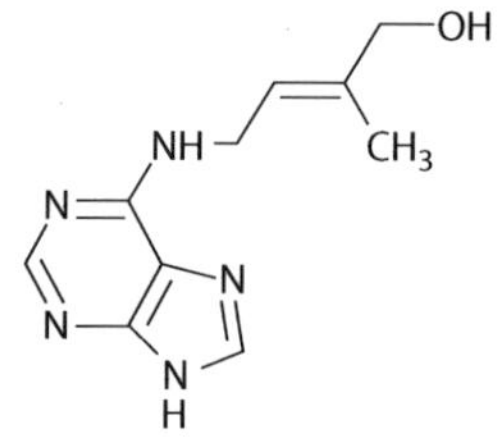

trans-Zeatin
6-(4-hydroxy-3-methylbut-2-enylamino)purine

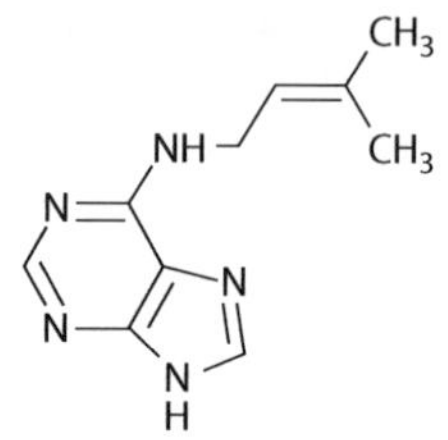

2-iP
N^6-(2-isopentyl)adenine

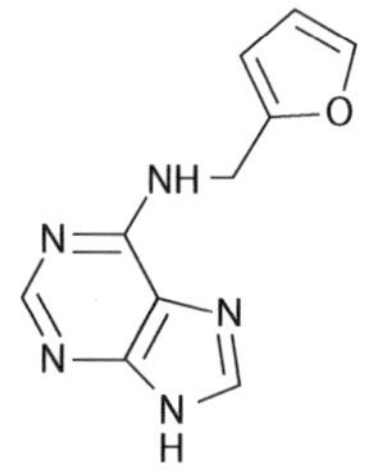

Kinetin (Kin)
6-furfurylaminopurine

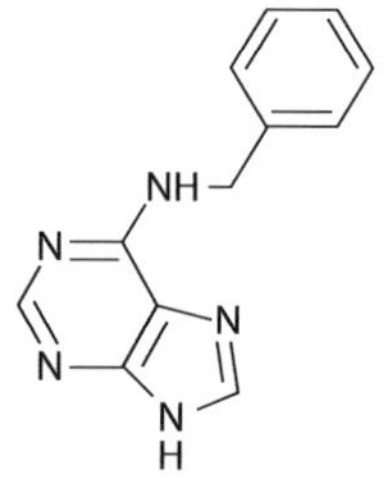

Benzyladenine (BA)
6-benzylaminopurine

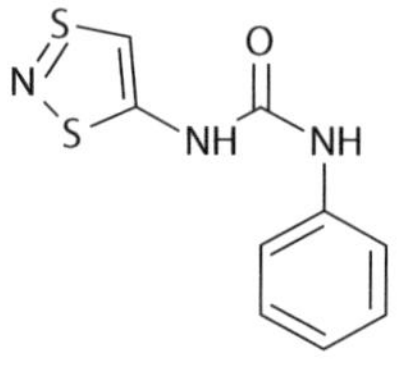

Thidiazuron (TDZ)
N-phenyl-N′-1,2,3-thiadizol-5-ylurea

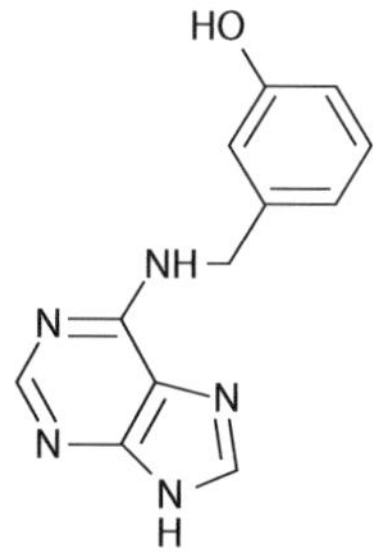

Meta-topolin (MT)
N^6-(*meta*-hydroxybenzyl)adenine

FIGURE 4.5 Cytokinins most commonly used in tissue culture, including trans-zeatin (ZEA), kinetin or 6-furfurylaminopurine (KIN), 6-benzylaminopurine or benzyladenine (BA), thidiazuron (TDZ), and a relative newcomer to the kinetin group, metatopolin (mT).

Probably the two most commonly used cytokinins are BA and KIN. These cytokinins have been proven to be effective in stimulating shoot proliferation and callus formation in a wide range of woody and herbaceous species. Naturally occurring cytokinins commonly used in tissue culture are IPA and ZEA, isolated from *Zea mays* kernals (Letham, 1963). Synthetic cytokinins include 6-furfurylaminopurine or KIN, and BA. The substituted phenylurea, TDZ, originally used as a defoliant for cotton, has proven to be a potent cytokinin exhibiting activity at concentrations as low as 10 pM (Preece et al., 1991) relative to the effective range of amino purine cytokinins (1–10 μM). Although it is different in structure from either the auxins or the more familiar cytokinins, TDZ has effects similar to both cytokinins and auxins (induction of callus, formation of shoots, somatic embryogenesis, androgenesis, and gynogenesis) and has the ability to work alone or with other PGRs to stimulate regeneration in vitro (Murthy et al., 1998). It appears to be particularly valuable for the micropropagation of recalcitrant species and may act via modulation of endogenous hormones (Murthy et al., 1998). Many species have proven responsive to TDZ in terms of embryogenesis, including tobacco, peanut, geranium, chickpea, neem, and St. John's wort (Murthy et al., 1998). TDZ in combination with IPA induced the development of floral structures from stamen explants in *Rhododendron* (Shevade and Preece, 1993).

An aromatic cytokinin isolated from poplar leaves, meta-topolin (mT; Strnad et al., 1997), has been shown to be more active than either ZEA or BA in promoting shoot formation of sugar beet in tissue culture (Kubalakova and Strnad, 1992) and both good shoot production and better rooting of *Spathiphyllum floribundum* (Werbrouck et al., 1996). At concentrations as high as 0.5 mg/L or higher, mT has been shown to stimulate callus formation at the base of the plantlets of potato (Baroja-Fernandez et al., 2002). With *Musa*, mT at 22.2 μM/L inhibited formation of roots but did not cause callus formation at the base of shoots (Escalona et al., 2003); however, the lowest concentration (1.33 μM/L) resulted in longer shoots and more roots than did the corresponding concentration of BA. At 5 mg/L, mT was also the most effective cytokinin tested for in vitro shoot multiplication of Jonagold apple (Magyar-Tabori et al., 2002). For turmeric micropropagation, mT was more effective than BA, resulting in the development of healthy shoots with broad leaves and thin short roots (Salvi et al., 2002).

GIBBERELLINS

Gibberellins in whole plants exert powerful effects on stem elongation and sex expression. They also play a dominant role in dormancy and germination. Commercially, gibberellins are used to increase the size of seedless grapes. Rice seedlings infected with *Gibberella fujikoroi*, the causal fungus of "foolish rice seedling disease" or *bakanae*, first caught the interest of a Japanese scientist (Kurosawa, 1926), and gibberellin was first isolated in 1935 from the fungus (Yabuta and Sumiki, 1938). Gibberellins typically do not play a large role in regulation of in vitro development, although they have been used to enhance shoot elongation before rooting or to stimulate the conversion of buds into shoots (Gaba, 2005), but they may interfere with bud initiation, reduce root formation, and interfere with embryogenesis. GA_3 has been shown to have a positive effect on rooting of Japanese plum in vitro (Rosati et al., 1980). In suspension culture, an auxin-independent green clone of spinach and an auxin-dependent line of rose had enhanced expansion growth by 10^{-11} to 10^{-6} M GA_3 (Fry and Street, 1980).

ABSCISIC ACID

Abscisic acid (ABA) was originally identified by Addicott and his colleagues studying abscission in cotton, and one of the compounds isolated was called "abscisin II." Wareing, studying bud dormancy, discovered the same compound and called it "dormin." The name abscisic acid was the compromise name settled on for the phytohormone (Salisbury and Ross, 1985). In whole plants, ABA plays a role in bud and seed dormancy, stomatal control, leaf abscission, and senescence. In tissue culture systems, ABA is generally positive in effect at relatively low concentrations, but high concentrations inhibit both callus growth and formation of buds, roots, and embryos (Gaspar et al., 1996). ABA is most often used to enhance the maturation and normal growth of nonzygotic (somatic) embryos. In the pursuit of embryogenesis in cereals such as wheat, enhanced embryogenesis occurred when low concentrations of abscisic acid were included in the medium (Brown et al., 1989). When germplasm preservation is desired, proliferation of shoots can be prevented by the addition of abscisic acid (Singha and Powell, 1978). For cell cultures, ABA has been used to increase freezing tolerance (Gaspar et al., 1996). ABA has also been useful in pretreating somatic embryos of carrot to increase the survival rate of embryos encapsulated with 5% polyethylene oxide and then dried to a constant weight (Kitto and Janick, 1985).

ETHYLENE

Ethylene in whole plants plays a role in ripening, senescence, abscission, and stress. Plants respond to exposure to ethylene with the classic "triple response," which includes epinasty, swollen stems, and loss of gravitropism. Ethylene, at high-enough concentrations, can inhibit the growth and development of tissues in vitro. This may be exacerbated by some culture vessels that restrict gas exchange. Sensitive species may exhibit reduced stem elongation and swollen stems, restricted leaf growth or crinkly leaves, premature leaf senescence, and breaking of apical dominance (Gaba, 2005). Ethylene may also have a beneficial effect in tissue culture; ethylene in combination with cytokinins stimulated indole alkaloid accumulation in cell suspension cultures of Madagascar periwinkle (Yahia et al., 1998).

OTHER REGULATORS OF PLANT GROWTH USEFUL IN TISSUE CULTURE

Polyamines are organic molecules with two or more amino groups that play a role in many processes in whole plants, including cell division and morphogenesis. They tend to be found in higher concentrations in juvenile tissues rather than mature ones (Rey et al., 1994). Common polyamines, putrescine, spermidine, and spermine enhanced shoot elongation by 83% and number of buds by 41% in cultures of hazelnut (Nas, 2004). The enhancement of morphogenesis by polyamines may be through their abatement of ethylene production (Bais et al., 2000). Putrescine (0.4 mM) caused an increase in protocorm-like bodies in *Dendrobium* (Saiprasad et al., 2004), and spermidine (0.2 mM) was useful in stimulating gametic embryogenesis in anther cultures of *Citrus clementina* (Chiancone et al., 2006). Polyamines may also have a role in the initiation phase of embryogenesis as spermidine increased the production of somatic embryos of *Panax ginseng* (Monteiro et al., 2002), and putrescine enhanced embryogenesis and plant regeneration of upland cotton (Sakhanokho et al., 2005). Both putrescine and spermidine together at 0.5 mM, combined with 2,4-D, stimulated the recalcitrant oat cultivar "Tibor" to produce the same number of mature and immature embryos as did the highly regenerable line GP-1 (Figure 4.6). This is very desirable as Tibor is a high-protein, low-fiber, hullless oat with excellent agronomic characteristics.

Jasmonic acid is a fragrant essential oil, and in whole plants, it and its volatile methyl ester are involved in wound and pathogen responses, formation of stress proteins, and ethylene synthesis (Gaspar et al., 1996). Jasmonates stimulate bulb formation in vitro (Gaspar et al., 1996), retard callus formation, inhibit root formation, and promote bud formation in potato meristems (Ravnikar and Gogala, 1990). Jasmonates inhibit shoot formation, causing fewer and shorter shoots from *Pinus radiata* cotyledons (Gaspar et al., 1996). Jasmonates may play a role in enhancing secondary plant products. Methyl jasmonate increased taxane accumulation in callus culture of *Taxus* x *media* (Furmanowa et al., 1997) and saponin content in bioreactor culture of ginseng (Kim et al., 2003).

Brassinosteroids are steroid-class compounds that promote cell expansion and elongation (Clouse and Sasse, 1998). Their effects on stem elongation and cell division were first discovered in organic extracts of pollen of *Brassica napus* or rapeseed (Grove et al., 1979), and hence their name. Various brassinosteroids have been shown to improve rooting efficiency and survival of Norway spruce (Ronsch et al., 1993), increase cell division of *Petunia* leaf protoplasts (Oh and Clouse, 1998), and substitute for cytokinin in *Arabidopsis* callus and suspension cells (Hu et al., 2000). The brassinosteroid, 24-epibrassinolide was used successfully to initiate protocorm-like bodies and successful regeneration of *Cymbidium elegans* (Malabadi and Nataraja, 2007). Brassinosteroids act synergistically with auxins and, overall, play a role in improving efficiency of regeneration.

Salicylic acid is a phenolic phytohormone involved in plant growth regulation and development (Raskin, 1992), but it is widely known for its role in defense mechanisms against biotic and abiotic stresses. One of the defense mechanisms is systemic acquired resistance (Metraux et al., 1990), which is a whole-plant response to a localized attack by a pathogen and results in the induction of pathogenesis-related proteins (Malamy et al., 1990). In vitro, salicylic acid has been used to enhance regeneration of *Coffea arabica* (Quiroz-Figueroa and Mendez-Zeel, 2001), *Astragalus adsurgens* (Luo et al., 2001), and *Avena nuda* (Hao et al., 2006). Salicylic acid (0.5 mM) had a beneficial effect on meristems of two Hibiscus species exposed to various salt concentrations and improved shoot growth and multiplication, root formation and elongation, plant survival rate, and proline accumulation (Sakhanokho and Kelley, 2009).

Oligosaccharins are cell-wall fragments that are involved with plant responses to disease and insect attack, but may also be involved with other development and morphogenic responses in whole plants. In tissue culture, they can promote callus proliferation in conjunction with auxin, enhance

FIGURE 4.6 Somatic embryogenesis of oat cultivar "Tibor," a high-protein, low-fiber, hullless oat cultivar that is recalcitrant to regeneration (top). Incubation with 9 μM 2,4-D plus 0.5 mM putrescine and 0.5 mM spermidine induced a number of mature and immature embryos equal to the highly regenerable cultivar GP-1 on comparable 2,4-D medium containing spermidine. Proliferation of shoots from Tibor embryogenic cultures in response to medium containing 0.2 mg/L NAA and 0.5 mg/L BA (bottom). (From Kelley, R. Y., A. E. Zipf, D. E. Wesenberg, and G. C. Sharma. 2002. *In Vitro Cell, Dev. Biol.*–Plant 38:508–512. Copyright 2002 by the Society for In Vitro Biology, formerly the Tissue Culture Association. Reproduced with permission of the copyright owner.)

root formation of wheat embryos in the absence of auxin, control tobacco organogenesis, and inhibit auxin-stimulated rooting in tobacco leaf explants and auxin-dependent embryogenesis is carrots (Tran Thanh Van et al., 1985; Tran Thanh Van and Trinh, 1990). Their effects may be due in part to their acting as antiauxins (Gaspar et al., 1996).

Triazoles, such as paclobutrazol and uniconazole, are growth retardants that inhibit the biosynthesis of gibberellins (Sankhla and Davis, 1999). They have proven useful in alleviating some of the desiccation associated with transfer to soil of chrysanthemum (Smith et al., 1990) and grapevine

(Smith et al., 1992). With *P. serotina*, PBZ reduced water loss and enhanced survival of rooted plantlets after transfer to the greenhouse (Eliasson et al., 1994). Another triazole, triadimefon, improved hardening of Stage III plantlets of banana. Treated shoots were turgid and healthy as opposed to control plants that wilted and died (Murali and Duncan, 1995). For endangered species *Symonanthus bancroftii*, PBZ improved tolerance to desiccation after in vitro rooted shoots were transferred to soil and increased plant survival from 50% with no PBZ to 90% (Panaia et al., 2000). Inclusion of PBZ in the rooting medium of *Chrysanthemum* resulted in plants that grew taller and flowered earlier after one month in the greenhouse (Kucharska and Orlikowska, 2008). Triazoles may have their greatest utility in enhancing acclimatization and post-transplant survival of micro-shoots after rooting.

Coconut water (coconut liquid endosperm) was first used as a component of plant tissue culture medium by Overbeek et al. (1941) for its ability to promote growth. In addition to the sugars, sugar alcohols, lipids, amino acids, nitrogenous compounds, organic acids, and various enzymes (Tulecke et al., 1961), coconut water contains several classes of PGRs, including auxin, cytokinins, GAs, and ABA (Ma et al., 2008). Indole-3-acetic acid is the auxin in coconut water, and GA_1 and GA_3 are the gibberellins present. The cytokinin in greatest concentration is trans-zeatin riboside, followed by trans-zeatin *O*-glucoside and dihydrozeatin *O*-glucoside (Ma et al., 2008). Coconut water has been used to promote the formation of protocorm-like bodies in orchid culture (Huan et al., 2004) and in the tissue culture of Chinese medicinal herbs (Yong et al., 2009). The growth stimulation effects of coconut water cannot completely be replaced solely by substituting a definite cytokinin, as the effects may be due to synergy among all of the components of coconut water.

PROCESSES THAT REQUIRE ENDOGENOUSLY SUPPLIED GROWTH REGULATORS' ESTABLISHMENT IN CULTURE AND CALLUS FORMATION

Auxins are the growth regulators most often used to initiate callus from explants, and the most-often used auxin for this is undoubtedly 2,4-D. Once callus is initiated by 2,4-D, maintaining it on the same medium may cause genetic variability, so sometimes callus is transferred to medium containing an alternative auxin such as NAA or IAA (Machakova et al., 2008).

Sometimes when plants are established in tissue culture, the explants will exude brownish substances into the medium, and the explants will often undergo necrosis soon after. These substances typically consist of tannins and phenolics. Inclusion of a low dose (0.1 μm) NAA might help prevent this (Wainwright and Flegmann, 1985), as will prompt reculturing to fresh medium. If callus is desired to subsequently be used for regeneration, use of TDZ as the cytokinin to stimulate adventitious shoot formation may be advantageous under these conditions because TDZ has been reported to reduce the activity of enzymes related to oxidative stress (Tang and Newton, 2005).

SUSPENSION AND PROTOPLAST CULTURE

In cell suspensions, auxins tend to favor cell dispersion, whereas cytokinins tend to favor cell aggregation (Machakova et al., 2008). Higher auxin levels may also promote embryogenesis and tend to inhibit formation of chlorophyll promoted by cytokinins. Maintenance of suspension cultures is usually sustained with 2,4-D, often with cytokinins, but 2,4-D may be replaced by NAA or IBA when the suspension is to be used for subsequent morphogenesis (Machakova et al., 2008).

Usually, the presence of both auxin and cytokinin in the culture is required to stimulate the division and further growth of protoplasts. Protoplasts are plant cells whose cell wall has been removed enzymatically. Protoplasts of different genotypes can be induced to fuse and then develop into normal plants. TDZ was more effective than either BA or ZEA in sustaining division of apple leaf protoplasts (Wallin and Johansson, 1989) and more effective than KIN or ZEA for growth of willow protoplasts (Vahala and Eriksson, 1991).

Axillary Shoot Proliferation and Adventitious Shoot Organogenesis

Adventitious shoot proliferation can occur directly on the explants or on callus proliferated from the explants. Another route to obtain shoot proliferation is from the release of apical dominance and growth of axillary buds or shoots. Cytokinins (most often KIN or BA) are the predominate growth regulators for the promotion of shoot proliferation or the development of adventitious shoots, but sometimes the response can be enhanced by the addition of low concentrations of an auxin or even the addition of another cytokinin. For example, when BA plus IBA or NAA were added to a medium containing TDZ, increased shoot proliferation was obtained with *Robinia*, *Sorbus*, and *Tilia* (Chalupa, 1987). Different species may also have a preference for certain cytokinins and respond to one or more but not all, with respect to shoot culture. For example, with *Castanea*, BA promoted axillary bud proliferation, but KIN was ineffective and ZEA only poorly effective (Vieitez and Vieitez, 1980). Sometimes a short duration of exposure to a cytokinin and auxin followed by incubation without growth regulators is effective as in the case of chrysanthemum exposed to a two-week pulse of BA and IAA (Figure 4.7). Finding the best cytokinin for shoot proliferation may require trial and error.

TDZ has been particularly effective for use with woody plants (Briggs et al., 1988) and at much lower concentrations than those used for other adenine-type cytokinins. TDZ has been more effective than purine-type cytokinins for adventitious shoot regeneration of woody species, including *Actinidia*, *Celtis*, *Cydonia*, *Fraxinus*, *Lonicera*, *Malus*, *Picea*, *Prunus*, *Pseudotsuga*, *Pyrus*, *Rhododendron*, *Rubus*, and *Ulmus* (Lu, 1993). Some disadvantages in its use that impact the complete development of shoots into finished plants include poor elongation and distortion of shoots as in the case of *Hibiscus acetosella* in response to TDZ (0.25 mg/L) and BA (1 mg/L), which only occurred at concentrations of TDZ greater than or equal to 0.25 mg/L (Figure 4.8). In response to 0.1 mg/L TDZ, *H. acetosella* exhibited normal shoot formation (Figure 4.9). TDZ at 1.0 mg/L even induced leaf drop and stunting. Another anomalous consequence of TDZ is the occurrence of fasciation as in the case of silver maple (Figure 4.10; Heutteman and Preece, 1993). Fasciated shoots are flat shoots appearing to consist of several shoots fused together. Shoots proliferated in response to TDZ may also exhibit poor rooting (Huetteman and Preece, 1993), which could be caused by using unnecessarily high concentrations of TDZ or because TDZ tends to persist in cultured tissues (Murthy et al., 1998). The poor elongation of shoots can be ameliorated by transfer to media containing lower concentrations of TDZ or, in some cases, a much less potent cytokinin.

FIGURE 4.7 Multiple adventitious shoots formed directly on leaves of chrysanthemum. Shoot formation was induced by a pulse treatment for two weeks with BA and IAA, followed by incubation without growth regulators. (Courtesy of R.N. Trigiano, University of Tennessee.)

FIGURE 4.8 Multiple shoot formation in a green variant of *Hibiscus acetosella* from meristem explants in response to TDZ (0.25 mg/L) and BA (1 mg/L). At the right, the clump of shoots are separated to better show individual shoot characteristics. At concentrations greater than or equal to 0.25 mg/L TDZ, "distorted" plantlets appear, and with TDZ concentrations of 1 mg/L, stunting and leaf drop also occurs (Courtesy of Hamidou Sakhanokho).

FIGURE 4.9 Normal appearance of multiple shoots of *Hibiscus acetocella* induced in response to TDZ (0.1 mg/L) and BA (1 mg/L). At the right, the clump of shoots are separated to better show individual shoot characteristics. (Courtesy of Hamidou Sakhanokho.)

In another case, the inhibition of shoot elongation in *Liquidambar* in response to TDZ was remedied by culturing on a medium containing NAA and BA in addition to TDZ (Kim et al., 1997). Keeping explants on TDZ-containing medium for less than 8 weeks is also recommended (Lu, 1993). TDZ stimulates both axillary shoot proliferation and adventitious shoot formation, so careful selection of the concentration of TDZ may be necessary to minimize variation (Heutteman and Preece, 1993). Shoots of *Juglans nigra* exhibited fasciation in response to 0.1 μM TDZ on 0.2% Gelrite medium (Figure 4.11 top; Bosela and Michler, 2008), but fasciated shoots were also found in *J. nigra* on 25 μM BA and 2 g/L Gelrite (Figure 4.11 bottom).

Under high concentrations of cytokinins, or particularly with certain cytokinins such as TDZ, hyperhydricity of the tissues occurs. Hyperhydric shoots appear water soaked and translucent with thickened stems, shorter internodes, and elongated malformed and brittle leaves. The glassy appearance of shoots of *Juglans nigra* cultured on medium containing 0.05 μM TDZ and 2 g/L Gelrite is easily seen in Figure 4.12 (Bosela and Michler, 2008). These shoots are not desirable

FIGURE 4.10 Fasciation of microshoots of silver maple on medium containing 10 nM TDZ. (From Heutteman, C.A. and Preece, J.E. 1993. *Plant Cell Tiss. Org. Cult.* 33:105–119. With permission.)

for proliferation because of deficient cell wall deposition, altered cuticle and wax formation, and even apex necrosis. Sometimes, both hyperhydric shoots and normal shoots can be seen in one culture as in the case of shoots of *Populus tremuloides* x P. *tremula*, Cl. 18-59-S-11, in response to 0.01 μM TDZ (Figure 4.13; Bosela, 2009). At one time, the term *vitrification* was used to describe the phenomenon of hyperhydricity, but the latter term was deemed more appropriate (Debergh et al., 1992), and vitrification is now used in conjunction with changes occurring to plant tissues during cryopreservation.

Organogenesis—Adventitious Root Formation

The formation of roots is stimulated by auxins, and the most often used auxins for this purpose are IBA and NAA. Cytokinins tend to inhibit root formation, and some cytokinins such as TDZ tend to have persistent negative effects on root formation on shoots. For some plants, roots are easily induced merely by transferring in vitro shoots on regeneration medium to one that contains no PGRs. These include *Nicotiana* species, cucumber, squash, *Acacia*, and some *Rosa* (Gaba, 2005). The type of auxin used to stimulate shoot proliferation prior to rooting in vitro also affects how easily roots are induced. Shoots of American yellowwood, stimulated in response to BA, rooted easily with IBA in vitro (Figure 4.14). Too high a concentration of auxin used to induce rooting may result in callus formation and poor root induction/quality. If a high concentration of auxin is needed to induce roots but is detrimental to further root development, an auxin pulse of limited duration can be used. Sometimes mixtures of more than one auxin at lower concentrations is effective in initiation of roots where a single auxin is ineffective as in the case of olive, *Eucalyptus*, *Hevea*, and *Vitis* (Gaba, 2005).

FIGURE 4.11 Fasciated black walnut (*Juglans nigra*) shoots induced on medium containing 0.1 μM TDZ and 0.2% (2 g/L) Gelrite (top). Fasciated black walnut shoots induced on medium containing 25 μM BA and 2 g/L Gelrite (bottom). (Top figure from Bosela and Michler, 2008. *In Vitro Cell. Dev. Biol.—Plant* 44:316–329. Copyright 2008 by the Society for In Vitro Biology, formerly the Tissue Culture Association. Reproduced with permission of the copyright owner.)

Somatic (Nonzygotic) Embryogenesis

Somatic embryos undergo several developmental stages closely paralleling sexual embryo development in vivo. In the case of dicots, embryogenesis proceeds through recognizable globular, heart-shaped, torpedo, and, finally, cotyledonary stages. Monocots undergo globular, scutellar, and then coleoptilar stages. Conifers progress through globular, early cotyledonary, and late cotyledonary stages. The plant growth regulator requirements for induction of somatic embryos can differ from that of later stages of maturation and germination, which may require a medium not containing auxin, possibly to establish a polar auxin gradient (Jiminez, 2001). Somatic embryogenesis can occur directly on the differentiated explant tissue or can proceed through an intermediary callus or suspension step. Generally, auxins, or higher auxin to cytokinin ratios, promote the development of somatic embryos.

Raemakers et al. (1995) conducted a review of growth regulators used to induce embryogenesis. Of the 65 dicot species reviewed, 17 could develop embryos on medium containing no growth regulators, 29 species required auxin, and 25 required cytokinin supplementation. In order of frequency,

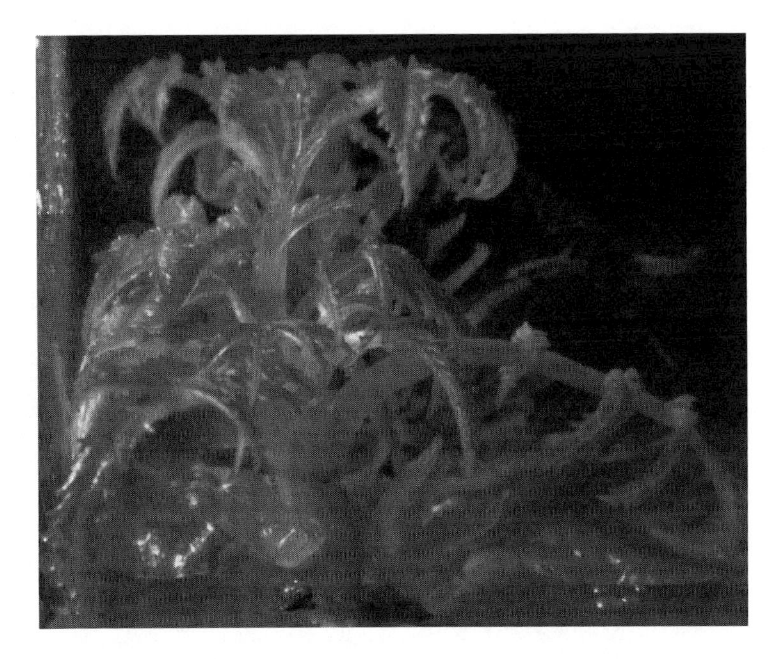

FIGURE 4.12 Glassy translucent appearance of shoots of black walnut induced on medium containing 0.05 μM TDZ and 2 g/L Gelrite. (Courtesy of M.J. Bosela and C.H. Michler.)

FIGURE 4.13 Both hyperhydric and normal shoots of *Populus tremuloides* x P. *tremula* Cl. 18-59-S-11 in response to 0.01 μM TDZ. (Courtesy of M.J. Bosela and C.H. Michler.)

the auxins used were 2,4-D (almost half), NAA, IAA, IBA, PIC, and DIC. Of the cytokinins, BA was used in more than half of the species, followed by KIN and, in a smaller number of species, ZEA and TDZ. With daylily, a combination of 0.1 mg/L TDZ plus 2 mg/L 2,4-D was effective in inducing embryogenesis (Figure 4.15). The concentrations of the PGRs used may be less important than the sensitivity of the tissues (Jiminez, 2001).

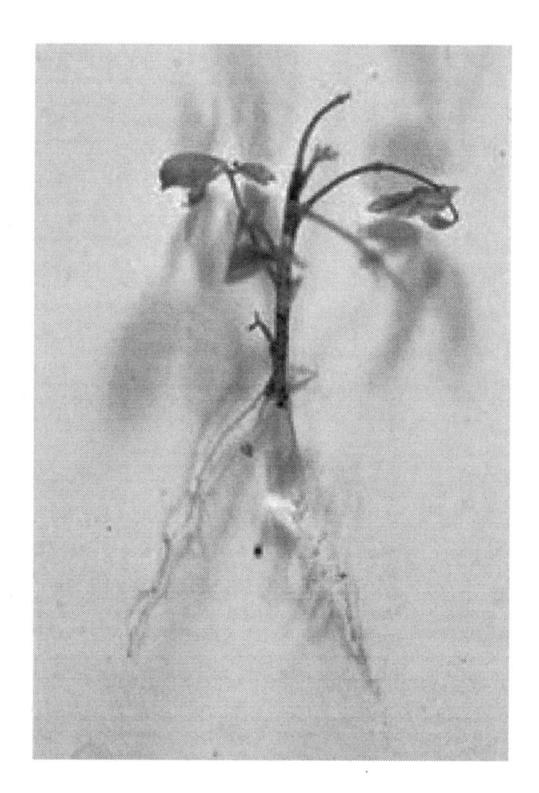

FIGURE 4.14 A rooted shoot of American yellowwood. The shoot was induced to elongate from an axillary bud using BA; rooting was achieved with IBA, an auxin used almost exclusively to root woody shoots. (Courtesy of R.N. Trigiano, University of Tennessee.)

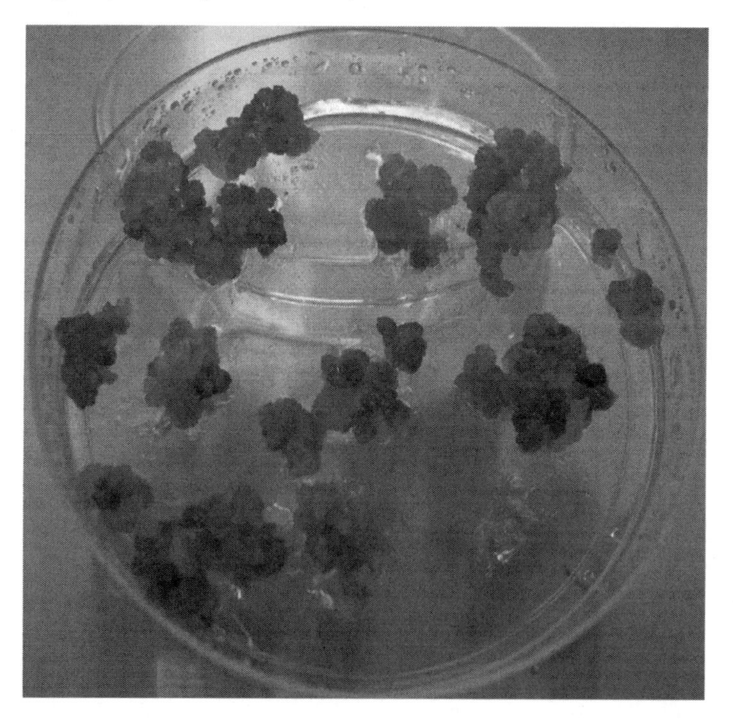

FIGURE 4.15 Somatic embryogenesis of daylily (*Hemerocallis littorea*) in response to 0.1 mg/L TDZ and 2 mg/L 2,4-D. (Courtesy of Hamidou Sakhanokho.)

The auxin of choice for induction of embryos from callus of cereals, particularly rice, wheat, and corn, has been 2,4-D (Bhaskaran and Smith, 1990). In some cases, other herbicidal auxins have been used successfully to induce embryogenic callus, for example, DIC with maize (Duncan et al., 1985) and orchardgrass (Figure 4.16, top), PIC combined with a low concentration of KIN with finger millet (Eapen and George, 1989), DIC or DIC with 2,4-D with barley, DIC or PIC plus 2,4-D in triticale, and DIC or PIC plus 2,4-D in wheat (Przetakiewicz et al., 2003). While many dicots can be induced to form somatic embryos in the presence of auxin alone (usually 2,4-D) as in the case of chrysanthemum (Figure 4.16, bottom), conifers usually require an auxin coupled with a cytokinin. For example, somatic embryos have been induced from *Picea abies* on media containing 2,4-D and BA (Chalupa, 1985; Hakman et al., 1985), NAA and BA (von Arnold, 1987; von Arnold and Hakman, 1988), and 2,4-D and BA and KIN (Boulay et al., 1988). The herbicidal auxin, PIC, was

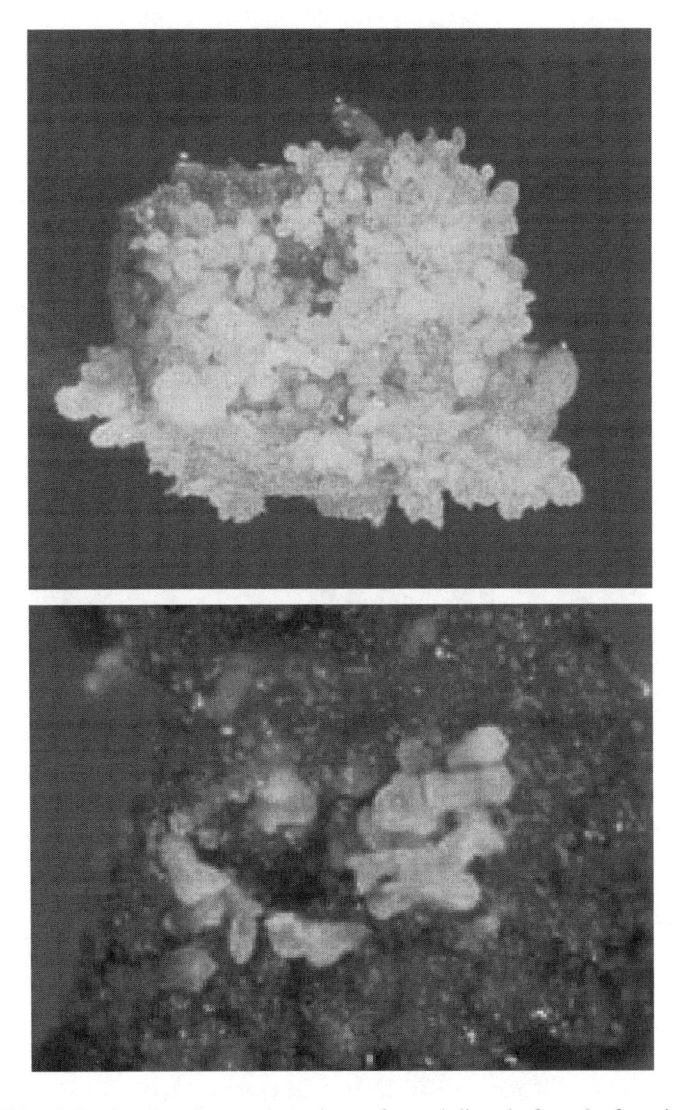

FIGURE 4.16 White, fully developed somatic embryos formed directly from leaf sections of orchardgrass (*Dactylis glomerata*) induced by the auxin-like herbicide dicamba (top). Globular and heart-shaped somatic embryos formed directly on leaves of chrysanthemum after incubation on a medium containing the auxin-like herbicide 2,4-D (bottom). A complex light treatment was also necessary to induce somatic embryogenesis. (Courtesy of R.N. Trigiano, University of Tennessee.)

more effective than 2,4-D for induction of embryogenesis in *Haworthia* and *Gasteria* (Figure 4.17; Beyl and Sharma, 1983).

Just as anomalies may occur in shoot development as a result of some growth regulators or their concentrations, aberrations in embryogenesis may also occur, including fasciation of cotyledons such as these for willow oak somatic embryos induced from mature embryos in response to 1 mM BA and 5 μM NAA (Figure 4.18; Geneve et al., 2003). In the case of eastern redbud, fasciated funnel-shaped cotyledons developed on somatic embryos induced from immature zygotic embryos in response to 5μM 2,4-D (Figure 4.19, top). A longitudinal cross section through the malformed cotyledon of the developing embryo indicates the lack of an apical meristem (Figure 4.19, bottom; Geneve and Kester, 1990).

Maturation of conifer somatic embryos is enhanced by the addition of ABA (Becwar et al., 1987; Boulay et al., 1988; Hakman and von Arnold, 1988; Misra and Green, 1990), but mature embryos are typically transferred to a medium lacking growth regulators to allow them to continue to develop into plants. ABA may act to inhibit precocious germination of somatic embryos as in the case of grape (Rajasekaran et al., 1982). In the case of carrot, the number of embryos obtained per explants was dependent on the ABA concentration (Nishiwaki et al., 2000). Gibberellins have been reported to inhibit embryogenesis or development of the somatic embryo in several species (Jiminez, 2001).

FIGURE 4.17 Callus (top) and subsequent embryo formation (bottom) from *Haworthia* leaf cross-section explants in response to PIC. PIC (0.5–3 mg/L) outperformed 2,4-D in both earliness of embryo induction and yield of embryos. (Top figure from Beyl, C. A. and G. C. Sharma. 1983. *Plant Cell, Tissue, and Organ Culture* 2:123–132. With permission.)

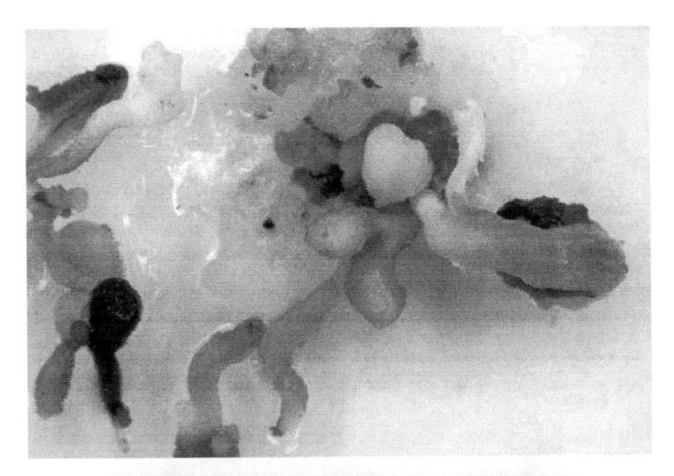

FIGURE 4.18 Fasciated cotyledons of Willow Oak (*Quercus phellos*) somatic embryos induced from immature zygotic embryos with 1 mM BA and 5 μM NAA (top and bottom). (From Geneve, R.L. et al. 2003. *Comb. Proc. Intl. Plant Prop. Soc.* 53:570–572. With permission.)

SUMMING IT UP

Of all the many factors that influence the success of micropropagation protocols, the choice of plant growth regulator and concentration has the most profound effect. Many other factors influence the successful establishment and proliferation of callus, shoots, roots, and embryos in culture, including the age, state, and part of the plant from which the original explants were taken. The genotype and environmental conditions of the mother plant also play a role. These factors in turn can influence the sensitivity of the tissue to the selected growth regulator or concentration. The explant itself may have a profound effect on the efficacy of the growth regulator, depending upon its sensitivity or other biochemical processes such as destruction or inactivation. Exogenously applied growth regulators may also interact with endogenous hormone levels, causing unanticipated consequences. The medium may also play a role as a sink into which endogenous growth regulators may diffuse, or as a medium allowing contact and diffusion of exogenous growth regulators into the tissue. Choice of growth regulator and concentration are not the only questions facing the researcher. Timing, duration of exposure, and subsequent culture conditions also impact success.

FIGURE 4.19 Fasciated funnel-shaped cotyledons of eastern redbud (*Cercis canadensis*) somatic embryos induced from immature zygotic embryos with 5 μM 2,4-D (top) and a longitudinal section through a developing somatic embryo showing malformed cotyledons and lack of an apical meristem (bottom). (From Geneve, R.S. and S.T. Kester. 1990. *Plant Cell Tiss. Org. Cult.* 22:71–76. With permission.)

Further complicating the decisions that the researcher must make are the interactions that exist among growth regulators of different classes, and between growth regulators and other metabolites within the plant. All in all, the field of growth regulation of plants during micropropagation is one that will provide interesting challenges for many researchers in the decades to come.

REFERENCES

Amasino, R. 2005. 1955: Kinetin arrives. The 50th anniversary of a new plant hormone. *Plant Physiol.* 138:1177–1184.

Bais, H.P., G. Sudha, and G. A. Ravishankar. 2000. Putrescine and silver nitrate influence shoot multiplication, in vitro flowering and endogenous titers of polyamines in *Cichorium intybus* L. cv. Lucknow local. *J. Plant Growth Regul.* 19:238–248.

Baroja-Fernandez, E., H. Aguirreola, J. Martinkova, J. Hanua, and M. Strnad. 2002. Aromatic cytokinins in micropropagated potato plants. *Plant Physiol. Biochem.* 40:217–227.

Becwar M.R., T. L. Noland, and S. R. Wann. 1987. Somatic embryo development and plant regeneration from embryogenic Norway spruce callus. *Tappi J.* 70:155–160.

Beyl, C. A. and G. C. Sharma. 1983. Picloram induced somatic embryogenesis in *Gasteria* and *Haworthia*. *Plant Cell Tiss. Org. Cult.* 2:123–132.

Bhaskaran, S. and R. H. Smith. 1990. Regeneration in cereal tissue culture: A review. *Crop Sci.* 30:1328–1336.

Bosela, M. J. 2009. Effects of β-lactans and auxins on indirect shoot regeneration in aspen tissue cultures. *Plant Cell Tiss. Org. Cult.* 98:249–261.

Bosela, M. J. and C. H. Michler. 2008. Media effects on black walnut (*Juglans nigra* L.) shoot culture growth in vitro: Evaluation of multiple nutrient formulations and cytokinin types. *In Vitro Cell. Dev. Biol.—Plant* 44:316–329.

Boulay, M.P., P.K. Gupta, P. Krogstrup, and D.J. Durzan. 1988. Development of somatic embryos from cell suspension cultures of Norway spruce (*Picea abies* Karst.). *Plant Cell Rep.* 7: 134–137.

Briggs, B. A., S. M. McCulloch, and L. A. Edick. 1988. Micropropagation of azaleas using TDZ. *Acta Hort.* 226:205–208.

Brown, C., F. J. Brooks, D. Pearson, and R. J. Mathias. 1989. Control of embryogenesis and organogenesis in immature wheat embryo callus using increased medium osmolarity and abscisic acid. *J. Plant Physiol.* 133:727–733.

Chalupa, V. 1985. Somatic embryogenesis and plantlet regeneration from cultured immature and mature embryos of *Picea abies* (L.) Karst. *Commun. Inst. Forest. Cechosl.* 14:57–63.

Chalupa, V. 1987. Effect of benzylaminopurine and thidiazuron on in vitro shoot proliferation of *Tilia cordata* Mill., *Sorbus aucuparia* L. and *Robinia pseudoacacia* L. *Biol. Plant (Praha)* 29:425–429.

Chiancone, B., A. Tassoni, N. Bagni, and M. A. Germana. 2006. Effect of polyamines on in vitro culture of *Citrus clementina* Hort. Ex Tan. *Plant Cell Tiss. Org. Cult.* 87:145–153.

Clouse, S. D. and J. M. Sasse. 1998. Brassinosteroids: Essential regulators of plant growth and development. *Ann. Rev. Plant Physiol. Plant Mol. Biol.* 49:427–451.

Darwin, C. R. 1880. *The Power of Movement in Plants*. London: John Murray.

Debergh, P., J. Aitken-Christie, D. Cohen, B. Grout, S. von Arnold, R. Zimmerman, and M. Ziv. 1992. Reconsideration of the term "vitrification'" as used in micropropagation. *Plant Cell Tiss. Org. Cult.* 30:135–140.

Duncan, D. R., M. E. Williams, B.E. Zehr, and J. M. Widholm. 1985. The production of callus capable of plant regeneration from immature embryos of numerous *Zea mays* (L.) genotypes. *Planta* 165:322–332.

Eapen, S. and L. George. 1989. High frequency plant regeneration through somatic embryogenesis in finger millet (*Eleusine coracana* (L.) Gaertn). *Plant Sci.* 61:127–130.

Eliasson, M. K., C. A. Beyl, and P. A. Barker. 1994. In vitro responses and acclimatization of *Prunus serotina* with paclobutrazol. *J. Plant Growth Regul.* 13:137–142.

Escalona, M., I. Cejas, J. González-Olmedo, I. Capote, S. Roels, M. J. Cañal, R. Rodríguez, J. Sandoval, and P. Debergh. 2003. The effect of meta-topolin on plantain propagation using a temporary immersion bioreactor. *InfoMusa* 12(2):28–30.

Fry, S. C. and H. E. Street. 1980. Gibberellin-sensitive suspension cultures. *Plant Physiol.* 65:472–477.

Furmanowa, M., K. Glowniak, K. Syklowska-Baranek, G. Zgorka, and A. Jozefczyk. 1997. Effect of picloram and methyl jasmonate on growth and taxane accumulation in callus culture of *Taxus* x *media* var. Hatfieldii. *Plant Cell Tiss. Org. Cult.* 49:75–79.

Gaba, V. P. 2005. PGRs in plant tissue culture and development. In *Plant, Tissue Culture and Development*, R. N. Trigano and D. J. Gray, eds., CRC Press, Boca Raton, FL, pp. 87–99.

Gaspar, T., C. Kevers, C. Penel, H. Greppin, D. M. Reid, and T. A. Thorpe. 1996. Plant hormones and PGRs in plant tissue culture. *In Vitro Cell. Dev. Biol. Plant* 32:272–289.

Geneve, R. S. and S. T. Kester. 1990. The initiation of somatic embryos and adventitious roots from developing zygotic embryo explants of *Cercis canadensis* L. cultured in vitro. *Plant Cell Tiss. Org. Cult.* 22:71–76.

Geneve, R. L., S. T. Kester, C. Edwards, and S. Wells. 2003. Somatic embryogenesis and callus induction in willow oak. *Combined Proceedings International Plant Propagators' Society* 53:570–572.

Grove, M. D., G. F. Spencer, W. K. Rohwedder, N. B. Mandava, J. F. Worley, J. D. Warthen, G. L. Steffens, J. L. Flippen-Anderson, and J. C. Cook. 1979. Brassinolide, a plant growth-promoting steroid isolated from *Brassica napus* pollen. *Nature* 281:216–217.

Guan, H. Y., P. Huisman, and G.-J. De Klerk. 1997. Rooting of apple stem slices in vitro is affected by indoleacetic-acid depletion of the medium. *Angewandte Botanik* 71:80–84.

Hakman, I., L. C. Fowke, S. Von Arnold, and T. Eriksson. 1985. The development of somatic embryos in tissue cultures initiated from immature embryos of *Picea abies* (Norway Spruce). *Plant Sci.* 38:53–59.

Hakman, I. and S. von Arnold. 1988. Somatic embryogenesis and plant regeneration from suspension cultures of *Picea glauca* (white spruce). *Physiol. Plant.* 72:579–587.

Hao, L., L. Zhou, X. Xu, J. Cao, and T. Xi. 2006. The role of salicylic acid and carrot embryogenic callus extracts in somatic embryogenesis of naked oat (*Avena nuda*). *Plant Cell Tiss. Org. Cult.* 85:109–113.

Heutteman, C. A. and J. E. Preece. 1993. Thidiazuron: A potent cytokinin for woody plant tissue culture. *Plant Cell Tiss. Org. Cult.* 33:105–119.

Hu, Y. X., F. Bao, and X. Y. Li. 2000. Promotive effect of brassinosteroids on cell division involves a distinct Cyc D3-induction pathway in Arabidopsis. *Plant J.* 24:693–701.

Huan. L. V. T., T. Takamura, and M. Tanaka. 2004. Callus formation and plant regeneration from callus through somatic embryo structures in *Cymbidium orchid. Plant Sci.* 166:1443–1449.

Jiminez, V. 2001. Regulation of in vitro somatic embryogenesis with emphasis on the role of endogenous hormones. *Rev. Bras. Fisiol. Veg.* 13:196–223.

Keim, P., J. Erion, and J. E. Fox. 1981. In *Metabolism and Molecular Activities of Cytokinins*, J. Guern and C. Peaud-Lenoel, eds., Springer, Berlin, p. 179.

Kelley, R. Y., A. E. Zipf, D. E. Wesenberg, and G. C. Sharma. 2002. Putrescine-enhanced somatic embryos and plant numbers from elite oat (*Avena* spp. L.) and reciprocal crosses. *In Vitro Cell, Dev. Biol.—Plant* 38:508–512.

Kim, M. K., H. E. Sommer, B. C. Bongarten, and S. A. Merkle. 1997. High-frequency induction of adventitious shoots from hypocotyls segments of *Liquidambar styraciflua* L. by thidiazuron. *Plant Cell Rep.* 16:536–540.

Kim. Y.-S., D. Chakrabarty, E.-J. Hahn, and K.-Y. Paek. 2003. Methyl jasmonate increases saponin content in bioreactor culture of ginseng (*Panax ginseng* C. A. Meyer) adventitious roots. *ISHS Acta Horticulturae* 625:289–292.

Kitto, S. L. and J. Janick. 1985. Hardening treatments increase survival of synthetically-coated asexual embryos of carrot. *J. Amer. Soc. Hort. Sci.* 110:283–286.

Kubalakova, M. and M. Strnad. 1992. The effects of aromatic cytokinins (populins) on micropropagation and regeneration of sugar beet in vitro. *Biologia Plantarum* 34:578.

Kucharska, D. and T. Orlikowska. 2008. The influence of paclobutrazol in the rooting medium on the quality of Chrysanthemum vitroplants. *J. Fruit Ornamental Plant Res.* 16:417–424.

Kurosawa E. 1926. Experimental studies on the nature of the substance secreted by the "bakanae" fungus. *Nat HistSoc Formosa*. 16, 213–227.

Letham, D. S. 1963. Zeatin, a factor inducing cell division isolated from *Zea mays*. *Life Sci.* 2:569–573.

Lu, Chin-Yi. 1993. The use of thidiazuron in tissue culture. *In Vitro Cell. Dev. Biol.* 29P:92–96.

Luo, J.-P., S.-T Jiang, and L.-J Pan. 2001. Enhanced somatic embryogenesis by salicylic acid of *Astragalus adsurgens* Pall.: Relationship with H_2O_2 production and H_2O_2-metabolizing enzyme activities. *Plant Sci.* 161:125–132.

Ma, Z., L. Ge, A. S. Y. Lee, J. W. H. Yong, S. N. Tan, and E. S. Ong. 2008. Simultaneous analysis of different classes of phytohormones in coconut (*Cocos nucifera* L.) water using high-performance liquid chromatography and liquid chromatography–tandem mass spectrometry after solid-phase extraction. *Anal. Chim. Acta* 610:274–281.

Machakova, I., E. Zazimalova, and E. F. George. 2008. PGRs I: Introduction: Auxins, their analogues and inhibitors. In *Plant Propagation by Tissue Culture*, 3rd edition, E. F. George, M. A. Hall, and G. De Klerk, eds., Springer, Berlin, pp. 175–204.

Magyar-Tabori, K., J. Dobranszki, and E. Jambor-Benczur. 2002. High in vitro shoot proliferation in the apple cultivar Jonagold induced by benzyladenine analogues. *Acta Agron. Hungarica* 50:191–195.

Malabadi, R. B. and K. Nataraja. 2007. Brassinosteroids influence in vitro regeneration using shoot tip sections of *Cymbidium elegans* Lindl. *Asian J. Plant Sci.* 6:308–313.

Malamy, J., J. P. Carr, D. F. Klessig, and I. Raskin. 1990. Salicylic acid: A likely endogenous signal in the resistance response of tobacco to viral infection. *Science* 250:1001–1004.

Metraux, J. P., H. Signer, J. Ryals, E. Ward, M. Wyss-Benz, J. Gaudin, K. Raschdorf, E. Schmid, W. Blum, and B. Inverardi. 1990. Increase in salicylic acid at the onset of systemic acquired resistance in cucumber. *Science* 250:1004–1006.

Miller, C. O., F. Skoog, M. H. von Saltza, and F. M. Strong. 1955. Kinetin, a cell division factor from deoxyribonucleic acid. *J. Am. Chem. Soc.* 77:1392.

Misra, S. and M. J. Green. 1990. Developmental gene expression in conifer embryogenesis and germination. I. Seed proteins and protein body composition of mature embryo and megagametophyte of white spruce (*Picea glauca* (Moench) Voss.). *Plant Sci.* 68:163–173.

Monteiro, M., C. Kevers, J. Dommes, and T. Gaspar. 2002. A specific role for spermidine in the initiation phase of somatic embryogenesis in *Panax ginseng* CA Meyer. *Plant Cell Tiss. Org. Cult.* 68:225–232.

Murali, T. P. and E. J. Duncan. 1995. The effects of in vitro hardening using triazoles on growth and acclimatization of banana. *Scientia Horticulturae* 64:243–251.

Murthy, B. N. S., S. J. Murch, and P. K. Saxena. 1998. Thidiazuron: A potent regulator of in vitro plant morphogenesis. *In Vitro Cell Dev. Biol. Plant* 34:267–275.

Nas, M. N. 2004. Inclusion of polyamines in the medium improves shoot elongation in hazelnut (*Corylus avellana* L.) micropropagation. *Turk. J. Agric. For.* 28:189–194.

Nishiwaki, M., K. Fujino, Y. Koda, K. Masuda, and Y. Kikuta. 2000. Somatic embryogenesis induced by the simple application of abscisic acid to carrot (*Daucus carota* L.) seedlings in culture. *Planta* 211:756–759.

Oh, M. H. and S. D. Clouse. 1998. Brassinolide affects the rate of cell division in isolated leaf protoplasts of Petunia hybrid. *Plant Cell Rep.* 17:921–924.

Panaia, M., T. Senaratna, E. Bunn, K. W. Dixon, and S. Sivasithamparam. 2000. Micropropagation of the critically endangered Western Australian species, *Symonanthus bancroftii* (F. Muell.) L. Haegi (Solanaceae). *Plant Cell Tiss. Org. Cult.* 63:23–29.

Preece, J.E., C.A. Heutteman, and W.C. Ashbey. 1991. Micro and cutting propagation of silver maple. I. Results with adult and juvenile propagates. *J. Amer. Soc. Hort. Sci.* 116:142–148.

Przetakiewicz, A., W. Orczyk, and A. Nadolska-Orczyk. 2003. The effect of auxin on plant regeneration of wheat, barley and triticale. *Plant Cell Tiss. Org. Cult.* 73:245–256.

Quiroz-Figueroa, F. and M. Mendez-Zeel. 2001. Picomolar concentration of salicylates induce cellular growth and enhance somatic embryogenesis in *Coffea arabica* tissue culture. *Plant Cell Rep.* 20:679–689.

Raemakers, C. J. J. M., E. Jacobsen, and R. G. F. Visser. 1995. Secondary somatic embryogenesis and applications in plant breeding. *Euphytica* 81:93–107.

Rajasekaran, K., J. Vine, and M. G. Mullins. 1982. Dormancy in somatic embryos and seeds of *Vitis*: Changes in endogenous abscisic acid during embryogeny and germination. *Planta* 154:139–144.

Raskin, I. 1992. Role of salicylic acid in plants. *Ann. Rev. Plant Physiol. Mol. Biol.* 43:439–463.

Ravnikar M. and N. Gogala. 1990. Regulation of potato meristem development by jasmonic acid in vitro. *Plant Growth Regul.* 9:233–236.

Rey, M., C. Diaz-Sala, and R. Rodriguez. 1994. Polyamines as markers for juvenility in filbert. *Acta Hort.* 351:233–237.

Ronsch, H., G. Adam, J. Matschke, and G. Schachler. 1993. Influence of (22*S*, 23*S*)-homobrassinolide on rooting capacity and survival of adult Norway spruce cuttings. *Tree Physiol.* 12:71–80.

Rosati, P., G. Marino, and C. Swierczewski. 1980. In vitro propagation of Japanese plum (*Prunus salicina* Lindl. cv. Calita). *J. Amer. Soc. Hort. Sci.* 105:126–129.

Saiprasad, G. V. S., P. Raghuveer, S. Khetarpal, and R. Chandra. 2003. Effect of various polyamines on production of protocorm-like bodies in orchid- *Dendrobium* "Sonia." *Scientia Horticulturae* 100:161–168.

Sakhanokho, H. and R. Y. Kelley. 2009. Influence of salicylic acid on in vitro propagation and salt tolerance in *Hibiscus acetosella* and *Hibiscus moscheutos* (cv Luna Red). *Afr. J. Biotechnol.* 8:1474–1481.

Sakhanokho, H., P. Ozias-Akins, O. L. May, and P. W. Chee. 2005. Putrescine enhances somatic embryogenesis and plant regeneration in upland cotton. *Plant Cell Tiss. Org. Cult.* 81:91–95.

Salisbury, F. B. and C. W. Ross. 1985. *Plant Physiology,* 3rd ed., Belmont, CA: Wadsworth Publishing.

Salvi, N. D., L. George, and S. Eapen. 2002. Micropropagation and field evaluation of micropropagated plants of turmeric. *Plant Cell Tiss. Org. Cult.* 68:143–151.

Sankhla, N. and T.D. Davis, 1999. Use of biosynthesis inhibitors for controlling woody plant growth. *Proc. Plant Growth Regul. Soc. Amer.* 26: 97–104.

Shevade, A. and J. E. Preece. 1993. In vitro shoot and floral organogenesis from stamen explants from a Rhododendron PJM group clone. *Scientia Horticulturae* 56:163–170.

Singha, S. and L. E. Powell. 1978. Response of apple buds culture in vitro to abscisic acid. *J. Amer. Soc. Hort. Sci.* 103:620–622.

Skoog F. and C. O. Miller. 1957. Chemical regulation of growth and organ formation in plant tissue cultured in vitro. *Symp Soc Exp Biol* XI:118–131.

Smith, E. F., A. V. Roberts, and J. Mottley. 1990. The preparation in vitro of chrysanthemum for transplantation to soil. 2. Improved resistance to desiccation conferred by paclobutrazol. *Plant Cell Tiss. Org. Cult.* 21:133–140.

Smith, E. F., I. Gribaudo, A. V. Roberts, and J. Mottley. 1992. Paclobutrazol and reduced humidity improve resistance to wilting of micropropagated grapevine. *HortScience* 27:111–113.

Strnad, M., J. Hanus, T. Vanek, M. Kaminek, J. A. Ballantine, B. Fussell, and D. E. Hanke. 1997. Meta-topolin, a highly active aromatic cytokinin from poplar leaves (*Populus* x *canadensis* Moench., cv. Robusta). *Phytochemistry* 45:213–218.

Tang, W., and R. J. Newton. 2005. Peroxidase and catalase activities are involved in direct adventitious shoot formation induced by thidiazuron in eastern white pine (*Pinus strobus* L.) zygotic embryos. *Plant Physiol. Biochem.* 43:760–769.

Tran Thanh Van, K., P. Toubart, A. Cousson A. G. Darvill, D. J. Gollin, P. Chelf, and P. Albersheim. 1985. Manipulation of morphogenic pathways of tobacco explants by oligosaccharins. *Nature* 314:615–617.

Tran Thanh Van, K. and T. H. Trinh. 1990. Organogenic differentiation. In S. S. Bhojwani, ed., *Plant Tissue Culture: Applications and Limitations.* Elsevier Science, Amsterdam, pp. 34–53.

Tulecke, W., L. Weinstein, A. Rutner, and H. Laurencot. 1961. The biochemical composition of coconut water (coconut milk) as related to its use in plant tissue culture. *Contrib. Boyce Thompson Inst.* 21:115–128.

Vahala, T. and T. Eriksson. 1991. Callus production from willow (*Salix viminalis* L.) protoplasts. *Plant Cell Tiss. Org. Cult.* 27:243–248.

Van Overbeek, J., M. E. Conklin, and A. F. Blakeslee. 1941. Factors in coconut milk essential for growth and development of very young *Datura* embryos. *Science* 94:350–351.

Vieitez, A. M. and M. L. Vieitez. 1980. Culture of chestnut shoots from buds in vitro. *Plant Physiol.* 55:83–84.

Von Arnold, S. 1987. Improved efficiency of somatic embryogenesis in mature embryos of *Picea abies*. *J. Plant Physiol.* 128:233–244.

Von Arnold, S. and I. Hakman. 1988. Regulation of somatic embryo development in *Picea abies* by abscisic acid (ABA). *J. Plant Physiol.* 132:164–169.

Wainwright, H. and A. W. Flegmann. 1985. The micropropagation of gooseberry: In vitro proliferation and in vivo establishment. *J. Hort. Sci.* 60:485–491.

Wallin, A. and L. Johansson. 1989. Plant regeneration from leaf mesophyll protoplasts of in vitro cultured shoots of a columnar apple. *J. Plant Physiol.* 135:565–570.

Went, F. W. 1926. On growth-accelerating substances in the coleoptiles of *Avena sativa*. *Proc. Kon. Ned. Akad. Wet.* 30:10–19.

Werbrouck, S., M. Strnad, H. Van Onckelen, and P. Debergh. 1996. Meta-topolin, an alternative to benzyladenine in tissue culture? *Physiol. Plantarum* 98:291.

Yabuta, T. and Sumiki, Y. 1938. On the crystal of gibberellin, a substance to promote plant growth. *J Agric Chem Soc Japan*. 14. 1526

Yahia, A., C. Kevers, T. Gaspar, J-C. Chenieux, M. Rideau, and J. Creche. 1998. Cytokinins and ethylene stimulate indole alkaloid accumulation in cell suspension cultures of *Catharanthus roseus* by two distinct mechanisms. *Plant Science* 133:9–15.

Yong, J. W. H., L. Ge, Y. F. Ng, and S. N. Tan. 2009. The chemical composition and biological properties of coconut (*Cocos nucifera* L.) water. *Molecules* 14:5144–5164.

5 Elements of In Vitro Research

Michael E. Compton

CONCEPTS

- Research is important for discovery and learning. Scientific methods are used to test new ideas and concepts in a controlled environment. Statistics is used to objectively evaluate data that are gathered and to determine the effectiveness of tested treatments.
- The most commonly used hypothesis for in vitro research is the null hypothesis. This popular hypothesis states that all treatments will elicit a similar response on all plant tissues.
- An experimental unit (EU) is the smallest unit that receives a single treatment—the culture vessel, in most tissue-culture experiments. The observational unit (OU) is the object being observed or measured. However, an explant may be considered as both EU and OU in cases where one explant is cultured per vessel.
- The completely randomized, randomized complete block, incomplete block, and split plot are experimental designs commonly used for in vitro research. These designs differ in their randomization schemes and experimental requirements.
- Many researchers find it beneficial to culture multiple explants in a culture vessel. This conserves resources and time but leads to a situation in which multiple measurements are made for each replicate (e.g., culture vessel). Statistically, this is referred to as subsampling or repeated measures.
- The nature of the data influences the type of statistical procedure that should be used. Most observations made in plant tissue-culture experiments result in data that can be classified as continuous measurements (shoot length, root length, or embryo size); as counts (number of organs per explant or callus [shoots, embryos, or protocorn-like bodies]); as binomial (response or no response, expressed as percentages); or as multinomial (shoot, root, callus, or no response). The statistical methods used to evaluate the treatment effects vary for each type of data observation.

WHY CONDUCT RESEARCH?

Research is important for discovery, learning, and the development of new and efficient practices. While some advances such as the discovery of penicillin, polycarbonates, and cellophane have resulted serendipitously, most discoveries are made using a methodical experimental approach in which treatments are devised and tested in controlled conditions. During these experiments, data are gathered and evaluated using statistical analysis to evaluate the effectiveness of each treatment at correcting the problem.

This chapter is meant to be an introduction to the basic elements of plant tissue culture research. More specifically, it aims to demonstrate the use of various experimental designs and statistical methods used to analyze and interpret data as well as present results using tables and graphs.

METHODOLOGIES ASSOCIATED WITH SCIENTIFIC RESEARCH

ESTABLISHING THE HYPOTHESIS AND EXPERIMENTAL OBJECTIVES

Research begins by reviewing what is known about the problem. We read books and published research articles and combine that information with the knowledge that we have gained from prior experiences. We use this base to formulate a hypothesis and experimental objectives that can be tested objectively.

The most commonly used hypothesis employed in in vitro research is the "null hypothesis." This popular hypothesis states that all treatments will elicit a similar response on all plant tissues. The application of the null hypothesis can be illustrated by using a situation in which different cytokinins [zeatin, zeatin riboside, 2ip, benzyladenine (BAP), kinetin, and thidiazuron (TDZ)] were tested to evaluate their ability to stimulate adventitious shoot organogenesis in petunia (*Petunia x hybrida*) leaf explants. In this case the null hypothesis would read, *adventitious shoot regeneration in petunia is similar among leaf explants incubated in medium containing equal concentrations of zeatin, zeatin riboside, 2iP, BAP, Kinetin, and TDZ*. Rejection of the null hypothesis through statistical analysis implies that the experimental tissues reacted differently to the imposed treatments. Statistical procedures used to test research hypotheses will be discussed later in this chapter.

SELECTING THE MAIN COMPONENTS AND EXPERIMENTAL DESIGN

The main components of the experiment can be selected after the hypothesis and experimental objectives are outlined. This includes selection of treatments and supportive materials required for the experiment. Treatments are conditions characteristic of the population of interest that can be measured and tested experimentally (Lentner and Bishop, 1986). Examples of treatments used for in vitro research include culture vessels and their contents (medium salt formulations, type or concentration of plant growth regulators), environmental conditions outside the culture vessel (temperature, light source, quality, intensity, or photoperiod), or factors inherent to the explant [genotype or stock plant physiology (explant size or age, stock plant age or maturation stage, preconditioning, etc.)]. Researchers may choose to examine one treatment factor at a time or test several factors simultaneously (e.g., stock plant genotype and plant growth regulator concentration). The latter is often popular in in vitro studies because many factors often interact with each other. In addition, time and materials can be conserved by conducting one experiment rather than two (Compton, 1994). It is important that treatments and treatment levels be chosen according to the experimental objectives and evoke a wide range of responses. This is important so that an optimum level can be easily identified.

Supportive materials are items necessary for the experiment but are not experimental treatments. In plant tissue culture studies, these materials may be culture vessels, medium components, growth regulator type and concentration, state of medium solidification, growth environment (light photoperiod and intensity), or any item important for satisfactory explant growth. Supportive materials should not interfere with treatment effects on plant tissues. Knowledge of previous experiments, successful or not, provides beneficial information when selecting treatments and supportive experimental materials.

Experimental units, observational units, and the experimental design are chosen once treatments and supportive materials are selected. By definition, an experimental unit (EU) is the smallest unit that receives a single treatment. This is the culture vessel in most tissue-culture experiments. The observational unit (OU) is the object being observed or measured. Explants are the most common OU in tissue-culture studies because most people are interested in discovering treatments that optimize their development and performance. However, an explant may be considered as both the EU and OU in cases in which there is one explant per culture vessel (Compton and Mize, 1999). The number of EUs required per treatment is determined once the EUs and OUs are selected.

Replication occurs when more than one EU is evaluated for each treatment and is necessary to estimate treatment effects and experimental error. The number of replicates per treatment is based on the expected degree of variability and the extent to which the researcher wishes to detect treatment effects (Zar, 1984). At least three replicates per treatment are required to calculate means and measure error. Most researchers rely on prior experience, review of literature, and speculation to determine the number of replicates to be used. Insufficient replication can reduce the ability of statistical procedures to detect treatment differences (Lentner and Bishop, 1984). Kempthorne (1973) outlines how information from previous studies can be used to estimate the number of replicates required for an experiment.

Observational units are randomly assigned to replicate EUs and then transferred to experimental conditions using a specific experimental design. An experimental design is a method used to randomly arrange replicate EUs in a manner that limits experimenter bias and the effects of nontreatment factors on the OUs. The most commonly used experimental designs for in vitro research are the completely randomized, randomized complete block, incomplete block, and split plot designs. These designs differ in their randomization schemes and experimental requirements. When selecting an experimental design, it is important to examine the supporting materials and population of interest for possible variation and choose a design that is simple to employ, efficient in regard to measuring treatment effects and residuals, and matches the experimental objectives (Compton and Mize, 1999).

The completely randomized (CR) design is most commonly used for in vitro research because it is easy to use and has a randomization scheme with no set pattern. This allows researchers to maximize replicate numbers and utilize either equal or unequal treatment replication (Little and Hills, 1978). This is important because unequal replicate numbers often occur in plant tissue culture studies due to contamination or death of explants. Randomization of the CR design can be illustrated using an example in which the effectiveness of six types of cytokinins on adventitious shoot organogenesis from leaf explants of *Petunia x hybrid* was examined (Figure 5.1A). In this experiment, individual explants were cultured in test tubes containing 20 mL of MS medium supplemented with 4 μM of either zeatin, zeatin riboside, 2iP, BA, kinetin, or TDZ. Ten vessels were established for each treatment. Statistically speaking, the CR design is most efficient because it allows researchers to maximize degrees of freedom (df) for error, which is important for detecting treatment differences during statistical analysis (Little and Hills, 1978). However, this design has some problems. One pitfall of the CR design is that the EUs must be homogeneous for data analysis to be effective. Situations in which EUs or OUs are highly heterogeneous increases variability in the experiment and decreases the likelihood that treatment differences will be detected during statistical analysis (Little and Hills, 1978).

Unevenness or inconsistency among environmental factors in the growth room, differences among stock plants or leaves from which explants are obtained, variability among technicians as well as differences among batches of *Agrobacterium*, or bombardments using a particle accelerator are a few nontreatment factors that introduce variability into the experiment. Experimental designs that arrange replicate EUs into similar units (i.e., blocks) are used when there is a high degree of variability among the EUs. When using blocking, only factors that are measurable can be blocked.

The randomized complete block (RCB) design assembles replicates of each treatment into blocks. This creates a high level of uniformity among treatments within a block and reduces nontreatment variation (Little and Hills, 1978). It is important that all explants in a block are as uniform as possible for the design to be effective. Explants from the same leaf or stock plant often have similar regeneration competence, and each leaf can constitute a block. The petunia study could be used to illustrate the RCB design by using explants from one leaf to establish replicates of each treatment in a block (Figure 5.1B). Using the RCB in this situation would maximize uniformity within blocks by utilizing the uniformity present in each leaf.

Situations often exist in which there is too little source material to produce the number of explants required to establish complete blocks, each containing one replicate of each treatment. In these cases, the incomplete block (IB) design may be the best choice (Compton and Mize, 1999). As in the case of the RCB design, this design uses blocking to maintain homogeneity, but differs from the RCB design in that not all treatments are present in a block. The petunia example could be modified to illustrate use of the IB design by limiting the number of explants obtained from each leaf to four.

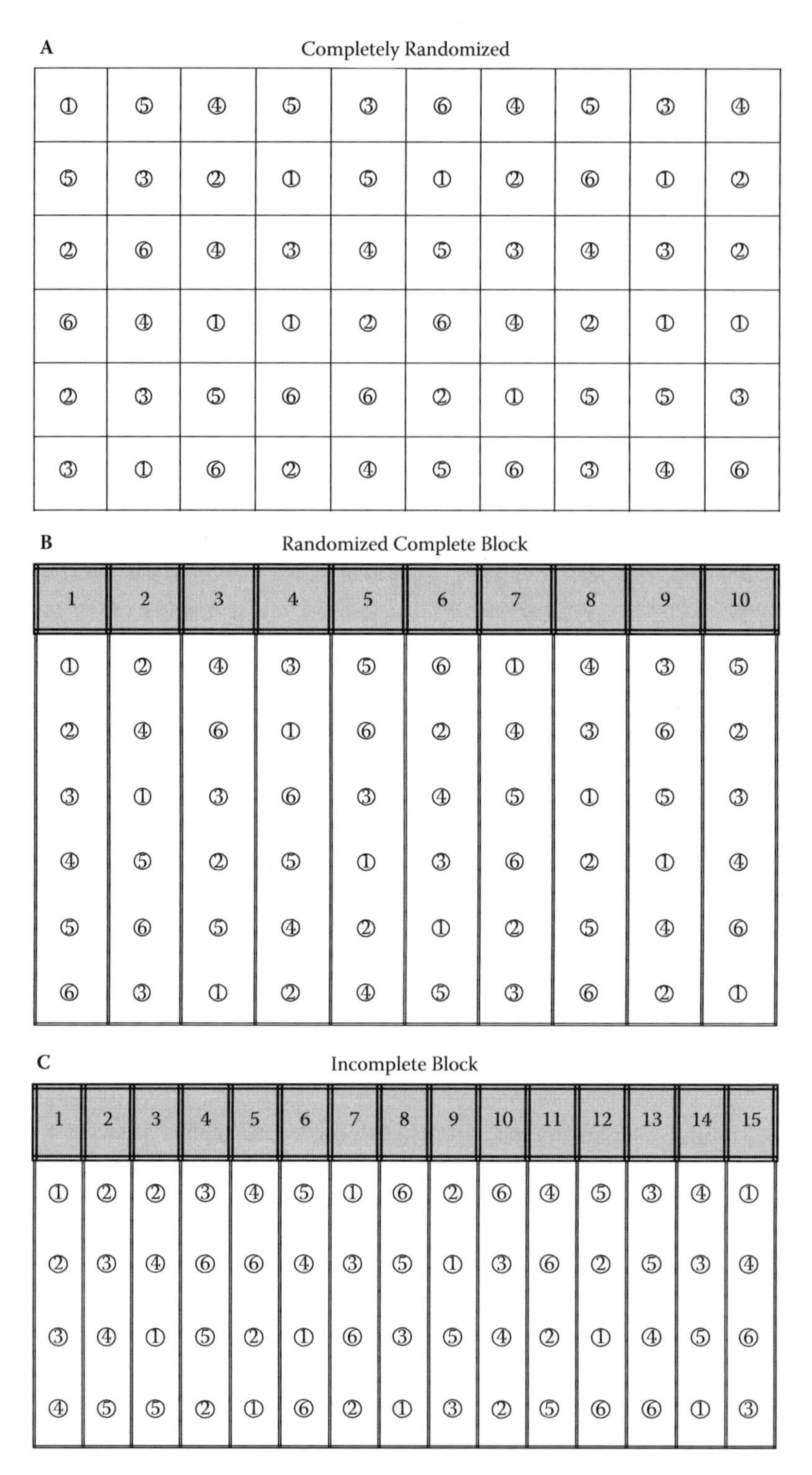

FIGURE 5.1 Randomization schemes for a completely randomized (A), randomized complete block (B), and incomplete block (C) designs with six treatments. Each treatment was replicated 10 times in all designs. *Note:* The number of blocks and the number of times that any two treatments appear in the same block for the IB design were determined using the formula kb = tr and $\lambda(t - 1) = r(k - 1)$.

In this case there would not be enough leaf material to establish complete blocks with all six treatments represented. Therefore, 15 incomplete blocks would be required to replicate each treatment 10 times (Figure 5.1C). Establishing the number of blocks, replicates, and number of times that any two treatments appear together in a block is determined using a special formula [$kb = tr$ and $\lambda(t\text{-}a) = r(k\text{-}1)$; see Lentner and Bishop (1986)]. Information regarding the use of IB designs in plant tissue culture was published by Kuklin et al. (1993). Although the IB is an excellent design for in vitro studies, an RCB design is often preferred when there is adequate explant material available to form complete blocks (Mize et al., 1999).

Another design that utilizes blocking is the split plot (SP). This design is best used when two treatment factors with different degrees of variation are examined simultaneously (Little and Hills, 1978). The SP design uses two separate levels of randomization for each factor. The treatment factor with the greatest degree of variation is assigned to large plots called main plots, and the treatment factor with less variation is assigned to subplots. Main plots can be completely randomized or arranged in blocks. It is common that the youngest fully expanded leaves are most competent for adventitious shoot organogenesis. Given this circumstance, researchers must often use leaves from several different stock plants to obtain enough explant material for experiments. This often introduces additional variability into the experiment. In the petunia example, the researcher believes that stock plants from which leaves are harvested for explantation, as well as cytokinin type, influence adventitious shoot regeneration. Based on this hypothesis, an experiment was designed in which the influences of stock plants and cytokinin type on shoot regeneration were examined simultaneously. Because greater variation was expected to be introduced into the experiment from the five stock plants than from cytokinin type, the stock plant source was assigned to main plots, and cytokinin type to subplots. Stock plants were replicated three times with main plots completely randomized and subplot treatments randomized within each main plot, resulting in 15 replicates of each cytokinin type (Figure 5.2). Although the SP is the most efficient experimental design to use when examining two treatment factors with different degrees of influence on explant response, the setup, data recording, as well as data analysis and interpretation are more complicated than for the other designs.

It is impossible to be absolutely sure if blocking will be required when designing experiments. The need for blocking can be determined by conducting a small preliminary experiment and testing for blocking when conducting the statistical analysis. Kempthorne (1973) outlines a method for determining if blocking is necessary.

Plant Genotype	1_1	3_1	4_1	2_1	5_1	....	4_3	3_3	2_3	5_3	1_3
	①	④	②	③	⑤		④	①	⑥	②	③
	③	⑥	⑤	⑥	①		③	④	①	⑥	⑤
Cytokinin Type	⑤	②	③	①	④		⑤	②	③	④	⑥
	②	⑤	⑥	④	②		①	③	④	①	①
	⑥	③	④	⑤	⑥		⑥	⑤	②	③	④
	④	①	①	②	③		②	⑥	⑤	⑤	②

FIGURE 5.2 Randomization scheme for a split-plot design in which main plots are completely randomized and subplots randomized within each main plant. Stock plants were assigned to main plots. Each stock plant (1–5) was replicated three times. Cytokinin types were assigned to subplots [zeatin (①), zeatin riboside (②), 2-Isopentenyl adenine (2iP) (③), BA (④), kinetin (⑤), and thidiazuron (TDZ) (⑥)].

Using Subsampling

Many researchers find it beneficial to culture multiple explants in a culture vessel. This conserves resources and time but leads to a situation in which multiple measurements are made for each replicate (e.g., culture vessel). Statistically, this is referred to as subsampling or repeated measures (Compton and Mize, 1999). Subsampling is not an experimental design but a modification that can be used with any of the designs discussed previously. It can be illustrated by modifying the petunia example so that two explants are cultured in petri dishes containing 25 mL of MS medium supplemented with 4 μM of zeatin, zeatin riboside, 2iP, BA, Kinetin, or TDZ. Subsampling may be used with any of the previously described experimental designs. Culturing multiple explants in one vessel not only saves resources and space but also maximizes the number of OUs and minimizes variation associated with explants by making multiple measurements per vessel (Mize and Winistorfer, 1982). One disadvantage of subsampling is that the statistical analysis is slightly more complicated than when subsampling is not used (Compton and Mize, 1999).

Determining How Many Times an Experiment Should Be Replicated

An experiment should be conducted more than once to confirm the validity of results. However, researchers often wonder how many times an experiment should be repeated. Most statisticians and researchers believe that an experiment must be conducted at least twice to validate the results (Lentner and Bishop, 1986). The number of times the experiment is to be conducted should be decided during planning and before the first time the experiment is conducted. Repeating an experiment does not guarantee a similar outcome even if the same stock plants are used. In addition, time can be introduced as a nontreatment factor if changes in stock plant physiology have occurred with increased stock plant age or time in the growing season, both of which can lead to differences in experimental results between runs of the experiment. Researchers generally have two options when differences between runs are obtained. The entire experiment can be repeated, which can be costly, or each run examined as a block or main plot. The best option would be to first look at the data from the two runs and determine if similar trends exist. If the trends are similar, the experiment should not be repeated, but the circumstances that caused the differences should be examined and explained. If the results obtained in the two runs contrast, the experiment should be repeated. If differences among runs persist, the possibility of seasonal effects on the results should be examined experimentally. A decision should be made during the planning phase regarding what will be done if differences between runs of the experiment are observed.

Choosing the Best Data and Planning the Statistical Analysis

The data recorded should help evaluate the treatment effects according to the experimental objectives. If inappropriate data are recorded, it is unlikely that the researcher will be able to assess explant response and address the experimental objectives. Therefore, it is important to discuss which observations should be recorded with colleagues and statisticians during planning.

Most researchers peruse relevant literature and combine the information gathered with previous experience to determine which observations to measure in their experiments. In the petunia illustrations used throughout this chapter, the number and percentage of explants producing shoots, number of shoots per explant (or all explants in a petri dish), and shoot length are observations that most researchers would record to evaluate explant competence. However, to get the most information from the project, the researcher may wish to make additional observations such as the number and percentage of explants that produced callus, the amount of callus produced per explant (weight or volume measurement), and ploidy level of regenerants as well as histological observations of explants periodically during the experiment to ascertain the mode of shoot regeneration (direct or indirect; see Chapter 7 on histology, and Chapters 19 through 21 on shoot regeneration).

Observations made during the experiment are recorded on data sheets to facilitate data processing. This activity can lead to mistakes if the sheets are not well organized. Designing data sheets the way observations will be entered into the computer helps to avoid mistakes when transcribing data. Many statistical software packages will organize and print data sheets for you (Compton, 2006). Data sheets should be designed during planning to ensure that the researcher is considering observations that measure treatment effects.

The nature of the data influences the type of statistical procedure that should be used. Most observations made in plant tissue-culture experiments result in data that can be classified as continuous (shoot length, root length, or embryo size), counts [number of organs per explant or callus (shoots, embryos, or protocorm-like bodies)], binomial [response or no response (e.g., percentages)], or multinomial (shoot, root, callus, or no response; Mize et al., 1999). Therefore, the statistical methods used to evaluate treatment effects vary for each type of data observation (Compton, 2006). Outlining the sources of variation and degrees of freedom during planning can help identify data analysis procedures that are most appropriate for the type of data recorded and the experimental objectives. Most researchers use analysis of variance (ANOVA) to evaluate data. While ANOVA is suited to continuous data, it is not suitable for analyzing count, binomial, or multinomial data without prior manipulation.

CONDUCTING THE EXPERIMENT

Experimentation may begin once planning is complete and thoroughly reviewed. Personal bias must be avoided during all phases of the experiment. When setting up an experiment, bias can be avoided by using the structure of the experimental design (Mize et al., 1999). Do not cut all explants for one treatment at a time. Instead, arrange the culture vessels according to the randomization scheme selected and place the explants in the vessel according to the design.

It is also important to avoid any circumstances that introduce error into the experiment. Many mistakes occur when researchers become fatigued. The time required for each phase of the experiment should be estimated during planning. Tasks that are taxing should be staggered to avoid worker strain. Mistakes may also occur when explants are transferred to fresh medium. Carefully arrange culture vessels (both new and currently used) in the laminar hood in a manner that reduces the opportunity for mistakes. Be careful to label each culture vessel with the correct information. If using codes, be sure to record keys to your codes in your laboratory notebook. There are many avenues for introducing operator error into your experiment. Take precautions necessary to encourage safe and accurate experimentation.

RECORDING DATA

As when starting an experiment, you should use the structure of the experiment when recording observations. For example, record observations of replicates in one block before moving to the next block. This helps to avoid bias. Be meticulous and take your time when recording data. Using sloppy technique introduces error and inconsistency. Take breaks to avoid fatigue when recording data. Many observations are stressful on the eyes, making it important to rest periodically.

Completed data sheets should be checked for errors. Erroneous data entries lead to incorrect treatment evaluation and interpretation (Mize et al., 1999). Make sure the data of each treatment fall within the expected boundaries. Data outside these boundaries are called outliers and may result from errors in reading, recording, or transcribing data, or may be values obtained from plant tissues with inherent problems or genetic mutations. Outliers are often discarded. However, researchers should exercise caution when deleting data, as personal bias may be unintentionally introduced, affecting the outcome of the experiment. Several statistical procedures can be used to detect outliers and are helpful to reduce bias when editing (Barnett and Lewis, 1989). Treatment codes should be examined while checking data. Do not assume that codes were entered correctly. Errors associated with entering codes can be examined by calculating the mean for the coded variables. In the petunia example, obtaining a mean of 3.5 for the cytokinin treatment codes (coded 1 through 6) indicates that the code was entered correctly for each observation. Editing data can be facilitated by printing data sheets and thoroughly reviewing them. This can be a time-consuming task but is worth the effort.

DATA ANALYSIS AND INTERPRETATION

Data must be statistically analyzed and interpreted as planned. As discussed earlier, the nature of the data influences the results and interpretation of the statistical analysis. Researchers should choose an analysis procedure best suited for the observations recorded and experimental objectives. When analyzing data, a general analysis is conducted first to determine if there are any differences between treatment levels. Further analysis of treatment means is conducted if differences are detected by the general analysis (Compton, 1994; Mize et al., 1999).

ANALYSIS OF CONTINUOUS DATA

Continuous data are normally distributed with treatments having similar variances (Mize et al., 1999). Because of this characteristic, ANOVA is well suited to analyzing this type of data. During ANOVA, the value of each observation is subtracted from the overall mean. Differences between these values are considered random error (residual) and used in the analysis (Mize et al., 1999). A model statement is created that identifies the dependent variables (observations recorded) and independent variables (treatments, treatment interactions) based on the experimental design used. The influence of independent variables on the dependent variables is tested according to the model statement. A summary table is generated during ANOVA that provides the results of the model tested (Lentner and Bishop, 1986). Information in the summary table includes the sources of variation, degrees of freedom (df), sums of squares (SS), mean square error (MSE), F-statistic (F), and estimates the probability (P) that a similar result would be obtained next time the experiment is conducted under similar conditions (Lentner and Bishop, 1986).

In the petunia regeneration example, the variable shoot length is considered continuous. Since a CR design was used in this study, cytokinin type and experimental error were identified as sources of variation (Table 5.1). According to the results, the length of regenerated shoots was influenced by the type of cytokinin used. This is determined by looking at the F and P values. In ANOVA, F (12.46) was obtained by dividing the MS for cytokinin (360.67) by the MSE (28.9537). The MS value reflects the degree of variation associated with the source. Therefore, the cytokinin MS reflects the amount of variation caused by cytokinin treatments. The MSE measures variation due to culture vessels and any other nontreatment sources. The *p* value (level of significance for the F) is obtained by comparing the calculated F against standardized values for ANOVA. The *p* values equal to or less than 0.05 are considered significant. This level is chosen by most researchers because 0.05 indicates that a similar result would be obtained 95% of the time that the experiment is repeated, signifying a high level of confidence for the outcome. Another reason for selecting 0.05 is that most journals require researchers to test at this level of significance.

ANOVA for experiments in which blocking is used must be conducted differently from when a CR design is used. When an RCB design is used, the block must be specified in the model statement (Compton, 1994). Comparing ANOVA summary tables for the RCB, IB, and CR designs illustrates the origin of df and SS for blocking (Table 5.1). The df and SS calculated for cytokinin is identical for all experimental designs. However, df and SS values for experimental error differ for the three randomization schemes. Adding the df for blocks to the df for experimental error in the RCB design results in the same value calculated for error df in the CR design. Compare experimental error rows

TABLE 5.1
Analysis of Variance Table Comparing Petunia Shoot Organogenesis Data (Shoot Length) Analyzed Using a Completely Randomized (CR), Randomized Complete Block (RCB), Incomplete Block (LB), and Completely Randomized Design with Subsampling (CR-SS)

Source	DF	SS	MS	F	p
Completely Randomized Design without Subsampling					
Cytokinin	5	1803.35	360.67	12.46	<0.000
Experimental error	54	1563.50	28.9537		
Total	59	3366.85			
Randomized Complete Block Design					
Cytokinin	5	1803.35	360.67	12.69	<0.000
Block	9	284.350	31.5944		
Experimental error	45	1279.15	28.4256		
Total	59	3366.85			
Incomplete Block Design					
Cytokinin	5	1803.35	360.67	67.20	<0.000
Incomplete blocks	14	490.10	63.928		
Experimental error	40	1073.40	26.835		
Total	59	3366.85			
Completely Randomized Design with Subsampling					
Cytokinin	5	1803.35	360.67	13.93	<0.000
Subsampling error	24	787.0	32.7917		
Experimental error	30	776.5	25.8833		
Total	59	3366.85			

Note: Explants were prepared from leaves of petunia plants grown in the greenhouse. Six cytokinin types were evaluated (zeatin, zeatin riboside, 2iP, BA, kinetin, and TDZ). Explants were either cultured in 25 × 150 mm test tubes (1 per tube; CR, RCB and IB) or 100 × 15 mm petri dishes (2 per dish; CR-SS) containing 25 mL of shoot regeneration medium. There were ten replicate test tubes (CR, RCB, and IB designs) or five petri dishes per treatment (CR-SS). Data are hypothetical.

in Table 5.1. The same is true for the SS values. Specifying blocks in the model statement instructs ANOVA to separate variation associated with blocks from the experimental error, resulting in a lower MSE and increased F for the treatment variable. This increases the sensitivity of ANOVA, increasing the likelihood that treatment differences will be detected by segregating variation not associated with the treatments.

Conducting ANOVA is slightly more complicated when an IB design is used. This is because two model statements must be generated. One statement is written similar to a CR design, while the second is written to generate a value for using incomplete blocks. The values calculated for treatment and incomplete blocks are used to construct an ANOVA table for calculating the F and associated *p* value for the treatment. In the petunia example, use of incomplete blocks removed variation from the experimental error more effectively than the RCB design (Table 5.1), possibly because the use of smaller blocks (four treatments per block) improved homogeneity within individual blocks.

ANOVA is preferred by most researchers because it is easy to perform, can be used to evaluate data obtained from virtually all experimental designs, and generates an MSE value that is considered to be the best estimate of experimental error (Mize et al., 1999). ANOVA provides a good overall analysis because it accurately measures how treatments relate to each other when used within the proper circumstances.

Count Data

Count data are not normally distributed because treatment variances are equal to the average response of the treatments (Zar, 1984). Because of their distribution, count data should be analyzed using Poisson regression (Mize et al., 1999). This statistical method is best suited for analyzing count data because it uses a logarithmic value of the mean counts, which normalizes the data during analysis. Poisson regression calculates a coefficient that is divided by the standard error to determine if the model is significant. In the petunia example, observations on the number of shoots per explant are considered count data. Poisson regression indicated that the cytokinin type influenced the number of shoots per explant ($p = 0.005$; see Table 5.2).

ANOVA is often used by researchers to analyze count data. Unfortunately, it is unreliable when observed count values are less than 10 (Mize et al., 1999). Since the number of organs regenerated by explants in adventitious organogenesis and embryogenesis studies is typically low (less than 10) use of Poisson regression is usually warranted. In experiments in which more than 10 organs are regenerated per explant (e.g., micropropagation studies using shoot tips), ANOVA and Poisson regression yield similar results (Mize et al., 1999). Zar (1984) believes that ANOVA should not be used to analyze count data regardless of how large values are. Therefore, it may be in the best interest of the researcher to transform count data in all situations.

Binomial Data

Response data (also known as binomial data) record observations related to the ability of explants to respond to a given treatment. These data are generated to evaluate the influence of treatments on regeneration competence, and have a binomial distribution that causes treatment variances to be dependent on explant success (Mize et al., 1999). Percentages calculated by these observations are often considered important because researchers are interested in identifying treatments that optimize explant response. The ability to optimize explant response can translate directly into increased profits for plant tissue-culture businesses, micropropagating house plants, or biotechnology researchers interested in using recombinant DNA technology to insert transgenes into plant tissues.

Most researchers use ANOVA to analyze response data. This is unfortunate because, as stated earlier, ANOVA assumes that data are normally distributed and treatment variances equal. Because treatment variances of response data are dependent on the success of responses, ANOVA results are unreliable. Logistic regression is the statistical procedure most suited to analyzing response data because the procedure does not produce separate estimates of experimental error but calculates a

TABLE 5.2
Poisson Regression Summary Table for Petunia Shoot Organogenesis Data (Number of Shoots per Explant)

Predictor Variables	Poisson Regression Coefficient (A)	Standard Error (B)	Poisson Regression Statistic (A/B)	*P*-Value
Constant	1.120	0.150	7.45	<0.0000
Cytokinin	0.101	0.036	2.79	0.0052

TABLE 5.3
Logistic Regression Summary Table for Petunia Shoot Organogenesis Data (Percentage of Explants with Shoots)

Predictor Variables	Logistic Regression Coefficient (A)	Standard Error (B)	Logistic Regression Statistic (A/B)	*P*-Value
Constant	1.1296	0.7435	1.74	0.0813
Cytokinin	0.0573	1.9580	0.29	0.7969

special coefficient and standard error value to arrive at a test statistic to determine statistical significance (Mize et al., 1999). In the petunia example, the percent regeneration observations were binomially distributed. Analyzing these data using logistic regression revealed that the percentage of explants that produced shoots was not influenced by the type of cytokinin used (P value = 0.7969; Table 5.3).

Before analyzing response data, it is important to determine if there are treatments in which the response did not vary. This usually occurs when all or none of the explants responded. These values should either be changed or deleted before analysis (Mize et al., 1999). If a decision is made to change the values, zeros should be changed to a slightly higher value (0.000001) and 100% values reduced slightly (0.999999). It is important to decide during planning if values are going to be altered or deleted. You should also indicate if treatments were dropped or values altered when writing reports and manuscripts.

Analyzing Experiments Using Subsampling

Most researchers culture several explants in a culture vessel (subsampling) to save time and resources, but analyze the data as if subsampling was not used. This action is incorrect because subsampling reduces variation introduced into the experiment by culturing multiple explants per vessel (Compton and Mize, 1999). Two methods can be used to analyze observations made when subsampling is used. Observations from each explant in a dish may be combined, resulting in one value per vessel. In this situation, the model statement used in the general analysis (ANOVA) would be written as if subsampling was not used (Lentner and Bishop, 1986). An alternative method would be to record observations from each explant in a culture vessel and enter them individually. However, the model statement should be written so that df and SS associated with subsampling is separated from the experimental error (Compton, 1994). This method is preferred by researchers wishing to document the variation associated with explants. If the subsampling error is not separated in this situation, the error df, SS, and MS will be inflated, resulting in a smaller F value for the treatments of interests, which reduces the likelihood that treatment differences will be detected. In the petunia regeneration example, 24 df and 787.0 SS were removed from the experimental error, which reduced the MSE from 28.9537 to 25.8833, resulting in a greater treatment F value (Table 5.1). Subsampling can be used with single factor and factorial experiments and any of the experimental designs discussed previously.

METHODS FOR COMPARING TREATMENT MEANS

Once a significant value is obtained in the preliminary statistical test (ANOVA, Poisson regression, or Logistic regression), the researcher must elucidate specific differences among treatments. This is not a problem when there are only two treatments because the general test alone is sufficient for determining statistical differences. However, most researchers test more than one treatment, requiring further analysis. The easiest way to compare treatment means is to rank them in ascending or descending order and pick the best treatment ignoring statistical treatment and account for variation among EUs and OUs. It is important to consider treatment variation when comparing means because the means alone may not accurately represent how the OUs respond to the treatments. There are many mean separation procedures that account for within-treatment variation, including the standard error of the mean (SEM), multiple comparison and multiple range tests, orthogonal contrasts, and trend analysis.

Standard Error of the Mean

The standard error of the mean (SEM) is obtained by dividing the sample standard deviation by the square root of the number of observations for that treatment (Zar, 1984). When most researchers use SEM values for mean separation purposes, they use treatment means along with their respective SEM and the difference between paired values calculated. Treatments are declared different if the collective values for the paired treatments do not overlap. Many researchers use SEM values to compare treatment means. Problems occur because most researchers use SEM values to compare the means of treatments ranked far apart. This use of SEM overestimates treatment differences and violates the assumption of ANOVA that treatment variances are equal (Mize et al., 1999). This does not produce useful results because SEM values increase with the numerical value of the data and do not accurately reflect population variance (i.e., treatments with large values have larger SEM than treatments with small values). When using SEM values to compare treatment means, one SEM value should be calculated from the MSE from ANOVA or SEM values obtained from Poisson or Logistic regression (Mize et al., 1999). This use of SEM is more likely to yield realistic results. However, mean comparison procedures that use sample variance often provide more useful information.

Multiple Comparison and Multiple Range Tests

Multiple comparison and multiple range tests are statistical procedures that use the sample variance in a formula to calculate a numerical value to compare treatment means. Means are ranked as when SEM is used and the difference between compared means calculated. The calculated value is compared with the critical value computed by the mean comparison statistic. If the difference between the compared means exceeds the computed statistical value, the treatments are considered statistically different. However, if the difference between treatment means is equal to or less than the computed statistical value, the treatments are considered similar (Mize et al., 1999). When presenting means in a table or graph, means designed as different are assigned different letters, whereas treatments declared similar are assigned the same letter.

Most researchers use the terms *multiple comparison* and *multiple range tests* interchangeably. However, the two differ in their method of making comparisons. Multiple comparison tests use the same critical value to compare adjacent and nonadjacent means, whereas multiple range tests employ different critical values to compare adjacent and nonadjacent means (Compton, 1994). Multiple range tests are considered to be more accurate than multiple comparison tests because different values are used to compare means ranked far apart, making it less likely that errors will be made when comparing distantly ranked means. Examples of multiple comparison tests are Bonferoni, Fisher's least significant difference (LSD), Scheffe's, Tukey's Honestly Significant Difference Test (Tukey's HSD), and Waller–Duncan K-ratio T test (Waller–Duncan). Examples of multiple range tests include Duncan's New Multiple Range Test (DNMRT), Ryan–Einot–Gabriel–Welsh Multiple F-test (REGWF), Ryan–Einot–Gabriel–Welsh Multiple Range Test (REGWQ), and Student–Newman–Kuels (SNK). The DNMRT is considered one of the best mean comparison tests to use. However, this procedure is not available in all computer software packages.

TABLE 5.4
Comparison of Results Obtained from Analyzing Treatment Means Using Least Significant Difference (LSD) and Tukey's Honestly Significant Difference (Tukey's HSD) Mean Comparison Tests. Data Were Analyzed Following ANOVA for a Single Factor Completely Randomized Design.

		Results of Mean Comparison Tests	
Cytokinin	**Shoot Length (mm)**	**LSD[a]**	**Tukey's[b]**
Zeatin	17.6	a	a
2iP	14.3	ab	ab
BA	11	bc	abc
Zeatin riboside	8.5	cd	bcd
Kinetin	4.1	de	cd
TDZ	1.8	e	d

[a] Means with the same letter are not significantly different according to LSD at 0.05.
[b] Means with the same letter are not significantly different according to Tukey's HSD at 0.05.

In the petunia example, Tukey's HSD and LSD were used to compare the effect of different cytokinin on elongation of regenerated shoots (Table 5.4). Normally, only one comparison test is used to compare treatment means. However, two are used here to demonstrate differences due to procedures. In this situation, a high treatment mean indicates a favorable response; therefore, means were ranked from highest to lowest and means of all treatments compared. Different results were obtained from the two tests. LSD sorted the means into five groups, while Tukey's HSD divided the treatment means into four groups. Which procedure is correct? LSD is considered by many statisticians to be the most liberal mean comparison procedure, that is, most likely to declare means different. This often leads researchers to declare means different that are similar. This is often true with means that are ranked far apart (e.g., highest versus lowest means). For this reason, LSD should not be used unless treatments are considered different in a general test (ANOVA, etc.). Tukey's HSD is considered more conservative than LSD and produces more reliable results.

Results of mean comparison tests should be presented in either tables or bar graphs since treatments are unrelated. If using bar graphs, letters assigned to specific treatments should be positioned above each bar (Figure 5.3).

Multiple comparison and multiple range tests should only be used when treatments are unrelated (Lentner and Bishop 1986). In plant tissue-culture studies, these are treatments in which different plant growth regulators, genotypes, culture vessels, medium solidification agents, etc., are tested. These procedures should not be used to compare means of treatments that are related, for example, different concentrations of the same plant growth regulator, antibiotic, medium solidification agent, activated charcoal, etc. (Compton, 1994; Mize et al., 1999).

Multiple comparison and multiple range tests can be used for count and binomial data. However, the data must be normalized first. Count data can be normalized using square root transformation (Compton, 1994). The arc sin transformation is typically used for binomial data. Once results of transformed data are calculated, the data are converted back to the original scale for presentation.

Orthogonal Contrasts

Multiple comparison procedures are most commonly used by researchers to detect differences between treatment means. However, there are mean comparison procedures that give more

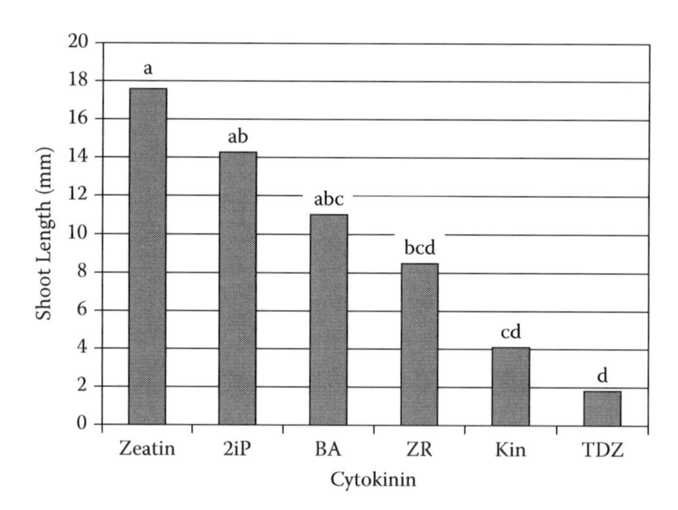

FIGURE 5.3 Effect of cytokinin type on elongation of adventitious shoots regenerated from leaf explants of Petunia x hybrida incubated in MS medium containing 4 μM of zeatin, 2-Isopentenyl adenine (2iP), benzyadenine (BA), zeatin riboside (ZR), Kinetin (Kin), or thidiazuron (TDZ).

meaningful results. Orthogonal contrast is a statistical procedure in which comparisons are made between treatments with similar characteristics (Compton, 1994; Mize et al., 1999). In plant tissue-culture studies, these may be plant growth regulators with similar activity, natural versus synthetic plant growth regulators, or DNA constructs with the same promoter. Orthogonal contrasts differ from mean comparison tests in that more than two treatments can be compared in one test. However, the number of comparisons that can be made are restricted to the number of degrees of freedom for the treatment variable. Contrast statements are usually performed as part of the ANOVA procedure (Compton, 1994). A contrast statement is written that specifies the treatments or group of treatments to be compared, and ANOVA calculates the degrees of freedom, SS, and MS for the comparison. An F-statistic is calculated by dividing the MS for the contrasts by the MSE and the level of significance (p-value) is determined. One degree of freedom is used for each comparison.

In the petunia example, five contrasts of interest were planned before the experiment [1—natural cytokinins (zeatin, zeatin riboside, and 2iP) versus BA, Kinetin, and TDZ; 2—cytokinins containing zeatin (zeatin and zeatin riboside) versus non-zeatin-containing cytokinins; 3—BA (as a control) versus all other cytokinins; 4—BA versus TDZ; and 5—TDZ versus all other cytokinins] and contrasts conducted for all dependent variables as part of ANOVA. Results of orthogonal contrasts are usually presented in a table containing ANOVA computations (Table 5.5). Means of each contrast may be placed in a table or presented in the text with the means and p-value given for each contrast. Examining results for shoot length, shoot elongation was promoted when plant growth regulators containing zeatin (Z and ZR) were used ($p = 0.039$). Likewise, natural cytokinins (Z, ZR, and 2iP) promoted shoot length more than synthetic formulations of BA, kinetin, and TDZ ($p = 0.0001$). Shoot elongation was inhibited ($p = 0.0008$) when TDZ (1.8 mm) was used compared to any of the other cytokinin formulations (11.1 mm).

Orthogonal contrasts can be used to analyze binomial or count data given that the data are transformed before analysis. Information gained from orthogonal contrasts is often much more useful than outcomes of multiple comparison or multiple range tests. Orthogonal contrasts are underutilized in plant tissue-culture studies possibly because researchers have not been exposed to the concept or because their usefulness is not completely understood. See Lentner and Bishop (1986), Little and Hills (1978), or Zar (1984) for more information regarding the use of orthogonal contrasts.

TABLE 5.5
Analysis of Variance Summary Table with Orthogonal Contrasts for Shoot Length from the Petunia Shoot Organogenesis Study

Source	DF	Mean Square	F-Value	*p* -Value
Cytokinin	5	1803.35	12.46	<0.00001
Experimental error	54	1563.50		
Corrected total	59			
Orthogonal contrasts				
zeatin and ZR versus 2iP, BA, kinetin, and TDZ	1	367.5	2.54	0.039
zeatin, ZR and 2iP versus BA, kinetin, and TDZ	1	920.42	6.36	0.0001
BA versus zeatin, ZR 2iP, kinetin, and TDZ	1	25.230	0.17	0.9711
BA versus TDZ	1	423.20	2.92	0.0209
TDZ versus zeatin, ZR 2iP, BA, and kinetin	1	720.75	4.98	0.0008

Note: Data were analyzed using ANOVA for a single factor completely randomized design.

Trend Analysis

Trend analysis is the most effective method for analyzing data of treatments consisting of various levels of a single factor (different concentrations of the same growth regulator) or increments in time (Kleinbaum et al., 1988). With these treatments, the primary objective is to identify a dose or time period that stimulates optimal explant response. Models identifying specific trends (linear, quadratic, cubic) are tested in a stepwise fashion from simplest (linear) to most complex (cubic in most cases). The procedure is stopped when a nonsignificant trend value is identified, and the last significant trend is then considered to best fit the data. Trend analysis uses SS, T, and R^2 values to indicate significant trends. Trends may be tested through regression analysis or polynomial contrasts statements in ANOVA (Compton, 1994).

Trend analysis can be illustrated by changing the cytokinin treatments evaluated in the petunia example from six cytokinin types to six concentrations of BA [0 (control), 2, 4, 6, 8, and 10 μM]. Data recorded included the percentage of explants that produced shoots, number of shoots per explant, and length of regenerated shoots. Since no shoots were produced by explants cultured in medium without BA, observations recorded for 0 BA were deleted before analysis. In addition, zero values from nonresponding explants were deleted from the remaining treatments because only shoot length values from responding explants were of interest. Based on information from responding explants, shoot length was dependent on BA concentration (Table 5.6), with the linear model (Y = 22.954 – 2.1014*BA) giving the best fit ($p < 0.00001$; $R^2 = 0.86$). Means of dosage treatments should be presented in graphs displaying the regression equation and fitted line, R^2, individual data values, and confidence intervals. Similar to mean comparison tests, confidence intervals of treatments that overlap are considered similar. In the petunia example, confidence intervals for 2 μM and 4 μM overlap (Figure 5.4). However, confidence intervals for 6 μM, 8 μM, and 10 μM do not overlap with 2 μM and 4 μM, and are considered significantly different from the former treatments. The conclusion would be that shoot elongation was inhibited with increasing concentration of BA above 4 μM.

Trend analysis and polynomial contrasts may be used for evaluating optimum treatment levels even if those levels were not directly tested but lie within treatment boundaries (Compton, 1994).

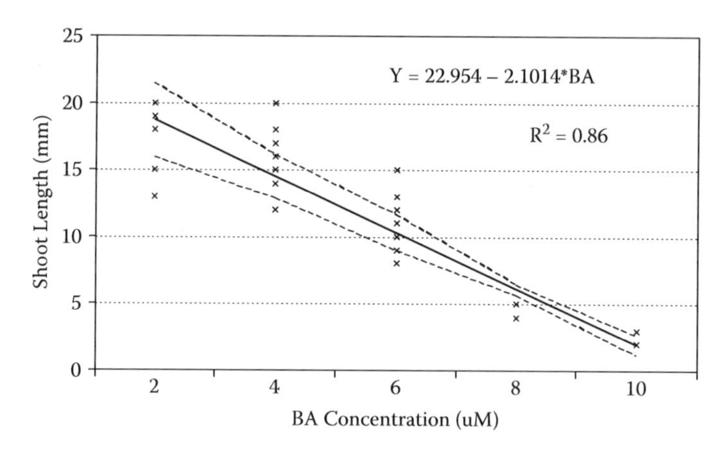

FIGURE 5.4 Influence of benzyadenine (BA) concentration on elongation of adventitious shoots regenerated from leaf explants of Petunia x hybrida incubated in MS medium containing 0, 2, 4, 6, 8, or 10 μM BA.

TABLE 5.6
Results of Regression Analysis Evaluating the Influence of Benzyladenine Concentration on the Elongation of Adventitious Shoots Regenerated from Petunia Leaf Explants

Predictor Variables	Coefficients (COEF)	Standard Error (SE)	COEF/SE	P-value
Constant	22.954	0.9164	25.05	<0.0001
BA	–2.1014	0.1366	–15.38	<0.0001
R^2	0.8616	EMS	4.8968	
Adjusted R^2	0.858	SD	2.2129	
	Regression Summary Table			
Source	DF	MS	F-value	P-value
Linear	1	1158.70	236.62	<0.0001
Quadratic	1	7.7861	1.616	0.485
Residual	37	4.81871		

Note: Explants were prepared from leaves of petunia plants grown in the greenhouse. The influence of various concentrations (0, 2, 4, 6, 8, and 10 μM) benzyladenine on the elongation of adventitious shoots was examined. Data from the control treatment (0 μM BA) were deleted before analysis because none of the explants responded. Likewise, data from non-responding explants incubated in media containing 2 to 10 μM BA were deleted before analysis. Data are hypothetical.

Predicted values can be obtained using information generated from the analyses, allowing researchers to estimate the effects of treatments that were not evaluated.

Trend analysis can be used to evaluate binomial and count data provided that the data are transformed prior to analysis. As with mean comparison procedures, data are converted back to the original scale for presentation.

CONCLUSIONS

It is important to remember that any well-conceived research project should be planned carefully before implementation begins. This helps to ensure that the experiments will answer the desired questions. Be sure to minimize bias, fatigue, and errors during the course of the experiments. These factors lead to untrue results that mislead researchers. Use the experimental design while conducting the experiment, and recording and analyzing data. This helps to provide an objective, nonbias way to accurately evaluate treatment differences. Used properly, statistics is a valuable tool to the scientific researcher, allowing investigators to declare experimental results with confidence.

REFERENCES

Barnett, V. and T. Lewis. 1984. *Outliers in Statistical Data*. John Wiley & Sons, Chichester, England.

Compton, M.E. 1994. Statistical methods suitable for the analysis of plant tissue culture data. *Plant Cell Tiss. Organ Cult.* 37: 217–242.

Compton, M.E. 2006. Statistics in plant biotechnology. pp. 145–163. In V.M. Loyola-Vargas, and F. Vázquez-Flota, eds. *Plant Cell Protocols*, Second edition. Humana Press, Totowa, NJ.

Compton, M.E. and C.W. Mize. 1999. Statistical considerations for in vitro research: I—Birth of an Idea to Collecting Data. *In Vitro Cell. Dev. Biol.—Plant* 35:115–121.

Kempthorne, O. 1973. *The Design and Analysis of Experiments*. Robert E. Krieger Publishing Co., Malabar, FL.

Kleinbaum, D.G., L.L. Kupper, and K.E. Muller. 1988. *Applied Regression Analysis and Other Multivariable Methods*, Second edition. PWS-Kent Publishing, Boston.

Kuklin, A.I., R. N. Trigiano, W.L. Sanders, and B.V. Conger. 1993. Incomplete block design in plant tissue culture research. *J. Plant Tiss. Cult. Meth.* 15:204–209.

Lentner, M. and T. Bishop. 1986. *Experimental Design and Analysis*. Valley Book Company, Blacksburg, VA.

Little, T.M. and F.J. Hills. 1978. *Agricultural Experimentation: Design and Analysis*. John Wiley & Sons, New York.

Mize, C.W. and Y.W. Chun. 1988. Analyzing treatment means in plant tissue culture research. *Plant Cell Tiss. Organ Cult.* 13:201–217.

Mize, C.W., K.J. Koehler, and M.E. Compton. 1999. Statistical considerations for in vitro research: II—Data to presentation. *In Vitro Cell. Dev. Biol.—Plant* 35:122–126.

Mize, C.W., and P.M. Winistorfer. 1982. Application of subsampling to improve precision. *Wood Sci.* 15:14–18.

Zar, J.H. 1984. *Biostatistical Analysis*, Second edition. Prentice-Hall, Englewood Cliffs, NJ.

6 Proper Use of Microscopes

David T. Webb

Although the compound microscope is the most commonly used biological instrument, it is often used improperly. This may not matter with very thin commercial slides at low-to-medium magnifications. However, proper alignment of the illumination system is essential for viewing thick sections, whole mounts, and, for highly magnified samples of cells, fungi and bacteria. It is also crucial for studying unstained specimens and for photomicroscopy.

This chapter was written as if you knew nothing about using a microscope. In some cases you may have learned bad practices that need to be corrected. It is also written for microscopes that have a field diaphragm, and a condenser that can be centered and focused to achieve Kohler illumination (Kohler, 1893). Many student scopes do not have these features because their condensers and field diaphragms are rudimentary. In the course of your careers you will encounter microscopes that have the ability to achieve Kohler illumination. At that point this article will be even more useful.

You will be using microscopes throughout your career. If you learn the simple lessons in this chapter, you will do much better work and see the exciting world of microscopy in a new light. The modified procedure we present was developed by the German scientist, August Kohler (1866–1948), and bears his name. Recently, his ideas were used to make the EM 910 Electron Microscope by Carl Zeiss. Thus, this procedure, which was introduced in 1893, has been of lasting value.

Microscopes are partly categorized by the number of oculars they contain. The first microscopes had one ocular and were called Monocular. Binocular scopes have two oculars, while Trinocular scopes have three. The third ocular is typically modified for the use of a camera.

THE COMPOUND MICROSCOPE

Because the optical units in a microscope are composed of many lenses, the term *compound microscope* is used. This is specifically applied to microscopes used to study thin sections with high power objectives (Figure 6.1). Dissecting or stereo microscopes (Figure 6.2) are used to examine larger, three-dimensional specimens at lower magnifications and with greater working distances. These also have compound lenses, but they are not generally called compound microscopes.

Both types of microscope can use transillumination (through illumination), in which light passes from the microscope base through the specimen and travels to your eyes through objectives and oculars. This requires a special transillumination base for stereo microscopes (Figure 6.2). Epiillumination (illumination from above) is typically used with stereo scopes but not with compound scopes. I am using a Zeiss Standard compound microscope and an American Optical stereo microscope to illustrate this chapter. Your microscopes may be somewhat different, but you should be able to transfer the terminology and procedures described herein to your instrument. We will examine compound microscopes first and deal with stereo scopes later.

Microscope Care and Handling

Please treat these instruments with care. Value what they can do and handle them with respect.

- Use two hands to carry microscopes.
- Place one hand on the arm, and one under the base. The microscope arm is the curved area that connects the Body to the Stage and Base (Figure 6.1 and Figure 6.2)

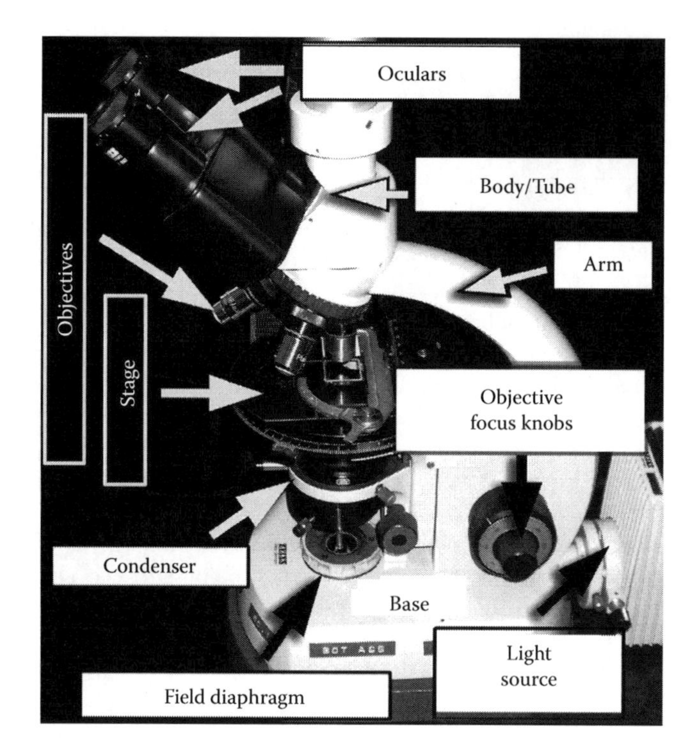

FIGURE 6.1 American Optical Standard Microscope showing major parts of a typical compound microscope.

- Do *not* carry scopes sideways or upside down, as the oculars and other parts will fall out.
- Use lens paper to clean all lenses on the compound scope before each lab session and especially after using immersion oil.
- Do not use any other kind of paper except lens paper to clean microscope lenses.
- Do not use liquids (except where specified) when cleaning the lenses.
- Always use the correct focusing technique to avoid contact between any objective and your slide.
- Use a plastic petri plate or some other surface to perform dissections and observations with a stereo microscope. This protects the stage plate.
- Turn off the light when not using scopes for long time periods to avoid overheating and to extend bulb life.
- Carefully place the power cord or any other cords out of the way at your workspace.
- Do the same when you return your microscope to its storage place.
- Always replace the cover on the microscope when you put it away.
- Report any problems immediately.
- Do not use the microscope if you cannot see your specimens clearly.

The Light Path

The light in a typical compound microscope follows the path below.

Light Source → Mirror → Field Diaphragm → Condenser → Stage → Specimen → Objective → Body/Tube → Ocular → Eye or Camera

- Locate the major parts of your microscope by referring to Figure 6.1.

There are various control knobs on the microscope that affect the light path. In addition, there are knobs for coarse and fine focusing, as well as knobs to move the stage.

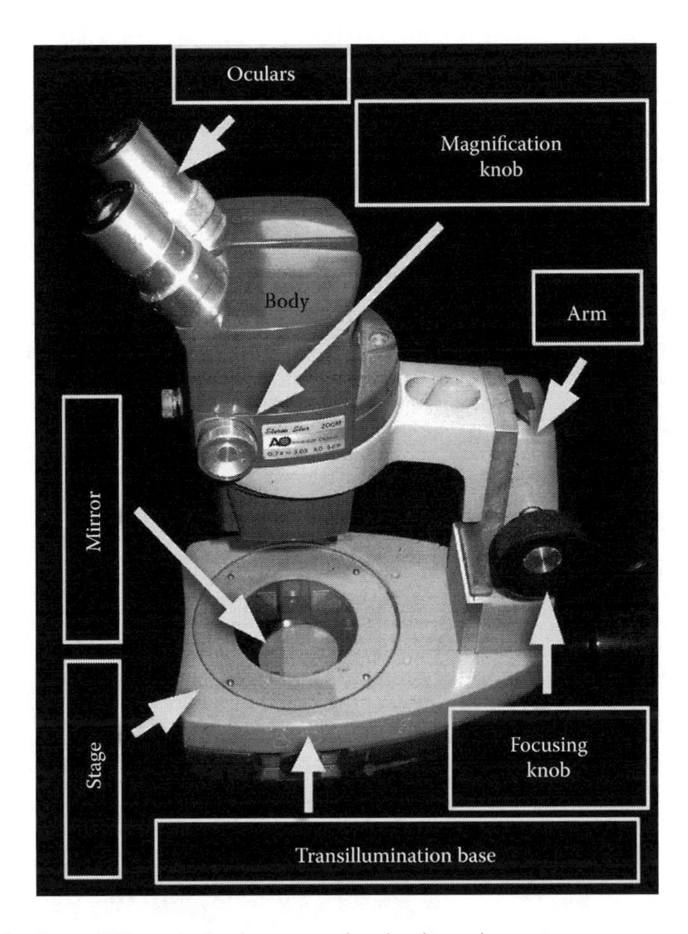

FIGURE 6.2 Zeiss Stereo (Dissecting) microscope showing its major parts.

Focusing the Objectives

The objectives are focused on the specimen by two sets of knobs that are located on both sides of your microscope (Figure 6.1). The large outer knob is for coarse focusing. This should be used at the lowest magnification when you first place a specimen on the stage. It should be used with caution at higher magnifications. The smaller, central knob is used for fine focusing.

- Locate the coarse and fine focusing knobs on each side of your scope.

Moving the Mechanical Stage

Light passes through the Stage Opening so that it can illuminate the specimen. The knobs that control the mechanical stage (Stage Transport Knobs) are usually on the left side of the microscope as it faces you. One of these moves the stage from side to side (X-axis), while the other moves the stage in and out (Y-axis). Most stages have X and Y scales. These allow you to record the precise location of objects that you may want to relocate without searching the entire specimen. Slides are held in place by a Mechanical Slide Holder.

- Locate the Stage Transport Knobs on your microscope.
- Locate the X and Y scales on your stage.
- Locate the Mechanical Slide Holder and explore its mode of action.

Using the Condenser

The condenser aligns and focuses light on the specimen. It may be equipped with a high-power condenser lens (Figure 6.3). This is used with 10 to 100 X objectives but is removed from the light path with low magnification objectives (4 to 5 X). The position of the High Power Condenser Lens (HPCL) may be controlled by a rotating knob. On some microscopes it is moved in and out of the light path by a push–pull plunger or a lever.

Failure to use this lens properly is the most common mistake that people make. The lens is typically left out of the light path at low magnification because it limits the field of illumination. A fully illuminated field is achieved with the HPCL out of the light path with low power objectives. There may not be a large penalty for examining commercial slides at higher magnification with the HPCL out, but there is a severe visual penalty at higher magnifications and with fresh mounts.

- Locate the High Power Condenser Lens on your microscope and determine how to move it into and out of the light path.

Condenser Centering Screws

A pair of screws, set apart by a 45° angle, is used to center the condenser. These are located along the back right and left sides of the condenser on the Zeiss Standard (Figure 6.3). However, they may be found near the front of the condenser with other scopes. In some cases there are four screws.

- Locate the Condenser Centering Screws on your microscope.

The condenser is used to focus light onto the specimen from below.

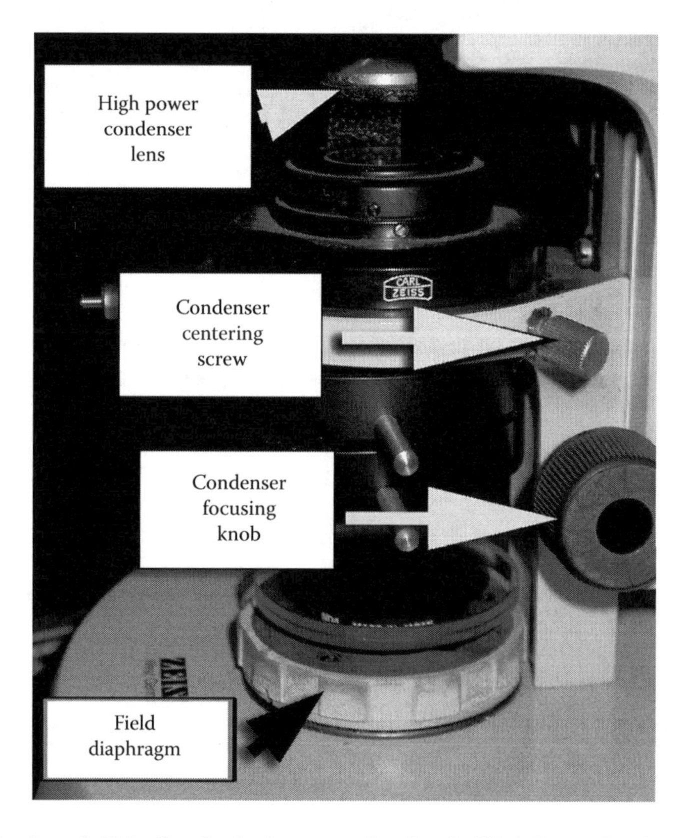

FIGURE 6.3 Side view of a Zeiss Standard microscope showing the High-Power Condenser Lens, Condenser Focusing Screws, and the Condenser Focusing Knob plus the Field Diaphragm.

This is an extremely important, but poorly understood function of the condenser. It is obvious that you must focus the objectives onto the specimen to see it clearly. It is less obvious that you need to focus the condenser on the specimen so that it is properly illuminated. Imagine that the condenser is a magnifying glass and you want to start a fire from straw. You need to move the magnifying glass to a position that produces the smallest focused beam of sunlight in order to start your fire. That is exactly the same concept you need to keep in mind when you focus the condenser. This is done by rotating the Condenser Focusing Knob, which is found on the right side of the condenser with the Zeiss Standard (Figure 6.3).

- Locate the Condenser Focusing Knob on your microscope.

Condenser (Aperture) Diaphragm

Finally, there is a lever that controls the diameter of the Condenser or Aperture Diaphragm (Figure 6.4). I like to refer to this as the Condenser Diaphragm to prevent confusion with the Field Diaphragm. However, it has typically been called the Aperture Diaphragm. Partially closing this iris improves contrast (the difference between light and dark), especially at intermediate and high magnifications. It also increases the depth of field, which is very small at high magnifications.

Do NOT use this to increase or decrease brightness! This is the second most frequent mistake that people make!

It is best to leave this completely open (rotated to the left on the Zeiss Standard) at the outset. Later, you will experiment with this to see its effects.

- Locate the Condenser (Aperture) Diaphragm Lever and manipulate it.
- Leave it in the open position for now.

Using the Field Diaphragm

The light source is usually housed in the base of the microscope. Light passes through the Field Diaphragm that also contains an iris (Figure 6.5). The size of its iris is controlled by rotating a knurled ring that is concentric with it. The Field Diaphragm controls the area or field of illumination.

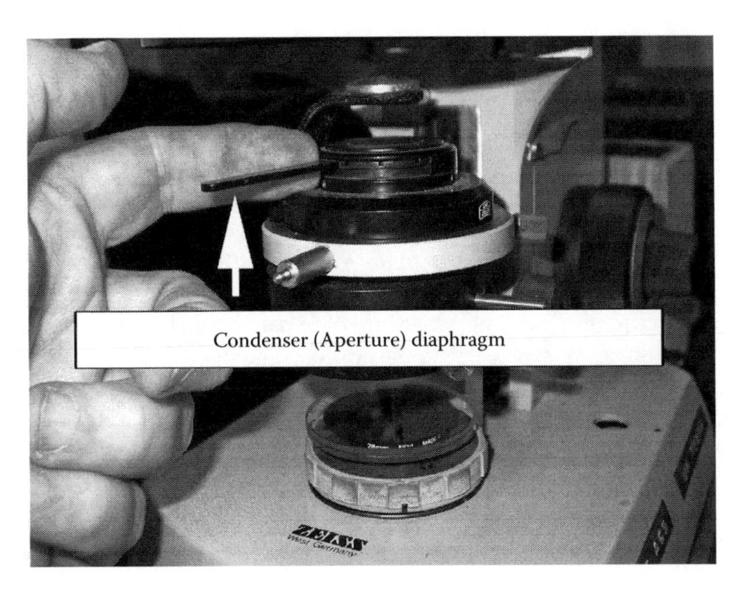

FIGURE 6.4 Lever that controls the Condenser (Aperture) Diaphragm. It is completely open in this position. Rotating the lever to the right closes the iris diaphragm inside.

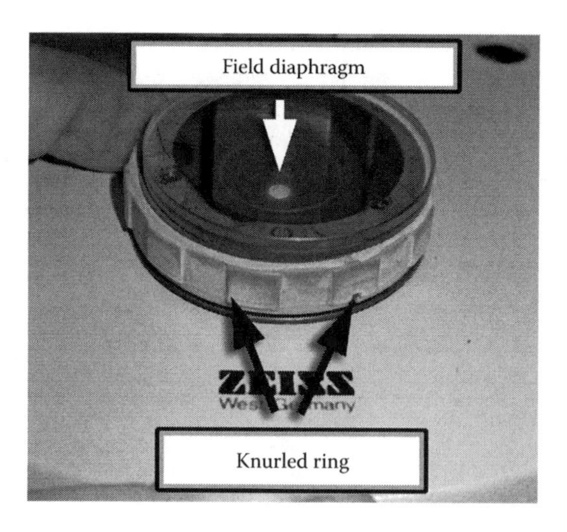

FIGURE 6.5 Field Diaphragm that is almost completely closed. Note the small illuminated polygon indicated by the white arrow. The knurled ring is used to open and close the iris inside.

- Locate the Field Diaphragm on your microscope.
- Manipulate the Knurled Ring to open and close the iris inside.
- Leave it fully open for now.

Using the Objectives

The magnification of an image is regulated by the objectives that are housed in a rotating nose piece (Figure 6.6). To change objectives, you rotate the nosepiece clockwise or counterclockwise. Be sure to know the correct rotation for your microscope.

Be sure that the low power objective is in place before you start using your scope AND when you are finished using it. This prevents damage to the objectives and your specimen.

- Always start viewing with the low power objective (4X or 5X).
- Do not start by swinging in the 10X to 100X objectives.
- Be sure that you rotate the nosepiece in the correct direction! You do not want to switch from 4X to 100X!
- Objectives may be damaged if they hit the specimen.
- You should focus on the sample with each objective before switching to the next.
- *Focus on the sample using the low power objective and rotate to the next lens (10X), refocus and repeat this until you reach the magnification you plan to use.*
- *It is vital to focus on your specimen with the 40X objective before switching to 100X.*

Very good quality microscopes have objectives that are parafocal. This means that you should be able to focus with one objective and switch to the next one and the one after that without refocusing. However, this rarely occurs, especially with student scopes. Special care must be used with fresh sections and whole mounts. These can be thick and irregular. Consequently, greater care must be taken when changing objectives. When in doubt, play it safe.

A typical microscope will have a series of objectives such as 5X, 10X, 20X, 40X, and 100X. The length of an objective is a rough indication of its magnification. However, the magnifications of microscopes are engraved on them (Figure 6.7). You should check to see exactly what your objective magnifications are.

- Check the magnification of your objectives.

FIGURE 6.6 Rotating nosepiece with objectives. In this case the nosepiece is rotated clockwise (left to right) to change objectives to higher magnifications.

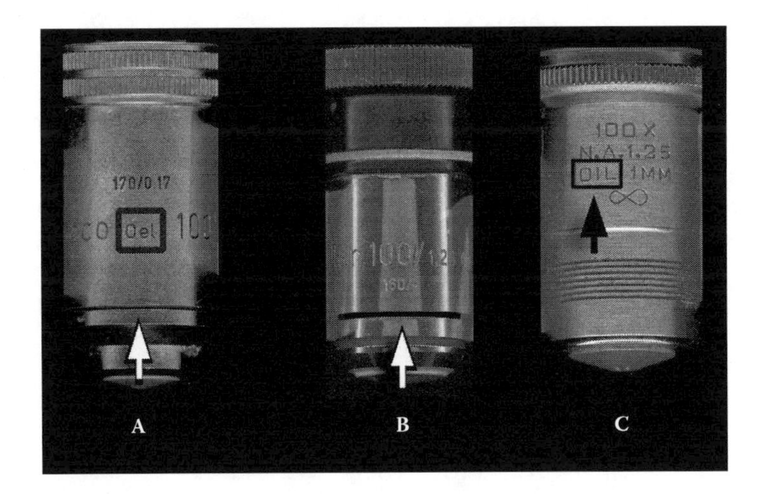

FIGURE 6.7 Oil Immersion Objectives from Leitz (A), Zeiss (B), and American Optical [AO] (C). Note the black lines (indicated by the white arrows) on A and B. This means that they are oil immersion objectives. The AO objective (C) lacks this line. However, it has the word *oil* inscribed on it as indicated by the black arrow and box. The word *oel* (oil) is inscribed on the Leitz objective (A). Note the markings N. A. 1.25 on objective C. This indicates a Numerical Aperture of 1.25. Immersion objectives typically have a *numerical aperture* greater than 1.0.

Most immersion objectives have a black line near their tip (Figure 6.7). However, this is not true for all manufacturers. The words *oil* or *oel* indicate that it is an oil immersion objective. Oil improves the image because it unites the cover slip and the objective. It replaces air with oil. Immersion oil has the same refractive index as glass. Thus, less light scattering and refraction occur when oil is used. It also protects the objective lens from scratching.

The markings HI (Homogenous Immersion) or Imm (Multiple Immersion) indicate oil or multiple medium immersion objectives, respectively. Water Immersion objectives are marked with the words *water*, *waser*, *water immersion*, or *WI*. Furthermore, an objective with a numerical aperture greater than 1.0 (Figure 6.7) is probably an immersion objective. Oil is the most common immersion medium, but water and glycerin can also be used with Imm objectives. Oil immersion objectives can be used without oil, but image quality will be degraded.

- Check your 100X objective to see its markings.

The procedure for properly using oil immersion objectives can be found in Procedure 6.1. Also see Peterson et al. (2008).

For optimal results, oil should also be placed between the uppermost condenser lens and the bottom of the slide. The condenser needs to have a Numerical Aperture greater than 1.0 in order for the oil to have a beneficial effect. This may be impractical for routine studies, but should be used for critical examinations and for the most detailed microphotography. However, most oil immersion objectives require you to partially close the Condenser (Aperture) Diaphragm. This effectively lowers the numerical aperture of the condenser such that adding oil is no longer beneficial.

Using the Oculars

The oculars must be adjusted to suit both of your eyes! You should be able to adjust the interpupillary distance between the two objectives. This means that you can move them to match the distance between your eyes. This is typically done by grasping the base of the ocular tubes and gently spreading them apart or drawing them together. In some cases there is a dial that you can use to move the oculars. Either one or both oculars may be movable. There should be a scale and a reference line or dot that allows you to record the best spacing for your eyes. You can readily readjust the oculars to your personal setting if they have been moved. Follow the steps in Procedure 6.2 to adjust the interpupillary distance of your oculars.

Your head position is very important! You need to find a comfortable distance for your eyes from the oculars in order to see things properly. This depends on the type and quality of your oculars and may require some experimentation. Most oculars can be used without glasses that are corrected for near- or farsighted eyes. Oculars are not compensated for astigmatism. If you have astigmatism, you need to use your eyeglasses or contact lenses. Oculars with eyeglasses engraved on them are suitable for use with glasses (Figure 6.8), but you are not required to wear glasses to use these.

Procedure 6.1
Using an Oil Immersion Objective

Step	Instructions and Comments
1	Locate the area of interest in the center of the field.
2	Focus on this with the 40X objective.
3	Raise the objective to its upper limit.
4	Swing the oil immersion objective into viewing position.
5	Place a drop of oil on the 100X oil immersion objective.
6	Place a small drop of oil on the cover slip.
7	Do NOT look through the oculars.
8	While observing from the side of the stage, focus your eyes on the objective and the specimen. Your eyes must be at the same height as the stage.
9	Lower the objective lens carefully (use the Coarse Focusing Knob) until it just touches the oil on the cover slip.
10	A light flash may be observed when the oil on the objective meets the oil on the slide.
11	Now look through the oculars and use the coarse focusing knob to bring the specimen into rough focus.
12	Use the Fine Focusing Knob to complete focusing on the sample.
13	Hereafter, avoid focusing down on the specimen with an oil immersion lens.
14	Change the focus so that the objective is traveling up, away from the slide.
15	If the image does not come into focus readily, carefully reverse the direction until it does.
16	When in doubt, STOP! Ask for HELP! The lens might be dirty or there may be some other problem.
17	IMPORTANT! Wipe the oil from the objective with Lens Paper when you are finished.
18	Clean the objective until no more oil is visible on the lens paper.
19	Wipe oil from the cover slip of the slide if it is to be saved.

Procedure 6.2
Adjusting the Interpupillary Distance of the Oculars

Step	Instructions and Comments
1	Position your head so that you can see through the oculars while focusing on a sample.
2	Focus on a specimen at 10X – 20X.
3	Grasp the base of the ocular tubes or use the dial located between the oculars.
4	Move the oculars so that they are as close together as possible.
5	Carefully move the oculars apart until you see only one image.

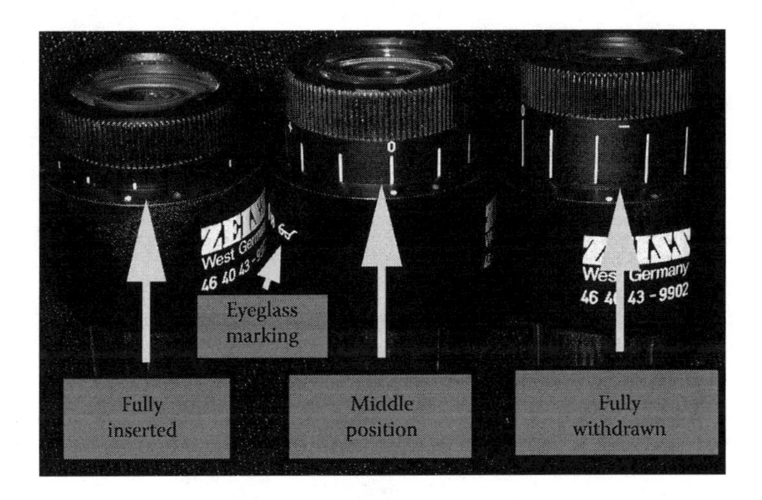

FIGURE 6.8 Zeiss oculars showing three focusing positions. These are focused by grasping the knurled ring at the top of the ocular and rotating it while holding on to the barrel below. The one on the left has been adjusted so that the rotating part of the ocular is fully inserted into the barrel. The one on the right shows the extreme opposite rotation. The one in the center is approximately in the middle. In this case the position of the ocular is indicated by the length of the white reference lines. In other cases there are numbers that can be used to designate the best focusing position for the ocular. Note the eyeglasses engraved on the middle objective. This indicates that these oculars were designed to be used with your glasses.

It is important that each ocular is in focus when you examine samples. Oculars that are capable of independent focusing will have a scale, a reference line, and a knurled ring on them (Figure 6.8). These markings may be on the ocular tube rather than on the oculars themselves. Follow procedure 6.3 to focus your oculars.

Kohler Illumination

The best resolution occurs when all elements of the microscope are in perfect alignment, and the iris diaphragms are properly adjusted to the best apertures for the objective you are using. On simple microscopes you may not be able to alter the alignment of the condenser, but on the Zeiss Standard and many other microscopes it is possible to do this and achieve "Kohler Illumination." This makes a significant difference for viewing unstained and lightly stained samples, especially at high magnifications.

Centering the Lamp Filament

The first step in this process involves centering the lamp filament. This may not be possible with your microscope. Furthermore, it is best done by someone who is very familiar with this process. Check the illuminator housing on the back of your microscope to see if there are any adjustable screws. This process will be described from a generic perspective so that you can attempt this with your scope.

- Turn on the microscope illuminator.
- Place a piece of thin paper over the Field Diaphragm.
- If the illumination is uneven, use the Lamp Centering Screws on the lamp housing or rotate the lamp to get uniform illumination.

Focusing on the Specimen for Kohler Illumination

The recommended procedure for focusing on a specimen as part of achieving Kohler illumination is given in Procedure 6.4. It is important to focus on a specimen before doing anything else! You may want to move the specimen out of the light path for the remaining steps the first few times you do this.

Focusing and Centering the Field Diaphragm

This is the heart of Kohler illumination. See Procedure 6.5 for the detailed steps in this process. Briefly, completely close the Field Diaphragm and use the Condenser Focusing Knob to focus the Field Diaphragm until you see that it is a small polygon (Figure 6.9A). Use the Condenser Centering Screws to center the image of the Field Diaphragm (Figure 6.9B). Partially open the Field Diaphragm and center it again (Figure 6.9C). Open the Field Diaphragm until the field is completely illuminated and stop.

Procedure 6.3
Focusing Oculars for Your Eyes

Step	Instructions and Comments
1	Before you make any adjustments, place a slide on the stage and focus on the central part of the specimen with a 10–20X objective.
2	Locate the oculars that can be focused independently. There may be only one, or there may be two.
3	Block one of the oculars.
4	Look through the other ocular with your matching eye (left eye ← → left ocular, or right eye ← → right ocular) and focus on a fine detail in the center of the specimen with the objective focusing knobs at the rear of the scope.
5	Switch to the other ocular that is capable of independent focusing, and look through it with the matching eye.
6	Do NOT look through the other ocular while you are doing this!
7	Rotate the knurled ring of the ocular to bring the fine detail into sharp focus. You will need to stabilize the barrel of the ocular with one hand while you turn the knurled ring with the other.
8	Check the first ocular to see that the image is still in focus with your other eye. Both oculars are now focused for your eyes.

Procedure 6.4
Focusing on a Specimen as Part of Kohler Illumination

Step	Instructions and Comments
1	Adjust the oculars so they have the correct interpupillary position and are in focus for your eyes.
2	Use a commercial slide with obvious, well-stained contents and move the specimen into the light path.
3	Focus on the specimen with your low power objective.
4	Rotate the 10X objective into viewing position.
5	Insert the high power condenser lens into the light path.
6	Watching from the side of the stage (NOT looking through the oculars), lower the 10X objective so that it comes close to the cover slip. Make a note on which direction lowers and retracts the objectives.
7	Look through the oculars and rotate the objective focusing knobs so the objective is retracted from the slide.
8	Stop when the sample comes into focus. Moving the mechanical stage during this process may help.

Adjusting the Condenser (Aperture) Diaphragm

When working with 10–100X objectives, it is important to adjust the Condenser (Aperture) Diaphragm (Figure 6.4). This is especially true for translucent structures. Closing this iris increases contrast. Thus, something indistinct becomes sharp, and something faint becomes dark. It also improves the depth of field, which is critically small at high magnifications. It is usually possible to close the iris and judge its effects subjectively. However, there is a "tried and true" procedure (Procedure 6.5) you should know.

In practice, you can experiment with this while viewing a specimen and adjust it without removing the ocular (Delly, 1988). In general, I slowly close this iris until I first see a perceptible change in the specimen, and I take a photograph. I close it some more and take another photo. I may repeat this for a third time. In reality, each specimen is different, and strict rules such as those outlined may not give the best results. Closing the Condenser (Aperture) Diaphragm also increases depth of field. Thus, more regions of a three-dimensional specimen will be in focus. However, if it is closed too much, a flat indistinct image results.

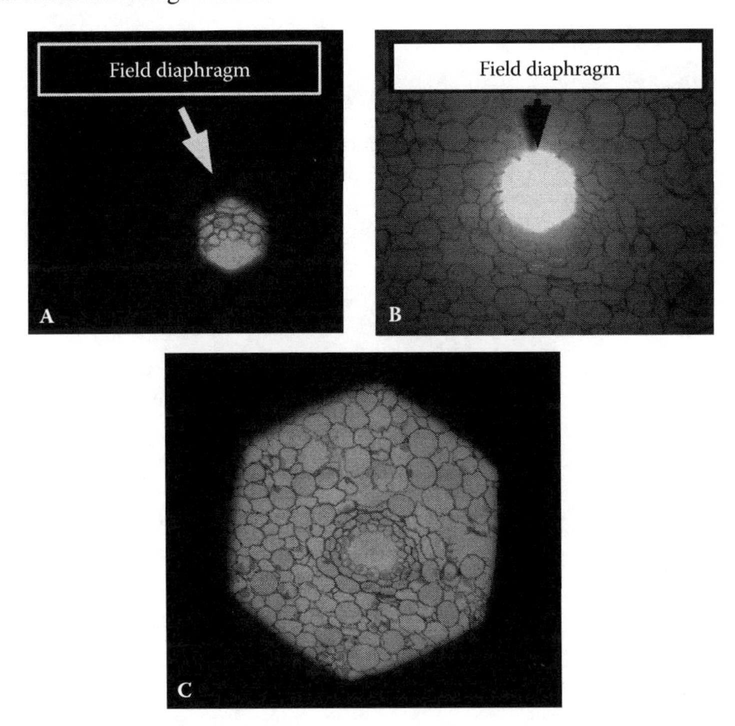

FIGURE 6.9 Focused (A) and Closed (B) Field Diaphragm, uncentered (A), and centered (B). Partly open and centered Field Diaphragm (C).

Procedure 6.5
Adjusting the Condenser (Aperture) Diaphragm

Step	Instructions and Comments
1	Complete all of the preceding operations.
2	Place a lightly stained specimen in the light path.
3	Focus on the specimen at 20–40X.
4	Remove one of the oculars and look directly down the tube at the light field.
5	Close the Condenser (Aperture) Diaphragm so that it occludes 1/4–1/3 of the area. This should give the best contrast.
6	Examine a specimen before and after adjusting this iris.
7	This should be done for each objective for critical viewing.

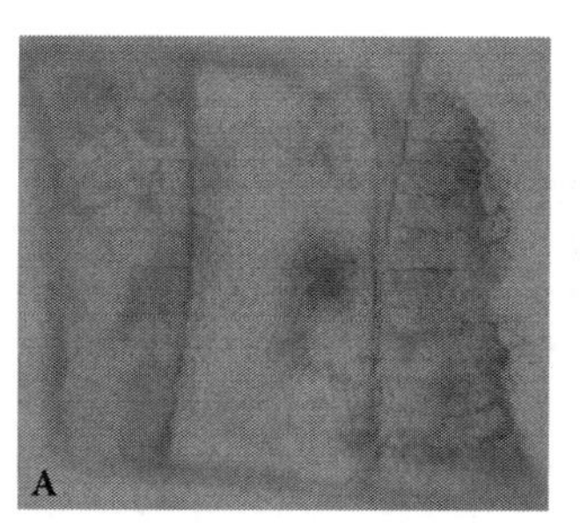

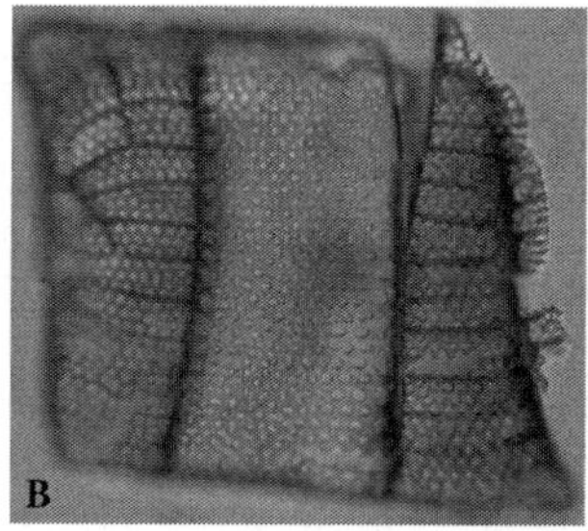

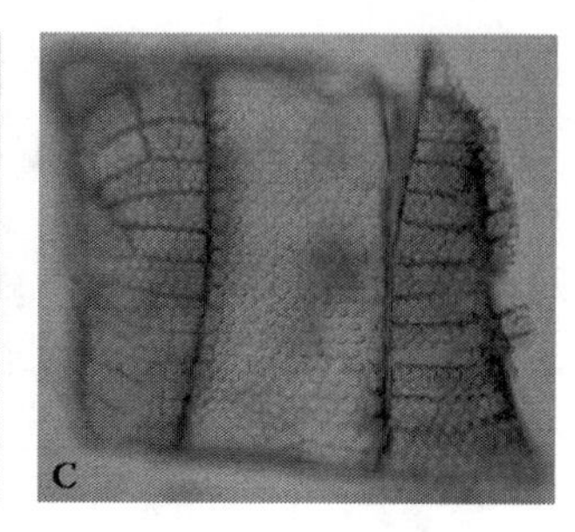

FIGURE 6.10 Diatom with Condenser (Aperture) Diaphragm completely open (A), completely closed (B), and partially closed (C). It is hard to see any details and most of the subject is out of focus due to the shallow depth of field in A. More details are visible due to increased contrast, and there is greater depth of field in B compared to A. More details are visible, and there is greater depth of field in C compared to A.

The examples in Figure 6.10 show a diatom frustule that is composed of silicon and is very translucent. There is little detail when the Condenser Diaphragm is wide open (Figure 6.10A). When it is fully closed (Figure 6.10B) the contrast and depth of field are greatly increased. When the iris is closed 25–30%, there is improved contrast and depth of field with less theoretical potential for aberrations (Figure 6.10C). In this case, Figure 6.10C should have been the best image, but Figure 6.10B appears to be the best.

- Experiment with the Condenser Diaphragm while viewing a lightly stained or unstained specimen.
- Once you have achieved what you think gives the best image, remove one of the oculars and see how much of the field is occluded by looking down the barrel. You may need to open and close the iris to see what is happening.

Throughout your career you will be using different stains to study their effects on fresh specimens. Experiment with the Condenser (Aperture) Diaphragm as you study these.

While these procedures may seem tedious, they will become routine as you progress in your work. Your results will be superior to that of others who do not know how to do this.

Adding Coverslips to Wet Mounts

It is essential that air bubbles be avoided when adding coverslips to wet mounts as they will interfere greatly with your observations. Follow the steps in Procedure 6.7 to avoid bubbles when adding a coverslip. Basically, you use forceps or a dissecting needle to slowly lower the coverslip onto the wet mount so that the specimen is covered without the formation of large air bubbles (Peterson et al., 2008).

Simple Measurements with a Compound Microscope

In most cases you CANNOT accurately determine the magnification of a compound microscope by multiplying the magnifications of the ocular and the objective. Microscope parts are not that precisely manufactured. Furthermore, the length of the microscope tube/body differs from one type of scope to another. This is especially true in photography because projection lenses of different magnifications may be used in place of oculars, and the total distance of the light path may be different from that used with the oculars. We will work through the procedure (Procedure 6.8) for calibrating an ocular micrometer that can be used to make measurements during observations.

Before we proceed, a quick review of the metric system will be helpful. A millimeter (mm) is 10^{-3} meters, while a micron (μ) is 10^{-6} meters. Consequently, 1 mm equals 1,000 μ, 0.1 mm equals 100 μ, and 0.01 mm equals 10 μ. A Stage Micrometer (Figure 6.11) is used to precisely determine magnification.

Procedure 6.6
Focusing and Centering the Condenser

Step	Instructions and Comments
1	Use the 10X objective to focus on the center of a specimen.
2	Reduce the illumination to a moderate level so that you do not hurt your eyes.
3	Check to see that the Condenser (Aperture) Diaphragm is open.
4	Check to see that the High Power Condenser Lens is in the light path.
5	Close the Field Diaphragm so that the circle of light becomes smaller.
6	Observe the Field Diaphragm through the oculars when it is being closed (Figure 6.9).
7	When the diaphragm is as small as possible, use the Condenser Focusing Knob (Figure 6.3) to make the "circle" of light as small as possible.
8	You should see that the Field Diaphragm is NOT circular in outline but has a polygonal shape (Figure 6.9). You may see a red or blue fringe as you bring the field diaphragm into focus. The best position is the one in between the red and blue fringes.
9	Use the Condenser Centering Screws to center the Field Diaphragm.
10	Open the Field Diaphragm by rotating its knurled ring. It may not be perfectly centered.
11	Perform final centering of the Field Diaphragm when it fills most of the field.
12	Expand the Field Diaphragm just beyond the field of view and stop!
13	Repeat this with the 20X and 40X objectives. For critical work, this should be done for each objective. This is especially important for taking photographs and for examining minute, translucent specimens such as fungi and bacteria.
14	This is difficult to do this with the 100X objective. However, if you achieve proper alignment with the 40X objective, the 100X will be similar.

Procedure 6.7
Proper Method for Adding a Coverslip to a Wet Mount

Step	Instructions and Comments
1	Place your sample in 2–3 drops of water or stain in the center of the slide.
2	Use a large (20 × 40 mm) coverslip.
3	Place one end of the coverslip on the slide at a 45° angle without touching the solution containing the specimens.
4	Steady this end with your thumb and index finger.
5	Grasp the other end of the coverslip with fine forceps.
6	Alternatively, rest it on a dissecting needle.
7	Slowly lower the forceps or needle until the coverslip touches the solution.
8	Continue until the forceps or needle touch the slide.
9	Release your grip on the forceps.
10	Slowly remove the forceps or needle by sliding them along the slide.
11	Remove excess solution by touching the side of a Kimwipe or paper towel near one of the coverslip edges.
12	Be careful to not sponge out your samples with the excess solution.
13	Slowly remove the Kimwipe so that you do not drag the coverslip over the slide.
14	If you have been using a stain that must be removed, add water to one end of the coverslip.
15	Withdraw the stain at the opposite end by blotting with a Kimwipe or paper towel.
16	Wipe excess fluid from the bottom of the slide, or it will stick on the stage and make slide transport difficult.
17	Excess fluids may damage the stage or other microscope parts.
18	Carefully place the slide into the slide holder on your stage.

Calibrating an Ocular Micrometer

The Stage Micrometer is the *known* in this process. It has finely etched distance calibrations on its surface. The largest dimensions from one end to the other are millimeters (Figure 6.11 A and B). Each millimeter (1,000 μ) is divided into 0.1 mm (100 μ) segments (Figure 6.11B). Each 0.1 mm segment is divided into 0.01 mm (10 μ) segments.

Procedure 6.8
Calibrating an Ocular Micrometer

Step	Instructions and Comments
1	Place the Stage Micrometer onto the stage of your microscope and move it into the light path.
2	Focus on its scale with the 4X Objective.
3	Move the Stage Micrometer so that some of the reference lines on it coincide with reference lines on the Ocular Micrometer.
4	Because the *distances* between the lines of the *Stage Micrometer* are *known*, you divide this known distance by the number of lines from the Ocular Micrometer.
5	This gives you the distance measured by the intervals of the Ocular Micrometer at that magnification.
6	Repeat this for each ocular on your microscope.

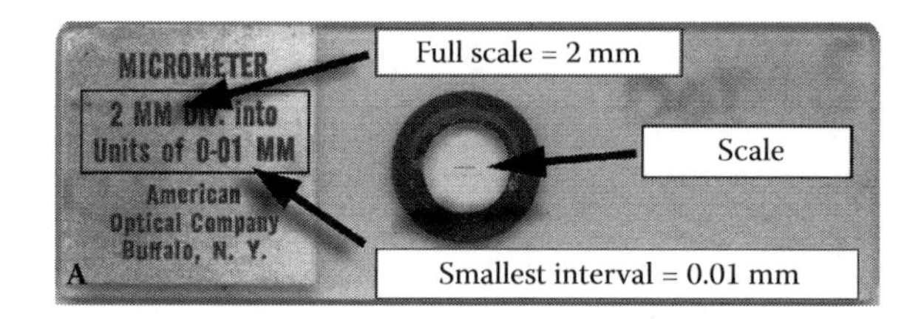

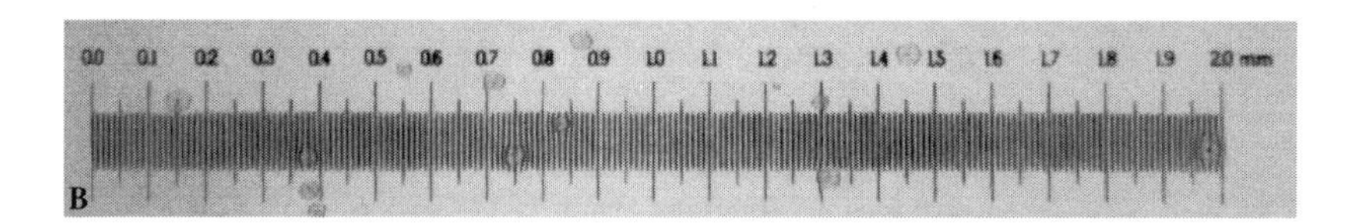

FIGURE 6.11 Stage Micrometer (A) and magnified scale from the Stage Micrometer (B). The total length of the scale is 2 mm (2,000 μ). It is divided into 20 intervals of 0.1 mm or 100 μ (B). Each of these is further divided into intervals of 0.01 mm or 10 μ. This is an old micrometer and there are some defective areas, but it is still usable.

The ocular micrometer has precisely etched lines engraved on it. However, due to differences in the optics of individual microscopes, ocular micrometers must be calibrated with a stage micrometer. The steps in this procedure are contained in Procedure 6.8. Briefly, the stage micrometer and the ocular micrometer are brought together under the microscope at 10X so that the intervals of the optical micrometer (unknown) are matched with intervals on the stage micrometer (known). The actual distance between units of the ocular micrometer is determined by dividing the known distance (stage micrometer) by the unknown units (ocular micrometer). This gives the actual distance for each unit on the ocular micrometer (Figure 6.12).

You need to record this value and repeat it for all of the objectives on your scope. If you transfer your ocular micrometer to another microscope, you must repeat this process for that scope.

In the example I have provided (Figure 6.12), two large units on the ocular micrometer equal 270 μ. Consequently, one large unit on the ocular micrometer equals 135 μ, and each small unit equals 13.5 μ. Thereafter, I could use the ocular micrometer to record the lengths of fungal hyphae and convert them to microns.

DISSECTING OR STEREO MICROSCOPES

Dissecting microscopes are also called *Stereo microscopes* because they contain two separate light paths that travel to different oculars. This results in three-dimensional images because specimens are seen from two different angles. This is a vital feature for viewing and dissecting 3-D subjects. Early dissecting scopes consisted of two monocular scopes bound together. Compound scopes use

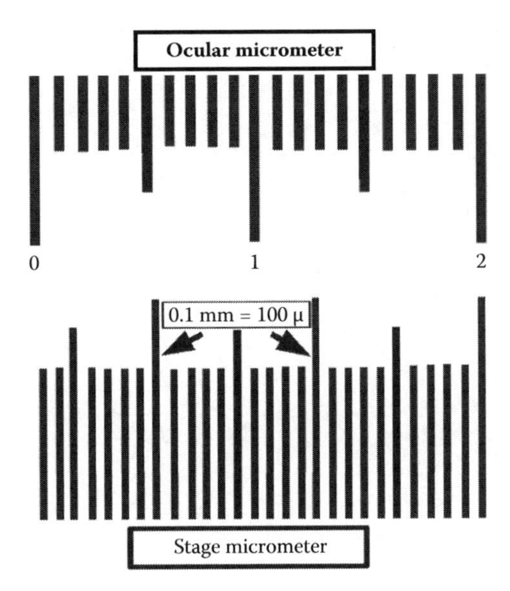

FIGURE 6.12 Diagram of a Stage Micrometer and Ocular Micrometer. In this case, two large units on the ocular micrometer = 270 μ. One large unit on the ocular micrometer = 135 μ, and one small unit = 13.5 μ.

only one light path that goes to both oculars. This produces a 2-D image. Dissecting scopes have many similarities to compound microscopes (Figure 6.1 and Figure 6.2).

The basic parts of a dissecting scope are Base, Stage, Arm, Focus Knob, Body, Magnification Knob, Oculars, and are shown in Figure 6.2.

- Locate the major parts of your dissecting scope by referring to Figure 6.2.

In most cases, Epiillumination (Figure 6.13) is used. There may be a "built in" illuminator that is located above and behind the microscope body. It may have a fixed angle of illumination or a knob that lets you vary the angle. The American Optical Stereo Star has a two port opening behind the body (Figure 6.13). This lets you insert an illuminator at two different angles. In all cases, the illuminator is positioned to avoid creating shadows during dissections. In many cases you will use a separate illuminator or pair of illuminators that can be positioned around the scope on your lab bench (Figure 6.14). There is often a flexible arm that can be used to vary the direction of illumination. This provides maximum flexibility for illuminating the sample.

Some dissecting scopes have a transillumination base. There is a mirror in the base that can direct light through the specimen. This is used to examine translucent specimens. It can provide an overview of a large, translucent sample that cannot be obtained with a compound scope. The mirror can be rotated using a knob on the base (Figure 6.13). This provides various angles of illumination. The mirror usually has a white opaque back that can provide diffuse reflected light. This can supplement epiillumination.

- Identify the types of illuminators available for your scopes.
- Explore their utility with various types of samples, illuminated at different angles.

Uniform, shadow-free illumination can be obtained from ring illuminators that can be mounted just below the Objective (Figure 6.14). Fluorescent ring illuminators are inexpensive, and they do not produce damaging heat. LED ring illuminators can produce more light and are also relatively "heat free." Fiber-optic ring illuminators are available but are significantly more expensive than other light sources. These also produce "cool" light because the heat is dissipated by the transformer. Separate fiber-optic light guides can be positioned independently to provide various angles of illumination (Figure 6.14). Shadow-free illuminators are extremely useful for most situations, but the ability to produce shadows

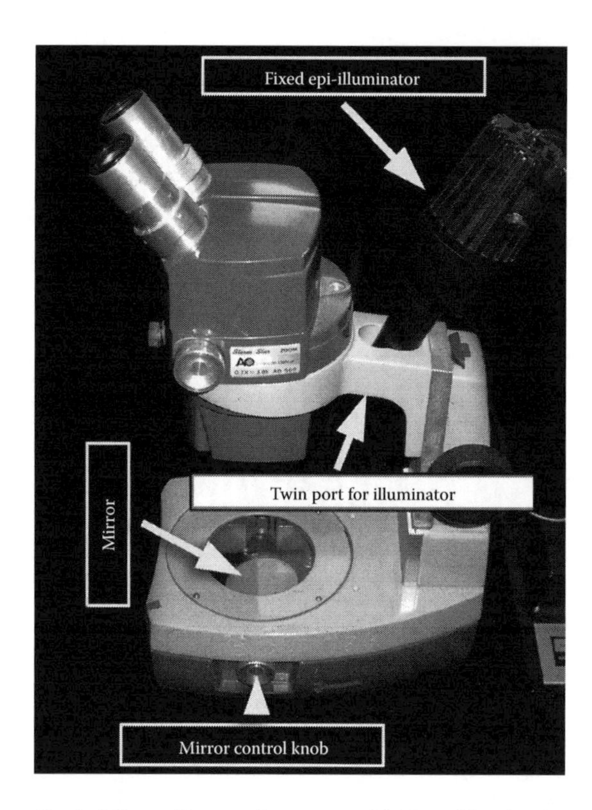

FIGURE 6.13 American Optical Stereo Zoom microscope with Transillumination Base, showing the mirror and its control knob plus an Illuminator placed in one of the Twin Ports that are designed to provide two angles for Epiillumination from behind the body of the microscope.

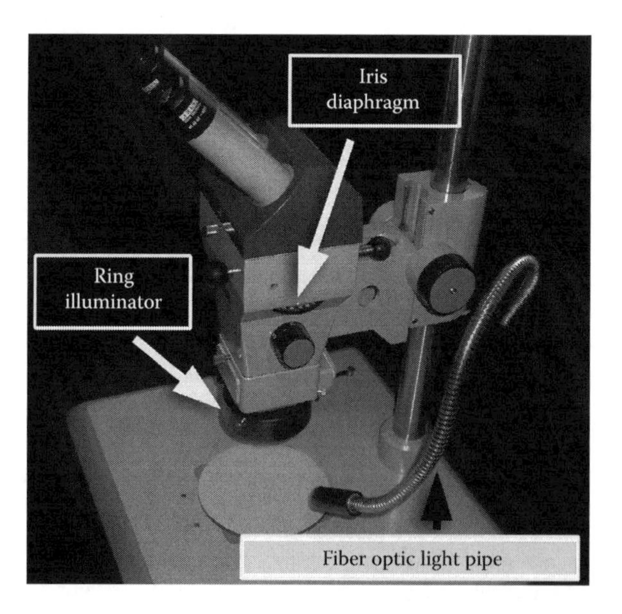

FIGURE 6.14 Zeiss Stereo Microscope with two types of illuminators. One is a Ring Illuminator that provides shadow-free light. The other is a fiber-optic illuminator that has two flexible Light Pipes. These can be adjusted to provide light from various angles. This scope also has an Iris Diaphragm that increases the depth of field and is especially useful for photography.

in a controlled fashion can be helpful in obtaining 3-D relief from your specimen. Cameras only use one of the tubes in a stereo scope. Consequently, pictures are only two dimensional.

The stage usually has a removable plate that is translucent, or white on one side and black on the other. This allows you to vary the background color, which can be important depending on the subject. Stage clips can usually be attached to the base. These are used for samples mounted on glass slides. There are mechanical stages for stereo scopes. These work like the same devices on compound scopes and are useful when fine adjustments are necessary. There are also clever devices that allow you to tilt your samples.

There is typically one focusing knob on each side of the arm. There is generally no fine focusing knob. Rotating the focusing knob moves the microscope body up and down.

Objectives are located near the base of the body. Magnification of the objectives is usually controlled by a knob located on the side or the top of the body (Figure 6.2). Older and less expensive microscopes contain a set of fixed objective lenses that are rotated into the light path by turning a similar knob. Most modern dissecting scopes have zoom objectives that can achieve continuous magnification over a range that is typically 1X to 3X. Auxiliary objective lenses can be fitted over the "built-in" objectives much like filters on a camera. These can increase or decrease the magnification range.

Some stereo scopes have an Iris Diaphragm inside the body (Figure 6.14). There is a dial or knurled ring that is used to control the opening of the iris. Closing this increases the depth of field and is very useful for photography (Gray, 1999).

It is important to experiment with the Iris Diaphragm to determine how much you should close its aperture, especially for photography. There are markings on the aperture control that signify the relative diameter of the iris. In general, the smaller the aperture, the greater is the depth of field. However, closing the aperture reduces the amount of light reaching the camera and can result in chromatic shifts. Furthermore, completely closing the aperture produces very poor photographs. Experiment with different aperture settings at a range of magnifications and record optimal combinations for photography. Spruce (*Picea glauca*) somatic embryos differentiating from an embryogenic callus are seen in Figure 6.15. The cotyledons and other details of these embryos were enhanced by using the iris diaphragm on a Nikon SMZ 10 stereo microscope.

The designation of Double Iris Diaphragm means that there is an iris in each of the two light paths. Closing the iris will diminish the amount of light that gets through but can also be useful during dissections. Illumination can be increased by adding extra light sources, especially if they are heat free. Iris diaphragms are usually not present in inexpensive microscopes.

The oculars are inserted into tubes that are attached to the body. These are similar to oculars used on a compound scope. They must be adjusted for your eyes following the steps in Procedures 6.2 and 6.3.

There may be an ocular micrometer in one of the oculars. To calibrate this, you may be able to use the same stage micrometer used to calibrate a compound scope. Otherwise, use an extremely accurate ruler like the one sold by Ted Pella Inc. (Product # 54480). Similar calibration aids may be available, but I am not aware of them. When in doubt, use the best ruler you can obtain.

TYPICAL STAINS FOR COMMERCIAL SLIDES OF PLANTS

Commercial slides are typically stained with Safranin O and Fast Green. Safranin O appears brilliant red in chromosomes, nuclei, and in lignified, suberized, or cutinized cell walls. Fast Green appears a brilliant green in cytoplasm and unlignified cellulosic cell walls. Fast Green turns blue in basic solutions and may appear blue to bluish-green in the stems and leaves of aquatic plants and most gymnosperms (Johansen, 1940).

FRESH SECTIONS OF PLANT ORGANS

There are several ways to generate fresh sections of plant organs. These include hand sectioning, or use of a hand microtome, or inexpensive sliding microtome, or a traditional sliding microtome.

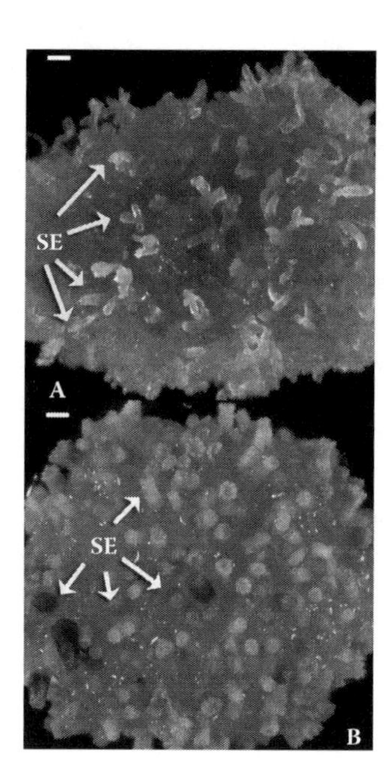

FIGURE 6.15 White Spruce (*Picea glauca*) embryogenic calli photographed with a Nikon SMZ-10 stereo microscope equipped with a Double Iris Diaphragm using epiillumination. Note the depth of field in each image. This was achieved by closing the iris diaphragm. A: Precotyledonary somatic embryos (SE), Scale Bar = 70μ. B: Cotyledonary somatic embryos (SE), Scale bar = 90μ.

Freezing microtome sections are nearly equivalent to fresh sections in many cases. This is a topic unto itself. However I will review the process of hand sectioning as this is a quick way to produce useful sections. Details are given in Procedure 6. 9.

The most important element in this process is the blade that you use. I use disposable freezing microtome blades, but you may not have these. Teflon-coated, stainless steel safety razor blades or injector razor blades work fairly well (Peterson et al., 2008). If you use a double-edged blade, be sure to put tape over one of the edges to prevent cutting your fingers. Use tape on your fingers as well. Single-edge utility razor blades give reasonable results, but they seem to lose their edge quickly. There is an inexpensive sliding microtome that is available from ScientificsOnline.com that surpasses most hand-sectioning attempts.

The ability to make free hand sections will allow you to quickly analyze plant specimens without resorting to laborious procedures. A tremendous amount of information can be derived from these. They have natural colors because they have not been extracted with organic or caustic chemicals. Sections need not be extremely thin to be of use. In addition, hand sections do not need to be complete or uniformly thin to be useful. They also provide 3-D information, which is not available with extremely thin sections. Your initial attempts will be frustrating; however, you will quickly become proficient.

The procedures that follow work well for stems, petioles, and roots, which are substantial organs. Leaves are more difficult to section because they are too flexible. One way to overcome this is to make a sandwich of three 5 × 10 mm leaf pieces containing the midrib with attached lamina. By squeezing these between your thumb and forefinger, reasonable leaf cross sections can be obtained. Another way to do this is to use artificial cork as a support medium. This is far superior to real cork. Hold a 5 × 10 mm leaf piece between two layers of artificial cork to make successful cross sections. You may want to trim excess "cork" away from the end of the cork so that you do not need to cut through too much of it. Real cork works, but it is difficult to use. Some callus is sturdy enough to section in this manner. You can purchase commercial pith, but it is expensive and delicate. Fresh carrot can be used, but you get a lot of debris from it. It is also rather slippery.

Procedure 6.9
Method for Making Hand Sections

Step	Instructions and Comments
1	Place a Band-Aid or adhesive tape on the thumb of your left hand. Have the cotton portion on the bottom of your thumb. The thumb is the backstop for this operation.
2	Place another Band-Aid or adhesive tape on the end of your index finger. The index finger will control the height of the specimen and thus its thickness.
3	Grasp the plant structure between your thumb and forefinger so that the top of the specimen extends above the level of your forefinger.
4	Take a razor blade in your right hand. Be sure that it is wet. Rest the blade on your forefinger and use a slicing motion to cut off the top of the specimen. This is a thick section that is designed to produce a flat surface.
5	Use a slicing motion that moves the blade away from your thumb in order to avoid cutting your thumb!
6	Raise the specimen slightly by manipulating it with your fingers and repeat the slicing motion. Make a lot of quick slices rather than a few slow and careful sections each time you elevate the specimen. Some will be thick and some will be thin.
7	Thin sections can often be obtained by pressing the blade down on your forefinger and then slicing through the specimen several times.
8	After several sections have accumulated on the blade, wash them off in a petri dish of water.
9	Keep on slicing until you have some thin sections. These will appear translucent when seen against a dark background. In most cases, the sections will have thin and thick regions. As long as part of the section is thin, you may be able to use it, and thick sections are frequently good for gross anatomy.
10	Sections can be removed with forceps or a wet artist's brush. The brush works best with delicate samples.
11	Place these in a drop of water or stain on a microscope slide. Sections will be released from the brush if you rotate it in the water on the slide.
12	It is a good idea to view unstained sections prior to staining. Proper use of the Condenser (Aperture) Diaphragm is important for viewing unstained samples.

IN VITRO SPECIMENS

Whole mounts are typically used to study cell and tissue-culture samples. This is easily accomplished for most suspension cultures by removing small samples with a pipette and placing a few drops on a microscope slide. Swirl the culture vessels before removing some liquid. This will give a more representative sample. If the culture is dilute, let it settle in the culture vessel and in the pipette before collecting and dispensing. A fair amount of cytological detail can be observed in unstained cells with Kohler illumination and the condenser iris diaphragm closed appropriately (Figure 6.16).

Small samples of loose, friable callus can be removed with a scalpel, forceps, or dissecting needle and placed in a drop of culture medium or water on a slide before adding a cover slip. Some careful pressure may be applied with the blunt end of a dissecting needle or similar instrument to disperse callus clumps. If the callus is compact and nodular, some dissection and dispersion under a stereo microscope may be necessary prior to microscopic observation. It is important to keep the sample submerged to avoid desiccation. Alternatively, such callus may be sectioned in a freezing microtome (O'Brien and McCully, 1981, Berlyn and Miksche, 1976) or through standard histological methods. With extremely nodular callus, free-hand sections can be used (Peterson et al., 2008). A support medium such as cork or artificial cork may be used to get good sections (O'Brien and McCully, 1981; Peterson et al., 2008).

STAINS FOR FRESH SPECIMENS

Virtually all water-soluble stains and some alcohol soluble stains can be used with small in vitro samples. Comprehensive information is contained in O'Brien and McCully (1981), Peterson et al. (2008), Ruzin (1999). I will report on three easily used stains.

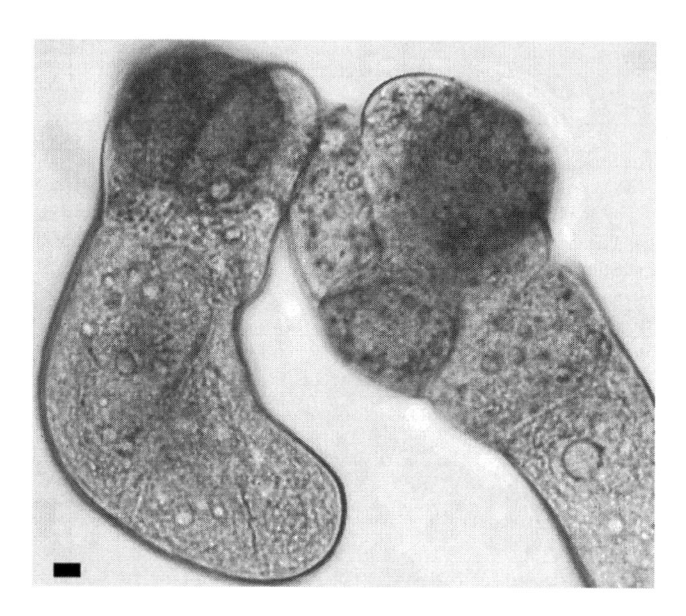

FIGURE 6.16 Unstained organized structures in Loblolly pine (*Pinus taeda*) suspension culture. Cytological details are visible after achieving Kohler illumination and adjusting the condenser (aperture) iris. Scale bar = 1μ.

Toluidine Blue O

Pectin-rich walls such as those found in parenchyma cells stain pink with Toluidine Blue. It stains lignified walls and phenol-containing vacuoles a blue to blue-green color (O'Brien and McCully, 1981; Peterson et al., 2008; Ruzin 1999). See Procedure 6.10 for details. Some walls, especially those of the phloem, may not stain. You need to act quickly because Toluidine Blue is fast acting, and overstaining may destroy its specificity (Figure 6.17A). It is wise to include several samples so that a range of staining intensities is present.

Toluidine Blue is one of the most frequently used stains for cell and tissue cultures because it is rapid and stains most cell components. Nuclei are densely stained. Consequently, meristematic cells and meristems are readily distinguished from more vacuolate cells (Figure 6.17A). It is impermanent, and photos must be taken to record the results.

Sudan

Sudan III and IV are alcohol-soluble stains (Peterson et al., 2008) that are used to stain lipids such as storage fats and oils in seeds and embryos (Figures 6.17B, 6.18). They also stain suberized and cutinized cell walls. When stained specimens are rinsed with water, oily and waxy materials that have taken up Sudan will remain red/orange, and other areas are colorless. Specimens are flooded with the stain prior to adding a cover slip. Alcohol evaporates rapidly, so add a coverslip right away. Staining time varies and must be determined empirically. Stain is removed by adding water to one side of the coverslip and a paper towel or other tissue to the opposite side until the stain is replaced by water. Red to orange areas indicate the presence of lipids (Figures 6.17B, 6.18).

Iodine Potassium Iodide (IKI)

IKI is water soluble and stains starch a blue-black to red-brown color (O'Brien and McCully, 1981; Peterson et al., 2008; Ruzin, 1999). It also imparts a golden color to cell wall cytoplasm and nuclei, although this incidental staining is not specific to any substance. IKI is very useful for staining cell

Procedure 6.10
Staining with Toluidine Blue

Step	Instructions and Comments
1	Add several sections to a drop of water on a slide.
2	Add 2–3 drops of Toluidine Blue to this.
3	Quickly add a coverslip.
4	Remove the excess stain by blotting with a Kimwipe. Wipe excess fluid from the bottom of the slide.
5	View right away.
6	This stain fades over an hour or two.
7	Caution: Toluidine Blue is hard to get out of clothing, so use it carefully and clean up any spills with lots of water.
8	In addition, it is mildly poisonous, so avoid getting it on your skin as much as possible.
9	Use latex or Nitex gloves to protect your hands.
10	Be sure to wash your hands well if they become stained. This applies to all of the stains you use.

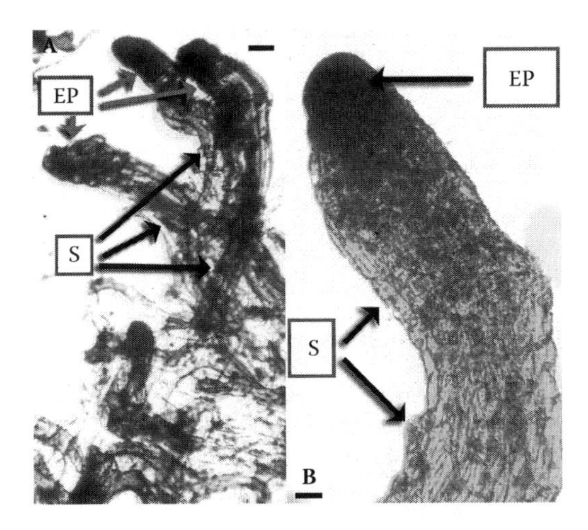

FIGURE 6.17 White Spruce (*Picea glauca*) precotyledonary somatic embryos. A: Stained with Toluidine Blue 0. Note the dense staining of the embryo proper (EP) cells due to their large nuclei and dense cytoplasm. The highly vacuolate suspensor (S) cells are lightly stained. Scale Bar = 18 μ. B: Stained with Sudan IV. The overall staining pattern is similar to A and shows that lipids are concentrated in the EP. Spruce and conifers in general store oil in their embryos and megagametophytes. Scale bar = 5 μ.

and tissue cultures. Densely cytoplasmic, meristematic cells concentrate the stain and are readily distinguished from highly vacuolate cells (Figures 6.19 and 6.20) and starch responds as expected.

POLARIZING FILTERS

Polarizing filters cause light to vibrate in one plane and thus produce "plane polarized light." Light traveling from a source vibrates in all possible planes. Imagine many radii emanating from a common center. These would represent the many vibrational planes of the light beam. A polarizer cuts out all but one of these. I think of polarizers as combs. A comb straightens tangled hair so that the strands are parallel to one another (Peterson et al., 2008).

If two polarizers are oriented parallel to one another, light will pass through them because the plane polarized light that passes through the first comb is parallel to the teeth in the second comb. However, if two polarizers are placed at an angle of 90° to each other, no light will pass through the second polarizer because the first polarizer eliminates all light that vibrates parallel to the teeth in

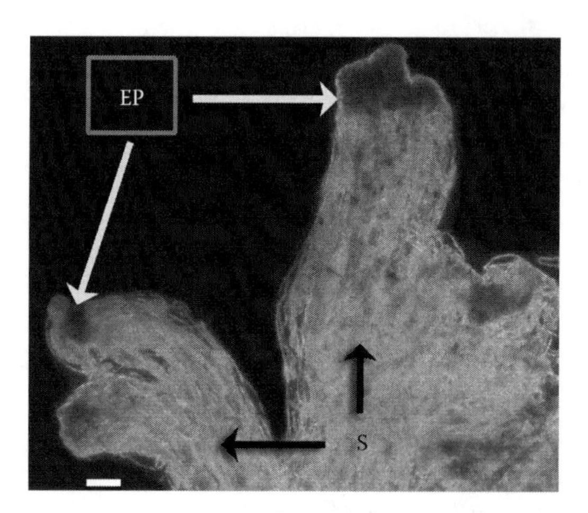

FIGURE 6.18 Precotyledonary White Spruce (*Picea glauca*) somatic embryos, stained with Sudan IV, and viewed with crossed Polarizing Filters. The latter are responsible for the black background due to the absence of birefringent objects. Suspensor (S) cell walls are highly birefringent and appear bright. Cells of the embryo proper (EP) are not birefringent due to the presence of oil, which was stained with Sudan IV, and thin cell walls. Scale bar = 12 μ.

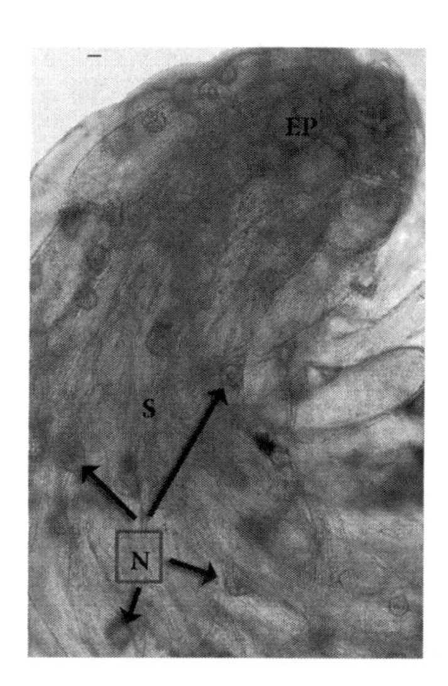

FIGURE 6.19 Precotyledonary Douglas Fir (*Pseudotsuga menziesii*) somatic embryo stained with IKI. Cells of the embryo proper (EP) have stained heavily due to their relatively large nuclei (N) and dense cytoplasm. More vacuolate suspensor (S) cells have stained lightly. Consequently, their nuclei are more prominent. Scale bar = 2 μ.

the second comb. You can verify this by holding one polarizer while looking at a light source other than the sun. Take a second polarizer in your other hand and superimpose it on the first. Turn either one until the light is completely blocked. You can do this with polarized sunglasses.

If a crystalline object, like a cell wall, is placed between crossed polarizers, it will depolarize the light that passes through the first polarizer (Berlyn and Miksche, 1976). This property is known as birefringence. Birefringent material will create light that vibrates in the same plane as the second polarizer, and it will be visible while all else will be dark. Cell walls, crystals, and starch grains are birefringent and become apparent using crossed polarizers (Figure 6.18). This works with unstained and some stained sections. Densely stained meristematic zones may not be birefringent at low power (Figure 6.18). Staining with IKI may destroy the birefringent properties of starch grains.

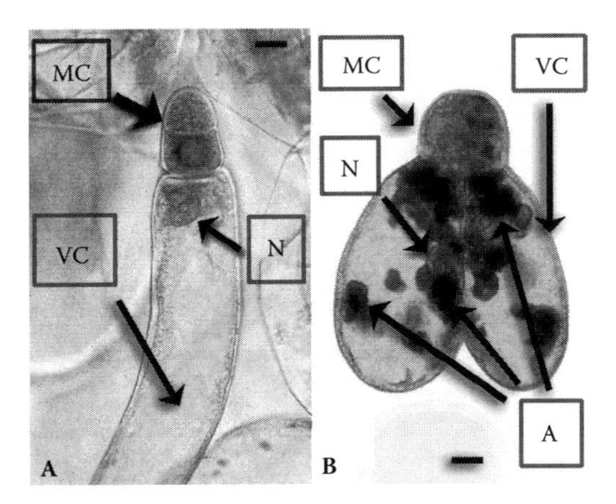

FIGURE 6.20 A, B: Organized structures in Douglas Fir (*Pseudotsuga menziesii*) suspension culture stained with IKI. Nuclei (N) have stained prominently in meristematic cells (MC) and vacuolate cells (VC) of A and B. MC are densely cytoplasmic and have accumulated stain in both cases. Amyloplasts (A) are absent in A but are prominently stained in VC of B. Scale bars = 1 μ.

Inexpensive polarizing filters and sheets of polarizing film can be purchased from Scientifics Online.com. Circular polarizing filters designed for cameras can be used. The polarizing filters for Nikon Coolpix 990, 995, and 4500 cameras work very well for digital photography. The same should be true of other camera filters. Use neutral gray circular polarizing filters.

Place one polarizing filter over the field diaphragm of a compound scope. Place another over, or inside, your ocular, or wear polarized sunglasses. This may look strange, but it really works. Focus on your samples as usual and rotate the polarizer over the field diaphragm. If your specimen has cells with thick walls, starch grains, or crystals, they will become bright while the background becomes dark. Intermediate effects are also possible and can reveal subtle features that are not visible otherwise. Most people open the condenser iris completely to let in as much light as possible. However, closing this iris can reduce some light scattering. Experiment with the condenser iris to find the right balance.

If you want to photograph your results, you will need to place one polarizer between the specimen and your camera. This can be done by cutting an appropriately sized piece of polarizing material from a larger filter sheet and placing it over the projection lens inside the microscope. Some dissecting scopes have internal polarizers to eliminate glare. An external filter can be cut from polarizing sheets and placed over the objectives to create the same effect.

REFERENCES

Berlyn, G. P. and J. P. Miksche. 1976. *Botanical Microtechnique and Cytochemistry*. The Iowa State University Press, Ames, IA.

Delly, J.G. 1988. *Photography through the Microscope*. The Eastman Kodak Company, Rochester, New York.

Feder, N. and T. P. O'Brien. 1968. Plant microtechnique: Some principles and new methods. *Amer. J. Bot*, Vol. 55, No. 1, 123–142.

Gray, D. J. 1999. Photographic methods for plant cell and tissue cultures. In R. N. Trigiano and D. J. Gray (Eds.), *Plant Tissue Culture Concepts and Laboratory Exercises*, 2nd ed., CRC Press, Boca Raton, FL.

Johansen, D. A. 1940. *Plant Microtechnique*. McGraw-Hill, New York.

Kohler, A. 1893. A new system of illumination for photomicrographic purposes. Royal Microscopical Society, Oxford, United Kingdom, Kohler Illumination Centenary (1994), pp. 1–5.

O'Brien, T. P. and M. E. McCully. 1981. *The Study of Plant Structure Principles and Selected Methods*. Termarcarphi, Melbourne, Australia.

Peterson, R. L., Peterson, C. A., and L. H. Melville. 2008. *Teaching Plant Anatomy through Creative Laboratory Exercises*. NRC Research Press, Ottawa, Canada.

Ruzin, S. E. 1999. *Plant Microtechnique and Microscopy*. Oxford University Press, New York.

IMPORTANT WEBSITES

A-Z Microscope Glossary of MicroscopeTerms (http://www.az-microscope.on.ca/glossaryofterms.htm).

Microscopy Primer (http://www.microscopy-uk.org.uk/index.html?http://www.microscopy-uk.org.uk/primer/index.htm).

Molecular Expressions Optical Microscopy Primer (http://micro.magnet.fsu.edu/primer/index.html).

Nikon MicroscopyU (http://www.microscopyu.com/).

7 Plant Histological Techniques

R. N. Trigiano, D. J. Gray, K. R. Malueg, K. A. Pickens, Z.-M. Cheng, and E. T. Graham

CONCEPTS

- Histological examination of plant tissues using paraffin or plastic media involves the following steps: fixing, dehydrating, infiltrating, embedding, sectioning, and staining of samples.
- Histological techniques can reveal structural details and developmental processes in plants.
- Histological techniques are useful for investigating/confirming various responses of plant tissues to in vitro manipulations.

Plant histology can be defined simply as the study of the microscopic structures or characteristics of cells and their assembly and arrangement into tissues and organs. Several histological techniques are commonly used for examining plant tissues, each providing somewhat similar gross information, but differing in resolution of details and the medium in which the samples are prepared. These techniques include those for bright field and fluorescent microscopy, in which specimens can be prepared for cutting into thick sections (10–40 μm) either without a stabilizing medium (fresh sections), in cryofluids (frozen sections), or embedded in paraffin-like materials, or in various formulations of plastic. Other techniques that employ electron microscopy either do not require a specialized embedding medium for specimen preparation (scanning electron microscopy) or have samples embedded in plastics (transmission electron microscopy) that can be cut into ultrathin sections (65–100 nm). An explanation of the details involved in all of these techniques is beyond the scope of this chapter.

One might reasonably ask why include a chapter on histology in a plant development and histology book? In many situations, histological techniques provide essential information that may not be evident by visual inspection alone. Much of our understanding of in vivo and in vitro developmental processes that are presented throughout this book resulted from detailed histological research. For example, somatic embryos (Chapter 22) can be produced on the surface of a leaf explant, but may be so morphologically aberrant as to be unrecognizable. Using histological techniques and scrutinizing anatomical features, the characteristics of somatic embryos can be more readily seen. Another example of using histological techniques is the investigation of the origin of specific structures, that is, adventitious shoots and roots (Chapter 21), embryos, etc., that develop in culture. Histological development can be studied over time by periodically sampling tissues, and/or the result can be examined in the mature structure. One must always be mindful that growth of tissues is dynamic and changes from moment to moment, whereas histological sections, for example, are static and fixed in time, and only present a very narrow glimpse of the developmental process. Nevertheless, the origin of tissues and organs may be convincingly inferred from serial observation of many samples and sections.

The field of plant histology and microtechniques is quite broad and cannot be described in one chapter. Therefore, the intent of this chapter is to briefly summarize the essentials and provide a "primer" of sorts for preparing specimens from plants and tissue cultures for histological examination using elementary paraffin and scanning electron microscopy techniques. The techniques presented herein are considerably easier and avoid many of the toxic materials found in more traditional protocols. For a broader view of general histological methodologies, students should consult one or more of the following references: Berlyn and Miksche (1976); Sass (1968); Jensen (1962); and Johansen (1940).

Please view the presentation on the DVD for a brief account of the paraffin histological process. The video depicts fixation, dehydration, embedding, traditional staining protocol, and cover glass mounting.

GENERAL CONSIDERATIONS

Equipment and Supplies

The equipment and materials needed to complete a paraffin histological study are as follows:

- Indelible marker or #2 lead pencil
- 500 or 1000 mL plastic beaker
- Aluminum pan
- Rotary microtome with disposable blade holder
- Warming oven (58°C–60°C)
- Two slide warming trays (40°C and 50°C)
- Water bath (50°C)
- Hot/stir plate
- 1% agar or agarose in a 150 mL beaker
- Disposable microtome blades
- Paraplast embedding medium
- Disposable plastic base molds and embedding rings
- 5 mL microbeakers
- Alcohol lamp and metal spatula
- Screw-capped Coplin jars
- Histochoice fixative (Amresco, Inc., Solon, Ohio) plus 20% ethanol
- 37% Formaldehyde solution and fume hood
- Disposable snap cap plastic 10 mL specimen vials
- Wooden applicator sticks
- Clean glass slides and cover glasses
- Camel or sable hair artist brush
- Test tube racks
- Eukitt or similar cover glass resin (EMS, Fort Washington, Pennsylvania)
- MicroClear (Micron Environmental Industries, Fairfax, Virginia)
- 30%, 50%, 75%, 95% (all aqueous), and 100% isopropanol
- Glacial acetic acid and sodium acetate
- Microbiology-grade agar
- 0.2% aqueous solution of Alcian blue 8GX (Sigma) 200 mg/100 mL of water
- Quick-mixed hematoxylin (Graham, 1991)—Stock A: 750 mL distilled water, 210 mL propylene glycol, 20 mL glacial acetic acid, 17.6 g aluminum sulfate and 0.2 g sodium iodate, Stock B: 100 mL propylene glycol and 10 g certified hematoxylin (Sigma). Stain is prepared by adding 2 mL of Stock B to 98 mL of Stock A.

GENERAL PROTOCOL FOR PARAFFIN STUDIES

The following procedure will work well with most tissues, including those from stock plants as well as those cultured in vitro. Moreover, it has several important advantages over more traditional paraffin protocols. First, it circumvents the use of toxic aldehydes (formaldehyde) and heavy metals (chromium and mercury) as fixatives. Second, it avoids the use of specimen/slide adhesive agents, such as

Haupt's, that require the use of formaldehyde and may produce staining artifacts. Third, it does not use toxic substances, such as xylene, for deparaffinizing sections. Fourth, sections are stained directly through the paraffin, greatly reducing the number of operational steps in the process. The foregoing advantages combine to make preparing tissue for histology and slides a safe and less intimidating process for both students and instructors alike. There are circumstances in which more traditional fixation and staining protocols are warranted, and one example is included in this chapter.

Fixation

The first step in the protocol is to identify typical specimens (i.e., embryos on leaf sections), remove them from the culture dishes, and place them in a small petri dish containing water. Most explants are too large to be adequately fixed and usually the subject (embryos) occupies a very small area. The specimen then should be carefully trimmed using a razor blade or a scalpel to include the subject and a small area of surrounding tissue. Five or fewer specimens are placed in about 5 mL of Histochoice, a nonaldehyde, nontoxic fixative amended with 20% ethanol, contained in each specimen vial. If possible, the open vials should be aspirated (-1 atm or less) for about 30 min to remove air from the samples and promote infiltration of the fixative. For very small or very delicate specimens, we suggest that fixation without aspiration be tried first. The vials are closed, and the tissue remains in the fixative at room temperature for a minimum of 24 h; samples may be stored at room temperature for weeks or months without degradation of structures. However, replace the original fluid with fresh fixative after 24 h. If long-term storage is needed, we suggest using screw-capped rather than snap-capped vials.

For some, a formaldehyde fixative may produce a better "fixation image," and we suggest that a 50% ethanol–formaldehyde–acetic acid fluid (FAA) can be used. It should be noted that 70% ethanol concentration can also be used. Mix the following ingredients under a fume hood according to Table 7.1 to make this fixative. Do not use glutaraldehyde fixation (or osmium tetraoxide postfixation) in paraffin protocols.

Dehydration and Infiltration

After fixation, the samples must be dehydrated and placed in a solvent compatible with paraffin. Traditionally, this has been accomplished with graded series of ethanol and transitioning to tertiary butyl alcohol. This process can be considerably shortened by using isopropanol to both dehydrate the tissue and dissolve the paraffin. Histochoice fixative can be removed from the vials with a long thin pipette and poured down the drain with plenty of water. Specimens are dehydrated using 30, 50, 75, 95, and three changes of 100% isopropanol—each for about 10–30 min. Note that if you are using 50% FAA as a fixative, start with 50% isopropanol and complete the dehydration process under a fume hood. After the last change of pure isopropanol, fill the vials about one-quarter full with fresh 100% isopropanol, add a few pellets of Paraplast embedding medium, loosely recap the vials, and

TABLE 7.1
Formulas for Aldehyde-Based Fixatives

Ingredient	50% FAA (mL)	70% FAA (mL)
Ethanol (95%)	53	74
Water	37	16
Formaldehyde (37%)	5	5
Glacial acetic acid	5	5
Total	100	100

incubate at 60°C. Periodically over the next several hours, swirl the contents, and add a few more pellets of Paraplast to each of the vials. At the end of the day, remove the caps from the vials to allow the isopropanol to evaporate. At this time, fill a 500 or 1000 mL plastic beaker with Paraplast pellets, place several pellets into a number of base molds contained in an aluminum pan, and incubate in the 60°C oven. The next morning all of the isopropanol should be dissipated, the specimens completely infiltrated with Paraplast, and the pellets in the beaker and base molds melted.

Casting Specimens into Blocks

Using a sharp razor blade, whittle a flattened paddle-shaped end on several wooden applicator sticks and store in a vial with the flattened end immersed in molten paraffin in the oven. Transfer a specimen into the molten paraffin contained in a base mold, and position it so that the sectioning plane of the specimen is parallel to the bottom of the base mold. For example, if a transverse section of a stem is required, the stem should be placed perpendicular to the base mold; the long axis of the stem should be "sticking up" straight up in the molten paraffin. If the stem is positioned flatly in the base mold, longitudinal sections will result. With more complex specimens, such as single somatic embryos or shoots growing from basal tissue, this orientation step becomes more critical. It is important to visualize the exact desired section plane at this step. Once the sample is oriented correctly, quickly transfer the base mold to a cool surface—an inverted aluminum pan containing ice works well. The surrounding paraffin in the bottom of the base mold will quickly become cloudy (congealed), and the specimen will be immobilized. Rapidly, attach an embedding ring to the base mold, and fill the apparatus with molten paraffin from the plastic beaker. This is a very vulnerable point in the process. If the immobilizing paraffin is allowed to congeal too far, the added molten paraffin will not bond with it, and the block will split apart. At this time, a label consisting of white paper and text identifying the specimen written in pencil (do not use ink) may be inserted partially into the molten paraffin. Place the casted blocks in the refrigerator, or float them in a large beaker containing ice water. The base mold may be removed after the block is hardened.

MICROTOMY AND MOUNTING SECTIONS ON SLIDES

A word of caution before beginning to section. Regardless of whether a stainless steel knife or disposable razor blade is used to cut sections, remember that they are both extremely sharp and cut flesh very easily. Exercise care while working with the microtome, and if the microtome is left unattended, place a note on the knife holder warning in bold print that a knife or blade has been installed. The rotary action arm should be kept locked at all times except when sectioning.

One of the advantages of using embedding molds is that the cutting face of the block is square. If the block were sectioned as is, the resulting "ribbon of very large sections" would be straight. However, seldom does the specimen occupy the entire block face. Therefore, it is often desirable to trim away some of the excess paraffin from around the sample. The horizontal edges of the block must be parallel with each other; or, another way of visualizing it is for the sides of the block to be of equal length. The block edges must also be kept parallel to the knife edge. If the ribbon of sections curves to the right or left, the side of the block opposite the direction of curvature is too long. Remove the block from the chuck, and retrim so that the sides are parallel and of equal length. Do not trim the block while it is mounted in the chuck over the knife!

Mount a new disposable blade in the blade holder; only a small portion of the blade should be clear of the carriage. Typically, new blades are coated with oil and may be cleaned with a small wad of tissue moistened with a little MicroClear; carefully wipe the blade perpendicular to the blade surface with the tissue and allow to completely dry. Set the microtome to cut 10–12 μm sections. Now mount the specimen block in the chuck, and orient it so the bottom of the cutting face is parallel to the edge of the blade. Position the specimen at the level of the knife, and carefully and gently slide the knife holder toward the block until they are very close. With the utmost care, turn the wheel clockwise, causing the specimen to advance 10 μm per revolution. The first sections may not include the entire block face; continue advancing the block until a complete section is obtained. Lock the advancing wheel into position, carefully remove the partial sections, and clean the knife as before. After the MicroClear has evaporated, continue to operate the microtome, and after five or more sections are produced, place a camel hair brush underneath the ribbon for support while it is being held away from the knife holder. A 30–50 section ribbon can be produced and easily manipulated. Lay the ribbon out on a clean piece of dark paper and, using a razor blade, divide it into segments (sections) that do not exceed about 75% the nonfrosted length of a glass microscope slide.

Modern Superfrost slides afford the following well-established advantages: (1) they can be used directly from the manufacturer without cleaning; (2) the glass surface bonds with tissue sections without an adhesive coating; and (3) the special labeling pen resists all common histological solvents (Graham and Trentham, 1998). If these are not unavailable in your laboratory, prewash glass microscope slides in a mild dishwashing detergent solution, rinse well with tap water, and rinse with distilled water. Allow the slides to dry in test tube racks with an aluminum foil base. Washed and dried slides can be stored indefinitely in slide boxes. Label the frosted portion of the slide with appropriate information using an indelible marker or lead pencil and cover specimen surface of the slide with distilled water. Moisten the tip of a brush, touch it to the surface of the first section to the left, and lift the entire ribbon off the paper. Now, touch the first section to the right onto the water, and lay the rest of the ribbon down so that air is not trapped underneath the sections. Place the slide on the 50°C warming table to expand the sections. After about 10 min, set the slide on a sponge, and slightly tilt it to drain off most of the water. Affix the sections to the glass by placing the slide on the 40°C warming table for 1–16 h or overnight. The sections are now ready to be stained.

STAINING SECTIONS

The staining procedure described in Procedure 7.1 is adapted from Graham and Joshi (1995).

An alternative to this staining process is offered by Graham and Trentham (1998). This procedure (Procedure 7.2) is very efficient and provides a triple stain in a single solution.

More traditional fixation and staining procedures are provided in Procedure 7.3.

Do not become discouraged if your results are less than perfect the first time—do not give up—try again! As with any other technique, skill develops with experience.v

Procedure 7.1

Staining through Paraffin with Alcian Blue and Hematoxylin

Step	Instructions and Comments
1	Preheat the 0.2% Alcian blue solution in a closed Coplin jar, and place the slides in the stain for 1 h at 50°C.
2	Gently rinse the slides in a stream of cold tap water, and carefully blot dry the remaining water droplets.
3	Transfer the slides to Coplin jars containing hematoxylin stain for 15 min at room temperature, rinse and dry as described before, and place on the 40°C slide table for a minimum of 30 min.
4	The paraffin in the sections is removed with three 5 min soaks in MicroClear followed by three 5 min changes of 100% isopropanol.
5	After the slides have air-dried, pool some resin (Eukitt) in the center of the slide, and gently lower a cover glass so that the resin spreads evenly over the surface without air bubbles. The resin may be cured overnight on a 40°C warming table.

Procedure 7.2
Triple Stain Using Alcian Blue, Safranin O, and Bismarck Brown

Step	Instructions and Comments
1	Prepare a 0.1 M acetic acid–sodium acetate buffer at pH 5.0. Dissolve 0.8 g of sodium acetate in 100 mL in a 250 mL beaker. The specific gravity of 100% glacial acetic acid is about 1.05; therefore, each liter contains 1050 g of acetic acid. To make a 0.1 M acetic acid solution, pipette 0.6 mL of glacial acetic acid into a 100 mL graduated cylinder, and then carefully bring the volume to 100 mL with water. Titrate the sodium acetate solution with 0.1 M acetic acid solution and bring to pH 5.0 using a pH meter. Use a magnetic stirrer while titrating.
2	Make three 1% stock solutions of Alcian blue 8 GX, safranin O, and Bismarck brown, using 0.1 g of dye powder dissolved in 100 mL of absolute (100%) alcohol.
3	To 100 mL of acetate buffer, add 5 mL of Alcian blue, 2 mL of safranin O, and 1 mL of Bismarck brown stock solutions.
4	Extract paraffin from the sections by moving slides through four baths of MicroClear, and then through five baths of 100% isopropanol to remove the paraffin solvent. Finally, place the slides on a rack to air-dry.
5	Immerse slides in the triple stain made in step 2 above. Coloring of tissue will be evident with 15 min, but optimum staining differs widely among various specimens. Progress of staining is monitored easily by rinsing a slide briefly with distilled water, blotting it dry with soft tissue paper, and observing under a compound microscope. The slide may be returned to the stain solution until satisfactory staining is achieved. We suggest reexamining the slide at 15 min intervals.
6	When staining is sufficient, dry slides for 1 h on a 40°C slide warmer, and affix a coverslip as described in Procedure 7.1, Step 5. The results of the triple stain are the following: polysaccharide in cells are blue; lignified walls, nuclei, and chloroplasts are red; and cuticle (if present) is brown or yellow-brown.

Procedure 7.3
Traditional Fixing and Staining Schedule

Step	Instructions and Comments
1	Fix tissue in 50% FAA (Table 7.1) for a minimum of 24 h. Caution: Do this operation under a fume hood, and avoid breathing vapors or contact with the skin.
2	Decant the 50% FAA from the specimen jars, and begin dehydration and paraffin infiltration of the samples with 50% isopropanol as described under the dehydration and infiltration section. Cut sections and mount on clean glass slides as previously described.
3	The following is a description of a triple stain and works well for most plant tissues. Place slides with sections in carrier, and immerse in the following solutions contained in Coplin jars for a minimum of 5 min each: 100% MicroClear, 100% MicroClear, 100% MicroClear, 50% MicroClear 50% absolute ethanol, 100% ethanol, 95% ethanol (may skip), 70% ethanol, and 50% ethanol.
4	Safranin O (4 g safranin O; 100 mL water; 100 mL 95% ethanol; 4 g sodium acetate) for 30 min to 2 days. It is very difficult, if not impossible, to overstain with safranin. Wash slides with several changes of water or until the water is no longer pink. Do not let running water directly contact slides. Immerse slides in crystal violet stain (1 g crystal violet; 100 mL of water) for 30–90 s. Very easy to overstain. Immediately wash in water as with safranin until water is no longer purple; water may remain slightly violet.
5	Dehydrate sections for 5 min each in the following solutions: 50% ethanol, 70% ethanol, 95% ethanol, and 100% ethanol. Some of the crystal violet "to leech" from the sections; this is normal. Place slides in fast green (1.2 g fast green; 80 mL methyl cellusolve; 80 mL methyl salicylate; 80 mL absolute ethanol) for 30–90 s. Caution: Easy to overstain. Immediately immerse slides in 100% ethanol for a few sections to remove most of excess stain, and then transfer to a new Coplin jar containing 100% ethanol for 5 min.
6	Transfer slides through the following solutions for a minimum of 5 min each: 50% MicroClear–50% absolute ethanol, 100% MicroClear, 100% MicroClear, and 100% MicroClear. Coverslips may now be added to the slides as described previously.

Immobilization of Specimens for Paraffin Sectioning

This sample preparation technique involves immobilization and precise orientation of specimens between two layers or a "sandwich" of agar. It offers several advantages over more conventional methods of preparation, especially when working with small and/or delicate tissues. Small samples, for example, embryos or callus, are often lost or damaged during the dehydration sequences when fluids are removed and replaced. By immobilizing the specimens in agar, it is nearly impossible to lose samples, and the agar minimizes damage from handling. However, the real advantage of the technique is that the sample can be oriented in the agar before fixation, eliminating the need to position it while casting the block.

This procedure is adapted primarily from Hock (1974). Prepare 1% water agar (1 g of microbiological-grade agar 100 mL of water in a tall 150 mL beaker) on a hot/stir plate and, after the agar is melted, store in a 50°C water bath. Dip a clean glass microscope slide into the molten agar, and then place it in a 100 × 15 mm petri dish containing moistened filter paper. The rest of the immobilization procedure is diagrammatically represented in Figures 7.1A–D. Select a specimen, and place it on top of the cool and hardened agar. With a Pasteur pipette, cover the sample with a small amount of molten agar. The position of the sample is now secure. Draw an arrow with pencil on a piece of white paper about the width of the sample. Using a stereomicroscope scope, if necessary, and fine forceps, place the arrow directly opposite or behind the desired cutting face of the tissue, and tack it into position using molten agar. Cover the entire preparation with a thin coating of molten agar. Once the agar has hardened, trim any excess agar away using a razor blade. The specimen can be removed from the slide by "slipping" the razor blade under the bottom of the agar sandwich and placing it in fixative. Follow the previously described protocols for fixation of tissue and infiltration with paraffin.

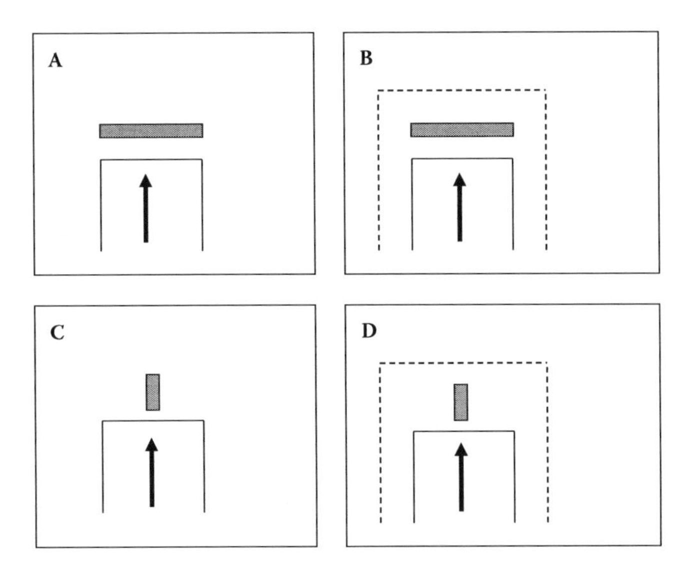

FIGURE 7.1 Immobilization. A. A specimen (stem) is placed on an agar-coated slide, then covered with molten agar. Since sectioning is orthogonal (right angles) to the direction of the arrow, this stem would be sectioned longitudinally or the length of the specimen. A white slip of paper with an arrow drawn in pencil is also affixed with agar. B. After the agar has hardened, the preparation may be trimmed and removed (dotted lines) from the glass slide. C. Stem oriented to yield cross sections upon microtomy—see 7.1A for details. D. Cross section orientation of sample—see 7.1B for details. Note: The subject of the sectioning may not be the explant, but rather what is growing from the explant, e.g. a non-zygotic embryo. Furthermore, the non-zygotic embryo may be oriented at some angle to the explant. Therefore, it may be necessary to orient the non-zygotic embryo so that it is sectioned longitudinally and the explant may be sectioned obliquely or cross sectionally. This is where the art in the science comes into play!

Instead of using an embedding mold, use a 5 mL plastic microbeaker to cast the specimen into a paraffin block. Draw a straight line on the bottom of the beaker with an indelible pen. Fill the container with molten paraffin, and transfer the sample to the beaker so that the sample is parallel, yet slightly behind, and the arrow is positioned orthogonally (at a right angle) to the line on the bottom (Figure 7.2). Place the microbeaker on a cold surface, and label with a paper slip as described before.

Dissection and Mounting of Specimens

The block of paraffin in which the specimen was cast cannot be mounted on the microtome. Therefore, it is necessary to dissect the specimen and a small amount of the surrounding block, and mount it on a block formed by an embedding mold. Usually, there are some old blocks around from previous projects, or newly cast ones may be used. Before the microbeaker is removed, take a sharp razor blade and cut through the beaker, and score the paraffin in the block underlying the line on the microbeaker. This will ensure that the cutting face is identified even if the specimen and arrow are not clearly visible in the block. Dissect the specimen from the block. Create a flat field on the face of the embedding mold block. Melt a thin layer of the embedding block using the flat side of a spatula heated with an alcohol burner. Quickly attach the side opposite of the cutting face of the specimen block to the embedding mold block. Reinforce the attachment by melting small chips of paraffin with a hot spatula tip along all four sides of the specimen block (Figure 7.3). The block may be sectioned and slides prepared as previously described.

Preparation of Specimens for Scanning Electron Microscopy (SEM)

The sections prepared for paraffin histology present a two-dimensional (flat) interior view of a small area of a much larger three-dimensional object. In contrast, SEM allows for three-dimensional topical or internal views of an entire specimen. Samples for SEM do not need to be embedded in paraffin nor do they need to be stained. Moreover, the resolving power of a scanning electron microscope is much greater than a compound light microscope. However, in many ways, preparation of samples for SEM is technically easier than for paraffin histology. (*Note:* Methodologies for critical point drying and sputter coating are not presented and should be performed by a competent SEM technician or other person familiar with these techniques.)

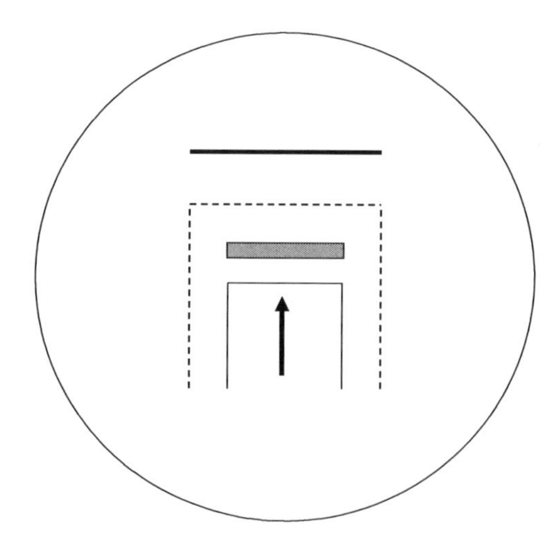

FIGURE 7.2 Top view of positioned specimen cast in a plastic microbeaker. The dark line at the top of the figure is made with a black Sharpie and helps orient the preparer after the wax in the beaker has hardened and become opaque.

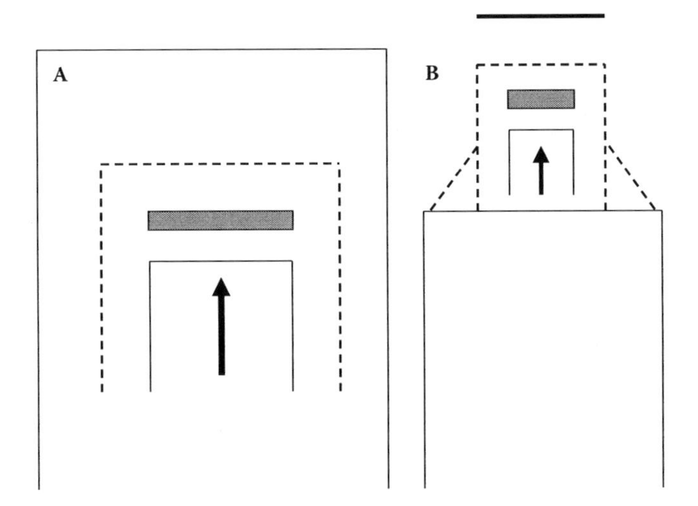

FIGURE 7.3 Mounting the dissected specimen (A) to a flat block cast using an embedding mold (B). Note the paraffin reinforcement along the sides of the specimen. The solid line at the top of the figure represents the knife blade.

The following materials will be needed to prepare samples for SEM:

- 0.05 M potassium phosphate buffer, pH 6.8–7.2—Prepare by dissolving 0.87 g K_2HPO_4 and 0.68 g KH_2PO_4 in 200 mL of water. These two solutions may be mixed 1:1 or titrated using a pH meter to achieve the desire pH
- 3% glutaraldehyde in 0.05 M potassium phosphate buffer, pH 6.8–7.2—Store in the refrigerator (4° C)
- 1% osmium tetraoxide in the same potassium phosphate buffer—Store in a shatter-proof bottle in the freezer. Postfixing with osmium tetraoxide is optional in SEM, although we believe that postfixed specimens produce superior results.
- Aluminum mounting stubs
- (All of the above may be purchased from Electron Microscopy Sciences, Fort Washington, Pennsylvania.)
- Double-sided sticky tape
- Graded series of ethanol or acetone (10%, 30%, 50%, 75%, 95%, and 100%)

(*Note:* Glutaraldehyde and osmium tetraoxide are extremely toxic and should only be handled wearing gloves and in a certified fume hood. For additional information, consult material safety data sheets.)

Follow the protocol outlined in Procedure 7.4 to prepare samples for SEM.

Chromosome Counting

Determination of the chromosome number of a plant species may be used in research in genetics, evolution, and taxonomy. The chromosome number is generally conserved within a species, and taxonomical assumptions can be made according to this number. Chromosome counting is also a necessary means of determining the ploidy level from plants derived from anther and ovule cultures or chromosomal doubling by mitotic inhibitors. This is a fun and easy exercise for students.

The equipment and materials needed for study areas follows:

- Onion bulb and moist sand in a large plastic beaker
- Acetocarmine stain (45 mL glacial acetic acid, 55 mL of distilled water, and 0.5 g carmine); combine in beaker covered with aluminum foil and gently boil solution for 5 min

Procedure 7.4
Preparation of Specimens for Scanning Electron Microscopy

Step	Instructions and Comments
1	Fix and aspirate small pieces of tissue in cold (4°C) phosphate-buffered (pH 6.8–7.2) solution of 3% glutaraldehyde for a least 24 h.
2	Dispose of the fixing solution in a waste bottle, and rinse specimens at least three times with cold (4°C) phosphate buffer (pH 6.8–7.2).
3	With great care and working in a fume hood, dispense enough of the cold, buffered osmium tetroxide (osmium) solution to cover the samples and recap the vials. After 2 h, dispose of the osmium in a properly labeled hazardous waste bottle and rinse the tissues with three changes of cold buffer. In many instances, the tissue will be blackened after postfixing with osmium; this is normal.
4	Dehydrate the tissue for 30 min in each member of the graded ethanol or acetone series except 100%. The final steps of dehydration are three changes of room temperature absolute ethanol or acetone over 30 min. The samples may be stored in absolute ethanol for a few days until they can be critical point dried by the SEM technician. (*Note:* We suggest that if the specimens cannot be dried within a few days, they be stored in 75% ethanol or acetone.)
5	Have an EM technician critically point dry the specimens. Place a piece of double-sided clear tape onto an aluminum stub, and mount the dried sample on the tape. The samples are ready to be gold-palladium coated and viewed with the aid of a technician.

Procedure 7.5
Staining Chromosomes in Onion Root Tip Squashes

Step	Instructions and Comments
1	Place an onion bulb with flat side down on moist sand contained in a large beaker. Bulb cultures may be incubated at room temperature on the lab bench.
2	New roots should emerge from the base of the bulb after about 10–14 days (dependent on temperature). No later than midmorning (10 AM), excise about 1–2 cm root tips, remove sand with water, and store in fixative for 2 h. Roots may be stained immediately in acetocarmine or stored in 70% ethanol until needed.
3	Trim roots to about 0.5 cm, and place in acetocarmine stain for 20–120 min. Transfer the tissue to a glass slide along with acetocarmine stain. Gently pass the slide through the flame from an alcohol lamp. (*Caution:* Do not allow the stain to boil away; add more to prevent drying.) Cover the preparation with a coverslip. Place several layers of tissue paper or blotting paper over the specimen, and squash the tissue flat by pressing evenly and firmly with the eraser end of a pencil. Try not to move the coverslip.
4	Examine the preparation with a high dry or an oil immersion lens for even spreads of chromosomes and mitotic figures. The preparation may be made semipermanent by sealing the edges of the coverslip with nail polish.

under a fume hood. Filter solution through filter paper (may require several days) and store in refrigerator. The stain should be dark red.
- Fixative (75 mL of absolute ethanol; 25 mL of glacial acetic acid) and 70% ethanol
- Microscope slides, coverslips, and nail polish (a wild or unusual color is fun and is typically inexpensive to purchase)
- Compound microscopes equipped with oil immersion lens
- Alcohol lamp, blotting paper, and pencil with rubber eraser

Follow the protocol listed in Procedure 7.5 to stain chromosomes of onion.

The biggest problem will be getting a thin mount of the meristem area, and some students will be frustrated with the process. Keep at it, try again, and from our experience about one in every five mounts is good for observing mitotic figures. Although most cells will be in interphase, students should be able to locate cells in prophase, metaphase, anaphase, and telophase.

REFERENCES

Berlyn, G. P. and J. P. Miksche. 1976. *Botanical Microtechnique and Cytochemistry*. Iowa State University Press, Ames, IA.

Graham, E. T. 1991. A quick-mixed aluminum hematoxylin stain. *Biotechnic Histochem.* 66:279–281.

Graham, E. T. and P. A. Joshi. 1995. Novel fixation of plant tissue, staining through paraffin with Alcian blue and hematoxylin, and improved slide assembly. *Biotechnic Histochem.* 70:263–266.

Graham, E.T. and W. R. Trentham. 1998. Staining paraffin extracted, alcohol rinsed and air dried plant tissue with an aqueous mixture of three dyes. *Biotechnic Histochem.* 73: 178-185.

Hock, H. S. 1974. Preparation of fungal hyphae grown on agar coated microscope slides for electron microscopy. *Stain Technol.* 49:318–320.

Jensen, W. A. 1962. *Botanical Histochemistry*. W. H. Freeman Co., San Francisco.

Johansen, D. A. 1940. *Plant Microtechnique*. McGraw-Hill, New York.

Sass, J. E. 1968. *Botanical Microtechnique*, Third edition. Iowa State University Press, Ames, IA.

8 A Brief Introduction to Plant Anatomy and Morphology

R. N. Trigiano, Jennifer A. Franklin, and D. J. Gray

CONCEPTS

- Plant cells can be classified into the following basic types: meristematic and their immediate derivatives, parenchyma, collenchyma, and sclerenchyma.
- Simple tissues consist of one cell type, whereas complex tissues contain more than one of the basic cell types.
- The four plant organs are roots, stems, leaves, and flowers.
- Generally, the organization of monocots and dicots are similar, but they each have distinctive arrangements of cells/tissues in their organs.
- The four units of a flower are sepals, petals, anthers, and pistils.
- The study of plant anatomy is useful for understanding plant development, determining the origin of cells and tissues in situ and in vitro, and identifying the mode of regenerated plants in culture.

This chapter explores some of the internal organization (cells, tissues, and organs) or anatomy of vascular plants. For simplicity, we have organized and illustrated the material by first looking at cell types and then comparing and contrasting the anatomy of tissues and organs of the monocotyledonous (monocot) and dicotyledonous (dicot) angiosperms, gymnosperms, and pteridophytes (ferns). For the purposes of this book, we will consider the following four organs: roots, stems, leaves, and reproductive structures. It is impossible to discuss adequately all the details of anatomy and development of these organs in this short chapter. Therefore, most treatments of cell types, tissue, and organs are described in broad, widespread terms and students are cautioned that many exceptions to our generalizations can be found. The relationship of anatomy to common forms and shapes, or morphology, of these organs will also be touched upon. Readers with greater interest in more exhaustive details of anatomy and development are directed to some botany and anatomy textbooks cited at the end of this chapter. Most of the material in this chapter is derived from Esau (1960) and Fahn (1990). Readers should also appreciate that while plant anatomy and morphology typically are studied through use of static materials, such as histological sections (Chapter 7), it is important to view these in the context of growing, changing three-dimensional organisms. In this way, a better understanding of plant development can be achieved.

Plants, similar to other complex organisms, are constructed of cells, the basic unit of life. However, unlike animal cells, plant cells are surrounded by a wall composed of structural polymers, which may include pectin, cellulose, lignin, and hemicellulose. The wall may be relatively thin and flexible as in many parenchyma cells or rather thick as in collenchyma and sclerenchyma cell types (see the following discussion). In fact, the structure of the plant cell wall imparts, to some degree, the function of the cell. Cells may have only a primary cell wall, which is more or less defined as the wall material deposited while the cell is increasing in size and having cellulose microfibrils that are laid down randomly or in more or less parallel orientations (Esau, 1960). The primary wall usually contains cellulose, hemicellulose, and pectic compounds which are referred to as the middle lamella.

The wall is flexible, and stretches as the cell grows, with the orientation of cellulose fibers determining the direction of growth and eventual cell shape. The middle lamella acts as "cement" between adjacent cells. Secondary walls found in some cells are deposited to the inside of the primary wall and middle lamella after the primary wall has been completed, and can be very thick. Lignin may then be deposited into primary and secondary cell walls, making the cell rigid. Secondary walls containing cellulose and hemicellulose and lignin may or may not be present (Esau, 1960).

CELL TYPES

Let us consider the basic cell types of plants before examining the internal arrangement of cells into tissues and, in turn, tissues into organs. For the purposes of this chapter, plant cells can be simply classified into the following types and their variations: (1) meristematic, (2) parenchyma, (3) collenchyma, and (4) sclerenchyma. Note that most references consider meristematic cells to be parenchyma.

1. Meristematic cells have the following characteristics:
   - Very thin-walled cells that undergo mitosis to increase the length (apical meristem) or thickness (lateral meristem) of the organ. The meristematic initials (stem cells; see Chapter 20) reproduce themselves as well as form new cells, termed derivatives, that increase the body of the plant. These derivative cells usually continue to divide several times before any significant differentiation into other cell types occurs. The initials and their derivative cells constitute the apical meristem (Esau, 1960), which can be found at shoot (Figure 8.5A) and root (Figure 8.1A) tips. The stem apical meristem of monocots and dicots may be divided further into the tunica and corpus. The tunica (coat) is one to several cell layers thick and divides only by anticlinal divisions to increase the surface area of the tip and surrounds the corpus (body), which consists of a number of cells that divide in different planes to increase the volume of the meristem. This two-part arrangement of the meristem is absent in roots. The apical meristem of ferns and most other seedless nonvascular plants consists of a single large, triangular cell from which the surrounding apical meristem cells are derived, and the tunica/corpus organization is not present. In gymnosperms, there is no clear outer layer of tunica, but zones can

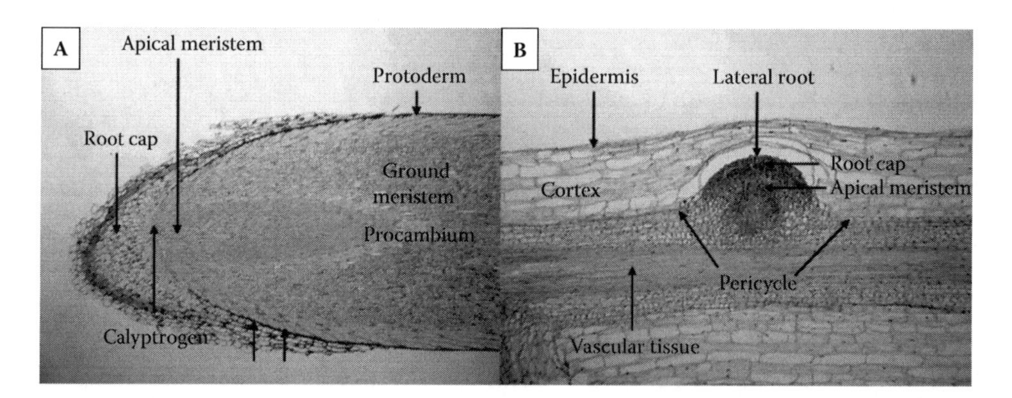

FIGURE 8.1 Root apical meristem and lateral root origin. A. Near median longitudinal section through a young corn (*Zea mays*) root. The calyptrogen, in this case, gives rise to the cells of the root cap. In other roots, including those of many dicots, the root cap is derived from the apical meristem. The three meristematic areas are the protoderm (gives rise to the epidermis), the ground meristem (cortex), and procambium (vascular tissue). Arrows indicate mucilaginous wall substance. (Slide courtesy of Carolina Biological Supply Co., Burlington, North Carolina.) B. A young lateral (secondary) bean (*Phaseolus vulgaris*) root that originated from the pericycle. Note as the lateral root develops, it pushes through and crushes the cortical and epidermal tissues of the primary root. The architecture of lateral roots is similar to that of primary roots.

be discerned within the apical meristem. An upper, lens-shaped region of large cells called the central mother cell zone and the surrounding initials derived from it divide infrequently. Cells on the periphery of this zone are active in cell division, and below the apex, the arrangement is similar to that of monocots and dicots. The shoot apical meristem can be divided into a peripheral zone, which gives rise to leaves, buds, and flowers (lateral organs), and the rib zone, which produces the stem tissues (see Chapter 20). Several cell layers distal to the apical meristem and in the rib zone, the procambium (vascular), ground meristem (cortex), and protoderm (epidermis) of the stem are differentiated and give rise to the primary tissues of the plant body.

- Primary growth of the plant is brought about by the activities of apical meristems and subsequent divisions and differentiation of the derivative cells into the tissues and organs of the plant. All tissues originating from a primary meristem are termed primary tissues, that is, primary xylem and epidermis. Most ferns, monocots, and some dicots complete their growth and development via primary growth only.
- Secondary growth exhibited by many dicots and gymnosperms is achieved through specialized lateral meristems. The vascular cambium (Figure 8.6A), located between the primary phloem and primary xylem, produces cells that differentiate into additional vascular tissue, that is, secondary xylem and phloem, which increases the girth of stems and roots. Another lateral meristem, the phellogen or cork cambium, is found near the exterior of stems and roots and arises in the primary cortex. It produces phellem and phelloderm cells that replace the epidermis and cortex, respectively, which is lost or crushed due to the expanding diameter of the root or stem. Collectively, this new tissue is called the periderm.

2. Parenchyma cells generally have the following characteristics:
   - They are typically nearly isodiametic (about as long as they are wide); however, cells may vary in shape, being elongated or even lobed.
   - The primary cell wall of this cell type is relatively thin and composed mainly of cellulose and hemicellulose with a layer of pectic substances, the middle lamella, on the exterior of the primary wall. Note that some parenchyma cells, especially in vascular tissue, may develop a secondary wall or become sclerified with lignin (see sclerenchyma).
   - Parenchyma cells always have nuclei and functioning protoplasts (cytoplasm).
   - These cells are generally considered to be relatively undifferentiated (compared to sclerenchyma) and capable of resuming meristematic activities by dedifferentiation. Indeed, this cell type and tissue is involved in the development of adventitious roots and shoots, wound healing, and other activities. Note that some parenchyma cells can be very differentiated and specialized in their function.
   - This cell type can be found through the body of the plant in primary and secondary tissues.

3. Collenchyma cells have the following characteristics:
   - They are typically more elongated than parenchyma cells and are specialized to function as mechanical support for the plant.
   - Collenchyma cells have soft, pliable, unevenly thickened primary walls composed mostly of cellulose with some pectin, but never lignin.
   - Collenchyma cells are similar to parenchyma cells in having nuclei and living protoplasm and are capable of dedifferentiation and meristematic activity.
   - This type of cell is generally found in young stems (Figure 8.6D) and leaf petioles and function as flexible supporting tissues.

4. Sclerenchyma cells have the following characteristics:
   - Sclerenchyma cells can be long and thin (fibers; Figure 8.6A) or isodiametic to elongated (sclerids) and are involved in mechanical support and water conduction. This cell type is widely distributed in the four major plant organs. Specific cells types include fibers, vessel elements, trachieds, and sclerids (various forms including astrosclerids [star-shaped], stone cells, etc.).
   - The hallmark of sclerenchyma cells is the deposition of a secondary wall on the interior of the primary wall. Proceeding from the outside of the cell inward, the wall layers encountered in these cells are the middle lamella, primary wall, and, lastly, the secondary wall. By definition, the secondary wall is laid down after growth of the cell ceases. The wall is composed primarily of cellulose arranged in parallel fibers and usually lacks pectic components. If a cell becomes lignified, deposition of lignin starts at the middle lamella and proceeds inward.
   - Although many sclerenchyma cells are dead (lack a protoplasm) at functional maturity, some types of cells may retain living protoplasm. However, the protoplasm of these cells appears to be physiologically nonfunctional or inactive (Esau, 1960).
   - Sclerenchyma cells are highly differentiated and are usually considered not to be capable of dedifferentiation and resumption of meristematic activity.

Cells are organized into simple tissues (one cell type) and complex tissues (more than one cell type). For example, young ground and pith tissue found in stems is a simple tissue made up of parenchyma cells (Esau, 1960), as is the mesophyll tissue in a leaf. Another example would be the collenchyma tissue found in the four corners of a mint stem (Figure 8.6D) or the "strings" in a celery stalk (petiole). An excellent example of a complex tissue is the secondary vascular tissues found in many dicots. This tissue contains representatives of both parenchyma and sclerenchyma cell types. Tissues, in turn, are organized into the four primary organs: roots, stems, leaves, and reproductive structures. The remainder of this chapter will be devoted to illustrating the general arrangement of cells and tissues within these organs and a brief account of seed anatomy. Anatomical studies of tissue culture regeneration of organs are included where applicable.

ROOTS

Roots serve as the primary water- and mineral-absorbing organ of plants. They also act to anchor the plant in the soil and may also function as storage organs and in vegetative (asexual) reproduction. Roots vary greatly in diameter, density of root hairs, and degree of lateral branching. Dicots and gymnosperms typically have a persistent taproot, and may exhibit secondary growth, whereas with many monocots, the taproot is ephemeral and is replaced with a fibrous root system consisting of many adventitious roots. The root morphology of ferns is fibrous and adventitious, similar to that of monocots, but the internal anatomy is similar to that of dicots.

While the general morphology of the root system is genetically determined, both overall appearance and internal anatomy can be modified by the growth environment. The diameter of primary roots ranges from 0.04–1 mm, with monocots often having roots that are smaller in diameter than those of dicots and gymnosperms. Very fine rootlike structures produced by the fern gametophyte are greatly elongated single cells, rather than true roots, and are referred to as rhizoids. Roots can contain a wide variety of pigments, and the color of roots ranges from white to brightly colored to nearly black. Young roots generally contain no pigments, and thus appear white. Exposure to light may result in a pink pigmentation. However in some species, the above-ground portions of roots contain chlorophyll, and so are green in color. While older roots often appear to be darker in color, pigmentation is not an accurate indicator of maturity or internal anatomy.

The anatomy of roots is extremely variable, but general models may be developed for both monocots and dicots. The anatomy of ferns and gymnosperms is similar to that of dicots. The apical meristem of most roots (Figure 8.1A) appears less conspicuous than shoot meristems (Figure 8.5A), which are arranged as a tunica and corpus. Ferns have a single large, triangular, meristematic apical cell rather than the multicellular meristem found in seed plants. In many plants, the root apical meristem gives rise to the cells of both the root cap and the primary meristems; however, in some grasses, a group of cells, the calyptrogen, produces the cells of the root cap (Figure 8.1A). In many instances, a quiescent zone is located in the apical meristem. This region exhibits low mitotic activity, but cell division typically resumes distal to this zone. The most noticeable differentiation of cells is in the vascular tissue: primary phloem differentiates first followed by primary xylem. Cells continue to mature by elongation and enlargement. Root hairs, extensions of epidermal cells that increase the surface area to absorb water and minerals, are typically first seen behind the zone of mitotic activity and mark the maturation of the first xylem elements. Root hair length and density are genetically determined and fairly consistent within a species, but is highly variable between species.

Lateral or secondary roots greatly increase the absorptive area of the root system and usually are initiated from the pericycle, a layer or layers of cells in the vascular tissue or stele, at some distance behind the apical meristem (Figure 8.1B). Several adjacent pericycle cells divide to form a root primordium, and continued divisions force the developing root through and crush the endodermis, cortex, and epidermis of the primary root. Vascular tissue within the lateral root is connected to similar elements in the primary or parent root by differentiation of pericycle cells (Esau, 1960). The number of lateral branches is influenced by water, nutrient availability, and biotic interactions. Genetic and hormonal control of lateral branch production is fairly well understood in dicots, whereas less is known concerning control of adventitious root proliferation in monocots (Osmont et al., 2007). Associations with ectomycorrhizal fungi often result in clusters of short lateral roots, which are covered with the fungal sheath. These are often visibly distinct, being dark in color, and occurring in clusters or in pairs (Figure 8.2). The pattern of branching is termed "root architecture." Lateral roots growing directly from the first, or primary, root are termed first-order lateral roots; laterals growing from those are termed second-order lateral roots, and so on.

Most of the same cells and tissues are present in monocots and dicots; however, the arrangement of these tissues is somewhat different. A summary of the primary differences between the two divisions of angiosperms is presented in Table 8.1. The most conspicuous difference between

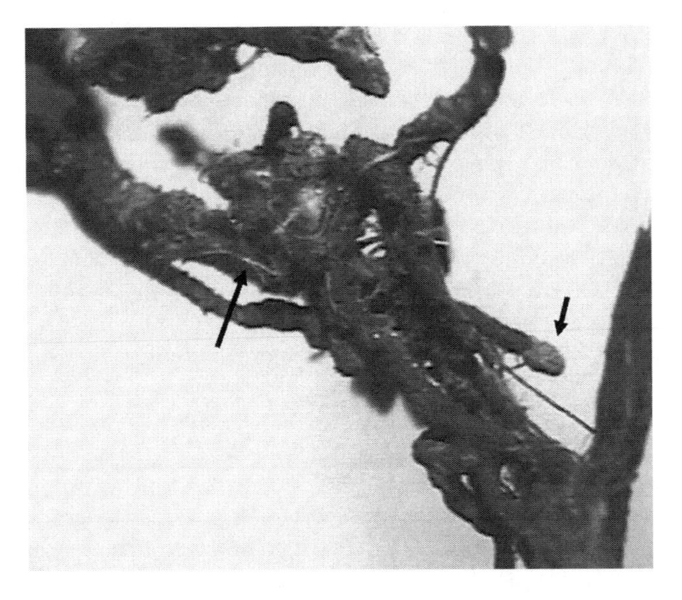

FIGURE 8.2 Ectomycorhizzae colonizing a pine (*Pinus echinata*) root. Mycorrhizal roots are surrounded by a mat of hyphae, with fungal threads visible on the surface. A newly colonized root tip appears lighter in color.

TABLE 8.1
A Summary Guide to Morphological and Anatomical Traits of Monocots and Dicots

Organ	Monocot	Dicot
Root	• Usually fibrous • Pith present • Lacks vascular and cork cambia or secondary growth	• Usually a taproot • Pith lacking • Vascular and cork cambia and secondary growth present • Primary xylem typically arranged in "arches"
Stem	• Pith lacking, but vascular bundles embedded in pith-like fundamental tissue • Vascular and cork cambia typically lacking; no secondary growth, although may have primary thickening meristems	• Pith present • Vascular and cork cambia typically present; secondary growth evident manifested as rings of vascular tissue
Leaf	• Typically blade-like with parallel venation • Leaf mesophyll generally undifferentiated into distinct layers	• Variously shaped with net venation • Mesophyll may be differentiated into spongy and palisade parenchyma layers
Flower and Seed	• Typically three-merous (flower parts in three or multiples of three) • Embryo has one cotyledon	• Typically four or five-merous (flower parts in four or five or multiples of four or five) • Embryo has two cotyledons

Note: Although this table provides very broad characterizations and contrasts of monocots and dicots, the reader is cautioned that there are many exceptions to these generalizations.

dicot and monocot root anatomy is the arrangement of the primary vascular tissue. Most dicot roots lack central pith tissue; instead, the core of the root is occupied by large metaxylem vessels (Figure 8.3A and B). Vascular tissue is contained in a single central area called the stele, with the primary xylem arranged in arches or arms, and the primary phloem located between the arms. Ferns generally have only two arms, termed diarch development. Gymnosperms and dicots often have 3 to 4 arms, triarch or tetrarch development, though some species have as many as 9. A vascular cambium is located between the primary xylem and phloem in dicot roots. In contrast, many primary monocot roots have pith in the center of the primary root surrounded by many arms of vascular tissue (Figure 8.3C and D). Dicot roots generally exhibit secondary growth via vascular cambia located between the primary xylem and phloem and by cork cambia, which forms in the cortex (see discussion in stems). Monocots and ferns lack lateral cambia or meristems.

Growth of plant tissues and organs in culture often presents anomalies, which are difficult to understand. One such instance was seen in a culture of redbud (*Cercis canadensis*) cotyledons on which adventitious roots had formed. At the tip of one such root, a green, multilobed tissue had grown (Figure 8.4 insert). Upon histological sectioning, the unrecognizable tissue was determined to be five fused somatic embryos (see Chapter 22) that had been initiated from the apical meristem of the adventitious root (Figure 8.4).

Some species have roots that are highly specialized for a specific function, and these may have an internal structure that is somewhat different than other portions of the root system. Dicots may have roots specialized for storage; these have a large diameter due to a proliferation of parenchyma cells within the secondary xylem and phloem, and may have several concentric layers of vascular cambia producing secondary vascular tissue. Aerial roots, or aerial portions of a root, generally lack root hairs, have smaller vessels than subterranean roots, and may be green due to the presence of chlorophyll in the cortex. Some roots, both above and below ground, have a cortex that appears "spongy" to facilitate the diffusion of gasses. Specialized sections of root may be produced to house associated organisms such as mycorrhizal fungi and certain types of bacteria. Specialized anatomy can also be found in parasitic roots.

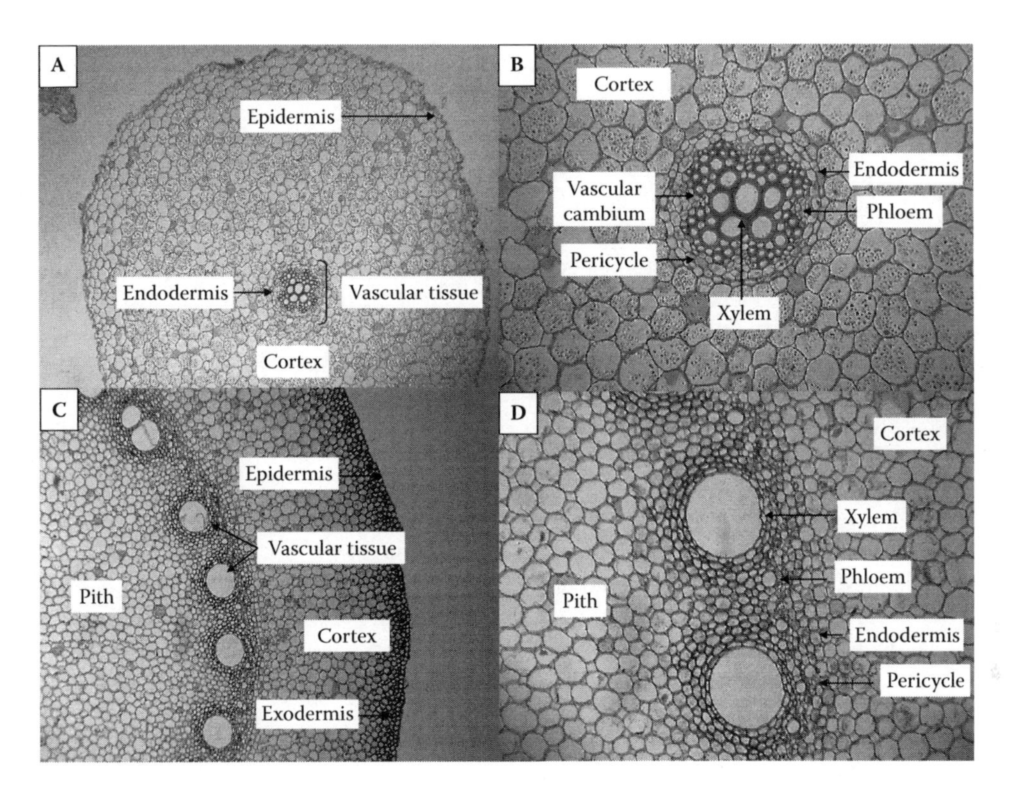

FIGURE 8.3 Dicot and monocot root structure. A. Cross section of a buttercup (*Ranunculus* species), a dicot root. Note the arrangement of tissues and the lack of pith in the center. B. Higher magnification of the central area shown in A. The "stele" includes all vascular tissue (xylem and phloem), the vascular cambium (discernible only by relative position in the young root), and the pericycle. It does not include the endodermis. Note the large xylem vessels in the center and the thickened walls (casparian strips) of the endodermal cells. C. Cross section of a large corn (*Zea mays*), a monocot root. Note the central core of pith and very large xylem vessels. D. Enlargement of the vascular area of the cross section shown in C. (All slides provided courtesy of Carolina Biological Supply Co., Burlington, North Carolina.)

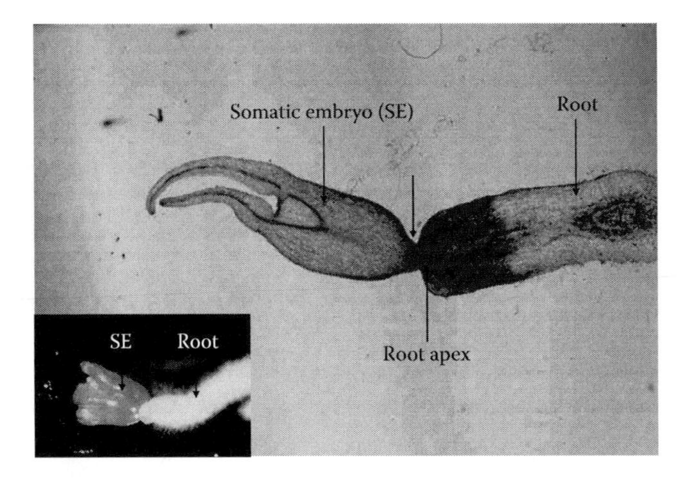

FIGURE 8.4 Somatic embryos of redbud (*Cercis canadensis*) that originated from root tissue. Near median longitudinal section through the somatic embryo and oblique section through the root, which developed from a callus culture (see insert). Close examination of serial sections revealed five distinct, but fused, somatic embryos from a common suspensor (arrow) that originated from the apical area of the root.

STEMS

As noted earlier, the shoot meristem (Figure 8.5A) of monocots and dicots is multicellular and organized into two parts, the tunica and the corpus, whereas that of gymnosperms has a zonal organization. As in roots, the shoot apex of ferns is a single apical cell. Just as in the root apical meristem, the primary meristems, the ground, the procambium and protoderm, are also found in the shoot tip. However, the shoot tip is more complex than the root apex as it differentiates leaves, axillary buds (Figure 8.5B), and flowers from the peripheral zone. The primary meristem at the end of each growing branch is the terminal bud. The stem of the plant is divided into nodes (leaves present) and internodes (leaves absent). There may be one, two, or several leaves at each node. Leaves are defined by the presence of an axillary bud (Figure 8.5B and insert) lying between the main stem and leaf petiole. The anatomy of axillary buds is equivalent to the original shoot tip. The bud has the same structure as that of the apical meristem and serves to initiate branch or lateral growth and/or flowers under the proper environmental conditions.

Stems are highly variable in form. The surface of the stem often has structures that reduce water loss and limit herbivory; trichomes and heavy cuticular waxes are present in many species. Trichomes may be a unicellular extension of the epidermis, or made up of several cells. Scales are papery and one cell thick, and bristles and thorns are rigid, multicellular outgrowths of the epidermis. Many young stems and some older stems contain chlorophyll, and so are green. Anthocyanins, which appear red or purple, are also common in stems. Green, photosynthesizing stems require a means of gas exchange, and this is provided by stomata in very young stems.

The arrangement of tissues in stems is variable and very different from the arrangement found in roots. Just as in root anatomy, dicot and monocot stems are generally different from one another. These differences are summarized in Table 8.1. Dicots typically have primary vascular bundles (fasciculars) arranged in a ring around the central pith (Figure 8.6A and B). The primary xylem is located toward the pith, while the primary phloem is toward and contiguous with the cortex. Note that in some instances the primary phloem may lie on either side of the primary xylem. This arrangement is common in ferns, which are similar in structure to dicots, although some primitive ferns lack pith. The fascicular may also have a cap of very conspicuous phloem fibers (Figure 8.6A). A portion of the vascular cambium (intrafascicular cambium) is located between the primary xylem and phloem, and another portion between (interfascicular cambium) the vascular bundles. The

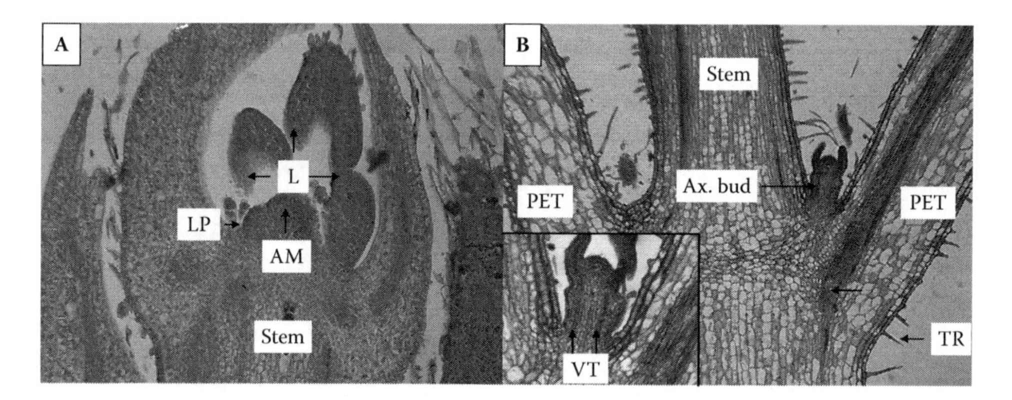

FIGURE 8.5 Apical and lateral (axillary) shoot meristems. A. Longitudinal section through the stem tip of bean (*Phaeseolus vulgaris*). The apical meristem (AM) is surrounded by a number of very young leaf primordia (LP) and more developed leaves (L). B. Longitudinal section through a node of catnip (*Nepeta cataria*). The axillary bud (Ax. bud) is located in the axil formed by the leaf petiole (PET) and the stem. Note the vascular tissue (VT) extending from the petiole and connecting to the vascular tissue of the stem (arrow). The apical meristem (AM) is dome-shaped, and the vascular tissue (VT) has differentiated (insert). TR = trichome.

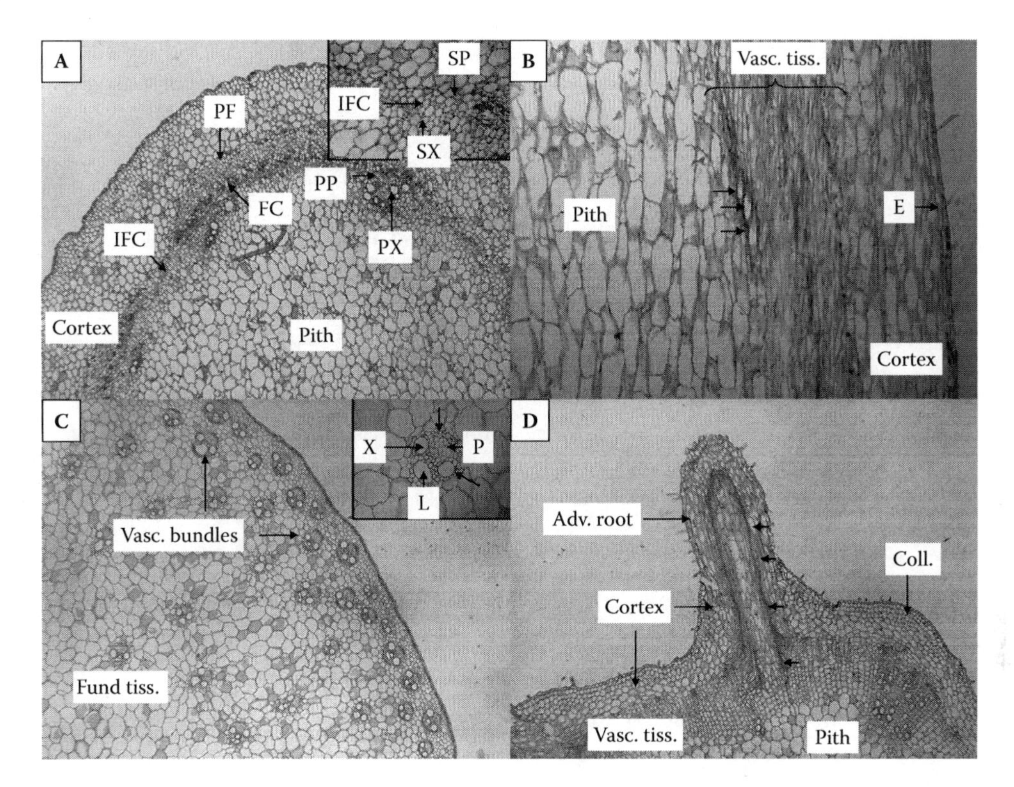

FIGURE 8.6 Anatomy of dicot and monocot stems. A. Cross section of a bean (*Phaseolus vulgaris*) stem. Note the prominent pith and the cap of phloem fibers (PF). Secondary xylem (SX) and phloem (SP) tissues, which originated from the interfascicular vascular cambium (IFC) and fascicular cambium (FC), are also common in dicot stems (insert). PP = primary phloem; PX = primary xylem. B. Longitudinal section of bean illustrates the arrangement of tissues in the stem. Arrows indicate xylem vessels with helical secondary wall patterns. E = epidermis. C. Cross section of an older corn (*Zea mays*) stem. Note that vascular bundles are scattered throughout the stem and "embedded" in fundamental tissue (Fund tiss.). The vascular bundles contain only primary xylem (X) and phloem (P) tissues (vascular cambium absent) with a prominent lacuna (L) or air space (insert). Arrows indicate sclerenchyma cells of the sheath. (Slide courtesy of Carolina Biological Supply Co., Burlington, North Carolina.) D. Cross section through the stem of catnip (*Nepeta cataria*) and a longitudinal section through an adventitious root. The root originated near primary phloem, and vascular tissue (arrows) has differentiated and connected to the vascular tissue (Vasc. tiss.) in the stem. Coll = collenchyma tissue.

activities of the vascular cambium produce secondary vascular tissues, which in due course surround the pith with a continuous ring of vascular tissue. The secondary vascular tissues eventually crush and obliterate the primary tissues, especially in perennials and woody plants. It is this continuous ring of vascular tissue that easily differentiates pith from cortex in the dicots.

Dicots also have another lateral or secondary meristem: phellogen or cork cambium. This cambium or meristem is formed in the cortex or phloem and is responsible for forming additional cortical cells (phelloderm) and replacement for the epidermis (periderm). Gas exchange is provided by lenticels, multicellular ruptures of the epidermis or periderm, in stems with secondary growth. These appear as dark, or more often light-colored, dots or patches on the stem.

In monocot stems the vascular bundles are scattered throughout the ground tissue or fundamental tissue, and as a result, a pith and cortex are not discernable (Figure 8.6C). The vascular bundles typically are surrounded by a sheath of sclerenchyma cells, which helps support the stem (Figure 8.6C insert). Additionally, many monocot stems, for example, corn, have an abundance of small vascular bundles located just under the epidermis, imparting stiffness to the stem and helping the plant to withstand environmental stresses. Monocots lack vascular cambia and therefore do not have secondary growth. If stems thicken, they do so through division of parenchyma cells in the ground tissue.

An important feature of stems, especially in the production of shoots in tissue culture or cuttings, is the ability to form adventitious roots (Figure 8.6D). These roots have their origin in parenchyma cells lying near the vascular (phloem) tissue, or from the interfascicular cambium, and grow similarly to the description provided for lateral roots. In many species, adventitious roots form most readily at nodes.

There are many modified stems among different plant taxa. Rhizomes are common in monocots, and are the primary stem form of ferns. These grow horizontally underground, generally have short internodes, and produce both roots and leaves. Stolons are similar to rhizomes, but are more often found in dicots, and generally have long internodes, grow horizontally above ground, and produce a new individual at each node. Tubers, corms, and bulbs are below-ground stems modified for starch storage. The term *tuber* refers to a swollen below-ground structure, which can be a root, or a modified stem as in a potato, that can produce new shoots from axillary buds at, sometimes inconspicuous, internodes.

LEAVES

Leaves are the primary photosynthesizing organs of vascular plants. Most angiosperm leaves are relatively thin and flat (large surface area compared to volume) and adapted for capturing light and facilitating gas/water exchange with the atmosphere. Fern leaves are similar, and are called fronds. The leaves of most gymnosperms are elongated; may be round, oval, or triangular in cross section; and are commonly referred to as needles. Other gymnosperms have small scale-like leaves that grow closely oppressed to the stem. There are, of course, many exceptions to these generalized statements, and leaves can exhibit extensive, often unusual, modifications depending on the environment coupled with genetic inputs and evolution.

Leaves are found at nodes of the stem and are variously arranged (phyllotaxis): alternate (one leaf at the node), opposite (two leaves at the node), or whorled (three or more leaves at the node). Some plants have leaves that are very difficult to recognize as leaves, whereas others are very obvious. Generally, leaves have three parts: blades or lamina, petioles, and stipules, although it is not unusual for many leaves to lack the petiole (sessile leaf) and/or the stipules. The blade may be simple with smooth or toothed margins, shallowly or deeply lobed, or compound with the leaf divided into leaflets (Figure 8.7). There are many arrangements found in compound leaves, variations of two common patterns: palmate, where leaflets are all connected at the top of the petiole, and pinnate, with leaflets forming two rows along the central midvein. All leaves have an axillary bud associated with them, and the bud is located in the axis formed by the petiole and stem. Compound leaves do not have buds at the base of the branch points or where leaflets join the common axis.

Many dicot and fern leaves exhibit net venation patterns with major and minor veins (Figure 8.8A), whereas monocots and gymnosperms typically have a parallel venation arrangement with most veins of more or less equal size (Figure 8.8B). Note that there are many exceptions to this general rule. Many gymnosperms have a single vascular bundle, or two bundles that run side-by-side, down the length of the needle. Anatomically, the vascular tissue in the node will exhibit a leaf gap, parenchyma tissue in the vascular cylinder of the stem where the leaf traces (vascular tissue connecting the leaf to the stem) are bent toward the leaf petiole (Esau, 1960).

Leaves typically do not exhibit secondary growth and therefore have only primary tissues. Leaf tissue can contain all three basic cell types discussed previously. The epidermis of leaves has many special features including a cuticle to prevent wall loss, various trichomes (hairs), guard cells and accessory cells to regulate water and gas exchange, and other cells with specialized functions (Figure 8.8A and B). Large bulliform cells in the epidermis of many grasses can "deflate," allowing the blade to fold or roll. A hypodermis beneath the epidermis of conifer needles protects inner tissue

FIGURE 8.7 Simple and compound leaves, with a petiole indicated by the arrow in each image. A. Simple leaves with an alternate leaf arrangement in privet (*Lingustrum japonicum*). B. Simple leaves with an alternate leaf arrangement in redbud (*Cercis canadensis*). C. Simple linear leaves of a monocot, a fescue grass (*Festuca* spp.). Note the horizontal stem, a rhizome (RH), from which the roots and vertical stem is growing. D. Pinnately compound leaves of *Cardamine hirsuta*. E. Palmately compound leaf of buckeye (*Aesculus flava*).

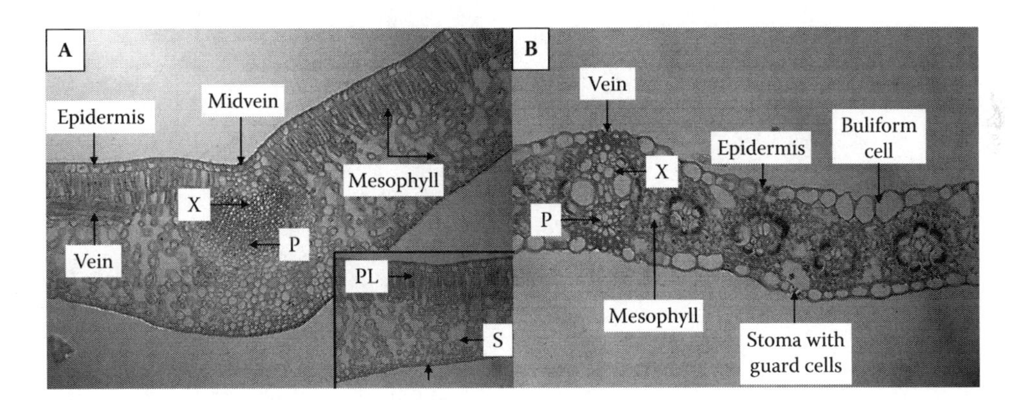

FIGURE 8.8A and B Leaf anatomy. A. Cross section of a dicot leaf (privet: *Ligustrum* species). A large midvein with xylem (X) toward the top (adaxial) surface and phloem (P) oriented toward the abaxial (bottom) surface is shown. The mesophyll is differentiated into palisade (PL) and spongy layers (S) with stomata and guard cells (arrow) on the bottom surface (insert). B. Cross section through a corn (*Zea mays*) leaf, a typical monocot. Notice that the epidermis contains specialized buliform cells and that the mesophyll tissue is not differentiated into layers. Stomata and guard cells are associated with large substomatal cavities (*). (Figures A and B are courtesy of Carolina Biological Supply Co., Burlington, North Carolina.)

from mechanical injury. The leaf mesophyll of many dicots and ferns is differentiated into one or more palisade layers toward the adaxial (top) surface of the leaf and the spongy mesophyll toward the abaxial (bottom) surface. The palisade cells are elongated (shoe box-shaped) and arranged "tightly" and orthogonally to the leaf surface. The cells of the spongy mesophyll tissue are isodiametric to slightly elongated and arranged "loosely" with large intercellular spaces (Figure 8.8A). Xylem tissue

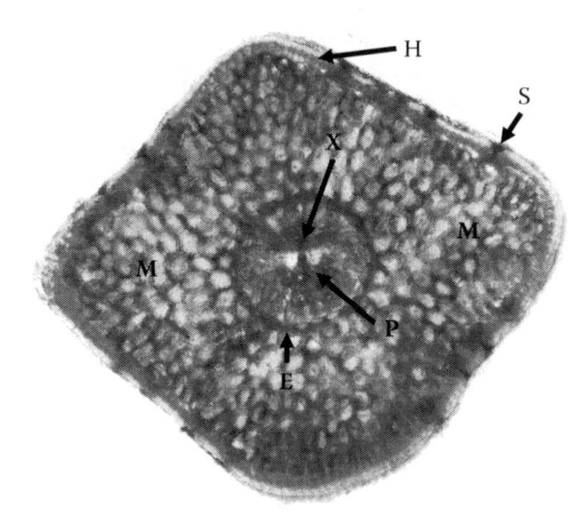

FIGURE 8.8C Anatomy of a spruce leaf. Cross section of a gymnosperm leaf (*Picea pungens*). The single vein of xylem (X) and phloem (P) are encircled by the endodermis (E). The mesophyll (M) is dense and undifferentiated. Note that immediately below the epidermis is a layer of sclerenchyma, the hypodermis (H), that contains channels for airflow from the stomatal opening (S) to mesophyll.

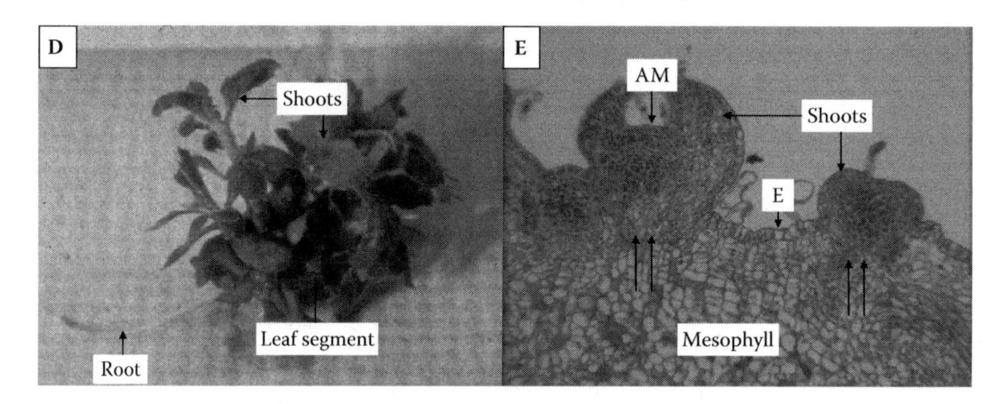

FIGURE 8.8D and E Tissue culture of chrysanthemum leaves. Adventitious shoots and roots forming directly from a mum leaf segment cultured on medium containing cytokinin. E. Cross section through the leaf segment and longitudinal sections through several adventitious shoots shown in Figure 8.8D. Note the lack of an intermediate callus and the vascular connections (arrow pairs) between the adventitious shoots and the leaf mesophyll. E = epidermis; AM = apical meristem.

is usually toward the top of the leaf and the phloem toward the bottom of the leaf. The mesophyll of monocot leaves and primitive ferns is typically not differentiated into layers and has only spongy parenchyma (Figure 8.8 B). Vascular tissue exists as bundles running through the mesophyll of the leaf. Mesophyll of gymnosperms is also undifferentiated, and is compact, often having highly invaginated cell walls. Large circular resin ducts may be found within the mesophyll. The vascular tissue is surrounded by transfusion tissue, made up of parenchyma and thin-walled tracheids, which in turn are surrounded by a suberized endodermis, similar to that of roots (Figure 8.8C). In many plants, the vascular tissue may be surrounded by a sheath of sclerenchyma cells. Vascular tissue of grasses is enclosed by a layer of parenchyma called bundle sheath cells, which in C_4 plants contain many large chloroplasts (Figure 8.8B).

Young, partially expanded leaves are a commonly explanted tissue for plant regeneration studies (Figure 8.8D). Shoots, roots, and somatic embryos arise from the explant, but it is sometimes difficult to visually determine one structure from another and from what tissue each regenerated structure had originated. A quick and simple anatomical study can help resolve these issues (Figure 8.8E).

In the case of regeneration from chrysanthemum leaf segment, the regenerated tissues were shoots originating from mesophyllic tissue. Not all cases are as clear-cut as this example, and a more extensive examination of serial sections of the tissue must be made in order to determine both the origin and identity of the regenerated tissue (somatic embryo versus shoot).

REPRODUCTIVE STRUCTURES

Flowers are organs unique to the angiosperms. They are incredibly diverse in morphology and range from very showy and conspicuous to very bland and almost imperceptible. They may occur singly or be arranged in different types of inflorescences (multiple flowers on a common axis). Flowers contain the sexual reproductive apparatus of plants and may have both female and male sexes in the same flower (monoecious: one house) or two separate flowers or plants (dioecious: two houses) for each of the sexes (pistillate and staminate flowers). Meiosis and alternation of generations occur in the ovule (female) and the anther (male). The male and female gametophytes are greatly reduced, consisting only of a few cells.

For the sake of simplicity, we will consider only perfect flowers (all flower parts including both male and female structures present). The four basic units in a flower are sepals, petals, stamens, and pistils (Figures 8.9A,B,C). The floral parts are arranged as whorls on the receptacle (Figure 8.9D). Working from the outside, inward at the base of the receptacle, the sepals, usually green, leaf-like structures, are the first whorl encountered, and these are collectively termed the calyx. The next layer is composed of the petals, which can be green, but are generally white or colored. The petals are usually larger than the sepals and are collectively termed the corolla. The petals and sepals taken together are also called the perianth. In some cases, when the sepals and petals are similar or almost indistinguishable, they are called tepals (Figure 8.9C). The stamens or the male portion of the flower are the next set of structures found in the flower. A stamen consists of the filament or stalk (Figures 8.9A,B) that terminates in the anther, which contains pollen (Figures 8.9C,D). The pistil or female structure, which is more or less flask-shaped, occupies the center of the flower and consists of a swollen base, the ovary, a smaller diameter stalk (style) that ends in a somewhat swollen tissue, the stigma (Figure 8.9B; stigma not shown). Dicots typically have sepals, petals, and stamens in whorls of four or five or multiples of four or five (Figure 8.9A), whereas monocots generally have these structures in groups of three or multiples of three (Figure 8.9C). Pistils may be composed of a single carpel (a structure analogous to a leaf rolled along the long axis bearing ovules on the inner surface) or of many carpels (Figure 8.9A). One or more ovules may be attached to a common surface or placenta in each carpel (Figure 8.9B). Microspore mother cells (2N) in the anther and one megaspore mother cell (2N) in each ovule undergo meiosis (a reduction division) followed by mitosis to form the male and female gametophyte, respectively. When mature, the male gametophyte consists of three cells or nuclei, whereas the female gametophyte typically has eight cells or nuclei. The gametophytes are often placed in tissue culture in hopes of obtaining haploid plants. One must be cautious since many diploid plants may arise from the tapetum, the inner layer of cells lining the anther (Figure 8.9D) or the nucellus, which is a maternal cell layer delimiting the female gametophyte.

Pollination occurs when pollen (containing the male gametophyte) is transported via wind, insect, or other vector to the stigma. If the pollen is compatible with the female tissue and the stigma is receptive, the pollen grain will germinate by a germ tube and grow downward through the style toward the female gametophyte in the ovule. The pollen tube contains two sperm nuclei (each N) of which one fuses with the polar nuclei to form the primary endosperm nucleus (3N or other polyploid number) while the other unites with the egg (N) to form the zygote (2N). This is called double fertilization. Thus, the sporophytic phase (2N) is restored. The primary endosperm nucleus divides and produces endosperm, and the zygote divides producing the suspensor and embryo. The embryo develops into a bipolar structure, possessing both shoot and root meristems and exhibiting bilateral symmetry. The dicot zygotic embryo has two cotyledons (Figure 8.10A), and the monocot embryo has one cotyledon (scutellum: Figure 8.10B). The integuments (seed coat) covering the ovule harden,

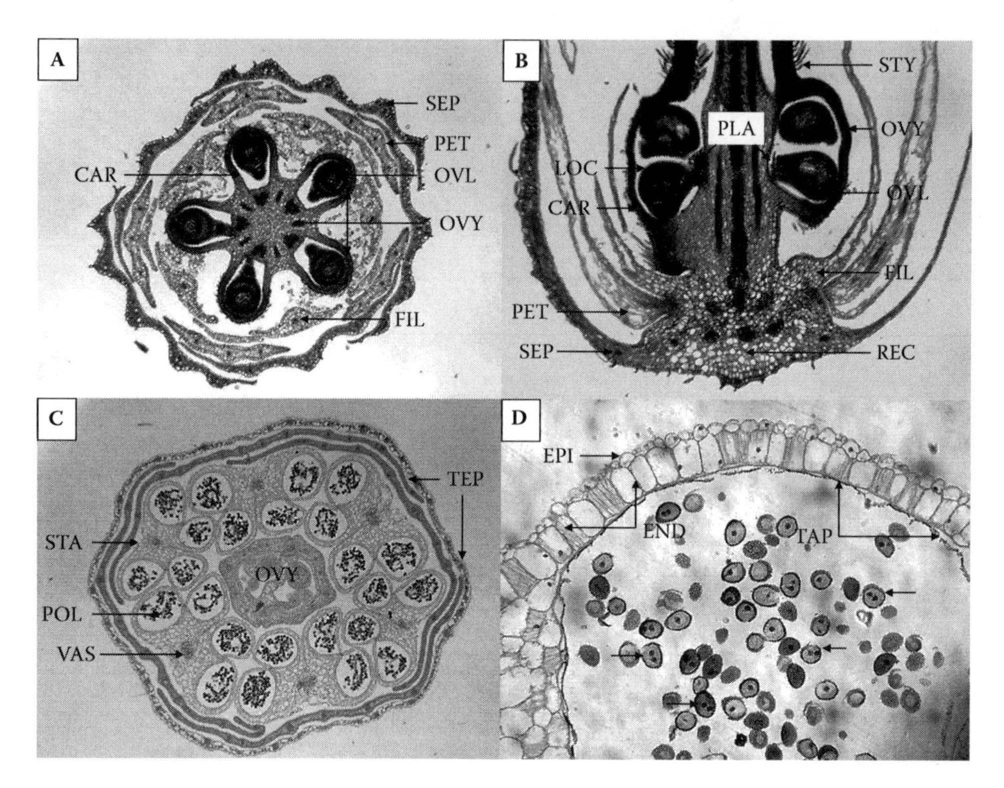

FIGURE 8.9 Flower anatomy. A. Cross section through a *Geranium* species (Stork's-bill) flower. Progressing from the outside inward, sepals (SEP), petals (PET), and filaments (FIL: the stalks of the anther) are borne. Occupying the center of the flower is the ovary (OVY), which is composed of five carpels (CAR) in which two ovules (OVL) are borne. (*Note:* Only a single ovule can be seen in each locule [space] with this section.) B. Longitudinal section through the Stork's-bill flower seen in Figure 8.9A. In this view, two carpels (CAR) and locules (LOC) of the ovary (OVY) are visible, but each contains only two ovules (OVL) of which only one will mature. Note the trichomes (hairs) on the style (STY) portion of the ovary. The sepals (SEP), petals (PET), and filaments (FIL) are inserted below the ovary on the receptacle (REC). C. Cross section of a Lily (*Lilium* species) flower. In this view, six stamens (STA) containing pollen (POL) are shown. The vascular tissue (VAS) in the filament and four lobes of each anther are evident. Sepals and petals are collectively termed tepals (TEP) for lily flowers. OVY = ovary. D. Higher magnification of an anther of lily. The anther wall is composed of the epidermis (EPI), the endothecium (END), and the tapetum (TAP), which in this case has degenerated and is only represented by the remnants of the cells. Many pollen grains are present, some in the binucleate stage (arrows). (Figures C and D courtesy of Carolina Biological Supply, Burlington, North Carolina.)

and the ovule is now a seed (Figure 8.10A). Endosperm may be completely absorbed by the embryo or may remain outside the embryo.

Immature zygotic embryos and/or embryo parts (hypocotyls and cotyledons) are often used in experiments in which somatic embryogenesis is the goal. These embryos are excellent sources of relatively undifferentiated, totipotent cells that are capable of producing somatic embryos and regenerating entire plants. In many cases, embryo tissues are exposed to high concentrations of synthetic auxins and somatic embryos are produced (Figure 8.10C). Histological sections of the materials provide evidence of somatic embryo formation (bipolar structures without vascular connections to the explant) and the origin of the somatic embryos (Figure 8.10D).

Gymnosperms produce seeds but do not flower, and with the exception of several nonconiferous gymnosperms, the cone is the primary reproductive structure. Microsporangiate (male) cones and megasporangiate (female) cones usually occur on the same plant. Each cone consists of an axis supporting a spiral of overlapping scales. On the female cone, two ovules develop on each scale within

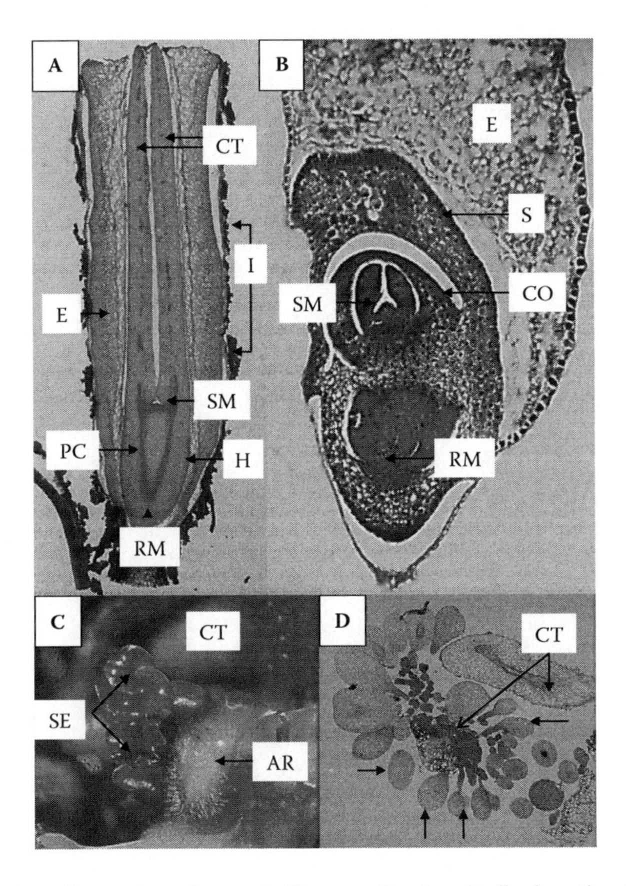

FIGURE 8.10 Zygotic and somatic embryos. A. Near median longitudinal section through an immature seed of the dicot, *Cercis canadensis* (redbud). Note the well-developed pair of cotyledons (CT), shoot meristem (SM), and hypocotyl (H); the root meristem (RM) is inconspicuous. E = endosperm, I = integument, and PC = provascular tissue. B. Near median longitudinal section through a seed of the monocot, *Dactylis glomerata* (orchardgrass). The shoot (SM) and root (RM) meristems, the single cotyledon (S = scutellum), and the first leaf (CO = coleoptile) are well developed. E = endosperm. (Photo courtesy of Springer-Verlag, McDaniel et al., 1982.) C. Somatic embryos (SE) of redbud (*Cercis canadensis*) that formed on a cotyledon (CT) cultured on medium containing 2,4-D. AR = adventitious root. D. Section through the cotyledon (CT) and somatic embryos (arrows) shown in Figure 8.10C. Note the well-developed suspensors on some of the somatic embryos.

which the megaspore mother cell undergoes meiosis to produce four megaspores (N). Only one of these develops, and over the course of a year it grows to thousands of cells in size before producing two archegonia at one end, each containing an egg cell. From the male cone, pollen production and gametophyte development occur in a manner similar to angiosperms. However, once the pollen tube germinates, one of the sperm fuses with the egg to produce the zygote (2N), and there is no endosperm produced. Here, the seed coat and thin layer of tissue to the inside of it originate from the female parent (2N), but the tissue that surrounds the zygote is megagametophyte (N). The embryo develops as in angiosperms, but with three to six cotyledons.

In ferns, no seeds are produced and the alternation of generations is more apparent, with the gametophyte being a free-living haploid (N) stage known as a prothallus. The prothallus is generally 2 to 20 mm in diameter and comprised entirely of parenchyma cells. These are variously shaped, but are often flat and heart-shaped, with thin rhizoids to serve the function of roots. They may contain chlorophyll and grow on the soil surface, or be colorless and dependant upon mycorrhizal associations. The prothallus produces cup-shaped archegonia (female) inside which a single egg is produced, and antheridia (male) that produce numerous motile sperm. A single prothallus may produce both structures, or they may occur on separate individuals. Water is required for the sperm to reach the egg and fertilization to occur. The resulting zygote grows into the familiar sporophyte generation (2N). When mature, clusters of sporangia called sori are produced on specialized fronds, or in rows or along the margins of regular fronds. Within the sporangia, the spore mother cell undergoes two meiotic divisions to create a tetrad of spores (N) that are shed and germinate to grow into the prothallus. Numerous variations on this cycle exist.

CONCLUSION

We have presented a very brief account of the anatomy of angiosperms, gymnosperms, and ferns, and we hope that we have whetted your appetite to further your study of this area of plant science. The study of plant anatomy allows us to understand more fully the process of plant development from the union of egg and sperm to formation of the zygotic embryo, to development of the various tissues in the embryo and the mature plant—fascinating processes. The study of anatomy also allows us to somewhat understand the origin and development of regenerated tissue from in vitro culture. Aspects of regeneration of plants in vitro are of keen interest to researchers conducting studies of genetic transformation (use of ballistics versus *Agrobacterium* mediated methods; see Chapter 33), evaluating the quality of regenerated plants, and determining the mode of regeneration (e.g., somatic embryogenesis versus shoot formation).

REFERENCES

Esau, K. 1960. *Anatomy of Seed Plants*. John Wiley & Sons, New York. 376 pp.

Fahn, A. 1990. *Plant Anatomy*. Fourth edition. Pergamon Press, New York. 588 pp.

Gifford, E. M., and R. H. Wetmore. 1957. Apical meristems of vegetative shoots and strobili in certain gymnosperms. *Proc. Natl. Acad. Sci. USA* 43:571–576.

Jones, D. L. 1987. *Encyclopaedia of Ferns*. Lothian Publishing Company, Melbourne. 433 pp.

McDaniel, J. K., B. V. Conger, and E. T. Graham. 1982. A histological study of tissue proliferation, embryogenesis, and organogenesis from tissue cultures of *Dactylis glomerata* L. *Protoplasma* 110:121–128.

Osmont, K. S., R. Sibout, and C. C. Hardtkey. 2007. Hidden branches: Developments in root system architecture. *Annu. Rev. Plant Biol.* 58:93–113.

Raven, P. H., R. F. Evert, and S. E. Eichhorn. 1999. *Biology of Plants*. Sixth edition. W.H. Freeman and Co., 944 pp.

Smucker, A. J. M. 1993. Soil environmental modifications of root dynamics and measurement. *Annual Review of Phytopathology* 31:191–213.

SUGGESTED READING

Dickison, W. C. 2000. *Integrative Plant Anatomy*. Academic Press, San Diego, CA.

Wilson, C.L., W. E. Loomis, and T. A. Steeves. 1971. *Botany*. Fifth edition. Holt, Rinehart and Winston, New York. 752 pp.

9 Seed Development and Germination

Feng Chen, Ruth C. Martin, Songquan Song, and Hiroyuki Nonogaki

CONCEPTS

- Prevention of water uptake. Prevention of water uptake by the seed coat is a common cause of seed dormancy. Without sufficient water, metabolic activities cannot be resumed and the embryo cannot expand, which results in the failure of radicle emergence.
- Mechanical constraint. The testa of some seeds is very rigid, which prevents the radicle from penetration. In some seeds, the endosperm also serves as a physical constraint.
- Interference with oxygen uptake. Oxygen is required for respiration. The low permeability of the testa to oxygen will limit the availability of oxygen, especially to the embryo.
- Presence of inhibitor. The testa and other surrounding tissues and the embryo may contain relatively high concentrations of growth inhibitors, with ABA being the most prominent one that can suppress germination of the embryo.

A seed is the fertilized and matured ovule of angiosperms and gymnosperms and represents a crucial stage in the life cycle of seed-bearing plants. Seeds of diverse plant species may display differences in size, shape, and color. Despite apparent morphological variations, most mature seeds consist of the following three major components: the embryo, the endosperm, and the testa (seed coat; Figure 9.1). The embryo consists of leaf, stem, and root precursor tissues. Growth of the embryo out of the seed leads to the formation of a seedling. The endosperm is the tissue that surrounds the embryo and provides nourishment for the development and growth of embryo and seedling. In some seeds, the endosperm is reduced to a few cell layers or may completely disappear at maturity. In these seeds, food reserves are stored in cotyledons. The testa surrounds the embryo and endosperm and provides a protective covering for the seed.

Seeds have a number of unique properties. Most seeds are highly tolerant to desiccation. Compared to the vegetative portions of plants, which are sessile, seeds are relatively mobile. They can be dispersed by animals, wind, or water. Additionally, seed germination is regulated by many internal and external factors, which usually optimize the distribution of germination over time in a population of seeds. These properties contribute significantly to the success of seed plants on earth. In addition to their role in plant reproduction, seeds are a vital component of the human diet. The majority of the world's food staple consists of seeds (e.g., cereal grains). Because of the importance of seeds as food and a genetic delivery system, seed biology has been an important area of both basic and applied plant science. In this chapter, we will describe the biological and physiological events that occur during seed development and germination as well as the dormancy mechanisms that regulate the germination process.

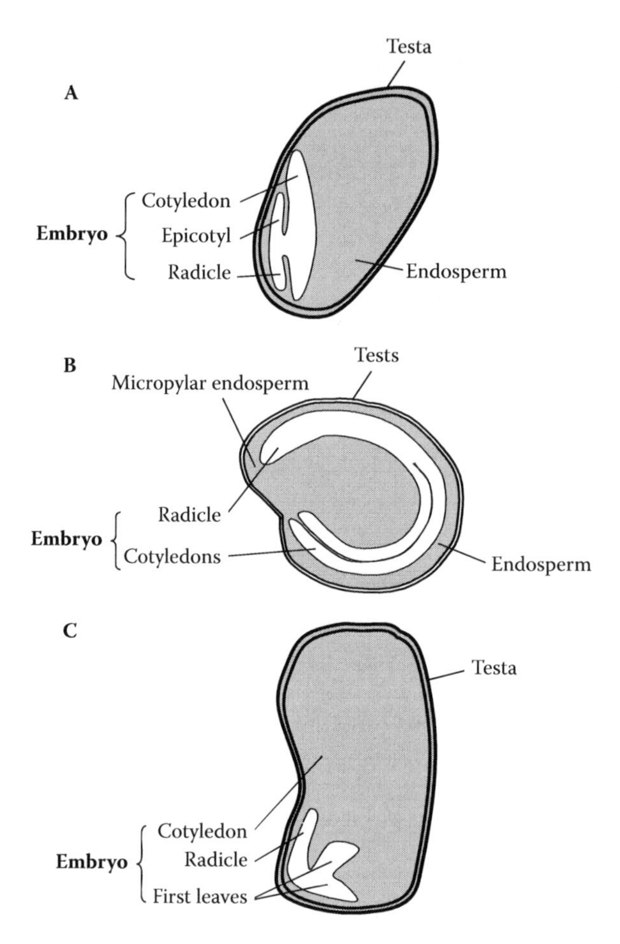

FIGURE 9.1 The major seed tissues in a monocot seed (A), a dicot seed with endosperm (B), and a dicot seed without endosperm (C).

SEED DEVELOPMENT

Formation of Seed

Seed formation is initiated shortly after pollination and fertilization. It is important to recall the mechanisms of fertilization in order to understand the developmental programs of seeds because seed tissues such as the embryo, endosperm, and testa have distinct origins that are associated with multiple events during fertilization. The embryo, which is a diploid ($2n$) tissue, results from the fusion between a gamete from the pollen and a nucleus in an egg cell. The triploid ($3n$) endosperm tissue is formed by the fusion of a second gamete from the pollen with two nuclei in the central cell. The testa is derived from integuments of the maternal tissues. A detailed description of double fertilization and early embryogenesis is provided in many plant biology textbooks and is not repeated here. This section focuses on events during the mid- to late stages of seed development.

The major events during early embryogenesis that involve cell division and differentiation are important in determining cell fates in the different domains of the globular, heart, and torpedo embryos (Figure 9.2). During the mid- and late stages of embryogenesis, seed reserves are deposited in the endosperm and translocated to the growing embryo. Some species such as castor bean and cereals contain a thick endosperm in mature seeds, while seeds of other species such as *Arabidopsis* and lettuce retain only a few cell layers of endosperm at maturity. Additionally, there are seeds from those species, including peas and beans, that do not retain endosperm at maturity, and these are

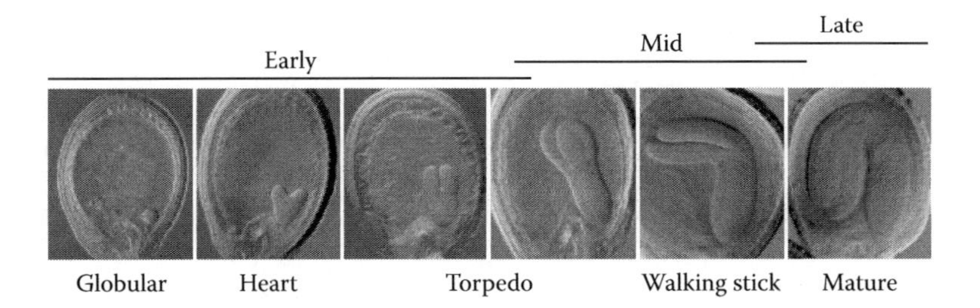

FIGURE 9.2 Developmental stages of *Arabidopsis* seeds. Seed development can be divided into early, mid- and late stages. These are not absolute definitions but represent overlapping stages. Early stages of development involve histodifferentiation, which is important for forming the plant body plan. Seed reserves accumulate during mid- to late stages. Differential interference contrast microscope images of the globular-, heart-, torpedo-, walking stick–, and mature stage embryos are shown. (Photo taken by Jessica Kristof and Jennifer Coppersmith.)

referred to as nonendospermic seeds. However, even in nonendospermic seeds or seeds with only trace endosperm at maturity, a substantial amount of endosperm is transiently formed during seed development but is digested during the completion of embryogenesis. While information concerning endosperm differentiation is limited compared to the information available for embryogenesis, recent molecular and genetic studies using the model plant *Arabidopsis* are providing valuable information on endosperm development.

There are two types of endosperm differentiation—cellular and nuclear. In cellular endosperm differentiation, nuclear division in endosperm cells is immediately followed by cellularization, which is similar to cell division in many other tissues. In contrast, during nuclear endosperm development, there is a delay in cell plate formation that results in cells with multiple nuclei: coenocytes. This is also called syncytial endosperm. The involvement of microtubules in cell plate formation after the syncytial stage of endosperm development has been characterized. Details of endosperm differentiation, however, are still unknown.

During seed development, the maternal integuments develop into the testa, which is important for protecting the embryo and endosperm from harsh environments. In *Arabidopsis* seed, the innermost layer of integument, or endothelium, accumulates proanthocyanidins (PAs) that are flavonoid pigments. The *transparent testa* (*tt*) mutants, which only accumulate low levels of PAs, lack seed dormancy, indicating that pigments in the testa are somehow involved in the imposition and/or maintenance of seed dormancy (Debeaujon et al., 2000).

Differentiation of the testa and endosperm starts shortly after fertilization, suggesting that fertilization triggers the development of these tissues. However, there are mutants that develop endosperm and testa without fertilization. In some mutants, nuclear endosperm differentiation occurs without fertilization and is followed by cellularization. A normal appearing testa including polygonal structures with a central elevation or columella is formed in *fis* mutants without fertilization. These observations indicate that the initial development of the seed-covering tissues is independent of embryogenesis, although the developing embryo can potentially affect the growth of the endosperm and testa. In addition to the genetic control of seed development mentioned above, environmental factors such as temperature, light, and soil nutrition can also affect seed development and have an effect on seed mass and quality.

Accumulation of Seed Reserves

Photosynthetic products, which are transported in the form of sucrose and amino acids, accumulate in developing seeds. Sucrose is transported from the maternal plant through the funiculus, converted to hexoses and absorbed by seeds. The products of photosynthesis may be transported in both an apoplastic (extracellular) and symplastic (intracellular) manner. Plasmodesmata, pores in cell walls between neighboring cells, may be involved in symplastic transport. The small molecules transported through the funiculus accumulate in developing seeds as macromolecules such as polysaccharides, lipids, and proteins.

The major carbohydrate storage reserve typically found in seeds is starch, which consists of amylose and amylopectin. Amylose is a linear α-1,4-glucan chain. Amylopectin is a glucan chain containing many branched amyloses through α-1,6 linkages. Amyloses and amylopectins, which eventually become large starch granules in cells, accumulate in the plastids of seed storage tissues such as the cereal endosperm and bean cotyledons. ADP-glucose pyrophosphorylase (AGP), which converts glucose-1-phospate to ADP-glucose, is a rate-limiting enzyme for starch synthesis and plays an important role during seed development. Maize *shrunken 2*, which has a mutation in this enzyme, accumulates less starch but more sucrose in developing seeds. The overaccumulation of sucrose attracts water due to osmotic potential. As a consequence, seeds become shrunken after maturation drying when water is eliminated from the seed, and the seed enters a metabolically quiescent state. The production of AGP is promoted by 3-phosphoglycerate (3-PGA), a product of photosynthesis. This means that as the supply of photosynthetic products increases, starch deposition is stimulated. This is an example of feed-forward regulation. When this enzyme converts glucose-1-phosphate to ADP-glucose using ATP, orthophosphate is released from ATP as a by-product. Orthophosphate exerts an inhibitory effect on AGP. That is, AGP is subject to feedback regulation. AGP consists of two small subunits and two large subunits, and the large subunits are sensitive to orthophosphate. The introduction of a mutation in the AGP large subunits in wheat makes the enzyme complex less sensitive to inhibition by orthophosphate and drastically increases seed weight. This type of transgenic approach for altering developing seeds to increase yield has potential for crop improvement.

β-1,4-mannose chains containing α-1,6-galactose or -glucose side branches such as galacto-gluco- and galactogluco-mannans represent another major type of carbohydrate found in seeds of endospermic legumes and other species such as tomato and coffee. The mannose main chains are degraded by β-mannanase, an endo-type enzyme, which means that the enzyme hydrolyzes internal bonds of the mannan polymer and produces end products with two (mannobiose) or three units of mannose (mannotriose). β-mannosidase, an exo-type enzyme, further digests these oligomannans to mannose.

Seeds of oil crops such as sunflower, canola, cotton, and soybean accumulate oils and proteins as major seed reserves. When cells of the seed storage tissues in these species are examined under a light or electron microscope, many small round oil bodies are observed surrounding the relatively larger protein storage vacuoles (Figure 9.3). The predominant oil in seeds is triacylglycerol (TAG), which is synthesized by the condensation of glycerol and three fatty acid molecules. TAGs containing saturated and unsaturated fatty acids accumulate in oil bodies, which appear as small (0.2–2.0 μm) particles with an electron-dense boundary and electron-opaque matrix when viewed with an electron microscope. The electron-dense surface is the half-unit membrane, a single layer of phospholipids which also contains specific proteins called oleosins. Oleosins have three different domains. The central portion of oleosin polypeptides is highly hydrophobic and is anchored to the TAG matrix inside an oil body and may be important in stabilizing the small particle. The polar N-terminal and amphipathic C-terminal portions of the polypeptides are located on the surface of the oil body facing the cytosol and thereby prevent fusion of oil bodies, which might otherwise happen because of their hydrophobic properties (Huang, 1994). Oleosins are not essential for oil synthesis. The size of oil bodies found in olive seeds is 0.5–2.0 μm, whereas the size of oil bodies in the olive fruit, which do not contain oleosins, is 10–20 μm. It is thought that the function of oleosins is to maintain the small size of oil bodies so that they are efficiently accessed and digested by lipase, a lipid-degrading enzyme, during reserve mobilization after germination. It is also possible that oleosins have a seed-specific physiological role in the control of dehydration/hydration. In *Brassica napus*, rapeseed, oleosins are induced concomitantly with storage proteins such as cruciferin and napin, all of which are regulated by abscisic acid (ABA; see the following text). The

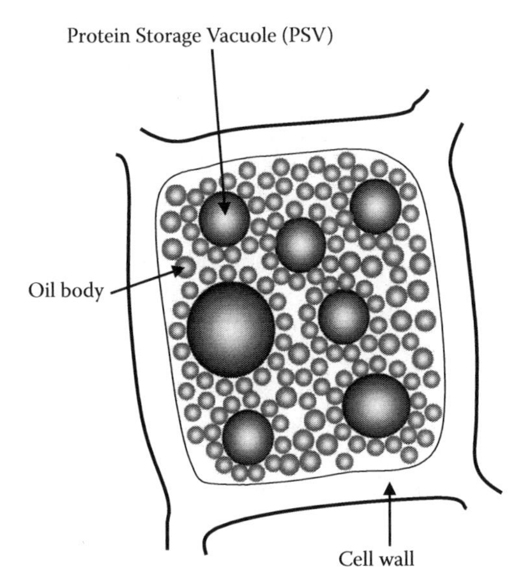

FIGURE 9.3 Schematic representation of a cell that would be found in seed storage tissues such as endosperm and cotyledons in oil- and protein-rich seeds. Storage proteins accumulate in protein storage vacuoles (PSVs), which are surrounded by small oil bodies containing specific proteins called oleosin (see text).

rate-limiting enzyme for TAG synthesis is diacylglycerol acyltransferase (DGAT). Mutations in this gene reduced enzyme activity and TAG content in seeds, whereas overexpression of this gene increased seed oil content and weight.

Amino acids are transported into seeds and synthesized as storage proteins. Seed proteins are generally classified into several categories depending on their solubility. Albumins (water soluble), globulins (salt soluble), glutelins (dilute acid/alkaline soluble), and prolamins (aqueous alcohol-soluble) are typical classes of proteins according to the Osborne classification system. Legume globulins and cereal glutelins and prolamins have been well characterized because these are edible proteins important in our food supply. Many seed proteins are present as homo- or heteropolymers with multiple subunits. The assembly of soybean 11S globulin (glycinin) subunits has been well characterized. The formation of an intramolecular disulfide bond and cleavage of polypeptides into distinct domains play an important role in the formation of different polymers. The glutelins predominant in wheat and rice are given specific names, glutenins and oryzenins, respectively. Likewise, maize and barley prolamins are called zein and hordein, respectively.

Seed storage proteins are synthesized in the endoplasmic reticulum (ER) and stored in protein storage vacuoles (PSV; Figure 9.3). There are two pathways in terms of protein synthesis and transport in the cells of seed storage tissues. In seeds of dicotyledonous (dicot) species, globulins are synthesized in the cisternal ER and transported to PSVs via the Golgi apparatus. In soybeans, the trans Golgi secrete dense vesicles containing globulins, which eventually fuse with PSVs. A similar pathway is observed for albumin transport in pumpkin seed cells. Prolamin synthesis in seeds of monocotyledonous (monocot) species occurs in the so-called modified ER. Prolamins directly accumulate in the lumen of rough ER (RER), which later creates a special budding structure called a protein body (PB). As this structure moves through the cytoplasm after it is detached from ER, it is taken into PSVs by autophagy.

In addition to the major storage proteins, seeds may accumulate toxic proteins called lectins. Plant hemagglutinin (PHA), which binds to specific oligosaccharides and precipitates blood, is a typical lectin. Lectins are thought to be important for seed survival against predators. Seeds also accumulate specific inhibitors to some digestive enzymes, which are also important in protecting seeds from predators. Bean seeds accumulate α-amylase inhibitors (α-AIs) that block the degradation of starch. When starch is incubated with insect α-amylase in a test tube, starch degradation is normally observed. However, when starch and α-amylase are co-incubated with α-AI, starch degradation is strongly inhibited. Interestingly, the α-AI does not inhibit α –amylases of plant origin. This inhibitor actually suppresses the growth of insect larvae. α-AI could be introduced into crops as a strategy for protecting developing seeds from insect damage. A potential problem with this application is that it has the potential to cause digestion problems in humans since we also need α-amylases to digest starches present in seed crops. There are some wild-type bean species that accumulate α-AIs that inhibit insect α-amylase but do not affect animal pancreatic α-amylase. The gene for this α-amylase inhibitor may be a useful candidate for protecting crops against insects with little effect on human consumption. Developing seeds accumulate inhibitors to proteinases as well.

Maturation Drying

One of the most characteristic events during seed development is spontaneous hydration or preprogrammed desiccation, which could be detrimental to other plant tissues. This is called maturation drying. Seeds of some tropical and aquatic plant species do not have desiccation tolerance and are termed recalcitrant seeds. Many other species have orthodox seeds, which are seeds that can survive under conditions in which seed water content is maintained at below 10%. Desiccation tolerance is also observed in some animals such as nematodes and tardigrades (water bear). The mechanisms of anhydrobiosis, or life without water, are not well understood. It is generally known that anhydrobiosis provides resistance to other stimuli in these creatures. Dry seeds are less sensitive to high temperatures as compared to seeds with a relatively higher water content.

It is important to understand that seeds are not capable of surviving desiccation at all developmental stages. Interestingly, seeds can survive desiccation before natural maturation drying, indicating that the physiological changes that confer desiccation tolerance to seeds occur at an earlier stage. Another important aspect of seed desiccation is that the reduction in water content can switch the seed from the developmental mode to the germinative mode. When fresh castor bean seeds are harvested prematurely, they do not germinate until 50 days after pollination. In contrast, when the seeds are harvested and subjected to desiccation, the seeds germinate as early as 25 days after pollination. This exemplifies the importance of desiccation in switching on germination potential. What is more significant in terms of understanding the mechanisms of desiccation tolerance is that castor bean seeds harvested 20 days after pollination cannot survive desiccation at all. This suggests that changes critical for desiccation tolerance occur during a narrow five-day window. One can imagine that specific protective substances may be synthesized in developing seeds during this narrow window. Sugars and proteins may function as protective agents in seeds. The damage caused by dehydration is thought to occur due to changes in the physical properties of membranes. Phospholipids, which are in the liquid crystalline state in the presence of water, turn to a gel state when tissues are dehydrated. Problems occur when membranes in the gel state absorb water. Sugars could protect membranes by inserting between the polar heads of phospholipids and maintaining the liquid-crystalline state of the membrane.

During maturation drying, the sugar and oligosaccharide content increases while the water content decreases, suggesting that these molecules may play an important role in desiccation tolerance. The major sugar and oligosaccharides that increase during maturation drying in developing seeds are sucrose and raffinoses. The raffinose family oligosaccharides (RFOs) include raffinose, stachyose, and verbascose, which are sucrose plus one, two, and three galactose residues, respectively. Sugars and RFOs may function to help maintain the glassy state of cells, which is referred to as vitrification, although the involvement of RFOs in the induction of desiccation tolerance is still unclear.

Developing seeds of various species accumulate high levels of a specific group of proteins at late stages of embryogenesis. These proteins are called late embryogenesis abundant (LEA) proteins.

LEA proteins generally have highly hydrophilic amino acids and are regulated by ABA (Galau et al., 1987). The hydrophilic nature of the proteins suggests that they attract or substitute water and protect cells from dehydration. The fact that these proteins are regulated by ABA also suggests that they may be involved in desiccation tolerance because ABA is known to be critical for desiccation tolerance. *Arabidopsis* mutants deficient in ABA synthesis or double mutants deficient in ABA and insensitive to ABA exhibit reduced or no desiccation tolerance. While the involvement of LEA proteins has been known for many years, the definite mechanisms of desiccation tolerance are still unknown.

SEED GERMINATION

During maturation drying, metabolism in seeds is gradually reduced and the embryo enters into a metabolically quiescent state. Seed germination is defined as the resumption of growth of the embryonic plant within the mature seed. Different criteria have been used to judge the completion of seed germination. Farmers consider a seed germinated only after the seedling emerges from the soil. Seed biologists, however, define the completion of seed germination by visible radicle emergence, and consider growth after radicle protrusion as a postgerminative event. The latter definition is used in this chapter.

IMBIBITION AND RESUMPTION OF METABOLISM AND OTHER CELLULAR ACTIVITIES

Seed germination is initiated with water uptake, which is also called imbibition. Mature seeds are often extremely dry and need to take in significant amounts of water before cellular metabolism and growth can resume. The permeability of the testa plays a pivotal role in the rate of imbibition by dry seeds. Testa is a dead tissue at maturity that contains lignified cells and a waxy layer and could limit water penetration. Water does not enter different parts of testa at the same rate. The micropylar region is the most water-permeable region of the testa in cereals and many other seeds. Water then spreads throughout the seed tissues. The uptake of water into cells can be facilitated by water channel proteins such as aquaporins that are present in the endosperm and embryo. During seed imbibition, swelling of the tissues mainly as a result of protein rehydration can lead to significant increases in the size of seeds.

Water uptake by a viable mature dry seed displays a triphasic pattern (Figure 9.4; Bewley, 1997). Phase I imbibition exhibits rapid initial water uptake. This is a physical process driven by the difference in water potential between the seed and its environment. Phase II is called the "plateau phase," in which seed water content remains constant or increases only slightly. After phase II, nondormant seeds then proceed to phase III. Similar to phase I, phase III is also marked by rapid water uptake. Unlike phase I, the increase in water content in phase III is mainly due to water absorption by the growing embryo emerging from seed.

The rapid initial imbibition during phase I can cause damage to seeds such as structural perturbations, especially in membranes. The damage to membranes leads to a rapid leakage of low molecular weight metabolites and solutes into the surrounding imbibing environment. In some seeds, such as corn, such damage is especially sensitive to low temperatures. Normally, the membrane structures are repaired within a short period of rehydration. At this time, leakage of cellular contents will be curtailed. Dead seeds can also take up water. Since tissues of dead seeds lose membrane integrity and cannot develop turgor, they often absorb more water than viable seeds, but do not enter phase III. Imbibed dormant seeds (see next section) remain in phase II until the dormant state is broken.

Within minutes of water entering the cells, cellular metabolic activities resume. Enzymes necessary for the initial metabolic activities are made during seed development and are present in dry mature seeds. Rehydration of seeds is sufficient to activate many metabolic events, with respiration

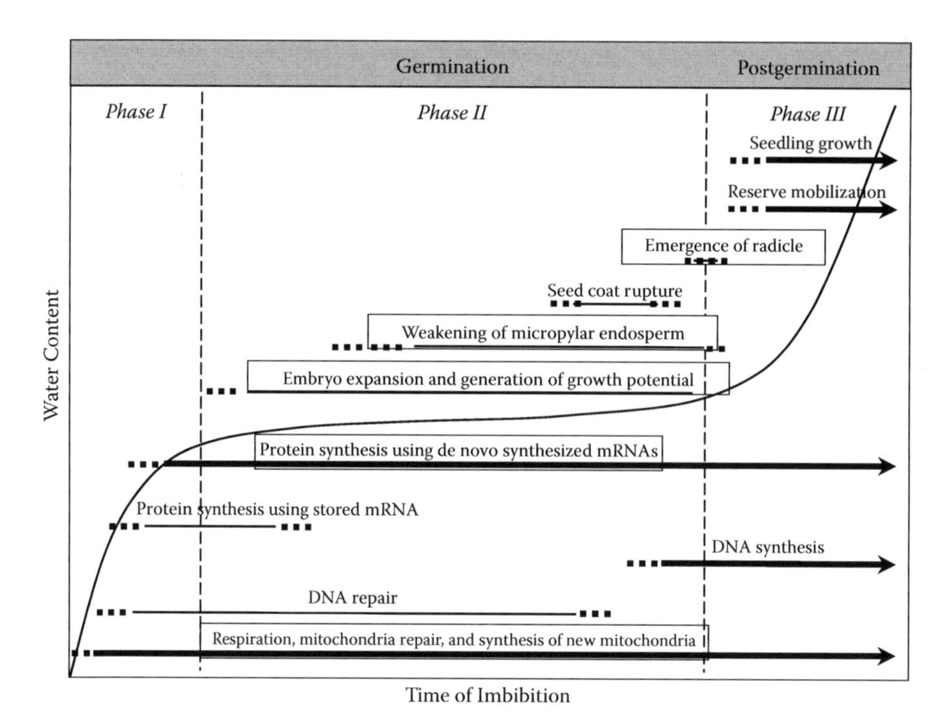

FIGURE 9.4 Time course of imbibition and some major events occurring during germination and postgermination (see text for explanation). (Modified from Bewley J.D. 1997. *Plant Cell* 9: 1055–1066.)

being one of the first activities to be resumed. As described in the seed development section, most plants store food such as starch, proteins, or oils when seeds are formed, which will be used to provide nourishment to the growing embryo inside the seed. The initial substrates for respiration are soluble sugars, such as sucrose. Mobilization of most reserves is a postgerminative event that occurs after the completion of seed germination. Oxygen consumption for respiration follows a pattern similar to water uptake. After a rapid initial increase in oxygen consumption, the rate of oxygen consumption declines and then increases again when the radicle penetrates the surrounding tissues. In some species, due to low levels of internal oxygen prior to radicle emergence, the major source of ATP is anaerobic respiration, which also leads to the production of ethanol.

Respiration occurs in the mitochondria. Activating the function of mitochondria is one of the earliest events that occurs during imbibition. Different patterns of mitochondrial development have been observed in seeds containing different types of reserves. Repair and activation of preexisting mitochondria predominate in starch-storing seeds such as mung bean. In contrast, new mitochondria are produced in oil-storing seeds such as soybean. The high-energy molecules ATP and NADPH produced via respiration are used to fuel essential cellular activities. The first synthetic activities are associated with the repair of structures, such as DNA, that are damaged during seed maturation drying and storage. Whereas some enzymes are produced during seed development, other enzymes are synthesized de novo. Some newly synthesized proteins are made from preexisting mRNAs and others are produced from newly transcribed mRNAs (Figure 9.4).

Completion of Germination

For most seeds, radicle protrusion through the structure surrounding the embryo marks the completion of germination, which is followed by the commencement of seeding growth. Seed germination may involve both cell expansion and division. For some species, however, only expansion of existing cells in the embryo occurs. Extension of the embryo is a turgor-driven process. Seeds display a wide range of embryo morphologies at seed maturity, from rudimentary to fully developed. In some seeds, such as celery, considerable embryo growth and development must occur within the seed prior to the completion of germination. In others, only expansion of the existing embryonic structures is required. In the latter case, DNA replication and cell cycle activity may commence prior to radicle emergence, but cell division in the meristem tissues is generally held until radicle emergence or later stages. Regardless, the embryo tissues must expand in order to protrude. Therefore, processes related to cell growth (embryo growth potential) are important for regulating seed germination.

In mature seeds, the testa protects the seed against harmful agents in the environment. As a result of this protective function, the testa may regulate seed germination through its mechanical strength, which restricts radicle protrusion. Testa rupture is associated, at least partly, with an increase of water uptake during imbibition that leads to embryo swelling and initial growth. Therefore, the extent of embryo swelling and/or testa strength during phase II of imbibition seems to be under physiological control. Testa mutants exhibiting reduced or enhanced dormancy characteristics demonstrate that the presence and/or properties of tissues external to the embryo can affect the timing and completion of germination.

In addition to testa, mature seeds of many plant species contain living endosperm that completely encloses the embryo. Examples of such seeds include tomato, tobacco, and pepper. Because of its complete enclosure of the embryo, the endosperm can also serve as a physical constraint, similar to that provided by the testa. This physical constraint has to be overcome in order for the radicle to emerge. The regulatory role of the endosperm on germination is mainly controlled by the micropylar endosperm (Figure 9.1). In tomato and other endospermic seeds, the micropylar endosperm forms a cone that encloses the radicle tip. It is therefore called the "endosperm cap." The endosperm cap can be distinguished anatomically from the lateral endosperm. For example, the endosperm cap of tomato seeds is composed of smaller cells, which contain thinner cell walls compared to those in the lateral endosperm. Dramatic changes are observed in the endosperm cap of germinating seeds. The inner surfaces of the endosperm cell walls are initially smooth, and become increasingly degraded following imbibition. These anatomical changes are associated with physical weakening of the endosperm cap tissue. Reduction in the physical force required to penetrate the endosperm cap has been observed in imbibing seeds of tomato, pepper, and coffee using radicle-sized probes and force analyzers. In all cases, the weakening of the endosperm cap is associated with radicle emergence, indicating that endosperm weakening is a prerequisite for successful germination of such seeds.

The balance between the restraints exerted by enclosing tissues and embryo growth potential determines when radicle emergence will occur in the end. Both embryo expansion and endosperm weakening probably involve cell wall modification. The biological processes that control cell wall properties have been suggested to be important for regulating seed germination. A number of cell wall hydrolases and the genes encoding them in the embryo and endosperm have been identified. The role of some of these cell wall hydrolases in germination have been demonstrated by genetic studies. Others were suggested based on the association between their accumulation and germination progression (Nonogaki et al., 2007).

Seed Dormancy

There are many occasions when seeds do not germinate, even when they are placed under favorable conditions, for example, a suitable temperature, an adequate water supply, and the normal gaseous environment. However, these viable seeds can be induced to germinate with various types of treatments. Such seeds are said to be dormant (Figure 9.5). Seed dormancy is very common in wild plants, where it may ensure the ability of a species to survive natural catastrophes,

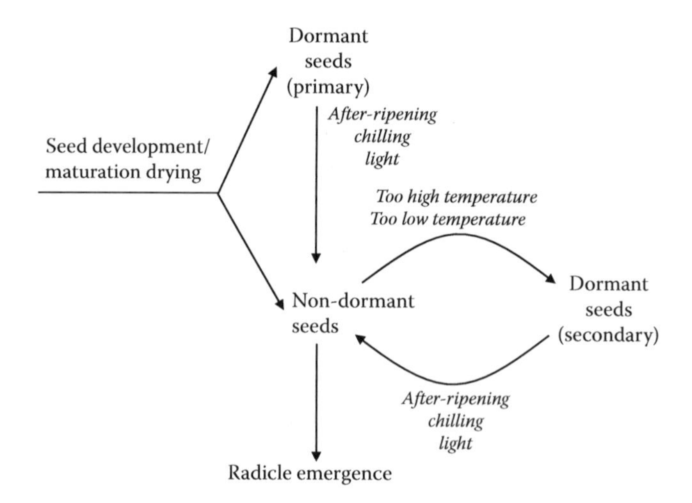

FIGURE 9.5 Relationships between seed dormancy and germination. "Primary" and "Secondary" in parentheses refer to primary dormancy and secondary dormancy, respectively.

decrease competition between individuals of the same species, or prevent germination out of season. In contrast, domesticated crops have been selected for fast and uniform germination followed by rapid seedling establishment to achieve good crop yield. However, lack of seed dormancy is not always desirable in agriculture because it may cause preharvest sprouting, a serious problem in cereal crops. Lack of dormancy has also been associated with reduced seed longevity.

Seed dormancy may be due to morphological and/or physiological properties of the seeds. Based on the causal mechanism, seed dormancy can be divided into five classes: morphological dormancy, physiological dormancy, morphophysiological dormancy, physical dormancy, and combinational dormancy (Baskin and Baskin, 2004). The system is hierarchical, each class of which can be further divided into levels and types.

Morphological dormancy (MD) is due to embryo immaturity; that is, the MD seeds contain underdeveloped or undifferentiated embryos. Embryos in MD seeds are not physiologically dormant. However, additional embryo growth after the seed is separated from the plant is necessary for the seed to germinate. Examples of species exhibiting MD include poppy and celery.

In contrast to MD, physiological dormancy (PD) is exhibited in seeds with fully developed and mature embryos. The presence of inhibitors is thought to be the main cause of PD. These inhibitors are present in the endosperm, cotyledons, or other food storage tissues. A deficiency of plant growth substances is also related to PD. The balance between inhibitors and promoters of germination determines whether a seed remains dormant or proceeds to germination. Dormancy in PD seeds can be released by treating seeds with plant hormones such as gibberellins and/or by altering environmental factors such as temperature and light. PD is the most common form of seed dormancy. It is found in seeds of gymnosperms and in all major angiosperm clades.

Seeds with morphophysiological dormancy (MPD) have an underdeveloped embryo with a physiological component of dormancy. Thus, they require a dormancy-breaking pretreatment in order to germinate. Because of the underdeveloped embryo, MPD seeds usually require a considerably longer period of time to release dormancy and complete germination than MD seeds.

Physical dormancy (PY) is caused by embryo coverings, including testa and/or endosperm. It is also called exogenous dormancy. The embryo coverings may impose dormancy by one or a combination of several mechanisms: restriction of water uptake, restriction of gas exchange, mechanical restriction of radicle protrusion, and presence of inhibitors in tissues. Dormancy of PY seeds may be released using scarification, a technique for breaking the seed coat using sand paper, hot water, acids, or other chemicals.

The fifth type of seed dormancy is called combinational dormancy, which is a combination of physical dormancy and physiological dormancy. In seeds with combinational dormancy (PY+PD), either PD or PY may be released first. Thus, release from PY and PD of PY+PD-dormant seeds appear to be independent events.

In addition to the cause of dormancy, the timing of dormancy onset has also been used to distinguish different types of seed dormancy. Primary dormancy indicates that seeds are dispersed from the parent plant in a dormant state. In secondary dormancy, seeds are released from the plant in a nondormant state but then become dormant under conditions unfavorable for germination. Secondary dormancy may develop as a result of prolonged inhibition of germination due to negative factors, such as ABA, unstable temperature, and high carbon dioxide. The occurrence of secondary dormancy is highly relevant to the behavior of seeds in the soil seed bank (seeds stored within the soil of many terrestrial ecosystems). It is a concept central to a phenomenon known as "dormancy cycling." The most important ecological significance of dormancy cycling is the prevention of germination of temperate annual species during short spells of favorable conditions within the unfavorable season (fall). The induction of secondary dormancy ensures that the seeds will only germinate prior to a season favorable for growth and reproduction. Although it may be assumed that the underlying mechanisms of secondary dormancy are the same as those of dormancy in general, some mechanisms, such as the physiological mechanisms of endogenous dormancy that control secondary dormancy, are largely unknown.

REGULATION OF SEED GERMINATION

Dormancy breakage may involve plant hormones as well as various types of environmental factors. These same factors also play important roles in regulating the rate of germination of nondormant seeds. In the remainder of this chapter, we will examine how these internal and external factors regulate seed dormancy release and germination.

Regulation of Germination by Plant Hormones

Primary seed dormancy is initiated during seed development, and ABA is believed to play a critical role in this process. ABA has been detected in all seed tissues. The levels of ABA increase during the first half of seed development and decline during the late maturation phase. As described in the seed development section, ABA is involved in regulating the synthesis and accumulation of storage proteins. Studies with ABA mutants have demonstrated that ABA is also important for the onset of seed dormancy. A transient increase in ABA in the embryo during seed development is required to induce primary dormancy. Dormancy release by fluridone (an ABA biosynthesis inhibitor) treatment provides further evidence that ABA inhibits seed germination in dormant seeds.

Gibberellin (GA) is another important plant hormone involved in many aspects of seed biology. Similar to ABA, GA is necessary for embryo growth during seed development. However, GA and ABA also antagonistically regulate the degree of seed dormancy during seed development. GA application can break dormancy in many seeds, especially those with physiological dormancy. In contrast to the inhibitory role of ABA, GA promotes seed germination. In mutant plants with a deficiency in GA biosynthesis, exogenous application of GA is required for seed germination. GA is believed to promote germination through multiple mechanisms, one of which involves inducing the production of hydrolytic enzymes that weaken the testa and endosperm cap and stimulate expansion of the embryo. Both the biosynthesis and signaling transduction pathways of ABA and GA are important for their roles in seed development, dormancy, and germination (Leubner-Metzger, 2005).

Regulation of Germination by Environmental Factors

Some dormant freshly harvested seeds slowly lose their dormancy during a period of several weeks to several months of dry storage at room temperature. This is called "after-ripening." After-ripening generally occurs in dry seeds or seeds below a specific water content. Therefore, the presence of water may prevent after-ripening. Nevertheless, a minimum water content is required for after-ripening. The process is delayed when seeds become too dry. The efficacy of after-ripening also depends on temperature. In general, the rate of after-ripening increases with temperature. Under natural conditions, the seeds of winter annuals may lose dormancy through after-ripening (dry and warm summer) and become germinable in the fall.

In addition to after-ripening, many seeds may lose dormancy through chilling, which is the exposure of imbibed dormant seeds to low temperatures for a certain period of time. In forestry nurseries, chilling is routinely carried out by overwintering seeds in shallow pits. Because the seeds are layered alternately with sand or soil, this process has frequently been described as "stratification." The significance of low-temperature control of seed dormancy release in nature is obvious. The dormancy of the hydrated seed is slowly broken over winter to prepare for germination during the correct season (usually in the spring) for subsequent growth.

Temperature also affects the rate of seed germination. Nondormant seeds of different types germinate over a wide range of temperatures. Germination responses to temperature can be characterized by what is called the cardinal temperatures containing minimum, maximum, and optimum temperatures. Minimum and maximum temperatures define limits of germination, and an optimum temperature is defined as the temperature at which seeds germinate with the greatest speed. Cardinal temperatures may vary with species. For many seeds, the optimum germination temperature is slightly above room temperature. Germination rates usually increase linearly between the minimum temperature and the optimum temperature. On the contrary, the germination rates decrease linearly between the optimum temperature and the maximum temperature.

Germination of seeds in many species is strictly regulated by light. Seeds of some plant species found in forests are also regulated by light. They will not germinate until an opening in the canopy allows them to receive sufficient light. Many seeds whose germination is regulated by light, for example, lettuce, have a thin seed coat that can be penetrated by light. In this case, the embryo can absorb light and respond biochemically. There are two interconvertible forms of phytochrome, a photoreceptor in plants. One form of phytochrome, named Pfr, is the form of phytochrome found in plant cells that are exposed to red or common white light. This form of phytochrome is biologically active and can initiate seed germination. The other form of phytochrome, named Pr, is formed when phytochrome is exposed to far-red light. This form is biologically inactive in terms of seed germination. Thus, if lettuce seeds are placed under far-red light (dark or buried in seed soil), they will not germinate. If placed under white or red light, even with a short exposure, lettuce seeds can germinate.

Water can affect both the rate of dormancy release and the rate of seed germination. Most seeds perform best when there is enough water to keep the seeds moistened but not completely soaked. When water availability is reduced, the germination rate is often reduced. If the water potential of the imbibing solution is sufficiently low, radicle emergence will not occur. Imbibing in a solution with a lower water potential can extend the duration of phase II imbibition. This is the basis of a seed treatment technique called seed priming, in which seeds are presoaked in osmotic solution before planting. Polyethylene glycerol is a commonly used osmotic priming material, which lowers the osmotic potential of the imbibing solution. When the presoaked seeds are dried and then reimbibed, they often display a faster and higher rate of germination.

REFERENCES

Baskin, J.M. and C.C. Baskin, 2004. A classification system for seed dormancy. *Seed Sci. Res.* 14: 1–16.
Bewley, J.D. (1997). Seed germination and dormancy. *Plant Cell* 9: 1055–1066.

Debeaujon, I., K.M., Leon-Kloosterziel, and M., Koornneef, 2000. Influence of the testa on seed dormancy, germination, and longevity in Arabidopsis. *Plant Physiol.* 122: 403–414.

Galau, G.A., N., Bijaisoradat, and D.W., Hughes, 1987. Accumulation kinetics of cotton late embryogenesis-abundant mRNAs and storage protein mRNAs: Coordinate regulation during embryogenesis and the role of abscisic acid. *Dev. Biol.* 123: 198–212.

Huang, A.H.C. 1994. Structure of plant seed oil bodies. *Curr. Opin. Struct. Biol.* 4: 493–498.

Leubner-Metzger, G. 2005. Hormonal interactions during seed dormancy release and germination. In A.S. Basra, (Ed.), *Handbook of Seed Science and Technology*. Haworth Press, New York.

Nonogaki, H., F., Chen, and K.J., Bradford, 2007. Mechanisms and genes involved in germination *sensu stricto*. pp 264–304. In K.J., Bradford, and H., Nonogaki, (Ed.), *Seed Development, Dormancy and Germination.* Blackwell Publishing, Oxford, U.K.

SUGGESTED READING

Baskin, C.C. and J.M. Baskin. 1998. *Seeds—Ecology, Biogeography, and Evolution of Dormancy and Germination.* Academic Press, San Diego, USA.

Bewley, J.D. 1997. Breaking down the walls—a role for endo-β-mannanase in release from seed dormancy? *Trends Plant Sci.* 2:464–469.

Bewley J.D. and M. Black. 1994. *Seeds: Physiology of Development and Germination.* Plenum Press, New York.

Chaudhury, A.M., L. Ming, C. Miller, S. Craig, E.S. Dennis, and W.J. Peacock. 1997. Fertilization-independent seed development in *Arabidopsis thaliana. Proc. Natl. Acad. Sci. USA* 94:4223–4228.

Chiwocha, S.D.S., A.J. Cutler, S.R. Abrams, S.J. Ambrose, J. Yang, A.R.S. Ross, and A.R. Kermode. 2005. The etr1-2 mutation in *Arabidopsis thaliana* affects the abscisic acid, auxin, cytokinin and gibberellin metabolic pathways during maintenance of seed dormancy, moist-chilling and germination. *Plant J.* 42:35–48.

Crowe, J.H., L.M. Crowe, J.F. Carpenter, and C.A. Wistrom. 1987. Stabilization of dry phospholipid-bilayers and proteins by sugars. *Biochem. J.* 242:1–10.

Finch-Savage, W.E., and G. Leubner-Metzger. 2006. Seed dormancy and the control of germination. *New Phytologist* 171:501–523.

Finkelstein, R., W. Reeves, T. Ariizumi, and C. Sreber. 2008. Molecular aspects of seed dormancy. *Annu. Rev. Plant. Biol.* 59:387–415.

Fischer-Iglesias, C. and G. Neuhaus. 2001. Zygotic embryogenesis—hormonal control of embryo development. Pp 223–247. In Bhojwani, S.S. and Soh, W.Y. (Eds.) *Current Trends in the Embryology of Angiosperms.* Kluwer Academic, Dordrecht.

Forbis, T.A., S.K. Floyd, and A. de Queiroz. 2002. The evolution of embryo size in angiosperms and other seed plants: Implications for the evolution of seed dormancy. *Evolution* 56:2112–2125.

Groot, S.P.C., B. Kieliszewska-Rokicka, E. Vermeer, and C.M. Karssen. 1988. Gibberellin-induced hydrolysis of endosperm cell walls in gibberellin-deficient tomato seeds prior to radicle protrusion. *Planta* 174:500–504.

Hallett, B.P. and J.D. Bewley. 2002. Membranes and seed dormancy: beyond the anaesthetic hypothesis. *Seed Sci. Res.* 12:69–82.

Hays, D.B., E.C. Yeung, and R.P. Pharis. 2002. The role of gibberellins in embryo axis development. *J. Exp. Bot.* 53:1747–1751.

Hilhorst, H.W.M. 1995. A critical update on seed dormancy. I. Primary dormancy. *Seed Sci. Res.* 5:61–73.

Holdsworth, R., W. Reeves, T. Ariizumi, and C. Steber. 2008. Molecular aspects of seed dormancy. *Annu. Rev. Plant Biol.* 59:387–415.

Homrichhausen, T.M., J.R. Hewitt, and H. Nonogaki. 2003. Endo-β-mannanase activity is associated with the completion of embryogenesis in imbibed carrot (*Daucus carota* L.) seeds. *Seed Sci. Res.* 13:219–227.

Kermode, A.R. and J.D. Bewley. 1985. The role of maturation drying in the transition from seed development to germination: I. Acquisition of desiccation-tolerance and germinability during development of *Ricinus communis* L. seeds. *J. Exp. Bot.* 36:1906–1915.

Manz, B., K. Müller, B. Kucera, F. Volke, and G. Leubner-Metzger. 2005. Water uptake and distribution in germinating tobacco seeds investigated in vivo by nuclear magnetic resonance imaging. *Plant Physiol.* 138:1538–1551.

Nambara, E. and A. Marion-Poll. 2003. ABA action and interactions in seeds. *Trends in Plant Sci.* 8:213–217.

Ogawa, M., A. Hanada, Y. Yamauchi, A. Kuwahara, Y. Kamiya, and S. Yamaguchi. 2003. Gibberellin biosynthesis and response during *Arabidopsis* seed germination. *Plant Cell* 15:1591–1604.

Richards, D.E., K.E. King, T. Aitali, and N.P. Harberd. 2001. How gibberellin regulates plant growth and development: A molecular genetic analysis of gibberellin signaling. *Annu. Rev. Plant Physiol. Plant Mol. Bio.* 52:67–88.

Singh, D.P., A.M. Jermakow, and S.M. Swain. 2002. Gibberellins are required for seed development and pollen tube growth in *Arabidopsis*. *Plant Cell* 14:3133–3147.

Steadman, K.J., A.D. Crawford, and R.S. Gallagher. 2003. Dormancy release in *Lolium rigidum* seeds is a function of thermal after-ripening time and seed water content. *Func. Plant Bio.* 30:345–352.

White, C.N., W.M. Proebsting, P. Hedden, and C.J. Rivin. 2000. Gibberellins and seed development in maize. I. Evidence that gibberellin/abscisic acid balance governs germination versus maturation pathways. *Plant Physiol.* 122:1081–1088.

White, C.N. and C.J. Rivin. 2000. Gibberellins and seed development in maize. II. Gibberellin synthesis inhibition enhances abscisic acid signaling in cultured embryos. *Plant Physiol.* 122:1089–1097.

10 Molecular Tools for Studying Plant Genetic Diversity

Timothy A. Rinehart, Xinwang Wang, R. N. Trigiano, Naomi R. Rowland, and Rene E. DeVries

CONCEPTS

- All living organisms contain DNA that can be assayed by a variety of methods to answer important questions about plant diversity.
- DNA fingerprinting techniques such as DAF, RAPD, and AFLP utilize arbitrary priming and do not require prior knowledge of the plant genome.
- SSR and SNP methods rely on DNA sequence data and provide more detailed information.

The ubiquitous nature of DNA is a central theme for all biology. The nucleus of each cell that makes up an organism contains genomic DNA, which is the blueprint for life. The differential expression of genes within each cell gives rise to different tissues, organs, and, ultimately, organisms. Changes in genomic DNA give rise to functional advantages that make some individuals more successful than others. Success, as measured by the ability to reproduce, dictates that organisms that accumulate useful mutations in their genomic DNA will be more likely to pass those changes on to future generations. Heritable mutations are the genetic variation that is visualized by molecular tools. In this chapter, we will discuss the many different forms of genetic variation, current molecular methods for characterizing genetic variation, and possible questions concerning plant diversity that can be answered with molecular tools.

The methods that we discuss rely heavily on understanding DNA, so it is worthwhile to review its structure. DNA is made up of four nucleotide bases: adenine (abbreviated A), cytosine (C), guanine (G), and thymine (T), which are covalently linked together by a sugar (deoxyribose)-phosphate backbone into a long polymer. Strands of DNA are linear and directional; that is, they are read from 3′ to 5′ along the sugar-phosphate backbone (Figure 10.1). Within the nucleus, two strands of DNA intertwine to form a double-helix structure where G bonds with C and A bonds with T to form base pairs. Complementary strands encode the same information, a redundancy that ensures fidelity during DNA replication. Under most conditions, complementary strands are annealed to each other via hydrogen bonding. During cell division, complementary DNA strands are split apart by DNA polymerase enzymes, and new DNA is synthesized from each of the template strands (Figure 10.1).

Genetic information is determined by the sequence of nucleotides. Regions of DNA that encode functional products are called genes and consist of trinucleotide units called codons, each of which codes for one of 22 possible amino acids. Genes are transcribed by cellular enzymes into a temporary nucleotide monomer called messenger RNA (mRNA). Proteins are produced by machinery that translates the codon sequence of the mRNA and assembles the corresponding amino acids into linear chains. As amino acids are linked together via peptide bonds, they often fold into complex structures that can be combined with other protein subunits, transported, embedded in cellular compartments, or modified to become functional units. Functional proteins are the cellular machinery that makes tissues, organs, and organisms what they are. The unidirectional flow of genetic information from DNA to mRNA to protein, often called the central paradigm of molecular biology, is

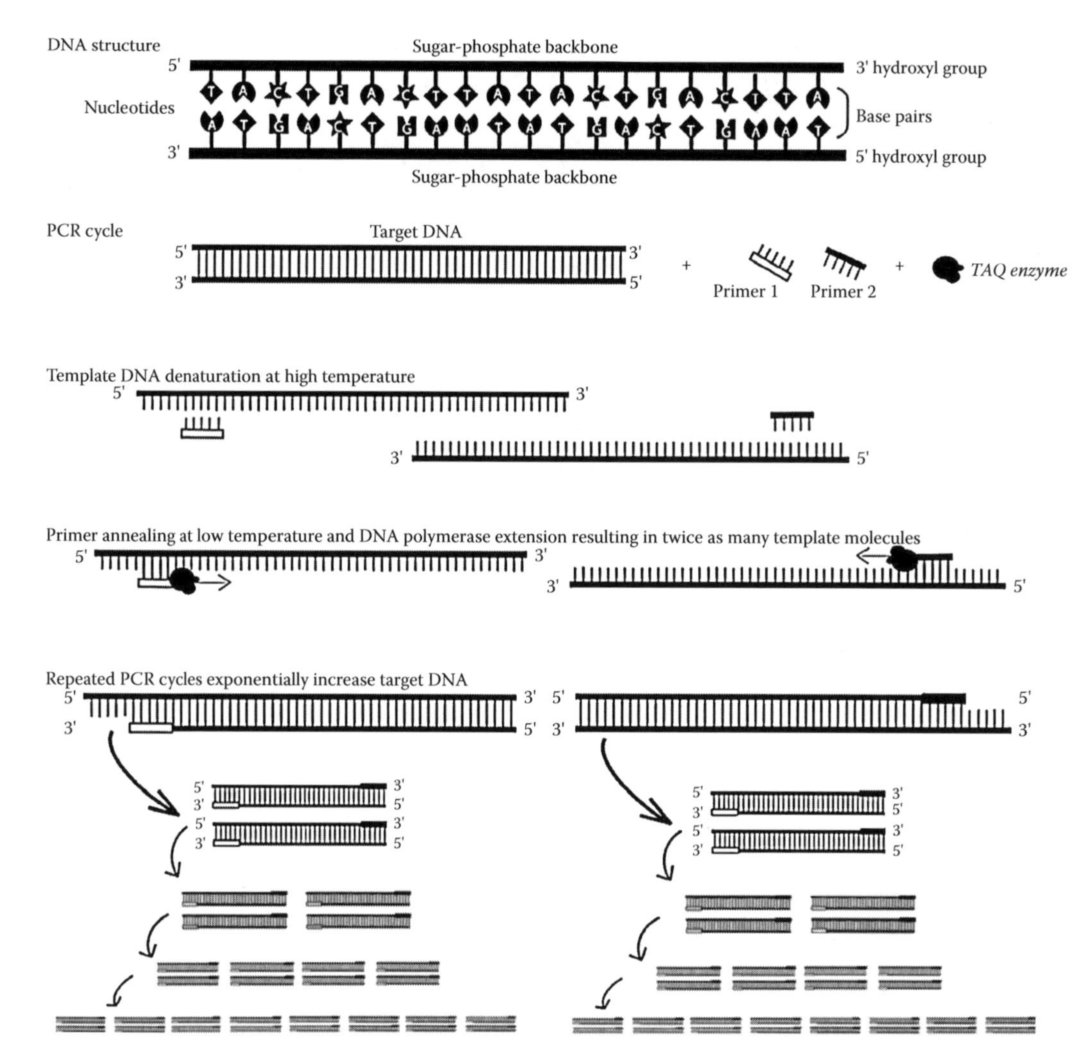

FIGURE 10.1 The complementary structure of DNA and PCR amplification of DNA fragments. Nucleotide bases are paired with each to form base pairs with A and T, and G and C as partners. Sugar phosphate backbone is directional, denoted by 5′ to 3′ end. Primers are short DNA fragments that anneal to specific regions of DNA according to the complementary base pair, which serves as a starting point for DNA polymerase to synthesize more DNA. Synthesis proceeds directionally from the 3′ starting point. When PCR primers are designed for both DNA strands facing each other, the resulting cycles of template DNA denaturation, primer annealing, and synthesis of new DNA results in exponential amplification of discreet regions of DNA defined by the 5′ ends of the primers.

critical for understanding how genetic variation in genomic DNA can produce phenotypes that are subjected to natural selection.

The type of genetic variation observed between individuals is highly dependent on where you look within the genome. Genes contain 64 possible combinations of three nucleotide codons that only correspond to 22 possible amino acids. Thus, there is considerable redundancy in codon usage. A single nucleotide mutation can change the genomic DNA sequence, but not necessarily the amino acid sequence of the protein. These types of mutations are called silent because the phenotypic effect is neither apparent nor subject to natural selection. Mutations that alter the codon so that a different amino acid is incorporated into the peptide sequence are called missense mutations. Mutations can also consist of nucleotide insertions or deletions. One or two nucleotide insertions or deletions can disrupt the trinucleotide codon sequence and are known as frameshift mutations. These mutations typically disrupt protein synthesis and are not observed as often as silent or missense mutations. This does not necessarily mean that frameshift mutations occur at a lower rate, just that they are less likely to be passed on due to their deleterious effects. Different regions of the genome demonstrate distinct patterns in the type of mutations they accumulate and the rate at which mutations are observed. Where to look for genetic variation and the types of variation observed plays an important role not only in how genetic data are analyzed but also the conclusions that can be made about plant diversity.

Not all genomic DNA codes for cellular machinery. Many DNA sequences are associated with packaging the genome into chromosomes and serve as recognition sites for DNA-binding proteins. Other DNA is only useful as spacing between genes or as a buffer against DNA loss. Noncoding regions, especially those with no apparent function at all, typically accumulate mutations at a higher rate. The types of mutations observed varies widely among noncoding sites but often includes large insertions and deletions that would not be tolerated in coding regions. Eukaryotic genomes also contain large amounts of mobile DNA, which appear to encode proteins solely for the purpose of copying and inserting additional copies of the DNA, the phenotypic effects of which may not be readily apparent. Regardless, all types of DNA play a vital role when assaying genetic diversity between individuals or populations. In this first part of the chapter, we will focus on methods that can be used to genetically identify and describe plant diversity. In the second half, we will focus on the types of questions that can be answered by a molecular approach to understanding plant genetic diversity.

PLANT GENETIC DIVERSITY AND IDENTIFICATION

DNA variation can be used to distinguish between all taxonomic levels of plants, including individuals, populations, strains, species, genera, and families. DNA fingerprinting has significantly accelerated the important task of plant identification. If a plant has been previously described using molecular tools, unknown plants can often be identified from tissue samples by laboratory testing. Plant systematics, or the study of plant diversity and relationships among groups of plants over time, utilizes many different DNA fingerprinting methods to identify and classify plants. Table 10.1 provides a brief description and comparison of the molecular techniques available. Understanding plant genetic diversity can also answer questions regarding population structure, modes of reproduction, modes of dispersal, and evolution.

Some of the most popular molecular tools to identify and genetically characterize plants are arbitrarily primed techniques such as DNA amplification fingerprinting (DAF), random amplified polymorphic DNA (RAPD), and amplified fragment length polymorphism (AFLP). These protocols do not require prior knowledge of the plant's genome. All three methods make use of polymerase chain reaction (PCR) to amplify large amounts of specific DNA from small amounts of total genomic DNA (Mullis and Faloona, 1987). PCR primers are short segments of human-made single-strand DNA that anneal to specific regions in the genome based on the complementary DNA sequence. DNA polymerase synthesizes new DNA using the genomic DNA as a template. When this reaction is repeated several times, the DNA fragment between the PCR primers is exponentially amplified (Figure 10.1). Thermal-stable DNA polymerase is used so that the template DNA can be melted apart (passive denaturation) with heat without destroying the function of the enzyme. A single PCR cycle consists of several seconds at high temperature to denature the DNA, followed by a low temperature to anneal the primers, and finally a period of optimum temperature for the DNA synthesis, usually 68–72° Celsius. PCR amplification of short regions, typically less than one kilobase (1000 nucleotides), is more robust than longer amplifications, which require specialized amplification protocols.

DAF and RAPD utilize arbitrary primers of short length, often 12 bases or less, that anneal throughout the genome (Welsh and McClelland, 1990; Williams et al., 1990). The sequence of the primers is random, and the probability of two primers annealing within 1.5 kilobases of each other can be calculated based on the size of the genome being assayed. However, genomes are not

TABLE 10.1
Summary of DNA Characterization Techniques

Technique	Prior Knowledge	Type of Information	Use/Comparison at Level	Difficulty/ Expense
DAF	None; PCR uses arbitrary primers	Anonymous bands or loci, dominant data	DNA profiling of individual isolates from the same population or same species	Easy and inexpensive
RAPD	None; PCR uses arbitrary primers	Anonymous bands or loci, dominant data	DNA profiling of individual isolates from the same population or same species	Easy and inexpensive
AFLP	None; PCR uses known primers annealing to linker sites after restriction digestion	Anonymous bands or loci, dominant and codominant data	DNA profiling of individual isolates from the same population, species, or genera	Moderately easy and inexpensive
SSR	DNA sequence of random SSR loci	Allele size variation of known loci, codominant data	DNA profiling of isolates from the same species, genera, or higher-order taxa	Moderately difficult, expensive to develop
DNA sequence	DNA sequence data of specific genes	DNA base mutations at known locus, codominant data	DNA profiling of isolates from the same species, genera, or higher-order taxa	Moderately difficult, expensive to develop
SNP	DNA sequence data of multiple genes	DNA base mutations at multiple loci, codominant data	DNA profiling of isolates from the same species, genera, or higher-order taxa	Moderately difficult, moderately expensive
Whole genome sequence	DNA sequence data of large genomic regions and/or chromosomes	DNA base mutations at multiple loci, codominant data	DNA profiling of isolates from the same genera or higher-order taxa	Difficult and very expensive

composed of random DNA sequences, so primer pairs must be empirically tested. The size separation of the PCR fragments can be done using either acrylamide or agarose gels or other inexpensive equipment, making optimization of DAF or RAPD affordable, especially since hundreds of discreet loci may be amplified during a single PCR amplification. The banding patterns produced indicate genetic similarities and differences between samples (Figure 10.2). Same-sized PCR fragments produced under identical cycling conditions indicate genetic similarity, while fragments that are different between samples, or polymorphic, suggest that a mutation disrupted the PCR amplification. Missing bands can be due to nucleotide changes in the primer annealing site, which would eliminate the production of the polymorphic fragment, or an insertion/deletion mutation in the region to be amplified, which would change the size of the fragment. PCR fragments that are unique to a group of samples suggest inheritance of that mutation and can be used to reconstruct the genetic relatedness of individuals and estimate genetic diversity within and between populations.

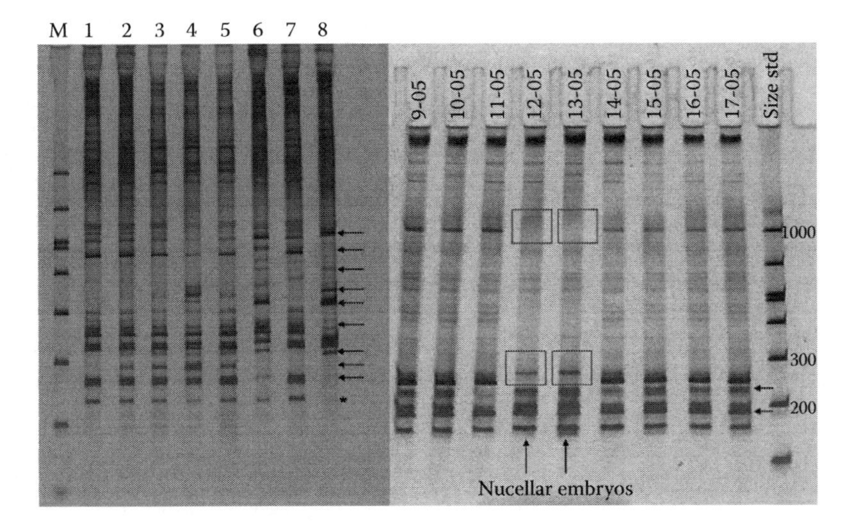

FIGURE 10.2 A. DAF analysis of turfgrass cultivars. DNA extracted from young leaves of several turfgrass cultivars was amplified by using the oligonucleotide primer GTAACGCC, generated fingerprints were separated using polyacrylamide gels, and fragments were stained with silver. Molecular markers are indicated by base pairs, and selected polymorphic bands are indicated by arrows. An "*" indicates a monomorphic band. B. DAF analysis of citrus seedlings using the foregoing procedure, except that GAGCCTGT was used as the oligonucleotide primer. Arrows indicate polymorphic loci, and boxes indicate determination of markers seedlings derived from parental nucellar (clonal) embryos.

Reproducibility is often cited as a disadvantage of DAF and RAPD analyses. This is understandable since slight differences in temperature or reagents might bias the amplification process toward or away from certain PCR fragments. When this bias is multiplied by the enormous number of fragments that can be produced, there is considerable justification for being cautious about DAF and RAPD results. However, the overall conclusions from DAF and RAPD analyses have proved reliable as long as the necessary controls are observed (Brown, 1996). Because of the low cost and virtually infinite combinations of primers and amplification conditions, researchers with enough dedication and time can compare an exhaustive number of loci between individuals. Randomly sampling genomic DNA includes comparisons between coding, noncoding, and mobile DNA, which increases the chances of finding a difference since some regions of the genome are more prone to mutation than others. Thus, RAPD potentially offers greater resolving power between highly related individuals, even when evaluating clonally reproduced plants such as tea cultivars (for example, see Singh et al., 2004).

AFLP is based on Restriction Fragment Length Polymorphism (RFLP), where genomic DNA is cut into small chunks by restriction endonuclease enzymes. The resulting DNA fragments are visualized by radiolabeled probes made from known genes (Vos et al., 1995). AFLP employs the same restriction endonucleases to chop up genomic DNA, but then utilizes PCR to selectively amplify hundreds of discreet fragments. Unlike DAF and RAPD, which uses random primers, AFLP primers are specific sequences designed to anneal to human-made linker DNA that is attached to the ends of the pieces of fragmented genome. Linkers are annealed to the overhangs left behind by the restriction endonuclease enzymes and attached by ligating them to the sugar-phosphate backbone. The resulting pool of DNA contains small fragments, typically 500–2000 base pairs, which can be PCR amplified using primers designed from the linker DNA sequence. Primers are radioactive or fluorescently labeled so that amplified fragments can be visualized on acrylamide gels, which can size-separate fragments that differ by a single nucleotide. Since there may be hundreds to thousands of amplifiable and detectable DNA, additional bases are sometimes included in the primer at the 3′ end to further reduce the complexity of the amplified DNA. The resulting DNA fingerprint consists of size-separated fragments for each sample that can be compared side by side. Similar to RAPD results, fragments that are present or absent in one sample but not the other suggest a genetic difference. Same-sized fragments produced by both organisms indicate genetic similarity. Researchers can choose from a number of restriction endonucleases and primer combinations, which potentially produce hundreds of fragments during a single PCR amplification. Thus, AFLP generates robust DNA fingerprints from loci across the entire genome but uses specific PCR primers, which increase reliability and repeatability. AFLP has been adapted to run on automated capillary array sequencing instruments using fluorescent labels for greater throughput and automated data analysis (Figure 10.3).

AFLP, DAF, and RAPD produce dominant markers. Data are tabulated by identifying DNA fragments that are present (1) or absent (0). The exact nature of the genetic mutation creating differences in the DNA fingerprints is unknown. Genetic similarity between samples and phylogenetic inference regarding shared ancestry are based solely on the mathematical frequency of DNA fragments, not biological models for genome evolution. Amplified DNA fragments that are unique to a specific individual or population can be purified, ligated into plasmid vectors, and cloned into *E. coli*. The plasmid DNA can then be sequenced to identify the underlying nature of the polymorphism. These sequenced regions are referred to as Sequence Characterized Regions (SCARs), which, once described, can be exploited as codominant markers. PCR primers are designed flanking the mutation such that the amplified DNA fragments can be visualized, either by size variation or DNA sequence differences depending on the nature of the mutation, and recorded as alleles specific to each sample. In a sexually reproducing diploid organism, researchers expect to see two alleles per sample, one corresponding to the paternal chromosome and another allele from the maternally contributed chromosome. Codominant data are not scored as a binary (absence or presence), but as allele variation. Identical alleles suggest shared ancestry, whereas different alleles indicate genetic divergence. Codominant data are considered more informative because every PCR amplification produces DNA fragments, and a technical failure during PCR amplification cannot be mistakenly scored as absence of a fragment (0).

One technique that generates codominant data is Simple Sequence Repeats, or SSR markers (Tautz, 1989). Eukaryotic genomes generally contain many diverse regions of repeated DNA.

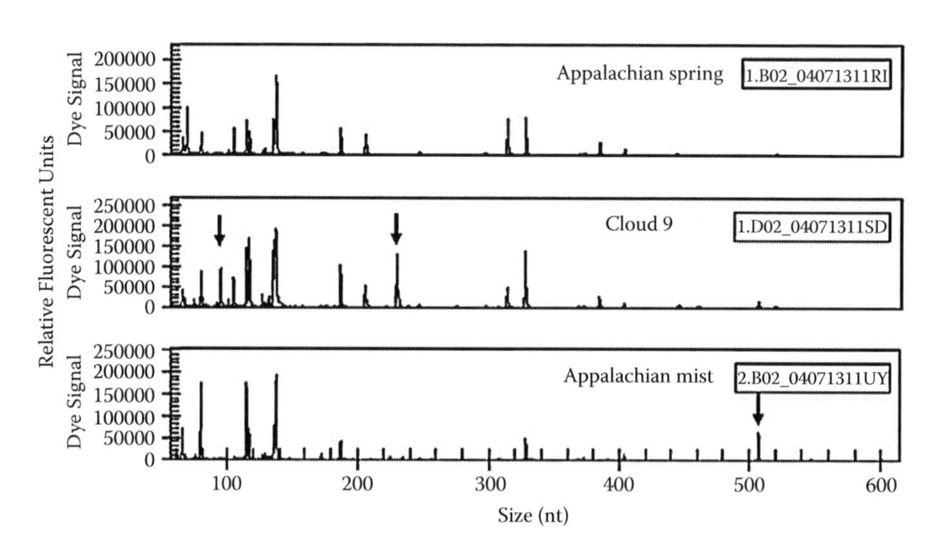

FIGURE 10.3 Amplified fragment length polymorphism (AFLP) fingerprints, or electropherograms, for *Cornus florida* "Appalachian Spring," "Cloud Nine," and "Appalachian Mist." Note the unique markers for "Cloud Nine" and "Appalachian Mist" (arrows). Size (nt) = size of fragment in nucleotides.

Generally, the repeat unit is 1 to 4 base pairs long and repeated 10 to 100 or more times. These repeats have a tendency to change in number when DNA is replicated due to a phenomenon known as DNA polymerase slippage. Slippage events that occur in germ cells (pollen or eggs) are passed on to progeny and become heritable mutations. PCR primers adjacent to an SSR region can amplify the repeat. Size differences in the repeat length can be visualized by radiolabeled or fluorescent molecules incorporated into the PCR products during amplification. Because they are uniformly spread around the genome and some mutate faster than others, SSR are robust molecular markers.

Polyploidy found in certain plants can produce numerous SSR alleles per amplification, making analysis of the data somewhat complex. SSR markers are particularly suited for diploid organisms that reproduce sexually and for evaluating genetic differences between species, genera, and higher-order taxa. The main disadvantage of the SSR technique is cost. In order to develop SSR markers, researchers must locate and sequence SSR regions before they can develop specific primers, although cross-amplification of SSR markers between genera has been documented for related plant taxa (Peakall et al., 1998). SSR data are reproducible and easily verified by sequencing the amplified products. There are also specific models for the evolution of SSR regions that can be used during data analysis for more accurate conclusions. For example, a trinucleotide repeat consists of three base pair units. Changes in allele size can be weighted during analysis such that a 15 base pair change is weighted more than a 3 base pair change, since it is likely that multiple slippage events, or more than one mutation, contributed to the 15 base pair variant.

The ultimate molecular tool for comparing genetic variation between plants is DNA sequencing the entire genome of each sample. Such an experiment would be cost prohibitive. Most DNA sequencing methods focus on only a few loci. DNA sequence variation can range from highly conserved to highly variable and researchers often use different regions of the genome to answer different questions. Conserved DNA sequences are more appropriate for evaluating higher-level relationships such as comparing genera or families, while more variable regions are appropriate for comparing individuals and populations. Public databases such as GenBank (www.ncbi.nih.nlm.gov) contain more than 100 gigabases of DNA sequence data and computational tools to search for analogous DNA sequences, or sequences that share a high level of similarity (Bensen et al., 2006). Gene sequences, particularly conserved gene sequences, from related taxa can be used to design PCR primers to amplify the same DNA regions in known and unknown plants. DNA sequences from the unknown plant can then be compared to related organisms in order to estimate genetic diversity and relatedness. Studies using conserved loci for species identification are also cataloged in GenBank, making it possible to classify unknown plants based solely on DNA sequence comparisons to previously classified organisms (Rinehart et al., 2006). This work builds upon the collective research of others with the expectation that researchers will add their own DNA sequences once studies are published. Genes commonly sequenced include elongation factor genes, tubulin genes, and other universally conserved eukaryotic sequences.

When more variation is desirable, which may be necessary for the identification of populations or species, non-coding regions of the genome are more useful since they accumulate mutations rapidly. The optimum scenario is a short hypervariable region sandwiched between conserved DNA sequences such that PCR primers can be designed to anneal to the conserved regions and amplify the more variable internal DNA. For example, Internal Transcribed Spacer (ITS) regions of ribosomal DNA (rDNA) are short sections of hypervariable DNA located adjacent to conserved 5.8s, SSU and LSU regions (White et al., 1990). PCR amplification is robust using universal primers, in part because the primers anneal to the conserved regions and because rDNA is repeated in tandem and found in high copy number per cell. ITS variants containing base changes, nucleotide insertions, and deletions are usually prevalent, sometimes even among individual samples in a population. Verified ITS sequence data for many plants are available publicly for comparison, which increases the validity of the results and reduces the labor involved.

There is a wide array of genes that could be PCR amplified and sequenced, but the main disadvantage is the cost, which is considerable as this requires genomic DNA extraction, PCR amplification, and then DNA sequencing for each sample to be genetically characterized. Moreover, valid conclusions require comparison of unknown DNA sequences to additional samples, preferably from related organisms or previously characterized reference plants. Unless these data exist already, researchers are obligated to repeat the DNA sequencing on many other related plants in order to align the DNA sequences and compare genetic variation. This can rapidly increase the number of samples and escalate costs, particularly if additional tissue samples must be collected from the wild. One advantage, aside from the detail inherent to DNA sequence variation, is that sophisticated models for DNA evolution can be incorporated into computer analyses of genetic diversity, making conclusions based on DNA sequence data statistically testable. Decisions to use DNA sequencing data to answer questions about plants diversity generally come down to how much is already known and how much effort is justified in acquiring DNA sequence data.

If large amounts of DNA sequence data are known or can be generated, then single base differences between samples can be measured. These single-nucleotide polymorphisms, or SNPs, can be tabulated for many different loci across the genome to generate high-resolution genetic characterization and organism identification. Once an SNP site has been identified as informative, or unique to a particular population or individual, it can be assayed in new or unknown samples, similar to how SCAR markers generate codominant data. SNP databases are powerful tools since they approximate the strengths of entire genome sequencing by focusing only on polymorphic base pairs in genes and ignoring noninformative DNA.

ANSWERING QUESTIONS ABOUT PLANT GENETIC DIVERSITY

Each individual plant contains a different mix of genes that are reshuffled every time a plant produces progeny through sexual reproduction. The "gene pool," or total number and variety of genes and alleles in a sexually reproducing population, determines what alleles are available for transmission to the next generation. This genetic diversity, or variation between individuals within species, is what allows populations to adapt to changes in climate and other local environmental conditions. Without genetic variation, there is little chance to evolve into new forms, populations, or species. Measures of genetic diversity allow researchers to quantify how many potential new forms, populations, or species exist for taxonomic analysis, conservation, breeding manipulation, or commercial exploitation. Estimating genetic diversity is an important first step in understanding plant germplasm resources.

Natural populations are the most basic source of genetic diversity, but controlled plant breeding of synthetic cultivars also produces new allele combinations, some of which may not be predicted by Hardy–Weinberg equilibrium equations, which are used to estimate the frequency of alleles in natural populations (Weinberg, 1908). Hardy–Weinberg is a mathematical model for the gene pool and is based on five factors: population size, migration between populations, mutation within a population, differences in fertility, and mating systems for a specific organism. We can calculate theoretical allele frequencies using the Hardy–Weinberg equations but, for our purposes, it is sufficient to understand that allele frequencies are generally balanced for large, random mating populations in nature where the effects of mutation, migration, and selection are minimal. In the altered framework of plant breeding, however, the directed evolution of small breeding populations is mainly due to the parents, the type of cross, and the artificial selection of desired traits.

Given what we discussed regarding factors that affect allele frequencies in populations, accurate estimates of genetic diversity using molecular markers requires a sampling strategy to minimize sample variance in diversity estimates. The number of individuals sampled in a given population is generally regarded as the most important variable when analyzing genetic diversity. The number of markers, population structure, mating systems, and population size or subpopulations should also be considered. Sampling strategy must take into account the level of genetic diversity being measured: individual genotypes, breeding lines, populations, species, or germplasm collections that include higher-order taxa.

Allele richness, or the number of genotypes present in a group of samples, is especially influenced by sampling errors. Rare alleles that occur in less than 5% of the genotypes are likely to be missed unless a sufficient number of random individuals are analyzed (Nei, 1987). Thus, using estimates of the total number of alleles as a measure of genetic diversity can be significantly impaired by sampling errors, including low number of samples or biased sample collection. Some of the variance due to small sample size can be overcome by using a larger set of molecular markers because the percentage of polymorphic loci can also be considered an estimate of genetic variation. In wild populations, Hardy–Weinberg frequencies can be tested and appropriate statistical tools applied to genetic diversity measures to estimate sample variance. Plants are either self-pollinating or cross-pollinating. While these naturally occurring reproductive systems can be artificially changed (self-pollinating species can be forced to outbreed, and vice versa), the underlying genetic structure that these reproductive systems impose cannot be changed. This "genetic architecture" must be considered when planning a sampling strategy since it necessarily affects the expected allele frequencies, and Hardy–Weinberg equilibrium only applies to randomly cross-pollinating populations.

Synthetic populations such as inbred lines or clonal accessions likely include genotypes not distributed in Hardy–Weinberg frequencies due to genetic linkage or selection. Where pedigrees are known, statistical tools have been developed to compensate. Sampling distributions are generally considered unknown when estimating diversity among species in germplasm collections. Regardless, some account of sample variance must be made when estimating genetic diversity since all individuals in a population or species are rarely available for analysis and the cost of analyzing everything would be prohibitive. Moreover, increasing sample size beyond what is necessary can lead to diminishing returns since the probability of observing a rare allele by adding an individual sample decreases greatly as sample size increases. The general rule of thumb is that sample sizes should be large enough to differentiate between groups of samples in order to accurately describe genetic variation within a group of samples. To that end, most sampling strategies include analyzing more than one group (i.e., subpopulations, populations, species, or genera). Refined models for determining sample size can be found in Baverstock and Moritz (1996), Crossa et al. (1993), Marshall and Brown (1975), Warburton et al. (2002), Weir (1996), and many other publications.

Heterozygosity measures are not as sensitive to sampling errors since these estimates do not rely on observing rare alleles. This is because heterozygosity measured in individuals can be used as estimates for the entire population. In other words, the fraction of loci in an individual that are observed as heterozygous is indicative of the fraction of individuals in the population that are heterozygous for that locus. Heterozygosity can also be averaged across all markers. Thus, the number of samples is not nearly as important as the number and distribution of markers when calculating heterozygosity. By comparing the observed (Ho) and expected (He) heterozygosity measures, researchers can estimate allele frequency, which is a useful measure of genetic diversity, especially if a population is at Hardy–Weinberg equilibrium.

Genetic distance is the "difference between two entities that can be described by allelic variation" (Nei, 1973). Entities can include individuals, populations, or species, and the differences can be quantified for any of the molecular markers outlined in the first half of this chapter. There are many ways to calculate distance measures. Some statistical calculations are specific to the molecular marker being used, while other calculations can be used for any marker system or even combined datasets from multiple marker systems and morphological data. The most common measures of genetic distance include Nei and Li (1979), Jaccard's coefficient (1908), and simple matching (Sokal and Michenener, 1958). All of these measures can be broadly applied to any marker data recorded in binary format. Genetic distances are based on the number of markers present in both individuals, the number absent in both individuals, and the number of markers present in one individual but not the other. While similar, all three measures use marker data differently. Jaccard's coefficient does not take into account markers that are absent between individuals, and calculations are based solely on shared markers between individuals. Nei and Li estimates of genetic distance are calculated from the proportion of shared markers divided by the average of the proportion of markers present in each individual. Simple matching calculations use both shared and nonshared markers equally when calculating genetic distance. When molecular marker data can be interpreted as codominant alleles, such as for SSRs, allele frequencies can be calculated directly, much as genetic distance estimates between two individuals is calculated from heterozygosity measures.

Genetic distance measures should be selected in accordance with the molecular marker system being used. For example, dominant and codominant markers are weighted differently when calculating distances with Nei and Li or Jaccard. Higher levels of missing data may also affect distance measures, especially if the presence of matching markers is equally weighted as the absence of matching markers (0-0), which also represents missing or failed marker generation. Jaccard (1908) explicitly ignores the absence of matching markers (0-0), so missing data would have minimal effect on genetic distance measures. Likewise, several genetic distance measures explicitly incorporate models for the molecular evolution of the marker system. For example, the observed DNA sequence variation in genes would be significantly different than DNA sequence variation in noncoding regions. Models for the accumulation of mutations in coding DNA take into account the 3-base codon structure, the biochemical bias in the occurrence of mutations (transitions versus transversion), and even the organelle where the DNA originated (mitochondrial, chloroplast, of nuclear genome). The level of genetic diversity being analyzed (individual, population, species) may also influence the selection of a particular genetic distance measure.

Genetic distances, often referred to as pairwise distances, are recorded in a matrix format with each individual compared to all other individuals in the sample (Figure 10.4). Often, researchers will employ more than one measure of genetic distance and look for correspondence between the resulting distance matrixes. Correspondence can be statistically tested using the Mantel test (Mantel, 1967) and may add validity to results when there is agreement between different distance measures. Similarly, Mantel tests can be applied to genetic distance matrices derived from different molecular marker systems that have been used on the same set of samples, such as AFLP and SSRs data from the same samples. Other statistical tests can be used to add credibility to genetic distance measures, such as Analysis of Molecular Variance (AMOVA), but there is not sufficient space to describe them in detail.

Genetic distance can be used to differentiate between groups of samples. Simple chi-square tests can be used to differentiate between subpopulations if heterozygosity and the number of alleles are relatively low. Genetic diversity estimates within and between population can also be statistically tested with Wright's (1951) F-statistic. F-coefficients are calculated for each individual marker and correlate with the allele frequencies within a group and between groups of samples. Multiple markers can be tested simultaneously using Gst and Dst statistics outlined in Nei (1973). Gst is a proportional measure, expressed as a percentage of the total genetic diversity that is due to differences between subgroups or subpopulations, while Dst estimates the genetic diversity among subpopulations.

Plant systematics uses genetic diversity measures to resolve taxonomic issues and accurately classify samples. Most of this work is based within a cladistic framework (also known as phylogenetics), which specifies a hierarchy based on shared ancestors. For example, genetic diversity may be used to support the relatedness between two groups of plants suggesting a recent common ancestor. The relatedness between these groups is directly proportional to the genetic distances between a most recent common ancestor (MRCA) and each group. Similarly, genetic distances may also be used to differentiate between species and assign taxonomic labels to unknown samples. Cladistic uses for molecular markers often involved cluster analysis of genetic distance measures to group-related samples with relatively high genetic similarity and separate group with high genetic divergence between clusters. Much like the statistical approach to differentiation, clustering analysis uses a genetic distance matrix created from one of the many distance measures noted earlier, which was derived from molecular marker data generated by one of the marker systems described earlier in the chapter. Computer software is necessary to complete the extensive analysis, and the resulting phylogenetic tree, or dendogram, should be considered a hypotheses of the evolutionary ancestry of the samples in the analysis (Figure 10.5). Despite advances in software and computational power, the

	#1	#2	#3	#4	#5	#6	#7	#8	#9	#10	#11	#12	#13	#14	#15	#16
#1 'All Summer Beauty'	0.000	0.231	0.603	0.372	0.449	0.372	0.013	0.474	0.716	0.737	0.628	0.792	0.756	0.625	0.724	0.513
#2 'David Ramsey'	0.231	0.000	0.590	0.436	0.474	0.410	0.224	0.500	0.743	0.750	0.628	0.792	0.756	0.639	0.697	0.526
#3 'Fuji Waterfall'	0.603	0.590	0.000	0.513	0.538	0.423	0.579	0.538	0.743	0.816	0.538	0.792	0.782	0.694	0.671	0.615
#4 'Kardinal'	0.372	0.436	0.513	0.000	0.372	0.346	0.355	0.462	0.649	0.684	0.564	0.750	0.705	0.569	0.632	0.462
#5 'Matilda Gutges'	0.449	0.474	0.538	0.372	0.000	0.269	0.434	0.474	0.770	0.750	0.628	0.806	0.833	0.722	0.724	0.526
#6 'Nigra'	0.372	0.410	0.423	0.346	0.269	0.000	0.355	0.397	0.703	0.776	0.551	0.819	0.833	0.667	0.671	0.449
#7 'Nikko Blue'	0.013	0.224	0.579	0.355	0.434	0.355	0.000	0.461	0.708	0.730	0.605	0.786	0.737	0.600	0.716	0.487
#8 'Veitchii'	0.474	0.500	0.538	0.462	0.474	0.397	0.461	0.000	0.527	0.724	0.551	0.722	0.731	0.597	0.579	0.449
#9 'Beni Gaku'	0.716	0.743	0.743	0.649	0.770	0.703	0.708	0.527	0.000	0.649	0.527	0.657	0.649	0.559	0.514	0.405
#10 'Blue Billow'	0.737	0.750	0.816	0.684	0.750	0.776	0.730	0.724	0.649	0.000	0.671	0.597	0.592	0.614	0.645	0.750
#11 'Coerulea'	0.628	0.628	0.538	0.564	0.628	0.551	0.605	0.551	0.527	0.671	0.000	0.667	0.679	0.375	0.355	0.564
#12 'Iyo Shibori'	0.792	0.792	0.792	0.750	0.806	0.819	0.786	0.722	0.657	0.597	0.667	0.000	0.472	0.588	0.625	0.736
#13 'Komachi'	0.756	0.756	0.782	0.705	0.833	0.833	0.737	0.731	0.649	0.592	0.679	0.472	0.000	0.611	0.684	0.731
#14 'Miranda'	0.625	0.639	0.694	0.569	0.722	0.667	0.600	0.597	0.559	0.614	0.375	0.588	0.611	0.000	0.529	0.611
#15 'Oamacha'	0.724	0.697	0.671	0.632	0.724	0.671	0.716	0.579	0.514	0.645	0.355	0.625	0.684	0.529	0.000	0.645
#16 'Preziosa'	0.513	0.526	0.615	0.462	0.526	0.449	0.487	0.449	0.405	0.750	0.564	0.736	0.731	0.611	0.645	0.000

FIGURE 10.4 Genetic distances calculated for *Hydrangea macrophylla* cultivars and arranged in a matrix so each individual can be compared to every other plant. Higher values indicate more differences in the SSR data, while lower values suggest less genetic divergence. When compared to themselves, the genetic distance should be zero, as seen diagonally from top left to bottom right in the matrix.

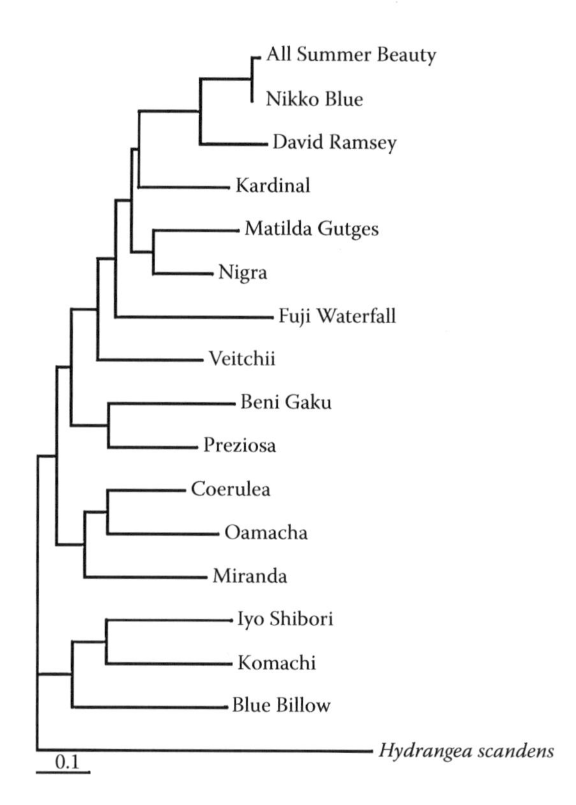

FIGURE 10.5 A tree dendogram was created from the genetic distance matrix in Figure 10.4. Samples with high similarity, or low distance numbers in the table, cluster when subjected to neighbor-joining analysis. Clustering of the top eight samples is expected since they belong to *Hydrangea macrophylla* subspecies *macrophylla* and the bottom eight samples are *H. macrophylla* sbsp *serrata*. An additional sample, *H. scandens*, was added as an outgroup to root the tree.

resulting trees are unlikely to be a perfect evolutionary tree, or historical ancestry, of all samples. The use of genetic distance methods has certainly improved the resolution of phylogenetic analysis and several algorithms, such as unweighted-pair group method (UPGMA) and neighbor-joining (NJ), are commonly accepted as robust measures of cluster-based relatedness.

It is worth noting that phylogenetic analyses can also be conducted on some molecular marker data without generating a genetic distance matrix. Maximum parsimony analysis of DNA sequence data is based on finding the smallest number of evolutionary steps to explain the differences between all samples. Samples are arbitrarily arranged in phylogenetic trees and "scored" based on DNA differences between samples using specific mutational constraints for DNA sequence evolution. The lowest-scoring trees are saved for further comparison, and trees with higher scores are discounted. Depending on the number of samples, the number of trees to be compared can be overwhelming, and finding a single most parsimonious tree is usually not possible. In an effort to reduce the amount of computation, maximum likelihood methods use probability statistics to infer the likelihood of phylogenetic trees and reduce the number of trees to be compared. Likelihood methods were recently improved by including Bayesian statistics to reduce the field of trees by assuming a starting distribution of possible trees. All of these methods are highly computational and require extensive data modeling, but the resulting trees can provide powerful insight into the genetic diversity within a group of samples.

Nonhierarchical clustering analyses do not produce trees. Results are typically viewed as scatter plots with genetic distances among individuals displayed as proportional distances between dots within the plot (Figure 10.6). Clustering of individuals within the plot indicates increased genetic

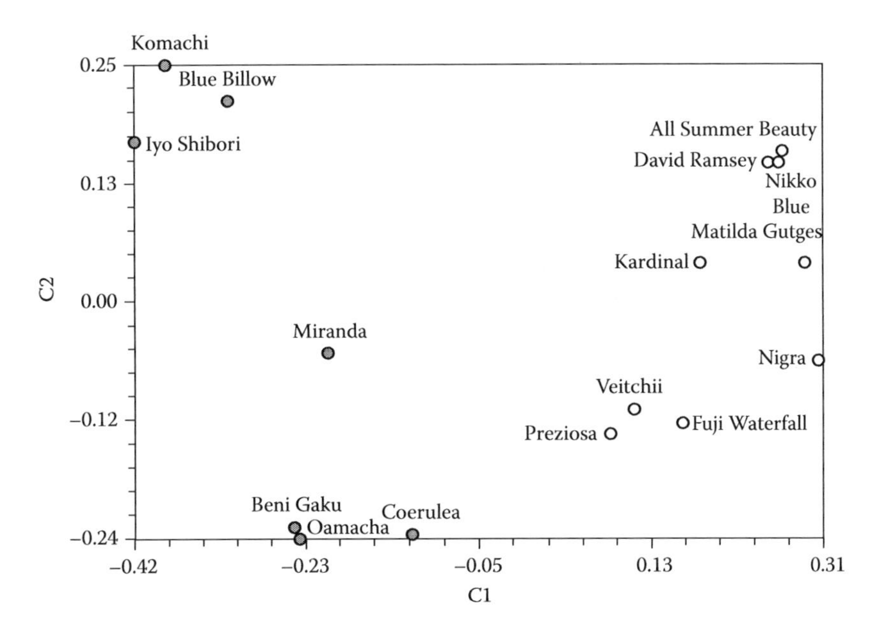

FIGURE 10.6 Clustering of subspecies is more visible in the principal coordinates analysis (PCoA) plot of the same genetic distance data. Samples with gray dots are *Hydrangea macrophylla* sbsp *serrata* (left side of plot), while samples identified as subspecies *macrophylla* are denoted by white dots (right side). The x-axis and y-axis represent 37% and 25% of the genetic diversity, respectively.

similarity. Data for these analyses, called principal component analysis or principal coordinate analysis, can also include other measures (morphological measures, other molecular markers, physiological ratings, etc.) as long as the data are scaled proportionally. Because each axis is measured in genetic diversity units, clusters can be deemed as significant based on user-specified cutoffs. Additional statistic tools are available to ascribe significance to genetic clusters such as multivariate analysis of variance, or MANOVA, which is often used to determine the optimum number of clusters. Under MANOVA, clusters are considered separate treatments, with each individual within a cluster representing a replicate treatment. Variance within a cluster should be less than the variance between clusters in order to be significant.

Similar statistical tools have been developed to test hierarchical clustering analyses. Bootstrap and Jackknife replication provide support for genetic relationship represented within and between branches on tree dendograms. Both statistics rely on resampling, or rerunning the tree-building analysis multiple times, and using each "independent observation" to calculate the frequency distribution of the sample hierarchy. Relationships between samples, as shown by branches and nodes on the tree, can be assigned bootstrap or jackknife support based on the number of times the same branching pattern is observed during resampling. Thus, a node connecting two specific samples that is observed in 90 out of 100 reiterations is assigned a bootstrap value of 90%. Bootstrap values less than 50% are generally ignored, and significance is probably not relevant until values are greater than 70%.

Cluster analysis is typically used with wild-collected plants but can be done on germplasm collections or breeding lines, albeit for different purposes. Genetic distance and clustering have been used to detect interspecific and intraspecific hybrids (whether naturally occurring or human-made), verify parentage of plants, maintain genetic diversity within collections without excessive duplication, unambiguously identify plants or cultivar, and legally protect plant patents. As biotechnology becomes less expensive and more widely available, molecular tools will likely become even more important techniques in understanding plant diversity, particularly since these methods require only small amounts of tissue. Advances in plant molecular genetics rely heavily on public sharing of genetic marker information and computer software. In this chapter, we covered common molecular markers and analyses being used to rapidly quantify genetic diversity, classify plants, and increase taxonomic understanding of the relationships between different plants.

REFERENCES

Baverstock, P.R. and C. Moritz. 1996. Project design. pp. 17–27. In D. M. Hillis et al. (Eds.), *Molecular Systematics*. Sinauer Associates, Sunderland, MA.

Benson, D.A., I. Karsch-Mizrachi, D.J. Lipman, J. Ostell, and D.L. Wheeler. 2006. GenBank. *Nucleic Acids Res*, 34: 16–20.

Brown, J.K.M. 1996. The choice of molecular marker methods for population genetic studies of plant pathogens. *New Phytologist*, 133: 183–195.

Crossa, J., C.M. Hernandez, P. Bretting, S.A. Eberhart, and S. Taba. 1993. Statistical genetic considerations for maintaining germplasm collections. *Theor. Appl. Genet*. 86: 673–678.

Jaccard, P. 1908. Nouvelles researches sur la distribution florale. *Bull. Soc. Vaudoise Sci. Natl*. 44: 223–270.

Mantel, N. 1967. The detection of disease clustering and a generalized regression approach. *Cancer Res*. 27: 209-220.

Marshall, D.R. and A.H.D. Brown. 1975. Optimum sampling strategies in genetic conservation. In O.H. Frankel and J.G. Hawkes (Eds.), *Crop Genetic Resources for Today and Tomorrow*. Cambridge Univ. Press, Cambridge, England, pp. 53–80.

Mullis, K.B., and F.A. Faloona. 1987. Specific synthesis of DNA in vitro via a polymerase-catalyzed chain reaction. *Methods Enzymol*. 155: 335–50.

Nei, M. 1973. Analysis of gene diversity in subdivided populations. *Proc. Natl. Acad. Sci. (USA)* 70: 3321–3323.

Nei, M. 1987. *Molecular Evolutionary Genetics*. Columbia University Press, New York.

Nei, M. and W. Li. 1979. Mathematical model for studying genetic variation in terms of restriction endonucleases. *Proc. Natl. Acad. Sci. (USA)* 76: 5269–5273.

Peakall, R., S. Gilmore, W. Keys, M. Morgante, and A. Rafalski. 1998. Cross-species amplification of soybean (Glycine max) simple sequence repeats (SSRs) within the genus and other legume genera: Implications for the transferability of SSRs in plants. *Mol. Biol. Evol*. 15: 1275–1287.

Rinehart, T.A., C. Copes, T. Toda, and M. Cubeta. 2006. Genetic characterization of binucleate *Rhizoctonia* species causing web blight on azalea in Mississippi and Alabama. *Plant Disease*, 91: 616–623.

Singh, M., J. Saroop, and B. Dhiman. 2004. Detection of intra-clonal genetic variability in vegetatively propagated tea using RAPD markers. *Biologia Plantarum*, 48: 113–115.

Sokal, R.R., and C.D. Michener. 1958. A statistical method for evaluating systematic relationships. *Univ. Kansas Sci. Bull*. 38: 1409–1438.

Tautz, D. 1989. Hypervariability of simple sequences as a general source for polymorphic DNA. *Nucleic Acids Res* 17: 6463–6471.

Vos, P., R. Hogers, M. Bleeker, M. Reijans, T. van de Lee, M. Hornes, A. Frijters, J. Pot, J. Peleman, and M. Kuiper. 1995. AFLP: A new technique for DNA fingerprinting. *Nucleic Acids Res*. 23: 4407–4414.

Warburton, M.L., X. Xianchun, J. Crossa, J. Franco, A.E. Melchinger, M. Frisch, M. Bohn, and D. Hoisington. 2002. Genetic characterization of CIMMYT inbred maize lines and open pollinated populations using large scale fingerprinting methods. *Crop Sci*. 42: 1832–1840.

Weinberg, W. 1908. Über den Nachweis der Vererbung beim Menschen. *Jahresh Verein f vaterl Naturk Württem* 64:368–382. On the demonstration of heredity in man. In Boyer, S.H. (Ed.), *Papers on Human Genetics*. Prentice-Hall, Englewood Cliffs, 1963, pp. 4–15.

Weir, B.S. 1996. Intraspecific differentiation. In D.M. Hillis et al. (Eds.), *Molecular Systematics*. 2nd edition, Sinauer Associates, Sunderland, MA, pp. 385–403.

Welsh, J., and M. McClelland. 1990. Fingerprinting genomes using PCR with arbitrary primers. *Nucleic Acids Res* 18: 7213–7218.

White, T.J., T. Bruns, S. Lee, and J. Taylor. 1990. Amplification and direct sequencing of fungal ribosomal RNA genes for phylogenetics. In M. A. Innis, D. H. Gelfand, J. J. Sninsky, and T. J. White, (Eds.), *PCR Protocols: A Guide to Methods and Applications*, Academic Press, San Diego, pp. 315–322.

Williams, J.G.K., A.R. Kubelik, K.J. Livak, J.A. Rafalski, and S.V. Tingey. 1990. DNA polymorphisms amplified by arbitrary primers are useful as genetic markers. *Nucleic Acids Res* 18: 6531–6535.

Wright, S. 1951. The genetical structure of populations. *Ann. Eugen*. 15:323–354.

11 Molecular Approaches to the Study of Plant Development

Albrecht G. von Arnim and Byung-Hoon Kim

CONCEPTS

- Many complex and long-standing problems in plant developmental biology are becoming understood through a multi-pronged combination of powerful experimental tools. These include molecular cloning and gene expression analysis, genetic reagents such as mutants, and cell biological and biochemical approaches.
- Forward genetics, the classical direction of molecular genetic investigation, dissects a biological process beginning with mutational analysis, followed by molecular cloning of the genes involved, and biochemical investigation of gene function. Reverse genetics, in contrast, begins with the gene as a molecular tool and seeks to identify a biological activity for this gene using transgenic approaches.
- The availability of the complete genome sequence and shared resources has fast-forwarded the developmental biology of *Arabidopsis* by facilitating gene cloning and gene knockout studies. Similar effects are imminent in other species whose genome sequences have become or are about to become available.

As we interact with plants on a daily basis, perhaps enjoying our backyard or a landscaped park, hiking across an alpine meadow, or harvesting the fruits of agricultural labor, we might ask ourselves: Why does a daffodil flower in the spring, and a dahlia in late summer? Why are the leaves on tobacco plants undivided (entire) and those on the related solanaceous species, tomato, are divided (compound)? What prevents the premature germination of corn kernels before they have been shed by the mother plant? These are just a few examples of the questions that have long inspired and fascinated plant developmental biologists. Each of them can be boiled down to one or more underlying conceptual events. For example, the induction of flowering by photoperiod and the induction of germination by stratification are both examples of developmental phase transitions. This and other concepts in plant development are summarized in Table 11.1.

Development is associated with an increase in biological complexity over time. Evidently, the competence for a specific developmental program is laid down in the DNA sequence, the genome, of a fertilized egg cell (zygote). However, how the genetic information is realized is a far more complicated question. In search of the mechanistic basis of developmental events certain underlying themes emerge repeatedly.

- Are there environmental factors that modulate a given genetically programmed event and how does this work?
- From where does a developing cell receive its cues? Is a particular developmental event informed by cell lineage information, that is, information handed down to the cells in question from its progenitors, or by positional information, that is, information transmitted to the cells from their current neighbors?

TABLE 11.1
Conceptual Events in Plant Development and Examples Thereof

Concept	Examples
Developmental phase transitions	Seed germination, flowering, transition from juvenile to adult
Pattern formation	Phyllotaxis, flower structure, arrangement of hairs on leaf surface
Histogenesis, including cell fate determination and cell differentiation	Formation of vasculature and wood, hairs, pericarp
Organogenesis	Emergence of leaves from the apical meristem, emergence of lateral roots from the pericycle, embryogenesis
Morphogenesis	Shapes of leaves and fruits

In summary, although this chapter focuses primarily on the molecular toolkit of plant developmental biology, we will also highlight the types of questions that are supposed to be answered using these tools.

ARABIDOPSIS

One critical event in plant developmental biology was the adoption of the small crucifer *Arabidopsis thaliana* as an experimental platform. *Arabidopsis*, a herbaceous (winter) annual with a generation time of about three months, is easily grown in the laboratory in either soil or defined growth media. It is self-compatible and produces an abundance of seeds. Its small genome (~125 million base pairs) is easily mutagenized. *Arabidopsis* is also very amenable to the addition of new genes (transformation) using the bacterium *Agrobacterium tumefaciens* as a vehicle. The bacteria are conveniently delivered to the *Arabidopsis* flowers by infiltration in the presence of detergent. Perhaps the main advantage of *Arabidopsis* lies in the fact that it has been adopted by hundreds of research groups around the world. The combined thrust has led to the first complete plant genome sequence, and numerous associated resources including full-length cDNA collections, insertional mutagenesis programs, bioinformatic resources, and the collegial sharing of reagents and tricks of the trade.

SCOPE OF THIS CHAPTER

This chapter is meant to deliver a concise introduction into the toolbox of molecular genetics in the context of plant developmental biology. The emphasis is on outlining the capabilities, limitations, and applications of these "molecular approaches" in the experimental model organism *A. thaliana* (Brassicaceae). We omitted some recent advances that rely on well-annotated full genome sequences; in turn, we included several techniques once popular in *Arabidopsis* and which still have merit in species where little genome sequence information is available. As background, a basic understanding of processes such as molecular cloning, gene structure and function, DNA sequencing, the polymerase chain reaction (PCR), genetic mapping, and Mendelian genetics can be obtained from introductory genetics textbooks.

GENE CHARACTERIZATION USING MUTANTS

Overview

Many aspects of plant development can be understood as a chain of events between signals (e.g., the hormone auxin), that are perceived via signal intermediates (e.g., auxin responsive genes) and resulting responses (e.g., growth regulation in axillary buds). In turn, the response often serves as a signal for a subsequent response. Signals, signal intermediates, and responses are either the products of genes or are dependent on gene products for their synthesis or perception. Therefore, much insightful information can be gained from carefully observing the phenotypic effects that result when the function of the underlying gene is altered by mutations.

Mutational analysis underlies many of the experiments in developmental biology. Mutations may either abolish gene function (null allele), compromise it (partial loss-of-function allele), or artificially enhance it (gain-of-function allele). Briefly, for a mutant screen, a family of plants whose genome has been exposed to a mutagen is screened for defects in a specific aspect of development, for example, pigmentation (Figure 11.1A) or the response to the hormone ethylene. Phenotypically abnormal plants (putative mutants) are identified (Figure 11.1B), and strains with single genetic defects are isolated via backcrossing (Figure 11.1D). Independent mutations are tested for allelism to distinguish mutations that are allelic (affect the same gene) from those affecting different genes (Figure 11.1C). As a result, one can already arrive at a minimal estimate of the number of genes involved in the developmental event under study. The precise phenotypic defects are examined at the physiological and anatomical levels, which often distinguishes pleiotropic mutations (those affecting many characters) from more specific ones (Figure 11.1E). Recessive mutations, usually suggesting reduced function of a gene, are distinguished from dominant ones, which may be due to gain of function, haploinsufficiency, or dominant negative loss of function. Much can be learned about the operational order of the genes involved by examining the phenotypes of plants with mutations in two genes at once (double mutants) for epistatic relationships between the alleles. The resulting data are often summarized in the form of a "genetic pathway," which is nothing but a set of hypotheses that proposes a hierarchy in which the gene products contribute to their overall function. For example, Figure 11.2 shows the famous ABC model, which addresses the specification of the four different whorls of a dicot flower as sepals, petals, stamens, and carpels (Coen and Meyerowitz, 1991). The ABC model as originally stated explains the determination of organ types on the basis of spatially overlapping activities of no more than three classes of genes. It was derived primarily from the phenotypic analysis of mutations in flowering genes, which caused characteristic transformations from one organ type to another (homeotic mutations).

MUTAGENS

Mutagens come in three types: chemical, physical, and biological. They differ in the spectrum of mutations caused, which has implications for the severity of phenotypes that can be expected as well as for the prospect of an eventual molecular cloning of the underlying gene (Table 11.2).

1. **Chemical mutagens.** The methylating chemical ethyl methane sulfonate (EMS) is widely used due to its moderate toxicity and high effectiveness in inducing multiple mutations per genome. Moreover, the mutations are usually single base substitutions. Thus, EMS is the mutagen of choice when mild, partial loss-of-function alleles are desired, although null-alleles ("knockouts") and gain-of function alleles will also be produced. Most other mutagens are more likely to cause complete knockouts.
2. **Physical mutagens.** Among the physical mutagens (radiation), fast neutrons and X-rays cause deletions, which are often knockouts. Deletion alleles can be very useful during map-based cloning of a mutated gene (see the following text), when the task at hand is to locate the gene within a larger chromosomal region containing, perhaps, several dozen genes. A deletion will cause a drastic change in the pattern of DNA fragments produced by restriction enzyme digestion (restriction fragment length polymorphism, or RFLP), whereas base substitutions usually do not. Probing for the mutated gene by Southern blotting will reveal a difference in the banding pattern between wild-type and mutant plants. This is a powerful way of distinguishing the gene of interest from surrounding genes.
3. **Biological mutagens.** Biological mutagens are fragments of DNA that insert themselves into the plant genome, either transposons or *Agrobacterium* T-DNA. Their insertion tends to cause a knockout of the affected gene (Figure 11.3A), although an insertion into a gene's

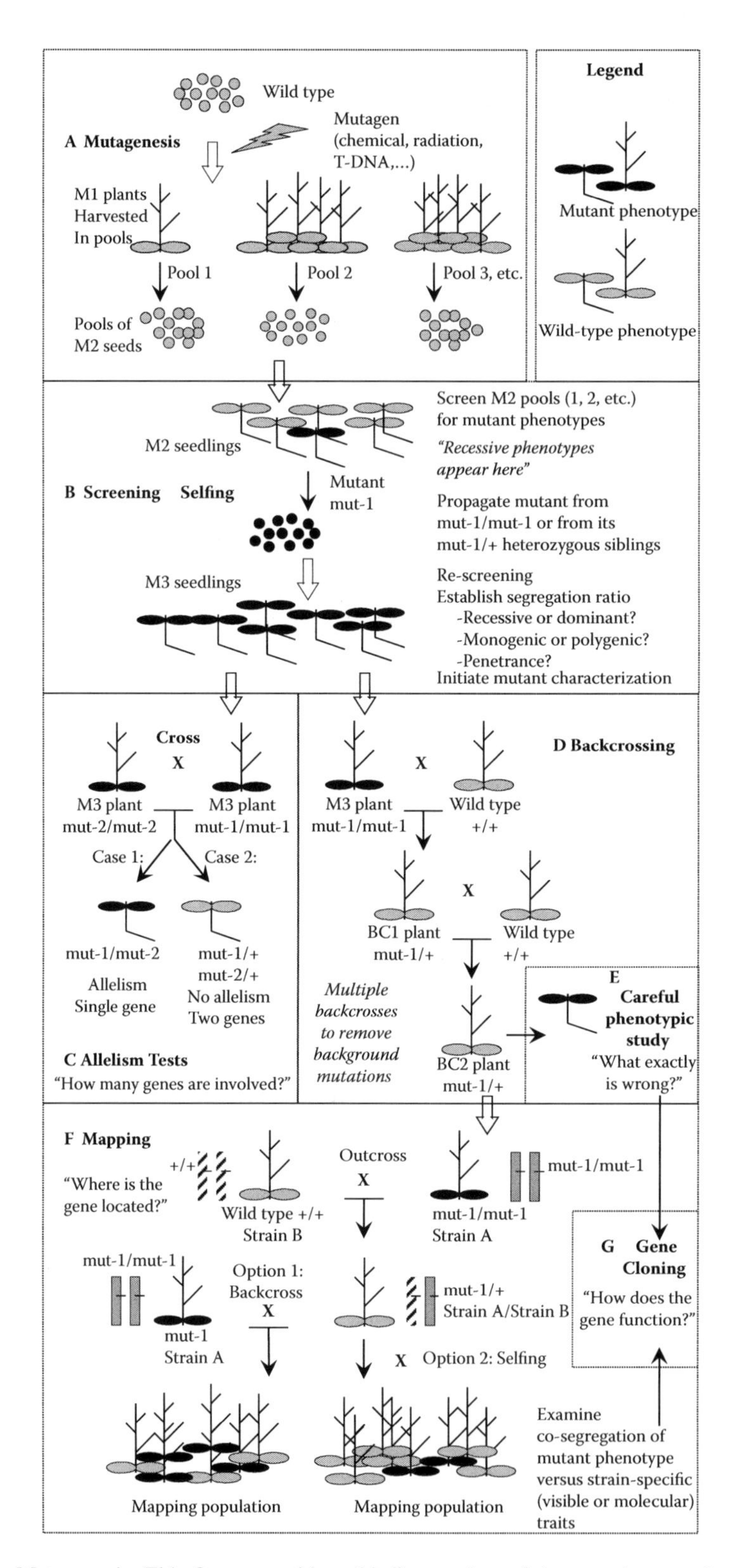

FIGURE 11.1 Mutagenesis. This figure provides a bird's-eye view of the genetic procedures and analyses conducted in order to identify mutations in a specific developmental program. For ease of illustration, the mutant phenotype shown here is excess pigmentation (for details, see text). +, wild-type allele.

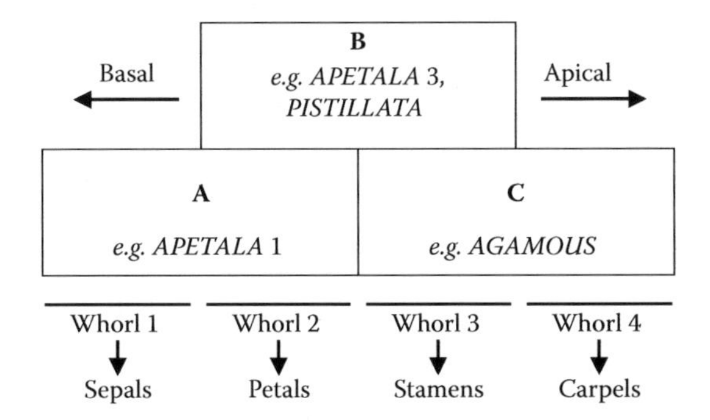

FIGURE 11.2 A simplified version of the ABC model for the specification of floral whorls. A, B, and C are three distinct gene activities that are expressed in a spatially overlapping fashion in basal, medial, and apical domains of the flower meristem. Each activity is provided by several genes, only a few of which are shown here. Thus, the combined presence of A and B activities specifies the formation of "petals" in whorl 2, etc.

TABLE 11.2
Mutagens, Their Effects and Implications for Molecular Analysis

Mutagen	Predominant Molecular Lesion	Implication for . . .	
		Typical Genetic Defect	Molecular Analysis
Ethyl methane sulfonate (EMS)	Base substitution (transition)	Reduced function, loss of function; efficient mutagen	
Fast neutron	Deletion (several kb)	Null-allele (knockout)	Molecular lesion usually reflected in RFLP
X-ray	Deletion, chromosome break	Broad spectrum	
Transposon (stable)	Insertion	Null-allele	Sequence tag
Transposon (unstable)	Small insertion (footprint)	Broad spectrum, including frameshift	
T-DNA (standard)	Insertion	Null-allele	Sequence tag
T-DNA (activation tagging)	Insertion	Gain-of-function; overexpression, ectopic expression	Sequence tag

regulatory region may result in a weak allele or even a gain-of-function allele. Although time-consuming to produce, insertional mutants have one powerful advantage: The DNA sequence of the insertion element is known. The insertion element thus provides a "tag," which can be used to obtain the unknown DNA sequence of the mutated gene.

One derivative of the T-DNA tagging strategy is "activation tagging." Here, the T-DNA contains a strong transcriptional enhancer sequence, which may lead to a dominant gain-of-function allele of the gene flanking the T-DNA, due to its overexpression or "ectopic" expression in the wrong cell type (Figure 11.3B).

Transposable elements are more efficient mutagens than T-DNA if the transposon catalyzes its own excision and reintegration. However, the resulting inconveniences are as follows: First, a transposon that re-excises from a tagged gene may cause a mutation by leaving behind no more than a small insertion ("footprint"), which is not sufficient to serve as a tag for molecular cloning. Second, if the transposon accumulates to a high copy number in the genome, it can be difficult to track down which specific copy of the transposon caused the mutant phenotype.

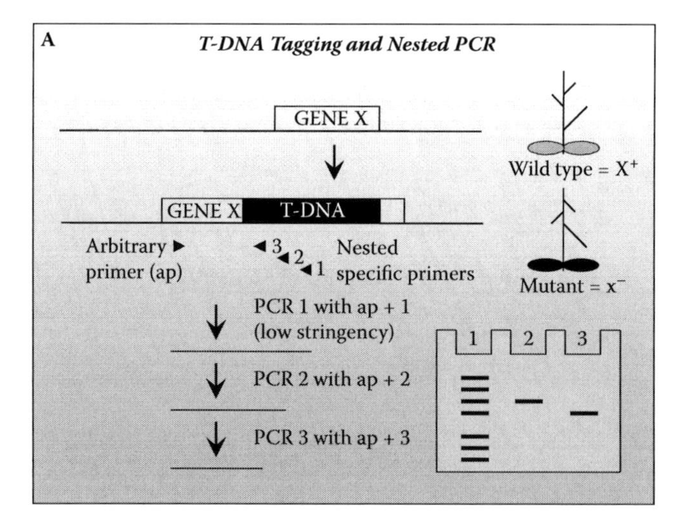

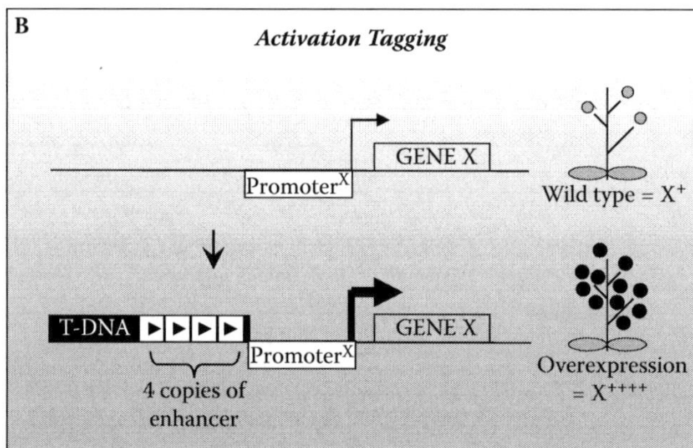

FIGURE 11.3 T-DNA tagging. Panel A illustrates how a T-DNA insertion into a postulated "Gene X" causes aberrant branching and pigmentation. Gene X can be identified through a stepwise PCR procedure using nested PCR primers. While round 1 results in numerous unspecific products, round 2 and round 3 amplify the desired product more specifically, as seen by the shift in the product size evident in the gel electrophoresis pattern. Panel B illustrates how a T-DNA carrying a transcriptional enhancer can boost the expression of a neighboring gene causing an overexpression phenotype, such as profuse flowering.

Building a Genetic Pathway from Mutants and Their Genetic Interactions

A successful mutant screen will usually yield a series of mutants in a number of genes that affect the process of interest. Once a set of mutants has been isolated, one can attempt to organize the underlying genes into a genetic hierarchy ("genetic pathway"), as exemplified by the ethylene signaling pathway illustrated in Figure 11.4A (Guo and Ecker, 2004). There are several genetic tools to this end:

1. **Pleiotropic versus specific effects.** The first order of the day is the careful observation of the various mutant phenotypes. Certain mutations will lead to very specific defects, whereas others may have broader consequences, referred to as pleiotropic effects. For example, mutations causing defects in the response to ethylene may display a defect in just one ethylene response or all ethylene responses. Mutations causing defects in root growth may affect the entire root or may affect just a single cell layer. Often, the more pleiotropic

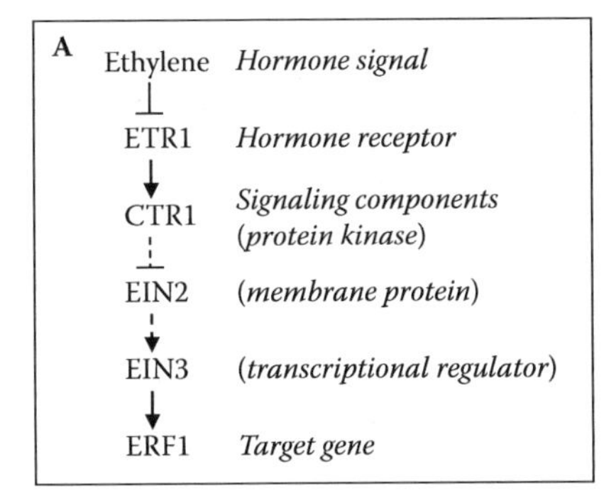

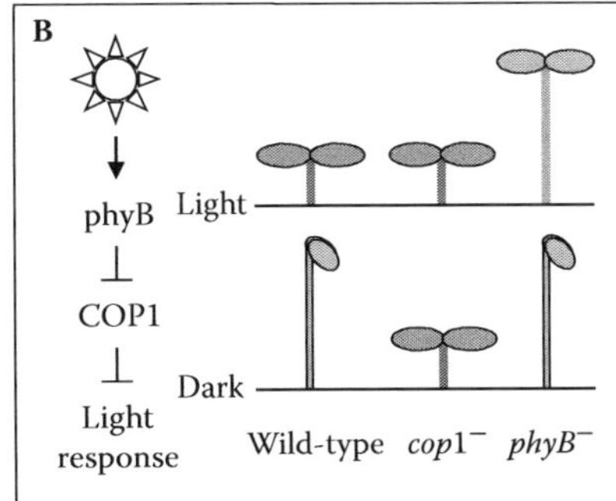

FIGURE 11.4 Genetic pathways. (A) A genetic pathway model for the mode of action of the hormone ethylene. The pathway was derived in large part from mutational analysis. Ethylene is perceived by the ethylene receptor, ETR1. ETR1 causes changes in gene expression through downstream signaling components that include CTR1, EIN2, and the transcriptional regulator EIN3. EIN3 binds to DNA sequences in target genes such as the *ERF1* gene. Blocked arrows indicate repression, and regular arrows symbolize activation. (B) A rudimentary genetic pathway for light signaling. In the dark, light responses are repressed by CONSTITUTIVE PHOTOMORPHOGENESIS1 (COP1). Light signals are perceived by photoreceptors, such as phytochrome B (phyB), which overcomes the inhibitory action of COP1. This model is supported by the phenotypes in light and darkness of seedlings mutated for COP1 and phyB.

mutation marks a gene that functions high up in the hierarchy. This is often referred to as "upstream in the genetic pathway," whereas the mutation with a more specific defect is said to mark a "downstream" gene.

2. **Activators versus repressors.** An important characteristic is whether the gene defined by mutation affects the process under study in a positive fashion, that is, as an activator, or in a negative fashion, as a repressor (Figure 11.4B). For instance, the "long hypocotyl" (*hy*) mutations of *Arabidopsis* mark positive regulators in the response of seedlings to light, given that an abnormally long hypocotyl is evidence of reduced sensitivity to light in the *hy* mutant. Not surprisingly, by the way, several *hy* mutations are due to lesions in phytochrome or cryptochrome photoreceptor genes. In contrast, the "constitutive photomorphogenesis" (*cop*) mutants mark negative regulators of the light response. The *cop* mutants appear like light-grown plants when germinated in darkness, due to a failure to repress the default light response pathway in darkness (Deng et al., 1991; Wei and Deng, 1992).

   When we make an inference about a wild-type allele based on its mutant phenotype, we must be certain that the mutation is showing us the loss-of-function phenotype of the gene in question. If the mutation is recessive, this assumption is usually justified.
3. **Allelic series.** For certain genes, the mutant phenotypes of individual alleles do not all look alike but form an allelic series. An allelic series can be very informative in defining the function of a specific gene at different developmental stages.
4. **Double mutant analysis.** One of the more powerful tools for defining the hierarchy among multiple genes is double mutant analysis. Let us assume we possess two loss-of-function mutants with opposite phenotypes. One mutant fails to carry out a specific response even when the appropriate stimulus is present (i.e., the wild-type allele is a positively acting gene); the second mutant carries out the specific response constitutively (i.e., marking a repressor gene). A suitable example is the already familiar *hy* and *cop* mutants, respectively. We can answer the question whether the *HY* or the *COP* gene functions higher in the pathway by examining the phenotype of a double mutant, for instance between *hy3* and *cop1*, which is generated by genetic crossing. If the *hy3/cop1* double mutant were to display the *hy* phenotype, then the *HY3* gene would be placed below the *COP1* gene in the hierarchy. In fact, however, the *hy3/cop1* double mutant displays the *cop1* phenotype, regardless of light conditions,

thus placing the *COP1* gene below ("downstream of") the *HY3* gene in the light signaling pathway. Consistent with its position upstream in the light signaling pathway, the *HY3* gene encodes the phytochrome B photoreceptor. COP1 encodes a nuclear protein that targets light regulatory factors for degradation. Its activity is inhibited by phytochrome B. Similar studies have contributed to the genetic models in Figures 11.2 and 11.4.

Although careful mutant characterization can in many cases define the relative position and functional interactions of distinct developmental regulator genes, there will often be genes that cannot be distinguished functionally. For example, the *CLAVATA1* and *CLAVATA3* genes both serve to restrict stem cell growth in the *Arabidopsis* shoot apical meristem (Sharma and Fletcher, 2002); thus, both *clv1* and *clv3* mutants share the phenotype of an enlarged shoot apex. In this case, only molecular analysis of the genes can distinguish their precise activities. In fact, as we will see, the two *CLAVATA* genes are not only expressed in different regions of the apical meristem, they also encode very different types of proteins. However, these two proteins function together as a cell-surface receptor (CLV1) and the cognate ligand molecule (CLV3), which explains the similar phenotypes of their mutants.

GENE ISOLATION: METHODS FOR CLONING GENES

Selecting the best among the numerous strategies for cloning a gene will depend on two things: What is my long-term objective and which tools and reagents do I have at my disposal. (1) Am I looking for the one-and-only gene that is defective in a specific, developmentally dysfunctional, mutant? In this case, a map-based ("positional") cloning strategy may be called for. (2) Are we looking for a gene responsible for a particular enzymatic reaction? This question might lend itself to functional complementation of a corresponding mutation in yeast. (3) Or am I generally interested in genes that are expressed in response to a particular stimulus, for example, cytokinin treatment? Subtractive hybridization could be used, although this kind of problem is now often approached using microarrays. (4) Finally, I might be interested in additional gene family members of other already known genes, for example, the cauliflower homolog of an *Arabidopsis* flowering gene. In this case, assuming sufficient genome sequence information is not available, a screen of a cDNA library by hybridization at low stringency may be the way to go. In principle, the aforementioned examples are representative of four general cases, the selection of genes based on

1. Mutant phenotype
2. Biochemical activity of the gene product
3. Expression pattern
4. DNA sequence

Cloning Genes Based on DNA Sequence Is Done by Hybridization

Molecular cloning based on DNA sequence is a straightforward method for isolating specific genes. Traditionally, the genome of interest is converted into a library of clones using a suitable cloning vector, usually bacteriophage lambda, and the library is screened by hybridization with an available fragment of DNA (the "probe") to identify a new gene that resembles the probe in its DNA sequence. Depending on the circumstances, the library is either a genomic library, representing chromosomal DNA, or a cDNA library, representing transcribed sequences (mRNAs). The stringency of hybridization should be adjusted depending on the sequence similarity.

If the amino acid sequence of a protein of interest is at least partially available, we can then reverse-translate a short amino acid sequence into a set of corresponding DNA sequences. After chemically synthesizing the DNA, we then proceed to isolate the gene of interest by hybridization. As protein sequencing technology by mass spectrometry improved dramatically in sensitivity and accessibility in recent years, cloning by reverse translation became very popular.

Library screening by hybridization is becoming all but obsolete in plants with a fully sequenced genome (*Arabidopsis*, rice, poplar, sorghum, maize, grape, and soon others). In many species that still lack a fully sequenced genome, many cDNA clones may be currently available. Full cDNA clones or at least partial cDNA clones, such as expressed sequence tags (ESTs), can be found in the Genbank database by sequence similarity search using a known gene sequence as a query (http://blast.ncbi.nlm.nih.gov/Blast.cgi). The cDNA clones found in this way can simply be ordered, since increasing numbers of molecular clones for individual genes have been cataloged and made available for distribution by genetic stock centers.

Cloning Genes Based on the Biochemical Activity of the Gene Product

The cloning strategies in this section have in common that they all rely on some form of biochemical assay for the gene product in question. In principle, a cDNA library is constructed in a gene expression vector. Clones are selected based on the biochemical activity of the expressed gene product, usually in a microorganism such as yeast or *E. coli*.

One elegant, powerful, and very popular cloning strategy in the category of activity-based methods is the yeast-two-hybrid system. We often wish to identify gene products that physically interact with an already known protein. In principle, the yeast-two-hybrid system consists of two expression plasmids and a reporter plasmid (Figure 11.5A). A physical interaction between two proteins X and Y is visualized by expressing protein X as a translational fusion protein with a specific DNA binding domain ("tagging" X with DBD), and tagging protein Y with a transcriptional activation domain (AD). DBD-X and AD-Y are then coexpressed in the same yeast cell. This yeast cell also possesses an inducible reporter gene, characterized by a promoter that

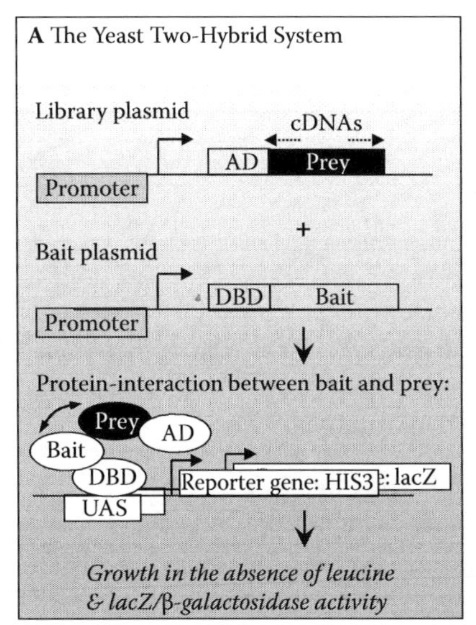

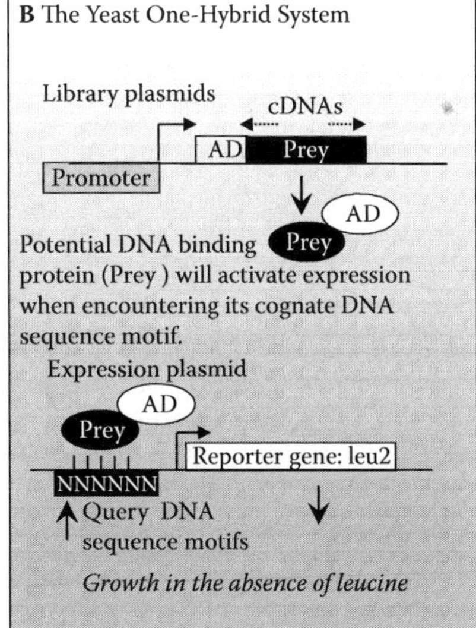

FIGURE 11.5 Library screening in yeast. (A) The yeast-two-hybrid system is a tool to identify proteins that interact with a given starter protein, the bait. The bait protein is expressed as a fusion to a DNA-binding domain (DBD). Plant cDNAs are selected from a library (black) constructed as a fusion to a transcriptional activation domain (AD). Prey cDNAs are selected based on their ability to interact with the bait (double-headed arrow), as evidenced by reporter gene activation (for details, see text). (B) The yeast-one-hybrid system is a strategy to identify the DNA-binding proteins of a given query promoter sequence using a cDNA library (black). The prey (library) proteins are expressed as a fusion to a transcriptional activation domain (AD). Only cells containing the query DNA-binding protein fused to the AD will show *leu2* reporter gene activation and, thus, growth in the absence of leucine.

contains a binding site for DBD. If X and Y interact by binding to each other, then AD-Y will activate transcription of the reporter gene because a transcriptional activation domain is recruited to the promoter by the DBD-X partner protein. Two reporter genes are widely used: the yeast *HIS3* selectable marker gene, which confers the ability to synthesize the amino acid histidine (histidine auxotrophy), and a colorimetric marker gene, such as *lacZ* (beta-galactosidase). To select new genes that interact with X, a cDNA library is made in the vector contributing the activation domain and candidate interactors of X are selected by looking for growth of yeast clones on medium lacking histidine.

A yeast one-hybrid system also exists (Figure 11.5B). To isolate new proteins and their genes that bind to a specific DNA sequence element, a cDNA library is expressed as a fusion to a transcriptional activation domain (AD), and the specific DNA sequence element is placed into the promoter region of a reporter plasmid expressing LEU2. Proteins that can bind to the DNA sequence element are selected by looking for growth of yeast clones on medium lacking leucine.

Another strategy for cloning a gene is by mutant rescue in yeast. Let us assume our charge consists of identifying a plant gene for a specific enzymatic step in amino acid biosynthesis. If a yeast loss-of-function mutation is available in the corresponding gene of yeast, then it may be possible to isolate the plant version of this gene as follows: A plant cDNA library is constructed in a yeast expression vector and is screened for cDNAs that rescue the yeast mutant defect. However, among plant developmental genes, few if any lend themselves to this approach because such genes are unlikely to be functionally conserved in yeast.

Cloning Genes Based on Gene Expression Characteristics

In order to identify genes based on their expression characteristics, say, their developmental timing, spatial pattern, or inducibility of expression, there are technically four categories of approaches:

1. Differential (subtractive) cloning or differential hybridization
2. Differential PCR amplification
3. Microarrays
4. Promoter trapping

The underlying assumption is that plant development is regulated by changes in the abundance of specific transcripts. Vice versa, a gene whose expression changes in response to a developmental stimulus might lead us to better understand developmental regulation. For instance, the gene *FLOWERING LOCUS T (FT)* functions as an integrator of various signals that induce flowering. FT expression is induced in response to long days with short nights in *Arabidopsis* and thus renders the transition to flowering dependent on day length (Imaizumi and Kay, 2006).

Differential (Subtractive) Cloning or Differential Hybridization

The first set of methods, now rarely used, begins by isolating mRNA transcripts from the two tissue samples to be compared, for example, shoot apices that have received a stimulus for flowering versus a control. During subtractive cloning, the population of transcripts is biochemically enriched for those that are present primarily in one sample and absent in the other. This can be accomplished by generating hybrid nucleic acids between the two populations and subtracting (removing) the hybrid, followed by selective cloning of the unique, nonhybrid cDNAs (Sagerström et al., 1997). For comparison, during differential hybridization, the library to be screened is left unselected. Instead, each set of mRNAs is converted into a set of cDNAs, and each set of cDNAs is used to probe the same library, for example, the library made from apices induced to flower. Clones are selected that hybridize preferentially to one of the probes, but not to the other.

Differential PCR Amplification

Both of the two methods in (1) are not only technically demanding but also suffer from the disadvantage that they tend to select for highly expressed genes, even though the most interesting developmental regulator genes are often expressed at low levels. Differential PCR amplification takes a stab at overcoming those problems.

Again, two different mRNA samples to be compared are isolated and converted into cDNA. In principle, two PCR primers of essentially arbitrary sequence are designed, and the cDNA then serves as the template for PCR. Because the primers are short and degenerate they amplify multiple cDNAs from each cDNA collection. Consequently, it is plausible that the resulting two sets of PCR products might differ in the abundance of a subset of products, which would reflect differentially expressed genes. The amplified fragment length polymorphism (cDNA-AFLP) method puts this idea into practice. cDNAs from two different samples are first digested into smaller fragments using a pair of restriction enzymes, then ligated to a short synthetic adapter of arbitrary DNA sequence. Because the base pair sequence of the adapter is known, the cDNA fragments are now PCR amplifiable with primers matching the adapters. Bulk PCR products are separated and distinguished according to their length by high-resolution gel electrophoresis. Again, products that accumulate to different levels when the two cDNA samples are compared may reflect differentially expressed genes. For practical reasons the adapter-modified cDNAs must be preselected during PCR to keep the total number of amplified fragments manageable. To this end, the primers contain an arbitrary 2-base extension at their 3′ end, which will allow amplification of only a subset (~1/16) of cDNA fragments.

Microarrays

With the advent of microarray-based expression profiling, a new way of identifying differentially expressed genes arose. A microarray ("DNA chip") is a device that measures the expression level of, in principle, all the genes in the genome of a plant. By probing the microarray with cDNA derived from two developmentally distinct plant samples, it is possible to identify, in principle, all the genes that are induced or repressed in one sample over the other.

There are two types of microarrays, two-color and single-color. For a two-color microarray, thousands of cDNAs or oligonucleotide sequences, each representing a single gene, are spotted on a conventional microscopy slide. Plant cDNAs from one treated sample and one control sample are each labeled with a different fluorescent dye (Cy3, green; Cy5, red). These two different cDNA populations are hybridized to the microarray slide simultaneously, which results in a range of color at a particular spot (gene) from green to red depending on the ratio between the red (Cy5)-labeled probe population and the green (Cy3)-labeled one. Genes that are overexpressed in the treated sample compared to the control will show up as green spots, and the genes that are underexpressed in the sample will show up as red spots.

In contrast, single-color microarrays are hybridized with just a single experimental sample, resulting in an absolute readout, not a relative ratio as for the two-color array, for the expression of each gene. Assignment of expression signals to individual genes is more robust than with the two-color array, because, according to the design implemented by Affymetrix Inc., each gene is represented by not just one but multiple oligonucleotide sequences, and closely related negative controls are also deposited on the chip. However, a pair of chips is needed to compare mRNA expression levels between a treated sample of interest and a reference sample.

In recent years, a large number of microarray datasets have been archived in various databases, such as GEO (http://www.ncbi.nlm.nih.gov/geo/) and ArrayExpress (http://www.ebi.ac.uk/microarray-as/ae/). Microarray meta-analysis Web sites such as the eFP browser (Geisler-Lee et al., 2007; Winter et al., 2007; http://www.bar.utoronto.ca/efp/cgi-bin/efpWeb.cgi) provide information about the spatial, temporal, and treatment-specific gene expression pattern based on microarray results. Such sites can be a source for identifying differentially expressed genes without having to do a single experiment.

Promoter Trapping

In contrast to the transcript-profiling methods covered earlier, promoter trapping is an insertional mutagenesis method designed to reveal the expression patterns of specific chromosomal genes *in situ*. To this end, a promoterless, that is, "silent," reporter gene, for example, β-glucuronidase (GUS), is integrated as a transgene into the plant genome. A collection of individual transgenic lines is created, each containing the reporter gene at exactly one random chromosomal location. Crucial to promoter trapping is the fact that by inserting into or next to an endogenous gene, the reporter gene may become expressed in a pattern similar to that of the endogenous gene, which is then said to be "trapped." For example, a promoter trap line in the *Arabidopsis FRUITFULL* gene was found by virtue of a characteristic GUS expression pattern in the floral meristem and the developing fruit (Gu et al., 1998). The *FRUITFULL* gene plays roles in floral induction and in coordinating fruit and seed development in *Arabidopsis*.

Two different reporters have proved to be most useful for promoter trapping. β-glucuronidase (GUS) is easily visualized by histochemical staining in situ. For comparison, green fluorescent protein (GFP) has the benefit of being visible in a live plant. Because GFP itself is not as sensitive a reporter as GUS, gene trapping with GFP is sometimes carried out with an amplification step (Figure 11.6). The promoter trapping cassette contains a transcriptional activator protein with sequence specific DNA binding activity, the yeast GAL4 protein. The expression pattern of GAL4 is visualized indirectly by operating in a plant line that harbors a GFP transgene with a GAL4 dependent promoter.

In principle, a promoter:reporter fusion with a defined expression pattern may serve as a starting point for subsequent mutagenesis to identify other genes that regulate its expression, thus moving one step forward in a developmental pathway. This particular application has often made use of a yet different reporter enzyme, the light-emitting firefly luciferase. For example, many genes regulating the circadian control of gene expression were discovered in this way (Millar et al., 1995). Although visualizing luciferase bioluminescence in situ requires a sensitive but expensive camera system, this disadvantage is outweighed by the advantages of real time in vivo and quantitative data acquisition using the light-emitting reporter.

Cloning Genes Based on Whole-Plant Mutant Phenotype

The most direct way of uncovering novel developmental regulator genes is to identify a mutant plant with a phenotypically informative defect in a single gene followed by molecular cloning of the

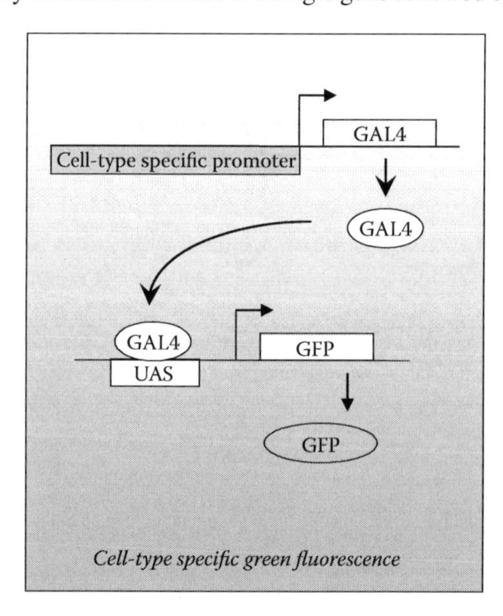

FIGURE 11.6 Promoter trapping by using GAL4 and GFP. The cell-type specific expression of the GAL4 transcriptional activator protein drives the cell-type specific accumulation of the green fluorescent protein (GFP) gene due to a GAL4 binding site (UAS) in the *GFP* promoter.

corresponding gene. Two rather different approaches to this problem are gene tagging (insertional mutagenesis) and map-based cloning.

1. **During gene tagging**, the mutation of interest is generated by insertion of a DNA element of known sequence, usually a transposon or a T-DNA (see above). Once this has been accomplished and established by confirming genetic linkage (cosegregation) between the mutant phenotype and a suitable marker gene on the insertional element, such as antibiotic resistance, the identification of the corresponding gene is often straightforward. Cloning a tagged gene hinges on isolating and sequencing the chromosomal DNA flanking the insertion element, for example, by nested PCR amplification (Figure 11.3A).
2. **Map-based cloning.** Alas, many mutations with interesting developmental phenotypes are not associated with a DNA sequence tag. In this case, the corresponding gene of interest may be identified by map-based cloning (positional cloning; Figure 11.7). In the first stage, the mutation is mapped to smaller and smaller genetic intervals, with the help of molecular markers (Figure 11.8). The smallest genetic interval that contains, among other genes, the gene of interest, is cloned in its entirety (Figure 11.9), and the gene of interest is identified from within the interval by preferably two independent means. First, cloned fragments are tested for their ability to rescue the mutant defect when reintroduced as a transgene into the mutant background (mutant rescue or complementation). Second, several independent

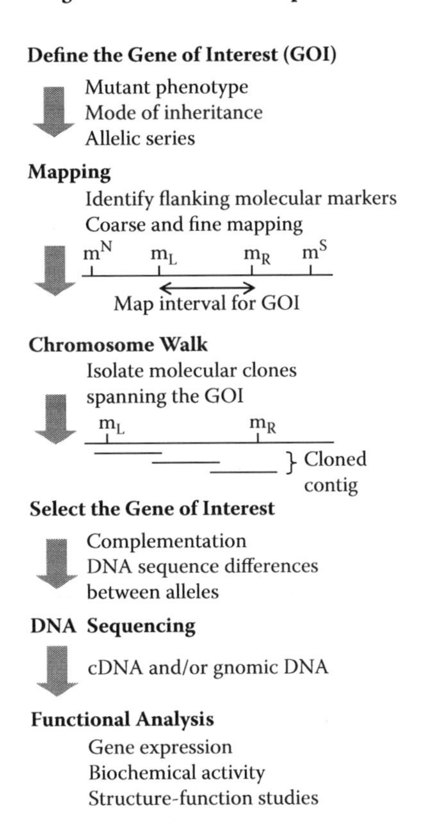

FIGURE 11.7 Cloning a gene based on its map position (positional cloning). Flowchart demonstrating the steps involved. For details, see Figures 10.8 and 10.9 and text. Bars indicate chromosomal regions with markers (m) interspersed.

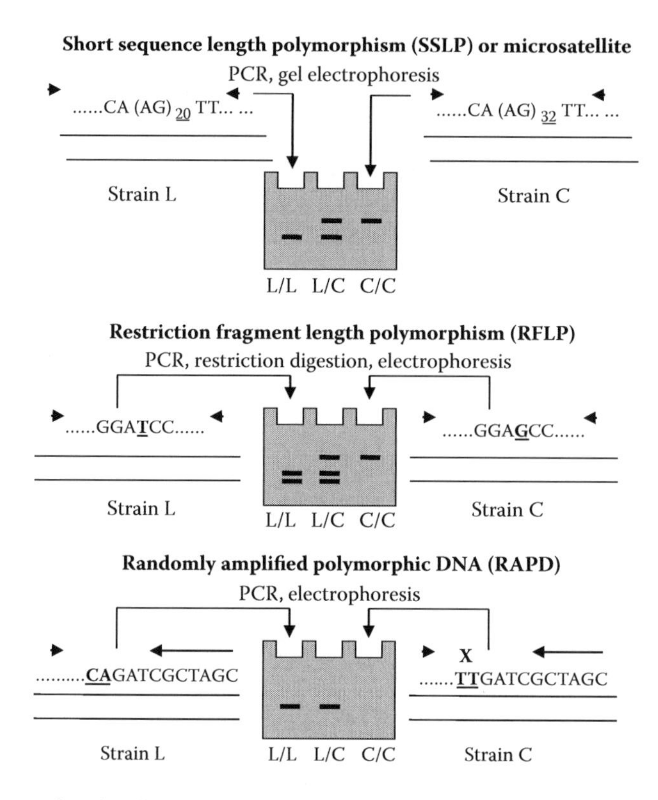

FIGURE 11.8 Types of molecular markers used in gene mapping. Each panel shows the DNA sequence difference (polymorphism) underlying the molecular marker for each of two strains of plants, L, and C. The sequence difference is highlighted in bold. Note that strains L and C are diploid, as indicated by the duplicated sequence in the SSLP panel. Short arrows symbolize PCR primers. The combed box represents an electrophoresis gel used to score the markers. L/L and C/C: Marker scored as homozygous, L/C: marker scored as heterozygous. SSLPs and RFLPs are scored with two different PCR primers that perfectly match the genomic sequence around the polymorphism. In contrast, RAPDs are scored with a single short primer. Small sequence differences between strains can interfere with efficient annealing of the primer, as shown here for strain C.

mutant alleles of the gene of interest are sequenced in order to demonstrate that each mutant allele contains a specific sequence alteration that is a plausible basis for the mutant defect (sequencing of an allelic series).

Map-based cloning relies on a dense map of genetic markers, most of which should be molecular markers, such as RFLPs and SSLPs, which can be scored painlessly using PCR in the large families necessary to establish tight linkage. Map-based cloning is also facilitated tremendously by shared genomic resources such as a completed genome sequence and large-insert genomic libraries. The portal for these resources for *Arabidopsis* is the TAIR database found at http://www.arabidopsis.org/.

METHODS FOR ANALYZING GENE FUNCTION

Once a developmentally important gene has been cloned, the immediate next question is: How does the gene function? Experimentally this question can be approached in many ways:

1. Domain structure—Does the predicted protein sequence contain clues?
2. Gene expression—Where and when is the gene expressed?

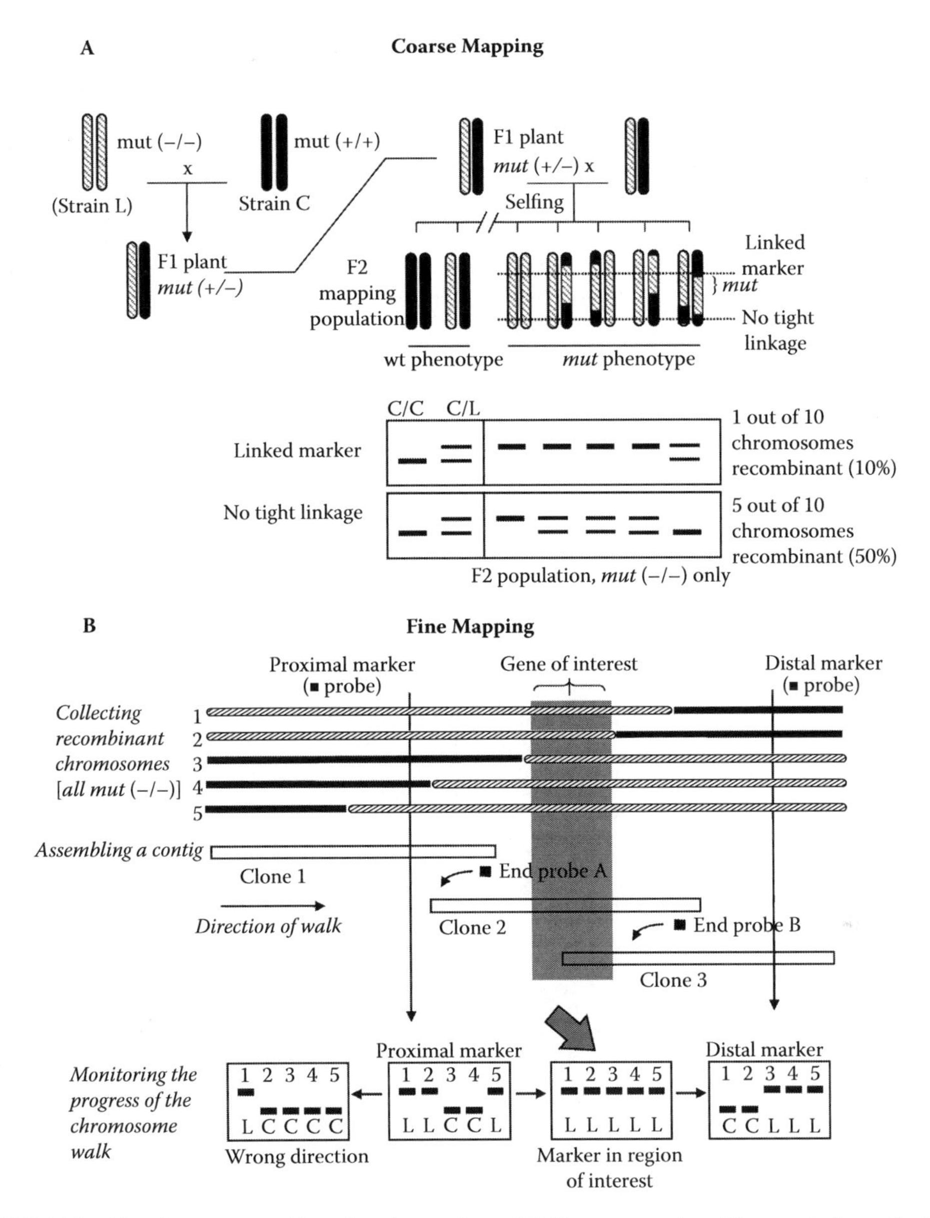

FIGURE 11.9 Mapping a gene with molecular markers. (A) Coarse mapping. The panel shows the three-generation crossing scheme to derive a mapping population (here: F2). A subset of F2 plants is shown to illustrate the principle of detecting linkage between the gene (*mut*) and two SSLP markers. The squared boxes represent gel electrophoresis patterns. Only one of the two markers shows clear linkage. Note that one of the ten chromosomes scored with the linked marker has a rare recombination event between *mut* (from strain L) and the marker (strain C). (B) Fine mapping and chromosome walk. After collecting a series of plants with recombinant chromosomes due to recombination events on either side of the gene of interest, the molecular markers are used to identify DNA clones near the region of interest (Clone 1 and Clone 3). Additional clones spanning the region of interest are derived, starting with probe A from Clone 1, by chromosome walking (Clone 2). The resulting set of clones is called a contig. Newly developed molecular markers (thick arrow) can be used to define the region of interest to a small interval, which must contain the gene of interest. Note that in a real map-based cloning project, much larger numbers of plants, chromosomes, and markers must be scored than shown here.

3. Biochemical activities—What are its enzymatic activities or partner proteins?
4. Protein localization—Where in the cell is the protein found?
5. Reverse genetics—What happens when the expression of the protein is modified?

Protein Domain Structure

Comparing the encoded protein sequence of the gene against other sequences deposited in the GenBank database often leads to a plausible hypothesis. For example, the gene product might appear to be a biosynthetic enzyme, a protein kinase, a transmembrane channel, a DNA binding protein with or without transcription activation domains, or a protein with interaction surfaces for other proteins. Does the protein consist of single or multiple domains and have other proteins with similar domain combinations been identified? Are there known orthologs of the protein in other species and is their function already understood? Answers to these questions can fast-forward the subsequent analysis tremendously. Nonetheless, we have to bear in mind, first, that any suggestion derived from sequence similarity is merely a hypothesis seeking experimental validation; and second, the protein sequence search may lead to no useful clues whatsoever.

Gene Expression Pattern

Much about gene function can be learned from the gene's spatial expression pattern and its dependence on experimental conditions. Genes involved in pattern specification, organogenesis, and cell differentiation are often expressed in a very specific cell type or in a restricted region of the plant. Conversely, genes involved in developmental phase transitions may follow a temporal pattern of induction or repression, whereas genes involved in a hormone response pathway may themselves be regulated by the same hormone. To visualize the gene expression pattern, the mRNA level of the gene is examined using RNA gel electrophoresis, followed by RNA hybridization ("Northern blotting"). Superior spatial resolution of expression is provided by in situ hybridization techniques. For example, the gene for the transcription factor *APETALA3*, which helps to specify the petals and stamen whorls in the flower (see Figure 11.2), is indeed expressed in a region of the early floral meristem that will later give rise to petals and stamens.

Another *in situ* gene expression technique makes use of fusing the 5′ upstream region of the gene (assumed to be the promoter) to a reporter gene, such as GUS. The resulting recombinant "promoter:GUS" fusion is introduced into the genome as a transgene and can be used to trace the expression of the corresponding gene by histochemical GUS staining. By crossing the transgene into specific mutant genetic backgrounds or by exposing it to specific environmental conditions, a researcher can gain insights into the regulation of the gene in question. For example, a GUS fusion to the auxin-responsive DNA sequence motif known as DR5 has become popular for reporting on endogenous auxin levels in plant tissues (Ulmasov et al., 1997).

By determining whether and how the expression of a gene depends on other genes or environmental parameters, one can establish a functional network around the gene of interest. Often, this is referred to as a "genetic pathway," with "upstream events," which regulate the gene, and "downstream events," by which the gene exerts its effects on plant development. These results should be reconciled with similar "pathway" results from double mutant analysis (see above).

Although the mRNA expression pattern is usually a reliable indicator for where the gene functions, this is not always true. Exceptions to the rule are among the most exciting recent discoveries because they suggest how genes can function over a physical distance of one or more cell layers. Plant cell differentiation is exquisitely sensitive to "position effects," that is, specific signals communicated by neighboring cells, but the underlying basis for such position effects has long been unclear. *SHORTROOT* is a regulatory gene for the radial pattern of cell types in the *Arabidopsis* root. In spite of its mutant phenotype, which affects mostly the peripheral endodermal cell layer of the root, *SHORTROOT* mRNA is found in the central stele and pericycle cells of the root (Nakajima et al., 2001). It turns out that the SHORTROOT protein is capable of migrating from the more central pericycle cells into neighboring endodermis cells, presumably via symplastic intercellular connections, the plasmodesmata.

Similar examples of mobile proteins have been discovered elsewhere. In the maize shoot apex, for instance, the Knotted1 protein is transported from the subepidermal layer, where its mRNA level is high, into the epidermal layer, where its mRNA level is negligible. Likewise, the CLAVATA3 protein appears to migrate from its point of synthesis in the central shoot apical meristem to more basal regions of the shoot apex, where it interacts with a cell surface receptor, CLAVATA1. CLAVATA3 is a small peptide and is thought to migrate through the extracellular cell wall space (Sharma and Fletcher, 2002).

Laser capture microdissection (LCM) is a method for isolating cells of interest from specific microscopic regions of tissue sections with the help of a tightly focused laser beam. DNA, RNA, and protein from a small region of tissues such as ovules or stamen abscission zones can be isolated and analyzed. In combination with microarray analysis, LCM provides a powerful tool to profile the gene expression pattern of specific cells types that would be difficult to dissect by other means (Cai and Lashbrook, 2006; Casson et al., 2005).

Biochemical Activities, Including Protein–DNA and Protein–Protein Interactions

A protein with a role in plant development can have any conceivable biochemical activity; it might function as a structural cell wall protein, a protein kinase, or a sequence-specific RNA-binding protein, to give just a few examples. Although the quest for the biochemical activity of a protein is a key to understanding its mode of action, unfortunately, there exists no systematic path that will guarantee a solution. Rather, defining this activity resembles a random walk, or searching for a needle in a haystack, and involves some trial and error. Nonetheless, two standard types of assays are applied routinely, assays for protein–DNA interactions and for protein–protein interactions.

1. Protein–DNA Interactions

If there is reason to believe that the protein might bind DNA, one can test whether the protein retards the electrophoretic mobility of DNA in a gel (gel-retardation assay). DNAse footprinting assays reveal the exact DNA sequence that is covered by the binding protein. As an alternative, a molecular genetic approach, the protein of interest is expressed in yeast as a fusion with a transcriptional activation domain in order to examine whether the protein can activate the promoter of a potential target gene (see "yeast one-hybrid system"; Figure 11.5B). One increasingly popular method to find target genes specifically for chromatin-associated proteins is to enrich the protein, together with its bound target DNA, from a nuclear extract using a specific antibody against the protein under scrutiny ("chromatin immunoprecipitation" or "ChIP"; Saleh et al., 2008). The enriched fraction is then tested by PCR for the presence of individual genes, if potential target DNA fragments are already known. In a less biased approach, the immunoprecipitated DNA fragments are cloned into a plasmid vector and sequenced to identify new target sites. High-throughput ("next-generation") DNA-sequencing techniques will likely play an important role in the effort to link the numerous chromatin-associated DNA binding proteins encoded in plant genomes with their presumptive target genes. Immunoprecipitated chromatin DNA is also being analyzed by hybridization to whole-genome microarrays ("ChIP-chip assay"; Lee et al., 2007).

2. Protein–Protein Interactions

Once a model for a genetic pathway is established, it is time to test it experimentally, for example, by addressing whether a genetic interaction between two genes is reflected in a direct biochemical interaction between their gene products. For proteins that appear to function by binding to other proteins, the yeast-two-hybrid cloning strategy outlined earlier is a powerful way to close in on the relevant cellular partners. Alternatively, one can develop a biochemical purification scheme for the protein at hand and determine the identity of any copurifying proteins. As hinted at earlier, identifying even small amounts (less than 1 μg) of copurifying proteins has become quite straightforward in recent years, driven by advances in mass spectrometric peptide sequencing. If a full genome sequence is also available, one can often identify the correct gene corresponding to the peptide sequence in silico, that is, by a search of the genome sequence database.

For example, the *Arabidopsis* repressor of light signaling, COP9, did not reveal any clues to its mode of action when the gene was first sequenced. However, the copurification of the entire COP9 associated protein complex, the eight-subunit COP9 signalosome complex led to hypotheses concerning the function of the complex (Chamovitz et al., 1996). Because the sequence of additional subunits resembled subunits of the proteasome proteolytic complex, it was hypothesized, and eventually confirmed, that COP9 and its partners are involved in regulating the degradation of light regulatory transcription factors.

Co-immunoprecipitation is another popular biochemical technique to examine whether two functionally related developmental regulators do in fact interact physically. Protein–protein interaction assays are, unfortunately, notorious for their false-positive results, requiring that results derived from one method be reproduced with an independent technique. In response to the potential for false-positive results under in vitro conditions, one recent trend has been to study protein interactions in vivo. To this end, specific optical probes, such as fluorescent proteins or luciferase, are attached to the two suspected interaction partners using recombinant techniques. An exquisitely distance-dependent physical phenomenon, fluorescence (or bioluminescence) resonance energy transfer (FRET/BRET) can be exploited to address whether two cellular proteins do in fact interact physically in a live cell (Subramanian et al., 2006).

Protein Localization

Many proteins contain one or more subcellular targeting signals that restrict where in the cell the protein will accumulate. Apart from their default location in the cytosol, proteins may be targeted to the chloroplast, to mitochondria, the nucleus, the endomembrane system, and numerous more specific subcellular addresses. Experiments to visualize the subcellular localization of a developmentally important gene product are informative for multiple reasons. First, the result will bolster or exclude certain hypotheses regarding its biological activity. For example, most proteins thought to bind DNA would be expected to reside in the nucleus. Second, the activity of many proteins is regulated at the level of subcellular localization. For instance, the *Arabidopsis* photoreceptor phytochrome is imported into the nucleus in a light-dependent fashion.

The cellular location of a gene product may be determined with the help of an antibody against the protein in conjunction with epifluorescence microscopy (Figure 11.10) or immunoelectron microscopy. Alternatively, especially when no specific antibody is available, the protein may be expressed transgenically as a fusion to a small peptide for which a specific antibody is commercially available (epitope tag). Antibodies are usually detected using a commercially available secondary antibody that carries a fluorescent or similar tag. An alternative powerful tag to reveal protein localization and its regulation is the green fluorescent protein (GFP) and its yellow, cyan, blue, and red siblings. For instance, it was demonstrated using fusion to GFP that the RGA protein, a negative regulator of gibberellic acid signaling, was excluded from the nuclei of *Arabidopsis* root cells and degraded in response to gibberellic acid (Dill et al., 2001). GFP can be visualized in live cells, thus opening the door to elegant real-time analysis of changes in protein localization in response to experimental treatments.

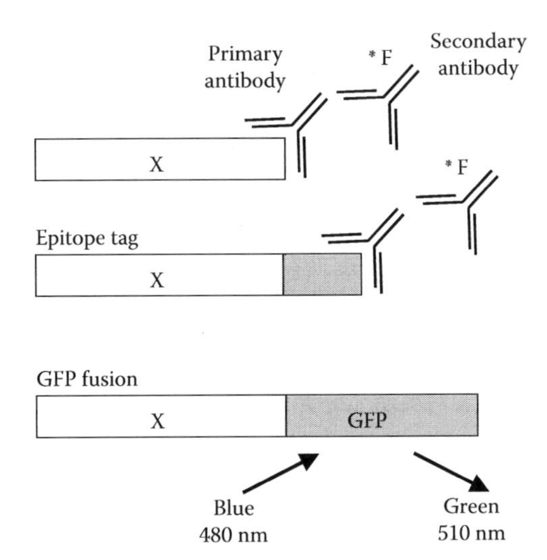

FIGURE 11.10 Determining the subcellular localization of a protein. Three approaches to visualizing the subcellular localization of protein X. Top: Indirect immunofluorescence using a primary antibody and a fluorescently labeled secondary antibody. Middle: A recombinant epitope tag (antibody-binding site) is introduced into the protein to be detected. Bottom: GFP-fusion proteins are detected by exciting their fluorescence at 480 nm and detecting fluorescence emission at 510 nm.

Reverse Genetics

Groundbreaking progress in developmental biology has been made based on the "forward genetic" paradigm. That is, phenotypic characterization of genetic lesions is followed by molecular cloning of the responsible gene and biochemical dissection of its activity. Reverse genetics makes use of the same tools, but in the opposite order. That is, reverse genetics seeks to ascribe a phenotype and thus a biological role to a gene that has been defined initially at the molecular level.

The typical starting point for this effort is as follows: Standard forward genetics may have identified a novel gene. A thorough search of the plant's genome then reveals that the novel gene is simply the first known member of a larger gene family of structurally and evolutionarily related genes. The question arises, then, as to what are the biological roles of the remaining gene family members. Do they function in the same developmental pathway or not?

In exceptional cases, the protein sequence per se is informative. For example, once two physiologically defined phytochrome photoreceptor genes, *PHYA* and *PHYB*, had been cloned, it became apparent that the *Arabidopsis* genome contains an additional three phytochrome-like genes, *PHYC*, *D*, and *E*. The notion that the *PHYC*, *D*, and *E* gene products are indeed photobiologically active phytochromes with special roles has since been confirmed.

However, more typically, protein sequence analysis will provide little, if any, clues with respect to the biological activity of otherwise anonymous gene family members. To illustrate this point, the *Arabidopsis* genome encodes a family of over 500 proteins that share the "F-box" sequence motif. The F-box implicates the proteins in the specification of protein turnover events. However, it is impossible to predict from sequence data alone which biological process might be affected by a given F-box protein, nor which specific proteins are the targets of a given F-box protein. To make progress, it is essential to generate or identify strains of plants in which the expression of the gene in question is altered, whether increased, decreased, abolished, or altered qualitatively. These strains can then be examined phenotypically for developmental abnormalities and can be subject to the molecular and biochemical gauntlets described earlier. Approaches to this effect are the following:

- Overexpression and ectopic expression
- Synthetic gain-of-function or loss-of-function alleles

- Loss of function by RNA interference (RNA silencing)
- Assembly of an allelic series by TILLING
- Gene knockouts

Overexpression and Ectopic Expression

Simple *overexpression* of a gene involves inserting additional copies of the gene into the genome, with the reasonable assumption that the additional copies are regulated in the same fashion as the original copy. By way of comparison, *ectopic expression* refers to the artificial (over)expression of the gene in cell types where the gene is normally silent. Ectopic expression can have major consequences if the gene is involved in pattern formation. For example, when proteins of the *Knotted* family of DNA-binding proteins, which are normally expressed exclusively in the shoot apex and the stem, are ectopically expressed in the leaf, striking abnormalities result. In maize, the result is outgrowth of extraneous tissue ("knots") on the leaf surface, whereas in tomato the leaves become excessively branched (more "compound"). These results provide evidence that the spatial restriction in the expression of *Knotted* in the stem contributes to cell fate specification in the stem versus the leaves.

Synthetic Alleles

Certain proteins lend themselves to the construction of synthetic gain- or loss-of-function alleles. As an example, the ethylene receptor, ETR1, is a member of a small gene family that includes the ERS1 and ERS2 (ethylene response sensor) proteins. The functionality of the ERS proteins was explored by introducing specific amino acid substitutions that were known to turn the ETR1 protein into a dominant gain-of-function allele. When introduced into *Arabidopsis*, the synthetic mutant alleles of ERS1 and ERS2 conferred the same ethylene insensitivity phenotype as did homologous changes in ETR1. This result indicates that ERS proteins are also ethylene receptors.

Aside from altering the protein sequence, one can create synthetic alleles by altering the regulation of gene expression. A trivial yet popular way of accomplishing this is by expressing a gene of interest under the control of a constitutive promoter such as the 35S promoter of cauliflower mosaic virus (CaMV 35S). More sophisticated methods involve inducible expression systems. A variety of designs have been described, including heavy metal inducibility, ethanol inducibility, and heat shock inducibility. Perhaps the most widely adopted design is based on the steroid hormone receptor, which provides inducibility of a target gene by externally added steroid hormones (Figure 11.11; Zuo et al., 2000).

RNA Interference (RNAi) and Related Gene-Silencing Tools

During RNAi, an RNA nuclease activity that is guided by a short double-stranded RNA molecule results in the sequence-specific degradation, or in some cases translational repression, of an mRNA. RNAi, formerly known as posttranscriptional gene silencing or cosuppression, is thought to represent a natural defense mechanism against invasive nucleic acids such as viruses or transposable elements. The effects of RNAi range from mild to severe, mimicking the phenotypic range of a conventional allelic series. For example, silencing of the *Arabidopsis* light regulatory *COP1* gene resulted in late phenotypes during the adult stage, which were mild, as well as early effects during the seedling stage, which may be lethal, as observed with conventional *cop1* mutant alleles.

To silence an RNA, a transgene is constructed that drives expression of an inverted repeat of the mRNA, with an intron separating the two repeats. Upon transcription, the intron is spliced out, and the RNA folds into a dsRNA hairpin structure, triggering the degradation of its target mRNA. Because double-stranded RNAs interfere with the expression of any mRNA that contains a short (~23 bp) stretch of sequence identity, long RNAi constructs can suffer from lack of specificity. In contrast, natural plant microRNAs are shorter and often include a few bases that are mismatched to their targets, which tends to reduce off-target effects. Using this principle, a technique called

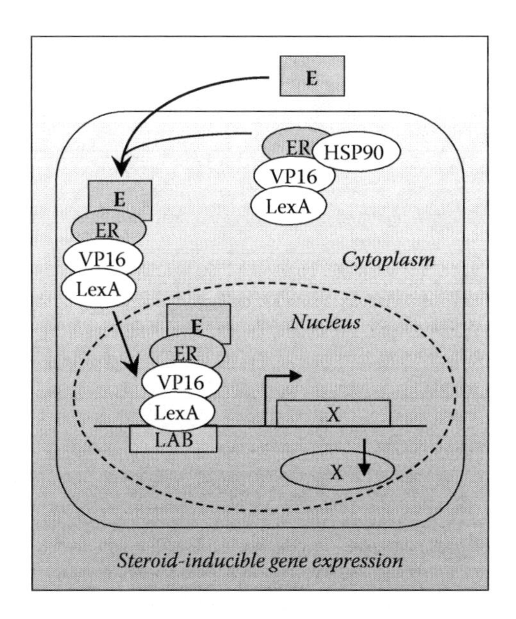

FIGURE 11.11 Steroid-inducible gene expression system. The gene (X) is inducible by steroid hormone, for example, estrogen (E). In the absence of steroid hormone, a chimeric transcriptional activator composed of the hormone-binding domain of the estrogen receptor (ER), the LexA DNA-binding domain, and the transactivating domain of VP16 is sequestered into an inactive form by the HSP90 protein. In the presence of hormone, this protein binds to a LexA-binding site (LAB) in the promoter of X and activates transcription.

artificial microRNAs (amiRNAs) has been developed to efficiently silence both single and multiple target genes. The design principles for amiRNAs have been generalized and integrated into a Web-based tool (Schwab et al., 2006; http://wmd.weigelworld.org).

Assembly of an Allelic Series by TILLING

Targeting Induced Local Lesions IN Genomes (TILLING) can be used for collecting a traditional allelic series of mutations in the gene of interest using a standard mutagen, such as EMS (Till et al., 2006). *Arabidopsis* is first mutagenized using EMS, which causes point mutations. Subsets of *Arabidopsis* plants are subjected to PCR amplification of the gene of interest. The bulk PCR product from individual subsets of plants is examined for point mutations by searching for mismatches with the wild-type allele using a high-throughput analytical biochemistry technique. If a point mutation is present in the subset, iteratively smaller subpools of the mutagenized population are screened for the same mutation until a single plant with this new mutant allele has been identified. The mutant may then be examined for a possible developmentally informative phenotype. The great advantage of the TILLING technique is that mild alleles can be found. Mild alleles are crucial to understand the *late* effects of genes that may be essential during early development.

Gene Knockouts

Last but not least, the first-best evidence for the function of an unknown gene comes from its "knockout" phenotype, that is, the phenotype of a plant in which this gene is missing. Other than in the mouse, there does not yet exist a reliable method for eliminating individual genes in the plant genome in a targeted fashion ("gene targeting"). However, at least in *Arabidopsis*, it is relatively easy to generate large collections of T-DNA insertion strains. As discussed for T-DNA tagging above, insertion of a T-DNA into a gene is likely to abolish the gene's function (Figure 11.3A). In recent years, major efforts have been expended in several countries to map the precise insertion point of large numbers of individual T-DNAs on a genomewide scale (Alonso et al., 2003; http://www.signal.salk.edu) and depositing the corresponding strains in genetic stock centers.

In summary, reverse genetic tools have brought the goal within reach of characterizing the function of each of the approximately 25,000 genes in the *Arabidopsis* genome, the first plant genome to be sequenced. Many of these genes will have roles associated with the development of *Arabidopsis*. This goal is being pursued under the auspices of the National Science Foundation in its *Arabidopsis* 2010 project.

REFERENCES

Alonso, J. M., A. N. Stepanova, T. J. Leisse, C. J. Kim, H. Chen, P. Shinn, D. K. Stevenson, J. Zimmerman, P. Barajas, R. Cheuk, C. Gadrinab, C. Heller, A. Jeske, E. Koesema, C. C. Meyers, H. Parker, L. Prednis, Y. Ansari, N. Choy, H. Deen, M. Geralt, N. Hazari, E. Hom, M. Karnes, C. Mulholland, R. Ndubaku, I. Schmidt, P. Guzman, L. Aguilar-Henonin, M. Schmid, D. Weigel, D. E. Carter, T. Marchand, E. Risseeuw, D. Brogden, A. Zeko, W. L. Crosby, C. C. Berry, and J. R. Ecker. 2003. Genome-wide insertional mutagenesis of *Arabidopsis thaliana*. *Science* 301: 653–657.

Cai, S. and C. C. Lashbrook. 2006. Laser capture microdissection of plant cells from tape-transferred paraffin sections promotes recovery of structurally intact RNA for global gene profiling. *Plant J.* 48:628–637.

Casson, S., M. Spencer, K. Walker, and K. Lindsey. 2005. Laser capture microdissection for the analysis of gene expression during embryogenesis of *Arabidopsis*. *Plant J.* 42:111–123.

Chamovitz, D. A., N. Wei, M. T. Osterlund, A. G. von Arnim, J. M. Staub, M. Matsui, and X. W. Deng. 1996. The COP9 complex, a novel multisubunit nuclear regulator involved in light control of a plant developmental switch. *Cell* 86:115–121.

Coen, E. S. and E. M. Meyerowitz. 1991. The war of the whorls: Genetic interactions controlling flower development. *Nature* 353:31–37.

Deng, X. W., T. Caspar, and P. H. Quail. 1991. cop1: A regulatory locus involved in light-controlled development and gene expression in *Arabidopsis*. *Genes Dev.* 5:1172–1182.

Dill, A., H. S. Jung, and T. P. Sun. 2001. The DELLA motif is essential for gibberellin-induced degradation of RGA. *Proc Natl Acad Sci (USA)* 98:14162–14167.

Geisler-Lee J., N. O'Toole, R. Ammar, N. J. Provart, A. H. Millar, and M. Geisler. 2007. A predicted interactome for *Arabidopsis*. *Plant Physiol.* 145:317–329.

Gu, Q., C. Ferrandiz, M. F. Yanofsky, and R. Martienssen. 1998. The FRUITFULL MADS-box gene mediates cell differentiation during *Arabidopsis* fruit development. *Development* 125: 1509–1517.

Guo, H. and J. R. Ecker. 2004. The ethylene signaling pathway: New insights. *Curr Opin Plant Biol.* 7:40–49.

Imaizumi, T. and S. A. Kay. 2006. Photoperiodic control of flowering: Not only by coincidence. *Trends Plant Sci.* 11:550–558. Erratum in: *Trends Plant Sci.* 2006 11:567.

Lee, J., K. He, V. Stolc, H. Lee, P. Figueroa, Y. Gao, W. Tongprasit, H. Zhao, I. Lee, and X. W. Deng. 2007. Analysis of transcription factor HY5 genomic binding sites revealed its hierarchical role in light regulation of development. *Plant Cell.* 19:731–749.

Millar, A. J., I. A. Carre, C. A. Strayer, N. H. Chua, and S. A. Kay. 1995. Circadian clock mutants in *Arabidopsis* identified by luciferase imaging. *Science* 267:1161–1163.

Nakajima, K., G. Sena, T. Nawy, and P. N. Benfey. 2001. Intercellular movement of the putative transcription factor SHR in root patterning. *Nature* 413:307–311.

Sagerström, C.G., B. I. Sun, and H. L. Sive. 1997. Subtractive cloning: Past, present, and future. *Annu Rev Biochem.* 66:751–783.

Saleh, A., R. Alvarez-Venegas, and Z. Avramova. 2008. An efficient chromatin immunoprecipitation (ChIP) protocol for studying histone modifications in *Arabidopsis* plants. *Nat Protoc.* 3:1018–1025.

Schwab, R., S. Ossowski, M. Riester, N. Warthmann, and D. Weigel. 2006. Highly specific gene silencing by artificial microRNAs in *Arabidopsis*. *Plant Cell.* 18:1121–1133.

Sharma, V. K. and J. C. Fletcher. 2002. Maintenance of shoot and floral meristem cell proliferation and fate. *Plant Physiol.* 129:31–39.

Subramanian, C., J. Woo, X. Cai, X. Xu, S. Servick, C. H. Johnson, A. Nebenführ, and A. G. von Arnim. 2006. A suite of tools and application notes for in vivo protein interaction assays using bioluminescence resonance energy transfer (BRET). *Plant J.* 48:138–152.

Till, B. J., T. Zerr, L. Comai, and S. Henikoff. 2006. A protocol for TILLING and Ecotilling in plants and animals. *Nat Protoc.* 1:2465–2477.

Ulmasov, T., J. Murfett, G. Hagen, and T. J. Guilfoyle. 1997. Aux/IAA proteins repress expression of reporter genes containing natural and highly active synthetic auxin response elements. *Plant Cell.* 9:1963–1971.

Wei, N. and X. W. Deng. 1992. COP9: A new genetic locus involved in light-regulated development and gene expression in *Arabidopsis*. *Plant Cell*. 4:1507–1518.

Winter, D., B. Vinegar, H. Nahal, R. Ammar, G. V. Wilson, and N. J. Provart. 2007. An "electronic fluorescent pictograph" browser for exploring and analyzing large-scale biological data sets. *PLoS ONE* 2:e718.

Zuo, J., Q. W. Niu, and N. H. Chua. 2000. Technical advance: An estrogen receptor-based transactivator XVE mediates highly inducible gene expression in transgenic plants. *Plant J*. 24:265–273.

Section III

Propagation and Development Concepts

12 Propagation by Shoot Culture

Michael Kane

CONCEPTS

- Shoot meristems retain the embryonic capacity for unlimited division.
- Smaller isolated meristem explants require more complex culture media for survival.
- Meristem and meristem-tip culture are methods for disease eradication.
- Shoot culture provides a means to multiply periclinal chimeras.
- Cytokinins disrupt apical dominance and enhance lateral shoot production.
- Increased auxin concentration increases rooting percentage and root number, but decreases root elongation.
- Negative carry-over effects of cytokinins used for Stage II shoot multiplication and auxins for Stage II rooting may affect ex vitro survival and plantlet growth.
- Low capacity for photosynthesis and poor control of water relations are the two principal causes of plant mortality during acclimatization to ex vitro conditions.

Micropropagation is defined as the true-to-type propagation of selected genotypes using in vitro culture techniques. Four basic methods are used to propagate plants in vitro. Depending on the species and cultural conditions, in vitro propagation can be achieved by the following: (1) enhanced axillary shoot proliferation (shoot culture); (2) node culture; (3) de novo formation of adventitious shoots through shoot organogenesis (Chapter 19); or (4) nonzygotic embryogenesis (Chapter 22). Currently, the most frequently used micropropagation method for commercial production utilizes enhanced axillary shoot proliferation from cultured meristems. This method provides genetic stability and is easily attainable for many plant species. Consequently, the shoot culture method has played an important role in development of a worldwide industry that produces more than 250 million plants yearly. Besides propagation, shoot meristems are cultured in vitro for two of the following other purposes: (1) production of pathogen-eradicated plants and (2) preservation of pathogen-eradicated germplasm (see Chapter 38). Concepts related to propagation by shoot and node culture will be discussed in this chapter.

SHOOT APICAL MERISTEMS

It is important to briefly review the general structure of shoot meristems. Shoot growth in mature plants is restricted to specialized regions that exhibit little differentiation and in which the cells retain the embryonic capacity for unlimited division. These regions, called *apical meristems*, are located in the apices of the main and lateral buds of the plant (Chapter 8). Cells derived from these apical meristems subsequently undergo differentiation to form the mature tissues of the plant body. Due to their highly organized structure, apical meristems tend to be genetically stable.

There are significant differences in the shape and size of shoot apices between different taxonomic plant groups (Fahn, 1974). A typical dicotyledon shoot apical meristem consists of a layered dome of actively dividing cells located at the extreme tip of a shoot and measures about 0.1 to 0.2 mm in diameter and 0.2 to 0.3 mm in length. The apical meristem has no connection to the vascular system of the stem. Below the apical meristem, localized areas of cell division and elongation represent sites of newly developing leaf primordia (Figure 12.1). Lateral buds, each containing an apical meristem, develop within the axils of the subtending leaves. In the intact plant, outgrowth of the lateral buds is usually inhibited by apical dominance of the terminal shoot tip. Organized shoot growth from the apical meristem of plants is potentially unlimited and is said to be *indeterminate*. However, shoot apical meristems may become committed to the formation of determinate organs such as flowers.

In Vitro Culture of Shoot Meristems

The recognized potential for unlimited shoot growth prompted early, but largely unsuccessful, attempts to aseptically culture isolated shoot meristems in the 1920s. By the middle 1940s, sustained growth and maintenance of organization of cultured shoot meristems through repeated subculture was achieved for several species. Ball (1946), however, provided the first detailed procedure for the isolation and production of plants from cultured shoot meristem tips and the successful transfer of rooted plantlets into soil. Ball is often called the *Father of Micropropagation* because his shoot tip culture procedure is the one most commonly used by commercial micropropagation laboratories today. Although these studies demonstrated the feasibility of regenerating shoots from cultured shoot tips, the procedures typically yielded unbranched shoots.

Several important findings facilitated application of in vitro culture techniques for large-scale clonal propagation from meristems. The discovery that virus-eradicated plants could be generated from cultured meristems led to the widespread application of the procedure for routine fungal and bacterial pathogen eradication as well (Morel and Martin, 1952; Styer and Chin, 1983). Demonstration of rapid production of orchids from cultured shoot tips supported the possibility of rapid clonal propagation in other crops (Morel, 1960, 1965). It should be noted that in vitro propagation in many orchids does not occur via axillary shoot proliferation; rather, the cultured meristems become disorganized and form spheroid protocorm-like bodies that are actually nonzygotic embryos (Chapter 22).

The final discovery was the elucidation of the role of cytokinins in the inhibition of apical dominance (Wickson and Thimann, 1958). This finding was eventually applied to enhance axillary shoot production in vitro. Application of this method was expedited by development of improved culture

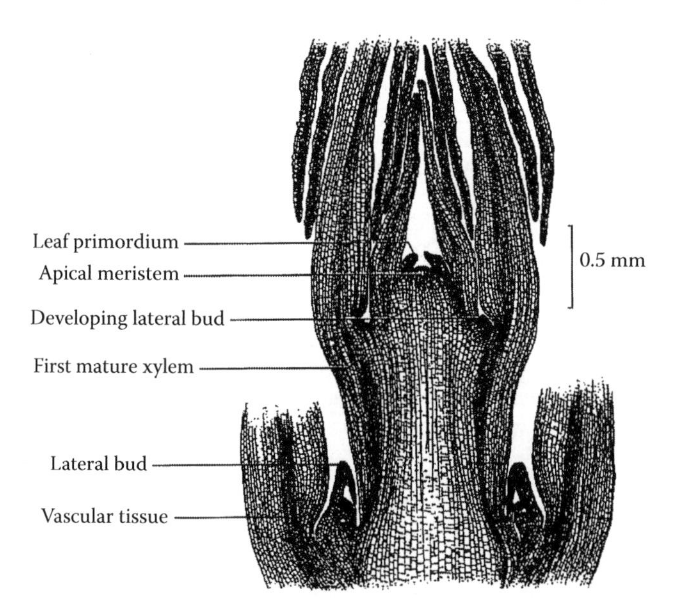

FIGURE 12.1 Diagrammatic representation of a dicotyledonous shoot tip. The shoot tip comprises the apical meristem, subtending leaf primordia, and lateral buds. (From Trigiano, R.N. and Gray, D.J., Eds., *Plant Tissue Culture Concepts and Laboratory Exercises, Second Edition*, CRC Press LLC, Boca Raton, FL, 2000.)

media that supported the propagation of a wide diversity of plant species (Murashige and Skoog, 1962; Lloyd and McCown, 1981).

Meristem and Meristem Tip Culture

Although not directly used for propagation, meristem and meristem tip culture will be briefly described since these procedures are used to generate pathogen-eradicated shoots that subsequently serve as propagules for in vitro propagation. Culture of the apical meristematic dome alone (Figure 12.2) from either terminal or lateral buds, for the purpose of pathogen elimination, is termed meristem culture. In reality, true meristem culture is rarely used because isolated apical meristems of many species exhibit both low survival rates and increased chance of genetic variability following callus formation and indirect shoot organogenesis.

Pathogen elimination can often be accomplished by culture of relatively larger (0.2–0.5 mm long) meristem tip explants excised from plants that have undergone thermo- or chemotherapy. The meristem tip comprises the apical meristem plus one or two subtending leaf primordia (Figure 12.2). This procedure is therefore termed *meristem tip culture.* Caution should be taken when interpreting much of the early published literature of successful "meristem" culture since, in many instances, meristem tip or even larger shoot tip explants were actually used. The term *meristemming* commonly used in the orchid literature is equally ambiguous.

Shoot and Node Culture

Although not the most efficient procedure, propagation from axillary shoots has proved to be a reliable method for the micropropagation of a large number of species (Kurtz et al., 1991). Depending

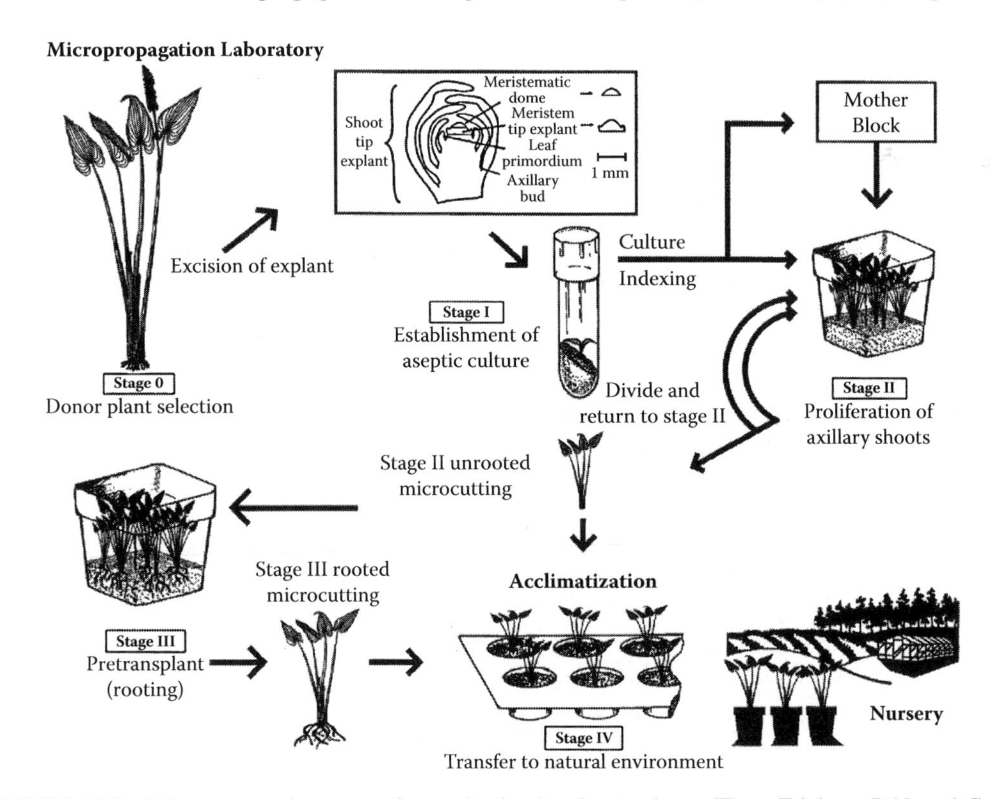

FIGURE 12.2 Micropropagation stages for production by shoot culture. (From Trigiano, R.N. and Gray, D.J., Eds., *Plant Tissue Culture Concepts and Laboratory Exercises, Second Edition*, CRC Press LLC, Boca Raton, FL, 2000.)

on the species, two methods, shoot and node culture, are used. Both methods rely on the stimulation of axillary shoot growth from lateral buds following disruption of apical dominance of the shoot apex. *Shoot culture* (shoot tip culture) refers to the in vitro propagation by *repeated* enhanced formation of axillary shoots from shoot tips or lateral buds cultured on medium supplemented with growth regulators, usually a cytokinin (George and Debergh, 2008). The axillary shoots produced are either subdivided into shoot tips and nodal segments that serve as secondary explants for further proliferation (see Figure 12.3a) or are treated as microcuttings for rooting.

When either verified pathogen-free stock plants are used or when pathogen elimination is not a concern, relatively larger (1–20 mm long) shoot tip or lateral bud primary explants (Figure 12.2) can be used for culture establishment and subsequent shoot culture. Advantages of using larger shoot tips include greater survival, more rapid growth responses, and the presence of more axillary buds. However, these larger explants are more difficult to completely surface sterilize and can potentially harbor undetected latent systemic microbial infection. Compared to other micropropagation methods, shoot cultures (1) provide reliable rates and consistency of multiplication following culture stabilization; (2) are less susceptible to genetic variation; and (3) may provide for clonal propagation of periclinal chimeras.

Node culture, a simplified form of shoot culture, is another method for production from preexisting meristems. Numerous plants such as potato (*Solanum tuberosum* L.) do not respond well to the cytokinin stimulation of axillary shoot proliferation observed in the micropropagation of many crops. Axillary shoot growth is promoted by the culture of either intact shoots (from meristem tip cultures) positioned horizontally on the medium (in vitro layering) or single or multiple node segments. Typically, single elongated unbranched shoots, comprising multiple nodes, are rapidly produced (see Figure 12.2). These shoots (microcuttings) are either rooted or acclimatized to ex vitro conditions or repeatedly subdivided into nodal cuttings to initiate additional cultures. In some species, modified shoot storage organs such as miniaturized tubers or corms may develop from

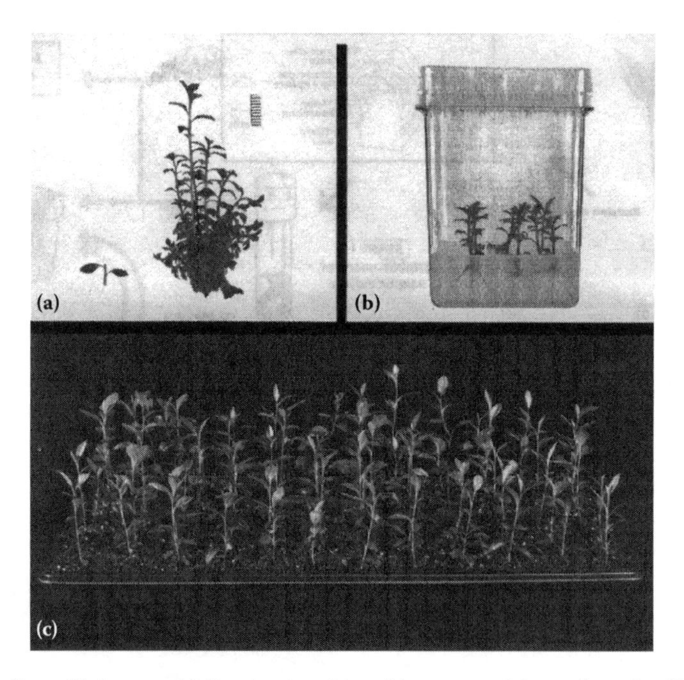

FIGURE 12.3 **a**. Stage II shoot multiplication is achieved by repeated formation of axillary shoot clusters from explants containing lateral buds (*Aronia arbutifolia* L.). Depending on the species, individual microcuttings or shoot clusters may be rooted and acclimatized ex vitro. Scale bar=1 cm. **b**. For maximum survival, Stage III rooting may be required prior to acclimatization to ex vitro conditions. **c**. Rooted and acclimatized Stage IV plantlets. (From Trigiano, R.N. and Gray, D.J., Eds., *Plant Tissue Culture Concepts and Laboratory Exercises, Second Edition*, CRC Press LLC, Boca Raton, FL, 2000.)

axillary shoots under inductive culture conditions. Although node culture is the simplest method, it is associated with the least genetic variation.

MICROPROPAGATION STAGES

Murashige (1974) originally described three basic stages (I–III) for successful micropropagation. Recognition of the contamination problems often associated with inoculation of primary explants prompted Debergh and Maene (1981) to include a Stage 0. This stage described specific cultural practices, which maintained the hygiene of stock plants and decreased the contamination frequency during initial establishment of primary explants. As a result of our increased information base, currently it is now agreed that there are five stages (Stage 0–IV) critical to successful micropropagation. These stages not only describe the procedural steps in the micropropagation process but also represent points at which the cultural environment is altered (Miller and Murashige, 1976). This system has been adopted by most commercial and research laboratories as it simplifies production scheduling, accounting, and cost analysis (Kurtz et al., 1991). Requirement for completion of each stage will depend on the plant material and specific method used. Diagrammatic representation of the micropropagation stages for propagation by shoot culture is provided in Figure 12.2.

STAGE O: DONOR PLANT SELECTION AND PREPARATION

Explant quality and subsequent responsiveness in vitro is significantly influenced by the phytosanitary, physiological state, and genotype of the donor plant (Debergh and Maene, 1981; Read, 1988; Valero-Aracama et al., 2008). Prior to culture establishment, careful attention is given to the selection and maintenance of the stock plants used as the source of explants. Stock plants are maintained in clean controlled conditions that allow active growth but reduce the probability of disease. Maintenance of specific-pathogen-tested stock plants under conditions of relatively lower humidity, use of drip irrigation, and antibiotic sprays have proved effective in reducing the contamination potential of candidate explants. Such practices also allow excision of relatively larger and more responsive explants, often without increased risks of contamination.

Numerous practices are also employed to increase explant responsiveness by modifying the physiological status of the stock plant. These include the following: (1) trimming to stimulate lateral shoot growth; (2) pretreatment sprays containing cytokinins or gibberellic acid; and (3) use of forcing solutions containing 2% sucrose and 200 mg/L 8-hydroxyquinoline citrate for induction of bud break and delivery of growth regulators to target explant tissues (Read, 1988). Currently, information on the effects of other factors such as stock plant nutrition, light, and temperature treatments on the subsequent in vitro performance of meristem explants is lacking.

STAGE I: ESTABLISHMENT OF ASEPTIC CULTURES

Initiation and aseptic establishment of pathogen-eradicated and responsive terminal or lateral shoot meristem explants is the goal of this stage. The primary explants obtained from the stock plants may consist of surface-sterilized shoot apical meristems or meristem tips for pathogen elimination or shoot tips from terminal or lateral buds (Figure 12.2).

The presence of microbial contaminants can adversely affect shoot survival, growth, and multiplication rate. Bacteria and fungal contaminants often persist within cultured tissues that visually appear contaminant free. Consequently, it is essential that Stage I cultures be indexed (screened) for the presence of internal microbial contaminants prior to serving as sources of shoot tip and nodal explants for Stage II multiplication (see Chapters 17 and 18).

The following factors may effect successful Stage I establishment of meristem explants: (1) explantation time; (2) position of the explant on the stem; (3) explant size; and (4) polyphenol oxidation. Time of explantation can significantly affect explant response in vitro. In deciduous woody perennials, shoot tips explants collected at various times during the spring growth flush may vary in their ability for shoot proliferation. Shoot tips collected during or at the end of the period of most rapid shoot elongation exhibited weak proliferation potential. Explants collected before or after this period are capable of strong shoot proliferation in vitro (Brand, 1993). Conversely, the best results are obtained with herbaceous perennials that form storage organs, such as tubers or corms, when explants are excised at the end of dormancy and after sprouting.

Explants also exhibit different capacities for establishment in vitro depending on their location on the donor plant. For example, survival and growth of terminal bud explants are typically greater than lateral bud explants. Often, similar lateral meristem explants from the top and bottom of a single shoot may respond differently in vitro. In woody plants exhibiting phasic development, juvenile explants typically are often more responsive than those obtained from the often-nonresponsive mature tissues of the same plant. Sources of juvenile explants include the following: (1) root suckers; (2) basal parts of mature trees; (3) stump sprouts; and (4) lateral shoots produced on heavily pruned plants.

The excision of primary explants often promotes the release of polyphenols and stimulates polyphenol oxidase activity within the damaged tissues. The polyphenol oxidation products often blacken the explant tissue and medium. Accumulation of these polyphenol oxidation products can eventually kill the explants. Procedures used to decrease tissue browning include the following: (1) use of liquid medium with frequent transfer; (2) adding antioxidants such as ascorbic acid or polyvinylpyrrolidone (PVP); (3) addition of activated charcoal; and (4) culture in reduced light or darkness.

There clearly is no one universal culture medium for establishment of all species. However, modifications to the Murashige and Skoog (Murashige and Skoog, 1962) basal medium formulation are most frequently used (see Chapter 3). Cytokinins or auxins are most frequently added to Stage I media to enhance explant survival and shoot development (Hu and Wang, 1983). The types and levels of growth regulators used in Stage I media are dependent on the species, genotype, and explant size.

Knowledge of the specific sites of hormone biosynthesis in intact plants provides insight into the relationship between explant size and dependence on exogenous growth regulators (Chapter 4) in the medium. Endogenous cytokinins and auxins are synthesized primarily in root tips and leaf primordia, respectively. Consequently, smaller explants, especially cultured apical meristem domes, exhibit greater dependence on medium supplementation with exogenous cytokinin and auxin for maximum shoot survival and development (Shabde and Murashige, 1977). Larger shoot tip explants usually do not require the addition of auxin in Stage I medium for establishment. Rapid adventitious rooting of shoot tip explants often provides a primary endogenous cytokinin source. Most Stage I media are agar-solidified and supplemented with at least a cytokinin (Wang and Charles, 1991). The most frequently used cytokinins are N^6-benzyladenine (BA), kinetin (Kin), and N^6-(2-isopentenyl)-adenine (2-iP). Due to its low cost and high effectiveness, the cytokinin BA is most widely used (Gaspar et al., 1996). Substituted urea compounds, such as thidiazuron, exhibit strong cytokinin-like activity and have been used to facilitate the shoot culture of recalcitrant woody species (Huetteman and Preece, 1993).

Many types of auxins are used. The naturally occurring auxin indole-3-acetic acid (IAA) is the least active, whereas the stronger and more stable compounds α-naphthalene acetic acid (NAA), a synthetic auxin, and indole-3-butyric acid (IBA), a naturally occurring auxin, are more often used. Stage I medium PGR levels and combinations that promote explant establishment and shoot growth but limit formation of callus and adventitious shoot formation are selected.

A commonly held misconception is that primary explants exhibit immediate and predictable growth responses following inoculation. For many species, particularly herbaceous and woody perennials, consistency in growth rate and shoot multiplication is achieved only after multiple subculture on Stage I medium. Physiological stabilization may require from 3 to 24 months and four to six subcultures. Failure to allow culture stabilization before transfer to a Stage II medium containing higher cytokinin levels may result in diminished shoot multiplication rates or production of undesirable basal callus and adventitious shoots. With some species, the time required for stabilization can be reduced by initial culture in liquid medium.

In many commercial laboratories, stabilized cultures, verified as being specific pathogen tested and free of cultivable contaminants, are often maintained on media that limit shoot production to maintain genetic stability. These cultures, called mother blocks, serve as sources of shoot tips or nodal segments for initiation of new Stage II cultures (Figure 12.2).

Stage II: Proliferation of Axillary Shoots

Stage II propagation is characterized by repeated enhanced formation of axillary shoots from shoot tips or lateral buds cultured on a medium supplemented with a relatively higher cytokinin level to disrupt apical dominance of the shoot tip. A subculture interval of 4 weeks with a three- to eightfold increase in shoot number is common for many crops propagated by shoot culture. Given these multiplication rates, conservatively, more than 4.3×10^7 shoots could be produced yearly from a single starting explant.

Stage II cultures are routinely subdivided into smaller clusters, individual shoot tips, or nodal segments that serve as propagules for further proliferation (Figure 12.3a). Additionally, axillary shoot clusters may be harvested as individual unrooted Stage II microcuttings or multiple shoot clusters for ex vitro rooting and acclimatization (Figure 12.2). Clearly, Stage II represents one of the most costly stages in the production process.

Both source and orientation of explants can affect Stage II axillary shoot proliferation. Subcultures inoculated with explants that had been shoot apices in the previous subculture often exhibit higher multiplication rates than lateral bud explants. Inverting shoot explants in the medium can double or triple the number of axillary shoots produced on vertically oriented explants per culture period in some species.

The number of subcultures possible before initiation of new Stage II cultures from the mother block is required depends on the species or cultivar and its inherent ability to maintain acceptable multiplication rates while exhibiting minimal genetic variation and off-types (Kurtz et al., 1991). Some species can be maintained with monthly subculture from 8 to 48 months in Stage II. In contrast, in some ferns (*Nephrolepsis*), as few as three subcultures may only be possible before the frequency of off-types increases to unacceptable levels.

Increased production of off-types is often attributed to production of adventitious shoots, particularly through an intermediary callus stage (Jain and De Klerk, 1998). Stage II cultures, originally regenerating from axillary shoots, often begin producing adventitious shoots at the base of axillary shoot clusters after a number of subcultures on the same medium. These so-called mixed cultures can develop without any morphological differences being apparent. Selecting only terminal shoots of axillary origin for subculture instead of shoot bases decreases the frequency of off-types, including the segregation of periclinal chimeras.

Selection of Stage II cytokinin type and concentration is made on the basis of shoot multiplication rate, shoot length, and frequency of genetic variation. Although shoot proliferation is enhanced at higher cytokinin concentrations, the shoots produced are usually smaller and may exhibit symptoms of hyperhydricity. Depending on the species, exogenous auxins may or may not enhance cytokinin-induced axillary shoot proliferation (Figure 12.4). Addition of auxin in the medium often mitigates the inhibitory effect of cytokinin on shoot elongation, thus increasing the number of usable shoots of sufficient length for rooting (Figure 12.5). This benefit must be weighed against the increased chance of callus formation.

The possibility of adverse carry-over effects on the survivability and rooting of plantlets in Stage IV should be evaluated when selecting a Stage II cytokinin. For example, with some plant species, the use of BA in Stage II can significantly reduce plantlet survival during acclimatization in Stage IV to as low as 10% (Griffis et al., 1981). Use of Kin or 2-iP instead of BA yields survival rates in excess of 90%. In some plants, the adverse effect of BA on Stage IV survival and rooting has been attributed to production of an inhibitory BA metabolite (Werbrouck et al., 1995). Substituting BA

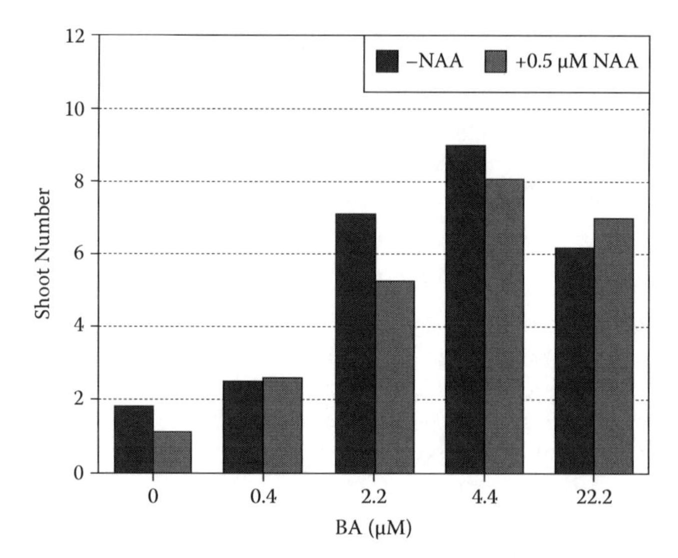

FIGURE 12.4 Effect of BA concentration on Stage II axillary shoot proliferation produced from two-node explants of *Aronia arbutifolia* after 28-day culture in presence and absence of 0.5 μM NAA.

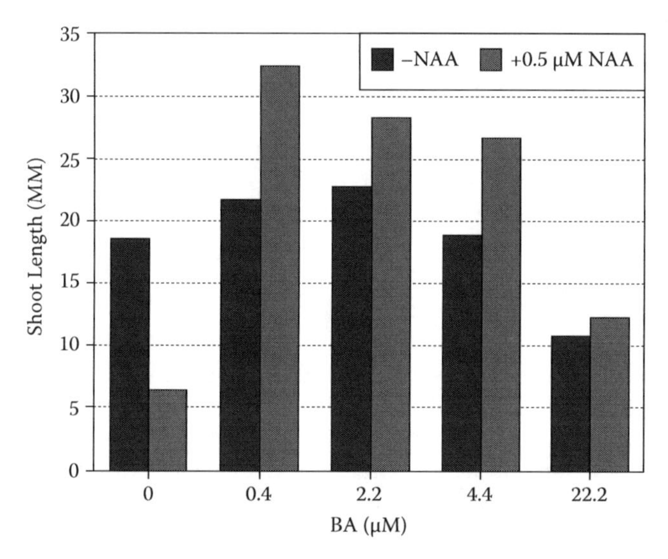

FIGURE 12.5 Inclusion of auxin may reduce the inhibitory effect of BA on axillary shoot elongation. Data shown for shoots generated from two-node explants of *Aronia arbutifolia* after 28-day culture in the presence and absence of 0.5 μM NAA.

with the BA-analog meta-topolin can significantly increase Stage IV survival of some plants and genotypes that display negative BA carry-over effects (Valero-Aracama et al., 2010). Conceivably, switching the Stage II cytokinin from BA may also eliminate the requirement for Stage III rooting in some species.

Stage III: Pretransplant (Rooting)

This step is characterized by preparation of Stage II shoots or shoot clusters for successful transfer to soil. The process may involve the following: (1) elongation of shoots prior to rooting; (2) rooting of individual shoots or shoot clumps; (3) fulfilling dormancy requirements of storage organs by cold treatment; or (4) prehardening cultures to increase survival. Where possible, commercial laboratories have developed procedures to transfer Stage II microcuttings to soil, thus bypassing Stage III rooting (Figure 12.2).

There are several reasons for eliminating Stage III rooting. Estimated costs for Stage III range from 35% to 75% of the total production costs. This reflects the significant input of labor and supplies required to complete Stage III rooting. Considerable cost savings can be realized if Stage III is eliminated. Furthermore, it is often observed that in vitro formed root systems are largely nonfunctional and die following transplanting. This results in a delay in transplant growth prior to production of new adventitious roots.

For various reasons, however, it may not always be feasible to transplant Stage II microcuttings directly to soil. Given the aforementioned limitations of Stage III rooting, Debergh and Maene (1981) proposed using Stage III solely to elongate Stage II shoots clusters prior to separation and rooting ex vitro. Elongated shoots may be further pretreated in an aqueous auxin solution prior to transplanting. Usually, Stage III rooting of herbaceous plants can be achieved on medium in the absence of auxins. However, with many woody species, the addition of an auxin (IBA or NAA) in Stage III medium is required to enhance adventitious rooting (Figure 12.2b). Optimal auxin concentration is determined based on percentage rooting, root number, and length (Table 12.1). It is critical that the roots not be allowed to elongate to prevent root damage during transplanting. Care must be taken when selecting an auxin. For example, use of NAA for Stage III rooting has been shown to decrease survival rates or suppress post-transplant growth (Conner and Thomas, 1981).

Stage IV: Transfer to Natural Environment

The ultimate success of shoot or node culture depends on the ability to transfer and reestablish vigorously growing plants from the in vitro to the ex vitro greenhouse environment (Figure 12.3c). This involves acclimatizing or hardening-off plantlets to conditions of significantly lower relative humidity and higher light intensity. Even when acclimatization procedures are carefully followed, poor survival rates are frequently encountered. Micropropagated plants are difficult to transplant for two primary reasons: (1) a heterotrophic mode of nutrition and (2) poor control of water loss.

Plants cultured in vitro in the presence of sucrose and under conditions of limited light and gas exchange exhibit no, or extremely reduced, capacities for photosynthesis. Reduced photosynthetic activity is often associated with low RubPcase activity (Valero-Aracama et al., 2006). During

TABLE 12.1
Addition of Auxin Promotes Stage III Rooting of Shoot Cuttings

Treatment IBA (μM)	% Rooting	Root Number	Root Length (mm)
0 (0)[a]	37	2.4	23.2
0.27 (0.05)	43	3.5	18.1
0.54 (0.1)	55	4.1	14.2
2.7 (0.5)	71	5.7	6.5
5.4 (1.0)	84	7.1	4.3

Source: Trigiano, R.N. and Gray, D.J., Eds., *Plant Tissue Culture Concepts and Laboratory Exercises, Second Edition*, CRC Press LLC, Boca Raton, FL, 2000.

Note: Effects of IBA on Stage III rooting of 10 mm shoots of *Aronia arbutifolia* after 4-week culture are shown.

[a] Concentrations in parentheses are mg/L.

acclimatization, there is a need for plants to rapidly transition from the heterotrophic to photoautotrophic state for survival (Preece and Sutter, 1991). Unfortunately, this transition is not immediate. For example, in cauliflower, no net increase in CO_2 uptake is achieved until 14 days after transplantation. This occurs only following development of new leaves since the leaves produced in vitro in the presence of sucrose never develop photosynthetic competency. Interestingly, before senescencing, these older leaves function as "lifeboats" by supplying stored carbohydrate to the developing and photosynthetically competent new leaves. This is not the rule with all micropropagated plants, since the leaves of some species become photosynthetic and persist after acclimatization.

A composite of anatomical and physiological features, characteristic of plants produced in vitro under 100% relative humidity, contribute to the limited capacity of micropropagated plants to regulate water loss immediately following transplanting. These features include reductions in leaf epicuticular wax, poorly differentiated mesophyll, abnormal stomate function, and poor vascular connection between shoots and roots.

To overcome these limitations, plantlets are transplanted into a well-drained "sterile" growing medium and maintained initially at high relative humidity and reduced light (40–160 $\mu mol \times m^{-2} \times s^{-1}$) at 20°–27°C. Relative humidity may be maintained with humidity tents, single tray propagation domes, intermittent misting, or fog systems. However, use of intermittent mist often results in slow plantlet growth following waterlogging of the medium and excessive leaching of nutrients. Transplants are acclimatized by gradually lowering the relative humidity over a 1–4 week period. Plants are gradually moved to higher light intensities to promote vigorous growth.

CONCLUSION

Propagation from preexisting meristems through shoot and node culture is the most reliable and widely used procedure. However, the need for multiple subcultures on different media makes shoot and node culture extremely labor intensive. Total labor costs, typically ranging from 50–70% of production costs, limit expansion of the micropropagation industry. Current application of the technology is restricted to high-value horticultural crops such as ornamental plants. Expansion of the industry to include production of vegetable, plantation, and forest crops depends on development of more efficient micropropagation systems. Cost-reduction strategies, including elimination of production steps and development of reliable automated micropropagation systems, will facilitate this expansion (Aitken-Christie et al., 1995).

REFERENCES

Aitken-Christie, J., T. Kozai, and S. Takayama. 1995. Automation in tissue culture—general introduction and overview. In *Automation and Environmental Control in Plant Tissue Culture*. J. Aitken-Christie, T. Kozai, and M.A.L. Smith, Eds., pp. 1–18. Kluwer Academic Publishers, Dordrecht.

Ball, E.A. 1946. Development in sterile culture of shoot tips and subjacent regions of *Tropaeolum majus* L. and of *Lupinus albus* L. *Amer. J. Bot.* 33:301–318.

Brand, M. 1993. Initiating cultures of *Halesia* and *Malus*: Influence of flushing stage and benzyladenine. *Plant Cell Tiss. Org. Cult.* 33:129–132.

Conner, A.J. and M.D. Thomas. 1981. Re-establishing plantlets from tissue culture: A review. *Comb. Proc. Intl. Plant Prop. Soc.* 31:342–357.

Debergh, P.C. and L.J. Maene. 1981. A scheme for commercial propagation of ornamental plants by tissue culture. *Sci. Hort.* 14:335–345.

Fahn, A. 1974. *Plant Anatomy*. Pergamon Press, New York.

Gaspar, T., C. Kevers, C. Penel, H. Greppin, D.M. Reid, and T. Thorpe. 1996. Plant hormones and plant growth regulators in plant tissue culture. *In Vitro Cell. Dev. Biol.—Plant* 32:272–289.

George, E.F. 1993. *Plant Propagation by Tissue Culture. Part 1. The Technology*. Exegetics, Ltd., London.

George, E.F. and P.C. Debergh. 2008. Micropropagation: uses and methods. In *Plant Propagation by Tissue Culture*. E.F. George, M.A. Hall, and G-J. DeKlerk (Eds.), pp. 29–64. Springer, The Netherlands.

Griffis, J.L., Jr., G. Hennen, and R.P. Oglesby. 1981. Establishing tissue-cultured plant in soil. *Comb. Proc. Intl. Plant Prop. Soc.* 33:618–622.
Heutteman, C.A. and J.E. Preece. 1993. Thidiazuron: A potent cytokinin for woody plant tissue culture. *Plant Cell Tiss. Org. Cult.* 33:105–119.
Hu, C.Y. and P.J. Wang. 1983. Meristem, shoot tip and bud cultures. In *Handbook of Plant Cell Culture. Vol. 1. Techniques for Propagation and Breeding*. D.F. Evans, W.R. Sharp, P.V. Ammirato, and Y. Yamada, Eds., pp. 177–227. Macmillan Publishing, New York.
Jain, S.M. and G.J. De Klerk. 1998. Somaclonal variation in breeding and propagation of ornamental crops. *Plant Tiss. Cult. Biotechnol.* 4:63–75.
Kurtz, S., R.D. Hartmann, and I.Y.E. Chu. 1991. Current methods of commercial micropropagation. In *Scale-up and Automation in Plant Propagation: Cell Culture and Somatic Cell Genetics of Plants*, Vol. 8, I.K. Vasil, Ed., pp. 7–34. Academic Press, San Diego, CA.
Lloyd, G. and B. McCown. 1981. Commercially-feasible micropropagation of Moutain laural, *Kalmia latifolia*, by use of shoot-tip culture. *Int. Plant Prop. Soc. Proc.* 30:421–427.
Miller, L.R. and T. Murashige. 1976. Tissue culture propagation of tropical foliage plants. *In Vitro* 12:797–813.
Morel, G. 1960. Producing virus-free *Cymbidium. Amer. Orchid Soc. Bull.* 29:495–497.
Morel, G. 1965. Clonal propagation of orchids by meristem culture. *Cymbidium Soc. News* 20:3–11.
Morel, G. and C. Martin. 1952. Guerison de Dahlias atteints d'une maladie a virus. *C.R. Acad. Sci. Ser. D.* 235:1324–1325.
Murashige, T. 1974. Plant propagation through tissue culture. *Annu. Rev. Plant Physiol.* 25:135–166.
Murashige, T. and F. Skoog. 1962. A revised medium for rapid growth and bioassays with tobacco tissue cultures. *Physiol. Plant.* 15:473–497.
Preece, J.E. and E.G. Sutter. 1991. Acclimatization of micropropagated plants to greenhouse and field. In *Micropropagation Technology and Application.*, P.C. Debergh and R.H. Zimmerman, Eds., pp. 71–93. Kluwer Academic Publishers, Boston.
Read, P.E. 1988. Stock plants influence micropropagation success. *Acta. Hort.* 226:41–52.
Shabde, M. and T. Murashige. 1977. Hormonal requirements of excised *Dianthus caryophyllus* L. shoot apical meristem in vitro. *Amer. J. Bot.* 64:443–448.
Styer, D.J. and C.K. Chin. 1983. Meristem and shoot-tip culture for propagation, pathogen elimination, and germplasm preservation. *Hort. Rev.* 5:221–277.
Valero-Aracama, C., M.E. Kane, S.B. Wilson, J.C. Vu, J. Anderson, and N.L. Philman. 2006. Photosynthetic and carbohydrate status of easy-and difficult-to-acclimatize sea oats (*Uniola paniculata* L.) genotypes during in vitro culture and ex vitro acclimatization. *In Vitro Cell. Dev. Biol.—Plant* 42:572–583.
Valero-Aracama, C., M.E. Kane, S.B. Wilson, and N.L. Philman. 2008. Comparative growth, morphology, and anatomy of easy- and difficult-to-acclimatize sea oats (*Uniola paniculata*) genotypes during in vitro culture and ex vitro acclimatization. *J. Amer. Hort. Sci.* 133:830–843.
Valero-Aracama, C., M.E. Kane, S.B. Wilson, and N.L. Philman. 2010. Substitution of benzyladenine with meta-topolin during shoot multiplication increases acclimatization of difficult- and easy-to-acclimatize sea oats (*Uniola paniculata* L.) genotypes. *Plant Growth Regul.* 60:43–49.
Wang, P.J. and A. Charles. 1991. Micropropagation through meristem culture. In *Biotechnology in Agriculture and Forestry, Vol. 17, High-Tech and Micropropagation* I. Y.P.S. Bajaj, Ed., pp. 32–52. Springer-Verlag, Berlin.
Werbrouck, S.P.O., B. van der Jeugt, W. Dewitte, E. Prinsen, H.A. Van Onckelen, and P.C. Debergh. 1995. The metabolism of benzyladenine in *Spathiphyllum floribundum* "Schott Petite" in relation to acclimatisation problems. *Plant Cell Rep.* 14:662–665.
Wickson, M. and K.V. Thimann. 1958. The antagonism of auxin and kinetin in apical dominance. *Physiol. Plant.* 11:62–74.

13 Micropropagation of *Syngonium* by Shoot Culture

Michael Kane

One of the most successful and reliable commercial applications of micropropagation has been the production of tropical foliage plants (house plants) using micropropagation (Henny et al., 1981; Chen et al., 2005). In fact, many of the foliage plants purchased in garden shops are produced using micropropagation procedures, especially shoot culture. Shoot culture is characterized by the establishment of Stage I cultures using isolated surface disinfested meristem tips, shoot tips, or lateral buds as primary explants (Chapter 12). Shoots that develop from these explants usually are first screened (indexed; see Chapters 17 and 18) for microbial contamination. Aseptic cultures, once physiologically adapted to culture, are then subdivided into nodal or shoot tip secondary explants and transferred to a medium supplemented with a cytokinin to promote axillary shoot proliferation. The axillary shoots produced are either subcultured for repeated proliferation (Stage II) or rooted as microcuttings in vitro (Stage III) or ex vitro (Stage IV) to produce plantlets (George, 1993).

Syngonium podophyllum Schott, commonly known as nephthytis or arrowhead vine, is one of the most commercially important tropical foliage plants propagated in vitro. Plants produce arrowhead-shaped variegated leaves in the juvenile state and climbing vines with 3-parted leaves when mature. Each year more than 6,000,000 plants of the various *Syngonium* cultivars are produced in the United States. Genetic instability of *Syngonium* in vitro has resulted in the selection of many new commercially valuable cultivars (Chen et al., 2006). Like other members of the Araceae, *Syngonium* is susceptible to infection by numerous systemic pathogens that negatively affect plant vigor and quality. However, use of meristem tip culture followed by shoot culture has enabled rapid production of highly branched specific pathogen eradicated plants. It is conservatively estimated that 5,000 syngoniums can be produced yearly from a single primary shoot tip explant (Miller and Murashige, 1976).

The primary objective of these laboratory exercises is to illustrate the sequential steps required for the micropropagation of *Syngonium* by shoot culture. Upon completion of the laboratory exercises students will be able to complete the following: (1) establish Stage I cultures using isolated meristem and shoot tip explants; (2) index Stage I cultures for cultivable bacterial and fungal contaminants; (3) rapidly propagate plants by axillary shoot proliferation (Stage II); and (4) successfully root and acclimatize microcuttings ex vitro (Stage IV). These exercises also offer the opportunity to exemplify several of the following important concepts: (1) the relationship between initial shoot tip explant size, response, and culture contamination; (2) the requirement for physiological adaptation of Stage I cultures prior to Stage II axillary shoot proliferation; and (3) direct rooting and acclimatization of microcuttings ex vitro. The two laboratory experiments outlined shortly include required information on the maintenance of stock plants, media preparation, culture procedures, and anticipated results. Refer to Figure 13.1 for a diagrammatic representation of the procedures used to complete both experiments.

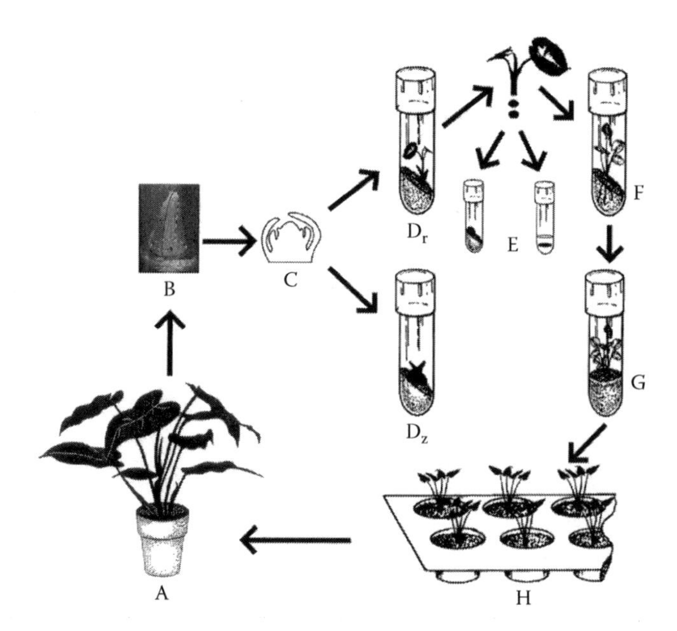

FIGURE 13.1 Diagrammatic representation of the *Syngonium* shoot culture laboratory exercise protocol. A. Nodal and terminal vine tip explants are obtained from potted plants and then surface disinfested. B. Lateral and terminal bud scales are aseptically removed by making shallow horizontal and vertical excisions (dotted lines). C. Successive inner leaf layers are removed to isolate the meristem and shoot tip explants. D. Primary explants are inoculated onto SEM (D_1) and SMM (D_2). Growth responses are compared after 6 weeks of culture. E. Thin basal stem cross sections, excised from 6-week-old shoots generated on SEM, are cultured in both liquid and agar-solidified Sterility Test Medium for 7 days to screen for the presence of cultivable contaminants. F. Shoots are transferred to SEM. G. Cultures screening positive for contaminants are discarded. Sterile cultures are transferred to SMM and cultured for 4 weeks. H. Microcuttings are rooted and acclimatized ex vitro (Stage IV). (From Trigiano, R.N. and Gray, D.J., Eds., *Plant Tissue Culture Concepts and Laboratory Exercises, Second Edition*, CRC Press LLC, Boca Raton, FL, 2000.)

GENERAL CONSIDERATIONS

Maintenance of Syngonium Stock Plants

Shoots tip explants excised from lateral buds and terminal shoot tips from *Syngonium* vines (Figure 13.2a) are placed in culture. It is recommended that vining stems be used because the lateral bud scales are usually larger and easier to remove without damaging the underlying meristem tips. However, if vines are not available, terminal shoot tips and lateral buds on nonvining plants can be used. Potted plants can be inexpensively purchased from most retail garden shops. Larger plants will produce vines more rapidly. Most *S. podophyllum* cultivars, such as "White Butterfly," produce climbing vines, and *S. wendlandii* Schott, another nephthytis species, is an especially fast vine producer. Plants can be grown in most soilless potting mix contained in 20-cm-diameter hanging baskets. Plants should be maintained in a shaded (70–80% light reduction) greenhouse or similar environment where they will receive about 150 $\mu mol \times m^{-2} \times sec^{-1}$ light at 22–30°C. Plants should be fertilized twice weekly (150 mg/L N) to promote vining. Depending on growing conditions and plant size, vines will be produced within 6–8 weeks. Two plants should produce sufficient numbers of explants for each student or team of students.

FIGURE 13.2 Shoot culture of *Syngonium*. a. *Syngonium* vine used as the source of excised meristem tip and shoot tip explants from surface disinfested nodal segments (lower center) and terminal tips (lower right). Scale bar = 1 cm. b. Exposed shoot meristem tip within a lateral bud prior to excision and inoculation. Distance between arrows is 3 mm. c. Left: Stage I shoot culture established from a primary shoot tip explant cultured on SEM for 6 weeks. Center: Inhibition of shoot growth and promotion of basal shoot organogenic callus following 6-week culture of primary shoot tip (arrow) explanted directly on SMM. Right: Typical shoot multiplication response after 4 weeks on SMM from a shoot explant initially established on SEM for 6 weeks. Scale bar = 1 cm. d. Rooted acclimatized plantlet 3 weeks following transfer to ex vitro conditions. (From Trigiano, R.N. and Gray, D.J., Eds., *Plant Tissue Culture Concepts and Laboratory Exercises, Second Edition*, CRC Press LLC, Boca Raton, FL, 2000.)

MATERIALS

The following are needed for each student or student team:

- *Syngonium* vines to provide 15 nodal explants (including terminal vine tip)
- 20% commercial bleach (containing 1.2% (v/v) sodium hypochlorite)
- Ten culture tubes each containing 12 mL *Syngonium* Stage I Establishment Medium (SEM)
- Ten culture tubes each containing 12 mL *Syngonium* Stage II Multiplication Medium (SMM)
- Incubator or a fluorescent light on a bench top providing approximately 30 µmol × m^{-2} × sec^{-1} for a 16-h photoperiod and 25°C

COMPOSITION AND PREPARATION OF THE MEDIA

Two culture media, SEM and SMM, are required. These media are identical to those described by Miller and Murashige (1976) with the exception that both media are solidified with Phytagel. The basal medium of both consists of Murashige and Skoog (1962) inorganic salts (see Chapter 2 for composition) supplemented with 1.25 mM (170 mg/L) $K_2PO_4 \times H_20$, 87.6 mM (30 g/L) sucrose, 0.55 mM (100 mg/L) myo-inositol, and 1.2 µM (0.4 mg/L) thiamine-HCL, solidified with 2 g/L Phytagel. SEM also contains 14.8 µM (3 mg/L) 2iP and 5.7 µM (1 mg/L) IAA, whereas SMM is supplemented with 98.4 µM (20 mg/L) 2iP alone. (The components to prepare these media can be purchased from *Phyto*Technology Laboratories, Shawnee Mission, Kansas.) Media are adjusted to pH 5.7 before adding Phytagel, heating and then dispensing as 12 mLvolumes into 150 × 25 mm culture tubes before autoclaving. The autoclaved media should be cooled and solidified at

45° slants. For Experiment 1, five culture tubes of each medium are prepared for each student or team of students. Additional SEM and SMM will be needed immediately before Experiment 2 is performed.

The liquid and agar-solidified Leifert and Waites Sterility Test Medium (STM) used in Experiment 2 is also available from *Phyto*Technology Laboratories (Product ID#: L476).

Solid STM is prepared by the addition of 10 g/L agar. Both media are prepared without pH adjustment and dispensed as 10-mL volumes into 20 mL glass scintillation vials. Culture tubes can be substituted. However, using relatively deeper 150 × 25 mm culture tubes will make it more difficult for the students to perform the stab-and-streak indexing procedure. At least six vials of each medium for each student should be prepared and dispensed immediately before Experiment 2 is performed. After autoclaving, the agar-supplemented STM is cooled and solidified as 45° slants.

EXERCISES

Experiment 1. Stage I Establishment of Shoot Tip Meristems

The purpose of this experiment is to familiarize students with the procedures used to successfully establish Stage I cultures from excised *Syngonium* meristem tip and shoot tip explants. The goal of Stage I is to establish axenic, responsive shoot cultures. Cultures are indexed for specific pathogens or microbial contamination and allowed to become physiologically adapted to in vitro culture conditions before being clonally multiplied by repeated proliferation of axillary shoots in Stage II (see Figure 12.2). Before proceeding, the structure of a generalized shoot apex should be briefly reviewed (Figure 12.1).

The size of the primary meristem explant will depend on the purpose of culture establishment. For example, when virus elimination is a goal, excision and culture of the apical meristematic dome alone (meristem culture) may be attempted. In reality, cultured apical meristems of many species exhibit low survival rates in vitro, and their use may increase the probability of genetic variability due to the tendency to form organogenetic callus (Styer and Chin, 1984). Consequently, relatively larger (0.2–1.0 mm) and more responsive meristem tip explants (meristem tip culture) are usually used. When pathogen eradication is not a concern, relatively larger (1–2 mm long) terminal or lateral bud shoot tip explants are used to establish cultures.

Follow the instructions in Procedure 13.1 to complete this experiment.

Anticipated Results

Syngonium primary explant size should significantly affect culture survival and responsiveness. Meristem tip explants about 2 mm in length or smaller usually exhibit phenolic browning and high mortality when cultured on either SEM or SMM. When cultured on SEM, larger explants enlarge rapidly and produce leaves and adventitious roots during the 6-week culture period (Figure 9.2c). However, cultures inoculated with larger explants may exhibit a greater frequency of visible contamination. It is important to note that the absence of visible contamination is not an assurance that a culture is aseptic or axenic.

It is often asked why primary explants are not directly inoculated onto Stage II multiplication medium to promote axillary shoot proliferation and save time. The first and most obvious reason for this is the danger of mass propagation of contaminated cultures. The second reason involves a requirement for the physiological adaptation of Stage I explants of some species to in vitro culture conditions. This is referred to as culture stabilization (see Chapter 12). Often culture stabilization must be completed before enhanced clonal multiplication at higher cytokinin levels is possible. For some plants, repeated subculture on Stage I medium is required before physiological stabilization is achieved.

The requirement for Stage I culture stabilization in *Sygonium* is exemplified by the poor response of the meristem tips explanted directly on SMM. Unlike explants cultured on SEM, shoot development from primary explants on SMM should be inhibited and the formation of shoot organogenetic

Procedure 13.1
Establishment of Stage I Cultures from *Syngonium* Meristem and Shoot Tip Explants

Step	Instructions and Comments
1	*Syngonium* vines consist of a terminal shoot apex and numerous lateral buds each enclosed by a sheathing petiole (Figure 13.2a). Remove the leaves with a scalpel to expose the terminal and lateral buds. Any adventitious roots should also be excised.
2	Cut the defoliated vines to yield 15 nodal segments, each about 1 cm long and consisting of a single bud. It can be noted that the smoothly cutinized epidermis of *Syngonium* should facilitate effective surface disinfestation of the explants. Nodes exhibiting bud break should not be used. Rinse nodal segments in tap water for 15 min and then surface disinfest in 20% commercial bleach with constant agitation for 10 min followed by three 5-min rinses in sterile deionized water.
3	In the transfer hood, under 15–30× magnification provided by a dissecting microscope, a 1–5 mm long meristem tip or shoot tip explant is excised from each surface-disinfested bud with a sterile scalpel. To accomplish this, remove the outer bud scales by making a very shallow, but continuous, incision around the base of the bud and then a continuous median longitudinal incision down the front and rear of the bud scale (see Figure 13.1b). Remove bud scales sections with forceps. The inner leaves are successively removed to expose the meristem tip (Figure 13.2b). Excisions should be performed in a sterile petri dish containing a small volume of sterile water to prevent tissue desiccation. The potential for cross contamination is decreased by using a clean petri dish for each nodal segment. Students should attempt to isolate meristem tip and shoot tip explants of various size ranging from 1–5 mm in length. Final explant length measurements are made using a thin plastic ruler positioned under the petri dish.
4	In the transfer hood, under 15–30× magnification provided by a dissecting microscope, a 1–5 mm long meristem tip or shoot tip explant is excised with a sterile scalpel from each surface-disinfested bud. To accomplish this, remove the outer bud scales by making a very shallow, but continuous, incision around the base of the bud and then a continuous median longitudinal incision down the front and rear of the bud scale (see Figure 13.1b). Remove bud scales sections with forceps. The inner leaves are successively removed to expose the meristem tip (Fig. 13.2b). Excisions should be performed in a sterile petri dish containing a small volume of sterile water to prevent tissue desiccation. The potential for cross contamination is decreased by using a clean petri dish for each nodal segment. Students should attempt to isolate meristem tip and shoot tip explants of various size ranging from 1–5 mm in length. Final explant length measurements are made using a thin plastic ruler positioned under the petri dish.
5	Make weekly observations and note differences in the growth responses of the various size explants cultured on SEM and SMM. After 6 weeks, determine the following: (1) percentage of cultures with visible contamination; (2) percentage of responsive explants on each medium; (3) the mean number of axillary shoots produced per responsive explant on each medium; and (4) presence of shoot organogenic callus at the base of each explant. Data should be collected without removing the plants from the culture vessels. Individual student or student team data should be compiled to provide more representative treatment responses.

callus should be promoted at the base of each explant (Figure 13.2c). This response is commercially undesirable since indirect shoot organogenesis may lead to genetic variability in the plants produced and will definitely result in separation of plant chimeras. It is interesting to note that both axillary shoot and adventitious shoot development may occur simultaneously in the same culture. These are often referred to as *mixed cultures*. Prevention of mixed cultures is usually controlled by the type, level, and combinations of plant growth regulators (PGRs) used.

Questions

- If your donor plant was virus infected, which explant tip source (lateral bud or vine tip) do you predict might yield the greatest number of virus-eradicated Stage I cultures?
- For commercial micropropagation, why is it unwise to attempt to save time by eliminating Stage I and immediately culturing surface disinfested primary explants (shoot tips) on SMM?
- In their study on *Syngonium* micropropagation, Miller and Murashige (1976) reported that both primary explant establishment and subsequent Stage II shoot multiplication was greatest when explants were cultured in liquid rather than on solid media. Give some possible explanations for these observations.

Experiment 2. Procedures for Stage I Cultures Indexing, Stage II Shoot Multiplication, and Stage IV Acclimatization

Rapid in vitro (contaminant-free) production of pathogen-eradicated plants is a fundamental goal of the micropropagation process. The surfaces of donor plants are normally colonized and/or inhabited with diverse bacteria and fungi, most of which are nonpathogenic, and surface disinfestion procedures are used to specifically eliminate this microflora. The presence of microbial contaminants in culture can adversely affects growth, multiplication rate, and survival of shoot cultures. Bacteria and fungal contaminants often persist within cultured shoots that visually appear contaminant-free (Knauss, 1976). Consequently, it is essential that Stage I shoot cultures be indexed (screened) for the presence of internal infections before being multiplied in Stage II (see Chapters 16 and 17).

The first objective of this experiment is to familiarize students with a general indexing procedure used to screen for cultivable bacteria and fungal contaminants in the 6-week-old Stage I *Syngonium* shoot cultures generated in Experiment 1. This indexing exercise can be completed during the same laboratory period that the final data collection for Experiment 1 is made. The second objective is to examine the response of secondary shoot explants excised from 6-week-old established and indexed Stage I shoot cultures following subculture on SMM for an additional 4 weeks. Lastly, procedures for successful rooting and Stage IV acclimatization of Stage II microcuttings will be examined.

Materials

The following supplies are needed for each student or student team:

- Six scintillation vials or 150 × 25 mm culture tubes containing liquid STM
- Six scintillation vials or 150 × 25 mm culture tubes containing agar-solidified STM (slants)
- Five culture tubes each containing 12 mL freshly prepared SEM
- Five culture tubes each containing 12 mL freshly prepared SMM
- Plug trays with soilless growing medium and propagation dome or mist bench

Follow the experimental protocol listed in Procedure 13.2 below to complete this experiment.

Anticipated Results

Rapid clouding of the inoculated liquid STM is a positive indication of the presence of cultivable contaminants in the tissue sample. The presence of contaminants on solid STM is confirmed by development of colonies on the surface of the medium and/or development of a halo in the medium where the tissue sample was stabbed. The uninoculated controls should not display microbial growth. Although STM will promote the growth of many microorganisms, it is important to note that growth of all bacteria and fungi will not be supported on this medium. Verified contaminated Stage I cultures are discarded.

After 4 weeks, each Stage II *Syngonium* culture should consist of three to five axillary shoots with no or minimal development of basal organogenic callus (Figure 13.2c). This is in contrast to the inhibition of shoot growth and promotion of organogenic callus observed on the primary explants inoculated directly onto SMM in Experiment 1. *Syngonium* microcuttings should readily root and become acclimatized to ex vitro conditions without the need for Stage III rooting. Rooted and vigorously growing acclimatized plantlets will be produced by week three (Figure 13.2d).

Questions

- Why is it important to both stab into and streak the Stage I tissue sample across the surface of the agar-solidified indexing medium?
- How is it possible to have a contaminated Stage I shoot culture that indexes negative for contamination?

Procedure 13.2
Stage I Cultures Indexing, Stage II Shoot Multiplication, and Stage IV Acclimatization

Step	Instructions and Comments
1	Each student or team should first consecutively number each of their five 6-week-old Stage I *Syngonium* cultures established on SEM in Experiment 1. One scintillation vial each of the liquid and solidified indexing medium should be labeled "Control." The other corresponding pairs of indexing media should be labeled "Syngonium #1–5."
2	Starting with culture #1, remove the shoot (see Figure 13.2c) from the culture vessel and place in a sterile petri dish. Trim roots from the shoot and make several thin cross sections of the stem base using a sterile scalpel. Gently crush the tissue cross sections with the forceps then inoculate the liquid STM in the scintillation vials with one or two tissue sections.
3	Using forceps, inoculate the solidified STM slant by gently stabbing the tissue partially into the medium at the base of the slant. Lift the tissue from the medium, drag it upward along the medium surface, and then position it at the top of the slant. Trim off leaves and transfer the remaining shoot onto freshly prepared SEM and label the tube "Syngonium #1."
4	Repeat the indexing procedure with the other Stage I *Syngonium* cultures. Stage I subcultures are maintained under the same culture conditions as described in Experiment 13.1.
5	In the event that all shoot cultures index negative for contamination, several visibly contaminated cultures should also be indexed. This can be accomplished, without removing the contaminated tissue from the tube, by dipping the end of the forceps into the contaminated medium and then inoculating the indexing media as described above. Cultures containing sporulating fungal contaminants should not be used. Vials or tubes containing inoculated indexing media are cultured at 25°C.
6	After 7 days, examine the indexing media for signs of microbial growth. Stage I shoot cultures that index positive for contamination should be discarded. Transfer the axenic Stage I cultures onto SMM and maintain cultures under the same culture conditions as described above.
7	After 4 weeks, Stage II shoot clusters are divided into single microcuttings and rooted directly in soilless growing medium contained in plug trays or any other small container placed under intermittent mist (5 s every 10 min) for 3 weeks under lightly shaded conditions. Alternatively, if a mist system is not available, plastic propagation domes can be placed over the plug trays to maintain humidity. After 10 days, the domes are gradually removed over a period of 7 days to acclimatize the plantlets to lower humidity.

- Provide a possible explanation for the differences in growth response of primary explants cultured first on SEM and then SMM verses those inoculated directly onto SMM.
- What are the two primary environmental factors that must be regulated during acclimatization?

REFERENCES

Chen, J., D.B. McConnell, R.J. Henny, and D.J. Norman. 2005. The foliage plant industry. *Hortic. Rev.* 31:47–112.

Chen, J., R.J. Henny, P.S. Devanand, and C.T. Chao. 2006. AFLP analysis of nephtytis (*Syngonium podophyllum* Schott) selected from somaclonal variants. *Plant Cell Rep*. 24:743–749.

George, E.F. 1993. *Plant Propagation by Tissue Culture. Part 1. The Technology*. Exegetics, Ltd., Edington.

Henny, R.J., J.F. Knauss, and A. Donnan, 1981. Foliage plant tissue culture. In *Foliage Plant Production*, J.N. Joiner (Ed.), pp. 137–178, Prentice-Hall, Englewood Cliffs, NJ.

Knauss, J. F. 1976. A tissue culture method for producing *Dieffenbachia picta* cv. Perfection free of fungi and bacteria. *Proc. Fla. State Hort. Soc.* 89:293–296.

Miller, L.R. and T. Murashige. 1976. Tissue culture propagation of tropical foliage plants. *In Vitro* 12:797–813.

Murashige, T. and F. Skoog. 1962. A revised medium for rapid growth and bioassays with tobacco cultures. *Physiol. Plant.* 15:473–497.

Styer, D.J. and C.K. Chin. 1984. Meristem and shoot-tip culture for propagation, pathogen elimination, and germplasm preservation. *Hort. Rev.* 5:221–277.

14 Micropropagation and In Vitro Flowering of Rose

Michael Kane, Timothy Johnson, and Philip Kauth

Roses (*Rosa* spp.) were among the first ornamental plants to be domesticated. In recent years there has been a significant increase in the demand for rose plants. In the United States alone, more than 1.2 billion blooms are sold annually. Roses are propagated for the following three distinct production sectors: greenhouse cut flower production, pot-grown flowering houseplant sales, and the home gardening market. Many rose cultivars are propagated asexually by budding or grafting of the desired scion cultivar on selected rootstock. Miniature (dwarf) roses are propagated on their own root systems. There have been numerous attempts to develop micropropagation protocols to produce roses that are true to type and superior to traditionally propagated rose plants in terms of price, quality, and growth performance (Dubois et al., 1988; Campos and Pais, 1990; Rout et al., 1999; Pati et al., 2006; Senapati and Rout, 2008).

Although commercial rose micropropagation has not been very successful in the United States, the technology has clearly facilitated early introduction of new cultivars to the consumer. In vitro techniques have also been used to introduce new genotypes by mutation breeding. One interesting response of in vitro cultured miniature rose is the production of flowers (Campos and Pais, 1990; Kane et al., 1991, 1992; Wang et al., 2002). In vitro flowering provides a unique system to also study factors controlling flower production and longevity (Kane et al., 1991; Van Staden and Dickens, 1992; Wang et al., 2002).

The purpose of this laboratory exercise is to illustrate the sequential steps required for the micropropagation of miniature rose by shoot culture (see Chapter 12). After completion of this laboratory exercise students will be able to do the following: (1) establish Stage I cultures using surface sterilized miniature rose nodal explants; (2) rapidly propagate plants by axillary shoot proliferation (Stage II); and (3) successfully root and acclimatize microcuttings ex vitro (Stage IV). This exercise also offers the opportunity for students to observe in vitro flowering. The laboratory exercise is outlined below including maintenance of stock plants, media preparation, procedures, and anticipated results.

GENERAL CONSIDERATIONS

Selection and Maintenance of Miniature Rose Cultivars

Many miniature rose cultivars are available and can be purchased as potted plants at most retail garden shops. Cultivars do vary with respect to initial culture establishment and capacity for in vitro flowering. Of 36 miniature rose cultivars screened, eight (22%) were unresponsive in vitro (chlorosis, leaf abscission, or severe hyperhydricity [Kane et al., 1992]). Of the remaining 28 responsive cultivars subcultured, 14 (50%) produced flower buds in vitro. Consequently, it is wise to screen several miniature rose cultivars as responses vary. Cultivars such as *Rosa* cv. *Royal Ruby*, *R.* cv. *Red Minimo*, *R.* cv. *Rosmarin*, *R.* cv. *Nancy Hall*, or *R.* cv. *Rosa Rouletti* are easy to establish in vitro and flower following subculture.

Regardless of the miniature rose cultivar chosen, purchase plants exhibiting new shoot growth if they are to be used immediately. Nodal explants taken from newly formed shoots can be more readily surface disinfested and are more responsive. If not used immediately, plants should be maintained in full sunlight (greenhouse in winter or outdoors in summer) and fertilized weekly (150 mg/L N). Three weeks before using donor plants for the exercise, plants should be trimmed to promote lateral shoot production. One plant in a 10-cm diameter pot should yield sufficient numbers of nodal explants for each student or team of students.

Materials

The following are needed for each student or student team:

- Potted miniature rose plants to provide 15 2-node explants per student
- 50% (v/v) ethanol
- 20% (v/v) commercial bleach (containing 1.2% sodium hypochlorite final concentration)
- Ten culture tubes each containing 12 mL Rose
- Establishment/Multiplication Medium (REM)
- Five GA-7 vessels containing 60 mL REM
- Incubator or a fluorescent light on a bench top providing approximately 30 $\mu mol \cdot m^{-2} \cdot s^{-1}$ for a 16-h photoperiod at 25°C

Composition and Preparation of the Medium

One advantage of this exercise is that the same culture medium can be used for both culture establishment (Stage I) and subsequent axillary shoot multiplication (Stage II) of miniature rose. REM consists of macro- and micronutrients and vitamins as described by Murashige and Skoog (1962; see Chapter 2 for composition) supplemented with 87.6 mM (30 g/L) sucrose, 2.5 μM (0.5 mg/L) BA, 0.6 μM (0.1 mg/L) IAA, 0.24 mM (50 mg/L) citric acid, and 0.28 mM (50 mg/L) ascorbic acid and solidified with 1.5 g/L Phytagel and 4.0 g/L agar. The Murashige and Skoog mineral salts (MS) and organics basal medium can be purchased prepackaged from PhytoTechnology Laboratories, Shawnee Mission, Kansas (Product ID#: M519). Medium is adjusted to pH 5.5 before addition of the gelling agents, heating, and dispensing as 12 mL volumes into 150 × 25 mm culture tubes and autoclaving. Ten culture tubes are prepared for each student or team of students.

Exercises

Exercise 1. Stage I Establishment and Stage II Shoot Multiplication from Nodal Explants

The goals of this two-part exercise is to familiarize students with procedures used to successfully establish Stage I miniature rose shoot cultures from 2-node explants (Figure 14.1a) and propagate shoots through production of axillary shoots (Stage II). The goal of Stage I is to establish axenic and responsive shoot cultures. Cultures are indexed for specific pathogens or cultivable microbial contamination and allowed to become physiologically adapted to in vitro culture conditions before being clonally multiplied by repeated proliferation of axillary shoots in Stage II (see Figure 12.2). Before proceeding, the structure of a generalized shoot apex should be briefly reviewed (Figure 12.1). Follow the instructions in Procedure 14.1 to complete this exercise.

Anticipated Results

Many miniature rose cultivars display vigorous growth when established in vitro, though the actual response is highly cultivar dependent. Axillary shoots should develop from the cultured primary nodal explants after two weeks of culture. By week four, 2–3 axillary shoots are produced from each nodal explant. In vitro flowering seldom occurs on shoots produced on primary nodal explants during the first culture period. This suggests that competency for flowering is acquired only following subculture.

FIGURE 14.1 Micropropagation and in vitro flowering of miniature rose. (a) Left: 2-node explants (left) are cut from young shoots of potted plants, and the leaf blades are removed prior to surface disinfestation. Right: petiole bases are removed after surface disinfestation to expose lateral buds before inoculation. Scale = 5 mm. (b) Production of axillary shoots and in vitro flowering of miniature rose following a 8-week subculture. (c) In-vitro-produced flowers are smaller and have fewer petals than those produced on plants grown in the greenhouse. Scale bar = 10 mm. (From Trigiano, R. N. and D. J. Gray, 2000. *Plant Tissue Culture Concepts and Laboratory Exercises, Second Edition*, CRC Press, LLC, Boca Raton, FL.)

Procedure 14.1
Stage I Establishment and Stage II Shoot Multiplication from Nodal Explants of Rose

Step	Instructions and Comments
1	Cut several newly produced shoots from the donor plant. Remove leaf blades with a scalpel leaving only a small basal section (ca. 3 mm) of each petiole. Cut shoots into 2-node explants about 15 mm in length (Figure 14.1a). It can be noted that the smoothly cutinized epidermis, typical of many miniature rose cultivars, should facilitate effective surface disinfestation of the explants. Nodes exhibiting bud break should not be used.
2	Rinse nodal segments in tap water for 15 min and then surface disinfest in 50% ethanol for 1 min followed by a 12-min agitated soak in 20% commercial bleach and three 5-min rinses in sterile deionized water.
3	In the transfer hood, trim off the bleached ends of each 2-node explant and the base of each petiole to expose the lateral bud (Figure 13.1a).
4	Transfer a single 2-node explant into each of 10 culture tubes containing REM. Partially embed the basal end vertically into the medium.
5	Label each tube and maintain in an incubator or under a fluorescent light on a bench top at 25°C and provided with approximately 30 μmol m^{-2} s^{-1} for 16 h. To prevent abnormal shoot development and premature leaf abscission, sealing films such as Parafilm should not be used.
6	After 4 weeks, students should determine (1) the percentage of visibly contaminated cultures, (2) mean number of axillary shoots produced per explants, and (3) percentage of flowering cultures. Data can be collected without removing the plants from the culture vessels.
7	In the transfer hood, Stage I shoot clusters are aseptically divided into defoliated 2-node secondary explants with five explants inoculated into each of five GA-7 vessels containing 60 mL freshly prepared REM. Cultures are maintained under the conditions described above and observed weekly.
8	After 7 to 8 weeks, students should determine (1) mean shoot number and length produced per explant and (2) percentage of explants producing flowers. The data collected by each student or student team at 4 and 7–8 weeks should be shared to provide more representative responses.

Nodal explants subcultured into GA-7 vessels should exhibit vigorous shoot development (Figure 14.1b). Shoots produced in larger vessels, such as Magenta GA-7 vessels, will exhibit a greater frequency of flowering. Floral buds are produced by week six and the flowers open by week eight (Figure 14.1b). Approximately 30% of the shoots produce open and very fragrant flowers. Flowers produced in vitro are much smaller and are composed of fewer petals than those produced

Procedure 14.2
Stage IV Rooting and Acclimatization

Step	Instructions and Comments
1	Rinse agar from Stage II shoot clusters produced in GA-7 vessels in exercise 1 and divide into single microcuttings (>8.0 mm long).
2	Place microcuttings directly into moistened soilless growing medium contained in plug trays or any other small container placed under intermittent mist (5 s every 10 min) for 3 weeks under lightly shaded conditions. Alternatively, if a mist system is not available, clear plastic propagation domes can be placed over the plug trays to maintain humidity. After 2 weeks, the domes are gradually removed over a period of 1 week to acclimatize the plantlets to lower humidity.
3	Evaluate microcutting survival and rooting 3 weeks posttransplant. Plants can be transferred to the greenhouse or laboratory.

on plants grown in the greenhouse (Figure 14.1c). Flowers become senescent within two weeks of opening. Recurrent flowering does not occur unless shoots are subcultured. Infrequently, vegetative shoots develop from the center of open flowers in vitro. If time constraints exist or demonstration of in vitro rose flowering is the only objective, previously established stock shoot cultures can be used as the source of explants.

Exercise 2. Stage IV Rooting and Acclimatization

Depending on the species, 35%–75% of the total cost to micropropagate a crop can be reduced by eliminating Stage III rooting (Figure 12.2). This reflects the significant input of labor and supplies required to complete this stage. The goal of this exercise is to demonstrate direct ex vitro rooting and acclimatization (Stage IV) using the Stage II miniature rose microcuttings produced in exercise 1. Follow the experimental protocol outlined in Procedure 14.2 to complete this experiment.

Anticipated Results

Miniature rose microcuttings should exhibit 100% ex vitro rooting and acclimatization by week three. Acclimatized plants should begin flowering again within four weeks posttransplant. Typically acclimatized micropropagated miniature roses are fuller and exhibit growth rates about twice that of plants produced by cutting propagation (Dubois et al., 1988).

Questions

- In terms of disease elimination, what is the disadvantage of using nodal explants?
- Why do shoots produced following subculture exhibit a greater capacity for flowering than those produced on primary explants?
- Which cultural factors could reduce the size of the flowers produced in vitro?
- Premature deterioration and/of wilting of cut rose flowers has been attributed to both bacterial growth in cut flower water and/or products exuded by the stem. How could in vitro flowering of rose be used to study this problem?

REFERENCES

Campos, P.S. and M.S.S. Pais. 1990. Mass propagation of the dwarf rose cultivar "Rosamini." *Scientia Hortic.* 43:321–330.

Dubois, L.A.M, J. Roggemans, G. Soyeurt, and D.P. De Vries. 1988. Comparison of the growth and development of dwarf rose cultivars propagated in vitro and in vivo by softwood cuttings. *Scientia Hortic.* 35:293–299.

Kane, M.E., F. Marousky, and N. Philman. 1991. Producing microorganism-free rose flowers: The tissue culture approach. *Amer. Rose Mag.* 31:16–17.

Kane, M.E., N.L. Philman, and F.J. Marousky. 1992. In vitro flowering of miniature rose cultivars. *HortScience* 27:206.

Murashige, T. and F. Skoog. 1962. A revised medium for rapid growth and bioassays with tobacco cultures. *Physiol. Plant.* 15:473–497.

Pati, P.K., S.P. Rath, M. Sharma, A. Sood, and P.S. Ahuja. 2006. In vitro propagation of rose—a review. *Biotechnol. Adv.* 24:94–114.

Rout, G.R., S. Samantaray, J. Mottley, and P. Das. 1999, Biotechnology of the rose: A review of recent progress. *Scientia Hortic.* 81:201–228.

Senapati, S.K. and G.R. Rout. 2008. Study of culture conditions for improved micropropagation of hybrid rose. *Hort. Sci.* 35:27–34.

Van Staden, J. and C.W.S. Dickens. 1992. In vitro induction of flowering and its relevance to micropropagation. In *Biotechnology in Agriculture and Forestry, Vol. 17. High Tech and Micropropagation 1.* Y.P.S. Bajaj, Ed., pp. 85–115. Springer-Verlag, Berlin.

Wang, G.Y., M.F. Yuan, and Y. Hong. 2002. In vitro flower induction in roses. *In Vitro Cell Dev. Biol.—Plant* 38:513–518.

15 Micropropagation of Potato by Node Culture and Microtuber Production

Michael Kane

The potato (*Solanum tuberosum* L.), a crop of worldwide importance, is an integral part of the diet of a large proportion of the population of the world. Potato production ranks fourth among major food crops and, unlike other major food crops, potatoes are vegetatively propagated and, unfortunately, susceptible to infection by many viral and bacterial pathogens. At least 23 viruses that decrease tuber quality and yield are known to infect potato. Reliance on vegetative propagation from field produced seed tubers results in multiplication and spread of infected tubers. Routine production of pathogen-eradicated seed tubers is necessary to maintain adequate yield in the field (Miller and Lipschultz, 1984). This is accomplished using meristem tip culture to produce specific pathogen-free tested (SPT) potato tuber seed stocks (Cassells and Long, 1982). Micropropagation techniques are subsequently used to clonally propagate these pathogen-eradicated stock plants (Hussey and Stacey, 1981; Dodds et al., 1992). For discussions concerning culture indexing for plant pathogens and microbial contaminants, see Chapters 17 and 18.

Two very simple, but reliable, micropropagation techniques can be demonstrated using potato. The first technique, called *single node* or *multiple node culture*, involves production of shoots from cultured single or multiple nodes (*in vitro layering*) positioned horizontally on the medium. In potato, the initial sources of these nodes are SPT shoots established in vitro from meristem tips (Figure 12.1). Interestingly, potato cannot be multiplied by using the shoot tip culture method (see Chapter 12) because axillary shoot proliferation is not usually promoted by the addition of cytokinin to the medium. Typically, single elongated unbranched shoots comprising multiple nodes are rapidly produced. These shoots (microcuttings) are either rooted and acclimatized to field conditions or are subdivided into single node cuttings to initiate additional cultures (Figure 15.1). Although not the most efficient method, node culture is the most reliable micropropagation procedure to produce plants that are true to type.

The second and perhaps more fascinating micropropagation technique involves production of miniaturized tubers (*microtubers*) on potato shoots cultured in vitro (Figure 15.1). Microtuber production is a very useful method to propagate and store valuable potato stocks and is adaptable to automated commercial propagation and large-scale mechanized field planting (McCown and Joyce, 1991; Donnelly et al., 2003). Depending on the cultivar, potato microtuber development may be promoted under short-day photoperiods and following addition of cytokinin and high levels of sucrose to the medium (Seabrook et al., 1993; Gopal et al., 1998). Under ideal planting and early growth conditions, field-planted microtubers perform as well as regular seed tubers (McCown and Wattimena, 1987).

The primary objective of this laboratory exercise is to illustrate micropropagation of potato by node culture and microtuber production. This exercise will familiarize students with the following concepts: (1) efficient shoot production is possible from propagules with preexisting shoot meristems in the absence of cytokinin; (2) potato shoot meristems that normally produce elongated

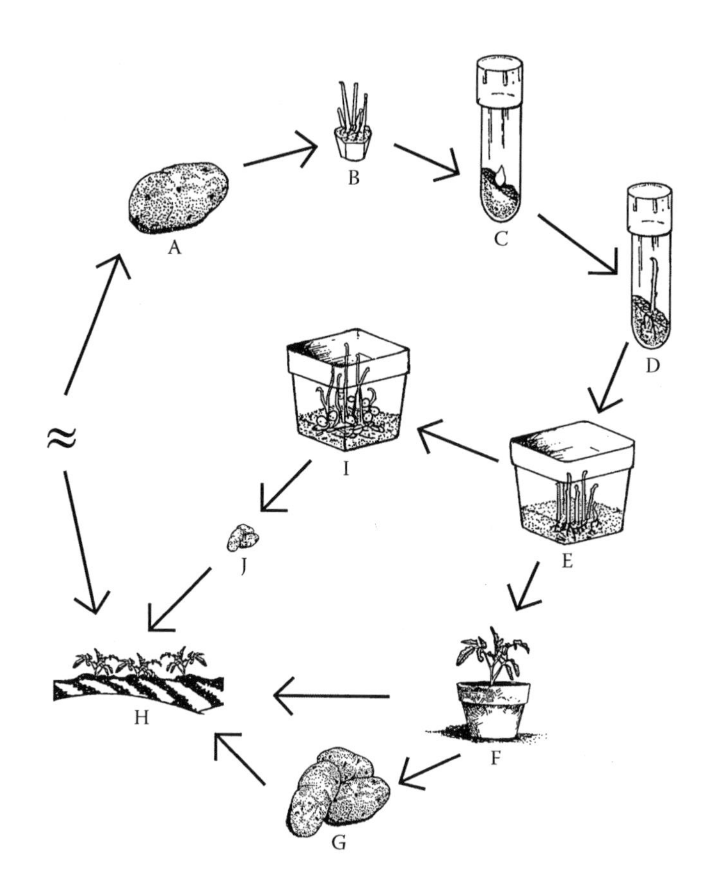

FIGURE 15.1 Diagrammatic representation of potato micropropagation by node culture and microtuber production. A. Tuber obtained from a multigenerational pathogen-infected field potato population. B. Meristem tip explants excised from sprouted tuber sections. C. Establishment of meristem tip in culture. D. Culture is SPT and indexed for other contaminants. E. Shoots determined to be specific pathogen free are propagated by node culture to increase numbers. F. Microcuttings are rooted and acclimatized ex vitro. G. Tubers produced ex vitro from STP plants are used as planting stock. H. Field planting of STP seed tuber. I. Shoots may be transferred to a microtuber induction medium and promotive conditions. J. Microtubers produced are stored or used as STP seed tubers for field planting. (From Trigiano, R.N. and Gray, D.J., Eds., *Plant Tissue Culture Concepts and Laboratory Exercises,* Second Edition, CRC Press LLC, Boca Raton, FL, 2000.)

shoots *in vitro* can be induced to form miniature tubers following alteration of environmental and cultural conditions; and (3) microtuberization is a valuable method for potato propagation, storage, and distribution. The laboratory exercise is detailed below, including establishment and maintenance of stock shoot cultures, media preparation, procedures, and anticipated results.

GENERAL CONSIDERATIONS

Establishment and Maintenance of Stock Potato Shoot Cultures

Students will use potato nodal cuttings and intact shoots from shoot cultures established from surface sterilized shoot tips excised from sprouted tuber sections. This eliminates the need for students to perform an often-unsuccessful surface sterilization step and decreases culture response time. Most potato cultivars (genotypes) can be induced to form tubers in vitro; however, the capacity for microtuberization is influenced by genotype. Potato cultivars such as “Kennebec,” “Superior,” “Red Pontiac,” or “Russet Burbank” have proven to be responsive (Leclerc et al., 1994) and are available in most produce stores. Only use fresh potato tubers that are free of decay.

Wash potato tubers under flowing tap water and then soak for 10 minutes in 20% (v/v) commercial bleach (1.2% NaOCl). Cut tubers into 5–10 pieces, each about 2.0 cm^2, and then soak for 1 h in 0.29 mM (100 mg/L) gibberellic acid (GA_3) to break bud dormancy and promote sprouting. Place treated tuber sections on moist vermiculite in an incubator maintained at approximately 25°C under a 16 h photoperiod provided by cool white fluorescent lamps at about 30 $\mu mol \times m^{-2} \times s^{-1}$. Excise sprouted shoots when they are approximately 3–5 cm in length. Large shoot tip explants 5–10 mm long should be used to initiate stock cultures. Rinse excised shoot tips in tap water for 10 min and then surface disinfest in 20% commercial bleach (1.2% sodium hypocholorite) with constant agitation for 10 min followed by three 5-min rinses in sterile distilled water. Trim away bleached tissues and inoculate shoot tip explants into separate 150 × 25 mm glass culture tubes containing 15 mL Murashige and Skoog (1962) basal salts (see Chapter 2 for composition) supplemented with 87.6 mM (30 g/L) sucrose, 0.56 mM (100 mg/L) myo-inositol, and 1.2 μM (0.4 mg/L) thiamine-HCL, solidified with 8 g/L agar. This medium can be purchased prepackaged without the sucrose and agar from PhytoTechnology Laboratories, Shawnee Mission, Kansas (Product ID#: M519). Cultures are maintained under the previously described culture conditions. Shoots should be about 4–6 cm long after 30–40 days culture. Subculture nodal cuttings from sterile cultures onto the same medium at 4–6 week intervals to generate additional shoot cultures.

MATERIALS

The following items are needed for each team of students:

- Twenty-four 5-week-old stock potato shoot cultures (in culture tubes)
- Long (29-cm) forceps
- Four GA-7 vessels each containing 60 mL Node Culture Medium (NCM)
- Eight GA-7 vessels each containing 60 mL Microtuber Induction Medium (MIM)
- Plug trays with soilless growing medium and propagation dome or mist bench
- Incubator or a fluorescent light on a bench top providing a 16 h photoperiod at approximately 30 $\mu mol \times m^{-2} \times s^{-1}$ and 20–22°C
- Incubator set to maintain 20–22°C and provide approximately 30 $\mu mol \times m^{-2} \times s^{-1}$ during an 8 h photoperiod (used for microtuberization experiment)

COMPOSITION AND PREPARATION OF THE EXERCISE MEDIA

Two culture media, NCM and MIM, are required for the laboratory exercise. Magenta GA-7 vessels (Magenta Corp., Chicago), each containing 60 mL medium, should be used if available (Figure 15.2). However, baby food jars containing 30 mL medium can be substituted. Each student team will require four GA-7 vessels containing NCM and eight GA-7 vessels containing MIM. The NCM is identical in composition to the stock shoot culture medium described earlier. The MIM contains the same components as the NCM except that the sucrose concentration is increased to 234 mM (80 g/L). Both media are adjusted to pH 5.7 before addition of agar and autoclaving

EXERCISES

EXPERIMENT 1: POTATO MICROPROPAGATION BY NODE CULTURE

The objective of this experiment is to demonstrate that node culture, the simplest micropropagation method, can be used to generate large numbers of plants in a short period of time. Each shoot produced in vitro consists of multiple nodes that potentially can be used as a propagule to produce another shoot. However, some potato cultivars may be less responsive or produce smaller shoots in vitro when single-node cuttings are used. During routine subculture of the stock shoot cultures it

FIGURE 15.2 Typical development of potato shoots from two-node cuttings cultured for 7 days in a Magenta GA-7 vessel containing NCM. Scale bar= 10 mm. Corresponds to part E of Figure 12.1. (From Trigiano, R.N. and Gray, D.J., Eds., *Plant Tissue Culture Concepts and Laboratory Exercises,* Second Edition. 2000. CRC Press LLC, Boca Raton, FL, 2000.)

Procedure 15.1
Micropropagation of Potato by Nodal Culture

Step	Instruction and Comments
1	Remove intact shoot from stock culture and place in a sterile petri dish. Add a small volume of sterile water to prevent tissue desiccation and subdivide into single-node cuttings using a sterile scalpel. Trim leaf blades from each nodal cutting, leaving only the petiole bases.
2	Orient four single-node cuttings horizontally onto NCM medium in each of the four GA-7 vessels. Place cultures in an incubator or on a laboratory bench top under fluorescent lighting providing approximately 30 μmol × m^{-2} × s^{-1} for a 16 h photoperiod and 20–22°C. The percent responsive nodal cuttings, average number of shoots, and total nodes produced per nodal cutting are determined after 3 weeks.
3	After data collection, the potato shoots produced may be treated as microcuttings and rooted directly in soilless growing medium contained in plug trays or any other small container and placed under intermittent mist (5 s every 10 min) for 2 weeks under lightly shaded conditions. Alternatively, if a mist system is not available, plastic humidity domes can be placed over the plug trays to maintain humidity. After 10–12 days, the domes are gradually removed over a period of 7 days to acclimatize the plantlets.

may be useful to compare the response of single-node and two-node cuttings. The typical response of two-node cuttings after only 7 days of culture is shown in Figure 15.2.

Follow the instructions in Procedure 15.1 to complete this experiment.

Anticipated Results

Shoot growth from the nodes should occur rapidly. Typically, an unbranched shoot approximately 5 cm long and consisting of three to five nodes, plus the shoot tip, will be produced from a nodal cutting after 3 weeks. Determine the average number of new nodes (plus the shoot tip) produced per nodal cutting after 3 weeks. Using this value, the students can then estimate the number of shoots generated from the original nodal cutting after one year. This is accomplished by calculating Y^x where Y = node number (including shoot tip) produced per nodal cutting after the 3 week culture cycle and X = 17.3, the number of culture cycles per year. This yearly production rate can be quite large and its calculation provides an opportunity to illustrate the power of this simple micropropagation technique. Potato microcuttings should quickly root and acclimatize to ex vitro (Stage IV) condition.

Questions

- Given a 3-week culture cycle and based upon your calculated average total nodes produced per nodal cutting, how many potato shoots can be produced from a single node cutting in a year?

- Based on the cellular level, why might node culture be considered the most reliable method to in vitro propagate plants that are true to type?

Experiment 2. Microtuberization of Potato

Follow the instructions in Procedure 15.2 to complete this experiment.

Anticipated Results

Potato tubers normally form underground but are produced in vitro from axillary meristems along the shoot (Figure 15.3). Usually only one microtuber fully develops on each shoot. During the tuberization process, there is apparently strong competition between axillary meristems along the same shoot such that the first meristem to tuberize inhibits tuberization of the other meristems (McCown and Joyce, 1991). This has been a factor limiting efficient microtuber production and makes it difficult to design experiments examining factors affecting microtuber production using multinode explants. However, it is important to point out to the students that each microtuber, being a modified stem, actually contains numerous shoot meristems, the so-called eyes of the potato. One alternative to increase efficiency is to induce microtuberization on single-node explants. Depending on the cultivar and inductive treatment, microtubers exhibit varying

Procedure 15.2
Microtuberization of Potato

Step	Instructions and Comments
1	Transfer four 4.0-cm-long microcuttings, with shoot tips intact, into each of the eight GA-7 vessels containing MIM. Position the microcuttings vertically by embedding the basal cut ends about 5 mm into the medium.
2	Place four inoculated culture vessels in an incubator set to provide 30 μmol × m^{-2} × s^{-1} fluorescent lighting for an 8 h photoperiod at 20–22°C and the remaining four culture vessels in an incubator set to maintain a 16-h photoperiod at the same light level and temperature. Usually microtuberization is promoted at temperatures ranging from 15–20°C. However, higher culture temperatures can be used if lower temperature control is not possible.
3	Depending on culture response, the numbers of microtubers produced per shoot and mean microtuber fresh weight can be determined after 6–10 weeks for both treatments. Class data can be compiled to provide greater replication.

FIGURE 15.3 Typical microtuber formation from cultured shoots 10 weeks after culture initiation. Note the presence of only one primary microtuber on each shoot. Scale bar = 10 mm. Corresponds to part I of Figure 12.1.

duration of dormancy up to 7 months that can be broken by storing at lower temperature (4–8°C). Depending on the cultivar used, microtubers produced under an 8-h photoperiod usually weigh more than those produced under a 16-h photoperiod. One variation to this experiment would be to compare the influence of sucrose concentration (0, 60, 120, 180, and 240 mM) on size and fresh weight of microtubers produced under 16 h and 8 h photoperiods or complete darkness.

Questions

- What are several advantages of producing potato in vitro via formation of microtubes?
- Provide a possible explanation why so few microtubers develop on each shoot microcutting.

REFERENCES

Cassells, A.C. and R.D. Long. 1982. The elimination of potato viruses X, Y, S and M in meristem and explant cultures of potato in the presence of Virazole. *Potato Res.* 25:165–173.

Dodds, J.H., D. Silva-Rodriguez, and P. Tovar. 1992. Micropropagation of Potato (*Solanum tuberosum* L.). In *Biotechnology in Agriculture and Forestry, Vol. 19. High Tech and Micropropagation III,* Y.P.S. Bajaj, Ed., pp. 91–106. Springer-Verlag, New York.

Donnelly, D.J., W.K. Coleman, and S.E. Coleman. 2003. Potato microtuber production and performance: A review. *Amer. J. Potato Res.* 80:103–115.

Gopal, J., J.L. Minocha, and H.S. Dhaliwal. 1998. Microtuberization in potato (*Solanum tuberosum* L.). *Plant Cell Rep.* 17:794–798.

Hussey, G. and N.J. Stacey. 1981. *In vitro* propagation of potato (*Solanum tuberosum* L.). *Ann. Bot.* 48:787–796.

Leclerc, Y., D.J. Donnelly, and J.E.A. Seabrook. 1994. Microtuberization of layered shoots and nodal cuttings of potato: The influence of growth regulators and incubation periods. *Plant Cell Tiss. Organ Cult.* 37:113–120.

McCown, B.H. and G.A. Wattimena. 1987. Field performance of micropropagated potato plants. In *Biotechnology in Agriculture and Forestry. Vol. 3*, Y.P.S. Bajaj, Ed., pp. 80–88. Springer-Verlag, New York.

McCown, B.H. and P.J. Joyce. 1991. Automated propagation of microtubers of potato. In *Scale-Up and Automation in Plant Propagation*, I.K. Vasil, Ed., pp. 95–109. Academic Press, San Diego, CA.

Miller, S.A. and L. Lipschultz. 1984. Potato. In *Handbook of Plant Cell Culture. Vol. 3. Crop Species*, P.V. Ammirato, D.A. Evans, W.R. Sharp, and Y. Yamada, Eds., pp. 291–326. Macmillan, New York.

Murashige, T. and F. Skoog. 1962. A revised medium for rapid growth and bioassays with tobacco cultures. *Physiol. Plant.* 15:473–497.

Seabrook, J.E.A., S. Coleman, and D. Levy. 1993. Effect of photoperiod on in vitro tuberization of potato (*Solanum tuberosum* L.). *Plant Cell Tiss. Organ Cult.* 34:43–51.

16 Commercial Laboratory Production

G. R. L. Suttle

CONCEPTS

- It is not enough to produce high-quality plant material; one must also produce it in a timely fashion, giving customers what they want, when they want it, at a price they can afford.
- Successful micropropagation on a commercial scale requires an understanding of two equally complex and dynamic factors—the plants and the marketplace.
- The primary method of increase in woody plant micropropagation is axillary shoot proliferation, although adventitious shoot proliferation may occur with some kinds of plants, such as species of the Ericaceae.
- The most obvious use for micropropagation is to get a jump-start on growing the newest and hottest items quickly. Micropropagation can cut 3 to 10 years off the time it takes to bulk-up new selections and get them to market.

Over the past 20 years, the horticultural nursery trade has found ever increasing ways to utilize micropropagation as a practical and cost effective production tool. This chapter will describe some of the basic steps involved in plant micropropagation, discuss some of the practical aspects of the commercial micropropagation business, and give some examples of how and why micropropagation is being used by growers today.

Many "scientific" papers have been published describing methods, formuli, and techniques for the successful micropropagation of a multitude of plant species. In theory, once one has the right "recipe" all that is needed is a well-equipped kitchen, some trucks to haul the goodies to market, and a good bank to hold all the money when it comes rolling back in. In reality, successful micropropagation on a commercial scale requires an understanding of two equally complex and dynamic factors—the plants and the marketplace. It is not enough to produce high quality, healthy plant material (a major challenge in itself); one must also produce it in a timely fashion, giving customers what they want, when they want it, for a price they can afford.

THE FACILITY

Well-designed laboratories generally have a good work flow allowing for the actual movement of supplies, people, cultures, and the finished product in an easy and logical pattern. Many small laboratories, unable to build a facility from scratch, use existing residential housing or mobile homes and remodel them to accommodate their needs. In order to keep airborne contaminants to a minimum, the ideal facility has very few entry sites (doors) and often has large areas designated as "clean rooms" with purified (HEPA filtered) air flowing under positive pressure. The basic areas in any laboratory are listed below (Figure 16.1).

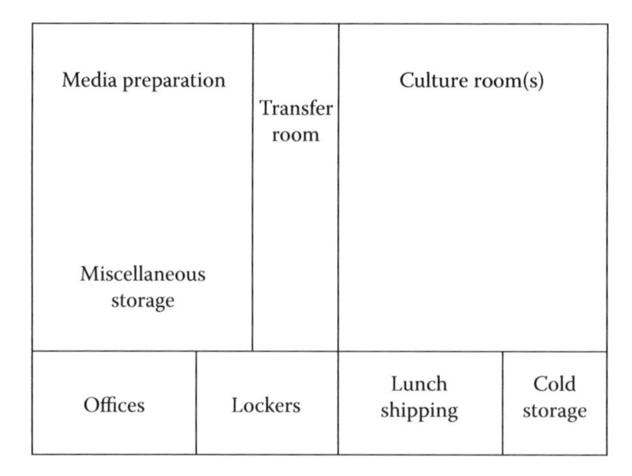

FIGURE 16.1 A design of a commercial tissue culture laboratory facility.

Media preparation/dishes. This area is set up similar to a kitchen with lots of counter space and shelves for chemicals and stock solutions, a pH meter, scales, autoclaves (usually more than one), media dispensing equipment, and a dishwashing area. Also important here is adequate storage space for empty culture vessels. Autoclaves create a tremendous heat load and usually are placed on an exterior wall with good clean ventilation.

Media storage. Once media is "cooked" (sterilized) and dispensed into culture vessels, it must be cooled down (again with good clean ventilation) and stored until needed. Often this area is combined either with the media preparation room or in a section of the transfer room.

Transfer room. This is the heart of any laboratory, filled with clean benches (laminar flow hoods) where all of the culture initiation and subsequent subculturing takes place. "Ripe" cultures (cultures with usuable shoots) are harvested and cultures replanted (transferred or reinitiated) to clean fresh media. Up to 80% of the labor force may be involved in this one room, so it must be a comfortable and pleasant working environment. Tools are sterilized at the hoods using glass bead sterilizers, bacticinerators, or alcohol and flame. Often transfer rooms tend to be cramped and the heat load generated by people, hoods, and sterilizers are compensated for with a good air conditioning system. People and plants suffer when it is hot and stuffy, and productivity declines.

Culture room(s). Usually the largest area of any commercial laboratory, where cultures are incubated on shelves under specific light and temperature regimes. Light quantity and quality can be key factors in culture success. Generally, regular or high output cool white fluorescent lighting is used, with higher intensities achieved by using more lights per square foot or by bringing cultures closer to the lights. Once again, heat buildup can be a major problem. Many laboratories remove the ballasts from the fluorescent light fixtures and mount them in an external area for easier cooling. Good air movement is critical in equalizing temperatures throughout the culture room. Temperature requirements are often crop specific, but generally ranging from 18 to 26°C. Different light or temperature regimes may require separate culture rooms for different crops. Usually air to the culture room is purified by HEPA filters.

Cold storage. Cold storage is an absolutely essential component of any commercial micropropagation facility growing woody plants. Cold storage (2–4°C) is used for the following purposes: (1) to maintain stock blocks in culture for plants that are produced on a seasonal basis, (2) to allow production to continue during the cold winter months by "banking" finished plant material ahead of time, and (3) to provide a chilling requirement for crops such as apple (*Malus*) and pear (*Pyrus*) species that respond in the greenhouse much more rapidly if given a "rest" or dormant period before out-planting.

Shipping/lunch area. Ripe product (either microcuttings or rooted plantlets) must be harvested, boxed, packed, and otherwise prepared for shipping. This activity requires sinks to rinse the plants and also should be near the cold storage unit for holding until shipping. This area may serve double duty as a lunch or break area.

Locker room. When first entering any commercial facility, employees usually must change shoes and may be required to wear additional clean clothing such as laboratory coats, hairnets, etc. into production areas. This locker area may also, as above, be incorporated with a lunch room and/or rest rooms.

Offices. Perhaps the least important place in a facility from an operations standpoint, but essential for running the minor details such as sales, production planning, accounting, and, of course, payroll, are the offices.

Miscellaneous storage. No laboratory ever has enough general storage space. Supplies, chemicals, shipping containers, extra equipment, spare parts, and items such as vessels used on a seasonal basis all require space. Logical locations for storage are near an external door (for receiving) or near the media preparation and shipping prep areas where supplies are most often used.

THE PROCESS

What to Grow

There are two different approaches in determining product focus. The first is to grow what you think customers might want (perhaps a novel plant) based on what you hope is good market research and analysis, and then working hard to convince potential customers that they really do want it. While this approach is the basis of our country's free enterprise system, it can be very risky if you happen to guess wrong. The most common reason for the failure of commercial laboratories over the past decade has been growing plants that the market either doesn't want or need. The second approach is to work with customers on more of a contractual basis, asking what their problems are, what they need, and when they need it, then going to work to try to solve the problem and fill the order. Arranging customers in advance is a much safer route to success considering the costs involved. Regardless of approach, every plant must be evaluated for suitability for micropropagation by asking the following questions:

1. *What are the minimum number of plants needed?* Most laboratories have a minimum requirement (5–10,000 or more), although smaller facilities may be willing to have lower minimums.
2. *How many customers are interested in a specific product?* The more the better, but sometimes exclusivity is important.
3. *How long is the product needed?* Is it just for mother block establishment, seasonal, or will it be needed on a continuous basis?
4. *How difficult is the plant to culture?* Is there existing literature or knowledge or will the technology need to be developed?
5. *Who "owns" the plant?* Are there patent or customer restrictions on marketing the plants?
6. *What is the estimated value of the plant once produced?* Can you charge enough to recover costs and make a profit?

All of these factors must be considered before starting any project. Often the price of a product cannot be defined until considerable research has been done on protocols for the given plant. Guarantees of success are impossible to make. The up-front cost of protocol development may or may not be shared with the customer. If exclusivity is needed, then the customer usually funds the work. Once a plant is deemed worthy for an attempt at micropropagation, the process of producing it may begin.

Stock Plant Preparation

Healthy well-labeled stock plants are essential. The most important job the customer must perform is to make absolutely sure that the stock wood sent to the laboratory is exactly the named material that will be wanted in large numbers when finished. This step sounds simple, but in reality, mistakes on labeling can and do happen far too frequently. This point cannot be overemphasized.

Ideally, 10–15 young, juvenile stock plants are placed under an intense fertilization and pesticide regime to provide rapid, healthy shoot growth. The plants are watered from the base to avoid getting moisture on the foliage, a breeding ground for contaminants. Sometimes there may be a single very old tree located thousands of miles away as the only source or a customer may have only a small amount of stock and does not want to risk sending it away. In these cases, wood must be collected periodically throughout the growing season and sent by overnight express.

Stage I: Culture Initiation

The process begins by carefully harvesting soft, active shoot growth from the stock plant. Usually 5–15 cm cuttings are snipped early in the morning while the plants are fully turgid. The cuttings are immediately placed in a cooler until disinfestation takes place. The leaves and most, if not all, of the petiole are removed, and each stem is cut into one to two nodal pieces.

The disinfestation or clean-up process usually involves several cold soapy water washes, then soaking the shoots in a 10–20% commercial bleach (sodium hypochlorite) solution along with a surfactant such as Tween 20 for 10–20 min followed by a series of sterile water rinses. Damaged tissue is trimmed away using aseptic techniques. The treated cuttings are then placed individually in test tubes filled with a nutrient gel (medium) composed of a "best guess" formula or range of formulae that may allow the buds to grow. The tubes are usually incubated in a culture room under 16 h light at 25°C.

Visual screening takes place every 3–5 days and the dead or dirty (visibly contaminated) cultures are discarded. Visibly clean, viable buds are transferred individually to fresh medium as new growth appears, usually 2–6 weeks after disinfestation. A portion of the stem piece is placed in a rich liquid or agar-based medium to detect bacterial contaminants that may be "invisible" or "unnoticable" to the naked eye. This contaminant screening process (also called indexing—see Chapters 17 and 18) allows us to discard any additional and unseen dirty material from the pool of viable shoots. This procedure may be repeated periodically; however and unfortunately, even this procedure does not catch everything and growing "clean" stock is a continuous struggle.

Culture initiation can be more of an art than a science. Often the concentration of bleach and/or time of disinfestation must be adjusted based on the woodiness, hairiness, and source of the stock tissue. Several variations may be tried if there is an abundance of source material, fine tuning the process based on previous efforts and the time of year. If materials are limited, the best guess is tried and hopefully will succeed.

Timing

The best success for getting plants into culture is typically realized by taking the first flushes of vegetative growth in the early spring. A very successful second option is to force dormant buds from sticks in the winter after the full chilling requirement has been satisfied. These sticks are placed in a "forcing solution" of 30 g/L Floralife Floral Preservative (Floralife, Inc., 120 Tower Drive, Burr Ridge, IL 60521), incubated at approximately 20°C and 10–12 h photoperiod (typical office conditions). Floral preservatives generally have sugar, an antimicrobial agent, and an acidifier to prolong bloom life. The dormant buds of many plant species will begin to sprout and grow under such conditions. Flower buds are removed and discarded when they appear to allow for better vegetative growth. When these soft shoots reach the length of 2–4 cm, they are easily removed from the stem and processed. They tend to be extremely clean so that surface disinfestation and establishment in culture are usually easier.

One theory explaining why such early growth seems to go into culture so well is that endogenous plant growth regulators (PGRs) and growth factors may have primed the plants to "spring" into rapid growth phase (inhibitory factors are at a low ebb). Fortunately, many times this new growth is very "clean"—it hasn't had time to acquire a load of contaminating organisms. Whatever the reason, early spring growth is without question usually the best starting material. By starting early in the growing season, it also gives more time to generate numbers and bring more cuttings into culture if needed.

Late summer or outdoor field grown stock posed particular problems from increased contamination and a lack of turgidity due to heat stress that causes plant tissues to be less tolerant of bleach disinfestation. Success rates for establishing material in Stage I can range from 0% to 95% due to all these factors and many more we do not yet comprehend. The ultimate goal of Stage I is to get the plant material clean and actively growing in culture. Once this is achieved, Stage II becomes the next challenge.

Stage II: Shoot Proliferation

The primary method of increase in woody plant micropropagation is by axillary shoot proliferation, although adventitious shoot proliferation may occur with some kinds of plants, such as species of the family Ericaceae. Axillary proliferation is preferred in order to avoid the potential of off-types (somaclonal variation—see Chapter 43) from occurring. Genetic mutation can occur by any vegetative propagation method, and if the characteristic is based on a relatively unstable chimera, such as some variegations are, micropropagation can enhance reversion. Micropropagators must continuously monitor the quality of plants they produce to ensure trueness. In our 15 years of experience of micropropagating trees, genetic mutation has not been even a minor problem. Useful strategies we employ to help reduce the potential of off-types are limiting the number of subcultures or length of time in culture, keeping PGRs levels in the culture medium as low as possible and always going back to the original stock source to start new lines.

Once a plant is clean and actively growing in culture, the goal of Stage II is to induce all (or at least some) of the axillary buds along the new, young, aseptically growing shoot to sprout and grow. In other words, we want the shoot to branch. Every 2–8 weeks (culture cycles vary depending upon the type of plant), we can subculture or harvest the new branches, "replant" or transfer them to fresh medium and wait another 2–8 weeks until each of these new branches sprout its own set of branches. There are rarely any roots involved during Stage II.

With woody plant micropropagation, there are two main methods of achieving multiplication of new axillary buds depending on the type of plant material involved. The first is by promoting many branches on a single stem piece resulting in a "clump" of new shoots. Species of apple and cherry (*Prunus*) would be two examples of plants best micropropagated by the clump method. During subculturing, the nice tall shoots are removed and either replanted in fresh multiplication medium to produce their own clumps or planted into a rooting medium or shipped out as microcuttings to be rooted in the greenhouse. The basal clump mass is left intact and replanted to yield another crop of branches. The clumps get thicker and thicker with new branches and can be used as a mini stock block. Eventually the clumps decline in quality and quantity, usually after 2–5 culture cycles, and must be discarded and replaced.

The second form of axillary shoot proliferation can be described as the single-node method. Plants exhibiting strong apical dominance such as lilac (*Syringa*) and maple (*Acer*) species are best micropropagated by single-node method. A single shoot grows up straight and tall, producing 2–10 sets of leaves. One or two-node sections are separated and replanted whereby each of these runs straight up again and is now ready to be chopped into nodal sections again. During any subculture, the nice terminals or tips can be harvested and planted into rooting medium or shipped out as microcuttings to be rooted in the greenhouse.

MEDIA FORMULI

The actual media formuli used during Stage I and II are often the same. There are 3 or 4 basic formuli that are adapted and modified as necessary, sometimes on a cultivar by cultivar basis. The proper ratios of the inorganic components in a formula actually are the main key to developing a successful protocol for a given plant. PGRs are also a key factor, but secondary to proper nutrition. It is not uncommon for a commercial laboratory to have over 50 modifications of basic media on file for multiplication alone and another 40 or so Stage III modifications. When a new plant is brought in, one takes an educated guess as to what might work based on past experience and literature searches, but usually modifications must be made by a systematic trial-and-error method of marching through the various components. In general, Stages I and II media usually have higher levels of salts, sugars, and cytokinins. Stage III media generally have lower levels of salts, sometimes lower sugars, and usually lack cytokinins, but include high levels of auxins.

STAGE III: ROOTING

When sufficient numbers of healthy shoots have been generated by one of the proliferation processes described above, healthy active terminal buds are selected either for in vitro rooting or for shipping directly to customers as microcuttings for rooting and acclimatization in the greenhouse. The decision whether or not to root in vitro is quite often made by the customer based on several factors. In vitro rooting adds about 30% to the final cost of the product and some crops, such as birch (*Betula*) and lilac species, root so well in the greenhouse as microcuttings that it simply does not make sense to root them in vitro. Other crops such as *Malus, Pyrus, Prunus,* and *Acer* are much easier to acclimatize as in vitro rooted plantlets. Perhaps the most important factor of all for the grower, other than survival, is that in vitro rooted plantlets require 2–5 weeks less time in the greenhouse to reach a size suitable for planting in the field. Many growers have limited greenhouse space and do not want to have to "baby-sit" tender microcuttings for such a long time. A grower may be able to process 2, 3, or 4 crops through his greenhouse in the same time it takes to raise one crop from microcuttings. Often, it pencils out as more economical to buy rooted plantlets.

STAGE IV: GREENHOUSE ACCLIMATIZATION

Fresh Stage II microcuttings (without roots) or Stage III in vitro rooted plantlets are both planted in any of several well-drained soilless potting mixes in the greenhouse under high humidity. Various types of fogs, tents, domes, and mist systems are used to wean the tender shoots into the real world by gradually reducing the humidity. The weaning process takes from 3 days to 6 weeks depending on what the plants are, the weather, and if they were rooted in vitro. Fertilization is not recommended until the plantlets become established and begin to grow. Some crops such as *Betula* can become stunted if overfertilized.

PRODUCTION SCHEDULING

When preparing an actual production schedule for any given plant, many factors play a role. How many are ordered and in what form (microcuttings or rooted plantlets)? When are they wanted? Can the order be prepared ahead of time (does it require or tolerate cold storage)? How does it multiply (nodal or clump)? What are the multiplication rates during build-up? What is the number of rootable or harvestable microcuttings (yield) and does it change over time? What rooting percentage can be expected? Can smaller batches be produced and stored or must it be produced in a lump sum? How much time is required between cycles (during build-up and during harvest cycles)? What are the estimated labor requirements for each cycle? Once these factors are determined, one works backwards from delivery date, calculating the number of cultures required at each of the various cycles along the way. Information on when each cycle must take place, what activities must be done at each step, and estimated labor required are all plotted on a production planning calendar. This planning process must be done for each cultivar on an individual basis since multiplication rates, yields, number ordered, and dates ordered will vary from plant to plant. Critical labor peaks must also be smoothed out as well as possible ahead of time. Simply put, not everything can come out at once.

One of the most frustrating factors in planning a production schedule is the very dynamic and very unpredictable nature of the plant material itself. Plants are living systems, and we simply don't control or even understand all the factors involved in in vitro propagation. Multiplication rates and harvestable yields, for example, can only be best guesstimated. Any plan has to be continuously updated and revised.

Seasonality

While a laboratory can pump out material on a year-round basis, anyone working on temperate deciduous crops knows that there is a season for everything. Plants can be fooled into thinking it is summertime in culture by continuously being subjected to long (16 h) days in the culture room, but as fall approaches, the rate of growth in the greenhouse slows considerably. To take maximum advantage of the growing season, customers want to get their product in the early spring through early summer. Use of cold storage to store crops ahead of time allows laboratory production of some plant material to take place in the winter, but much must come out fresh, making labor needs seasonal as well.

Reasons Why Micropropagation Is Used

While micropropagated material represents only a tiny percentage of the total number of plants produced in America today, there is no doubt that many of the largest and smallest nurseries in the United States view micropropagation as an essential tool that helps them maintain their competitive edge by growing better plants more efficiently. The remainder of this chapter will be devoted to a discussion of some of the advantages that in vitro micropropagation offers over conventional production systems.

New Introductions

Perhaps the most obvious use for micropropagation is to get a "jump start" on growing the newest and hottest items quickly. Micropropagation cuts 3–10 years off the time it takes to bulk-up new selections and get them to market. For example, micropropagation was used by one grower to establish layer beds of a new apple understock. He sold over one million rootstocks in the same amount of time his competition had bulked up to only a few thousand using conventional propagation methods. The market is always looking for something new and exciting. Nurseries on the forefront of introducing new plant material build their reputation as leaders in the industry, causing customers to come back year after year to find out what else is new.

In some cases, growers use micropropagated plant material to establish mother blocks and to fill in production shortages while mother blocks are too young to be productive. Conventional propagation methods such as hardwood or softwood cuttings, layer beds, or budding and grafting may take over long-term production needs once enough wood becomes available. In many other cases, micropropagation remains the method of choice for a variety of reasons.

Rapid Response to Market Demand

Large mother blocks or scion orchards are time consuming and expensive to establish and maintain. It is often difficult for growers to adjust quickly to the rise and fall in popularity of a given plant. With micropropagation, the stock block is maintained in a 10 × 10 foot cold storage unit. If a customer gets a call for an additional 10,000 liners of a particular blueberry, for example, they simply call up and ask when is the earliest they can take delivery on the additional microcuttings. They then add the time they need for the greenhouse growing and call their customer back with a delivery date.

Clean Plants

Micropropagation is inherently a cleaner system for producing plants compared to traditional production methods. Since the plants are grown in culture, diseases are not transmitted from the field into the greenhouse and on to subsequent generations. A single disease-free mother plant can theoretically produce unlimited disease-free daughter plants without the possibility of reinfection. Conversely, individual mother plants in a traditional virus-free cutting block must be tested again every year in order to maintain and ensure virus-free status (see Chapter 43). Testing fees add significantly to the expense of maintaining large mother blocks in the field.

Ease of Propagation

Bud incompatibility on budded or grafted stock, and poor rooting percentages with softwood or hardwood cuttings makes micropropagation the method of choice or the only option on many difficult to propagate plants such as *Syringa* and redbud (*Cercis*). Some red maple (*A. rubrum* L.) cultivars, such as "Karpick" and "Bowhall," are absolutely impossible to root from cuttings. Unreliable seedling availability and poor or unpredictable bud stands on *Betula*, *Tilia*, and *Morus* are problems growers are able to avoid by planting micropropagated material. Growing plants on their own roots offers major advantages for plants, such as contorted Filbert (*Corylus avellana* "contorta"), where suckering of understocks can be a major problem.

Sometimes micropropagated material provides the grower with a nucleus of "juvenile" material from which additional cuttings can be more easily rooted. Each year or two, the customer starts over with a fresh batch of starter material from the laboratory.

Speed

Research done at various universities indicate that it is possible to grow plants much more quickly to size than is traditionally seen in nurseries today, regardless of how the plants are propagated. Such rapid growth requires optimization of all growing conditions including fertilizer, light, and temperature. However, it is amazing the results that can be achieved with even modest adjustments of growing practices. Several field growers are now producing well-formed, small branched trees of *Prunus* "Kwanzan," *M. alba* "Chaparral" (mulberry), and others in 1 year instead of two. Blueberry plant production can be dramatically speeded up. Use of micropropagated cherry understock yields increased vigor and earlier fruit set (and earlier payback) for the orchardist.

Better Branching

Because the internode length is greatly reduced on micropropagated plants, there is generally more opportunity to develop a more full head on the growing plant. Indeed, this is also one of the reasons why survival is often greater. If something happens to destroy the terminal bud (for example, damage caused by freezing, hungry rabbits, or poor pruning), there are other buds below available to choose from. One grower accidentally sprayed a young block of *A. rubrum* liners twice with Surflan, an herbicide. The stems of the young micropropagated liners were girdled right at soil level. All was not lost, for the grower dug the soil out from around the base of the plants, removed the damaged tops, and the buds below the girdling all pushed out again. The grower lost some height on his crop (about 1–2 feet), but shorter plants are better than no plants at all. Having more buds to choose from also allows a grower to cut back closer to the ground, producing straighter trees.

The better branching is a real advantage when it comes to growing well-formed shrubs. *Hydrangea quercifolia* "Snow Queen" (PP4458) produced from rooted cuttings tends not to throw many branches at an early age, whereas micropropagated plants are easily developed into a bushy habit with routine pinching. The increased branching of blueberry plants is seen as a great advantage by some growers; others prefer the more traditional vase shape formed from rooted cuttings.

Greater Survival and Uniformity

While cultural practices play an important role in the ultimate performance of any block of plant material, the two key attributes most often given in describing micropropagated material are superior survival rates and much greater uniformity. Better survival is explained, in part, by heavier root systems and more buds to choose from on the top. The fact that the root-to-shoot ratio is more balanced from the very beginning and that the small plantlets are highly uniform right from the start helps to explain why subsequent growth is more reliable and consistent.

CONCLUSION

As the use of micropropagation becomes increasingly varied and widespread, growers will find new and ingenious ways to take advantage of its power. Even so, growers must weigh the pros and cons of this relatively new technique against the old methods and determine for themselves, on a case by case basis, whether it is worth the trouble to change. It must help them solve problems, find new markets, and save time or money. The successful micropropagator will be one who fills these needs and many more.

17 Detection and Elimination of Microbial Endophytes and Prevention of Contamination in Plant Tissue Culture

Alan C. Cassells

CONCEPTS

- Microbial endophytes are microorganisms that are introduced into culture in tissue explants.
- Endophytes may overrun the cultures or transmit disease in progeny plants.
- Endophytes may remain latent in the tissues and influence biochemical or transformation studies using the cultures.
- Latency (nongrowth) of endophytes may be influenced by the composition of the culture medium or by the environment.
- Some endophytes are nonculturable or fastidious, requiring special media.
- Endophytic plant pathogens can be detected and eliminated using established procedures.
- Nonpathogenic cultivable endophytes can be detected by culture indexing and DNA probes; elimination may be possible by incorporation of antibiotics into the culture medium.
- Laboratory contamination is caused by environmental and human-associated microorganisms and is controlled by good laboratory management practice.

Plant pathologists operate within a highly developed infrastructure to manage disease in the planting material, the growing crop, and in the harvested produce. There are professional societies and journals that publish descriptions of new diseases and of new methods of pathogen identification and disease control. National and international legislation exists to prevent pathogen spread and to ensure that methodology is standardized. The principle approach to disease management in crops is to plant "disease-free" material and to delay the entry of the pathogen into the crop by spraying with pesticides, etc. Fortunately, true seed is generally free of pathogens; however, for vegetatively propagated crops, there is a high risk of transmission of disease from the parent plant to the vegetative propagules, and so stock-plant health certification schemes have been established to provide healthy planting material for growers. The latter schemes are based on international guidelines for pathogen testing and pathogen elimination.

Both the nature and control of disease differ in tissue cultures compared with the field. Tissue culturists have the both the problems of (a) establishing pathogen-free aseptic cultures and (b) of maintaining aseptic cultures. Ideally cultures should be *axenic,* that is, free of all biological contamination, but this difficult to confirm and so the term *aseptic* (free of detectable contamination) is used. A major problem that plant tissue culturists face is that common cultivable environmental microorganisms (pathogenic and nonpathogenic) can grow rapidly on plant tissue culture media and overgrow the slower-growing plant cultures. The term *vitropaths* (vitropathogens) has been used to distinguish these culture contaminants from field pathogens (phytopathogens). Another distinction is that tissue culturists work with a very diverse range of mostly vegetatively propagated plants. These include major world crops but also minor horticultural crops and rare medicinal plants whose pathology is poorly documented and which may carry contaminating microorganisms from exotic locations. All in vitro cultures are susceptible to "disease" from microbial culture contaminants and to micro-arthropods infestation.

Chemical sterilization is used to eliminate surface contaminants in the preparation of explants for culture. Commonly, explant surface-sterilization treatments involve immersion in aqueous dilutions of ethanol and/or of hypochlorite with wetting agents. In intractable cases, alterative sterilants have been used (George, 1993; Hoffman et al., 1981). The primary source of culture contamination is the presence of microbial endophytes, that is, microorganisms that inhabit plant tissues (Bacon and White, 2000), in contrast to microbial epiphytes, which exist on the surfaces of plants (Andrews and Harris, 2000). For practical reasons, plant tissue culturists consider as "endophytes" contaminants that survive the procedures used to decontaminate (i.e., surface sterilize) the plant parts ("explants") used to initiate plant tissue cultures and that pass into the cultures in or on the explants.

Endophytes may:

- Be expressed, that is, grow readily on plant tissue culture medium, which is the response of many microbial contaminants;
- Be latent (unseen or unexpressed) in the plant tissue and transmitted subliminally on subculture, or be latent initially, but grow (be expressed) if the medium or environmental parameters are changed, causing culture losses or reducing quality in micropropagation (Table 17.1);
- Be fastidious, that is, only culturable on specialized media, or be nonculturable viruses and viroids, which may be transmitted on subculture (Table 17.2).

Some bacterial contaminants may be inhibited by the high osmotic potential or hormonal composition of Stage I culture medium, but may rapidly overrun the cultures when the hormonal composition

TABLE 17.1
Definitions of Culturability Used in the Text

Category	Remarks Regarding Culturability Murashige and Skoog (1962) Medium
Culturable	Microorganisms that grow on full strength M&S medium, that is, are visible ("expressed")
Latent	Microorganisms that are not expressed on M&S medium: (a) due to the high osmotic strength of medium or (b) are inhibited by components of the medium or (c) do not grow vigorously, or at all, at the pH of the medium or (d) do not grow at the temperature of the growthroom. Latent microorganisms may be expressed as the M&S medium components are diluted by tissue grow or on subculture to diluted M&S medium; or when the plant growth regulators in the medium are changed; or if the medium pH changes due to the influence of the plant tissue or metabolic activity of the contaminating organism. These microorganisms grow on common microbiological media.
Fastidious	The distinction between latent and fastidious microorganisms is that the latter do not grow on common microbiological media and they may have specific temperature requirements for optimal growth. Specialist media are available for mollicutes and phloem- and xylem-restricted bacteria and for other fastidious microorganisms. Detection is possible using bacterial DNA probes.
Nonculturable	Viruses and viroids are noncultivable. A majority of the bacteria in soils and many in biofilms/aggregates are described as viable but nonculturable (VBNC). VBNC bacteria can be detected using DNA probes.

Note: There is a distinction between fastidious organisms that have specialized media requirements and latent organisms that are suppressed by media and growth environmental factors.

TABLE 17.2
Categories of Tissue Culture Microbial Contaminants Referred to in the Text and Their General Growth Responses on the Commonly Used Murashige and Skoog (1962) Tissue Culture Medium with 6gl^{-1} Sucrose, Plant Growth Regulators pH 5.8 (George, 1993)

Category	Growth on Plant Tissue Culture Medium
Intracellular endophyte	
• Viruses	Nonculturable
• Viroids	Nonculturable
• Mollicutes	Fastidious/recalcitrant
• Xylem-restricted walled bacteria	Fastidious and possibly latent
• Phloem-restricted walled bacteria	Fastidious
• Symbionts/mutualists	Ectomycorrial fungi culturable; others nonculturable
Intercellular endophytes	
• Pathogenic fungi	Many culturable
• Nonpathogenic fungi	Mostly culturable
• Pathogenic bacteria	Mostly culturable
• Environmental bacteria	Culturable, possibly latent
Tissue culture contaminants	
• Environmental bacteria	Culturable
• Human-associated bacteria	Culturable
• Environmental yeasts	Culturable
• Human-associated yeasts	Culturable
• Mycelia fungi	Culturable

is altered or when the medium strength is reduced for root induction (Stage III; George, 1996). Loss of cultures in multiplication or release of pathogen-infected progeny plants has major implications in commercial micropropagation (Cassells, 1997). Contaminants may interfere with cultures used for experimental or industrial purposes such as secondary metabolite production where the contaminants may modify target compounds or contaminate products with microbial metabolites, which makes the product unfit for sale under health and safety legislation. Latent endophytic microbial contamination of plant tissues has long been a cause of concern to plant physiologists who have reported false experimental results due to the present of latent microbial contaminants (Holland and Polacco, 1994). Microbial contaminants also have the potential to interfere with genetic transformation experiments by, for example, transferring antibiotic resistance to *Agrobacterium fasciens* making it difficult to eliminate the latter from transformed tissues.

Endophytes are mainly bacteria, viruses, or viroids, although some fungal examples are known (Table 17.2). They may be characterized plant pathogens, symbiotic/mutualistic microorganisms, or environmental microorganisms including human pathogenic bacteria (Tyler and Triplett, 2008). Many plant pathogens tend to have fairly narrow host ranges, with some pathogen strains being restricted to specific host plant genotypes, which facilitates their detection/identification. There is extensive plant pathology literature and commercial diagnostic services are widely available (Table 17.3). In contrast, the plant may be a nonselective host for environmental microorganisms that can establish as facultative ("opportunistic") endophytes. Such environmental microorganisms may be nonpathogenic isolates of pathogenic species, epiphytes, and microorganisms with high local inoculum pressure, such as *Escherichia coli* in animal manure applied as a plant fertilizer. An ecological succession of plant epiphytic organisms may occur in which some are persistent

TABLE 17.3
Addresses of Two Search Engines, Which Provide Details of Consumables and Services for Plant Pathology, and for Environmental and Clinical Microbiology

Service	Website
Culture collections (some organizations may provide diagnostic services)	http://wdcm.nig.ac.jp/hpcc.html
Microbiological media, kits, and diagnostic service providers	http://www.rapidmicrobiology.com

and others transient in the season/crop development (Andrews and Harris, 2000). The epiphytic flora of cultivated plants may be disturbed by cultural practices and by the application of fungicides, fertilizers, and organism amendments such as compost. In summary, potentially endophytic contaminants of plant tissue cultures consist of host-specific and facultative plant pathogens, a few symbiotic/mutualistic microorganisms, and diverse common, mainly plant-associated, potentially cultivable environmental microorganisms.

In addition, aseptic cultures may become contaminated secondarily by cultivable laboratory microorganisms, mainly air-borne and human-associated bacteria and fungi and by microarthropods (Leifert and Cassells, 2001—see section on Laboratory Contamination Management).

ORIGIN OF PRIMARY EXPLANT MICROBIAL CONTAMINATION

Plants in the environment may be infected by plant pathogens and environmental microorganisms externally (epiphytically) or internally (endophytically; Figure 17.1). The term "hemi-endophyte" is also occasionally used for microorganisms in leaf glands bases, etc., which are intractable to surface sterilants. Epiphytic microorganisms in aggregates may also be resistant to surface sterilants (Andrews and Harris, 2000). With the exception of small intracellular microorganisms, namely, viruses and viroids, which may be systemic in plant tissues, larger viruses, for example, rhabdoviruses and some plant pathogenic bacteria, are restricted to the vascular system of the plant (Hull, 2001). Other pathogenic and nonpathogenic bacteria and fungi are restricted to the intercellular spaces that open up between cells with development of the tissues, for example, in the leaf spaces (see Chapter 8).

The cuticle, which overlays the epidermal cells of the leaf, is the primary physical barrier to the entry of organisms into the above ground tissues of plants. Epiphytes may enter the plant tissues locally, via wounds induced by wind and insects, or through natural openings such as stomata and lenticels or through sites of pathogen damage (Andrews and Harris, 2000; Bacon and White, 2000; Figures 17.1 and 17.2). These microorganisms may be localized in the airspaces of leaves or cavities in stems. Extensive microbial spread may occur in tissues in which spaces (cavities) develop between the cells as the tissues mature, for example, in stems and storage organs. Plant roots generally lack a cuticle, although in some plants the epidermal cells of the root may have suberized walls. The internal suberized endodermis is the barrier to soil microorganisms entering the roots and moving up in the vascular system into the stem. The region immediately behind the root tip where root hairs are formed is heavily colonized by soil microorganisms providing high inoculum pressure for the establishment of endophytes. The root environment (rhizosphere) is rich in microbial activity, and these microorganisms may colonize the intercellular spaces of the root cortex, in addition to the colonization by rhizobium of the roots of legumes and by mycorrhizal fungi of the roots of most plant species (Andrews and Harris, 2000). Natural openings/wounds caused by the emergence of lateral roots from the pericycle of the root, and damage caused by insects such as nematodes and fungi allow microorganisms to enter the root tissues (Figure 17.1).

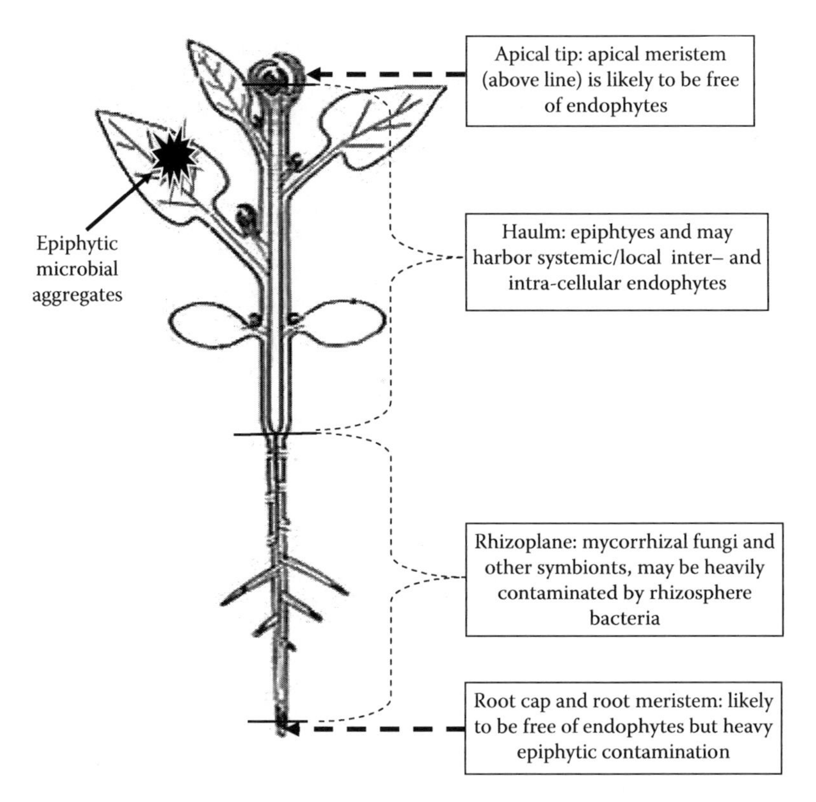

FIGURE 17.1 An illustration of a plant showing the roots (rhizoplane) and the haulm (phylloplane or leaves and stem); also indicated are the root and stem apical meristems. Plant anatomy in relation to tissue culture is discussed in Chapter 8.

The relationship between pathogenic organisms and their host plants is well characterized. The former can use natural openings, enzymes, and/or physical pressure to enter their host, and use enzymes and toxins to debilitate the host (Strange, 2003). The plant protects itself by preformed barriers, enzymes, and antimicrobial compounds and responds by forming antipathogenic metabolites and proteins. The dynamics of the interactions determine whether the pathogen is contained or disease develops (Strange, 2003). Pathogens can be inter- or intra-cellular, localized or systemic. On the other hand, the relationship between opportunistic colonizers and the plant's biotic stress defenses is unclear in most cases. There is some evidence that bacterial intercellular contamination of mature plant tissues may be widespread but little is known of the interaction between such tissue contaminants and the host cells. Some interactions may resemble elements of the interaction between pathogens and plants in that the contaminants may elicit host cell leakage and induce host biotic stress responses (Tyler and Triplett, 2008). The implication of the latter interaction is that, in some cases, expression of endophytic contaminants of plant tissues may be suppressed by a host defense response.

The abundance of root exudates and the emergence of secondary roots can result in the presence of high numbers of microorganisms on and in the root that are not killed by surface sterilization treatments prior to establishing root explants on a culture medium. Thus, roots and vegetative storage organs formed in the soil are avoided as sources of explants for the establishment of tissue cultures except where tissue-specific expression of a metabolite requires that they be used (Stafford and Warren, 1991). In these cases caution is urged and rigorous screening for endophytes is advised (see below). There is limited nutrient availability to epiphytes on the aerial parts of plants, hence lower contamination inoculum pressure on above ground parts, which makes aerial explants more suitable than root explants for the initiation of aseptic cultures. A factor that influences endophytic

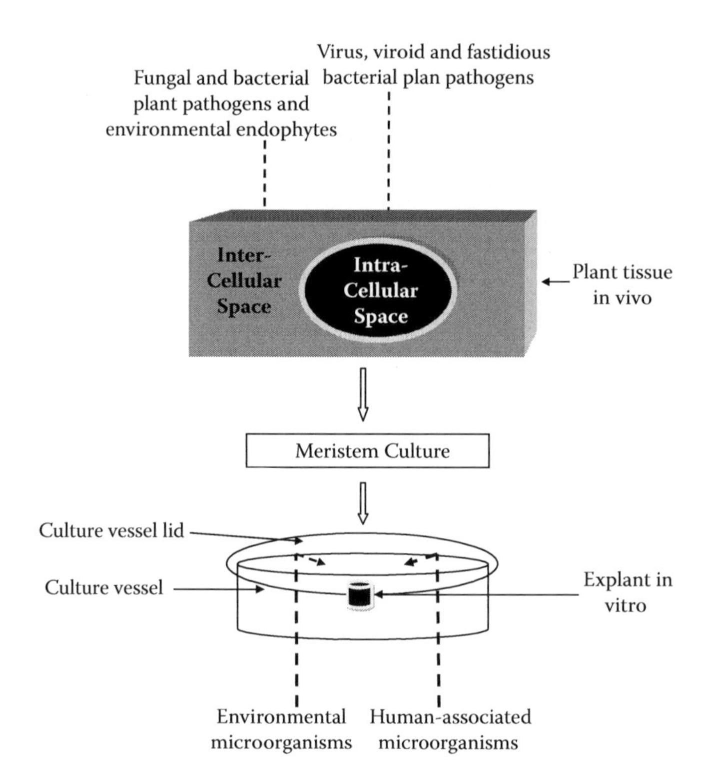

FIGURE 17.2 The general patterns of infection of plant tissues and cultures. Fungal and bacterial plant pathogens have mechanisms to directly enter plants through the cuticle or natural opening and are restricted to the intercellular spaces and may spread extensively within the plant. Environmental bacteria enter through natural openings or wound damage and are confined to intercellular spaces; their distribution in the plant may be localized. Viruses, viroids, and phloem- and xylem-inhabiting bacteria are introduced into the plant cells by vectors feeding on the plants where their distribution is intracellular and commonly systemic. Most pathogens and contaminants are eliminated in meristem culture (see text); however, aseptic cultures can be infected in the laboratory or growth room by cultivable environmental and human-associated microorganisms.

colonization of tissues by microorganisms is nutrient availability within plant tissues. It appears that there is adequate nutrient availability to support a wide range of endophytic colonists, and/or successful contaminants may be able to increase leakage of plant cell metabolites into the intercellular spaces (Tyler and Triplett, 2008).

For the above reasons, shoot tip explants should be the preferred explant for the establishment of aseptic pathogen-free cultures (Figure 17.1). Nodal explants carry the same high contamination risks as other explants from developed plant tissues (Figures 17.1 and 17.2). Finally, in spite of all the cautions about using leaf or stem explants to establish aseptic cultures, it is common practice to establish some ornamental plants, including *Begonia* spp. and African violet (*Saintpaulia ionantha*), from leaf and stem explants without reported contamination problems.

DETECTION OF PATHOGENIC AND NONPATHOGENIC ENDOPHYTES

General Considerations

As discussed above, the tasks facing a plant pathologist in selecting pathogen-free planting material differ from those of the tissue culturist seeing to establish aseptic cultures. Nevertheless, especially when dealing with micropropagation of pathogen-free stock, the strategies used by plant pathologists to screen for pathogens should be followed with the addition of rigorous culture-indexing for nonpathogenic endophytes. One should start by selecting vigorous plants of the species to be used, which should be examined carefully for signs of disease. Plant tissues showing signs of damage should be avoided as sources of explants to reduce the risk that the damaged tissues may harbor opportunistic endophytes. If the plants are suspected to be lacking in vigor or are showing symptoms of biotic (fungal, bacterial, viral, viroid diseases) or abiotic (nutrient deficiency, temperature) stress, the literature on the crop should be consulted. While some disease symptoms are distinctive and may allow disease identification, there may be overlap between the symptoms of different stresses, for example, between viruses and nutrient deficiencies and between bacterial and viral intracellular pathogens. The American Phytopathological Society (APS; http://www.apsnet.org) has published a very extensive range of compendia of major crop diseases and manuals on pathogen identification. Extensive information on crop diseases including early warning of emerging diseases and manuals for the production of healthy stock plants are available on the Web sites of the Food and Agriculture Organisation of the United Nations (FAO; http://www.fao.org), the North American Plant Protection Organisation (NAPPO; http://www.nappo.org), and the European and Mediterranean Plant Protection Organisation (EPPO; http://www.eppo.org).

Detection of Endophytic Intercellular Plant Pathogens

This group of contaminants includes fungal and bacterial plant pathogens that reside in the intercellular spaces. The detection/identification of fungal plant pathogens is based on the presence of their mycelium and reproductive structures on or in the diseased tissues. Many fungi may be endophytic at stages in their disease cycles, the most extreme example of which is the smut (*Ustilago*) disease of cereals that remains endophytic for most of its life cycle. Many fungal pathogens will be expressed when the explants is cultured on plant tissue culture or fungal media. When no fungal structures are associated with disease symptoms, the tissue should be plated on bacteriological media and, if microbial growth is detected, pure cultures should then be established (Lelliot and Stead, 1987) and, if required, the bacterium be identified as below. Bacterial plant pathogens may be latent in tissues until quorum sensing triggers pathogenesis (von Bodman et al., 2003). So, the advice is always to sample the explant tissues and plate on a range of bacteriological media to detect any latent bacterial contamination before embarking on tissues culture. Many soil bacteria are described as "viable, noncultivable" (VBNC; Table 17.1). It is not considered that this category posed a major contamination risk in tissue culture due to their extreme fastidious nutritional requirements (Andrews and Harris, 2000), which makes it unlikely that they will multiply in the tissues *in vitro*.

The isolation of fungal pathogens from damaged tissue can be problematic due to the presence of fast-growing secondary bacterial contaminants that may over-run the pathogen in culture. Commonly, a selective/semi-selective medium with inclusion of inhibitors/antibiotics is used to suppress bacteria in fungal isolation. This may be followed by the use of Koch's postulates to confirm that the fungal isolate is pathogenic to the host. For fungal plant pathogens, the characteristics of the mycelium and spores are used in identification (see Suggested Reading). There are service providers who will carry out fungal isolation and identification (Table 17.3). Cell and colony characteristics are used in bacterial identification (Schaad et al., 2001). Biochemical test kits are available for the identification of common environmental bacterial and yeast species but these are not useful for plant pathogen identification as the latter are not well represented on the databases (Tables 17.3 and 17.4). Increasingly, nucleic acid diagnostics are used in pathogen detection and identification, but the probes may not distinguish between plant pathogenic and nonpathogenic isolates. In the later case, confirmation of pathogenicity depends on applying Koch's postulates. Confirmation of Koch's postulates is the classic way of determining that a (culturable) isolate is the causal agent of disease. It involves isolating the pathogen in pure culture, inoculation of a healthy plant of the same crop cultivar with the pure isolate, observing the development of the same symptoms as in the original diseased sample, and reisolation and confirmation of the identity of the isolate (Strange, 2003).

Detection of Endophytic Intracellular Plant Pathogens

Intracellular plant pathogens include bacteria, viruses, and viroids. Intracellular bacteria include fastidious walled xylem- and phloem-located bacteria and fastidious wall-less phloem-located pathogens (mollicutes; Hull, 2001; Fletcher and Wayadande, 2002). As emphasized above, it is important to search for symptomless stock plants for tissue culture. But a caution: Many intracellular pathogens may be symptomless, at least at times in the season or under some cultivation conditions (Hull, 2001). Indeed, that a plant was infected with a virus is sometimes only inferred when progeny plants showing increased vigor compared with the parent (stock) plants are established from meristem cultures from which the pathogen has been eliminated subliminally (Hull, 2001). Pathogen-free plants, or more correctly, plants free from specified pathogens (see FAO, NAPPO, and EPPO certification schemes referred to above) may be obtained from official (National or State) crop certification schemes, from official germplasm collections, or from commercial companies (Hadidi et al., 1998). If certified stock is not available, the selected plants, including those provided by clients, should be indexed for latent viral and bacterial pathogens following the appropriate FAO guidelines for the crop. If no guidelines are available for the specific crop species, protocols should be developed based on the appropriate FAO guidelines for the habit of plant and suspected pathogen(s).

The detection/identification of intracellular pathogens is based on the following:

- *For viruses*: the mechanical inoculation of indicator plants, that is, species which give develop characteristic symptoms on infection with specific pathogen strains(s)
- *For all pathogens types*: by grafting to indicator species; by transmission via dodder or insect or other vectors to indicator plants; or by the use of serological or nucleic acid diagnostics

Mechanical transmission of viruses can be prevented by inhibitors in the sap of some species and by sap pH. Some viruses retain infectivity for only a short period in sap and many are susceptible to mechanical damage that renders them noninfectious (Hull, 2001).

Identification of plant pathogenic viruses is commonly based on the use of ELISA (enzyme-linked immunosorbent assay) kits using antibodies raised against the virus coat protein to achieve specificity, linked to an enzyme that amplifies the signal giving sensitivity (Clark and Adams, 1977). Kits are available from commercial plant diagnostics companies (Table 17.3). Increasingly, nucleic acid diagnostics services are also becoming available commercially (Table 17.4). These are based on the use of nucleic acid probes, which provide specificity, and the polymerase chain reaction, which provides sensitivity and appropriate detection by fluorescent or other staining (Schaad et al., 2003). Unlike serological diagnostics, which rely on variability in viral coat proteins to detect/differentiate specific pathogen isolates, nucleic acid probes can be developed to bind to both variable (e.g., viral coat gene) and conserved (e.g., viral helicase gene) regions of the genome (Hull, 2001). NA probes are the standard method for the detection of fastidious bacterial plant pathogens (Fletcher and Wayadande, 2002) and viroids that lack a protein coat (Flores et al., 2003).

While ELISA kits are commonly used in the tissue culture laboratory, nucleic acid detection is commonly carried out by a service provider (Table 17.3). The latter's protocol may involve sap extraction, partial purification, and pathogen nucleic acid-trapping being carried out in the tissue culture laboratory; the sample is then sent to the commercial laboratory for pathogen identification by probing and PCR. Test tube detection methods using fluorescent molecular beacons have also

TABLE 17.4
Diagnostics are Available Commercially for the Main Categories of Microbial Contaminants of Tissue Cultures. ELISA (Enzyme-Linked Immunosorbent Assay); NA (Nucleic Acid Probes); for Suppliers see Table 17.3

Organism	Commercial Diagnostics
Viruses	ELISA, NA probes
Viroid	NA probes
Bacteria	
• Plant pathogens	Morphology, NA probes
• Environmental	Biochemical test kits, FAP, NA probes
• Human-associated	Biochemical test kits, NA probes
• Fastidious	Specialized media for some, NA probes
VBNC	NA probes
Yeasts	Biochemical test kits, NA probes
Mycelial fungi	NA probes

been developed. It is possible to identify different virus/virus stains in the same extract using specific probes and tags (Schaad et al., 2003). Note that it is not uncommon for plants to be simultaneously infected by two (or more) viruses, and these viruses may interact to increase symptom severity or reduce symptoms, and to increase or decrease the relative titres (amounts) of the respective viruses (Hull, 2001). In general, the methods referred to above for pathogen detection are confirmatory; that is, they test the hypothesis that a specific pathogen is present. It is not practical to use them for screening for unknown viruses.

Detection of Nonpathogenic Endophytes

The approach to the detection of nonpathogenic endophytes (potential vitropaths) differs from that of detecting plant pathogens. In contrast to the use of specific diagnostics, traditionally the detection of nonpathogenic endophytes by tissue culturists has depended on the culture-indexing of explants, based on the assumption of endophyte culturability (George, 1993; see Chapter 18). In some cases, tissue culturists have taken tissue samples from stock plants for culture indexing before setting up cultures. In culture-indexing, explants or tissue samples, after surface sterilization, are normally placed on one or a few microbial media and the plates are incubated at approximately 20°C, then examined over an approximately 10-day time course. Many endophytes will grow on common bacteriological media appearing as cream, yellow, or pink colonies (Lelliot and Stead, 1987; Schaad et al., 2001). More than one contaminant may be present. To isolate pure cultures, the contaminants should be restreaked on fresh media and single spore colonies be used for identification and/or antibiotic testing, as appropriate (Barrett and Cassells, 1994). Some bacteria (i.e., fastidious bacteria) may have uncommon media requirements so it is advisable to use a range of bacteriological media, and some may have relatively high temperature optima so culture should be over a temperature range (Atlas, 2005; for background on the discovery and significance of nonpathogenic endophytes see Holland and Polacco, 1994; Andrews and Harris 2000; Bacon and White, 2000).

We are now moving into an era where commercial broad spectrum nucleic acid probes are becoming available, which can be used to screen for the presence of bacteria in general (Maidak et al., 1999) or members of plant virus families (Hull, 2001; http://www.agdia.com). While multiplexing (use of multiple probes) in PCR to detect more than one organism in a test is possible, spurious results increase with increasing numbers of probes. The emerging development of diagnostic microarrays should enable plant tissue culturists, in the future, to confirm in one test that their cultures are free of both specific pathogens and of bacteria, that is, not just aseptic but axenic (Boonham et al., 2007).

ELIMINATION OF ENDOPHYTES FROM STOCK PLANTS

There are three approaches to the elimination of the pathogens from plants, namely, thermotherapy, chemotherapy, and meristem culture with/without chemo- or thermo-therapy (George, 1993, 1996). Thermotherapy and meristem culture are used extensively to eliminate viruses and other systemic intracellular endophytes. The former depends on the selective killing effects of heat on pathogens and the resistance of the plant tissues, especially buds. There are published protocols that indicate that treatments of about 40°C for 4–6 weeks are widely used (George, 1993). It is very important to maintain the heat-treated plants in vector-proof conditions for 4–6 weeks after treatment with repeated indexing to confirm pathogen elimination before using them as sources of explants. The FAO, EPPO, and NAPPO (see above) have published protocols for thermotherapy in their certification schemes, which can be consulted and used as models (see also Hadidi et al., 1998).

Chemotherapy based on antiviral and antibacterial chemicals is not widely used for the treatment of plants in vivo (see use in vitro following), largely because of the cost and the difficulty of maintaining inhibitory concentrations in the plant. Attempts to use antibiotics to control intracellular bacterial infections in plants have generally resulted in symptom remission rather than pathogen elimination (Fletcher and Wayadande, 2002). Penicillins and other inhibitors of wall-forming bacteria are ineffective against the wall-less mollicutes where tetracyclines are used. First-generation antibiotics, formulated for application to plants, have been used to reduce the titre of bacteria in plants facilitating the elimination of contaminants from stem-tip explants; however, the application of antibiotics to plants is prohibited in many countries.

Meristem culture is used widely to eliminate plant pathogens and nonpathogenic endophytes, based on the hypothesis that the apex will be free of all microorganisms (see above; George, 1993; Figure 17.2). However, there is much ambiguity in the literature regarding the anatomy/morphology of the explants described as "shoot tip," apical meristem ("apical dome") by researchers (see Chapter 8). Arguably, the smallest explant capable of survival should be used to reduce the risk of carry-over of endophytes. Commonly, the explant used is the apical dome plus first pair of leaf primordia. The survival of small explants can be enhanced by micro-grafting to aseptic (in vitro) seedlings or microplants of the same species (George, 1993). The resulting shoot is excised and used for subsequent multiplication after appropriate indexing for pathogens and nonpathogenic endophytes.

In cases where the plant tissue is known to harbor cultivable endophytes—that is, where the primary explants shows signs of contamination—presoaking the explants in antibiotics, based on methods developed for seed treatments, may be attempted (Maude, 1996). Alternatively, antibiotics are incorporated into the plant tissue culture medium (see below).

LABORATORY AND GROWTHROOM CONTAMINATION MANAGEMENT

LABORATORY PRACTICE

Good laboratory practice is essential to avoid the risk of contamination of cultures by environmental microorganisms including human-associated microorganisms and by microarthropods (Leifert and Cassells, 2001). Contamination of cultures in the laboratory is due to poor aseptic technique or equipment failure. The contaminants encountered are generally common nonfastidious cultivable environmental bacteria and yeasts, and human-associated microorganisms, which grow on plant tissue culture media (Leifert and Cassells, 2001; Weller and Leifert, 1966). As mentioned above, it is unlikely, due to low inoculum pressure and absence of vectors, that plant pathogens will be introduced into cultures under laboratory conditions.

Good laboratory management practice is based on risk assessment using the hazard analysis critical control point (HACCP) strategy employed in the food and pharmaceutical industries (Leifert and Cassells, 2001). The basis of good laboratory practice is adequate staff-training in aseptic technique; the provision of personnel with appropriate footware, gloves, face-masks, and headwear; having a regular equipment servicing program; and scheduled monitoring of the laboratory and growth room atmosphere for the build-up of microbial spores. It is important to label cultures adequately such that they can be traced back to the individual who set them up and to the sequence in which they were set up. Frequently, a tissue culture batch is contaminated at a point in the production run where, for example, instrument sterilization fails or aseptic technique breaks down. Random sampling of a production run should be followed by examination of cultures up- and downstream from any contaminated culture to determine the full range of contamination. Cultures should be examined routinely by eye and hand lens at sub-culture for signs of microbial contamination. Examination is facilitated by the use of a dissecting microscope with background lighting. Detection of contamination can be facilitated by the use of nonopaque alternative gelling agents to agar, or by the use of good quality agar, which give clearer gels, with the caveat that the properties of the gelling agent may influence the tissue culture responses (George, 1993, 1996).

Occasionally, sometimes seasonally, a heavy build-up of airborne spores occurs in the laboratory, putting aseptic technique under severe pressure and resulting in heavy losses of cultures. Contamination can persist and may necessitate the pressurization of the laboratory or the use of an air filtration system for the laboratory. Autoclaves should be regularly serviced, and each batch from the autoclave monitored using autoclave tape. Airflow should be regularly monitored to ensure that the laminar flow cabinet HEPA filter is functioning correctly. Faulty instrument sterilization during subculture is a common cause of culture contamination. Bead sterilizers are preferred over traditional flame sterilization. Contamination biomarkers include atmospheric yeast, fungi, and bacteria; heat-resistant bacilli indicate autoclave failure and human-associated bacteria, and yeasts indicate poor aseptic technique (Leifert and Cassells, 2001).

Growthroom Management

Micropropagators routinely visually inspect cultures in the growthroom and prior to subculture for the visible present of microbial and microarthropod contamination (Figure 17.3). Where contamination is suspected, the cultures should be autoclaved unopened and disposed of appropriately.

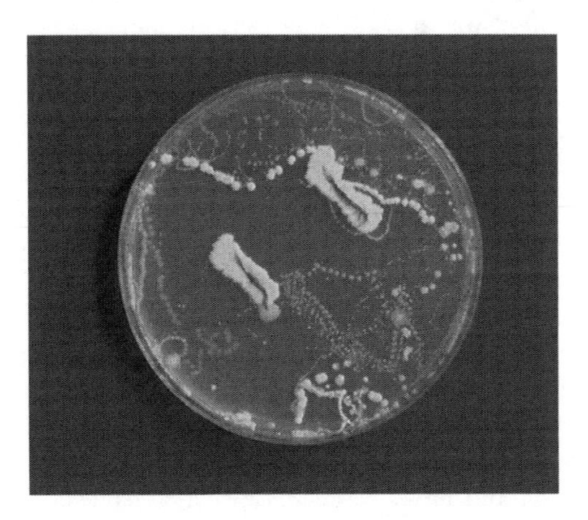

FIGURE 17.3 Trails of microbial colonies on the surface of the plant tissue culture representing the "foot prints" of microarthropods (mites) moving around the cultures.

Proprietary insecticides can be incorporated into the medium to control microarthropods (Pype et al., 1997), but infested cultures cannot be recovered due to the secondary microbial contamination. Infested cultures should be autoclaved unopened and disposed of appropriately. It is important to clear out microbially contaminated cultures and to avoid storing fungal cultures in the tissue culture laboratory and growthrooms as microarthropods may infest these and spread infection to clean cultures. If the growthroom shelves are of open mesh, these can be lined with transparent plastic sheeting to prevent micro-arthropods dropping down from upper shelves to infect cultures below. It may be necessary to fumigate infested growthrooms with an appropriate pesticide to control serious infestations. Fumigation should be carried out by accredited specialist companies.

In contrast to contamination introduced with the primary explants, which is due to association with pathogens of the crop and plant-associated microorganisms, laboratory/growthroom (secondary) contamination is caused by cultivable atmospheric environmental and human-associated microorganisms and microarthropods (mites; Figure 17.3; Pype et al., 1997). The time of expression of contamination after explants establishment and the pattern of contamination that is whether:

- Associated with the explants
- Within the medium
- Random on the surface of the medium
- In spore trails

provides information on the likely origin of the contamination (Leifert and Cassells, 2001).

It is common practice to dispense a medium into the culture vessels and to allow them to stand for a few days before use. Under these circumstances the presence of microbial colonies within the medium is an indication that the autoclaving of the medium was faulty. These colonies are likely due to contamination with *Bacillus* spp. from the atmosphere, which produce heat-resistant spores. Contamination on the surface, with blue moulds or pink or buff colored colonies, indicates bacterial or yeast contamination due to faulty dispensing equipment, poor operator technique, or that the HEPA filter in the laminar flow cabinet needs servicing or replacement (Leifert and Cassells, 2001).

The pattern of contamination in cultures in the growthroom also indicates the possible causes. Visible contamination around the explants indicates that it may have been contaminated due to failure to surface sterilize the explants (see comments re: hemi-endophytes and bacterial aggregates above); contact with contaminated instruments during dissection due to faulty instrument sterilization; due to poor aseptic technique by the culturist; or failure of the HEPA filter in the laminar flow cabinet (Leifert and Cassells, 2001). All cultures should be examined to see if this random contamination of cultures or associated with all the explants indicating batch contamination.

Sometimes a halo effect is seen around the ends or along the length of cut explants. This may be due to the presence of a latent or fastidious endophyte that is inhibited by the medium or by induction of a stress-response in the tissue, e.g., latex exudation or release of phenols; or may be due to natural exudation resulting from tissue damage caused at excision. The latter exudation may be transient but the exudate may block the uptake of nutrients and growth regulators by the explants. This stress response can be reduced by incorporation of antioxidants, such as vitamin C and phenol-binding agents, such as polyvinylpyrrolidone (PVP) in the medium (Menard et al., 1985); or the explants can be repositioned on the medium away from the exudates (George, 1996).

Contamination on the surface of the medium away from the explants may indicate poor aseptic technique or equipment failure (Leifert and Cassells, 2001), but it can also reflect failure of the culture vessel lid to exclude atmospheric contaminants. The atmospheric pressure in a culture vessel varies with the diurnal cycle. Pressure and humidity increase when the lights are on due to heating effects, whereas condensation and pressure decrease may occur when the vessels cool after the lights go off. Negative pressure in the vessels causes inward airflow. Contamination can occur in open-lidded vessels with no filters under these circumstances. Culture ventilation is necessary to facilitate gaseous exchange with the atmosphere and to allow ethylene to escape from the vessels. Problems due to contamination of ventilated cultures can be overcome by using gas permeable filters as lid seals or by replacing the lid or wrapping the open-lidded vessel in gas permeable plastic film (Cassells and O'Herlihy, 2003). There is a wide range of transparent plastic wrapping materials used in the food industry, which can be used instead of a conventional vessel lid or can be wrapped around an open-lidded vessel with the proviso that the film be adequately permeable to oxygen, carbon dioxide, and ethylene to allow equilibration with the ambient atmosphere. If replacing the lid with a plastic film, the permeability of the film to water vapor should be chosen such that the cultures do not dry out (Cassells and O'Herlihy, 2003).

Some tissue culturists use liquid culture systems ranging from conventional sized (ca. 500 ml) tissue culture containers to large industrial-scale bioreactors (fermenters). In these sealed systems, using either temporary immersion, misting with nutrients, or submersion with forced aeration, microbial filters are used to exclude contamination (Sharaf-Eldin and Weathers, 2006). "Cloudiness" in the medium may indicate contamination.

While vectors such as aphids, nematodes, and soil-borne fungi play an important part in the transmission of intracellular plant pathogens in the field and greenhouse (Hull, 2001; Fletcher and Wayadande, 2002), these vectors do not pose a contamination threat in the laboratory or growthroom. However, microarthropods can act as vectors of environmental microorganisms. Microarthropod numbers build up rapidly in contaminated cultures and then spread in the growthroom. Microarthropods may enter open (ventilated) vessels resulting in a characteristic pattern of contamination—trails of spore colonies across the surface of the medium and from explants to explants (Figure 17.3).

Elimination of Contaminants from Cultures

Fast-growing contaminants usually over-run cultures too rapidly to allow remedial action. However, it is possible to backtrack and to reestablish contaminated explants on antibiotic-containing medium. If contamination is detected, the contaminants should be streak-plated to obtain pure isolates, which can then be screened for antibiotic sensitivity. Cultivable contaminants are mycelia fungi, yeasts, and bacteria. Microscopic examination will distinguish between these. If the contaminant is a bacterium, the first step is Gram-staining to determine whether the isolate is Gram positive or Gram negative to determine an appropriate antibiotic to use (Barrett and Cassells, 1994). Additional staining, preliminary biochemical tests, and determination of growth characteristics can be carried out to identify the isolate further (Schaad et al., 2001).

In selecting antibiotics for use in plant tissue culture, it is important to avoid antibiotics that may cause mutation in the plant tissues, such as inhibitors of nucleic acid metabolism/protein synthesis or inhibit plant growth (Barrett and Cassells, 1994). Antibiotic sensitivity is determined on solid diagnostic sensitivity testing medium (DST) by inoculation of a plate with the isolate and by placing antibiotic discs, impregnated with different antibiotics and varying antibiotic concentration, on the medium. The minimum inhibitory concentration (MIC) is determined by inoculation with the isolate of serial dilutions of the antibiotic and the concentration at which bacterial growth is inhibited or killed is determined visually. Bacterial growth can also be determined spectrophotometrically. Antibiotics are usually used in the plant culture medium at 2–4×, the MIC. Most antibiotics are biostatic rather than biocidal and are not effective in eliminating fast-growing bacterial contaminants that can grow/are established on plant tissue culture media. Where a stock plant tissue is known to be contaminated, the freshly excised explant should be placed on the antibiotic-containing medium; the new growth, excised and transferred to fresh antibiotic-containing, repeated subculture, may be required to eliminate the contaminant as most antibiotics are bacteriostatic. The cultures should be tested after each passage on antibiotic to confirm elimination of the bacterium. Nystatin and other fungicides have occasionally been used to control fungal infections in vitro.

The antiviral compound Ribavirin (syn. Virazole) has been used to eliminate viruses from plant tissues in vitro; like most antibiotics this compound is virustatic rather than virucidal, and so only new tissue should be excised from infected explants and serial subcultured on ribavirin as required. Ribavirin is used at 10–100 μg/L in the plant culture medium. Ribavirin has cytokinin activity and so interaction with cytokinins in the plant tissue culture medium may influence its activity. Ideally, Ribavirin should be used in a simple hormone-free autotrophic medium (Cassells and O'Herlihy, 2003). In using Ribavirin nodal explants, or preferably apical explants, should be used and only the new growth excised for sub-cultured as the contaminant may be present, suppressed but viable, in the original tissues. It is inadvisable to attempt virus elimination by incorporation of antiviral chemicals into adventitious regeneration protocols as adventitious shoots may be genetically unstable, giving rise to mutant progeny (George, 1993).

Thermotherapy (see above) has also been used in vitro to eliminate pathogens from plant tissue cultures (Leonhardt et al., 1988).

CONCLUSIONS

Micropropagation is used extensively in the production of disease-free propagules (planting material) for vegetatively-propagated crops (George, 1993, 1996). The FAO guidelines under which micropropagated material is certified emphasize the importance of the methods used to detect specific pathogens and specify the plant material that should be screened. The FAO's concern is that the methods be sensitive and repeatable. It is recognized that pathogen levels may be below the threshold for detection in in vitro material. Consequently, the pathogen indexing is generally based on growing on the micropropagated plants in vector-free/pathogen-free environment before carrying out pathogen indexing to confirm that they are disease-free. Tissue culturists in many cases do not have greenhouses. They receive stock plants from their clients and return in vitro cultures to the clients for establishment, so they do not have access to mature plant tissue for pathogen indexing. Tissue culturists, in addition to managing cultivable contaminants as discussed previously, should recognize that pathogens/contaminants may be below the level of detection in in vitro material. It should also be recognized that some laboratories may use antibiotics to suppress contamination in micropropagation. As a precaution, when bringing stock cultures into the laboratory (as opposed to stock plants), it is important to inoculate tissue samples onto a range of microbial media and to incubate the cultures over a range of temperatures, that is, to culture-index them for latent contaminants.

The problems of culture contamination relate directly to the use, by approximately 90% of plant tissue culturists, of the culture medium developed by Murashige and Skoog (1962). This medium is a good substrate for environmental microorganisms that overgrow the plant tissues, directly or indirectly killing them. Other microbial contaminants may remain latent, suppressed by the high osmotic pressure, medium pH, or other components such as plant growth regulators, only to emerge and overrun the cultures when these factors are changed. (Photo-)autotropic culture (aseptic microhydroponics) on simple mineral medium has been advocated as an alternative to conventional heterotrophic culture and has been demonstrated to be applicable to some species (Cassells and O'Herlihy, 2003). However, this does not entirely solve the problem of environmental microbial contamination as the tissues release microbial nutrients into the medium. Cell and callus culture will continue to require heterotrophic media with consequent contamination risks.

The most exciting opportunity for disease and contamination management in vitro lies in the further development and commercialization of nucleic acid based diagnostics. Commercially available diagnostics developed for specific pathogens are of only limited application in micropropagation; however, broad spectrum diagnostics based on micro-arrays, capable of nonspecific detection of bacteria and bacteria-like and virus and virus-like contaminants, would have universal application in tissue-culture-contaminant management (Boonham et al., 2007). Advances in diagnostic are essential to the certification of plants from tissue culture (van der Linde, 2000).

REFERENCES

Andrews, J.H. and R.F. Harris. 2000. The ecology and biogeography of microorganisms on plant surfaces. *Annu. Rev. Phytopathol.* 38:145–180.

Atlas, R.M. 2005. *Handbook of Media for Environmental Microbiology*. CRC Press, Roca Baton, FL, 446 pp.

Bacon, C.W. and J.F. White Jr. 2000. *Microbial Endophytes*. CRC Press, Roca Baton, FL, 500 pp.

Barrett, C. and A. C. Cassells. 1994. An evaluation of antibiotics for the elimination of *Xanthomonas campestris* pv. pelargonii (Brown) from *Pelargonium* x *domesticum* cv. Grand Slam explants *in vitro. Plant Cell, Tiss. Org. Cult.* 36: 169–175.

Boonham, N., J. Tomlinson, and R. Mumford. 2007. Microarrays for rapid identification of plant viruses. *Annu. Rev. Phytopathol.* 45: 307–328.

Cassells, A.C. (Ed.). 1997. *Pathogen and Microbial Contamination Management in Micropropagation*. Kluwer, Dordrecht, p. 370.

Cassells, A.C. and E. O'Herlihy. 2003. Disease management of microplants: Good laboratory practice. *Acta Hortic.* 616: 105–114.

Clark, M.F. and A. N. Adams. 1977. Characteristics of the microplate method of enzyme-linked immunosorbent assay for the detection of plant viruses. *J. Gen. Virol.* 34: 475–483.

Fletcher J. and A. Wayadande. 2002. Fastidious vascular-colonizing bacteria. Plant Health Instructor. DOI: 10.1094/PHI-I-2002-1218-02. http://www.apsnet.org/education/IntroPlantPath/PathogenGroups/fastidious/default.htm.

Flores, R., C. Hernandez, A.E. Martinez de Alba, J-A. Daros, and F. Di Serio. 2003. Viroids and viroid-host interactions. *Annu. Rev. Phytopathol.* 43: 117–139.

George, E.F. 1993. *Plant Propagation by Tissue Culture, Part 1—The Technology.* Exegetics, Basingstoke.

George, E.F. 1996. *Plant Propagation by Tissue Culture, Part 1—In Practice.* Exegetics, Basingstoke.

Hadidi, A., R. K. Khetarpal, and H. Koganezawa (Eds.) 1998. *Plant Virus Disease Control.* APS Press, St. Paul, MO. 704 pp.

Hoffman, P.N., J.E. Death, and D. Coates, in C. H. Collins, M.C. Allwood, S. J. Bloomfield, and Fox, (Eds.). 1981. *Disinfectants: Their Use and Evaluation of Effectiveness*, Academic Press, London, pp. 77–83.

Holland, M.A. and J.C. Polacco. 1994. PPFMs and other covert contaminants: Is there more to plant physiology than just plant? *Annu. Rev. Plant Physiol.* 45:197–209.

Hull, R. 2001. *Matthew's Plant Virology*. Academic Press, New York. 400 pp.

Leifert, C. and A. C. Cassells. 2001. Microbial hazards in plant tissue and cell cultures. *In Vitro Cell. Devel. Biol.—Plant* 37, 133–138.

Lelliott R.A. and D. E. Stead. 1987. *Methods for the Diagnosis of Bacterial Diseases of Plants.* Blackwell Scientific Publications, Oxford.

Leonhardt, W., C. Wawrosch, A. Auer, and B. Kopp. 1988. Monitoring of virus diseases in Austrian grapevine varieties and virus elimination using in vitro thermotherapy. *Plant Cell Tiss. Org. Cult.* 52: 71–74.

Maidak, B.L., J.R. Cole, C.T. Parker Jr., G.M. Gerrity, and N. Larsen et al. 1999. A new version of the RDP (Ribosomal Database Project). *Nucl. Acid. Res.* 27:171–173.

Maude, R.B. 1996. *Seedborne Diseases and Their Control: Principles and Practices.* CAB International, Wallingford. 628 pp.

Menard, D., M. Coumans and T. H. Gaspar. 1985. Micropagation du Pelargonium a partir de meristems. *Meded. Fac. Landbouwett. Rijksuniv. Gent.* 50, 327–331.

Murashige, T. and F. Skoog. 1962. A revised medium for rapid growth and bioassays with tobacco cultures. *Physiol. Plant.* 15:473–497.

Pype, J., K. Everaert, and P. C. Debergh. 1997. Pp. 259–266. In Cassells, A.C. (Ed.), *Pathogen and Microbial Contamination Management in Micropropagation*, Kluwer Academic Publishers, Dordrecht.

Schaad, N.W., R.D. Frederick, J. Shaw, W.L. Schneider, R. Hickson, et al. 2003. Advances in molecular-based diagnostics in meeting crop biosecurity and phytosanitary issues. *Annu. Rev. Phytopathol.* 41: 305–324.

Schaad, N.W., J. B. Jones, and W. Chun. 2001. *Laboratory Guide for Identification of Plant Pathogenic Bacteria.* APS Press, St Louis.

Sharaf-Eldin, M.A. and P.J. Weathers. 2006. Movement and containment of microbial contamination in the nutrient mist bioreactor. *In vitro Cell. Devel. Biol.—Plant* 42:553–557.

Stafford, A. and G. Warren. 1991. *Plant Cell and Tissue Culture*. Open University Press, Milton Keynes.

Strange, R. 2001. *Essential Plant Pathology.* John Wiley & Sons, New York.

Strange, R. 2003. *Introduction to Plant Pathology.* John Wiley & Sons, New York.

Tyler, H.L. and E.W. Triplett. 2008. Plants as habitat for beneficial and/or human pathogenic bacteria. *Annu. Rev. Phytopathol.* 46: 73–73.
Van der Linde, P.C.G. 2000. Certified plants from tissue culture. *Acta Hortic.* 530: 93–101.
Von Bodman, S.B., W. Dietz Bauer, and D.L Coplin. 2003. Quorum sensing in plant pathogenic bacteria. *Annu. Rev. Phytopathol.* 41: 455–482.
Weller, R. and C. Leifert. 1966. Transmission of *Trichophyton interdigitale* via an intermediate plant host. *Br. J. Dermatol.* 135: 656–657.

SUGGESTED READING

Agrios, G.N. 2004. *Plant Pathology*, 5th Ed. Academic Press, New York, p. 952.
Barnett H.L. and B. B. Hunter. 1998. *Illustrated Genera of Imperfect Fungi*. 4th Ed., APS Press, St. Louis. 218 pp.
Cassells, A.C. and P. Gahan. 2006. *Dictionary of Plant Tissue Culture*. Haworth Press, New York, 267 pp.
Cassells, A.C. and V. Tahmatsidou. 1997. The influence of local plant growth conditions on non-fastidious bacterial contamination of meristem-tips of *Hydrangea* cultured in vitro. *Plant Cell Tiss. Org. Cult.* 47: 15–26.
Duggan, F.M. 2006. *The Identification of Fungi*. APS Press, St. Louis, MO. 218 pp.
Gregory, P.H. 1973. *The Microbiology of the Atmosphere*. Leonard Hill Books, Aylesbury, p. 377.
Kolozsvari Nagy, J., S. Sule, and J.P. Sampaio. 2005. Apple tissue culture contamination by *Rhodototula* spp: Identification and prevention. *In vitro Cell. Devel. Biol.—Plant* 41: 520–524.
Narayanasamy, P. 2001. *Plant Pathogen Detection and Disease Diagnosis*, 2nd Ed. Marcel Dekker, New York, 544 pp.
Walsh, C. 2003. *Antibiotics: Actions, Origins, Resistance.* ASM Press, Washington, DC. 272 pp.

18 Culture Indexing for Bacterial and Fungal Contaminants

Michael Kane, Philip Kauth, and Timothy Johnson

Rapid production of specific pathogen eradicated plants is a fundamental goal of the micropropagation process (Chapter 16). Whether for commercial in vitro propagation, or more fundamental metabolic, genetic, or morphogenetic research, it is desirable to establish and maintain plant cultures that are also free of nonpathogenic microbial contaminants (Cassells, 1997; Knauss and Knauss, 1979). The surfaces of plants are naturally populated with a diverse microflora of bacteria, fungi, yeast, and other organisms. A primary objective of Stage I is the elimination of this microflora and the subsequent establishment of aseptic cultures (Chapter 12). This is usually accomplished through surface disinfecting explants (e.g., meristem tips, shoot tips, stem or leaf tissue) with alcohol and/or sodium hypochlorite prior to culture inoculation. Once explants are established in vitro, it is essential that cultures be indexed (screened) for the presence of microbial contaminants. Contaminated cultures may exhibit no symptoms, variable growth, regeneration, reduced shoot proliferation, rooting, or poor survival (Leifert et al., 1989). Many culture contaminants are not pathogenic to plants under field conditions but become pathogenic in vitro often due to the release of toxic secondary metabolites into the medium (Leifert et al., 1994; Leifert and Cassells, 2001).

Research and commercial tissue culture laboratories often use only visual methods to index for culture contamination (the so-called *EBD* or eyeball determination method). There are several important reasons why screening only for visible microbial contamination is not adequate. Visible growth of microbial contaminants may be suppressed in plant culture media (Leifert and Waites, 1992). Although many fungal contaminants quickly become visible, endophytic bacterial infections may be latent and difficult to detect (Tanprasert and Reed, 1997; Thomas, 2004). Microbial contaminants such as *Methylobacteria* have evolved close biochemical relationships with the plant epidermis in vivo and will not grow independently on standard plant culture media (Holland and Polacco, 1994).

Indexing for microbial contaminants is usually accomplished by inoculating tissue sections or intact shoots into enriched-selection medium that will promote the visible growth of bacteria, filamentous fungi, yeast, or other contaminants (Chapter 17). Since secondary culture contamination can occur as a result of poor aseptic technique or contaminant vectors such as mites, cultures should be routinely reindexed. Management and quality assurance strategies such as Hazard Analysis Critical Control Points (HACCP) procedures have been adapted in commercial micropropagation to both limit contamination from all sources and increase plant quality (Leifert and Woodward, 1998).

Many procedures are available to screen for culture contaminants. Knauss (1976) screened for the presence of systemic bacteria and fungi by culturing 0.5–1.0 mm thick internodal and nodal stem cross sections from Stage I plantlets in four enriched microbiological media that promoted microbial growth. This procedure was effective in establishing cultures of foliage plants such as *Dieffenbachia* that were free of specific pathogens such as *Xanthomonas dieffenbachiae* and *Erwinia chrysanthemi*. Other procedures use a single indexing medium to screen for culture contamination in explants following surface disinfestation (Leifert et al., 1989; Viss et al., 1991). More recently, a systematic three-step indexing procedures involves the following: (1) visual examination of plant tissue cultures; (2) indexing of the medium of visually "sterile" cultures on bacteriological media; and (3) indexing using split tissues from different culture parts have been used to routinely detect covert and endophytic bacteria (Thomas, 2004).

The primary objective of this exercise is to illustrate indexing procedures for cultivable contaminates using a single indexing medium. After completing this exercise, students will be able to do the following: (1) discuss the importance and limitations of culture indexing; (2) select culture indexing media; (3) nondestructively excise plant culture tissue samples for indexing; (4) demonstrate the "stab and streak" method for tissue inoculation; and (5) reliably interpret indexing results.

GENERAL CONSIDERATIONS

Indexing Media Selection

Without specific knowledge of the actual microbial contaminants present, limitations in detecting contamination can occur if inappropriate indexing media are used. One approach is to use several highly enriched indexing media such as Sabouraud dextrose medium, Yeast extract dextrose broth, and AC broth (Knauss, 1976). Sabouraud dextrose medium is formulated for the growth of fungi including yeast, mold, and aciduric microorganisms. Yeast extract dextrose broth medium is used to stimulate bacterial growth, while AC broth is a sterility test medium for a variety of microorganisms. Contaminated cultures will index positive in all three media about 98% of the time (Kane, unpublished). These and many other sterility test media are manufactured by Difco Laboratories (Detroit, Michigan) and distributed through Fisher Scientific (Norcross, Georgia). For this laboratory exercise it is probably best to select a single broad-spectrum sterility test medium.

Several media have been formulated to also sustain the plant tissue being indexed. This increases efficiency by allowing tissues, such as shoot-tips, to first be placed in the selection medium and then transferred to fresh tissue culture medium only if they index negative. A Murashige and Skoog (1962) medium enriched with yeast extract, peptone, or glucose may serve as an adequate indexing medium (Tanprasert and Reed, 1997). A highly effective and recommended indexing medium is Leifert and Waites Sterility Test Medium (Leifert et al., 1989). This medium both sustains plant tissue and promotes growth of a broad range of latent bacterial contaminants with the exception of some *Xanthosomonas, Methylobacterium,* and *Hyphomycrobium* species and all mycoplasma-like organisms, and is available from *Phyto*Technology Laboratories, Shawnee Mission, Kansas (Product ID # 476).

Indexing Medium Preparation and Plant Cultures

For this exercise, both semisolid and liquid Leifert and Waites Sterility Test Medium will be used. The semisolid medium is prepared by the addition of 10 g/L agar. Medium is prepared without pH adjustment and dispensed as 10 mL volumes into 20 mL glass scintillation vials covered with autoclavable screw caps (Figure 18.1c). Scintillation vials are both shallow, which facilitates inoculation, and reusable. The cardboard tray that the vials are shipped in makes a convenient rack to hold one hundred vials during incubation. After autoclaving, the agar-supplemented medium is cooled and solidified as 45° slants (Figure 18.1c). Although indexing of Stage I shoot cultures is described in this exercise, almost any type of plant culture can provide tissue for indexing. This indexing exercise can be readily incorporated into any of the laboratory exercises outlined in this text that use a surface disinfecting procedure.

Exercise: Indexing Stage I Shoot Cultures

Materials

The following are needed for each student or student team:

- Five established Stage I shoot cultures
- Seven scintillation vials each of liquid and agar-solidified Leifert and Waites Sterility Test Medium
- Five culture vessels containing fresh Stage I medium

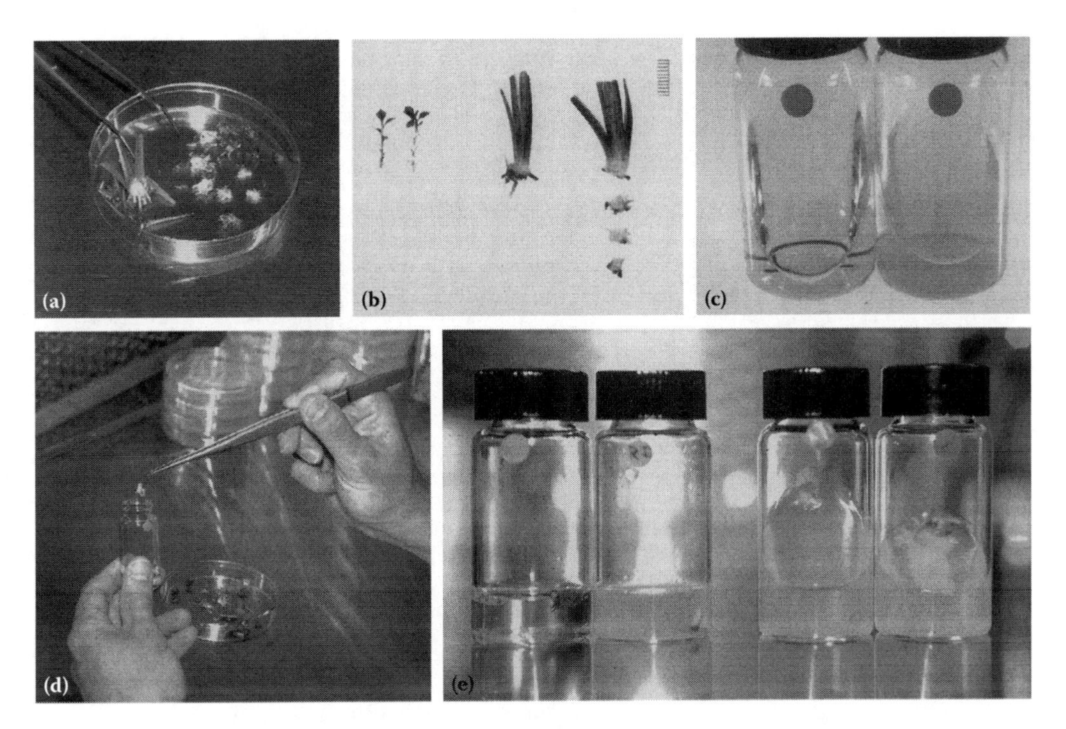

FIGURE 18.1 Culture indexing for cultivable contamination. (a) Aseptic technique is used to obtain tissue sections from the Stage I shoot culture. (b) Shoots with upright stems (left) are sectioned into nodal segments. The shoot tip is either cultured in the indexing medium or transferred to fresh Stage I medium. Cross-sectional stem disks are made from shoots with basal meristems (right). Scale bar = 1 cm. (c) Indexing medium is prepared as a liquid and agar-solidified slants. (d) Several tissue samples are directly placed in the liquid indexing medium. (e) Clouding of inoculated liquid indexing medium is a positive indication of the presence of cultivable contaminants in the tissue sample. Development of colonies on the surface of the medium and/or halo in the medium where the tissue sample was stabbed indicates culture contamination. (From Trigiano, R.N. and Gray, D.J., Eds., *Plant Tissue Culture Concepts and Laboratory Exercises, Second Edition*, CRC Press LLC, Boca Raton, FL, 2000.)

Follow the protocol in Procedure 44.1 to complete this experiment.

Anticipated Results

Rapid clouding of inoculated liquid indexing media is a positive indication of the presence of cultivable contaminants in a tissue sample (Figure 18.1e). The presence of contaminants on agar-solidified indexing media is usually confirmed by development of colonies on the surface of the medium and/or development of a halo in the medium where the tissue sample was stabbed (Figure 18.1e). Detection of slow-growing contaminants is usually facilitated on agar-solidified media. Students should make close observations for the presence of slowly developing colonies on the surface of the indexed tissues. Control vials should not display microbial growth. It should also be noted that systemic contamination does not occur uniformly throughout the plant and may be localized in specific regions or tissues that are not indexed. This results in false negative indexing. It is important that cultures be reindexed after a period of growth.

The occurrence of contaminated cultures provides an opportunity to discuss what to do with them. Options include redisinfestation and isolation of meristem tips or larger shoot tips, and the use of antibiotics (Falkiner, 1997).

Many indexing media will promote the growth of diverse microorganisms. However, growth of many microbial species will not be supported on these media. Because of this uncertainty, it is recommended that indexed negative plant cultures should not be described as being aseptic, sterile, or free of contaminants (Leifert et al., 1994). These cultures can only be described as being indexed

Procedure 18.1
Indexing Stage I Shoot Cultures

Step	Instructions and Comments
1	Aseptically remove shoots from culture and place in sterile petri plate (Figure 18.1a). If shoot consists of multiple nodes (Figure 18.1b), cut into single node segments, leaving the terminal apex intact. If shoots develop from a basal meristem (Figure 18.1b), make thin cross sectional disks at the base. Be sure not to cut through the basal meristem.
2	Slightly crush the node segments or stem disks and inoculate several segments into a vial containing liquid indexing medium (Figure 18.1d). Lightly swirl vial and replace cap but do not tighten completely.
3	Inoculate each agar-solidified slant by first gently stabbing the tissue into the medium at the base of the slant (try not to leave a bubble in the stabbed medium). Slowly lift tissue out of the medium and then lightly streak (drag) the tissue up along the medium surface, leaving it at the top of the slant (Figure 18.1e). This stabbing procedure inoculates the deeper regions in the medium where lower oxygen levels might promote the growth of some microbes.
4	Transfer remaining shoot-tip onto fresh Stage I medium (or inoculate it into the same vial inoculated in step #3). For reference, label both scintillation vials with the same culture number as the transferred shoot culture.
5	Repeat steps 1–5 for the four remaining Stage I cultures. Do not inoculate the remaining vial of liquid and semi-solid medium; label each as "Control."
6	Inoculated indexing medium and control vials are maintained in the dark at 22–30°C for 3 weeks and then screened for visible contamination. Autoclave vials with contamination before disposing.

negative or free of detectable (cultivable) contaminants if such statements are qualified by a description of the microbiological indexing methods used (Leifert et al., 1994).

Questions

- Why is it important to stab and then streak the Stage I tissue sample across the surface of the agar-solidified indexing medium?
- How is it possible to have a contaminated Stage I shoot culture that indexes negative for contamination?
- Why is it recommended to index a culture more than once?

REFERENCES

Cassells, A.C. 1997. Pathogen and microbial contamination management in micropropagation—an overview. In *Pathogen and Microbial Contamination Management in Micropropagation*, A.C. Cassells, Ed., pp. 1–13. Kluwer Academic Publishers, Dordrecht.

Falkiner, F. 1997. Antibiotics in plant tissue culture and micropropagation—what are we aiming at? In *Pathogen and Microbial Contamination Management in Micropropagation*, A.C. Cassells, Ed., pp. 155–160. Kluwer Academic Publishers, Dordrecht.

Holland, M.A. and J.C. Polacco. 1994. PPFMs and other covert contaminants: Is there more to plant physiology than just plant? *Annu. Rev. Plant Physiol. Mol. Biol.* 45:197–209.

Knauss, J.F. 1976. A tissue culture method for producing *Dieffenbachia picta* cv. "Perfection" free of fungi and bacteria. *Proc. Fla. State Hort. Soc.* 89:293–296.

Knauss, J.F. and M.E. Knauss. 1979. Contamination of plant tissue cultures. *Proc. Fla. State Hort. Soc.* 92:341–343.

Leifert, C. and W.M. Waites. 1992. Bacterial growth in plant tissue culture media. *J. Appl. Bacteriol.* 72:460–466.

Leifert, C. and S. Woodward. 1998. Laboratory contamination management: The requirement for microbiological quality assurance. *Plant Cell Tiss. Org. Cult.* 52:83–88.

Leifert, C. and A.C. Cassells. 2001. Microbial hazards in plant tissue cultures. *In Vitro Cell. Dev. Biol.—Plant* 37:133–138.

Leifert, C., W.M. Waites, and J.R. Nicholas. 1989. Bacterial contaminants of micropropagagated plant cultures. *J. Appl. Bacteriol.* 67:353–361.

Leifert, C., C. E. Morris, and W. M. Waites. 1994. Ecology of microbial saprophytes and pathogens in tissue culture and field grown plants: Reasons for contamination problems in vitro. *Crit. Rev. Plant Sci.* 13:139–183.

Murashige, T. and F. Skoog. 1962. A revised medium for rapid growth and bioassays with tobacco cultures. *Physiol. Plant.* 15:473–497.

Tanprasert, P. and B. Reed. 1997. Detection and identification of bacterial contaminants of strawberry runner explants. In *Pathogen and Microbial Contamination Management in Micropropagation*, A.C. Cassells, (Ed.), pp. 139–143. Kluwer Academic Publishers, Dordrecht.

Thomas, P. 2004. A three-step screening procedure for detection of covert and endophytic bacteria in plant tissue cultures. *Current Sci.* 87:67–72.

Viss, P.R., E.M. Brooks, and J.A. Driver. 1991. A simplified method for the control of bacterial contamination in woody plant tissue culture. *In Vitro Cell Dev. Biol.* 27P:42.

19 Propagation from Nonmeristematic Tissues—Organogenesis

Robert L. Geneve

CONCEPTS

- Adventitious shoots and roots are the most common forms of adventitious organ formation.
- Vegetative propagation via organogenesis is an important reproduction method used naturally by plants. Certain plants have evolved unique alternative life cycles that use organogenesis rather than seed propagation to exploit unique habitats.
- Clonal propagation via organogenesis is also an important alternative to seed propagation in agricultural, horticultural, and forestry production systems.
- Organogenesis in tissue culture can be exploited in novel ways for plant improvement including the induction of somaclonal variation and as an important step in the induction and recovery of genetically transformed plants.
- Organogenesis progresses through dedifferentiation and redifferentiation phases. During these phases, plant cells become competent to respond to inductive signals and determined for organ fate (i.e., an adventitious root or shoot).
- The ratio of auxin to cytokinin is the critical factor determining whether morphogenesis will result in either an adventitious shoot or root in competent explants.
- In addition to hormone signals, tissue maturity of the explant donor can be critical to the explant's competency to respond to morphogenetic signals to form adventitious organs.
- In the modern age of bioinformatics, the most important model systems for studying adventitious organ formation will have tissue that is easily genetically transformed, have available or easily induced mutations, and significant genome sequence information.

Nonmeristematic or adventitious organogenesis is de novo organ formation from differentiated plant tissues. Organs produced include shoots, roots, geophytic structures (i.e., bulbs, tubers), and flowers. All these organs are integral to either reproductive or asexual propagation. It is interesting to consider adventitious plant organogenesis in the current context and discussion surrounding stem cell research in mammals. Current evidence suggests that only a few cells in mammals possess the ability to reverse their cell fate and differentiate into a new organ. In mammals, this ability appears to be primal and resides in embryonic tissue prior to significant tissue differentiation. In contrast, both embryonic and relatively mature cell lines from many diverse plant tissues retain the ability to alter their cell fate and differentiate into novel plant organs. The precocious nature of plant cells was suggested in the 1800s in the concept of totipotency (Gautheret, 1983). Totipotency suggests that a plant cell retains the genetic architecture to regenerate an entire plant. The quest for experimental evidence for totipotency led to the advent of plant tissue culture.

Experiments toward demonstrating totipotency were first initiated by Gottlieb Haberlandt (1902). He attempted to grow isolated plant cells in a nutrient solution developed 40 years earlier by Johann Knop for hydroponic plant growth. The cells would increase in size but they failed to divide and multiply. However, he observed that slices of potato tuber that contained a vascular bundle would show some cell division. He concluded that there was a substance in the cut potato tissue that induced cell division and called this substance the "wound hormone."

Roger Gautheret (1934) grew callus from cambial tissue isolated from several woody plants including willow (*Salix*), sycamore (*Platanus*), and elder (*Sambucus*). These were the first sustained dividing callus cultures, but they would only survive about 6 months on his new agar-based nutrient medium. He rightfully concluded that something was missing from the culture medium that would sustain unlimited cell growth in tissue culture. Auxin was isolated as the first plant hormone in the late 1930s. Soon after, auxin was added to tobacco (*Nicotiana*) callus cultures with the result that they exhibited sustained cell division and growth (Nobecourt, 1939; White, 1939).

In 1948, Folke Skoog and his colleagues discovered kinetin as a substance that could induce organ formation in callus cultures of tobacco. This new class of plant hormone was called *cytokinin* because of its ability to induce cell division. It was later established that the ratio of cytokinin and auxin determined whether tobacco callus would make roots or shoots (Skoog and Miller, 1957). This is the fundamental basis for all subsequent tissue culture systems.

In 1954, Hildebrandt and his colleagues, still on the quest to demonstrate totipotency, were able to produce proliferating callus cultures from single cells (Muir et al., 1954). However, it wasn't until 1965 that Vasil and Hildebrandt (1965) were finally able to produce a complete plant derived from a single tobacco cell. Thus, totipotency was shown to exist in plant cells.

SIGNIFICANCE OF ORGANOGENESIS TO PROPAGATION

Vegetative propagation via organogenesis is an important reproduction method used in nature by plants. Plants have evolved unique alternative life cycles that bypass typical seed production in favor of clonal organogenesis reproduction systems. This may seem counter intuitive because sexual propagation leads to a greater genetic diversity in offspring. These sexual offspring have a higher potential to adapt to new or changing environments compared to clonal plants. However, investing in clonal reproduction seems to increase the likelihood that a species can colonize specific and often unique environmental niches. Vegetative reproductive systems are incredibly unique and their diversity is listed in Table 19.1. These reproductive structures can be aerial or subterranean modifications of stems, leaves, inflorescence, and roots. They range from basal stem modifications that slowly colonize local areas (like offset production in geophytes) to the production of detachable vegetative plantlets (like aerial bulbils) that develop on leaves or flower stems that can be disseminated much like true seed (Figure 19.1). When the clonal progeny is very adaptive to a particular environment, it becomes the dominant reproduction unit (Eriksson, 1993). An extreme example can be seen in aspen (*Populus tremuloides*) populations that can have over 1,000 clonal offspring (ramets) from a single dominant mother plant (genet) covering over 30 acres (Kemperman and Barnes, 1976).

Clonal propagation via organogenesis is also an important alternative to seed propagation in agricultural, horticultural, and forestry production systems (Hartmann et al., 2002). Cutting propagation is an important commercial technique that relies on adventitious root formation on stems or adventitious shoots on roots. Clonal propagation of monocots is often from adventitious shoot production, and the majority of their root systems are formed adventitiously on stems. Tubers are stem tissue modified as storage organs and often form adventitiously on modified stems called *stolons*. Potatoes propagated from tubers may represent the oldest example of clonal reproduction of a food crop. An extension of these propagation systems is found in tissue culture via micropropagation systems. Since the pioneering work by Skoog and Miller (1957) and Murashige and Skoog (1962) micropropagation has been adopted as an alternative propagation method in almost all phases of food, fiber, and ornamental plant production systems. Since the mid-1900s, there have been reports of thousands of plants that can be successfully regenerated from tissue culture.

TABLE 19.1
Clonal Propagation Strategies in Nature That Use Adventitious Organ Formation

	Representative plants	
Clonal reproduction strategies	**Common name**	**Latin name**
1. Modified stems (basal)	Tulip	*Tulipa*
a. Bulbs (offsets)	Lily	*Lillium*
b. Tubers	Potato	*Solanum tuberosum*
	Wind flower	*Anemone*
c. Corms (cormels)	Crocus	*Crocus*
	Gayfeather	*Liatris*
d. Stolons (runners)	Strawberry	*Fragaria*
	Red stem dogwood	*Cornus stolonifera*
2. Modified stems (above ground)	European bittercress	*Dentaria bulbifera*
a. Arial bulbs (bulbils)	Lily	*Lillium*
b. Arial tubers (tubercles)	Meadow saxifrage	*Saxifraga granulata*
	Hardy begonia	*Begonia evansonia*
	Winged yam	*Dioscorea alata*
3. Plantlets on leaves	Mother of thousands	*Kalanchoe daigremontiana*
	Piggy-back plant	*Tolmiea menzeisii*
	Hen and chick fern	*Asplenium bulbiferum*
	Walking fern	*Camptosorus rhizophyllus*
4. Plantlets on roots	Pawpaw	*Asimina triloba*
	Blackberry	*Rubus*
5. Plantlets on inflorescence	Century plant	*Agave americana*
	Wild garlic	*Allium vineale*
	Spiderplant	*Chlorophytum comosum*
	Orchard grass	*Dactylis glomerata*
	Meadow grass	*Poa bulbosa*
6. Roots on stems	English ivy	*Hedera helix*
	Figs	*Ficus*

SIGNIFICANCE OF ORGANOGENESIS TO PLANT IMPROVEMENT

In addition to micropropagation, organogenesis in tissue culture has been exploited in novel ways for plant improvement including the induction of somaclonal variants, chimera separation, di-haploid formation, and as an intermediate step in the recovery of genetically transformed plants.

Somaclonal variation is the general term used to describe genetic mutations that occur during the tissue culture process. An in-depth discussion of somaclonal variation can be found in Chapter 43. It is being noted here because it is generally accepted that plants regenerated from adventitious shoots have a greater chance of showing somaclonal variation compared to micropropagation systems using only axillary shoot proliferation (Tripepi, 1997). In a clonal propagation system, variation in the regenerants is obviously a concern and needs to be continually monitored during commercial micropropagation. However, somaclonal variation can be a benefit for selection of new cultivars with abiotic or biotic stress tolerance as well as unique ornamental characteristics (Bouman and de Klerk, 1997; Jain, 2001).

One type of genetic variation that consistently shows up in nature as well as during tissue culture is chimeras. A chimera is generated when one of the histogen layers in the meristem has a different genetic make-up compared to the other layers (Marcotrigiano and Bernatzky, 1995). The most commonly selected chimera is one where one of the layers fails to produce chlorophyll resulting in a green and white variegated leaf form. Other chimeras can exhibit different ploidy levels within layers, different pigment production, or epidermal thorns. It is also common for chimeras to form

FIGURE 19.1 Comparison of in vivo and in vitro organogenesis. 1. Shoot organogenesis on intact leaves in begonia (*Begonia hispida*) (1a) versus in vitro shoot induction in poplar leaf discs (1b). 2. Adventitious root formation in screwpine (*Pandanus*) in the form of brace roots (2a) versus adventitious roots on *Arabidopsis* hypocotyl cuttings (2b). 3. Aerial bublils (3a) versus in vitro formed bulblets (3b) in lily.

during attempts at genetic transformation. In many of these cases, the researcher or breeder will want to apply adventitious shoot regeneration protocols to separate the chimeral layers to produce a plant with identical genetics in each of the meristem's histogen layers. A good example of this type of chimeral separation can be found in thornless mutants of plants such as blackberry (*Rubus;* McPheeters and Skirvin, 1983) and rose (*Rosa;* Canli and Skirvin, 2008). Thornless plants that are chimeral, epidermal mutations can be unstable often reverting back to thorny phenotypes. One method to stabilize these plants is to regenerate them from adventitious shoots that initiate exclusively from all thornless cell types. The resultant plants are nonchimeral and stable for thornlessness.

Traditionally breeding relies on backcrosses and self-crosses to stabilize genetics for seed produced plants (Anderson, 2005). The objective is to create isogenic or near isogenic lines for controlled pollinated seed crops or for use as inbred hybrid parents. This is a time-consuming process for the plant breeder requiring many generations of selection that can be circumvented by employing di-haploid tissue culture protocols. This process involves the generation of haploid plants via anther culture followed by chemical treatments like colchicine to reestablish the diploid status of the tissue culture regenerants. Haploid culture requires either an organogenesis or somatic embryogenesis step for regeneration as an initial step in this di-haploid process.

Finally, it should be noted that except for the seed transformation systems, genetic transformation relies on tissue culture as an intermediate step to recover genetically altered transformants. Because of the nature of biolistic and *Agrobacterium*-mediated transformation, adventitious organogenesis is the most common method for regenerating and screening potential genetic transformants for the traits of interest.

CONCEPTUAL BASIS FOR ADVENTITIOUS PLANT ORGANOGENESIS

There are two general phases of organogenesis—dedifferentiation and redifferentiation (Figure 19.2). During the dedifferentiation phase, the plant cell must reverse its cell state and become "competent" to express its organogenic potential. There are two patterns for the early stage of dedifferentiation. In the direct pattern, specific cells in the explant directly become competent for organ formation. In the indirect pattern, there is an intervening cell division (callus) stage in which these new cells become competent. Competent cells are then able to respond to an induction treatment and progress toward the attainment of an organ specific fate. Once this organ specific fate is acquired the cells are considered "determined" to move into the redifferentiation phase of organogenesis and produce the new organ.

The concepts of competency and determination for plant organogenesis were first suggested in a series of papers by Christianson and Warnick (1983, 1984, 1985). They used transfer experiments with leaf explants of field bindweed (*Convolvulus arvensis*) to establish critical time periods for shoot or root organogenesis. They first established auxin and cytokinin ratios to produce shoot-inducing (SIM), root-inducing (RIM), or callus-inducing (CIM) media. They then moved explants from one medium to another at different times to establish periods when the tissue was competent to respond to root or shoot induction treatments. They also established the time when explants became determined for organ formation. Determination is the phase where the tissue proceeds to organ morphogenesis, independent of the type of induction signal. For example, an explant placed on CIM for 3 days, then moved to SIM and subsequently producing shoots, would be considered competent to respond to SIM induction by the third day of culture. Similarly, an explant placed on SIM for 10 days and then moved to CIM or RIM would be considered determined by 10 days if they formed shoots.

Similar experimental approaches have been established for other species such as tobacco (Dhaliwal et al., 2003) and *Arabidopsis* (Valvekens et al., 1988) that support the dedifferentiation–redifferentiation model. The recent use of *Arabidopsis* as a model system for organogenesis has dramatically advanced our insight into the gene activity during these phases of organogenesis (Zhang and Lemaux, 2004; Zhao et al., 2008). The molecular dissection of organogenesis will be further discussed in detail in Chapter 20.

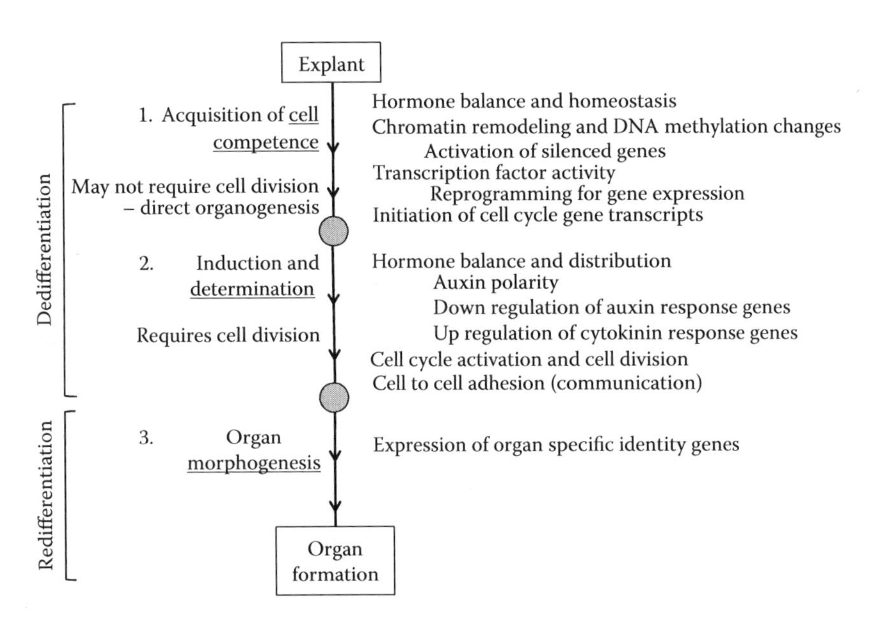

FIGURE 19.2 A scheme to describe events leading to adventitious organ formation.

Hormone Balance and Distribution

Since the classic experiments by Skoog and Miller (1957), it has been recognized that in tissue culture, the ratio of auxin to cytokinin drives morphogenesis toward either root or shoot organogenesis. This ratio is still a major driving force during organogenesis, and we now recognize that this balance establishes regulatory networks initiated to coordinate different steps in the differentiation pathway (Ramirez-Parra et al., 2005). It has also become apparent that the auxin to cytokinin ratio is established by hormone-initiated cross-talk and that other hormones (i.e., ethylene, gibberellin, abscisic acid, brassinosteroids, and polyamines) impact organogenesis by influencing factors within the regulatory network.

One interesting question about tissue culture organogenesis is why do only selective cells within an explant respond to hormone signals to form organs? It must be recognized that explants are relatively complex tissues made up of a plurality of cell types. Therefore, it is reasonable to expect that not all cells within an explant would have the same auxin to cytokinin ratio after exogenous hormone application, nor would each cell have the same competency for organ formation.

The movement and distribution of auxin has been shown to be important for cell attainment of organogenic competency (Vieten et al., 2007). Auxin is unique among the hormones because it only moves in a polar direction. Thus, the movement from cell to cell can be under the regulatory control of efflux carriers. Establishment of an auxin gradient within a tissue has been shown to be critical for pattern organ formation at the whole-plant level during embryo and shoot formation (Vieten et al., 2007). Stated another way, asymmetric distribution of auxin between cells is essential for directing organogenesis. Impeding auxin distribution either chemically or genetically disrupts normal organ formation (Benkova et al., 2003). This asymmetric distribution has been elegantly demonstrated using a green fluorescent protein reporter gene fused to a PIN1 promoter sequence (Gordon et al., 2007). PIN1 codes for an auxin efflux carrier protein and its expression would indicate a cell's ability to transport auxin. It was observed that this reporter construct was more highly expressed in cells that were to become adventitious shoot meristems. This suggests one physiological reason why only certain cells within an explant become organogenic. Only certain cells along the auxin gradient will have an internal concentration to complement the cytokinin concentration (and possibly other factors) to establish the hormone balance required for organogenesis.

For cytokinin, an asymmetric concentration gradient may not be as important as the cell's ability to respond to the cytokinin signal. The expression of a cytokinin-regulated response gene (ARR5) has been followed in *Arabidopsis* during shoot organogenesis (Che et al., 2002; Gordon et al., 2007). This gene is expressed during the early stages of dedifferentiation with its highest levels around 6 days, which is the time of cell determination for shoot formation. At this time, the level of gene transcription declines in those cells that will become shoot meristems setting up an asymmetric signally gradient among different cells within the explant.

Ethylene production is increased in most tissue culture systems studied (Biddington, 1992) and appears to play a modulating role in adventitious organ formation. Increased ethylene production is most likely the result on upregulation of ethylene biosynthesis genes induced by both auxin and cytokinin commonly used in tissue culture induction media. Numerous studies have shown that ethylene can have a detrimental or promotive effect on shoot and root organogenesis (Kumar et al., 1998; Mudge, 1988). Most of these studies have used exogenous chemical application to increase or decrease ethylene levels that could simply have unrelated pharmacological effects, but genetic studies have yielded similar contradictory results. For example, tomato ethylene perception mutants showed reduced adventitious root formation on stem cuttings (Clark et al., 1999) and leaf discs (Kittrel et al., 2006), while exogenous ethylene application decreased auxin-induced root formation (Coleman et al., 1980). Using gene expression profiling in *Arabidopsis*, Che et al. (2006) observed increases in numerous ethylene-related genes for biosynthesis and signal transduction during shoot organogenesis. Studies using ethylene insensitive and ethylene constitutive response mutants in *Arabidopsis* indicate that ethylene enhances shoot organogenesis via increased activity in the ethylene signal transduction pathway (Chatfield and Raizada, 2008) possibly enhancing tissue sensitivity to auxin (Chilley et al., 2006). However, exogenous ethylene application as well as increased-production mutants (*eto1*) have reduced capacity for shoot production. It was suggested that this apparent contradiction may occur when excessive ethylene levels increase senescence-related responses in explants.

Tissue Compentency

Another fundamental question about tissue culture organogenesis is whether within a given tissue there are cells with a greater competency to form organs or a particular organ (i.e., roots or shoots). There are systems that suggest both cases. For example, in tobacco leaf disc explants, shoots developed from the cells along the cut edge, while root primordia initiated from internal cells (Attfield and Evans, 1991). Dhaliwal et al. (2003), using transfer treatments between SIM and RIM, clearly showed that meristematic centers that initiated at these two sites in tobacco leaf discs could not be interconverted from a shoot to root meristem or vice versa. In contrast, field bindweed (*Convulvulus*) root explants initiated both roots or shoots internally from cells associated with the protoxylem (Bonnett and Torrey, 1966) and petunia (*Petunia inflata*) leaf explants produced both organs along the edge (Handro et al., 1973).

Differential regeneration capacity for adventitious shoots on different explants from the same plant has been documented numerous times. For example, in St. John's wort (*Hypericum perforatum* L.) root explants were more responsive compared to leaf explants treated with the same growth regulator combinations (Zobayed and Saxena, 2003). For leaf sections compared to petiole explants, petioles produced more adventitious shoots compared to leaf sections in sugar beet (*Beta vulgaris* L.), pathos (*Epipremnum aureum* Linden and Andre), and petasites (*Petasites hybridus* Gaertn., Mey. and Scherb.) (Qu et al., 2002; Wildi et al., 1998; Zhang et al., 2001).

This differential response seen between tissue sources may be due to tissue competency, which in turn could be due to different levels of inductive factors (morphogens) preexisting in the tissue. This can be illustrated in a comparison of seedling Kentucky coffeetree (*Gymnocladus dioicus*) petiole versus root explants (Geneve, 2005). Root explants have the capacity to form adventitious shoots on the basal medium (no exogenous growth regulators), while petiole explants required exogenous cytokinin treatment to become morphogenic (Figure 19.3). Given the importance of the cytokinin to auxin balance for shoot organogenesis discussed earlier, this suggests that root explants have a critical innate cytokinin signal while petiole explants do not.

Tissue maturity of the explant donor is also critical to the explant's competency to respond to morphogenetic signals to form adventitious organs (Hackett, 1985). A plant's life cycle proceeds from seedling through vegetative into reproductive phases that have distinct characteristics (Hartmann et al., 2010). This progression from juvenile to mature phase is a form of ontogenetic aging and the consequences of phase change can have profound effects on regeneration capacity (Hackett and Murray, 1997). Phase change is critical to understanding flowering and organogenic competency and has been extensively reviewed (Poethig, 1990, 2003). It is a large body of information that is beyond the scope of this chapter, but a case study from pawpaw (*Asimina triloba*) will serve to illustrate the key points for the relationship between plant maturity and adventitious regeneration capacity (Geneve et al., 2003). In attempts to establish micropropagation systems, three initial explant sources representing different maturity states were compared for shoot regeneration. These included stem explants taken from 12-week-old seedlings, rejuvenated shoots derived adventitiously from roots cuttings, and mature shoots from plants that had reached a flowering stage. Seedling explants established at 100% and developed faster than other explants. Explants from rejuvenated stems established at a slower rate but produced shoots in over 40% of the cultures after 8 weeks. Of the 551 mature explants cultured only 4% survived in culture and none produced shoots after

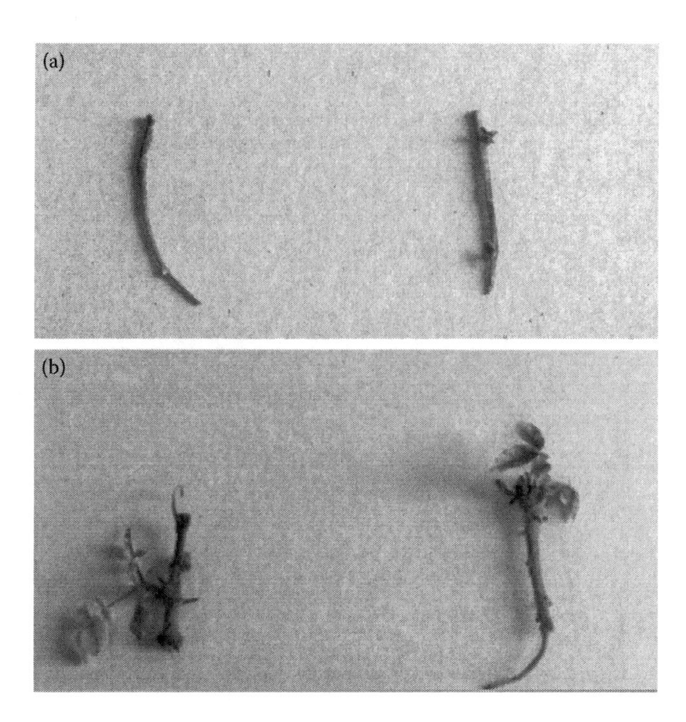

FIGURE 19.3 Shoot organogenesis in Kentucky coffeetree (*Gymnocladus*) isolated petioles (a) and roots (b) on a growth regulator free medium. Under these conditions, only root explants form adventitious shoots, while petiole explants would only form shoots when exposed to cytokinin.

8 weeks in culture. Similar results were obtained in attempts to root pawpaw cuttings. Pawpaw cuttings taken from seedlings of various ages demonstrate the significant impact of juvenility on rooting competency. Seedlings up to 2 months old showed a high capacity for adventitious root formation. Cuttings treated with auxin rooted at 75% and averaged approximately two roots. Seedlings beyond 2 months old showed a reduced capacity to form roots, and seedlings completely lost rooting capacity after 1 year.

The ability to form adventitious roots on cuttings has become diagnostic for plant maturity status in many woody perennials (Hackett, 1985; Greenwood and Hutchinson, 1993). As woody perennial plants transition from the juvenile to mature phase, there can be a dramatic loss in rooting competency. During this process, direct and indirect patterns of adventitious root formation can be recognized as illustrated for radiata pine (*Pinus radiata*; Smith and Thorpe, 1975) and English ivy (*Hedera helix*; Geneve et al., 1988). Juvenile cuttings tend to be easy-to-root and form root primordia directly from existing phloem parenchyma cells. As plant's mature, these cells lose their competency to directly form adventitious roots, but respond to auxin to first produce callus cells that may gain competency for root formation in an indirect pattern (Figure 19.4). Using English ivy debladed petiole explants as a model, it was demonstrated that juvenile phase explants had innate, competent root-forming cells as they responded to only a single 24-hour dose of auxin to directly form adventitious roots (Geneve, 1991). These cells lost that competency if cultured on an auxin-free medium for 15 days prior to auxin exposure. These explants reverted from a direct to an indirect pattern where roots only formed from callus cells rather than existing phloem parenchyma cells. However, if the base of the petiole was re-cut on day 15 prior to auxin exposure, these petioles again rooted in a direct pattern. This suggests that wounding may be part of an initial signal for these cells to directly respond to auxin for rooting.

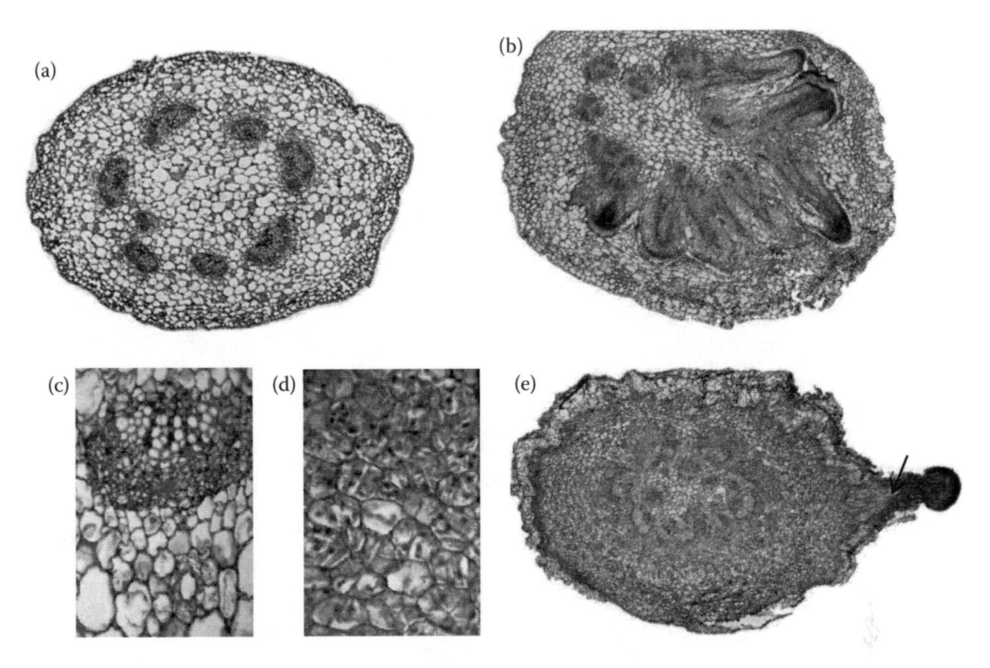

FIGURE 19.4 A comparison of adventitious root formation in isolated de-bladed petioles of English ivy (*Hedera*). (a) Basic cross-sectional anatomy prior to auxin treatment. (b) Direct adventitious root formation in juvenile phase explants. Note the origin of the roots is associated with a vascular bundle, specifically phloem parenchyma tissue. (c) Close-up showing initial polar cell divisions in phloem parenchyma that directly respond to auxin for root formation. (d) Close-up of nondirected cell division in the cortex of mature phase explants. (e) Indirect adventitious root formation in mature phase explants. Note root origin is from callus cells with no direct connection to the vascular bundle. (From Geneve, R.L., Hackett, W.P., and Swanson, B.T. 1988. *J. Amer. Soc. Hor. Sci.* 113:630–635. With permission.)

Novel Regeneration Systems

A number of novel systems have been developed that are derivations of traditional leaf, petiole, root, and stem explants. These include thin cell layers, thin sections, micro-cross sections, nodules, and bulb scale bioreactors. Each of these systems has a common goal for reducing the size of the explant to maximize the ratio of responding and nonresponding cells in order to reduce the dilution effect during biochemical or molecular extractions. They also strive for quick responses and the plasticity to respond to different morphogens to initiate multiple organs.

Thin cell layer explants are epidermal peels or thin transverse sections that are only a few cell layers thick (Nhut et al., 2003; Teixeira da Silva, 2003). Depending on the species, location of extraction from the donor plant, and the growth regulator used, thin cell layer explants can initiate callus, shoots, roots, flowers, or somatic embryos. These were first developed from flowering pedicels of tobacco (Tran Thanh Van, 1973) where all types of vegetative organs as well as flowers could be initiated. In this system, tobacco thin cell layers can produce up to 50 flowers or 800 shoots per explant. The thin cell layer system has also been a good target for *Agrobacterium* or particle biolistic transformation (Teixeira da Silva, 2003). It also has advantages for reduced chimerism in transformed regenerants because of the simplified explant and small number of cells responding to form the adventitious organ. There is now an extensive list of plants amenable for adventitious organ production via thin cell layers. A brief sample includes tobacco, lily (*Lillium*), chrysanthemum (*Dendranthema grandiflora*), petunia (Petunia hybrid), amaranth (*Amaranthus edulis*), sugar beet (*Beta vulgaris*), canola (*Brassica napus*), bean (*Phaseolus vulgaris*), corn (*Zea mays*), orange

TABLE 19.2
Model Systems for Studying Adventitious Organ Formation

Species	Genetic characteristics					
	Mutants	Genome	Transformable	Explant type	Organs initiated	References
Arabidopsis	Yes	Sequenced	Yes	Root	[z]S, R	Zhao et al., 2008
				Leaf	S, R	Gordon et al., 2007
				Hypocotyl	R	Sorin et al., 2005
				Thin layer	S, R	Falasca 2004
Tobacco	Yes	Partial sequence; ESTs, BACs	Yes	Leaf	S, R	Dhaliwal et al., 2003
				Thin layer	S, R, F	Tran Thanh Van, 1973
Tomato	Yes	Partial sequence; ESTs, BACs	Yes	Leaf	S, R	Bhatia et al., 2005
				Thin layer	S, R, F	Compton and Veilleux, 1991
Brassica	Yes	Partial sequence; ESTs, BACs	Yes	Leaf	S	Akasaka-Kennedy et al., 2005
				Thin layer	S	Ben Ghnaya et al., 2008
Rice	Yes	Sequenced	Yes	Seedling	R	Lui et al., 2005
Corn	Yes	Partial sequence; BACs	Yes	Apical meristems	S	Zhang et al., 1998
				Seedling stem	R	
Poplar	No	Sequenced	Yes	Leaf	S, R	Tsai, 1994
				Root	S, R	Jehan et al., 1994
				Stem cutting	R	Kohler et al., 2004
				Nodules	S, R	McCown et al., 1998
				Stem internodes micro-section	S	Lee-Stadelmann et al., 1989
Medicago	Yes	Sequenced	Yes	Leaf	S, R	Nolan et al., 2003
				Cotyledon	S	Voisey et al., 1994
Pinus taeda	No	*Partial sequence; ESTs*	Yes	Hypocotyl	R	Brinker et al., 2004
				Seedling	S	Tang et al., 2001
Convulvulus	No	No	No	Leaf	S, R	Christianson and Warnick, 1983
				Root	S, R	Bonnett and Torrey, 1966
Mung bean	No	No	No	Hypocotyl	R	Blazich and Heuser, 1979
Hedera	No	No	No	Petiole	R	Geneve et al., 1988

[z] Shoots (S), roots (R), and flowers (F).

(*Citrus* and *Poncirus*), banana (*Musa*), tomato (*Solanum esculentum*), pine (*Pinus radiata*), rose (*Rosa*), and poplar (*Populus;* Nhut et al., 2003; Teixeira da Silva, 2003 and references within). Recently, Falasca et al. (2004) developed a thin cell layer technique in *Arabidopsis* for adventitious root formation. This opens an exciting area for utilizing the extensive genetic toolbox *Arabidopsis* brings to studying this basic organogenic process.

Thin and micro-cross stem sections are similar to thin cell layer explants but are cut as cross-sections and include more cell layers within the stem. The additional explant size is often advantageous for woody perennial crops such as apple (Sedira et al., 2007) and poplar (Lee-Stadelman et al., 1989). Thin stem sections have also proved to be efficient systems for rapid mass propagation of orchids and geophytes.

Liquid culture systems employed as bioreactors have been used to produce mass quantities of meristemoids and bulb clusters. Meristemoids formed in woody perennials have been termed *nodules*. They have a distinctive anatomical structure and show plasticity to form roots or shoots once stabilized on a solid medium (McCown et al., 1991). Bulb and tuber clusters formed in bioreactors can be used for mass propagation in such plants as lily (*Lilium*), *Nerine*, *Gladiolus*, and potato (*Solanum;* Ascough and Fennell, 2004; Paek et al., 2001).

Model Regeneration Systems

Traditional model systems used to study organogenesis (i.e., pea [*Pisum*], mung bean [*Vigna*], sunflower [*Helianthus*], and tobacco) were selected based on their ease of adventitious organ initiation and experimental manipulation rather than their molecular genetic characteristics. Current day model systems should ideally consist of a relatively simple explant with a high ratio of responding to nonresponding cells, tissue that is easily genetically transformed and available, or easily induced mutations, and significant genome sequence information. Identification of explants with organ plasticity would be advantageous as well as access to libraries for microarray and microRNA analysis. Use of microdissection technology would be useful for targeting responding and non-responding cells during organogenesis allowing for larger initial explant selection or even whole plants. A list of model systems is provided in Table 19.2. One additional feature that would be important for a model system is availability of tissue sources with distinct orders of ontogenetic maturity to study epigenetic influences on organogenic competency.

REFERENCES

Akasaka-Kennedy, Y., Yoshida, H., and Takahata, Y. 2005. Efficient plant regeneration from leaves of rapeseed (*Brassica napus* L.): The influence of AgNO3 and genotype. *Plant Cell Rep.* 24:649–654.

Anderson, N.O. 2005. Breeding flower seed crops. In *Flower Seeds: Biology and Technology,* (Ed.) McDonald, M.B., and F.Y. Kwong. CAB International, London. 55–86.

Ascough, G.D. and Fennell, C.W. 2004. The regulation of plant growth and development in liquid culture. *S. Afr. J. Bot.* 70:181–190.

Attfield, E.M. and Evans, P.K. 1991. Stages in the initiation of root and shoot organogenesis in cultured leaf explants of *Nicotiana tabacum* cv Xanthi. *J. Exp. Bot.* 42:59–63.

Ben Ghnaya, A., Charles, G., and Branchard, M. 2008. Rapid shoot regeneration from thin cell layer explants excised from petioles and hypocotyls in four cultivars of *Brassica napus* L. *Plant Cell Tiss. Org. Cult.* 92:25–30.

Benkova, E., Michniewicz, M., Sauer, M., Teichmann, T., Seifertova, D., Jurgens, G., and Friml, J. 2003. Local, efflux-dependent auxin gradients as a common module for plant organ formation. *Cell* 115:591–602.

Bhatia, P., Ashwath, N., and Midmore, D.J. 2005. Effects of genotype, explant orientation, and wounding on shoot regeneration in tomato. *In Vitro Cell. Dev. Biol.—Plant* 41:457–464.

Blazich, F.A. and Heuser, C.W. 1979. Mung bean rooting bioassay—re-examination. *J. Amer. Soc. Hort. Sci.* 104:117–120.

Bonnett, H.T. and Torrey, J.G. 1966. Comparative anatomy of endogenous bud and lateral root formation in *Convolvulus arvensis* roots cultured in vitro. *Amer. J. Bot.* 53:496–501.

Bouman, H. and de Klerk G. 1997. Somaclonal variation. In *Biotechnology of Ornamental Plants.* (Ed.) Geneve, R. L., J. Preece, and S. Merkle. CAB International, London. 165–184.

Brinker, M., van Zyl, L., Liu, W.B., Craig, D., Sederoff, R.R., Clapham, D.H., and von Arnold, S. 2004. Microarray analyses of gene expression during adventitious root development in *Pinus contorta*. *Plant Physiol.* 135:1526–1539.

Canli, F.A. and Skirvin, R.M. 2008. In vitro separation of a rose chimera. *Plant Cell Tiss. Org. Cult.* 95:353–361.

Chatfield, S.P. and Raizada, M.N. 2008. Ethylene and shoot regeneration: Hookless1 modulates de novo shoot organogenesis in *Arabidopsis thaliana*. *Plant Cell Rep.* 27:655–666.

Che, P., Gingerich, D.J., Lall, S., and Howell, S.H. 2002. Global and hormone-induced gene expression changes during shoot development in *Arabidopsis*. *Plant Cell* 14:2771–2785.

Che, P., Lall, S., Nettleton, D., and Howell, S.H. 2006. Gene expression programs during shoot, root, and callus development in *Arabidopsis* tissue culture. *Plant Physiol.* 141:620–637.

Chilley, P.M., Casson, S.A., Tarkowski, P., Hawkins, N., Wang, K.L., Hussey, P.J., Beale, M., Ecker, J.R., Sandberg, G.K., and Lindsey, K. 2006. The POLARIS peptide of *Arabidopsis* regulates auxin transport and root growth via effects on ethylene signaling. *Plant Cell* 14:1705–1721.

Christianson, M.L. and Warnick, D.A. 1983. Competence and determination in the process of *in vitro* shoot organogenesis. *Dev. Biol.* 95:288–293.

Christianson, M.L. and Warnick, D.A. 1984. Phenocritical times in the process of *in vitro* shoot organogenesis. *Dev. Biol.* 101:382–390.

Christianson, M.L. and Warnick, D.A. 1985. Temporal requirement for phytohormone balance in the control of organogenesis *in vitro*. *Dev. Biol.* 112:494–497.

Clark, D.G., Gubrium, E.K., Barrett, J.E., Nell, T.A., and Klee, H.J. 1999. Root formation in ethylene-insensitive plants. *Plant Physiol.* 121:53–60.

Coleman, W.K., Huxter, T.J., Reid, D.M., and Thorpe, T.A. 1980. Ethylene as an endogenous inhibitor of root regeneration in tomato leaf discs cultured in vitro. *Physiol. Plant.* 48:519–525.

Compton, M.E. and Veilleux, R.E. 1991. Shoot, root and flower morphogenesis on tomato inflorescence explants. *Plant Cell Tiss. Org. Cult.* 24:223–231.

Dhaliwal, H.S., Ramesar-Fortner, N.S., Yeung, E.C., and Thorpe, T.A. 2003. Competence, determination, and meristemoid plasticity in tobacco organogenesis in vitro. *Can. J. Bot.* 81:611–621.

Eriksson, O. 1993. Dynamics of genets in clonal plants. *Trends Ecol. Evol.* 8:313–316.

Falasca, G., Zaghi, D., Possenti, M., and Altamura, M.M. 2004. Adventitious root formation in *Arabidopsis thaliana* thin cell layers. *Plant Cell Reports* 23:17–25.

Gautheret, R.J. 1934. The culture of cambial tissue. *Comptes Rendus Hebdomadaires Des Seances De L Academie Des Sciences* 198:2195–2196.

Gautheret, R.J. 1983. Plant tissue culture—a history. *Botanical Magazine* 96:393–410.

Geneve, R.L. 1991. Patterns of adventitious root formation in English ivy. *J. Plant Growth Reg.* 10:215–220.

Geneve, R.L. 2005. Comparative adventitious shoot induction in Kentucky coffeetree root and petiole explants treated with thidiazuron and benzylaminopurine. *In Vitro Cell. Dev. Biol—Plant* 41:489–493.

Geneve, R.L., Hackett, W.P., and Swanson, B.T. 1988. Adventitious root initiation in de-bladed petioles from the juvenile and mature phases of English ivy. *J. Amer. Soc. Hor. Sci.* 113:630–635.

Geneve, R.L., Pomper, K.W., Kester, S.T., Egilla, J.N., Finneseth, C.L.H., Crabtree, S.B., and Layne, D.R. 2003. Propagation of pawpaw—A review. *Horttechnology* 13:428–433.

Gordon, S.P., Heisler, M.G., Reddy, G.V., Ohno, C., Das, P., and Meyerowitz, E.M. 2007. Pattern formation during *de novo* assembly of the *Arabidopsis* shoot meristem. *Development* 134:3539–3548.

Greenwood, M.S. and Hutchison, K.W. 1993. Maturation as a developmental process. In *Clonal Forestry I: Genetics and Biotechnology.* (Ed.) M.R. Ahuja and W.J. Libby. Springer Verlag. 14–33.

Haberlandt, G. 1902. Kulturversuche mit isolierten Pflanzenzellen. Sitzungsber. *Akademie der Wissenschaften in Wien. Mathematisch-Naturwissenschaftliche* 111:69–92.

Hackett, W.P. 1985. Juvenility, maturation, and rejuvenation in woody plants. *Hort. Rev.* 7: 109–155.

Hackett, W.P. and Murray, J. R. 1997. Approaches to understanding maturation or phase change. In *Biotechnology of Ornamental Plants*. (Ed.) Geneve, R. L., J. Preece, and S. Merkle. CAB International, London. 73–86.

Handro, W., Rao, P.S., and Harada, H. 1973. Histological study of development of buds, roots, and embryos in organ-cultures of *Petunia inflata*. *Ann. Bot.* 37:817–22.

Hartmann, H.T., Kester D.E., Davies, Jr., F. T., and Geneve, R.L. 2002. *Hartmann and Kester's Plant Propagation: Principles and Practices.* Prentice-Hall, Inc., Upper Saddle River, N. J. Seventh edition.

Jain, S.M., 2001. Tissue culture-derived variation in crop improvement. *Euphytica* 118:153–166.

Jehan, H., Brown, S., Marie, D., Noin, M., Prouteau, M., and Chriqui, D. 1994. Ontogeny and ploidy level of plantlets regenerated from *Populous trichocarpa x deltoides* cv Hunnegem root, leaf and stem explants. *J. Plant Physiol.* 144:576–585.

Kemperman, J.A. and Barnes, B.V. 1976. Clone size in American aspens. *Can. J. Bot.* 54:2603–2607.

Kittrel, K., S.T. Kester, and R.L. Geneve. 2006. Adventitious root formation in tomato hormone mutants. *Comb. Proc. Intl. Plant Propagator's Soc.* 56:453–457.

Leestadelmann, O.Y., Lee, S.W., Hackett, W.P., and Read, P.E. 1989. The formation of adventitious buds invitro on micro-cross sections of hybrid *Populus* leaf midveins. *Plant Sci.* 61:263–272.

Lian, M.L., Chakrabarty, D., and Paek, K. 2003a. Bulblet formation from bulb scale segments of *Lilium* using bioreactor system. *Biologia Plantarum* 46:199–203.

Lian, M.L., Chakrabarty, D., and Paek, K.Y. 2003b. Growth of *Lilium oriental* hybrid "Casablanca" bulblet using bioreactor culture. *Scientia Horticulturae* 97:41–48.

Marcotrigiano, M. and Bernatzky, R. 1995. Arrangement of cell-layers in the shoot apical meristems of periclinal chimeras influences cell fate. *Plant J.* 7:193–202.

McCown, B.H., McCabe, D.E., Russell, D.R., Robison, D.J., Barton, K.A., and Raffa, K.F. 1991. Stable transformation of populus and incorporation of pest resistance by electric-discharge particle-acceleration. *Plant Cell Rep.* 9:590–594.

McPheeters, K. and Skirvin, R.M. 1983. Histogenic layer manipulation in chimeral thornless evergreen trailing blackberry. *Euphytica* 32:351–360.

Mudge, K.W. 1988. Effect of ethylene on rooting. In *Adventitious Root Formation in Cuttings.* (Ed.) T.D. Davis, B.E., Haissig, N. Sankhla. Dioscorides Press, Portland, OR. 150–161.

Muir, W.H., Hildebrandt, A.C., and Riker, A.J. 1954. Plant tissue cultures produced from single isolated cells. *Science* 119:877–878.

Murashige, T. and Skoog, F. 1962. A revised medium for rapid growth and bioassays with tobacco tissue cultures. *Physiologia Plantarum* 15:473—&.

Nhut, D.T., Da Silva, J.A.T., and Aswath, C.R. 2003. The importance of the explant on regeneration in thin cell layer technology. *In Vitro Cell. Dev. Biol.—Plant* 39:266–276.

Nobecourt, P. 1939. The life expectancy and augmentation in the volume of vegetal tissue cultures. *Comptes Rendus Des Seances De La Societe De Biologie Et De Ses Filiales* 130:1270–1271.

Nolan, K.E., Irwanto, R.R., and Rose, R.J. 2003. Auxin up-regulates MtSERK1 expression in both *Medicago truncatula* root-forming and embryogenic cultures. *Plant Physiol.* 133:218–230.

Ozawa, S., Yasutani, I., Fukuda, H., Komamine, A., and Sugiyama, M. 1998. Organogenic responses in tissue culture of srd mutants of *Arabidopsis thaliana. Development* 125:135–142.

Paek, K.Y., Hahn, E.J., and Son, S.H. 2001. Application of bioreactors for large-scale micropropagation systems of plants. *In Vitro Cell. Dev. Biol.—Plant* 37:149–157.

Peer, K.R. and Greenwood, M.S. 2001. Maturation, topophysis and other factors in relation to rooting in *Larix. Tree Physiol.* 21:267–272.

Poethig, R.S. 1990. Phase-change and the regulation of shoot morphogenesis in plants. *Science* 250:923–930.

Poethig, R.S. 2003. Phase change and the regulation of developmental timing in plants. *Science* 301:334–336.

Qu, L.P., Chen, J.J., Henny, R.J., Huang, Y.F., Caldwell, R.D., and Robinson, C.A. 2002. Thidiazuron promotes adventitious shoot regeneration from pothos (*Epipremnum aureum*) leaf and petiole explants. *In Vitro Cell. Dev. Biol.—Plant* 38:268–271.

Ramirez-Parra, E., Desvoyes, B., and Gutierrez, C. 2005. Balance between cell division and differentiation during plant development. *International J. Dev. Biol.* 49:467–477.

Sedira, M., Welander, M., and Geier, T. 2007. Influence of IBA and aphidicolin on DNA synthesis and adventitious root regeneration from Malus "Jork 9" stem discs. *Plant Cell Rep.* 26:539–545.

Skoog, F., and Miller, C.O. 1957. Chemical regulation of growth and organ formation in plant tissues cultured in vitro. *Symp. Soc. Exp. Biol.* 11:118–31.

Smith, D.R. and Thorpe, T.A. 1975. Root initiation in cuttings of *Pinus radiata* seedlings. 1. Developmental sequence. *J. Exp. Bot.* 26:184–94.

Sorin, C., Bussell, J.D., Camus, I., Ljung, K., Kowalczyk, M., Geiss, G., McKhann, H., Garcion, C., Vaucheret, H., Sandberg, G., and Bellini, C. 2005. Auxin and light control of adventitious rooting in *Arabidopsis* require ARGONAUTE1. *Plant Cell* 17:1343–1359.

Tang, W. 2001. In vitro regeneration of loblolly pine and random amplified polymorphic DNA analyses of regenerated plantlets. *Plant Cell Rep.* 20:163–168.

Teixeira da Silva, J.A., 2003. Thin cell layer technology in ornamental plant micropropagation and biotechnology. *Afr. J. Biotechnol.* 2:683–691.

Tran Thanh Van, M. 1973. In vitro control of de novo flower, bud, root and callus differentiation from excised epidermal tissues. *Nature* 245:44–45.

Tripepi, R.R. 1997. Adventitious shoot regeneration. In *Biotechnology of Ornamental Plants*. (Ed.) Geneve, R. L., J. Preece and S. Merkle. CAB International, London. 45–72.

Tsai, C.J., Podila, G.K., and Chiang, V.L. 1994. *Agrobacterium*-mediated transformation of quaking aspen (*Populus tremuloides*) and regeneration of transgenic plants. *Plant Cell Rep.* 14:94–97.

Valledor, L., Hasbun, R., Meijon, M., Rodriguez, J.L., Santamaria, E., Viejo, M., Berdasco, M., Feito, I., Fraga, M.F., Canal, M.J., and Rodriguez, R. 2007. Involvement of DNA methylation in tree development and micropropagation. *Plant Cell Tiss. Org. Cult.* 91:75–86.

Valvekens, D., Vanmontagu, M., and Vanlijsebettens, M. 1988. *Agrobacterium tumefaciens* mediated transformation of *Arabidopsis thaliana* root explants by using kanamycin selection. *Proc. Nat. Acad. Sci.* 85:5536–5540.

Vasil, V. and Hildebrandt, 1965. Differentiation of tobacco plants from single isolated cells in microcultures. *Science* 150:889–890.

Vieten, A., Sauer, M., Brewer, P.B. and Friml, J. 2007. Molecular and cellular aspects of auxin-transport mediated development. *Trends Plant Sci.* 12:160–168.

Voisey, C.R., White, D.W.R., Dudas, B., Appleby, R.D., Ealing, P.M., and Scott, A.G., 1994. *Agrobacterium* mediated transformation of white clover using direct shoot organogenesis. *Plant Cell Rep.* 13:309–314.

White, P.R. 1939. Potentially unlimited growth of excised plant callus in an artificial medium. *American J. Bot.* 26:59–64.

Wildi, E., Schaffner, W., and Buter, K.B. 1998. In vitro propagation of *Petasites hybridus* (Asteraceae) from leaf and petiole explants and from inflorescence buds. *Plant Cell Rep.* 18:336–340.

Zhang, C.L., Chen, D.F., Elliott, M.C., and Slater, A. 2001. Thidiazuron-induced organogenesis and somatic embryogenesis in sugar beet (*Beta vulgaris* L.). *In Vitro Cell. Dev. Biol.—Plant* 37:305–310.

Zhang, S.B. and Lemaux, P.G. 2004. Molecular analysis of in vitro shoot organogenesis. *Critical Reviews in Plant Sci.* 23:325–335.

Zhang, S.B., Williams-Carrier, R., Jackson, D., and Lemaux, P.G. 1998. Expression of CDC2Zm and KNOTTED1 during in-vitro axillary shoot meristem proliferation and adventitious shoot meristem formation in maize (*Zea mays* L.) and barley (*Hordeum vulgare* L.). *Planta* 204:542–549.

Zhao, X.Y., Su, Y.H., Cheng, Z.J., and Zhang, X.S. 2008. Cell fate switch during *in vitro* plant organogenesis. *J. Integ. Plant Biol.* 50:816–824.

Zobayed, S.M.A. and Saxena, P.K. 2003. In vitro-grown roots: A superior explant for prolific shoot regeneration of St. John's wort (*Hypericum perforatum* L. cv "New Stem") in a temporary immersion bioreactor. *Plant Sci.* 165:463–470.

20 Developing a Molecular Understanding of In Vitro and In Planta Shoot Organogenesis

Ling Meng, Shibo Zhang, and Peggy G. Lemaux*

CONCEPTS

- During in vitro micropropagation, shoot organogenesis initiates from differentiated somatic cells.
- Shoot development depends on correct cell-fate determination from the shoot apical meristem.
- Phytohormones play essential roles in controlling shoot organogenesis and development.
- In vitro shoot organogenesis and development is a complex, well-coordinated process, involving certain identified genes and pathways.
- Other, as yet unidentified, molecular events trigger dedifferentiation and act as developmental switches for shoot development.

Both somatic embryogenesis and shoot organogenesis can be involved in the process of in vitro culturing of plant tissue. During shoot organogenesis an adventitious shoot forms, followed by the development of adventitious roots from the shoot, resulting in an entire plant. The process of in vitro shoot induction and development via organogenesis has been reviewed in detail from the perspective of developmental biology (Hick, 1994) and from a physiological, biochemical, and molecular viewpoint (Zhang and Lemaux, 2004b). In this review, we will update a previous review (Zhang and Lemaux, 2004a) with more recent molecular details of in vitro shoot organogenesis.

In planta shoot development from an embryo differs from the process of in vitro shoot organogenesis since in the latter case the shoot meristem initiates from differentiated somatic cell(s), not from embryonic cell(s) as it does during embryogenesis. The process of in vitro shoot organogenesis usually involves three main stages: response of somatic cells to exogenous hormones, cell division of responding cell(s), and initiation and development of new shoots either directly from the newly dividing cell(s) or indirectly through a callus phase. When somatic cells respond appropriately to exogenously applied plant hormone(s), they can activate or accelerate the timing of the cell cycle, resulting in reprogramming of cells that assume the de novo developmental fate of shoot organogenesis.

Cytokinins (CKs), sometimes in concert with auxins, are the most efficient plant growth regulators that induce in vitro shoot organogenesis. Genetic and molecular analyses led to the identification in *Arabidopsis thaliana* (*Arabidopsis*) of important genes involved in the CK signal transduction pathway (Haberer and Kieber, 2002). The *Arabidopsis* histidine protein kinases (AHKs) serve as CK receptors and the histidine phosphotransfer proteins (AHPs) transmit the signal from AHKs to nuclear response regulators (ARRs), which activate or repress transcription. A few key regulatory genes have been shown to be involved in the plant cell cycle, such as p^{34}cdc2 and cyclin genes

* Dedicated to the memory of Shibo Zhang, who continues to inspire our interest in this topic.

(Shaul, 1996). During shoot meristem development, several other regulatory genes were identified (e.g., maize *KNOTTED1 (KN1),* Vollbrecht et al., 1991; *Arabidopsis SHOOT MERISTEMLESS* (*STM,* Long et al., 1996; *WUSCHEL* (*WUS*), Laux, et al. 1996, Mayer et al., 1998; and *CLAVATA 1-3* (*CLV1-3*), Fletcher et al., 1999). Studies suggest that interactions at the molecular level exist among these three processes: CK reception, cell cycle, and shoot meristem development (Riou-Khamlichi et al., 1999).

Identification of these genes suggests that they could be used as molecular markers to develop a better understanding of in vitro shoot organogenesis. Such approaches have also been used to identify genes more directly involved during in vitro shoot organogenesis. The current molecular understanding of in planta and in vitro shoot meristem development, cell cycle regulation, and CK signal transduction will be described in this review.

MOLECULAR ANALYSIS OF IN PLANTA SHOOT MERISTEM DEVELOPMENT

During zygotic embryogenesis the shoot meristem initiates and develops within the embryo that derives from the zygote. Because of its position at the growing tip, the initial shoot meristem becomes the shoot apical meristem (SAM), which generates all stems, leaves, and lateral shoot meristems during the entire process of shoot development (Kwiatkowska, 2008). The SAM, which is located at the shoot apex above the youngest leaf primordium, provides cells for new organ initiation from a pool of cells that undergo continuous renewal by cell division.

Based on histological and anatomical analyses, the SAM can be subdivided into three distinct radial layers (L1, L2, and L3, Figure 20.1A) and zones [peripheral (PZ), central (CZ), and rib (RZ), Figure 20.1B]. L1 and L2 cells form the tunica, dividing exclusively anticlinally to maintain the tunica layer. L3 cells, constituting the corpus tissue, divide both anticlinally and periclinally. The L1 layer gives rise to the epidermis, L2 to mesophyll cells that provide germ line cells that form gametes, and L3 cells to central tissues of the leaf and stem. Lateral organs originate from the PZ; stem tissue derives from the RZ. The CZ holds the stem cell pool that continuously replenishes itself and cells in the PZ and RZ (Kwiatkowska, 2008). The size and the number of stem cells in the SAM appear generally constant, suggesting that the continuous process of cell division and differentiation of daughter cells into organ primordia appear to be well balanced.

Plant shoot development from the SAM depends on correct cell-fate determination and maintenance of a proper stem cell pool and specification of organ (leaf) founder cells from the PZ where lateral organs arise. Cell fate is thought to be largely determined by the spatial position of cues (pattern formation) within a morphogenetic field during critical time periods—not by cell lineage via inherited cues or gene expression (Scheres, 2001). Spatial specification of organ-founder and stem cells is thus critical to proper shoot meristem activity and development.

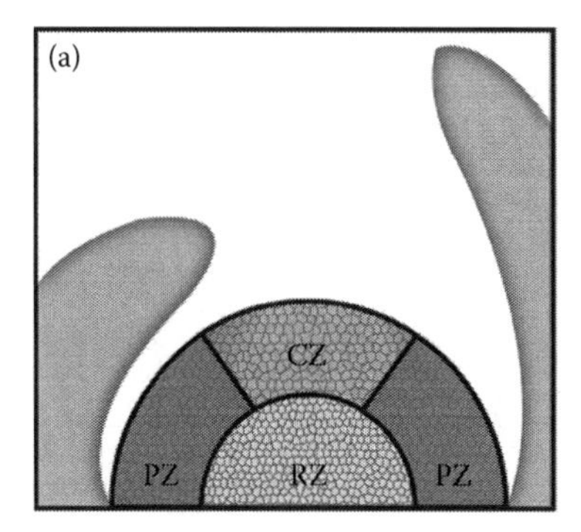

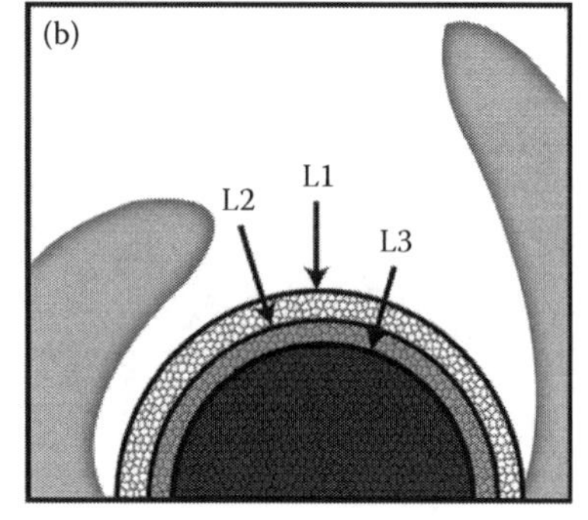

FIGURE 20.1 Diagrams of the shoot apical meristem (SAM) (a) showing the central zone (CZ), peripheral zone (PZ), and rib zone (RZ) and (b) showing the L1, L2, and L3 layers.

Genetic Control of SAM Patterning and Development

Intercellular communication is critical in controlling SAM organization and maintenance, as well as cell-fate specification (Williams and Fletcher, 2005). Genetic and molecular analyses of shoot meristem development in maize, *Arabidopsis*, and other plants have led to identification of genes critical to this process.

KNOX I and Asymmetric Leaves 1 and 2 Regulate Shoot Meristem Development

The gene for the maize homeodomain protein, Knotted 1(*KN1*) that causes "knots" when misexpressed in leaves (Vollbrecht et al., 1991), is expressed exclusively in the SAM and required for its activity and development (Jackson et al., 1994). Expression patterns of *KN1*-homologs in other cereals, like barley, are consistent with this pattern (Figure 20.2). The *KNOX* (knotted-like homeobox) gene family, found in all plant species, is divided into two classes based on expression patterns. Class I KNOX (*KNOX I*) genes are specifically expressed in the SAM and promote meristem activity, whereas class II *KNOX* genes are more widely expressed in plant tissues (Kerstetter et al., 1994). Relevant to shoot organogenesis, ectopic expression of maize *KN1* in tobacco, or its orthologs in *Arabidopsis*, consistently causes adventitious shoot formation in leaf tissues (Lincoln et al., 1994, Sinha et al., 1993). Misexpression in the barley awn induces ectopic meristems that form inflorescence-like structures on the awn (Williams-Carrier et al., 1997)—a phenotype similar to that in the dominant gain-of-function hooded mutant that is caused by misexpression of *HvKNOX3*.

Four KNOX I protein family members were found in *Arabidopsis*: SHOOT MERISTEMLESS (STM), BREVIPEDICELLUS (BP), KN1-like in *A. thaliana* 2 (KNAT2) and KNAT6. *STM*, the first *KNOX* gene shown to have recessive loss-of-function phenotypes, is most similar to maize *KN1* in terms of expression pattern and function (Long et al., 1996). *STM* expression, required to maintain all shoot meristems, is restricted to the SAM, starting in the embryo in a single apical cell at the late globular stage (Long and Barton, 1998). Loss-of-function mutations in *STM* result in premature termination of the SAM (Long et al., 1996). STM and BP act together to maintain SAM activity by preventing meristematic cells from developing organ (leaf)-specific cell fates (Byrne et al., 2002). KNAT2 and KNAT6, which share higher sequence similarity to one another than to STM or BP, do not share identical functions; *KNAT6*, but not *KNAT2*, regulates SAM activity and organ separation in the embryo. Mutations in *KNAT2* and *KNAT6* result in antagonistic interactions with BP; expression of both increases in *bp* mutants, suggesting BP represses expression of *KNAT2* and *KNAT6* (Ragni et al., 2008).

Correct cell fate determination in the SAM depends on mutual expression/repression between *Asymmetric Leaves 1* (*AS1*) and *AS2* and *KNOX I* genes. The latter, which promote meristem activity,

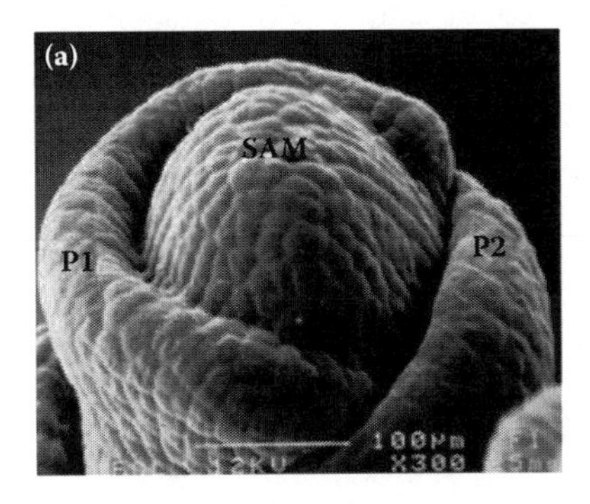

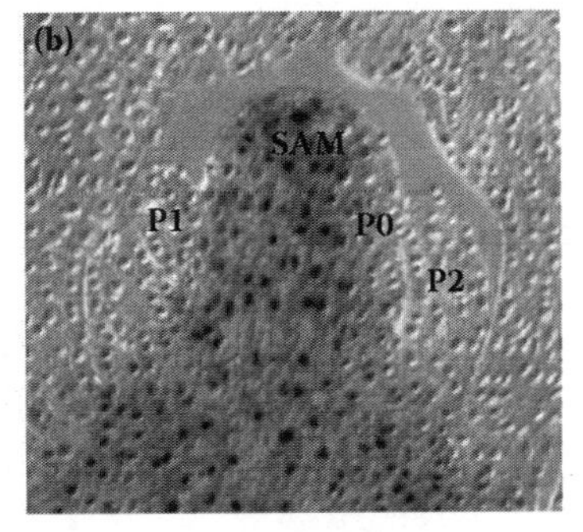

FIGURE 20.2 Expression of *KNOTTED1*-homolog in a shoot apical meristem (SAM) of barley. (a) Scanning electron micrograph of a barley shoot apex including the SAM and two leaf primordia (P1, P2). (b) Immunolocalization of *KNOTTED1*-homolog in barley shoot apex using KNOTTED1 antibody, showing expression of *KNOTTED1*-homolog only in the SAM, not in leaf primordia (P0, P1, and P2). (From Trigiano, R. N. and D. J. Gray, Eds., *Plant Tissue Culture and Development,* CRC Press LLC, Boca Raton, FL, 2004.)

are expressed in the SAM but are down-regulated in founder cells at leaf initiation (Jackson et al., 1994). *AS1* encodes a myb domain transcription factor closely related to *PHANTASTICA* (*PHAN*) in *Antirrhinum* and *ROUGH SHEATH2* (*RS2*) in maize. All three genes, which are expressed in leaf founder cells, influence leaf fate. AS1 represses *KNAT1* and *KNAT2* expression and is itself negatively regulated by STM. In loss-of-function *as1* mutants occasional lobes appear on the leaves similar to those in 35S:BP plants (Byrne et al., 2000, Ori et al., 2000). *BP*, *KNAT2*, and *KNAT6*, but not *STM*, are ectopically expressed in leaves of *as1* mutants. *AS1* expression expands into the SAM in *stm* mutants, consistent with its restricting *AS1* expression.

AS2, which encodes a LATERAL ORGAN BOUNDARIES (LOB) domain transcription factor, is expressed mainly in adaxial domains of leaf primordia. Phenotypes of *as2* mutants are similar to those of *as1*, with rumpled, slightly lobed leaves. AS2 is believed to control formation of symmetric flat leaf lamina and establish a prominent midvein and other patterns of venation (Iwakawa et al., 2002); it also appears to restrict *KNOX I* expression. Expression from *BP*, *KNAT2*, and *KNAT6*, but not *STM*, is ectopic in *as2* mutants. AS1, acting with AS2, directly binds to *BP* and *KNAT2* promoters and may form a repressive chromatin complex (Guo et al., 2008). This mutual interaction between AS1, AS2, and KNOX I defines a mechanism for differentiating stem and organ founder cells within the SAM, demonstrating that genes expressed in organ primordia interact with meristematic genes to regulate shoot morphogenesis (Byrne et al., 2000, 2002). *BLADE-ON-PETIOLE1* (*BOP1*) encodes a transcription factor involved in regulating leaf differentiation; BOP1 is thought to down-regulate *KNOX I* expression during leaf formation. *KNAT1*, *KNAT2*, and *KNAT6* are expressed ectopically in leaves of the *bop1*-1 mutant. BOP1 has synergistic effects with ASI or AS2 in *bop1/as1* or *bop1/as2* double mutants, and with STM in *bop1/stm-1* double mutants suggesting it promotes leaf development via *KNOX1* repression (Ha et al., 2004).

Second Pathway-Regulating SAM Activity

Local signaling to maintain stem cells in the SAM in Arabidopsis involves the CLAVATA-WUSCHEL (CLV-WUS) signal transduction pathway. WUS, distantly related to STM homeodomain transcription factors, acts non-cell autonomously to promote stem cell fate. It is first expressed at the 16-cell stage in subepidermal cells of the SAM but ultimately becomes limited to a few central cells of the RZ in the organizing center (OC; Mayer et al., 1998). In *wus* mutants, shoot meristem development stops after the first few leaves, resulting in bushy plants (Laux et al., 1996). Ectopic *WUS* expression in roots induces shoot meristems and organs, such as leaves, in the root (Gallois et al., 2004). Thus, WUS can establish stem cells with shoot meristem identity and can maintain stem cells in the CZ of the SAM. WUS and STM are activated independently in *Arabidopsis*; however, expression of *STM* is absent in *wus* mutants and vice versa (Lenhard et al., 2002).

WUS induces expression from *CLAVATA3 (CLV3)*, which encodes a small extracellular signal ligand (Fletcher et al., 1999) that is present in stem cells of the CZ. CLV3 is secreted from L1 and L2 toward interior stem cells of the CZ to directly interact (Ogawa et al., 2008) with CLAVATA 1 (CLV1), a leucine-rich repeat (LRR) receptor kinase (Clark et al., 1997) that forms a heterodimeric complex with CLAVATA 2 (CLV2) to suppress *WUS* expression. This feedback regulation allows the SAM to balance stem cell division in the CZ with cell differentiation in the PZ.

CLV3 is expressed in L1 and L2 layers, CLV1 only in L3 of the CZ (Clark et al., 1997), whereas CLV2 transcripts are detected in most plant tissues (Jeong et al., 1999). Loss-of-function CLV1, CLV2, or CLV3 mutations result in expanded *WUS* expression, enlarged shoot and floral meristems, and flowers that contain extra organs (Clark et al., 1997). Conversely, over-expression of *CLV3* causes loss of *WUS* expression and premature shoot and floral meristem termination, showing that interaction between secreted CLV3 and CLV1/CLV2 receptor complex limits the size of the *WUS* expression domain and thus the meristem. The CLV3/WUS-signaling pathway forms a spatial, negative feedback loop that controls stem cell accumulation in shoot and floral meristems (for review, see Williams and Fletcher, 2005).

More than 40 CLV3/ESR (Embryo Surrounding Region)-related (CLE) proteins have been identified in plants and plant-parasitic nematodes. The *CLE* family encodes small, putative signal ligands with a conserved 14-amino acid motif at the C-terminus and a signal peptide at the N-terminus (Cock and Mccormick, 2001). A predominant ectopic over-expression phenotype for some *CLE* genes is premature SAM termination (Strabala et al., 2006). Of note, a 12-amino-acid dodeca-modified peptide derived from the CLE motif was observed in transgenic *Arabidopsis callus* over-expressing *CLV3* (Kondo et al., 2006). When used in culture media, the synthetic 12 or 14 amino acid peptides from several CLE motifs phenocopied their gain-of-function phenotypes, and in some cases rescued *clv3* loss-of-function phenotypes (Fiers et al., 2005). Together, these results suggest that CLE peptides negatively regulate meristem activity; however, there is no direct evidence for a CLV3 peptide–ligand interaction with the CLV1/CLV2 receptor complex (Ogawa et al., 2008).

Regulators of SAM Activity via Interaction with WUS-Signaling Pathway

Besides *CLV* genes, several signaling components, which regulate the SAM via interaction with the WUS pathway, were identified. ULTRAPETALA1 (ULT1), a DNA binding-domain transcription factor found in all meristems and developing stamens, carpels, and ovules, restricts SAM size by repressing *WUS* expression. ULT1 appears to have WUS-independent functions in maintaining SAM activity as an *ult1/wus* double mutant shows additive phenotypes (Williams and Fletcher, 2005). HANABA TARANU (HAN), a GATA-3-like transcription factor, is also a WUS-signaling pathway component. *HAN*, transcribed at the boundaries between the meristem and its newly initiated organ primordia, and in vascular tissues, establishes organ boundaries in shoots and controls the number and correct positioning of WUS-expressing cells.

The BARD1 (BRCA1-associated RING domain 1) protein, containing a RING domain and two tandem BRCA1 C-terminal domains that function in phosphorylation-dependent protein–protein interactions, interacts directly with the *WUS* promoter region. A *bard1* mutant has severe SAM defects, increased *WUS* expression and overexpression of *BARD1*, which causes a *WUS* mutant phenotype. BARD1 is thought to be responsible for limiting *WUS* expression and regulating SAM organization and maintenance (Han et al., 2008).

Two widely expressed homeodomain finger proteins OBERON1 (OBE1) and OBE2 function redundantly to allow plant cells to acquire meristematic activity via the WUSCHEL–CLAVATA pathway. In *obe1/obe2* double mutants, the shoot meristem prematurely terminates with a dramatic reduction in *CLV3* and *WUS* expression. OBE1 and OBE2 are believed to be responsible for cells reaching a state that leads to establishment and maintenance of the meristem (Saiga et al., 2008). Class III homeodomain-leucine zipper (HD-ZIP III) transcription factors also appear involved in SAM maintenance by modulating *WUS* transcription levels in the OC. *Arabidopsis* contains five *HDZIP III* genes, *CORONA/ATHB15* (*CNA*), *REVOLUTA/INTERFASCICULAR FIBERLESS1* (*REV*), *PHABULOSA* (*PHB*), *PHAVOLUTA* (*PHV*), and *ATHB8*. Although *cna* single mutants exhibit subtle defects in meristem development, *clv/cna* double mutants develop a massively enlarged and disorganized SAM and ectopic expression of *WUS* and *CLV3*. Thus, CNA is believed to regulate stem cell identity in a pathway parallel to that of CLVs (Williams and Fletcher, 2005). The triple mutant, *can/phb/phv*, re-creates the *clv* enlarged-SAM phenotype. Thus, the CNAs appear to regulate SAM patterning and meristem function in a complex pathway.

Expression from *WUS* is regulated epigenetically by chromatin remodeling. Loss-of-function mutants in *FASCIATA1* (*FAS1*) or *FAS2*, encoding two subunits of the chromatin assembly factor 1 complex that function by regulating *WUS* expression, have a flat, enlarged, disorganized SAM (Williams and Fletcher, 2005). In *FAS* mutants, the *WUS* expression domain expands laterally, but not uniformly, resulting in a varied pattern. WUS may also be a direct target of the chromatin remodeling factor, SPLAYED (SYD), which interacts with the *WUS* promoter, giving evidence that *WUS* expression is regulated at the chromatin level. Mutants in *SYD* with reduced transcription levels from *WUS* and *CLV3* prematurely terminate the SAM.

Besides the KNOX I-AS1 and WUS-CLV pathways, other genes are involved in shoot meristem development and maintenance. *CUP-SHAPED COTYLEDON1* (*CUC1*) and *CUC2* are essential for meristem establishment and in forming boundaries between meristems and adjacent organs. Both *CUC1* and *CUC2*, which are expressed in a strip of cells at the apex of the globular-stage embryo, are involved in embryonic SAM formation and cotyledon separation (Aida et al., 1997; Takada et al., 2001). Seedlings of the *cuc1/cuc2* double mutant lack an embryonic SAM, and the two cotyledons are fused along both edges to form a cup-shaped structure (Aida et al., 1997). *CUC1* and *CUC2*, thought to function upstream of *STM*, regulate SAM formation through transcriptional activation of *STM* (for review, Fletcher and Meyerowitz, 2000).

In summary, in planta and in vitro shoot meristem development and maintenance involve a large number of genes that orchestrate a complicated and well-regulated process. Certain aspects of the molecular mechanisms involved have been discovered; however, full elucidation of all of the pathways still requires additional research.

Phytohormones Regulate SAM Patterning and Development

In addition to genetic factors, phytohormones (e.g., auxins and CKs) play major roles in controlling plant development. During in vitro culture an excess of CK over auxin promotes shoot formation in callus, while auxin excess induces root formation (Skoog and Miller, 1957). Evidence suggests that differentially distributed hormone activities across the SAM seem to be linked to basic aspects of meristem activity and development (Veit, 2009). Relatively high levels of auxin and gibberellic acid (GA) are closely associated with initiation and outgrowth of lateral organs in the PZ, whereas high levels of CK in the CZ can be linked to maintenance of a reservoir of undetermined cells that enable indeterminate growth.

High CK activity promotes meristem function. CKs play crucial roles in plant development, (i.e., promoting cell division, shoot meristem initiation, root and vasculature patterning, chloroplast biogenesis, and photomorphogenesis). Besides inducing shoot meristems from callus during in vitro culture, high levels of CK relative to auxin in the shoot apex is required for SAM function. This was first noted in the *Arabidopsis amp1* (altered meristem program1) mutants, which have higher than normal CK levels, an enlarged SAM and increased cell proliferation and cyclin *cycD3* expression (Helliwell et al., 2001). Another clue to role of CK in shoot meristem development is the rice gene, *Lonely Guy (LOG)*, which encodes a novel CK-activating enzyme, the final step in bioactive CK synthesis; loss-of-function *LOG* causes premature termination of the shoot meristem. *LOG* is expressed specifically in the shoot apex, suggesting specific activation of CKs as a mechanism to regulate SAM function (Kurakawa et al., 2007). Other studies suggest that decreasing CK by over-expressing CK oxidases reduces meristem size and, sometimes, causes meristem termination (Werner et al., 2003).

CK response involves multiple steps involving two-component circuitry through a histidine- and aspartate-phosphorelay-signaling pathway. Two-component AHKs contain a variable CK sensor and a conserved histidine kinase domain that acts as a CK receptor. AHPs act as signaling shuttles between CK receptors and downstream nuclear responses and are translocated to the nucleus upon phosphorylation by an activated AHK, thus relaying the phosphate to ARRs, which regulate a transcriptional network that controls plant responses (Kakimoto, 2003).

Using activation-tagging that causes overexpression of a tagged gene, a putative CK receptor, cytokinin-independent1 (CKI1), was identified (for review, Kakimoto, 2003). *cki1* encodes a protein similar to two-component regulators, having histidine kinase and receiver domains, thus resembling a CK receptor. Over-expression of *cki1* results in a response independent of CK—elevated cell division and greening—suggesting that CKI1 is involved in CK signaling, perhaps as a CK receptor. Subsequently CYTOKININ RESPONSE 1 (CRE1), also known as AHK4 or WOL (WOODEN LOG) (Ueguchi et al., 2001), was identified (Inoue et al., 2001) and confirmed as a receptor because it binds CK directly (Yamada et al., 2001). Cells from a *cre1* mutant displayed reduced sensitivity to CK in vitro. Analyses in yeast and *E. coli* indicate that CK can act as a ligand for CRE1/AHK4 (Inoue et al., 2001; Yamada et al., 2001). CRE1/AHK4 and two homologues, AHK2 and AHK3, share a CK-binding domain with varying numbers of transmembrane segments. In protoplasts transient over-expression of any of the receptors increased CK sensitivity (Hwang and Sheen, 2001), suggesting partially redundant functions (Higuchi et al., 2004; Riefler et al., 2006).

Six AHPs have been identified in *Arabidopsis*. AHP1 and AHP2 translocate to the nucleus after CK treatment (Hwang and Sheen, 2001). Loss-of-function mutation analysis revealed strong functional redundancy of AHPs; single *ahp* mutants were indistinguishable in CK response from wild-type, while higher order mutants displayed variable reduced sensitivity to CK. A quintuple *ahp1, 2, 3, 4, 5* mutant showed the most apparent reduction in CK sensitivity (Hutchison et al., 2006). AHP6, a pseudophosphotransfer protein lacking the conserved His required for phosphotransfer, negatively regulates CK, likely by competing with functional AHPs for AHK interaction. Alternatively, the presence of AHP6 in a specific spatial domain, may limit CK activity, contributing to fine-tuning cell differentiation boundaries (Mähönen et al., 2006).

ARRs, which share a receiver domain with conserved Asp residues, are classified based on structure and CK-inducible expression patterns into two subtypes, A and B (Imamura et al., 1999). Upon activation, nuclear B-types function as transcriptional factors, activating target gene transcription, including A-types (Hwang and Sheen, 2001). Conversely, CK-inducible A-types repress CK signaling, and thus may act as negative regulators of CK response in a negative-feedback loop. These studies suggest a multi-step phosphorelay pathway is involved in CK signal transduction: CRE1/AHK4/WOL→AHPs→ARRs→response.

A close correlation exists between CK response and SAM activity. Some type A ARRs (e.g., ARR7) negatively regulate SAM activity. Over-expression of *arr7* causes reduced meristem activities, resulting in *wus*-like phenotypes (Leibfried et al., 2005). Similarly, loss-of-function mutants of the maize phyllotaxis-controlling gene, *abph1*, encoding a CK-inducible type A ARR, have increased meristem size and altered phyllotaxy patterns. Since *abph1* is expressed specifically in the youngest leaf primordia, it was proposed that it function in limiting SAM size through negative regulation of CK signaling (Giulini et al., 2004). Moreover, loss-of-function in the three CK receptor genes, *cre/ahk4/wol, ahk2*, and *ahk3*, resulted in reduction in shoot meristem size and cell proliferation (Higuchi et al., 2004).

Based on the sum of all these studies, it seems that higher CK activity relative to auxin in the CZ is required for optimum SAM function.

High auxin and gibberellic acid activities facilitate lateral organogenesis. Auxin, being central to many aspects of plant development, is produced primarily in young leaves and primordia, and is transported to sites where it functions. It controls plant development through gradients and maxima established by auxin efflux carriers, i.e., the PIN (PINFORMED) family that drives polar auxin transport that maximizes in leaf primordia on the flank of the SAM, thus aiding lateral organ formation (Gälweiler et al., 1998). Loss-of-function mutations of *PIN1* or chemical treatment with an auxin transport inhibitor that causes loss of auxin, blocks lateral organ formation (Benkova et al., 2003; Scanlon, 2003). The block can be overcome by exogenous auxin, which restores leaf formation at the application site (Reinhardt et al., 2000; Reinhardt et al., 2003). Altering polar distribution by exogenous auxin application to *Brassica juncea* embryos resulted in cotyledon fusion and failure of further organ formation (Hadfi et al., 1998).

SAMs have a gradient of auxin concentration with high levels at the PZ, decreasing toward the CZ. Differential auxin distribution is regulated by the influence of CK on cell-to-cell polar auxin transport by altering expression of several PIN proteins (Pernisová et al., 2009). Another possibility is that auxin rapidly decreases CK biosynthesis in the SAM, which may locally affect the CK gradient from the CZ to the PZ (Nordstrom et al., 2004). High auxin-to-CK ratios, not absolute auxin concentrations, might trigger lateral organ initiation since their primordia initiate only from the PZ, not the CZ, even though auxin is present in the CZ (Treml et al., 2005). Since exogenous auxin application at the CZ does not induce lateral organ initiation (Reinhardt et al., 2003), additional factors appear to operate.

Specific binding of Aux/IAA proteins to the transcription factor, ARF (auxin response factor), prevents ARFs from activating transcription of auxin-inducible genes, including Aux/IAA. Targeting Aux/IAA proteins for degradation in an auxin-dependent manner through ubiquitin-mediated protein degradation results in activation of an auxin response (Gray et al., 2001). Dominant negative, loss-of-function, mutant alleles of the co-repressor, TOPLESS (PTL), cause the embryonic SAM to become root-like, suggesting PTL represses root development in apical regions of the developing embryo (Long et al., 2006). Defects may also result from limiting ARF5/MP activities in the CZ, not by altering auxin distribution (Szemenyei et al., 2008).

Gibberellic acid (GA) also regulates many processes in plant development, including leaf initiation and outgrowth and morphogenesis (Fleet and Sun, 2005). Similar to auxin, GA levels are high in leaf primordia and young leaves, but low in the SAM (Hay et al., 2004). Tight spatial control of GA accumulation appears crucial for establishing boundaries between meristematic and leaf founder cells (Sakamoto et al., 2001); repression of GA in the SAM maintains meristem function (Hay et al., 2004). Synthesis and degradation of GA are both controlled by KNOX I and auxin; auxin promotes GA synthesis in decapitated stems (Ross et al., 2000; Wolbang and Ross, 2001), while KNOX I is required to maintain low GA levels in the SAM (Jasinski et al., 2005).

Expression from *KNOX I* in the SAM excludes expression of the GA biosynthetic gene, *GA20ox,* the rate-limiting step in GA biosynthesis (Hay et al., 2004). In tobacco and potato, *KNOX I* genes bind to the *GA20ox* promoter, restricting its expression in the meristem periphery (Chen et al., 2004; Sakamoto et al., 2001). *KNOX I* genes further restrict GA synthesis by promoting, in a ring surrounding the SAM, transcription of *GA2ox2*, which mediates GA inactivation (Jasinski et al., 2005; Sakamoto et al., 2001). Maize KN1 directly regulates expression from *GA2ox1* by binding to an intron (Bolduc and Hake, 2009). No such restriction of GA occurs in lateral organ primordia, where *KNOX I* genes are partially repressed by high auxin activities (Hay et al., 2006). Expression from *GA2ox1* is likely regulated by other factors since expression only partially overlaps with that of *STM*. A mutant of SPINDLY (SPY), another negative regulator of GA, is insensitive to exogenous CK application but has a constitutive GA signaling response (Greenboim-Wainberg et al., 2005).

Cross-Talk between KNOX I/WUS and the Phytohormone Pathway

Study of mechanisms underlying the role of KNOX and WUS pathways and phytohormone activities in maintaining SAM function show that *KNOX I* and *WUS* genes directly link to phytohormone pathways. This results from a local CK activity gradient, high auxin at the CZ decreasing toward the PZ. High auxin relative to CK and GA represses *KNOX I* and *CUC* expression and CK activity, resulting in specification of lateral organ primordia.

KNOX I Is a Major Activator of Meristematic Activity

Direct Molecular evidence for positive regulation of CK biosynthesis by *KNOX I* is that misexpression of STM results in a rapid increase in expression from *ISOPENTENYLTRANSFERASE7* (*IPT7*), which encodes an enzyme involved in CK biosynthesis and accumulation. A severe mutant of the CK receptor, AHK4/CRE1/ WOL, resulted in a weak *stm* allele (Jasinski et al., 2005); exogenous application of CK or expression of bacterial IPT in the SAM partially restores meristem activity (Yanai et al., 2005). Thus, STM promotes SAM activity at least partially by activating CK biosynthesis in the SAM, but high CK activity also results in increased *KNAT1* and *STM* transcript levels (Rupp et al., 1999).

WUS Regulates CK Signaling Response

A direct link was identified between WUS and CK signaling that induces stem cell identity. WUS directly represses expression of several CK-inducible type A *ARR*s, which act in a negative feedback loop of CK signaling in the SAM (Leibfried et al., 2005). *ARR7* expression is limited specifically to the PZ; WUS was shown to bind upstream of the *ARR7* transcription start site. Ectopic activation of *ARR7* causes a *wus* loss-of-function phenotype. A mutant *arr7* allele causes formation of aberrant SAMs; a loss-of-function mutation in a maize *ARR* homolog resulted in enlarged meristems.

Negative Feedback Loop between Auxin Signals and CUC and KNOXI in the SAM

PIN, *KNOX I*, and *CUC* are expressed in mutually exclusive domains in the SAM (Heisler et al., 2005). Loss of maximum auxin levels in shoot tips is consistent with loss of organogenic capacity that is caused at least partially by ectopic expression from *KNOX I* genes. Organ initiation is partially restored in *pin1* mutants by loss of BP function (Section 1.A; Hay et al., 2006). High auxin activity in lateral organ primordia, together with AS1, represses *CUC* and *KNOX1* expression, facilitating lateral organogenesis (Hay et al., 2006; Heisler et al., 2005). KNOX I proteins might also inhibit auxin transport (Treml et al., 2005). Ectopic *KNOX I* expression in leaves alters auxin transport and activity gradients, dramatically altering leaf shape (Hay et al., 2006), suggesting KNOX I and auxin may function in a feedback loop that reinforces a developmental boundary between meristem and leaf primordia. Combined regulation of KNOX1 and auxin in boundary delimitation was also suggested by analysis of the boundary-expressed LOB domain protein JAGGED LATERAL ORGANS (JLO), which activates *STM* and *BP* expression and represses expression of *PIN* auxin efflux transporters (Borghi et al., 2007).

MOLECULAR ANALYSIS OF PLANT CELL DIVISION

The mitotic cell division cycle consists of four sequential phases: gap phase 1 (G_1), DNA synthetic phase (S), gap phase 2 (G_2), and mitosis (M). G_1 intercedes between M and entry into the next S phase; G_2 separates S from M. Cells in G_2 contain double the genetic material compared to cells in G_1. The gaps are regulatory points for cell division cycle controls that respond to variable internal developmental and external environmental factors (Dewitte and Murray, 2003). An alternative cell cycle that occurs in some plant cells, the endocycle (DNA replication without mitosis) involves repeated S and G phases without subsequent mitosis, resulting in endopolyploidy. For example, *Arabidopsis* trichomes have 32C DNA content versus 2C for diploid cells (Melaragno et al., 1993). Cell division has been suggested to be a principal determinant of meristem activity and overall growth rate; modulation of plant cell growth rate is mainly achieved through regulation of G_1-to-S phase transition (Cockcroft et al., 2000).

CYCLIN-DEPENDENT KINASE COMPLEXES

Large functional similarities exist between plants and animals in the core molecular mechanisms controlling cell cycle, although plants have some unique features (e.g., cell walls and continuous generation of new organs). As with all eukaryotes, the cell division cycle in plants is directly controlled by Ser/Thr kinases (i.e., cyclin-dependent kinase [CDK] complexes), which contain a catalytic CDK subunit, a regulatory cyclin subunit—either activating or inhibiting—and scaffolding proteins. Cell status is controlled by overall levels of CDK activity. The activity of the catalytic CDK subunit, which depends on binding of regulatory proteins or cyclins, is responsible for recognizing and phosphorylating a target motif present in substrate proteins; regulatory cyclins can discriminate between distinct protein substrates. Different CDK-cyclin complexes phosphorylate numerous substrates at the G_1-to-S and G_2-to-M transitions, triggering DNA replication and mitosis, respectively, to regulate the cell division cycle (Dewitte and Murray, 2003; Menges et al., 2005; Shaul, 1996).

CYCLIN-DEPENDENT KINASES (CDKs)

The first *CDK* gene, *cdc2*, was identified in *Schizosaccharomyces pombe* (Hindley and Phear, 1984). Plant homologues were found in pea (Feiler and Jacobs, 1990), alfalfa (Hirt et al., 1992), maize (Colasanti et al., 1993), rice (Hashimoto et al., 1992), petunia (Bergounioux et al., 1992), soybean (Miao et al., 1993), *Arabidopsis* (Ferreira et al., 1991), and *Antirrhinum* (Fobert et al., 1994). Partial complementation of yeast cell-cycle mutants proved functional equivalence of those from alfalfa, maize, rice, and soybean.

Expression of the first *Arabidopsis* CDK isolated, *cdc2aAt* (*CDKA; 1*), correlated with cell division and competence for cell division (Ferreira et al., 1991; Hemerly et al., 1993; Martinez et al., 1992). Expression from *cdc2aAt* was in root and shoot apices, young developing leaves, and at the root-shoot junction where adventitious roots initiate; it was nearly undetectable in fully expanded leaves. In the SAM, *cdc2aAt* expression corresponds to patterns of mitotic activity. Unlike the single *cdc2* in yeast, plants have numerous distinct *CDK*s that function in different temporal and spatial phases. The 29-member *Arabidopsis* CDK family is classified into six groups, A to F, in addition to numerous CDK-like proteins (Menges et al., 2005). Four of the six groups function in cell cycle regulation.

A-type CDKs, which contain a conserved PSTAIRE motif in their cyclin-binding domain, is required for cyclin binding (Dewitte and Murray, 2003; Inzé and De Veylder, 2006). A-type *CDK*s, expressed nearly constitutively throughout the cell cycle, peak at the G_1-to-S and G_2-to-M transitions, suggesting they function by interacting with distinct cyclins in different cell cycle phases. B-type *CDK*s, unique to plants and expressed only from S through M phases or during G_2 to M phase transition, are activated by binding with mitotic cyclins. Two subgroups were identified: CDKB1 present from S-phase to early M and CDKB2 from the G_2 to M boundary (Menges et al., 2005). Cell division-dependent *CDKB2* expression is required for SAM activity (Andersen et al., 2008). Function of C-type CDKs, similar to two human proteins, and E-types, unique to plants, is largely unknown.

CDK activity is regulated by reversible phosphorylation, performed by CDK-activating kinases (CAKs); D- and F- type CDKs act as CAKs, inducing a conformational change that allows proper substrate recognition. *Arabidopsis* has four CAKs in two distinct groups. (*i*) Cyclin H-dependent D-type CDKs (CDKD) bind cyclin H, phosphorylate and activate both CDKs and the C-terminal (CTD) tail of RNA polymerase II. (*ii*) Plant-specific cyclin H-independent F-type CDK (CDKF), specific for CDKs only (Shimotohno et al., 2004; Yamaguchi et al., 2003).

Activity of CAKs correlates with the cell division cycle. Increased expression of rice nuclear *CDKD1*, which expresses preferentially in S-phase, promotes S-phase progression and overall growth rate of suspension cells (Fabian-Marwedel et al., 2002). Overexpression of rice *CDKD2* in tobacco leaf explants caused callus formation without CK addition; callus induction depended on CDK activation (Yamaguchi et al., 2003). Decreasing *CDKF* (*CAK1At*) expression caused gradual reduction in CDK activity, arrest of cell division and premature differentiation of root meristems in transformed plants (Umeda et al., 2000), indicating CAKs play important roles in determining growth rates and differentiation.

Cyclins

Plant cyclin families are complex with orthologs of most major mammalian cyclin groups and some unique members. *Arabidopsis* has at least 50 cyclins of ten types (Wang et al., 2004) and, except for a few A-, B-, D-, and H-type cyclins, the majority are poorly understood. In general, A and B types, required for cells to enter mitosis, are termed mitotic cyclins. A-type cyclins regulate the S-to-M phase transition; B-type cyclins function in both G_2-to-M transition and intra-M-phase control. C-, D-, and E-types are G_1-to-S phase cyclins (Lew et al., 1991).

D-type cyclins (CYCDs), the expression of which mainly correlates with the proliferative status of cells, appear to be the most important regulators of cell division as they aid exit from the quiescent state (G_0) and reentry into the cell cycle (Potuschak and Doerner, 2001). CYCD levels respond to endogenous and extracellular signals (Hu et al., 2000; Meijer and Murray, 2000; Richard et al., 2002; Riou-Khamlichi et al., 1999). If such signals are removed, D-type cyclin levels decline rapidly, resulting in cells remaining blocked in G_1 (Diehl et al., 1997). CYCDs play key roles in G_1-to-S phase transition and *Arabidopsis* has 10 *CYCD* genes (Wang et al., 2004). It appears D-type cyclins may regulate G_2-to-M transition and some D-types appear to act as key triggers in hormonal response. Overexpression of *CYCD3;1* induces calli formation in the absence of CK (Riou-Khamlichi et al., 1999) and expression levels of *CYCD3;1* were rate-limiting for cell division in calli induced by CK.

Transcripts for *cyc1At* are restricted primarily to the root apical meristem (Hemerly et al., 1993). When the *cyc1At* promoter was fused to GUS, a close correlation between GUS and mitotic activity was observed in SAMs, developing flowers and embryos (Ferreira et al., 1994). During de-differentiation of mesophyll protoplasts, *cyc1At*-driven GUS expression was induced only when cell division occurred after treatment with appropriate combinations of auxins and CKs. Various cyclins are expressed during the cell cycle with each cyclin perhaps playing a different role during cell division (Fobert et al., 1994; Hirt et al., 1992).

Based on studies of *cyc1At* and *cdc2aAt* in *Arabidopsis,* a simplified model was proposed for the role of cyclins during plant development and dedifferentiation (Shaul, 1996). Cell cycle genes are highly expressed in young, dividing tissues; however, during differentiation, reduction in *cdc2aAt*, but not elimination, expression occurs, followed by cessation of *cyc1At* expression. In differentiated tissues, *cdc2aAt* expression levels reflect division competency. Dedifferentiation and re-acquisition of division competency requires *cdc2aAt* activation. D-type cyclins mediate exit from G_0 and re-entry into the cell cycle. Compared to A and B types, expression of D types is not restricted to dividing cells; it is expressed earlier in the cell cycle.

Regulation of Cell Cycle during Plant Development

To maintain normal organization and activity of a meristem during plant development, cell division must be tightly controlled by machinery that regulates the cell cycle and is linked directly to SAM activity.

Genetic Factors Regulate Plant Cell Cycle and Meristem Activity

Cell cycle regulation involves transcriptional activation of D-type cyclin genes. *AINTEGUMENTA* (*ANT*) encodes an AP2-domain transcription factor found only in plants that is expressed on flanks of the SAM at sites of developing leaf primordia. Loss of function *ant* mutants form smaller leaves. Plants overexpressing ANT form larger leaves and this correlates with induction of cell cycle genes and promotion of cell proliferation (Mizukami, 2000).

In plants, Myb proteins control G_2-to-M transition by activating or repressing transcription. MYB3R1 and MYB3R4 play partially redundant roles in positively regulating cell division. Double mutants of *myb3r1/myb3r4* often fail to complete cell division, which correlates with transcript reduction from several G_2-to-M phase-specific genes (Haga et al., 2007). A Myb-related cell-division-cycle transcription factor, AtCDC5, may be essential for G_2-to-M transition, and may regulate SAM activity by controlling expression from *STM* and *WUS* (Lin et al., 2007).

Arabidopsis PASTICCINO (*PAS*) encodes protein phosphatases, possibly involved in cell division regulation in the SAM in response to phytohormones (Faure et al., 1998, Harrar et al., 2003). Cells in *pas* mutants are more competent for cell division, demonstrated in the CK hypersensitive response and the ectopic cell division in the SAM. Expression of *CDKA* and *CYCB1* and certain *KNOX I* genes, *STM*, *KNAT2*, and *KNAT6*, are upregulated in *pas* mutants.

Exogenous Signals Regulate the Plant Cell Division Cycle

Upstream signaling components of the cell cycle are less well characterized but appear to involve exogenous signals like phytohormones (particularly, CK and auxin), sugars, and stress. *CycD3*, elevated in a mutant with high CK levels, was rapidly induced by CK application in both cell cultures and whole plants. Constitutive expression of *CycD3* in transgenic plants leads to induction and maintenance of cell division without exogenous CK. CK activates cell division through induction of *CycD3* at the G_1-to-S transition (Riou-Khamlichi et al., 1999). Auxin is also thought to promote cell cycle activity by triggering degradation of inhibitory proteins that induce expression of genes involved in G_1-to-S and G_2-to-M transitions (Blilou et al., 2002; Hartig and Beck, 2006).

Using a *cdc2aAt* promoter-*GUS* fusion, *GUS* expression in mesophyll protoplasts from leaves was studied during dedifferentiation (Hemerly et al., 1993). Cultivation of protoplasts in the presence of auxin or CK caused *cdc2aAt*-mediated induction of *GUS* expression. Despite lack of cell division, division competence was likely due to hormonal effects on endogenous *cdc2aAt*. Consistent with this, there was rapid induction of *cdc2aAt* promoter-driven *GUS* expression around damaged surfaces of wounded transgenic *Arabidopsis* leaves.

Possible links have been identified between CK and cell division, possibly in the G_2-to-M phase. In tobacco protoplasts CK controls the cell cycle at mitosis by stimulating Tyr dephosphorylation and activation of the p^{34}cdc2-like H1 histone kinase (Zhang et al., 1996). CK also regulates G_1-to-S transition, mediated by CycD3 (Riou-Khamlichi et al., 1999). Application of CK to both seedlings and in vitro cultured cells resulted in increased steady-state levels of *CycD3* mRNA. With regard to in vitro plant development, leaf explants over-expressing *CycD3* form healthy green calli without CK, in contrast to wild-type explants that form such calli only with addition of CK, suggesting that *CycD3* over-expression bypasses CK in activating the cell cycle; CK appears to regulate the cell cycle through CycD3.

CK transport through tissues may also play a role in the in vitro CK response since cells responding during in vitro shoot organogenesis often are not in direct contact with hormones in the media. During in vitro culture of vegetative shoots of maize and barley on a medium with a high CK to auxin ratio, responding cells were in axillary shoot meristematic domes (Zhang et al., 1998) or nodal regions (Zhang et al., 2002), neither of which were in direct contact with hormone-containing medium, while cells at the cut edge of the stem in direct contact with the medium did not respond. The possible role of a CK transporter is further supported by studies of its role during plant growth. CK exists in the xylem sap while the root tip is its major site of biosynthesis, suggesting that CK is transported through the xylem to aerial parts of a plant. Identification of a putative CK transporter, AtPUP1, was based on functional complementation of a yeast mutant deficient in adenine uptake (Gillissen et al., 2000).

MOLECULAR ANALYSIS OF SHOOT ORGANOGENESIS IN VITRO

Use of Specific Genes as Molecular Markers

Historically, study of in vitro shoot organogenesis was based on morphological or physiological observations. With cloned genes available, molecular markers can now be used to develop a better molecular understanding of in vitro shoot organogenesis. The first example of this approach used expression analysis of maize KN1 and its homolog in barley during in vitro axillary shoot meristematic cell proliferation and adventitious shoot meristem formation (Zhang et al., 1998). Vegetative shoot segments from germinated maize and barley seedlings were cultured in vitro on a high CK, low auxin medium. Within weeks, cell division in the cultured axillary shoot meristems changed from a well-regulated state to a proliferating state, with the small meristematic dome becoming enlarged. Adventitious meristems (ADMs) arose directly from cells in the enlarged dome (Figure 20.3a). Expression of *KN1* was maintained in shoot meristematic cells during in vitro cell proliferation of axillary shoot meristems (Figure 20.3b). ADMs appear to derive directly from *KN1*-expressing shoot meristematic cells. Thus, *KN1* can detect in vitro shoot meristem formation that appears to follow paths similar to in planta shoot meristem development.

More recently, *CUC2* and *WUS* were shown to be the earliest molecular markers for in vitro shoot organogenesis. Both are expressed in a small number of progenitor cells, deemed competent for new shoot meristem initiation (Gordon et al., 2007). Dynamic fine-tuning of *CUC2* and *WUS* expression and a local CK and auxin gradient feedback loop have been suggested to lead

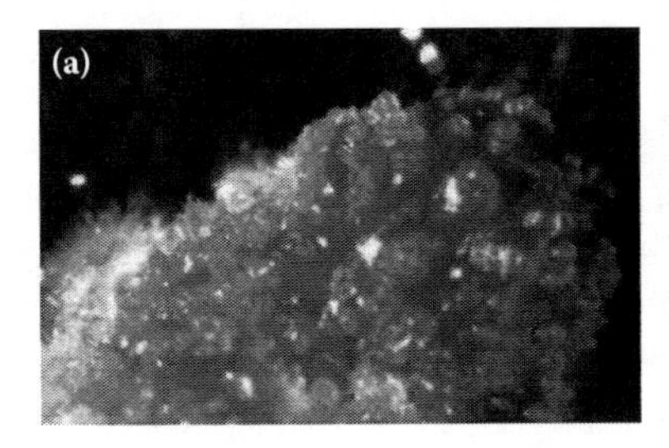

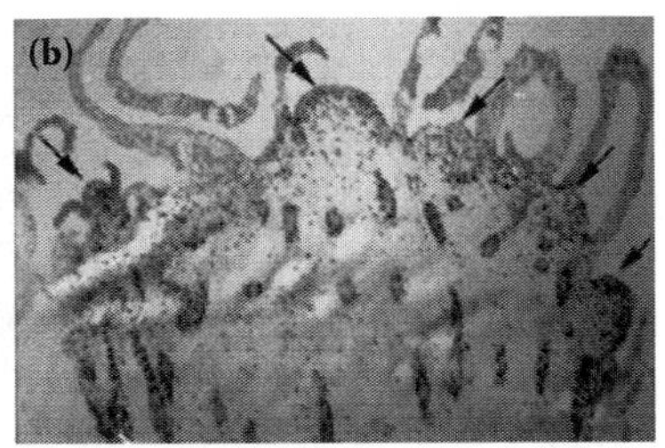

FIGURE 20.3 Expression of *KNOTTED1* in the in vitro–generated adventitious shoot meristems (ADMs) of maize. (a) Multiple ADMs induced from an enlarged shoot meristematic dome in vitro. (b) Expression of *KNOTTED1* (arrow) in the in vitro ADMs. (From Trigiano, R. N. and D. J. Gray, Eds., *Plant Tissue Culture and Development*, CRC Press LLC, Boca Raton, FL, 2004.)

to a self-organizing cell identity partition that gradually establishes shoot meristem cell niches within callus tissue. Subsequent patterning and development of the shoot meristems, once shoot premeristems are initiated, involves local activation of genes expressed in early shoot meristems, (e.g., *PIN1, STM, REV, FIL, ATML1,* and *CLV3*), thought to be largely autonomous.

CUC1 and *CUC2* genes are required for embryonic SAM formation and cotyledon separation (Aida et al., 1997). Mutations in *CUC1* and *CUC2* or *CUC1* alone reduce initiation efficiency of adventitious shoots during in vitro culture (Aida et al., 1997; Daimon et al., 2003; Takada et al., 2001), while overexpression increases adventitious shoot formation on calli; CUC1 and CUC2 activate *STM* expression in calli (Daimon et al., 2003). *CUC2* expression marks a small number of progenitor cells that proliferate to form a relatively homogeneous cell mass that later patterns into a de novo shoot meristem.

WUS expression in calli is essential to initiate shoot meristem progenitor cell identity; continuous *WUS* expression is required for further shoot meristem development. A strong *wus* mutant (*wus-1*) generated only 5% as many shoots as wild-type (Gordon et al., 2007). Ectopic *WUS* expression induces shoot generation in roots (Gallois et al., 2004). Moreover, a mutant in which *WUS* expression is elevated causes somatic embryo formation in a range of tissues and organs (Zuo et al., 2002), suggesting that WUS is sufficient to establish shoot meristem identity. Therefore, *CUC2* and *WUS* can be used as molecular markers for shoot premeristem progenitor cells, whereas *PIN1, STM, REV, FIL, ATML1,* and *CLV3*, can be used to identify developing shoot meristems during in vitro shoot organogenesis.

Using genes as molecular markers provides new insights into in planta and in vitro plant development. They will continue to increase our understanding of these complex processes as additional genes and networks are identified and characterized. These studies continue to validate the hypothesis that de novo organogenesis involves the following three steps: acquisition of competence, shoot induction, and organogenesis determination (Christianson and Warnick, 1985).

Identification of Genes Involved in In Vitro Shoot Organogenesis

As described earlier in this review many regulatory factors and pathways are involved in controlling formation and development of the SAM in planta and in vitro. Besides the key pathways of CUC1/CUC2-STM and WUS-CLV3, other important regulators have also been identified by genetic analyses of SAM-defective mutants. These studies reveal a complex and elegant regulation network involved in embryonic formation of the SAM and in adventitious shoot organogenesis.

Genetic and molecular efforts are now used to identify genes, which might either regulate or be involved directly in shoot organogenesis in vitro. For example, a novel MADS box cDNA, *PkMADS1*, was isolated from a cDNA library of leaf explants from a woody tree species, *Paulownia kawakamii*, which undergoes adventitious shoot formation (Prakash and Kumar, 2002). Deduced amino acid sequence of its MADS domain revealed 90% homology to the *Arabidopsis*

AGL24 (AGAMOUS-like) protein (Hartmann et al., 2000) and to the STMADS16 protein in potato (Carmona et al., 1998). Expression from *PkMADS1* was not detected in callus cultures of *Paulownia*, but was in shoot-forming cultures. In planta, *PkMADS1* transcripts were found only in shoot apices, not in root apices, flowers, or leaf explants. Plants with antisense knockouts of the gene had stunted shoots, altered phyllotaxy, and in some cases lacked SAM development. Shoot regeneration from leaf explants of antisense plants was reduced ten-fold versus wild-type or *PkMADS1* over-expressing plants, suggesting *PkMADS1* expression is essential for in vitro and in planta shoot formation.

Because CKs are the most efficient growth regulators inducing in vitro shoot organogenesis, genes involved in CK metabolism or signal transduction are likely to affect shoot organogenesis. When *ESR1* (*ENHANCER OF SHOOT REGENERATION1*), which allows for CK-independent shoot induction (Banno et al., 2001), is over-expressed, this putative transcription factor greatly enhances shoot regeneration efficiency following CK addition to root explants, coupled with a reduction in the optimal CK concentration needed (Buttner and Singh 1997; Fujimoto et al., 2000). Also wild-type root explants, cultured on CK-containing shoot-induction medium, had transiently elevated *ESR1* transcript levels, which occurred before *STM* expression but after acquisition of competence for shoot regeneration. These results suggest that *ESR1* may be a downstream effecter for CK that enhances shoot regeneration initiation after acquisition of competence for shoot organogenesis. Interestingly, inducible overexpression from *ESR1* leads to CK-independent shoot regeneration on roots. Constitutive *ESR1* overexpression, however, results in dark green calli with no shoot development (Banno et al., 2001). These results suggest an *ESR1* gradient and/or other factors may be necessary for shoot development during in vitro organogenesis.

Besides ESR1, a homologue, ESR2, is involved, and even more active than ESR1, in promoting shoot regeneration in tissue culture. Overexpression of *ESR2* results in CK-independent shoot regeneration from *cre1/ahk4/wol* mutant roots and rescues *cre1/ahk4/wol* mutants, suggesting ESR2 is an important regulator in the CK response. ESR2 plays a role in shoot regeneration through transcriptional regulation of *CUC1* based on an *ESR2* knockdown that downregulates *CUC1* expression and phenocopies the *cuc1* mutant (Aida et al., 1997). ESR2 also activates *CYCD1;1*, *AHP6* and *CUC1* expression (Ikeda et al., 2006).

Two genes, *RGD3* (*ROOT GRORWTH DEFECTIVE3*) and *RID3* (*ROOT INITIATION DEFECTIVE3*), were shown to control cell division in the SAM during shoot organogenesis from hypocotyl explants during in vitro culture. RID3, a negative regulator, and RGD3, a positive regulator, of the CUC-STM pathway participate in the proper control of cell division in the SAM. *RGD3* is expressed in the developing SAM, while *RID3* is expressed outside the SAM in the early stages of shoot regeneration (Tamaki et al., 2009).

SUMMARY

Molecular and genetic analyses of in planta and in vitro shoot organogenesis to date lend support to the developmental model proposed by Christianson and Warnick (1985), namely that shoot organogenesis is divided into three phases: competence acquisition, that is, the stage before cells are induced to develop into a shoot; shoot induction; and shoot development. Expression of genes involved in the CK signal transduction pathway (e.g., *CKI1* and *CRE1*) might be indicators of cell competence to respond to CK, normally required for shoot organogenesis. The best candidates as indicators of cell division competence are likely *CDC2a*-and cyclin-type genes, induced during shoot organogenesis from stem tissues (Boucheron et al., 2002). For the shoot induction phase, maize *KN1*-type homologues appear to be the most definitive molecular markers to identify early stages of induction of in vitro shoot organogenesis.

CK likely acts through activation of *CycD-3* type genes to initiate or accelerate the cell cycle; however, the molecular connection between CK action and CycD-3 activation has not yet been fully elucidated. Although molecular insights already exist as to why a higher ratio of CK to auxin favors shoot organogenesis, the precise molecular pathway still is not known. Future study of genes and regulatory pathways involved in dedifferentiation of somatic cells in vitro, coupled with the elucidation of the role of chromatin structure changes during the course of dedifferentiation, will provide the insights needed to understand fully in vitro shoot organogenesis and the developmental flexibility of the plant cell.

ACKNOWLEDGMENTS

We dedicate this review to the memory of Dr. Shibo Zhang, whose life goal was to understand the molecular pathways involved in in vitro shoot development. Although his death in a tragic accident prevented completion of his dream, we are sure that he would be pleased at the progress made in this field. We are grateful to Barbara Alonso for preparing Figure 20.1 and formatting the reference list. We attempted to cite all pertinent references but apologize for any work inadvertently missed.

REFERENCES

Aida, M., T. Ishida, H. Fukaki, H. Fujisawa, M. Tasaka. 1997. Genes involved in organ separation in *Arabidopsis:* an analysis of the cup-shaped cotyledon mutant. *Plant Cell* 9:841–857.

Andersen, S.U., Buechel, S., Zhao, Z., Ljung, K., Novak, O., Busch W., Schuster, C., Lohmann, J.U. 2008. Requirement of B2-type cyclin-dependent kinases for meristem integrity in *Arabidopsis thaliana. Plant Cell* 20:88–100.

Banno, H., Ikeda, Y., Niu, Q.-W., Chua, N.-H. 2001. Overexpression of *Arabidopsis* ESR1 induces initiationof shoot regeneration. *Plant Cell* 13:2609–2618.

Benkova, E., Michniewicz, M., Sauer, M., Teichmann, T., Seifertova, D., Jürgens, G., Friml. J. 2003. Local, efflux-dependent auxin gradients as a common module for plant organ formation. *Cell* 115:591–602.

Bergounioux, C., Perennes, C., Hemerly, A. S., Qin, L. X., Sarda, C., Inze, D., Gadal, P. 1992. A *Cdc2* Gene of petunia-hybrida is differentially expressed in leaves protoplasts and during various cell cycle phases. *Plant Mol. Biol.* 20:1121–1130.

Blilou, I., Frugier, F., Folmer, S., Serralbo, O., Willemsen, V.,Wolkenfelt, H., Eloy, N.B., Ferreira, P.C., Weisbeck, P., Scheres, B. 2002. The *Arabidopsis* HOBBIT gene encodes a CDC27 homolog that links the plant cell cycle to progression of cell differentiation. *Genes Dev.* 16:566–575.

Bolduc, N., Hake, S. 2009. The maize transcription factor KNOTTED1 directly regulates the gibberellin catabolism gene *ga2ox1 Plant Cell* 21:1647–1658.

Borghi, L., Bureau, M., Simon, R. 2007. Arabidopsis JAGGED LATERAL ORGANS is expressed in boundaries and coordinates KNOX and PIN activity. *Plant Cell* 19:1795–1808.

Boucheron, E., Guivarc'h, A., Azmi, A., Dewitte, W., Van Onckelen, H., Chriqui, D. 2002. Competency of *Nicotiana tabacum* L. stem tissues to dedifferentiate is associated with differential levels of cell cycle gene expression and endogenous cytokinins. *Planta* 215:267–78.

Buttner, M., Singh, K.B. 1997. *Arabidopsis thaliana* ethylene-responsive element binding protein (AtEBP), an ethylene-inducible, GCC box DNA-binding protein interacts with an ocs element binding protein. *Proc. Natl. Acad. Sci. USA* 94:5961–5966.

Byrne, M.E., Barley, R., Curtis, M., Arroyo, J. M., Dunham, M., Hudson, A. Martienssen, R.A. 2000. Asymmetric leaves1 mediates leaf patterning and stem cell function in Arabidopsis. *Nature* 408:967–971.

Byrne, M.E., Simorowski, J., Martienssen, R.A. 2002. ASYMMETRIC LEAVES1 reveals knox gene redundancy in *Arabidopsis*. *Development* 129:1957–1965.

Carmona, M.J., Ortega, N., Garcia-Maroto, F. 1998. Isolation and molecular characterization of a new vegetative MADS-boxgene from *Solanum tuberosum* L. *Planta* 207:181–188.

Chen, H., Banerjee, A.K., Hannapel, D.J. 2004. The tandem complex of BEL and KNOX partners is required for transcriptional repression of *ga20ox1*. *Plant J.* 38:276–284.

Christianson, M.L., Warnick, D.A. 1985. Temporal requirement for phytohormone balance in the control of organogenesis in vitro. *Dev. Biol.* 12:494–497.

Clark, S.E., Williams, R.W., Meyerowitz, E.M. 1997. Control of shoot and floral meristem size in *Arabidopsis* by a putative receptor-kinase encoded by the *CLAVATA1* gene. *Cell* 89:575–585.

Cock, J.M., Mccormick, S. 2001. A large family of genes that share homology with CLAVATA3. *Plant Physiol.* 126: 939–942.

Cockcroft, C.E., Den Boer, B.G.W., Healy, J.M.S., Murray, J.A.H. 2000. Cyclin D control of growth rate in plants. *Nature* 405:575–579.

Colasanti, J., Cho, S.-O., Wick, S., Sundaresan, V. 1993. Localization of the functional p34-cdc2 homolog of maize in root tip and stomatal complex cells: Association with predicted division sites. *Plant Cell* 5: 1101–1111.

Daimon, Y., Takabe, K., Tasaka, M. 2003. The CUP-SHAPED COTYLEDON genes promote adventitious shoot formation on calli. *Plant Cell Physiol.* 44:113–121.

Dewitte, W., Murray, J.A. 2003. The plant cell cycle. *Annu. Rev. Plant Biol.* 54:235–264.

Diehl, J.A., Zindy, F., Sherr, C.J. 1997. Inhibition of cyclin D1 phosphorylation on threonine-286 prevents its rapid degradation via the ubiquitin-proteasome pathway. *Genes Dev.* 11:957–972.

Fabian-Marwedel, T., Umeda, M., Sauter, M. 2002. The rice cyclin-dependent-kinase—activating kinase R2 regulates S-phase progression. *Plant Cell* 14:197–210.

Faure, J.D., Vittorioso, P., Santoni, V., Fraisier, V., Prinsen, E., Barlier, I., Van Onckelen, H., Caboche, M., Bellini, C. 1998. The *PASTICCINO* genes of *Arabidopsis thaliana* are involved in the control of cell division and differentiation. *Development* 125:909–918.

Feiler, H.S., Jacobs, T.W. 1990. Cell division in higher plants: A *cdc2* gene, its 34-kDa product, and histone H1 kinase activity in pea. *Proc. Natl. Acad. Sci. USA* 87:5397–5401.

Ferreira, P.C.G., Hemerly, A.S., Engler, J.D.A., Montagu, M.V., Engler, G., Inze, D. 1994. Developmental expression of the *Arabidopsis* cyclin gene cyc1At. *Plant Cell* 6:1763–1774.

Ferreira, P.C.G., Hemerly, A.S., Villarroel, R., Van Montagu, M., Inze, D. 1991. The *Arabidopsis* functional homolog of the p34cdc2 protein kinase. *Plant Cell* 3:531–540.

Fiers, M., Golemiec, E., Xu, J., Van Der Geest, L., Heidstra, R., Stiekema, W., Liu, C.M. 2005. The 14-amino acid CLV3, CLE19, and CLE40 peptides trigger consumption of the root meristem in *Arabidopsis* through a CLAVATA2-dependent pathway. *Plant Cell* 17:2542–2553.

Fleet, C.M., Sun, T.P.A. 2005. DELLAcate balance: the role of gibberellin in plant morphogenesis. *Curr. Opin. Plant Biol.* 8:77–85.

Fletcher, J.C., Brand, U., Running, M.P., Simon, R., Meyerowitz, E.M. 1999. Communication of cell fate decisions by CLAVATA3 in *Arabidopsis* shoot meristems. *Science* 283:1911–1914.

Fletcher, J.C., Meyerowitz, E.M. 2000. Cell Signaling within the shoot meristem. *Curr. Opin. Plant Biol.* 3:23–30.

Fobert, P.R., Coen, E.S., Murphy, G.J.P., Doonan, J.H. 1994. Patterns of cell division revealed by transcriptional regulation of genes during the cell cycle in plants. *EMBO J.* 13:616–624.

Fujimoto, S.Y., Ohta, M., Usui, A., Shinshi, H., Ohme-Takagi, M. 2000. *Arabidopsis* ethylene-responsive element binding factors act as transcriptional activators or repressors of GCC box-mediated gene expression. *Plant Cell* 12:393–404.

Gallois, J.-L., Nora, F.R., Mizukami, Y., Sablowski, R. 2004. WUSCHEL induces shoot stem cell activity and developmental plasticity in the root meristem. *Genes Dev.* 18:375–380.

Gälweiler, L., Guan, C., Muller, A., Wisman, E., Mendgen, K., Yephremov, A., Palme, K. 1998. Regulation of polar auxin transport by AtPIN1 in *Arabidopsis* vascular tissue. *Science* 282:2226–2230.

Gillissen, B., Bürklea, L., Andréb, B., Kühna, C., Rentscha, D., Brandla, B., Frommera, W. B. 2000. A new family of high-affinity transporters for adenine, cytosine, and purine derivatives in *Arabidopsis*. *Plant Cell* 12:291–300.

Giulini, A., Wang, J., Jackson, D. 2004. Control of phyllotaxy by the cytokinin-inducible response regulator homologue ABPHYL1. *Nature* 430:1031–1034.

Gordon, S.P., Heisler, M.G., Reddy, G.V., Ohno, C., Das, P., Meyerowitz, E.M. 2007. Pattern formation during de novo assembly of the *Arabidopsis* shoot meristem. *Development* 134:3539–3548.

Gray, W.M., Kepinski, S., Rouse, D., Leyser, O., Estelle, M. 2001. Auxin regulates SCFTIR1-dependent degradation of AUX/IAA proteins. *Nature* 414:271–276.

Greenboim-Wainberg, Y., Maymon, I., Borochov, R., Alvarez, J., Olszewski, N., Ori, N., Eshed, Y., Weiss, D. 2005. Cross talk between gibberellin and cytokinin: the *Arabidopsis* GA response inhibitor SPINDLY plays a positive role in cytokinin signaling. *Plant Cell* 17:92–102.

Guo, M., Thomas, J., Collins, G., Timmermans, M.C. 2008. Direct repression of KNOX loci by the ASYMMETRIC LEAVES1 complex of *Arabidopsis*. *Plant Cell* 20:48–58.

Ha, C.M., Jun, J.H., Nam, H.G., Fletcher, J.C. 2004. *BLADE-ON-PETIOLE1* encodes a BTB/POZ domain protein required for leaf morphogenesis in *Arabidopsis thaliana*. *Plant Cell Physiol.* 45:1361–1370.

Haberer, G., Kieber, J.J. 2002. Cytokinins. New insights into a classic phytohormone. *Plant Physiol.* 128:354–362.

Hadfi, K., Speth, V., Neuhaus, G. 1998. Auxin-induced developmental patterns in *Brassica juncea* embryos. *Development* 125:879–887.

Haga, N., Kato, K., Murase, M., Araki, S., Kubo, M., Demura, T., Suzuki, K., Muller, I., Voss, U., Jurgens, G, Ito, M. 2007. R1R2R3-Myb proteins positively regulate cytokinesis through activation of *KNOLLE* transcription in *Arabidopsis thaliana. Development* 134:1101–1110.

Han, P., Li, Q., Zhu, Y.X. 2008. Mutation of *Arabidopsis* BARD1 causes meristem defects by failing to confine WUSCHEL expression to the organizing center. *Plant Cell* 20:1482–1493.

Harrar, Y., Bellec, Y., Bellini, C., Faure, J.D. 2003. Hormonal control of cell proliferation requires *PASTICCINO* genes. *Plant Physiol.* 132:1217–1227.

Hartig, K., Beck, E. 2006. Crosstalk between auxin, cytokinins, and sugars in the plant cell cycle. *Plant Biol.* 8:389–396.

Hartmann, U., Hoehmann, S., Nettesheim, K., Wisman, E., Saedler, H., Huijser, P. 2000. Molecular cloning of SVP: A negative regulator of the floral transition in *Arabidopsis. Plant J.* 21:351–360.

Hashimoto, J., Hirabayashi, T., Hayano, Y., Hata, S., Ohashi, Y., Suzuka, I., Utsugi, T., Toh, E.A., Kikuchi, Y. 1992. Isolation and characterization of cDNA clones encoding cdc2 homologues from *Oryza sativa*: A functional homologue and cognate variants. *Mol. Gen. Genet.* 233:10–16.

Hay, A., Barkoulas, M., Tsiantis, M. 2004. PINning down the connections: transcription factors and hormones in leaf morphogenesis. *Curr. Opin. Plant Biol.* 7:575–581.

Hay, A., Barkoulas, M., Tsiantis, M. 2006. ASYMMETRIC LEAVES1 and auxin activities converge to repress *BREVIPEDICELLUS* expression and promote leaf development in Arabidopsis. *Development* 133:3955–3961.

Heisler, M.G., Ohno, C., Das, P., Sieber, P., Reddy, G.V., Long, J.A., Meyerowitz, E.M. 2005. Patterns of auxin transport and gene expression during primordium development revealed by live imaging of the *Arabidopsis* inflorescence meristem. *Curr. Biol.* 15:1899–1911.

Helliwell, C.A., Chin-Atkins, A.N., Wilson, I.W., Chapple, R., Dennis, E.S., Chaudhury, A. 2001. The *Arabidopsis* AMP1 gene encodes a putative glutamate carboxypeptidase. *Plant Cell* 13:2115–2125.

Hemerly, A.S., Ferreira, P., De Almeida Engler, J., Van Montagu, M., Engler, G., Inze, D. 1993. Cdc2a Expression in *Arabidopsis* is linked with competence for cell division. *Plant Cell* 5:1711–1723.

Hick, G.S. 1994. Shoot induction and organogenesis *in vitro*: A developmental perspective. *In Vitro Cell. Dev. Biol.—Plant* 30:10–15.

Higuchi, M., Pischke, M.S., Mähönen, A.P., Miyawaki, K., Hashimoto,Y., Seki, M., Kobayashi, M., Shinozaki, K., Kato, T., Tabata, S., Helariutta, Y., Sussman, M.R., Kakimoto, T. 2004. In planta functions of the *Arabidopsis* cytokinin receptor family. *Proc. Natl. Acad. Sci. USA* 101:8821–8826.

Hindley, J., Phear, G.A. 1984. Sequence of the cell division gene *CDC2* from *Schizosaccharomyces pombe*: Patterns of splicing and homology to protein kinases. *Gene* 31:129–134.

Hirt, H., Mink, M., Pfosser, M., Bogre, L., Gyorgyey, J., Jonak, C., Gartner, A., Dudits, D., Heberle-Bors, E. 1992. Alfalfa cyclins differential expression during the cell cycle and in plant organs. *Plant Cell* 4:1531–1538.

Hu, Y., Bao, F., Li, J. 2000. Promotive effect of brassinosteroids on cell division involves a distinct CycD3-induction pathway in *Arabidopsis. Plant J.* 24:693–701.

Hutchison, C.E., Li, J., Argueso, C., Gonzalez, M., Lee, E., Lewis, M.W., Maxwell, B.B., Perdue, T.D., Schaller, G.E., Alonso, J.M., Ecker, J.R., Kieber, J.J. 2006. The *Arabidopsis* histidine phosphotransfer proteins are redundant positive regulators of cytokinin signaling. *Plant Cell* 18:3073–3087.

Hwang, I., Sheen, J. 2001. Two-component circuitry in *Arabidopsis* cytokinin signal transduction. *Nature* 413:383–289.

Ikeda, Y., Banno, H., Niu, Q.W., Howell, S.H., Chua, N.H. 2006. The *ENHANCER OF SHOOT REGENERATION 2* gene in *Arabidopsis* regulates *CUP-SHAPED COTYLEDON 1* at the transcriptional level and controls cotyledon development. *Plant Cell Physiol.* 47:1443–1456.

Imamura, A., Hanaki, N., Nakamura, A., Suzuki, T., Taniguchi, M., Kiba, T., Ueguchi, C., Sugiyama, T., Mizuno, T. 1999. Compilation and characterization of *Arabiopsis thaliana* response regulators implicated in His-Asp phosphorelay signal transduction. *Plant Cell Physiol.* 40:733–742.

Inoue, T., Higuchi, M., Hashimoto, Y., Seki, M., Kobayashi, M., Kato, T., Tabata, S., Shinozaki, K., Kakimoto, T. 2001. Identification of CRE1 as a cytokinin receptor from *Arabidopsis. Nature* 409:1060–1063.

Inzé, D., De Veylder, L. 2006. Cell cycle regulation in plant development. *Ann. Rev. Genet.* 40:77–105.

Iwakawa, H., Ueno, Y., Semiarti, E., Onouchi, H., Kojima, S., Tsukaya, H., Hasebe, M., Soma, T., Ikezaki, M., Machida, C., Machida, Y. 2002. The *ASYMMETRIC LEAVES2* gene of *Arabidopsis thaliana*, required for formation of a symmetric flat leaf lamina, encodes a member of a novel family of proteins characterized by cysteine repeats and a leucine zipper. *Plant Cell Physiol.* 43:467–478.

Jackson, D., Veit, B., Hake, S. 1994. Expression of maize KNOTTED1 related homeobox genes in the shoot apical meristem predicts patterns of morphogenesis in the vegetative shoot. *Development* 120:405–413.

Jasinski, S., Piazza, P., Craft, J., Hay, A., Woolley, L., Rieu, I., Phillips, A., Hedden, P., Tsiantis, M. 2005. KNOX action in *Arabidopsis* is mediated by coordinate regulation of cytokinin and gibberellin activities. *Curr. Biol.* 15 1560–1565.

Jeong, S., Trotochaud, A.E., Clark, S.E. 1999. The *Arabidopsis* CLAVATA2 gene encodes a receptor-like protein required for the stability of the CLAVATA1 receptor-like kinase. *Plant Cell* 11:1925–1934.

Kakimoto, T. 2003. Perception and signal transduction of cytokinins. *Annu. Rev. Plant Biol.* 54 605–627.

Kerstetter, R., Vollbrecht, E., Lowe, B., Veit, B., Yamaguchi, J., Hake, S. 1994. Sequence analysis and expression patterns divide the maize *knotted1*-like homeobox genes into two classes. *Plant Cell* 6:1877–1887.

Kondo, T., Sawa, S., Kinoshita, A., Mizuno, S., Kakimoto, T., Fukuda, H., Sakagami, Y. 2006. A plant peptide encoded by CLV3 identified by in situ MALDI-TOF MS analysis. *Science* 313:845–848.

Kurakawa, T., Ueda, N., Maekawa, M., Kobayashi, K., Kojima, M., Nagato, Y., Sakakibara, H., Kyozuka, J. 2007. Direct control of shoot meristem activity by a cytokinin-activating enzyme. *Nature* 445:652–655.

Kwiatkowska, D. 2008. Flowering and apical meristem growth dynamics. *J. Exp. Bot.* 59:187–201.

Laux, T., Mayer, K.F.X., Berger, J., Jürgens, G. 1996. The WUSCHEL gene is required for shoot and floral meristem integrity in *Arabidopsis*. *Development* 122:87–96.

Leibfried, A., to, J.P.C., Busch. W., Stehling, S., Kehle, A., Demar, M., Kieber, J.J., Lohmann, J.U. 2005. WUSCHEL controls meristem function by direct regulation of cytokinin-inducible response regulators. *Nature* 438:1172–1175.

Lenhard, M., Juergens, G., Laux, T. 2002. The WUSCHEL and SHOOTMERISTEMLESS genes fulfill complementary roles in *Arabidopsis* shoot meristem regulation. *Development* 129:3195–3206.

Lew, D.J., Dulic, V., Reed, S.I. 1991. Isolation of three novel human cyclins by rescue of G1 cyclin (Cln) function in yeast. *Cell* 66:1197–1206.

Lin, Z., Yin, K., Zhu, D., Chen, Z.L., Gu, H., Qu L.-J. 2007. AtCDC5 regulates G2 to M transition of cell cycle and is critical for the function of *Arabidopsis* shoot apical meristems. *Cell Res.* 17:815–828.

Lincoln, C., Long, J., Yamaguchi, J., Serikawa, K., Hake, S. 1994. A knotted1-like homeobox gene in *Arabidopsis* is expressed in the vegetative meristem and dramatically alters leaf morphology when overexpressed in transgenic plants. *Plant Cell* 6:1859–1876.

Long, J.A., Barton, M.K. 1998. The development of apical embryonic pattern in *Arabidopsis*. *Development* 125:3027–3035.

Long, J.A., Moan, E.I., Medford, J.I., Barton, M.K. 1996. A member of the KNOTTED class of homeodomain proteins encoded by the STM gene of *Arabidopsis*. *Nature* 379:66–69.

Long, J.A., Ohno, C., Smith, Z.R., Meyerowitz, E.M. 2006. TOPLESS regulates apical embryonic fate in *Arabidopsis*. *Science* 312:1520–1523.

Mähönen, A.P., Bishopp, A., Higuchi, M., Nieminen, K.M., Kinoshita, K., Törmäkangas, K., Ikeda, Y., Oka, A., Kakimoto, T., Helariutta, Y. 2006. Cytokinin signaling and its inhibitor AHP6 regulate cell fate during vascular development. *Science* 311:94–98.

Martinez, M.C., Jorgensen, J.E., Lawton, M.A., Lamb, C.J., Doerner, P.W. 1992. Spatial pattern of cdc2 expression in relation to meristem activity and cell proliferation during plant development. *Proc. Nat. Acad. Sci. USA* 89:7360–7364.

Mayer, K.F.X., Schoof, H., Haecker, A., Lenhard, M., Jürgens, G., Laux, T. 1998. Role of WUSCHEL in regulating stem cell fate in the *Arabidopsis* shoot meristem. *Cell* 95:805.

Meijer, M., Murray, J.A.H. 2000. The role and regulation of D-type cyclins in the plant cell cycle. *Plant Mol. Biol.* 43:621–633.

Melaragno, J., Mehrota, B., Coleman, A. 1993. Relationship between endoploidy and cell size in epidermal tissue of *Arabidopsis*. *Plant Cell* 5:1661–1668.

Menges, M., De Jager, S.M., Gruissem, W., Murray, J.A. 2005. Global analysis of the core cell cycle regulators of *Arabidopsis* identifies novel genes, reveals multiple and highly specific profiles of expression and provides a coherent model for plant cell cycle control. *Plant J.* 41:546–566.

Miao, G.-H., Hong, Z., Verma, D.P.S. 1993. Two functional soybean genes encoding p34-cdc2 protein kinases are regulated by different plant developmental pathways. *Proc. Natl. Acad. Sci. USA* 90:943–947.

Mizukami, Y.A.F. and R.L. Fischer. 2000. Plant organ size control: *AINTEGUMENTA* regulates growth and cell numbers during organogenesis. *Proc. Natl. Acad. Sci. USA* 97:942–947.

Nordstrom, A., Tarkowski, P., Tarkowska, D., Norbaek, R., Astot, C.,Dolezal, K., Sandberg, G. 2004. Auxin regulation of cytokinin biosynthesis in *Arabidopsis thaliana*: a factor of potential importance for auxin-cytokinin-regulated development. *Proc. Natl. Acad. Sci. USA* 101:8039–8044.

Ogawa, M., Shinohara, H., Sakagami, Y., Matsubayashi, Y. 2008. *Arabidopsis* CLV3 peptide directly binds CLV1 ectodomain. *Science* 319:294.

Ori, N., Eshed, Y., Chuck, G., Bowman, J.L., Hake, S. 2000. Mechanisms that control knox gene expression in the *Arabidopsis* shoot. *Development* 127:5523–5532.

Pernisová, M., Klíma, P., Horák, J., Válková, M., Malbeck, J., Souek, P., Reichman, P., Hoyerová, K., Dubová, J., Friml, J., Zažímalová, E., Hejátko, J. 2009. Cytokinins modulate auxin-induced organogenesis in plants via regulation of the auxin efflux. *Proc. Natl. Acad. Sci. USA* 106:3609–3614.

Potuschak, T., Doerner, P. 2001. Cell cycle controls: genome-wide analysis in Arabidopsis. *Curr. Opin. Plant Biol.* 4:501–506.

Prakash, A.P., Kumar, P.P. 2002. PkMADS1 is a novel MADS box gene regulating adventitious shoot induction and vegetative shoot development in *Paulownia kawakamii*. *Plant J.* 29:141–151.

Ragni, L., Belles-Boix, E., Gunl, M., Pautot, V. 2008. Interaction of *KNAT6* and *KNAT2* with *BREVIPEDICELLUS* and *PENNYWISE* in *Arabidopsis* inflorescences. *Plant Cell* 20:888–900.

Reinhardt, D., Mandel, T., Kuhlemeier, C. 2000. Auxin regulates the initiation and radial position of plant lateral organs. *Plant Cell* 12:507–518.

Reinhardt, D., Pesce, E., Stieger, P., Mandel, T., Baltensperger, K., Bennett, M., Traas, J., Friml, J., Kuhlemeier, C. 2003. Regulation of phyllotaxis by polar auxin transport. *Nature* 426:255–260.

Richard, C., Lescot, M., Inzé, D., De Veylder, L. 2002. Effect of auxin, cytokinin, and sucrose on cell cycle gene expression in *Arabidopsis thaliana* cell suspension cultures. *Plant Cell, Tiss. Org. Cult.* 69:167–176.

Riefler, M., Novak, O., Strnad, M., Schmülling, T. 2006. *Arabidopsis* cytokinin receptor mutants reveal functions in shoot growth, leaf senescence, seed size, germination, root development, and cytokinin metabolism. *Plant Cell* 18:40–54.

Riou-Khamlichi, C., Huntley, R., Jacqmard, A., Murray, J.A.H. 1999. Cytokinin activation of *Arabidopsis* cell division through a D-type cyclin. *Science* 283:1541–1544.

Ross, J.J., O'neill, D.P., Smith, J.J., Kerckhoffs, L.H.J., Elliott, R.C. 2000. Evidence that auxin promotes gibberellin A_1 biosynthesis in pea. *Plant J.* 21:547–552.

Rupp, H.-M., Frank, M., Werner, T., Strnad, M., Schmuelling, T. 1999. Increased steady state mRNA levels of the *STM* and *KNAT1* homeobox genes in cytokinin overproducing *Arabidopsis thaliana* indicate a role for cytokinins in the shoot apical meristem. *Plant J.* 18:557–563.

Saiga, S., Furumizu, C., Yokoyama, R., Kurata, T., Sato, S., Kato, T., Tabata, S., Suzuki, M., Komeda, Y. 2008. The *Arabidopsis OBERON1* and *OBERON2* genes encode plant homeodomain finger proteins and are required for apical meristem maintenance. *Development* 135:1751–1759.

Sakamoto, T., Kobayashi, M., Itoh, H., Tagiri, A., Kayano, T., Tanaka, H., Iwahori, S., Matsuoka, M. 2001. Expression of a gibberellin 2-oxidase gene around the shoot apex is related to phase transition in rice. *Plant Physiol.* 125:1508–1516.

Scanlon, M.J. 2003. The polar auxin transport inhibitor *N*-1-naphthylphthalamic acid disrupts leaf initiation, KNOX protein regulation, and formation of leaf margins in maize. *Plant Physiol.* 133:597–605.

Scheres, B. 2001. Plant cell identity. The role of position and lineage. *Plant Physiol.* 125:112–114.

Shaul, O. 1996. Regulation of cell division in *Arabidopsis*. *Crit. Rev. Plant Sci.* 15:97–112.

Shimotohno, A., Umeda-Hara, C., Bisova, K., Uchimiya, H., Umeda, M. 2004. The plant-specific kinase CDKF;1 is involved in activating phosphorylation of cyclin-dependent kinase-activating kinases in *Arabidopsis*. *Plant Cell* 16:2954–2966.

Sinha, N.R., Williams, R.E., Hake, S. 1993. Overexpression of the maize homeo box gene, *KNOTTED-1*, causes a switch from determinate to indeterminate cell fates. *Genes Dev.* 7:787–795.

Skoog, F., Miller, C.O. 1957. Chemical regulation of growth and organ formation in plant tissues cultured in vitro. *Symp. Soc. Exp. Biol.* 11:118–140.

Strabala, T.J., O'donnell, P.J., Smit, A.M., Ampomah-Dwamena, C., Martin, E.J., Netzler, N., Nieuwenhuizen, N.J., Quinn, B.D., Foote, H.C., Hudson, K.R. 2006. Gain-of-function phenotypes of many CLAVATA3/ESR genes, including four new family members, correlate with tandem variations in the conserved CLAVATA3/ESR domain. *Plant Physiol.* 140:1331–1344.

Szemenyei, H., Hannon, M., Long, J.A. 2008. TOPLESS mediates auxindependent transcriptional repression during *Arabidopsis* embryogenesis. *Science* 19:1384–1386.

Takada, S., Hibara, K.-I., Ishida, T., Tasaka, M. 2001. The CUP-SHAPED COTYLEDON1 gene of *Arabidopsis* regulates shoot apical meristem formation. *Development* 128:1127–1135.

Tamaki, H., Konishi, M., Daimon, Y., Aida, M., Tasaka, M., Sugiyama, M. 2009. Identification of novel meristem factors involved in shoot regeneration through the analysis of temperature-sensitive mutants of *Arabidopsis*. *Plant J.* 57:1027–1039.

Treml, B.S., Winderl, S., Radykewicz, R., Herz, M., Schweizer, G., Hutzler, P., Glawischnig, E., Ruiz, R.A. 2005. The gene *ENHANCER OF PINOID* controls cotyledon development in the *Arabidopsis* embryo. *Development* 132:4063–4074.

Ueguchi, C., Koizumi, H., Suzuki, T., Mizuno, T. 2001. Novel family of sensor histidine kinase genes in *Arabidopsis thaliana*. *Plant Cell Physiol.* 42:231–235.

Umeda, M., Umeda-Hara, C., Uchimiya, H. 2000. A cyclin-dependent kinase-activating kinase regulates differentiation of root initial cells in *Arabidopsis*. *Proc. Natl. Acad. Sci. USA* 97:13396–13400.

Veit, B. 2009. Hormone mediated regulation of the shoot apical meristem. *Plant Mol. Biol.* 69:397–408.

Vollbrecht, E., Veit, B., Sinha, N., Hake, S. 1991. The developmental gene *Knotted-1* is a member of a maize homeobox gene family. *Nature* 350:241–243.

Wang, G., Kong, H., Sun, Y., Zhang, X., Zhang, W., Altman, N., Depamphilis, C.W., Ma, H. 2004. Genome-wide analysis of the cyclin family in *Arabidopsis* and comparative phylogenetic analysis of plant cyclin-like proteins. *Plant Physiol.* 135:1084–1099.

Werner, T., Motyka, V., Laucou, V., Smets, R., Van Onckelen, H., Schmulling, T. 2003. Cytokinin-deficient transgenic *Arabidopsis* plants show multiple developmental alterations indicating opposite functions of cytokinins in the regulation of shoot and root meristem activity. *Plant Cell* 15:2532–2350.

Williams, L., Fletcher, J.C. 2005. Stem cell regulation in the *Arabidopsis* shoot apical meristem. *Curr. Opin. Plant Biol.* 8:582–586.

Williams-Carrier, R.E., Lie, Y.S., Hake, S., Lemaux, P.G. 1997. Ectopic expression of the maize *kn1* gene phenocopies the *Hooded* mutant of barley. *Development* 124:3737–3745.

Wolbang, C.M., Ross, J.J. 2001. Auxin promotes gibberellin biosynthesis in decapitated tobacco plants. *Planta* 214:153–157.

Yamada, H., Suzuki, T., Terada, K., Takei, K., Ishikawa, K., Miwa, K., Yamashino, T., Mizuno, T. 2001. The *Arabidopsis* AHK4 histidine kinase is a cytokinin-binding receptor that transduces cytokinin signals across the membrane. *Plant Cell Physiol.* 42:1017–1023.

Yamaguchi, M., Kato, H., Yoshida, S., Yamamura, S., Uchimiya, H., Umeda, M. 2003. Control of *in vitro* organogenesis by cyclin-dependent kinase activities in plants. *Proc. Natl. Acad. Sci. USA* 100:8019–8023.

Yanai, O., Shani, E., Dolezal, K., Tarkowski, P., Sablowski, R., Sandberg, G., Samach, A., Ori, N. 2005. *Arabidopsis* KNOX1 proteins activate cytokinin biosynthesis. *Curr. Biol.* 15:1566–1571.

Zhang, K., Letham, D.S., John, P.C. 1996. Cytokinin controls the cell cycle at mitosis by stimulating the tyrosine dephosphorylation and activation of p34cdc2-like H1 histone kinase. *Planta* 200:2–12.

Zhang, S., Lemaux, P.G. 2004a. Molecular Analysis of In Vitro Organogenesis. *Critical Reviews of Plant Science* 23:325–335.

Zhang, S., Lemaux, P.G. .2004b. Molecular Aspects of In Vitro Shoot Organogenesis. In Trigiano, R. N., Gray, D.J. (Ed.), *Plant Development and Biotechnology.* CRC Press, Boca Raton, FL.

Zhang, S., Williams-Carrier, R., Jackson, D., Lemaux, P.G. 1998. CDC2Zm and KN1 expression during adventitious shoot meristem formation from in vitro—proliferating axillary shoot meristems in maize and barley. *Planta* 204:542–549.

Zhang, S., Wong, L., Meng, L., Lemaux, P.G. 2002. Similarity of expression patterns of *knotted1* and *ZmLEC1* during somatic and zygotic embryogenesis in maize (*Zea mays* L.). *Planta* 215:191–194.

Zuo, J., Niu, Q.-W., Frugis, G., Chua, N.-H. 2002. The *WUSCHEL* gene promotes vegetative-to-embryonic transition in *Arabidopsis*. *Plant J.* 30:349–359.

21 Direct Shoot Organogenesis from Leaf Explants of Chrysanthemum and African Violets

R. N. Trigiano, L. M. Vito, M. T. Windham, Sarah Boggess, and Denita Hadziabdic

Chrysanthemums or "mums" (*Dendranthema grandiflora* Tzvelev. synonymous with *Chrysanthemum morifolium* Ramat.) are among the most popular floricultural species and grown as garden and pot crops or as cut flowers. Mums originated in Asia and are now a major floricultural crop in many countries throughout the world. The majority of the hundreds of different cultivars are propagated commercially by stem cuttings, but many have been successfully micropropagated also by adventitious shoot formation (organogenesis) from a variety of tissue and callus cultures (e.g., Lu et al., 1990; Hossain et al., 2007). Some of these regeneration protocols can be used in eliminating viruses from cultivars (meristem tip cultures), rapid increase of the number of plants from unique sports, recovery of mutations induced by either chemical or irradiation processes, and genetic transformation of selected genotypes for flower color and disease resistance and salt tolerance. Although infrequently encountered in exercises for plant tissue culture classes, chrysanthemums can be used to demonstrate many methodologies and concepts in regeneration protocols that are based on direct shoot organogenesis.

Since their discovery in the Usambara Mountains of eastern Africa in 1892, African violets (*Saintpaulia ionantha* Wendl.) have become one of the most commonly cultivated houseplants in the world. Commercially, new cultivars are developed through sports or through traditional plant breeding techniques and are then multiplied via leaf cuttings. More recently, micropropagation is being used to produce a large number of true-to-type plants in a short period of time. African violets are an excellent choice for demonstrating organogenesis techniques because plant material is readily available and micropropagation is straightforward and dependable. In fact, most cultivars of African violets readily produce shoots in vitro.

The following laboratory exercises illustrate the technique of direct shoot organogenesis using leaf sections of chrysanthemum and African violets. Specifically, the laboratory experiences will familiarize students with the following concepts: (1) differences in the ability of genotypes to regenerate shoots and roots; (2) effects of plant growth regulator (PGR) combinations on shoot and root formation; and (3) pulse treatments for initiation and growth of shoots. The chrysanthemum exercises are adapted primarily from a publication by Trigiano and May (1994).

GENERAL CONSIDERATIONS

CHRYSANTHEMUM

All of the cultivars of chrysanthemum mentioned in the experiments may be obtained from Yoder Brothers, Barberton, Ohio, as unrooted or rooted cuttings and have been evaluated for their responses in culture. We suggest growing the following cultivars for the exercises: "Little Rock," "Regal Jamestown," "Spirit Lake," "Tahoe," and "Vail." However, if laboratory or greenhouse space is limited, grow only "Spirit Lake" and "Vail." Plant 3–5 cuttings in each of at least ten 10-cm-diameter plastic pots containing any soilless media 3 months prior to the beginning of the exercises. This should provide adequate materials for classes of about 15–20 students. A good estimate of the number of plants needed per class is to provide a minimum of one plant per student or student team per exercise. The conditions for growing the plants used for explants is very important—if possible, cultivate the plants in the laboratory or growth room illuminated with about 100 $\mu mol \cdot m^{-2}.s^{-1}$ fluorescent and incandescent light for 16 h per day at 22°–25°C. Alternately, plants grown in the greenhouse under shade and with cooling will provide suitable materials. Supplemental lighting in the greenhouse will be necessary to maintain long days and inhibit flowering if the plants are grown in the spring and fall. Pinch longer shoots often to encourage branching and initiation of new leaves. Fertilize regularly (twice weekly with 300 ppm N) to maintain vigorous growth (see May and Trigiano [1991] for details of growing conditions). Not getting the foliage wet with water will help eliminate most contaminating organisms and make surface disinfestation much more effective.

African Violets

Just about any commercially available African violet should respond in culture. The largest selection of cultivars is available online from Optimara through their distributor (www.selectivegardener.com). A large selection of African violets can usually be found at any local gardening center store or large grocery store and are generally available as finished plants in 4" pots. Allow one plant per student or student team. The plants should be maintained under the cleanest conditions possible in the laboratory or growth room illuminated with approximately 70 $\mu mol \cdot m^{-2}.s^{-1}$ fluorescent light for 16 h per day at 22°–25°C. Watering from the bottom is recommended for African violets to prevent leaf distortion, and a commercially available African violet fertilizer should be added to the water monthly.

EXPLANT PREPARATION AND BASAL MEDIUM FOR CHRYSANTHEMUM AND AFRICAN VIOLETS

Chrysanthemum

For satisfactory and reproducible results, select only young, partially expanded, light green, 2–5 cm long leaves for explant tissue. Surface disinfest whole leaves (do not remove the petiole) using 5% commercial bleach (0.26% NaOCl) for 5–10 min (or 10% bleach for material grown in the greenhouse) with constant agitation followed by three rinses with sterile distilled water. Place a single leaf with the top surface up in a sterile 100 × 15 mm plastic petri dish and drain excess water by tilting and gently tapping the dish on the flow hood table. The top surface of the leaf is easily identified by its darker color and abundant trichomes. Excise four or five ½–1 cm^2 sections from the midrib area of the leaf using a number 10 scalpel blade. Place the sections with the abaxial surface (underside) of the leaf in contact with the agar medium (Figure 21.1).

The basal culture medium is composed of Murashige and Skoog (MS; 1962) basal salts (see Chapter 2 for composition) amended with 30 g (88 mM) sucrose, 100 mg (0.06 mM) myo-inositol, 5 mg (2.9 μM) thiamine-HCl, pH 5.7, and 8 g of phytoagar per liter. After autoclaving, the medium may be poured into 60 15-mm (or other sized) plastic petri dishes. Typically, one or two leaf sections are placed in each of the dishes (Figure 21.1). PGR requirements will be provided for each variation of the experiment. We usually prepare at least four dishes of each treatment for each student or student team.

FIGURE 21.1 Two chrysanthemum leaf segments placed with the lower leaf surface on the growth medium.

African violet

Select healthy, undamaged leaves 4–6 cm in length. While this size is preferred, all sizes (ages) of leaves are responsive in tissue culture to some degree. Explant preparation is the same as for chrysanthemum, with the only exception being that the entire leaf can be used to provide explant material (Figure 21.2). Basal medium is also the same as described above for chrysanthemum.

Experiment 1. Differences between Genotypes of Chrysanthemums and of African Violets in the Ability to Regenerate Shoots and Roots

An important basic concept in the tissue culture of plants is that not all genotypes (in this case, genotypes are identified as cultivars) will respond equally under similar cultural conditions. In fact, there are some cultivars of chrysanthemum that will not produce either shoots or roots in vitro using any treatment, while some African violet cultivars require little or no treatment to produce shoots. This experiment was developed to illustrate this concept and is probably the easiest of the following exercises to conduct with large classes. We suggest that the exercise be completed by teams of two or three students. This experiment also provides the basic protocol for the other two experiments in this chapter. Experiment 1 will require approximately 12 weeks for completion. Please view the video mum experiments on the accompanying DVD before completing this experiment. It depicts all of the procedures for 22.1 Shoots from Chrysanthemum Leaf Explants.

Materials

The following items are needed for each student or team of students:

Chrysanthemum

- 100 mL of 5% commercial bleach per cultivar
- Four 60 × 15 mm petri dishes containing MS + 0.25 mg/L (1 µM) BA + 2 mg/L (11.5) µM IAA per cultivar
- Four 60 × 15 mm petri dishes containing MS medium with 11.5 µM IAA only
- Four 60 × 15 mm petri dishes containing MS medium with 1.0 µM BA only
- Sixteen 60 × 15 mm petri dishes containing MS medium without PGRs per cultivar

FIGURE 21.2 African violet leaf dissected for explants. The entire leaf excluding the petiole may be used in the experiments.

- Two GA7 Magenta culture vessels or other similar size vessel (baby food jars) containing 75 mL MS medium without PGRs
- Incubator providing approximately 25 μmol.m^{-2}.s^{-1} and 25°C
- Cell packs with soilless medium, and mist bed or plastic wrap or plastic bag to construct a moist chamber for high humidity

African Violet

- 250 mL of 5% commercial bleach per cultivar
- Four 60 × 15 mm petri dishes containing MS + 1.0 mg/L (4.4 μM) BA + 1 mg/L (5.75) μM IAA per cultivar.
- Four 60 × 15 mm petri dishes containing MS medium with 5.75 μM IAA only
- Four 60 × 15 mm petri dishes containing MS medium with 4.4 μM BA only
- Sixteen 60 × 15 mm petri dishes containing MS medium without PGRs per cultivar
- Two GA7 Magenta culture vessels or other similar size vessel (baby food jars) containing 75 mL MS medium without PGRs
- Incubator providing approximately 25 μmol.m^{-2}.s^{-1} and 25°C
- Cell packs with soilless medium or Jiffy-7 peat pellets and plastic containers or plastic bags to construct a moist chamber for high humidity

Follow the protocol outlined in Procedure 21.1 for chrysanthemum and Procedure 21.2 for African violet to complete these experiments.

Anticipated Results

Chrysanthemum

After the first 2 weeks of culture, leaf explants exposed to IAA and BA should have distorted shapes, and sparse, light green crystalline callus may have formed around the cut edges of all cultivars. Darker green meristematic zones or small buds may also be apparent on "Little Rock," "Tahoe," and "Vail" explants. This ends the initiation phase of the exercise Stages 0 (selection of materials) and I (establishment of cultures) of traditional micropropagation.

Three weeks after transfer to medium without PGRs, shoots and some roots should be formed on leaf sections of "Little Rock," "Tahoe," and "Vail" that were initially on cultured

Procedure 21.1
Differences Between Chrysanthemum Genotypes in the Ability to Regenerate Shoots and Roots

Step	Instruction and Comments
1	Surface disinfest leaves of the different chrysanthemum cultivars with 5% Clorox laboratory grown plants (10% for greenhouse grown plants) for 5–10 min and rinse three times with sterile distilled water.
2	Dissect leaves as described previously under explant preparation and place two sections in each 60 × 15 mm petri dish containing MS medium supplemented with 11.5 μM IAA and 1.0 μM BA; IAA alone, BA alone, or MS basal medium lacking PGRs. This is the shoot initiation phase. The experiment should be considered a randomized complete block design with four treatments and replications.
3	Incubate dishes at 25°C with 25 $\mu mol \cdot m^{-2} s^{-1}$ of fluorescent light for 2 weeks.
4	Transfer all explants to MS basal medium without PGRs and incubate under the same conditions as stated in Step #3 for 3 weeks. This is the shoot elongation phase.
5	The number of shoots and/or roots within a petri dish, an experimental unit, should be counted after 3 weeks. Student teams should share and analyze data (standard deviation or ANOVA—see Chapter 5).
6	Excise shoots greater than 2 cm and remove any large leaves that are positioned near the cut surface. Insert the cut end of the shoots into MS basal medium without PGRs contained in GA-7 Magenta vessels. Three rows of five shoots can be place in each vessel and incubated as in Step #3 for the root initiation phase.
7	After 3 weeks, gently wash the agar from the roots of plants with running tap water. Remove large basal leaves and transfer the plantlets to soilless medium in cell packs or individual 1 × 1 cells and place under intermittent mist for 2–3 weeks (acclimatization phase). Alternative, plastic wrap or a bag may be used to build a "tent" around the tray of cell packs. After a week, make several holes in the "tent" with a sharpened pencil. Add a few holes every 2–3 days. The plants should be acclimatized to ambient conditions in 2–3 weeks and now can be grown in the greenhouse or laboratory.

Procedure 21.2
The Ability of African Violet Genotypes to Regenerate Shoots and Roots

Step	Instruction and Comments
1	Surface disinfest leaves of the different African violet cultivars with 5% Clorox for 5–10 min and rinse three times with sterile distilled water.
2	Dissect leaves as described previously under explant preparation and place two sections in each 60 × 15 mm petri dish containing MS medium supplemented with 4.9 μM IAA (1 mg/L) and 4.4 μM BA (1 mg/L); IAA alone; BA alone; or MS basal medium lacking PGRs. This is the shoot initiation phase. The experiment should be considered a randomized complete block design with four treatments and replications.
3	Incubate dishes at 25°C with 25 $\mu mol \cdot m^{-2} \cdot s^{-1}$ of fluorescent light for 2 weeks.
4	Transfer all explants to MS basal medium without PGRs and incubate under the same conditions as stated in Step #3 for 3 weeks. This is the shoot elongation phase.
5	The number of shoots and/or roots within a petri dish, an experimental unit, should be counted after 3 weeks. Student teams should share and analyze data (standard deviation or ANOVA—see Chapter 5).
6	Excise shoots greater than 2 cm and remove any large leaves that are positioned near the cut surface. Insert the cut end of the shoots into MS basal medium without PGRs contained in GA-7 Magenta vessels. Three rows of five shoots can be placed in each vessel and incubated as in Step #3 for the root initiation phase.
7	After 3 weeks, gently wash the agar from the roots of plants with running tap water. Remove large basal leaves and transfer the plantlets to soilless medium in cell packs or hydrated Jiffy-7 peat pellets and place in a solid bottom container. Cover the container with a lid or clear plastic bag (acclimatization phase). As the peat pellet or soil dries, rehydrate by pouring tepid water into the container. African violet leaves are very sensitive to water and become damaged if allowed to get wet. After a week, loosen the cover and remove gradually over a week, or if using plastic bags, make several holes in the bag with a sharpened pencil. Add a few holes every 2–3 days. The plants should be acclimatized to ambient conditions in 2–3 weeks and now can be grown in the greenhouse or laboratory.

medium containing PGRs. In this case, shoots and roots are formed directly from mesophyll cells without an intervening callus phase or direct organogenesis. Shoots should have formed primarily on the cut margins of the explant (Figure 21.3). The most shoots (and least number of roots) should be on "Vail" explants and decreasing numbers on "Little Rock" and "Tahoe." "Regal Jamestown" and "Spirit Lake" explants may have produced a few roots and maybe an occasional shoot (Figure 21.4). None of the explants of any of the cultivars should have produced appreciable amounts of callus, and those initially incubated on either only IAA, BA, or medium lacking PGRs should have formed very few shoots or roots: "Regal Jamestown" and "Spirit Lake" may have formed some roots. This is the "shoot elongation" phase of the experiment or Stage II of micropropagation.

FIGURE 21.3 Two chrysanthemum "Vail" leaf segments after five weeks in culture. Notice the number of direct (no callus) adventitious shoots that have formed on the cut surface of the leaf piece.

FIGURE 21.4 Roots form on leaf segments of chrysanthemum "Regal Jamestown" after 5 weeks in culture.

Shoots that were excised from leaves (Figures 21.5A and B) should have elongated to almost twice the original length and produced roots (Stage III of micropropagation) in the Magenta vessels containing MS medium without PGRs (Figure 21.5C). Typically, 3–5 adventitious roots are formed per shoot. These shoots are easily acclimatized to normal growing conditions within three weeks using the described procedures (Stage IV of micropropagation) and grown to flower (Figure 21.5D).

African Violet

After the first two weeks of culture, results for African violets should be similar to those of chrysanthemum. In addition, some cultivars will have raised or bumpy areas across the entire explant surface. Three weeks after transfer to medium without PGRs, shoots and some roots should be formed on leaf sections of all cultivars that were initially cultured on medium containing PGRs (Figure 21.6).

Shoots that were excised from leaves should have elongated to almost twice the original length and produced roots (Stage III of micropropagation) in the Magenta GA7 vessels containing MS medium without PGRs (Figure 21.7). Some cultivars will also produce new shoots or plantlets and will need to be carefully separated when washing roots prior to transfer to peat pellets (Figure 21.8A). These shoots are easily acclimatized to normal growing conditions within 3 weeks using the described procedures (Figure 21.8B; Stage IV of micropropagation).

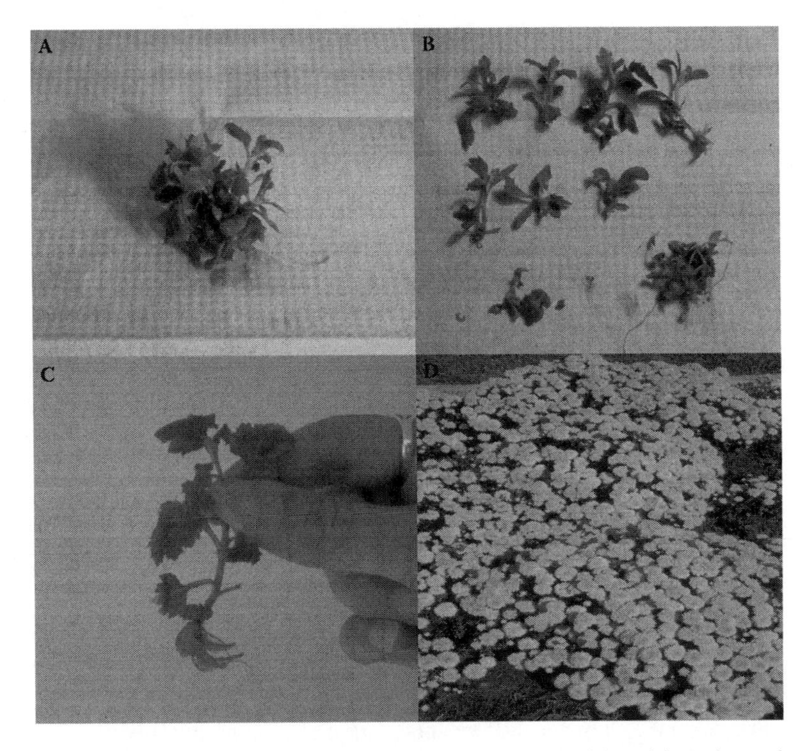

FIGURE 21.5 All figures are of the chrysanthemum cultivar Iridon. A. Adventitious shoots developed along margins of a leaf explant. B. Adventitious shoots excised from the leaf margin. C. Adventitious roots formed from the base of the shoot explanted onto MS medium without plant growth regulators. Notice that the shoot has elongated significantly. D. Plants regenerated from excised leaves and allowed to flower. The flower color is the same as the original parent plant.

FIGURE 21.6 Multiple shoots formed directly from African violet leaf explants. Note that roots may also form.

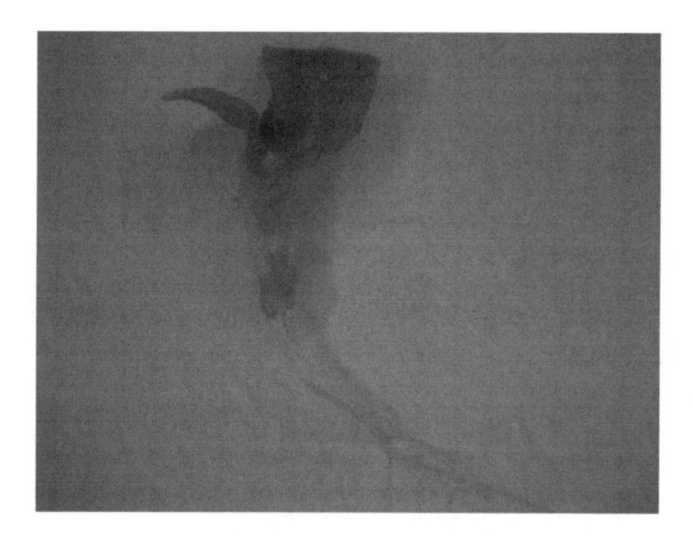

FIGURE 21.7 Rooted shoot of African violet.

Questions

- What purpose do the treatments containing individual PGRs (IAA or BA only) and without any PGRs serve in the initiation phase of the experiment? Why are they necessary in the design of the experiment?
- Why is it necessary to transfer the explants from initiation medium (with PGRs) to shoot elongation medium (without PGRs)?
- Why can't the plants be transferred directly from the Magenta vessels to normal growing conditions found in the laboratory or greenhouse?

FIGURE 21.8 Initial acclimatization of African violet rooted shoots. A. Shoots placed in peat pellets for acclimatization. Peat pellets are contained in a plastic box to maintain high humidity. B. Acclimatization of African violet. Note the different stages of growth proceeding from the bottom to the top of the lighted growth rack. Box depicted in Figure 21.7A is on the lower rack and "finished" plants are on the top of the rack.

Experiment 2. The Effects of PGRs on Formation of Shoots and Roots from Leaf Sections of Chrysanthemum

One of the first steps in developing a micropropagation protocol of any plant is to determine the type (or combination) and concentration of PGRs that will provide the desired response—in our case, shoot formation. Leaf tissue of some cultivars of chrysanthemum is very responsive to both the type and concentration of cytokinin in the initiation medium.

This experiment uses the cultivars "Regal Jamestown" and "Vail" or if time and/or materials are limited, the experiment may be completed with "Vail" only. Prepare several modifications of MS medium, each containing 11.5 μM IAA combined with the following concentrations and types of cytokinin or cytokinin-like compounds: 0.01, 0.1, 1.0, and 5.0 μM kinetin, 2iP, BA, and thidiazuron (TDZ). There are a total of 17 treatments including IAA only.

Since there are 17 treatments, it is impossible to include all treatments in every block without using at least four or five different leaves. The inherent variability for the ability to form shoots between different leaves from the same stock plant would probably necessitate increasing the replications to some very large number in order to detect significant differences between treatments. Therefore, from our experience, this experiment works best when designed as a randomized incomplete block design with eight replications (see Chapter 5). Each incomplete block is composed of four entries or the number of explants obtained from one leaf, thus limiting the variability within an individual block to that of an individual leaf. As matter of fact, explants from within an individual leaf are incredibly uniform in their ability to produce shoots. A more complete treatment of this experimental design with chrysanthemum may be found in Kuklin et al. (1993). A design for this exercise can be found in Table 21.1.

An associated exercise to demonstrate that explants from an individual leaf are equally capable of forming shoots is easily prepared and executed. Culture the four midvein explants from 10 leaves on MS medium supplemented with 1.0 μM BA and 11.5 μM for 2 weeks, then transfer to MS medium without PGRs for three weeks. Simply compare the number of shoots produced by each explant of an individual leaf—they should be more or less equal.

TABLE 21.1
An Incomplete Block Design with 17 Treatments and Eight Replications

Block	Treatment Number				Block	Treatment Number			
1	8	17	6	1	18	8	9	7	14
2	1	13	11	17	19	8	10	12	2
3	8	13	7	17	20	2	7	14	6
4	15	8	3	17	21	3	14	14	4
5	1	2	17	9	22	7	5	12	10
6	11	10	1	5	23	6	15	3	9
7	16	17	12	4	24	12	7	6	2
8	9	12	15	11	25	3	4	11	12
9	11	12	3	9	26	14	10	13	15
10	13	1	11	4	27	9	5	15	14
11	9	14	2	1	28	10	13	16	9
12	15	7	4	13	29	8	16	2	14
13	15	1	11	12	30	5	6	4	16
14	10	2	14	17	31	13	5	2	16
15	16	6	13	3	32	7	15	4	6
16	16	11	7	5	33	5	16	10	8
17	11	5	17	3	34	6	2	8	10

Materials

For each cultivar in the experiment:

- Eight 60 ×15 mm petri dishes for each treatment
- Eight 60 × 15 mm petri dishes containing MS medium without PGRs

Follow the protocols outlined in Procedure 21.3.

Anticipated Results

After incubation for 2 weeks on shoot initiation medium and an additional 3 weeks on shoot elongation medium, shoots and roots should be formed. Most treatments, except IAA only, will support some shoot formation on explants of "Vail"; only roots should form on explants of "Regal Jamestown" regardless of the combination of PGRs employed. For a pictorial representation of this concept, see Trigiano and May (2000). Shoots produced on medium containing higher concentrations of TDZ may not elongate after transfer to medium without PGRs. Also, most shoots produced on these treatments will exhibit hyperhydricity. Students should determine which treatment produced (statistically) the most shoots and/or roots.

Questions

- Which PGRs would you use to initiate shoots for a regeneration protocol and why?
- Are there other factors not considered by this experiment that might influence the number of shoots produced per experimental unit?
- What advantages does an incomplete block design offer over other designs such as completely randomized or randomized complete block designs?
- Some treatments allowed for the production of many shoots; however, a similar or greater number of roots were also formed. What are some of the disadvantages of forming roots during the shoots initiation and elongation phases of micropropagation of chrysanthemum? Are there any advantages?

Procedure 21.3
The Effects of PGRs in the Formation of Shoots and Roots from Leaf Explants of Chrysanthemum

Step	Instructions and Comments
1	Treatments (PGR type and concentration) should be represented by a number ranging from 1 to 17. Depending on the size of the class, each student or team should be assigned to complete specific blocks, not treatments, as shown in Table 21.1.
2	Surface disinfest chrysanthemum leaves as described in Procedure 21.1 and dissect leaves into four mid-vein sections as previously described under explant preparation and basal medium. One leaf piece should be placed in each of four 60 mm petri dishes (treatments) contained within each incomplete block.
3	Incubate the cultures at 25°C with 25 $\mu mol \cdot m^{-2} \cdot s^{-1}$ of fluorescent light for 2 weeks and then transfer to medium without PGRs for an additional 3 weeks under the same conditions.
4	Count the number of shoots and roots at the end of 5 weeks after the initiation of the experiment. Analyze data for significant differences between the means for each of the treatments.

Experiment 3. Pulse Treatment with PGRs for Initiation and Growth of Shoots from Chrysanthemum and African Violet Leaf Sections

Most adventitious shoot micropropagation protocols are based on a treatment of explants with PGRs for a specific period of time (initiation period) followed by incubation on medium that either lacks or has reduced concentrations of PGRs. We have used this technique in each of the previous two exercises and such treatments are usually referred to as "pulse" treatments. The practical reasons for utilizing "pulse" treatments are easily demonstrated using the chrysanthemum cultivar "Vail" or any African violet cultivar, and varying the time the explants are incubated on initiation medium. The objective of this experiment is to determine the optimum time that explants should remain on initiation medium to form the maximum number of shoots.

Materials

For each student or student team:

Chrysanthemum

- Sixteen 60 × 15 mm petri dishes containing MS medium with 11.5 μM IAA + 1.0 μM BA
- Twenty 60 ×15 mm petri dishes containing MS medium without PGRs

African Violet

- Sixteen 60 × 15 mm petri dishes containing MS medium with 5.75 μM IAA + 4.4 μM BA
- Twenty 60 × 15 mm petri dishes containing MS medium without PGRs

Follow the protocol outlined in Procedure 21.4 for chrysanthemum and 21.5 for African violet.

Anticipated Results

Chrysanthemum

Few, if any, shoots should form on explants from treatments 0, 1, 3, and 5 days, whereas 10–17 days of incubation on medium with PGRs should foster the development of the greatest number of shoots. Fewer shoots are formed on explants from the 21-day treatment, and although some shoots develop on explants from the 35-day treatment, they usually are stunted and/or exhibit hyperhydricity, a water-soaked appearance.

African Violet

As with chrysanthemum, few shoots should form on explants from treatments 0, 1, 3, and 5 days. The greatest number of shoots is formed on explants exposed to PGRs for 10–21 days. Beyond

Procedure 21.4
Pulse Treatment with PGRs for Initiation and Growth of Shoots from Chrysanthemum Leaf Explants

Step	Instructions and Comments
1	Surface disinfest chrysanthemum leaves (see Procedure 21.1) and dissect leaves as described under explant preparation.
2	Place two leaf sections into each 60 × 15 mm petri dish containing MS medium augmented with 1.0 μM BA + 11.5 μM IAA or MS medium without PGRs (for day 0).
3	The treatments are arranged in a completely randomized design, 0, 1, 3, 5, 7, 10, 14, 21, and 35 days of culture on initiation medium followed by MS medium without PGRs for the remainder of the 35-day period. The 0-day treatment should be initiated and transferred to MS medium without PGRs after 14 days. All cultures should be incubated at 25°C with 25 $\mu mol \cdot m^{-2} \cdot s^{-1}$ of fluorescent light for the entire 35 days.
4	Count the number of shoots after 35 days and compare the means of the treatments.

Procedure 21.5
Pulse Treatment with PGRs for Initiation and Growth of Shoots from African Violet Leaf Explants

Step	Instructions and Comments
1	Surface disinfest African violet leaves (see Procedure 21.1b) and dissect leaves as described under explant preparation.
2	Place two leaf sections into each 60 × 15 mm petri dish containing MS medium augmented with 4.4 μM BA + 5.75 μM IAA or MS medium without PGRs (for day 0).
3	The treatments are arranged in a completely randomized design, 0, 1, 3, 5, 7, 10, 14, 21, and 35 days of culture on initiation medium followed by MS medium without PGRs for the remainder of the 35-day period. The 0-day treatment should be initiated and transferred to MS medium without PGRs after 14 days. All cultures should be incubated at 25°C with 25 $\mu mol \cdot m^{-2} \cdot s^{-1}$ of fluorescent light for the entire 35 days.
4	Count the number of shoots after 35 days and compare the means of the treatments.

21 days, most cultivars produce such a large number of roots that it is difficult to harvest shoots. These shoots also exhibit abnormal growth and hyperhydricity.

Questions

- What is hyperhydricity and why did it occur with shoots produced on the 35-day treatment in this experiment?
- Why were shoots not formed from explants included in the 0-, 1-, 3-, and 5-day treatments?

REFERENCES

Hossain, Z., A. K. A. Mandal, S. K. Datta, and A. K. Biswas. 2007. Development of NaCl-tolerant line in *Chrysanthemum morifolium* Ramat. through shoot organogenesis of selected callus line. *J. Biotechnol.* 129:658–667.

Kuklin, A. I., R. N. Trigiano, W. L. Sanders, and B. V. Conger. 1993. Incomplete block design in plant tissue culture research. *J. Tiss. Cult. Meth.* 15:204–209.

Lu, C-Y., G. Nugent, and T. Wardley. 1990. Efficient, direct plant regeneration from stem segments of chrysanthemum (*Chrysanthemum morifolium* Ramat. cv. Royal Purple). *Plant Cell Rep.* 8:733–736.

May, R.A. and R. N. Trigiano. 1991. Somatic embryogenesis and plant regeneration from leaves of *Dendranthema grandiflora*. *J. Amer. Soc. Hort. Sci.* 116:366–371.

Murashige, T. and F. Skoog. 1962. A revised medium for rapid growth and bioassays with tobacco tissue cultures. *Physiol. Plant.* 15:473–497.

Trigiano, R. N. and R. A. May. 1994. Laboratory exercises illustrating organogenesis and transformation using chrysanthemum cultivars. *HortTechnology* 4: 325–327.

Trigiano, R.N. and R. A. May. 2000. Direct shoot organogenesis from leaf explants of chrysanthemum. Pp. 139–147 in Trigiano, R.N., and D.J. Gray (Eds.), *Plant Tissue Culture Concepts and Laboratory Exercises*, Second Edition, CRC Press LLC, Boca Raton, FL, 454 pp.

22 Propagation from Nonmeristematic Tissues—Nonzygotic Embryogenesis

D. J. Gray

CONCEPTS

- Nonzygotic embryogenesis occurs among widely disparate cell and tissue types and may be regarded to be a universal capability of higher plants.
- Nonzygotic embryos appear to arise from single cells, not as a result of sexual reproduction, but otherwise are identical to their zygotic counterparts.
- Nonzygotic embryogenesis conclusively demonstrates cellular totipotency and represents a highly efficient method of plant propagation.
- Embryogenic cell cultures are widely used for genetic manipulation such as in vitro selection and transgenic technologies.

An embryo can be defined as the earliest recognizable multicellular stage of an individual that occurs before it has developed the structures or organs characteristic of a given species. In most organisms, embryos are morphologically distinct entities that function as an intermediate stage in the transition between the gametophytic to sporophytic life cycle (Figure 22.1). For example, in higher plants, we are most familiar with embryos that develop within seeds; such embryos usually arise from gametic fusion products (zygotes) following sexual reproduction and are termed *zygotic embryos*, although seed-borne embryos also can develop apomictically (i.e., without sexual reproduction). However, plants are unique in that morphologically and functionally correct nonzygotic embryos also can arise from widely disparate cell and tissue types at a number of different points from both the gametophytic and sporophytic phases of the life cycle (Figure 22.2).

The first demonstration that plants could produce nonzygotic embryos in vitro was published in 1958 by Steward et al. Subsequently, Reinert (1959) observed bipolar embryos to differentiate in a culture of carrot roots after transfer from one medium to another. While carrot was the first species in which in vitro nonzygotic embryogenesis was reported, in subsequent years many species of angiosperms and gymnosperms have been added to the list of successes. In fact, demonstrations of nonzygotic embryogenesis are so widespread that it may be regarded to be a universal capability of higher plants.

A plethora of terminology has arisen to designate nonzygotic embryos. Such embryos originally were termed "embryoids" to denote perceived significant differences from zygotic embryos and, unfortunately, this term persists in some literature. However, differences in embryogenic cell origins notwithstanding, distinctions between zygotic and nonzygotic embryos become blurred as our understanding of embryo development increases. Nonzygotic embryos are now shown to be functionally equivalent to zygotic embryos and the suffix "oid" should be dropped. Other terms for nonzygotic embryos are based primarily upon differences in their specific sites of origin and often are interchangeable, leading to some inconsistencies in the literature. For example, nonzygotic embryos can arise from plant vegetative cells, reproductive tissues, zygotic embryos, or callus cells

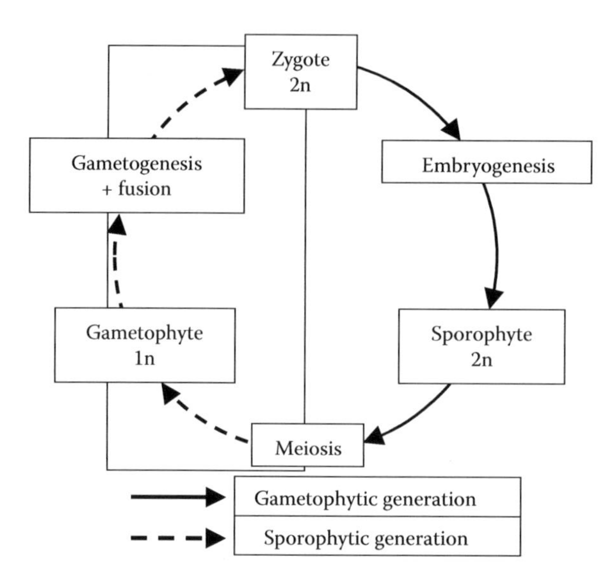

FIGURE 22.1 Typical angiosperm life cycle showing the natural role of embryogenesis in sporophyte development.

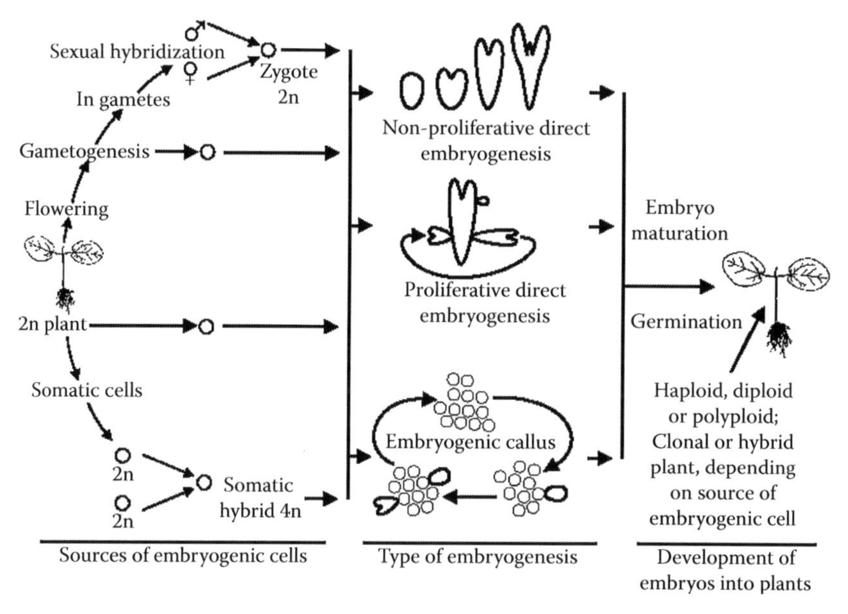

FIGURE 22.2 Sources of embryogenic cells, types of embryogenesis, and plants obtained via in vitro culture of higher plants.

derived from any of these. Thus, "somatic embryos" grow from somatic cells; "haploid or pollen embryos" are derived from pollen grains or microspore mother cells; "nucellar embryos" are formed from nonzygotic nucellar seed tissue; "direct secondary embryos" develop from previously formed embryos, etc.

Most of these embryo-types share the common trait of being able to be manipulated via in vitro culture. Such embryogenic culture systems form the basis for many biotechnological approaches to plant improvement, since they allow not only clonal plant propagation but also specific and directed changes to be introduced into desirable, elite individuals by genetic engineering of somatic cells.

Modified individual cells and/or embryos then can be efficiently multiplied in vitro to very high numbers prior to plant development. This approach to genetic improvement bypasses the unwanted consequences of sexual reproduction (mass genetic recombination and required cycles of selection) inherent to conventional breeding technology.

Because of the many commonalities exhibited by these embryo types as well as the various methods available to study and manipulate them, for convenience, this chapter will only utilize the term "nonzygotic embryo" to designate the embryos that develop through in vitro culture, regardless of origin. Thus, the developmental processes and experimental procedures described will tend to be applicable to many of the embryo types mentioned above.

IN VIVO VERSUS IN VITRO GROWTH CONDITIONS

In seeds, nutritive tissues directly encase the developing embryo. Early in embryogenesis, nutritive substances enter the zygotic embryo either through the suspensor or the embryo body (Figure 22.3A). The relative reliance of the developing embryo on obtaining nutrition via the suspensor versus endosperm differs greatly depending on the species. In contrast, nonzygotic embryos develop naked, not encased in seed endosperm and, as such, are not subjected to specialized and highly regulated nutritional regimes (Gray and Purohit, 1991). Often, a suspensor is the only link between the embryo and growth medium (Figure 22.3B). This demonstrates that the suspensor can serve as the pathway for all needed nutrition and that endosperm is not absolutely necessary for embryogenesis and germination to occur.

COMMON ATTRIBUTES OF EMBRYOGENIC CULTURE PROTOCOLS

Given the range of different explants, culture conditions, and media used to initiate and maintain embryogenic cultures, certain generalities can be made regarding basic methodology. The use of specific plant growth regulators (PGRs), which are required in most instances, will be discussed below with regard to embryogenic cell initiation.

The choice of genotype and explant often is crucial in obtaining an embryogenic response (see Gray, 1990; Gray and Meredith, 1992, for reviews). For example, with corn, only a limited range of genotypes are capable of producing embryogenic cultures and, with few exceptions, only immature embryo explants can be used. With corn, the age of immature embryos also is important, since embryos that are too young do not survive culture and those that are too old do not produce embryogenic callus. Interestingly, most species tested to date can be induced to produce embryogenic cultures from at least

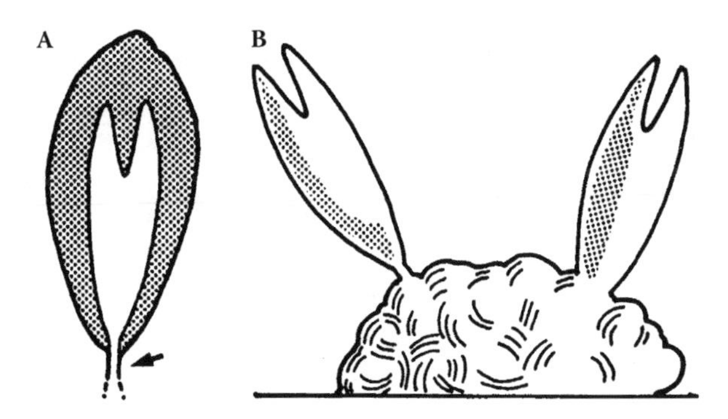

FIGURE 22.3 Comparison of typical zygotic embryo development in seed (A) with nonzygotic embryo development from callus (B). Zygotic embryos typically develop within nutritive seed tissues (shaded area) and are connected to the mother plant by a suspensor (arrow). In contrast, nonzygotic embryos often develop perched above subtending tissue; the only source of nutrition is the narrowed suspensor. (From Gray, D.J. and A. Purohit, 1991. *Crit. Rev. Plant Sci.* 10:33–61.)

some genotypes and tissues. Additionally, the embryogenic response is heritable, such that it can be bred from embryogenic lines into nonembryogenic lines via sexual hybridization, albeit with a certain degree of difficulty and transfer of other, often undesirable, traits via sexual hybridization.

Environmental requirements for optimum culture growth often are quite specific, but not unusual. Cultures may require growth in either dark or light or a combination of both over time, the optimization of which can be experimentally quantified. Culture in dark may be necessary in order to prevent the triggering effect of light on many plant biological processes that may adversely affect the growth of embryogenic cell populations. Culture in dark suppresses unwanted tissue differentiation in explant tissue, for example, by limiting the development of plastids into chloroplasts. Similarly, dark conditions may inhibit precocious germination of young embryos. Temperature requirements also are specific, but tend to be in the range of "room temperature" (i.e., 23°–27°C). Despite the similarity in culture requirements, improvements in nonzygotic embryo development and maturation increasingly are being obtained by optimization of medium composition.

ORIGIN OF NONZYGOTIC EMBRYOS

It is generally accepted that, like zygotic embryos, nonzygotic embryos arise from a single cell, in contrast to budding from a cell mass. This distinction is important in considering efficient genetic engineering, since modification of a single embryogenic cell might eventually result in a modified plant, compared to genetic modification of a cell within a bud, which would result in a chimeric plant.

Nonzygotic embryos growing from isolated cells, such as microspores or protoplasts, clearly develop from single cells. However, the origin of nonzygotic embryos that develop from complex intact primary explant tissue or callus is more difficult to resolve, since the action of microscopic single cells cannot be readily followed. Nonzygotic embryos often develop with a well-defined suspensor apparatus identical to that of zygotic embryos, which somewhat suggests a single cell origin, whereas others can develop with a broad basal attachment, suggesting a multicellular budding phenomenon (Figure 22.4).

INITIATION OF EMBRYOGENIC CELLS

In complex explants, nonzygotic embryos typically can be initiated only from the more juvenile or meristematic tissues. For example, immature zygotic embryos or zygotic embryo cotyledons and hypocotyls dissected from ungerminated seeds are commonly used explants. The young leaves, shoot tips, or even roots of established plants sometimes are used to initiate embryogenic cultures. However, the explant response is highly genotype dependent, so that, for any given species, only a certain type or range of explant can be used to initiate embryogenic cultures.

There are several pathways by which nonzygotic plant cells become embryo initials (Christensen, 1985). In instances where the explant consists of undifferentiated embryonic tissue, such as an immature zygotic embryo, initiation and maintenance of an embryogenic callus is akin to culturing and propagating a preexisting proembryonal complex. Thus, the embryogenic cells present in explant tissue prior to culture initiation are simply propagated and otherwise manipulated in vitro. However, in many instances, embryogenic cells are induced from nonembryogenic cells; this represents a dramatic change in their presumptive fate. The shift in developmental pattern involves a dedifferentiation away from the cell's "normal" fate followed by redetermination toward an embryogenic cell-type. For example, cells in leaf explants, which normally would develop into constituents of relatively short-lived parenchymatous tissue, instead become embryogenic under certain conditions. This is a pivotal change in development, since cells that normally would be capable of only a few divisions, at most, before senescence instead become redirected to become totipotent and capable of possibly unlimited divisions. Such embryogenic cells become immortal in the sense that they reinstate the germ line by being capable of developing into mature, reproductive individuals.

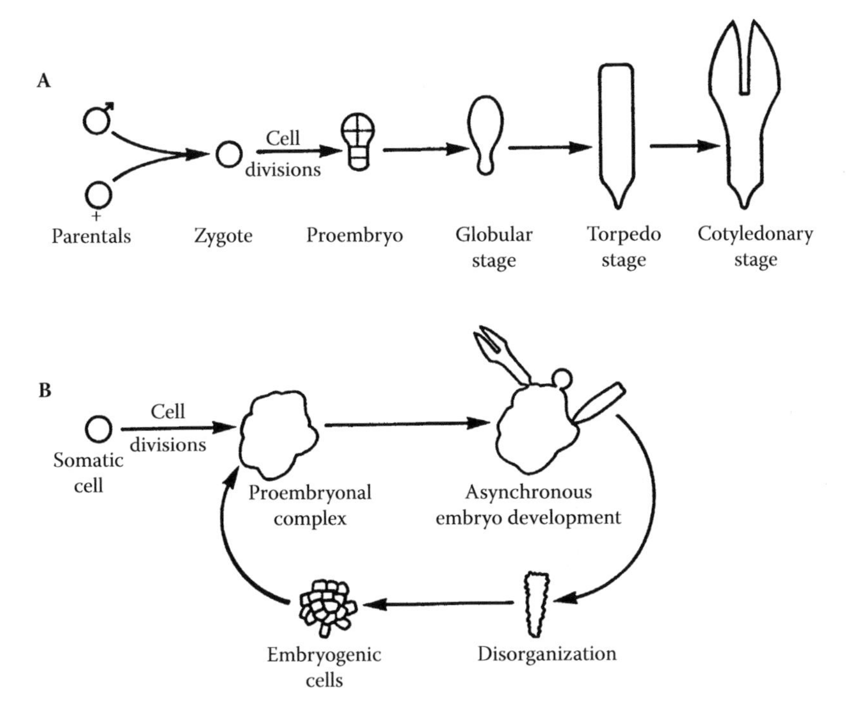

FIGURE 22.4 Comparison of typical zygotic embryogenesis occurring within seeds (A) with proliferative nonzygotic embryogenesis occurring in vitro (B). Zygotic embryogenesis is characterized as being very regulated, such that embryos synchronously pass through distinct developmental stages, whereas nonzygotic embryogenesis often is non-uniform, with many stages present at a given time. Nonzygotic embryos may bypass maturation and become disorganized, adding to the proembryonic tissue mass. (From Gray, D.J. and A. Purohit, 1991. *Crit. Rev. Plant Sci.* 10:33–61.)

The fact that isolated somatic cells can develop normally into embryos irrevocably demonstrates that the developmental program for embryogenesis is contained within and controlled by the cell itself and not by external factors. The exact nature of all triggering mechanism(s) of embryogenesis, whether it be a physical, biochemical, or genetic event, is becoming better understood through research, particularly with molecular genetic tools (see Chapter 11). Early attempts to identify genes that became activated as a direct consequence of embryogenic cell induction failed due to inadequate understanding of the basic induction phenomenon, which resulted in faulty experimental designs. Since the cell cultures used were already induced, any genes critical to the induction step already had been activated. Hence, the genes that were observed in such studies actually were those activated later in embryogenesis such that they were controlling down-stream developmental processes and were not related to the triggering event. Despite the technical difficulty of identifying critical induction-controlling genes, a number of other genes and their products have been identified during nonzygotic embryo development (e.g., Zimmerman, 1993). In some instances, the genetic mechanisms resolved for nonzygotic embryos can be related directly to that of the corresponding zygotic embryo.

INDUCTIVE PLANT GROWTH REGULATORS (PGRs)

In practice, the initiation of embryogenic cells requires in vitro culture of the appropriate explant on or in a medium that contains specific PGRs (Chapter 4). In fact, a preponderance of reports of embryogenic culture initiation employed a very narrow range of PGRs that typically are added to culture medium. Synthetic auxins, notably 2,4-dichlorophenoxyacetic acid (2,4-D), are added to medium in the predominance of reported protocols. Similar auxins used include dicamba, indolebutyric acid (IBA), naphthoxyacetic acid (NOA), picloram, and others. In addition, several weaker auxins—such as indoleacetic acid (IAA), a natural plant hormone, and napthyleneacetic acid (NAA)—have been utilized in a few culture systems. Auxins serve to induce the formation of embryogenic cells, possibly by initiating differential gene activation, as noted above, and also appear to promote increase of embryogenic cell populations through repetitive cell division, while simultaneously suppressing cell differentiation and growth into embryos. However, an auxin often is not required in instances where the explant consists of preexisting embryogenic cells, as described above, possibly because a discrete induction step then is not required.

In addition to auxin-like PGRs, cytokinins also are required to induce embryogenesis in many dicotyledonous species. In a few instances, only a cytokinin is required to cause embryogenic cultures to develop. The most commonly used cytokinin is benzyladenine (BA), but others such as thidiazuron (TDZ) and kinetin, and the natural cytokinin, zeatin, also are utilized.

Actual PGR concentration is important for an optimum response, since concentrations that are too low may not trigger the inductive event and concentrations that are too high, particularly when considering phenoxy-auxins, may become toxic. Typically, following the induction period, resulting culture material is transferred to medium lacking PGRs, which removes the auxin-induced suppression of embryo development and allows embryogenesis to occur; however, not all culture systems require this two-step procedure. For example, as mentioned above, certain species do not require any induction step at all and others become induced and undergo complete embryogenesis in the continuous presence of auxin. Examples of different PGR regimes used in culture systems are illustrated via the laboratory exercises provided after this chapter.

EMBRYO DEVELOPMENT

The physical, observable transition from a nonembryogenic to an embryogenic cell may occur when the progenitor cell undergoes an unequal division, resulting in a larger vacuolate cell and a smaller, densely cytoplasmic (embryogenic) cell. Embryogenic cells are readily distinguished by their small size, isodiametric shape, and densely cytoplasmic appearance (Figure 22.5). This type of unequal cell division is identical to that observed in zygotes and may be an early indication of developmental polarity. The embryogenic cell then either continues to divide irregularly to form a proembryonal complex, or divides in a highly organized manner to form a somatic embryo (Figures 22.2 and 22.4). However, an often undesirable difference exhibited by nonzygotic embryos is that they frequently deviate from the normal pattern of development either by producing callus, undergoing direct secondary embryogenesis, or germinating precociously. This tendency toward erratic development clearly is due to environmental factors as discussed in the coming text.

Zygotic and nonzygotic embryos share the same gross pattern of development, with both typically passing through globular, scutellar, and coleoptilar stages for monocots, or globular, heart, torpedo, and cotyledonary stages for dicots and conifers. Generally, the anatomy and morphology of well-developed nonzygotic embryos is faithful to the corresponding zygotic embryo-type such that they easily can be identified by eye or with the aid of a stereomicroscope (Figure 22.6). For example, nonzygotic embryos of grass and cereal species typically possess a scutellum, coleoptile, and embryo axis, which are distinctive embryonic organs of monocots (Figure 22.7A). Embryos of dicotyledonous species have a distinct hypocotyl and cotyledons (usually two; Figure 22.7B); those of conifers also exhibit a hypocotyl and numerous cotyledons (Figure 22.7C), the number of which is similar to that of the given species.

Embryo development occurs through an exceptionally organized sequence of cell division, enlargement, and differentiation. During early development the embryo assumes a clavate, globular shape and remains essentially an undifferentiated but organized mass of dividing cells with a well-defined epidermis. Subsequent heart through early torpedo stages are characterized by cell differentiation and polarized growth, notably elongation and initiation of rudimentary cotyledons

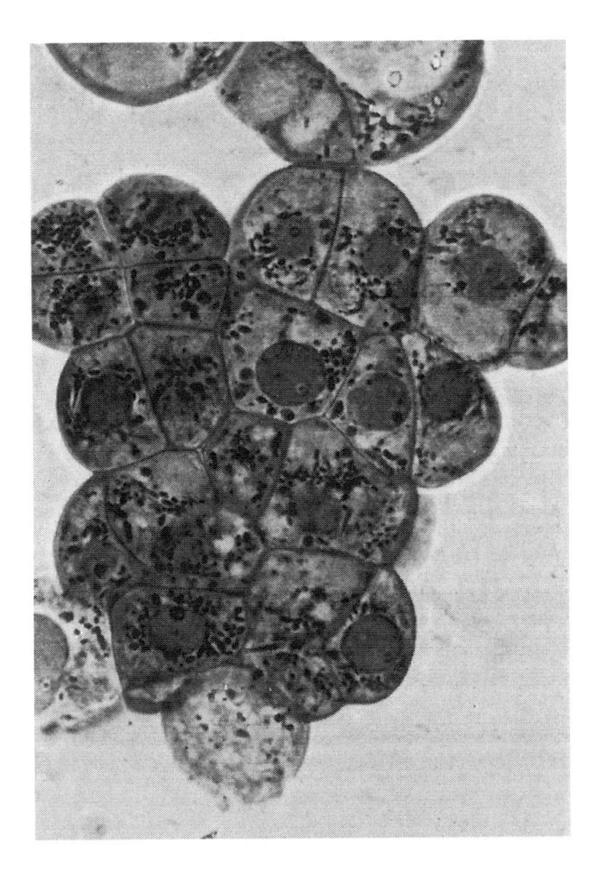

FIGURE 22.5 Typical embryogenic cells of grape. Note the three-cell stage embryo in the upper center part of the cell mass. (From Gray, D.J., *Somatic Embryogenesis in Woody Plants*, Kluwer, 1995. Reprinted with permission from Kluwer Academic Publishers.)

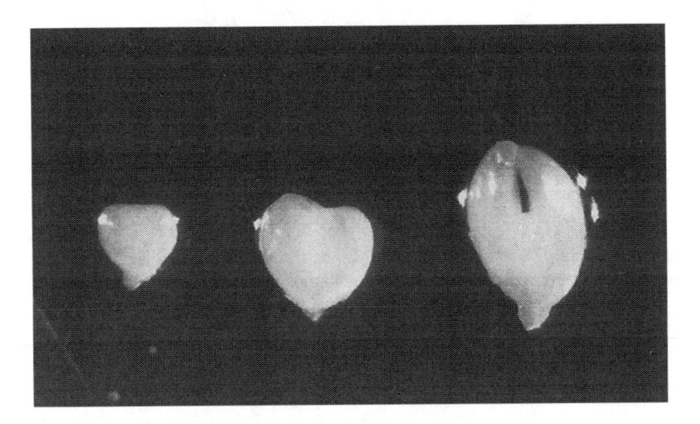

FIGURE 22.6 Somatic embryos of grape at heart and early cotyledonary stages, illustrating morphologies highly faithful to those of zygotic embryos.

in dicots (Figure 22.6) and development of the scutellum with initiation of the coleoptilar notch in poaceous monocots. At the same time, obvious tissue differentiation begins with the development of embryonic vasculature (Figure 22.8) and accumulation of intracellular storage substances. Final stages of development toward maturation are distinguished by overall enlargement, increase in cotyledon size in dicots (Figure 22.9), and coleoptilar enlargement in monocots. At the same time,

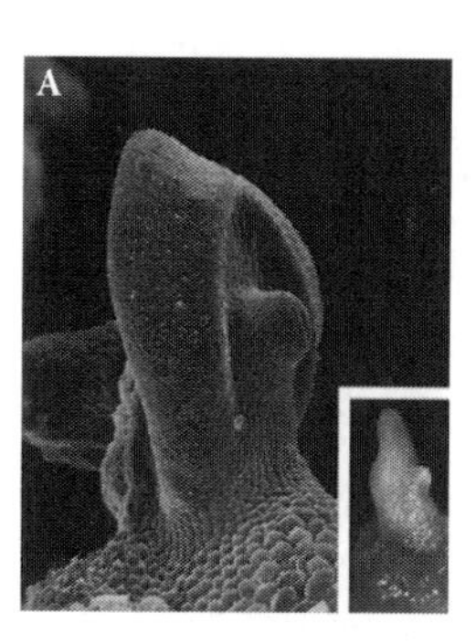

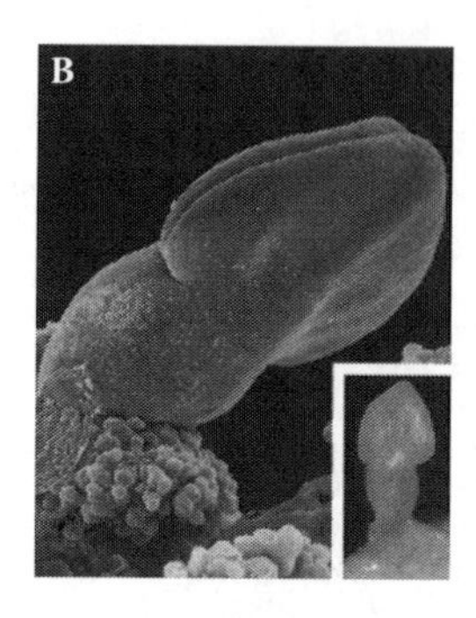

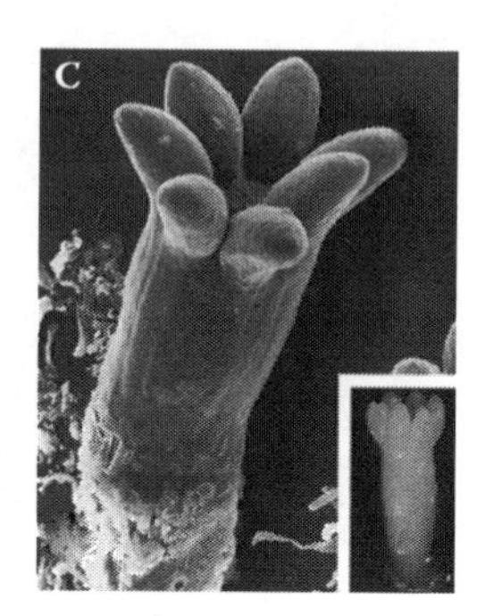

FIGURE 22.7 Comparison of monocot (A), dicot (B), and gymnosperm (C) nonzygotic embryos. Scanning electron micrographs, with corresponding stereomicrographs (inset). A. Somatic embryo of orchardgrass growing from an embryogenic callus. Note the protruding coleoptile and notch, through which the first leaves will emerge after germination. The prominent large flattened body of the embryo is the scutellum. (From Gray, D.J., et al., 1984. *Protoplasma*. 122:196–202.) B. Somatic embryo of grape growing from an embryogenic callus. Note two flattened cotyledons and subtending hypocotyl. (From Gray,D.J.,1995. *Somatic Embryogenesis in Woody Plants*, Kluwer. Reprinted with permission from Kluwer Academic Publishers.) C. Somatic embryo of Norway Spruce. Note multiple cotyledons and elongated hypocotyl. (From Fowke, L.C., et al.,1994. *Plant Cell Rep*. 13:612–618. With permission.)

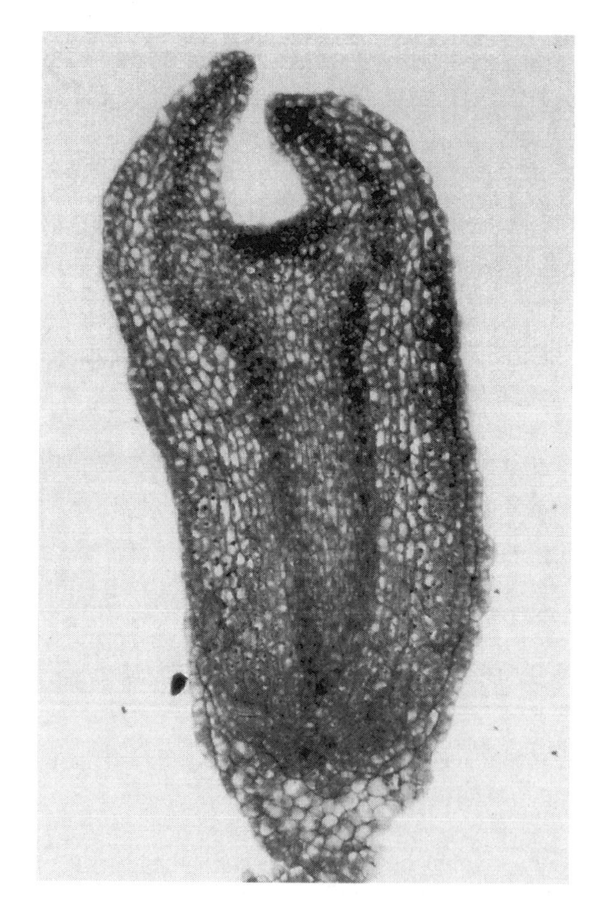

FIGURE 22.8 Longitudinal section through a grape somatic embryo showing typical vascular system. (From Gray, D. J., 1995. *Somatic Embryogenesis in Woody Plants*, Kluwer. Reprinted with permission from Kluwer Academic Publisher.)

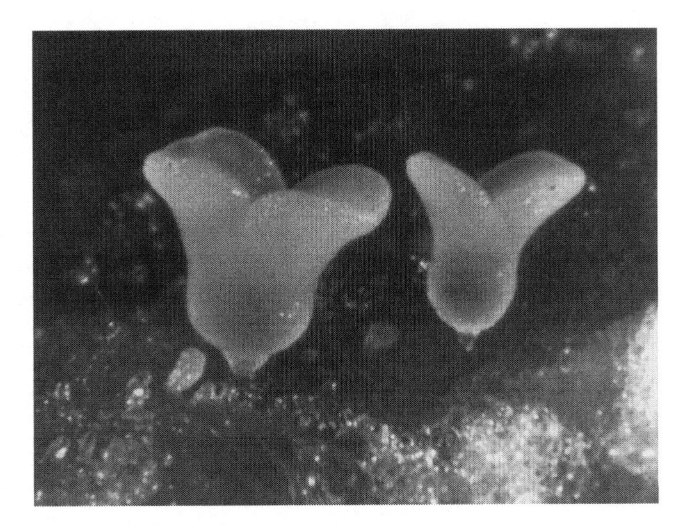

FIGURE 22.9 Cotyledonary-stage somatic embryos of cantaloupe growing from a cultured cotyledon. Note that the fine suspensor is the only connection to explant. (From Gray, D.J., et al., 1993. *J. Amer. Soc. Hort. Sci.* 118:425–432.)

the embryonic axis becomes increasingly developed. In dicots, the root apical meristem becomes well-established, embedded in tissue located above the suspensor apparatus and at the base of the hypocotyl, whereas the shoot apical meristem develops externally between the cotyledons. In monocots, the embryo axis develops laterally and parallel to the scutellum. The root apical meristem is embedded, and the shoot apical meristem develops externally, but is protected by the coleoptile. All of the events described above occur in concert with each other in a manner that is essentially identical to that of zygotic embryos. But, for a number of reasons, as described below, nonzygotic embryos often differ somewhat from their zygotic counterparts in morphology and/or performance.

An obvious difference in gross morphology between nonzygotic embryos growing in vitro and zygotic embryos in seeds is simply caused by the physical constraint on zygotic embryos imposed by the developing seed coat, often causing them to become compressed and/or flattened into a shape and size distinct for a given species or variety. This is made apparent by comparing the morphology of seed-borne zygotic embryos with corresponding nonzygotic embryos of a given species (Figure 22.10). Zygotic embryos excised from seed typically exhibit a highly compressed shape because the embryos become highly flattened during development. In contrast, nonzygotic embryos tend to be larger and have wider hypocotyls and fleshier cotyledons. It is possible that pressure exerted by the seed coat contributes to other aspects of embryo development that are lacking during nonzygotic embryogenesis (Gray and Purohit, 1991).

In addition to differences in development related to simple physical constraints, for several reasons, significantly more instances of abnormal development are known to occur during nonzygotic embryogenesis when compared to zygotic embryogenesis. For example, as mentioned above, species that normally produce zygotic embryos with suspensors often produce clusters of nonzygotic embryos from a proembryonal cell complex (Figure 22.4). This basic change in developmental pattern is likely due to differences between the seed and in vitro environments, since immature zygotic embryos dissected from seeds and cultured also often develop abnormally (Gray and Purohit, 1991).

Nonzygotic embryos growing in mass from proembryonal complexes tend to develop asynchronously so that several stages are present in cultures at any given time (Figures 22.2 and 22.4). Nonzygotic embryos initiated over time are subjected to different nutrient regimes as medium becomes depleted then replenished between and during subcultures. This leads to differences in

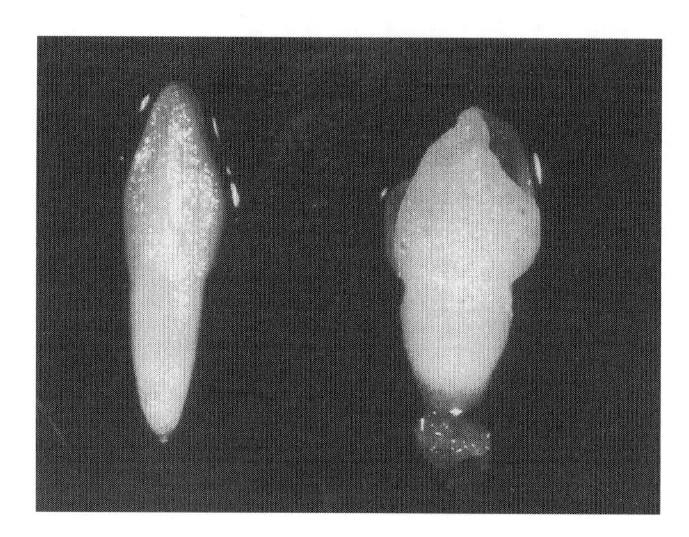

FIGURE 22.10 Comparison of grape zygotic embryo, compressed and flattened by development within a seed (left), with grape somatic embryo, which is not flattened (right). (From Gray, D.J. and A. Purohit, 1991. *Crit. Rev. Plant Sci.* 10:33–61.)

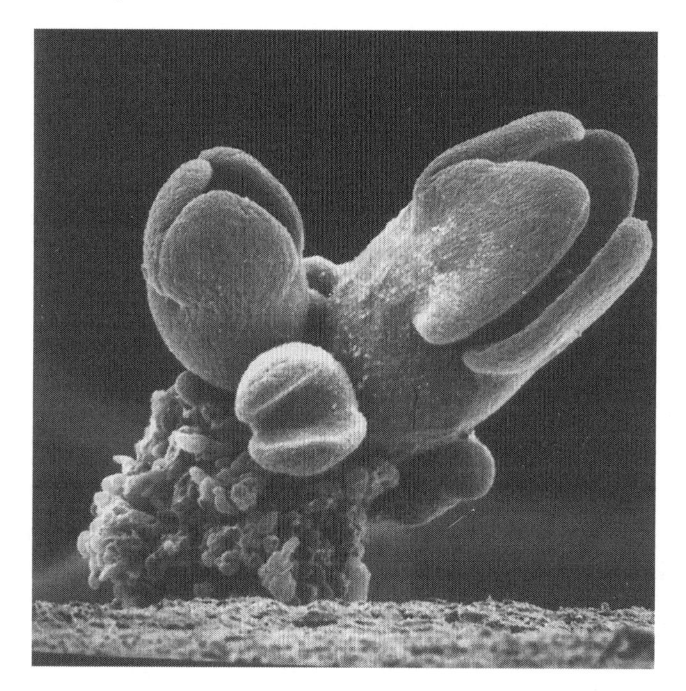

FIGURE 22.11 Asynchronous and abnormal somatic embryo development from embryogenic callus of grape. Increasing levels of development (clockwise from lower right globular embryo) are evident in a single group of embryos. In addition, the normal cotyledon number of two (lower center embryo) contrasts with supernumery cotyledon development of three and four on the two upper embryos. (From Gray, D.J. and J.A. Mortensen, 1987. *Plant Cell, Tissue and Organ Cult.* 9:73–80. Reprinted with permission from Kluwer Academic Publishers.)

development even among embryos from a single culture. With such variable and unregulated environmental conditions, nonzygotic embryos often bypass maturation altogether, becoming disorganized, forming new embryogenic cells and contributing to asynchrony (Figure 22.4). Nonzygotic embryos also often exhibit structural anomalies such as extra cotyledons (Figure 22.11) and poorly developed apical meristems.

EMBRYO MATURATION

Maturation is the terminal event of embryogenesis and is characterized by the attainment of mature embryo morphology, accumulation of storage carbohydrates, lipids and proteins (see prior section regarding in vivo versus in vitro growth conditions), reduction in water content, and, often, a gradual decline or cessation of metabolism. Nonzygotic embryos typically do not mature properly when compared to zygotic embryos. In fact, in the preponderance of instances, rapid growth continues to occur, leading to precocious germination.

Although complete maturation is not absolutely necessary in order to obtain plants from nonzygotic embryos, it is required to achieve high rates of plant recovery. As such, factors that influence and/or enhance nonzygotic embryo maturation have been explored. A number of culture medium components have been shown to promote maturation. In particular high-sucrose levels (9–12%), timed pulses of abscisic acid (ABA), a naturally occurring PGR, and polyethylene glycol (PEG), an osmotically active compound, in conjunction with certain amino acids, notably glutamine, have been remarkably successful. For example, an 8-week treatment with ABA and PEG caused white spruce nonzygotic embryos to accumulate more lipids and to better resemble zygotic embryos morphologically (Attree and Fowke, 1993). Similarly, a short pulse of ABA in conjunction with glutamine resulted in alfalfa nonzygotic embryos that could withstand dehydration as mentioned in following text (Senaratna et al., 1989).

QUIESCENCE AND DORMANCY

Perhaps the most obvious developmental difference between zygotic and nonzygotic embryos is that the latter lacks a quiescent resting phase. By comparison, zygotic embryos of many species begin a resting period during seed maturation. Zygotic embryo quiescence, in conjunction with the protective and nutritive tissues that comprise a seed, is the major factor allowing seeds to be conveniently stored and be useful in agriculture. Thus, the ability to withstand dehydration appears to be a normal step in embryo development.

However, nonzygotic embryos tend either to grow and germinate without normal maturation, become disorganized into embryogenic tissue, or die. They rarely enter a resting stage. Although quiescence and dormancy have been documented in somatic embryos, it is possible that dormancy occurs and is problematic in instances where plants cannot be obtained from well-developed embryos (see Gray, 1986; Gray and Purohit, 1991, for reviews).

EMBRYO GERMINATION AND PLANT DEVELOPMENT

Obtaining plants from nonzygotic embryos often is more difficult than would be expected. The early literature concerning first reports of nonzygotic embryogenesis for a given species or cultivar often did not include information on plant recovery. When plants were obtained, the recovery rate was very low or not reported, suggesting that a majority of nonzygotic embryos were too abnormal to germinate. Typically, plant recovery from nonzygotic embryos ranges from 0%–50%. This is very low compared to zygotic embryos in commercial seed, for which germination and plant development typically exceeds 90% in soil. In all but a few instances, nonzygotic embryos develop very poorly if at all when planted in soil like seed. However, research been conducted to determine conditions that raise plant recovery rates.

Progress in culture methodology has resulted in better development and germination of nonzygotic embryos. As mentioned above regarding maturation, pulse treatments with various amino acids, osmotica and PGRs, particularly ABA, have resulted in nonzygotic embryos with better maturation, including the ability to be dehydrated and stored like seeds, as well as improved germination characteristics (Attree and Fowke, 1993). This demonstrates that careful attention to culture conditions and nutrition, especially with regard to timed pulses of certain physical and chemical factors, results in plant recovery rates from nonzygotic embryos equivalent to those of zygotic embryos. In general, it can be concluded that conditions favoring embryo maturation also favor the recovery of plants.

After the event of germination, the embryo begins development into a plant. Typically, the storage reserves present (lipids, proteins, and/or starch—depending on the species) become depleted concomitant with the start of increased mitotic activity in shoot and root meristems. Eventually, a young photosynthetically competent plant develops (Figure 22.12), which then can be gradually acclimated to ambient conditions.

USES FOR EMBRYOGENIC CULTURES

Embryogenic cultures systems are used for a number of purposes:

1. They constitute an important tool for the study of plant development, both due to the unique convenience of in vitro culture over in planta growth and the contrasts that can be drawn from differences between nonzygotic embryogenesis in vitro and zygotic embryogenesis in seeds.
2. Embryogenic cultures are an efficient vehicle for genetic engineering, since embryogenic cultures often produce many embryos per volume of cell mass and isolated genes integrated into single embryogenic cells can become incorporated into the genome of the plant that ultimately develops.
3. Many nonzygotic embryos can be produced from a single desirable plant for efficient clonal propagation.

Certain potential uses of embryogenic systems as research tools have been described or suggested throughout this chapter and an in depth discussion of genetic engineering applications is presented in Chapter 33. The use of nonzygotic embryogenesis as a cloning vehicle is discussed in more detail in the following text.

SYNTHETIC SEED TECHNOLOGY

Research has been conducted with the goal of developing nonzygotic embryogenesis into a commercially useful method of plant propagation. The technology that has emerged is termed *synthetic* (or artificial) *seed technology*. A synthetic seed is defined as a somatic embryo that is engineered to be of practical use in commercial plant production (Gray and Purohit, 1991). Applications for

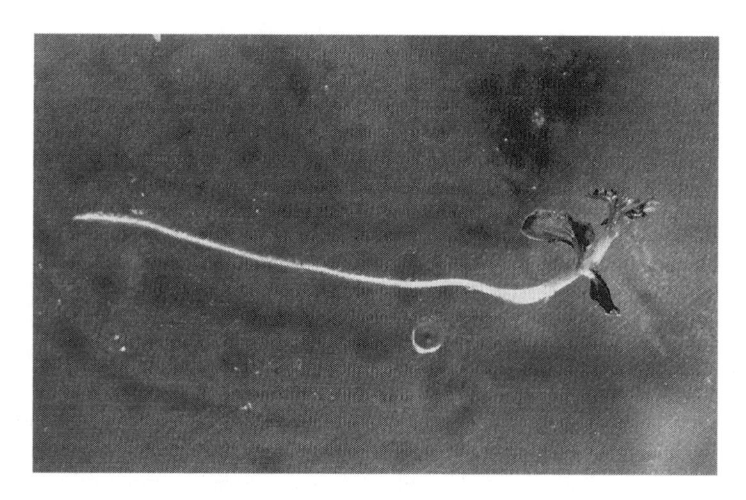

FIGURE 22.12 Plant development from a germinated melon somatic embryo. Note shoot and root development. (From Gray, D.J., et al., 1993. *J. Amer. Soc. Hort. Soc.* 118:425–432.)

synthetic seed vary, depending on the relative sophistication of existing production systems for a given crop and the opportunities for improvement. Whether or not a cost advantage results from synthetic seed will ultimately determine its commercial use. For seed-propagated agronomic crops, quiescent, nonzygotic embryos produced in bioreactors and encapsulated in synthetic seed coats will be necessary. Certain vegetable crops with expensive-to-produce seeds, such as seedless watermelon, are attractive candidates for synthetic seed technology, since the per-plant cost might be reduced.

Similarly, conifers, which are difficult to improve by breeding due to a long life cycle, would benefit from application of synthetic seed technology if elite individuals could be cloned and planted with seed-efficiency (Farnum et al., 1983). For vegetatively propagated crops, particularly those with a high per-plant value, naked, hand-manipulated, nonquiescent embryos may be cost effective. For example, many ornamental crops are painstakingly commercially micropropagated in tissue culture via adventitious bud proliferation with the major expense being the labor needed for multiple culture and rooting steps. Substitution of embryogenic culture systems for such crops would greatly reduce labor costs since mass produced somatic embryos easily could be hand selected and placed directly into planting flats and rooted (Gray and Purohit, 1991).

Ongoing efforts to develop synthetic seed technology constitutes one practical area of nonzygotic embryogenesis research. The potential benefits of using "clonal seeds" in agricultural production is great enough to stimulate continued investigation of factors regulating nonzygotic embryo maturation.

CONCLUSION

In vitro embryogenesis from nonzygotic cells must be regarded as a universal property of higher plants, albeit one that generally does not occur in nature. When induced from isolated cells, nonzygotic embryogenesis conclusively demonstrates not only cellular totipotency but also the presence of a highly conserved developmental program that is innate to a wide range of cell types. For the student, nonzygotic embryogenesis represents an opportunity to conveniently study in a culture vessel one of the more developmentally complex aspects of the plant life cycle.

Embryogenic cell culture allows the entire genome of a plant to be manipulated with microbiological techniques. Once the cells are proliferated, genetically engineered, and/or otherwise managed, intact plants can be reconstituted. From such simple manipulation come numerous possibilities for better understanding and ultimately improving our use of plants.

REFERENCES

Attree, S.M. and L.C. Fowke. 1993. Embryogeny of gymnosperms: Advances in synthetic seed technology of conifers. *Plant Cell Tiss. Org. Cult.* 35:1–35.

Christensen, M.L. 1985. An embryogenic culture of soybean: Towards a general theory of embryogenesis. In *Tissue Culture in Forestry and Agriculture*, R.R. Henke, K.W. Hughes, M.J. Constantine, and A. Hollaender, Eds., pp. 83–103, Plenum, New York.

Farnum, P., R. Timmis, and J.L. Kulp. 1983. Biotechnology of forest yield. *Science* 219:694–702.

Fowke, L. C., S.M. Attree, and P.J. Rennie. 1994. Scanning electron microscopy of hydrated and desiccated mature somatic embryos and zygotic embryos of white spruce (*Picea glauca* [Moench] Voss.). *Plant Cell Rep.* 13:612–618.

Gray, D.J. 1986. Quiescence in monocotyledonous and dicotyledonous somatic embryos induced by dehydration. In *Proc. Symp. Synthetic Seed Technology for the Mass Cloning of Crop Plants: Problems and Perspectives. HortScience* 22:810–814.

Gray, D.J. 1990. Somatic embryogenesis and cell culture in the Poaceae. In *Biotechnology in Tall Fescue Improvement*, M.J. Kasperbauer, Ed., CRC Press, Boca Raton, FL, 1990, 25–57.

Gray, D.J. 1995. Somatic embryogenesis in grape. In *Somatic Embryogenesis in Woody Plants*, pp. 191–217. S.M. Jain, P.K. Gupta, and R.J. Newton, Eds., Kluwer Academic Publishers, Dordrecht, The Netherlands.

Gray, D.J., B.V. Conger, and G.E. Hanning, 1984. Somatic embryogenesis in suspension and suspension-derived callus cultures of *Dactylis glomerata*. *Protoplasma* 122:196–202.

Gray, D.J., D.W. McColley, and M.E. Compton. 1993. High-frequency somatic embryogenesis from quiescent seed cotyledons of *Cucumis melo* cultivars. *J. Amer. Soc. Hort. Sci.* 118:425–432.

Gray, D.J. and C.P. Meredith. 1992. Grape. In *Biotechnology of Perennial Fruit Crops*. F.A. Hammerschlag and R.E. Litz, Eds., CAB International, Wallingford, UK.

Gray, D.J. and J.A. Mortensen. 1987. Initiation and maintenance of long term somatic embryogenesis from anthers and ovaries of *Vitis longii* "Microsperma." *Plant Cell Tiss. Org. Cult.* 9:73–80.

Gray, D.J. and A. Purohit. 1991. Somatic embryogenesis and the development of synthetic seed technology. *Crit. Rev. Plant Sci.* 10:33–61.

Reinert, J. 1959. Ueber die Kontrolle der Morphogenese und die Induktion von Adventiveembryonen an Gewebekulturen aus Karotten. *Planta* 53:318–333.

Senaratna, T., B.D. McKersie, and S.R. Bowley. 1989. Desiccation tolerance of alfalfa (*Medicago sativa* L.) somatic embryos—influence of abscisic acid, stress pretreatments and drying rates. *Plant Sci.* 65:253–259.

Steward, F.C., M.O. Mapes, and K. Mears. 1958. Growth and organized development of cultured cells. II. Organization in cultures grown from freely suspended cells. *Amer. J. Bot.* 45:705–708.

Zimmerman, L.J. 1993. Somatic embryogenesis: A model for early development in higher plants. *Plant Cell* 5:1411–1423.

23 Developmental and Molecular Aspects of Nonzygotic (Somatic) Embryogenesis

Xiyan Yang and Xianlong Zhang

CONCEPTS

- Somatic embryogenesis is the developmental process by which somatic cells, under suitable induction conditions, undergo restructuring through the embryogenic pathway to generate embryogenic cells. These cells then go through a series of morphological and biochemical changes that result in the formation of a somatic embryo and the generation of new plants.
- Somatic embryos differentiate from the explant with or without an intervening embryogenic callus phase, which were regarded as indirect somatic embryogenesis (ISE) or direct somatic embryogenesis (DSE).
- Somatic or nonzygoticembryos can originate from a single cell or from multiple cells.
- The embryogenesis pathway of somatic embryo can be divided into the following stages: globular-shaped, heart-shaped, torpedo-shaped, and cotyledonal stages.
- The genes identified during somatic embryogenesis include genes responsible for cell cycle and cell wall, hormone-responsive genes, genes involved in signal transduction pathways, and transcription factors.
- The auxin surges occurred during somatic embryogenesis, with the consequent accumulation of numerous mRNAs, which resulted in the isolation of several corresponding gene classes in plants. These include GH3s, PINs, auxin/indoleacetic acid (Aux/IAAs), and auxin response factors (ARFs), as well as small auxin-up RNAs (SAURs).
- Extracellular protein can serve as markers for somatic embryogenesis that offers the possibility of determining the embryogenic potential of plant cells in culture long before any morphological changes have taken place.

The life cycle of higher plants includes the following two generations: a haploid or gametophytic generation, and a diploid or sporophytic generation. The sporophytic generation begins with the double fertilization process that results in the formation of a zygote embryo and an endosperm nucleus. The embryogenesis pathway, which represents the critical transition during zygote embryo development, can be divided into the following stages: globular-shaped, heart-shaped, torpedo-shaped, and cotyledonal stages in dicots; globular scutellar and coleoptilar stages in monocots; and globular, early cotyledonary, and late cotyledonary embryos in conifers. Completion of the stages results in the generation of a new plant. Alternatively, the plant can be derived from a single somatic cell or a group of somatic cells. This regeneration process, which differs from the zygote embryo pathway, is known as somatic embryogenesis or nonzygotic embryogenesis. Here after we will refer to the process as *somatic embryogenesis.*

Somatic embryogenesis is the developmental process by which somatic cells, under suitable induction conditions, undergo restructuring through the embryogenic pathway to generate embryogenic cells. These cells then go through a series of morphological and biochemical changes that result in the formation of a somatic embryo and the generation of new plants. Somatic embryos resemble zygotic embryos and undergo almost the same developmental stages. However, in contrast to the development of a zygote embryo, the observable process and the capacity to form embryos from a diverse set of tissues have permitted somatic embryogenesis to serve as a model system for the study of morphological, physiological, molecular, and biochemical events occurring during the onset and development of embryogenesis in higher plants.

SCOPE OF THIS CHAPTER

This chapter attempts to provide a simple introduction on developmental and molecular aspects of somatic embryogenesis. Somatic embryogenesis represents a unique developmental pathway of characteristic events, during which cells dedifferentiate, activate their cell division, and reprogram their physiology, metabolism, and gene expression patterns. An understanding of the embryogenic initiation, the origin of the somatic embryo, is critical to scientific and biotechnological applications. Cell tracking could be successfully applied to determine the fate of embryogenic cells. Identification of hormone-inducible genes has yielded clues to the hormonal control of gene expression during embryogenic development. Characterization of genes taking part in signal transduction pathways has generated great interest in the switching of several signal cascades during somatic embryogenesis, and identification of genes encoding transcription factors could be used to regulate embryogenic development. Meanwhile, protein markers are useful probes for defining embryogenic potential and for marking different phases in plant development.

DEVELOPMENTAL PATHWAYS OF SOMATIC EMBRYOGENESIS

Overview

Somatic embryogenesis in higher plants has been studied intensively during the past century. Studies using light and electron microscopy have provided detailed descriptions of the morphological and anatomical changes that characterize embryonic development. We can understand this process from the following points:

- Embryogenic initiation
- Direct and indirect somatic embryogenesis
- The origin of somatic embryos
- Developmental stages of somatic embryos
- Tracking of somatic embryogenesis

Embryogenic Initiation

Generally speaking, early somatic embryogenesis involves differentiated somatic cells acquiring embryogenic competence and proliferating as embryogenic cells. Initiation of the embryogenic pathway is restricted to certain responsive cells that have the potential to activate genes involved in generating embryogenic cells. Once these genes are activated, an embryogenic gene expression program replaces the established gene expression pattern in the explant tissue (Yang and Zhang, 2010). Determining specific physical and chemical factors that switch on the embryogenic pathway of development is a key step in embryogenic induction.

Plant growth regulators (PGRs; Chapter 4) and stresses play a central role in mediating the signal transduction cascade leading to the reprogramming of gene expression, followed by a series of cell divisions that induce either unorganized callus growth or polarized growth leading to somatic embryogenesis (Quiroz-Figueroa et al., 2006). Activation of auxin responses may be a key event in cellular adaptation and genetic, metabolic, and physiological reprogramming, leading to the embryogenic competence of somatic plant cells. Other than auxin being a main inducer, somatic embryogenesis also responds to other growth regulators such as cytokinin or abscisic acid (ABA) either being present or absent.

Once embryogenic cells have been formed, they continue to proliferate, forming proembryogenic masses (PEMs). Auxin is required for proliferation of PEMs but is inhibitory for the development of somatic embryos. The degree of embryo differentiation that takes place in the presence of auxin varies in different species. Embryogenic cultures of some species and some genotypes can be subcultured for a prolonged period on medium containing PGRs and still retain their full embryogenic potential; however, in most crops, the embryogenic potential decreases with prolonged culture and is eventually lost (van Arnold et al., 2002).

Direct and Indirect Somatic Embryogenesis

1. *Direct somatic embryogenesis.* Somatic embryos differentiate directly from the explant without an intervening embryogenic callus phase (Figure 23.1A). According to previous hypothesis, proembryogenic competent cells are already present—in DSE, a minimal proliferation or reprogramming of unorganized tissue precedes embryo formation.
2. *Indirect somatic embryogenesis.* Somatic embryos differentiate indirectly after an embryogenic callus phase (Figure 23.1B). In ISE a major reprogramming is required to get proliferated calli with embryogenic ability before embryo formation.

The main factors involved in each case depend on the source and physiological state of the explant employed and the type and concentration of the growth regulators. Explants from which direct embryogenesis is most likely to occur include microspores (microsporogenesis), ovules, zygotic and somatic embryos, and seedlings. ISE has been most extensively studied and used in transformation and somatic hybridization. During ISE, both embryogenic and nonembryogenic calli are present. It is usually easy to distinguish between embryogenic and nonembryogenic calli on the basis of morphology and color. Embryogenic calli show nodular features and a smooth surface (Figure 23.2A). And the embryogenic cells are characterized generally as small and isodiametric in shape. These cells have large and densely staining nuclei and nucleoli and are densely cytoplasmic, with small vacuoles, thick cell walls, and a higher metabolic activity (Figure 23.2C); while nonembryogenic calli are rough, friable, and translucent (Figure 23.2B); and the nonembryogenic cells are of different sizes and shapes, with little cell content (Figure 23.2D–E). The features of embryogenic and nonembryogenic calli vary in some degree depending on different plant species.

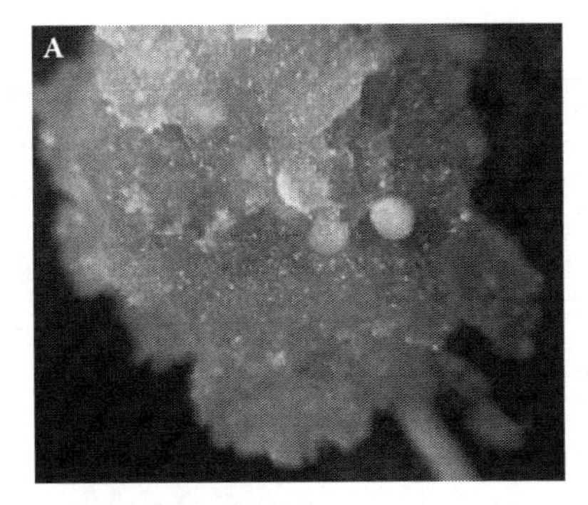

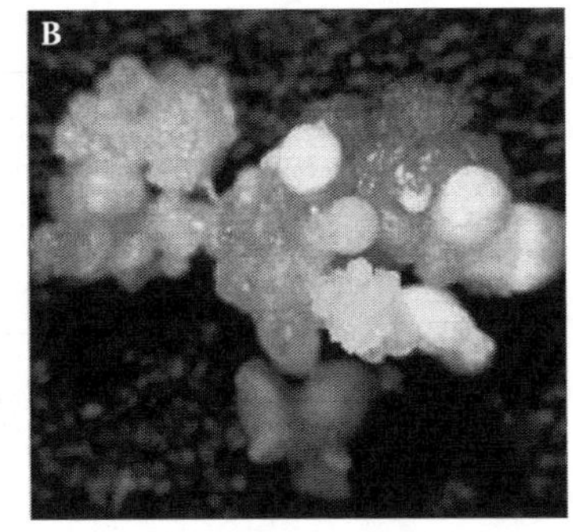

FIGURE 23.1 Schematic representation of direct (A) and indirect (B) somatic embryogenesis.

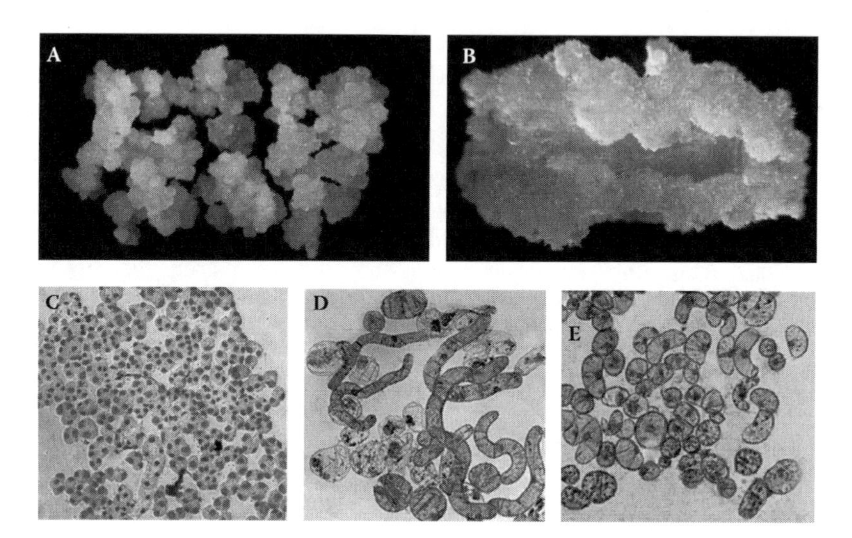

FIGURE 23.2 The morphology of embryogenic/noembryogenic calli and cells. (A) Embryogenic calli of *Gossypium hirsutum* cv. YZ1. (B) Noembryogenic calli of *G. hirsutum* cv. YZ1.D (C) Embryogenic cells of *G. hirsutum* cv. YZ1. (D–E) Noembryogenic cells of *G. hirsutum* cv. YZ1.

The Origin of Somatic Embryos

Somatic embryos generally originate in one of the two ways: unicellular or multicellular. The question of a single- or multiple-cell origin for somatic embryos is directly related to coordinated behavior of neighboring cells as a morphogenetic group.

1. *Unicellular origin.* When embryos have a unicellular origin, coordinated cell divisions are seen and the embryo is sometimes connected to the maternal tissue by a suspensor-like structure (Figure 23.3A).
2. *Multicellular origin.* Embryos with a multicellular origin, however, are initially observed as a protuberance that lacks coordinated cell divisions and the embryos in contact with the basal area are typically fused to the maternal tissue (Figure 23.3B).

Another controversial point is that the somatic embryos originate from a superficial/epidermal cell or subepidermal cells. When somatic embryo originates from a superficial cell, it possibly indicated a unicellular origin, or from subepidermal cells, representing a multicellular origin.

Developmental Stages of Somatic Embryos

Conceptually the development of the zygotic embryo can be divided into two stages: a first morphogenetic stage in which the basic structure of the embryo is established and a second metabolic stage characterized by biochemical activities that prepare the embryo for quiescence (Quiroz-Figueroa et al., 2006). Additionally, during the morphogenetic stage, zygotic embryo development in dicots can be divided into four general sequential stages: globular-shaped, heart-shaped, torpedo-shaped, and cotyledonal. As the zygotic counterpart, somatic embryos pass through such morphological stages (Zimmerman, 1993). The PEMs organize into a typical globular embryo, which then progresses through heart-shaped and torpedo-shaped stages. At the cotyledon stage, instead of entering developmental arrest, the embryos initiate a shoot meristem and seedling growth (Figure 23.4).

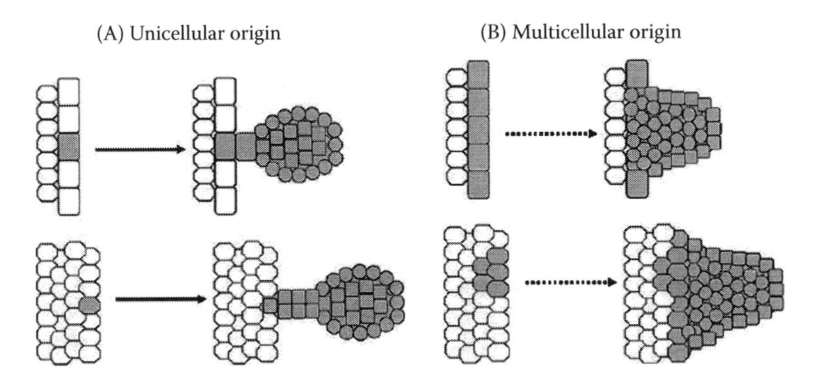

FIGURE 23.3 Unicellular (A) or multicellular (B) origin of somatic embryogenesis. (A) The embryo is connected to the maternal tissue by a suspensor-like structure. (B) The embryo fused to the maternal tissue by its basal part. (From Quiroz-Figueroa, F. R., et al. 2006. *Plant Cell Tiss. Org. Cult.* 86:285–301. With permission.)

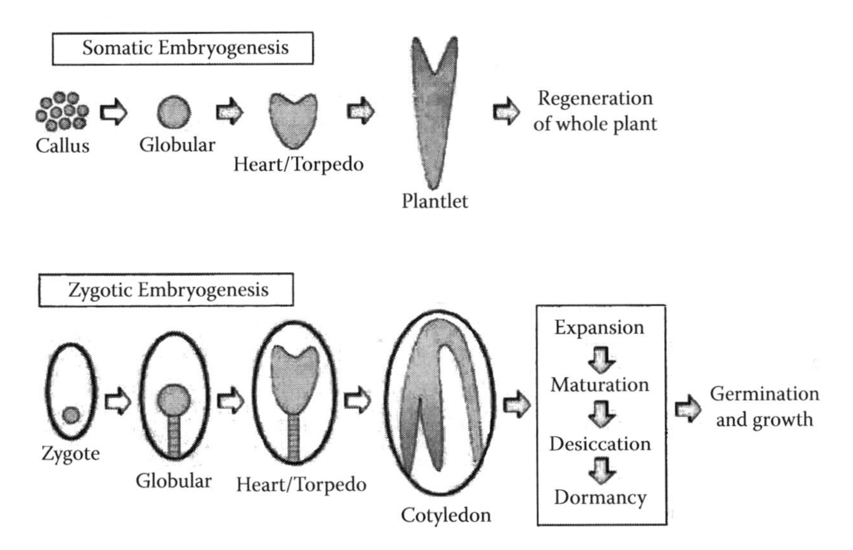

FIGURE 23.4 A comparison of somatic and zygotic embryogenesis. Morphologically and developmentally, somatic embryos and zygotic embryos are most similar from the globular stage through the torpedo stage. Somatic embryos do not experience desiccation or dormancy, but rather continue to grow into fully differentiated plantlets. (From Zimmerman, J. L. 1993. *Plant Cell* 5:1411–1423. With permission.)

Tracking of Somatic Embryogenesis

Ideally, the potential of somatic cells' transition to embryos is facilitated by observing events occurring during embryogenesis through the construction of a fate map. This map includes an adequate number of morphological and molecular markers to specify distinct developmental stages within the whole process. Moreover, once constructed, a fate map showing the correct progression of somatic embryogenesis would facilitate further analyses of specification, induction, and patterning of the embryonic tissues and organs. Construction of a fate map in somatic embryogenesis can be based on two alternative approaches, one using synchronous cell division systems and the other using time-lapse tracking of the development of individual protoplasts, cells, and multicellular structures. Time-lapse tracking usually yields more consistent data, for which there is no need to use drugs that affect the cell cycle and centrifugation treatments that otherwise might interfere with embryonic development. Furthermore, this technique starts with individual cells or cell aggregates that can be preselected based on certain criteria, making it possible to perform simultaneous analysis of dynamics and distribution of injected molecular probes conjugated with low molecular weight fluorochromes or fused to green fluorescent protein.

Time-Lapse Tracking of Somatic Embryogenesis in *Daucus carota*

Daucus carota, which has the best understanding of developmental pathways and molecular mechanisms of somatic embryogenesis, is the first angiosperm species in which both approaches have been successfully employed. By using time-lapse tracking, Toonen et al. (1994) showed that single suspension cells competent to form embryogenic cells have variable morphology. Based on the morphology, single cells isolated from embryogenic suspensions were classified into five types: oval vacuolated cells, elongated vacuolated cells, spherical vacuolated cells, spherical cytoplasm-rich cells, and irregular-shaped cells. It was found that all cell types could develop into somatic embryos, although with different frequencies. Embryo formation, in all cases, proceeded through the same sequence of stages (state-0, -1, and -2 cell clusters). Interestingly, depending on the initial cell type, embryogenesis could occur via three different pathways distinguished by a lack or presence of geometrical symmetry of state-0, -1, and -2 cell clusters (van Arnold et al., 2002). Oval vacuolated and elongated vacuolated cells developed into somatic embryos via asymmetrical cell clusters. Spherical cells, either vacuolated or cytoplasm rich, developed via symmetrical cell clusters into somatic embryos. Irregular-shaped cells first developed aberrantly shaped cell clusters, which then transformed into somatic embryos. These observations implied that organized growth and polarity are not always the case during somatic embryogenesis and suggested that somatic embryogenesis of auxin-induced PEMs is intermediate between unorganized growth and conservative embryonic pattern formation.

Time-Lapse Tracking of Somatic Embryogenesis in *Picea abies*

In gymnosperm, *Picea abies* was employed as a model plant to analyze the developmental pathway of somatic embryogenesis (Filonova et al., 2000a). Somatic embryogenesis in *Picea abies* involves two broad phases, which are each divided into specific developmental stages. Two types of cells, highly vacuolated cells and rounded cytoplasm-rich cells, are present in *Picea abies* embryogenic cell lines. However, neither of them can alone develop into somatic embryos. Thus, cell tracking begins with cell aggregates composed of a compact clump of densely cytoplasmic cells adjacent to a single vacuolated cell, referred to as PEM I cell aggregates, and the first phase is represented by proliferating PEMs. When a PEM I cell aggregate forms an additional vacuolated cell, it progresses to stage PEM II. Successively, PEM II enlarges in size by producing more cells of both types, while maintaining a bipolar pattern. At stage PEM III, an enlarged clump of densely cytoplasmic cells appears loose rather than compact, with disturbed polarity. Histological analysis has shown that PEMs lack a distinct embryonal mass, protoderm-like layer, and embryonal tube cells (Filonova et al., 2000a). The second phase encompasses development of somatic embryos, which arise from PEM III, and then proceed through the same typical sequence of stages as zygotic embryogenesis of Pinaceae. The tracking is terminated when fully mature somatic embryos resembling their zygotic counterparts have formed. Auxins and cytokinins are necessary during the first phase to maintain PEM proliferation, whereas embryo formation from PEM III is triggered by the withdrawal of PGRs. Once early somatic embryos have formed, their further development to mature forms requires ABA (Figure 23.5).

GENE EXPRESSION ANALYSIS DURING SOMATIC EMBRYOGENESIS

Overview

During somatic embryogenesis, somatic cells are induced to form totipotent embryogenic cells capable of regenerating into complete plants. Such developmental switching involves a series of events

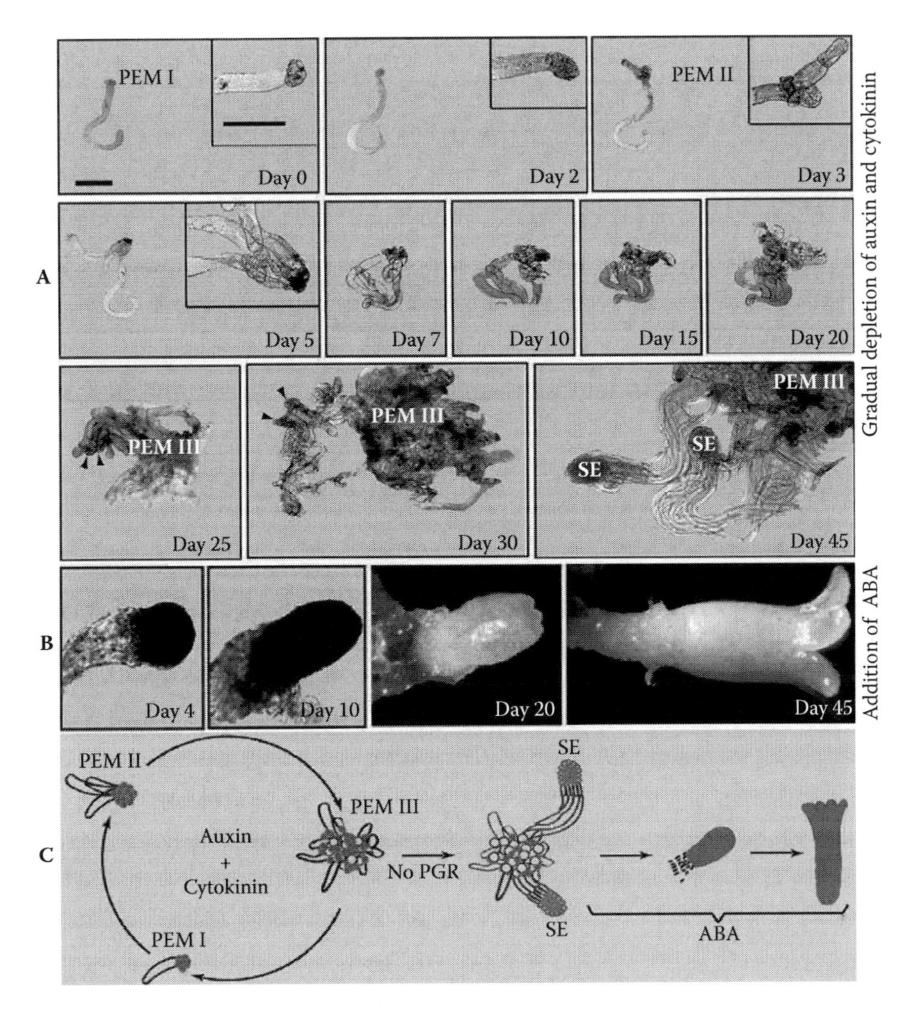

FIGURE 23.5 Time-lapse tracking the developmental pathway of somatic embryogenesis in *Norway spruce*. (A) The tracking was performed in a thin layer of agarose, under gradual depletion of auxin and cytokinin. Microscopic images were recorded at day 0, 2, 3, 5, 7, 10, 15, 20, 25, 30, and 45. (B) Somatic embryo developed by addition of ABA. (Figures A and B from Filonova, L. H. et al. 2000a. *J. Exp. Bot.* 51:249–264. With permission.) (C) The model of somatic embryogenesis in Norway spruce. Proliferation of PEMs is stimulated by auxin and cytokinin. An individual PEM should pass through a series of three characteristic stages (I, II, and III) to transdifferentiate to somatic embryos (SE). ABA is necessary to promote further development of somatic embryos through late embryogeny to mature forms. (Figure C from Filonova, L. H. et al. 2000b. *J. Cell Sci.* 113:4399–4411. With permission.)

associated with the molecular recognition of internal signals and external stimuli. The perception of response to these events sets off various signal cascades, and the downstream pathways followed during the transition of single cells to somatic embryos eventually result in specific gene expression and somatic embryogenesis. It is believed that analyses of gene expression during somatic embryogenesis can provide information for better understanding of this process.

Various types of somatic embryo-specific genes that have been identified (Figure 23.6) including the following:

- Genes responsible for cell cycle and cell wall
- Hormone-responsive genes
- Signal transduction pathway in somatic embryogenesis
- Transcription factors involved in somatic embryogenesis

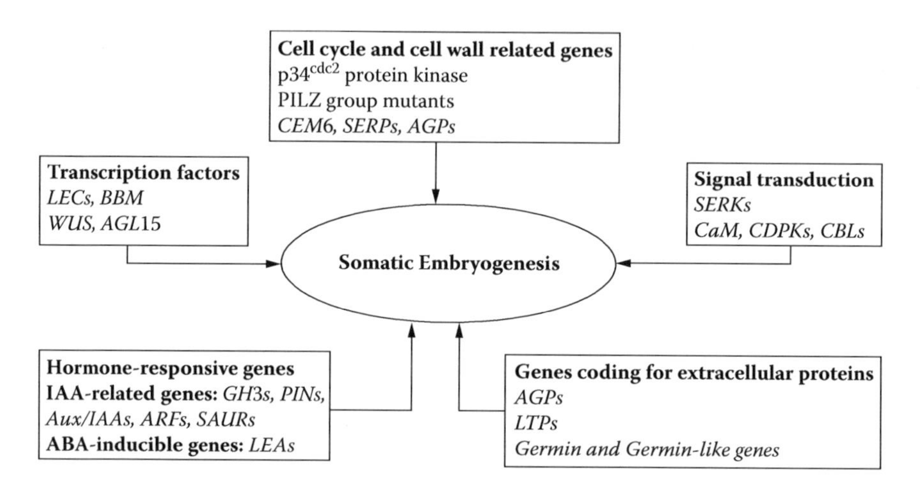

FIGURE 23.6 Various genes related to somatic embryogenesis in higher plants.

Genes Responsible for Cell Cycle and Cell Wall

Somatic embryogenesis requires strict spatio-temporal control over cell division and elongation. The polarity within the embryo is established through the precisely controlled cell division pattern of embryogenic cells and elongation of supporting suspensor-like and callus cells. Thus, the initiation of DNA synthesis and cell division are key events during initial as well as later steps of somatic embryogenesis in vitro. Additionally the molecular characterization of somatic embryogenesis can also be based on genes with cell cycle–dependent expression (Yang and Zhang, 2010). Somatic embryogenesis also depends on accumulation of cell wall and regulation of signal molecules bound to the cell wall. These processes may result in essential alterations in the expression of a set of genes responsible for cell cycle and cell wall, and subsequently cause coordinated changes in cellular functions at various stages of somatic embryogenesis.

Cell Cycle Genes

The old models for the molecular mechanisms of cell cycle control in eukaryotes describe $p34^{cdc2}$ protein kinase as a key regulatory element with a complex, phosphorylation-dependent interaction with cyclins (Yang and Zhang, 2010). The $p34^{cdc2}$ kinase phosphorylates many different substrates. Different regulatory pathways were suggested based on alteration of the active and inactive forms of the maturation promoting factor in embryogenic and somatic cells (Murray and Kirschner, 1989). Later, various components of this control system were shown to be present in plant cells. Further evidence for the existence of a plant cdc2 protein kinase has been obtained by isolation of a cDNA clone that appears to code for a *Medicago sativa* homolog of the *cdc2* gene. This cDNA (*cdc2Ms*) encodes a protein with the characteristic amino acid sequence elements required for cdc2 kinase function. This function was proved by complementation of a temperature-sensitive cdc2 mutant in fission yeast by the plant cDNA (*cdc2Ms*) (Hirt et al., 1991). Also, the transcript levels of *cdc2MS* were found to be higher in *M. sativa* shoots and auxin-induced suspension cultures. The housekeeping proteins Elongation factor 1, actin and the translationally controlled tumor protein homolog have already been reported as being specifically overexpressed during plant somatic embryogenesis (Chugh and Khurana, 2002).

Several *Arabidospis* mutants of the *PILZ* group show very small embryos consisting of only one (*porzino*) or a few large cells (*champignon, pfifferling, hallimasch*) having enlarged nuclei and severe microtubule and cytokinesis defects. Spindles are generally absent from mitotic nuclei and interphase cells have no cortical microtubules in these mutants, suggesting that products of these four mutant genes might be involved in regulating microtubular organization required for proper mitosis and/or cytokinesis (Mayer et al., 1999). Later, all these genes together with another gene,

KIESEL, which showed a weakened embryo-lethal phenotype, were cloned and were revealed to encode tubulin-folding cofactors and related G-protein *Arl2* involved in the formation of tubulin heterodimers (Steinborn et al., 2002). Additionally, *CHAMPIGNON* was found to be identical to *TITAN1*. Mutant embryos of the *PILZ* group lack microtubules while in the *KIESEL* mutant they are disorganized; actin seems to be present, although it is also disorganized, appearing as patchy structures rather than a fine meshwork.

Cell-Wall-Related Genes

Along with increased activity of the genes involved in cell cycle regulation, there is enhanced cell wall synthesis in embryogenic tissues and somatic embryos. Changes in the expression of actin and tubulin genes have been demonstrated during embryogenesis as enhanced cell wall and membrane formation result in an increase in the expression of these genes as well. CEM6 was found to be specific to the preglobular and globular stages of somatic embryo formation in *Daucus carota*, and its protein sequence characteristics suggested an important role as a cell wall protein in embryogenesis and as a specific marker for embryogenic cells (Sato et al., 1995). Aquea and Arce-Johnson (2008) found that SEPR1 and SEPR43, homologous to α-d-galactosidase and myo-inositol oxygenase, respectively, are up-regulated, whereas SEPR91, SEPR110, and SEPR114, which encode pectinesterase family proteins, are down-regulated. These SEPRs are involved in modification of the cell wall, and the identification of the related genes supports the idea that specific alterations in wall composition of the embryogenic cells may be important for proper embryo development.

During the past few years, the cell wall was found to participate in embryogenesis by its involvement in signal transduction and the formation of tensions influencing the cell shape and division plane. During embryogenesis specific rebuilding of the cell wall takes place. The cell wall also mediates the cell–cell (apoplastic) and cell-to-cell (symplastic) information flow. Embryogenic cells show specific structural features related to the composition of their cell wall arrangements. For example, the extracellular matrix surface network, which represents a thin outer cell wall layer, can be considered a specific structural marker for embryogenic cells in diverse plant species. This layer is composed of both arabinogalactan proteins (AGPs) and pectins. AGPs serve not only as specific molecular markers for cells having embryogenic competence, but they also play an important role in intracellular and intercellular signaling, and participate in apoptotic events during embryogenic development (Samaj et al., 2005).

Hormone-Responsive Genes

In many plant species, somatic cells of plant explants can be induced to proliferate and produce somatic embryos in suitable media supplemented with different PGRs. After proliferation in the presence of PGRs, two main cell types can be distinguished: nonembryogenic cells and embryogenic cells; somatic embryos develop from clusters of embryogenic cells. Some hormones are known to have a primary role in somatic embryogenesis but may cross-talk with others.

Auxin-Related Genes

For *in vitro* culture conditions, the strategy most often used to elicit somatic embryogenesis is to expose excised plant tissue to a high concentration of auxin, and auxin has emerged as one of the efficient initiators of somatic embryogenesis. Molecular approaches investigating the role of auxin signaling in somatic embryogenesis showed that auxin-induced growth and development involves changes in gene expression. The auxin surges occurred during somatic embryogenesis, with the consequent accumulation of numerous mRNAs, which resulted in the isolation of several corresponding gene classes in plants. These include *GH3s*, *PINs*, auxin/indoleacetic acid (*Aux/IAAs*), and auxin response factors (*ARFs*), as well as small auxin-up RNAs (*SAURs*) (Yang and Zhang, 2010).

1. GH3. The *GH3* gene is one of several sequences that were recovered in a differential hybridization screen of auxin-induced cDNA sequences derived from auxin-treated *Glycine max* excised hypocotyl sections. Expression of the *GH3* gene has been shown to be rapidly and specifically induced by the application of auxin.
2. PINs. The *Arabidopsis PIN* gene family consists of eight members and their polarity rearrangements define one of the earliest events in the regulation of different patterning and organogenesis processes. During the earliest developmental stages, *PIN1* is first expressed in proembryogenic cells in a nonpolar manner and then becomes polarized to the basal side of provascular cells once the early globular stage is reached.
3. Aux/IAAs and ARFs. The *Aux/IAA* genes are rapidly induced by auxin exposure. Many *Aux/IAA* genes have isolated from *Glycine max*, *Pisum sativum*, *Arabidopsis*, *Mung bean* and *Oryza sativa*. The Aux/IAA proteins consist of four highly conserved domains. Domains III and IV mediate homodimerization and heterodimerization among the Aux/IAAs and ARFs (Quint and Gray, 2006). Increased transcription of *Aux/IAA* genes after an auxin stimulus is likely to be mediated by ARF proteins via AuxREs in Aux/IAA promoter regions. Mutant analyses of several *Aux/IAA* genes have demonstrated that they play a central role in regulating plant growth and development, including embryonic patterning.
4. SAURs. Small RNAs have the potential to either counteract or reinforce the auxin signal by altering the stability of particular mRNAs in the auxin pathway. *pJCW1* and *pJCW2*, two members of *SAUR* family genes, were first identified in *Glycine max*. When used as probes, these clones indicate that auxin specifically induces accumulation of mRNAs that hybridize with these sequences (Hagen et al., 1984). Such auxin-responsive cDNAs can thus serve as an effective tool for screening the embryogenic potential of embryogenic and nonembryogenic lines.

ABA-Inducible Genes

The role of ABA in embryo development and maturation has been demonstrated in zygote embryos and somatic embryos. The ABA-regulated gene expression program includes transcriptional and posttranscriptional events, such as transcript processing, mRNA stability, translational control, and protein metabolism. Somatic embryo developmental stages are characterized by the accumulation of distinct sets of mRNAs and corresponding proteins in the somatic embryo. The accumulation of storage proteins, which are considered markers of the maturation phase, is followed by the accumulation of late embryogenesis abundant (LEA) proteins, some of which have been identified to be components of the ABA-inducible systems. *LEA* genes are abundantly expressed in late zygotic embryogenesis in many plant species, including *Gossypium hirsutum*, *Hordeum uhulgare*, *Oryza sativa*, *Brassica napus*, and *Triticum aestivum*. The expression pattern in embryogenesis and their ABA inducibility have led to the suggestion that they play a role in protecting the embryo during desiccation in zygotic embryos. High levels of LEA transcripts accumulate during embryogenesis, and several cDNAs of embryo-specific/embryogenic cell proteins have been isolated and characterized: *DcECP31*, *DcECP40*, and *DcECP63* from *Daucus carota* and *AtECP31* and *AtECP63* from *Arabidopsis* (Chugh and Khurana, 2002).

Signal Transduction Pathway in Somatic Embryogenesis

The acquisition of embryogeny followed by the dramatic transition from somatic cells to somatic embryos, coupled with establishment of body plan and embryo maturation, involves molecular events encompassing not only differential gene expression but also various signal transduction pathways for activating or repressing numerous genes sets. Species that easily form somatic embryos such as *Arabidopsis*, *Daucus carota*, *Picea glauca*, *Medicago sativa*, and *G. max* have been used for the identification and characterization of signal molecules that are important for the induction and maintenance of embryonic development.

Somatic Embryogenesis Receptor-Like Kinases (SERKs)

Various kinases have been identified in somatic embryogenesis. These kinases are often activated via autophosphorylation and transduce the signal from the cell membrane to the action site, thus regulating the successive downstream transducers in the signal transduction pathway. SERKs constitute a special subgroup of receptor protein kinases and are associated with the process of somatic embryogenesis. The first SERK gene (*DcSERK*) was isolated from *D. carota* on the basis of its expression in embryogenic competent cells of suspension cultures up to the globular-shaped stage of embryogenesis. This expression pattern appears consistent with the hypothesis that the SERK proteins may act as transmembrane receptors for signals in the culture medium and may trigger embryogenesis. However, genetic or biochemical evidence is not yet available to demonstrate the indispensable role of *DcSERK* in developing somatic embryos (Schmidt et al., 1997).

Ectopic overexpression of the *Arabidopsis* ortholog of SERK (*AtSERK1*) enhanced the ability of suspension cells to undergo somatic embryogenesis. Further research demonstrated its subcellular location as a membrane protein and phosporylation/dephosphorylation activities with KAPP as well as the precise spatial localization of gene expression during somatic embryogenesis (Salaj et al., 2008). Besides *AtSERK1*, four additional putative *SERK* genes were characterized in the *Arabidopsis* genome, with their phosphorylation sites identified by proteomic analysis. Putative *SERK* genes were also identified from other plant species including *Zea mays*, *Medicago truncatula*, *Helianthus annuus*, *Theobroma cacao*, *Oryza sativa*, *Triticum aestivum*, and *Cocos nucigera*.

Calmodulin-Mediated Signal Transduction

Calcium is a key regulator of various cellular and physiological processes in higher plants. Work on the *D. carota* system has shown that Ca^{2+} enhances embryogenic frequency, and its absence arrests somatic embryo formation. The enhancement of somatic embryogenesis by calcium and the presence of vacular Ca^{2+} as the first signal that allows the recognition of embryogenic cells are well documented. These observations are suggestive of an intermediary role for Ca^{2+} during plant embryogenesis. Cellular calcium signals are detected and transmitted by sensor molecules. The following three major classes of Ca^{2+} sensors have been identified in plants (Figure 23.7):

1. Calmodulin (CaM). CaM comprise the best of the well-known Ca^{2+} sensors, which are ubiquitous Ca^{2+}-binding proteins highly conserved in eukaryotes. CaM is generally localized in the meristematic regions of developing embryos and is also found in embryogenic cell cultures, with its transcript increased somewhat in globular-shaped and heart-shaped-stage embryos compared with low levels in the undifferentiated callus. In cultures of *Saccharum officenarum,* beginning with undifferentiated cells and continuing through the embryogenic cultures and the somatic embryo development stages, CaM expression was specific to the embryogenic stage compared with the nonembryogenic stage. An increase of CaM expression seemed to be related to the stages in which increased protein turnover in systems undergoing rapid cell division occurred, and it was suggested that spatial regulation of CaM may be important for regulating the embryogenic program (Suprasanna et al., 2004).
2. The calcium-dependent protein kinase (CDPK). CDPK, which represents a second class of Ca^{2+} sensor, contains a C-terminal CaM-like domain that can directly bind Ca^{2+}, thereby making CaM unnecessary for its activation. This activity was first purified from *Glycine max* and later identified and/or purified from a host of other plant species. CDPKs are now implicated as playing intermediary and regulatory roles in a variety of developmental and metabolic processes. Anil and Rao (2000) studied the possible involvement of Ca^{2+}-mediated signaling in the induction/regulation of somatic embryogenesis from

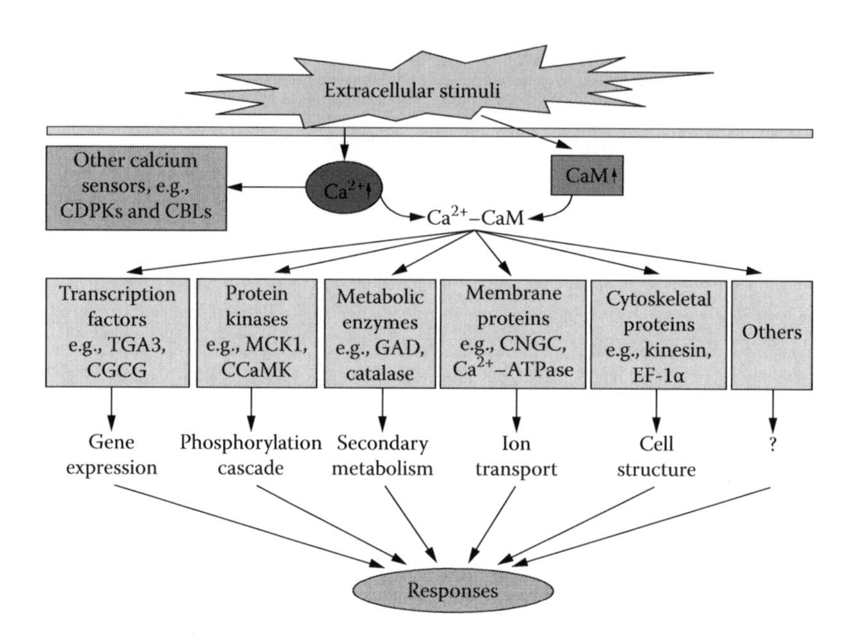

FIGURE 23.7 Model of Ca^{2+}/calmodulin-mediated network in plants. Ca^{2+} signal changes are triggered by environmental, hormonal, or developmental signals. The calcium signatures are decoded by calcium sensors, such as calmodulin (CaM), calcium-dependent protein kinase (CDPK), and calcineurin-B like protein (CBL). Expressions of some CaM genes are also induced by these signals. The activated Ca^{2+}/CaM complex binds to numerous target proteins and modulates their activities. Those target proteins include transcription factors, protein kinases, metabolic enzymes, ion channels and transporters, and cytoskeleton proteins. Finally, the Ca^{2+}/CaM-mediated signal network results in physiological responses, such as cell growth or differentiation, stress tolerance or growth arrest, and cell death. (From Yang, T. B., and B. W. Poovaiah. 2003. *Trends Plant Sci.* 8:505–512. With permission.)

proembryogenic cells of *Santalum album*, and the results indicated that blocking the CDPK-involved signaling pathway inhibits embryogenesis.

3. Calcineurin B-like proteins (CBLs). Unlike CaM genes, CBLs have been previously identified only in higher plants, suggesting that CBLs may function in plant-specific signaling processes. In plants, the CBL family represents a unique group of calcium sensors and plays a key role in decoding calcium transients by specifically interacting with and regulating a family of protein kinases (CIPKs). Several CBL proteins appear to be targeted to the plasma membrane by means of dual lipid modification by myristoylation and S-acylation. In addition, CBL/CIPK complexes have been identified in other cellular localizations, suggesting that this network may confer spatial specificity in Ca^{2+} signaling. Molecular genetic analyses of loss-of-function mutants have implicated several CBL proteins and CIPKs as important components of abiotic stress responses, hormone reactions, and ion transport processes. The isolation of an EST homology to *CIPK9* during initial cellular dedifferentiation in *G. hirsutum* might indicate its functions in early stage of somatic embryogenesis (Zhu et al., 2008).

Transcription Factors Involved in Somatic Embryogenesis

LEC Genes

In higher plants, embryogenesis has been largely characterized at the morphological and physiological levels. Also, the genetic network that controls embryogenic processes is becoming better understood because of the identification of several genes that play regulatory roles either in specific phases of embryogenesis or during the whole process. The most well-known are the *LEC* genes. The first class of *LEC* genes, exemplified by *Arabidopsis LEC1* and *L1L*, encode HAP3-related transcription factors, and the second class of genes, which includes *Arabidopsis LEC2*, *FUS3*, and *ABI3*, as well as maize Viviparous1 (*Vp1*), encode B3 domain transcription factors. Both gene classes encode regulatory proteins involved in embryogenesis and are essential for induction of somatic embryo development (Figure 23.8). Mutational analyses showed that the *LEC* genes function early in embryogenesis to maintain suspensor cell fate and specify cotyledon identity. Late in embryogenesis, the *LEC* genes are required for initiating and/or maintaining maturation phase and repressing precocious germination.

1. LEC1. *LEC1* gene has been proposed as a key regulator for embryonic identity in *Arabidopsis*. Ectopic expression of *LEC1* in vegetative cells of *Arabidopsis* leads to severely abnormal plant growth and development, with occasional formation of somatic embryo-like structures (Lotan et al., 1998).
2. LEC2. *LEC2* encodes a transcription factor containing a B3 domain unique to several other plant transcription factors. Overexpression of *LEC2* leads to formation of somatic embryos as well as formation of calli and cotyledon-like and leaf-like structures. Likewise, ectopic expression of the B3 domain transcription factor FUS3 in the L1 layer via the ML1 promoter causes lateral organs to develop with embryonic features rather than as postgerminative vegetative tissue (Gazzarrini et al., 2004).

More recently, *LEC* genes were found to directly interact with hormone-response genes. For example, the penetrance of the turnip (*tnp*) mutant phenotype of *Arabidopsis*, a gain-of-function mutant of *LEC1*, is strongly enhanced or antagonized in the presence of exogenous auxin and cytokinins, respectively (Casson and Lindsey, 2006). The VP1/ABI3-LIKE (VAL) proteins are a class of B3 domain proteins with a possible chromatin-related repressor function. Double *val1 val2* or *val1 val3* mutants and single *val1* mutants treated with gibberellin acid (GA) biosynthesis inhibitors develop embryonic characteristics on seedlings, a phenotype that correlates with derepression of *LEC1* and other genes encoding embryo-expressed B3 domain factors (Suzuki et al., 2007).

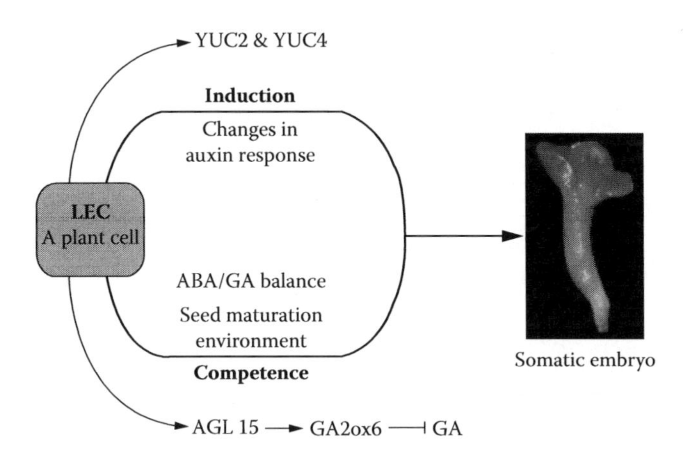

FIGURE 23.8 Model to explain the role of LEC transcription factors (TFs) in somatic embryogenesis (SE). The initiation of SE requires (i) an induction signal that causes a cell to change its identity and become embryonic and (ii) a cell that is competent to respond to the inductive signal. In many forms of SE, the induction signal is exogenously applied auxin. Ectopic LEC expression induces SE in the absence of exogenous auxin. It was proposed that the induction signal is an increase in endogenous auxin level caused by LEC2-mediated activation of YUC2 and YUC4 genes that encode auxin biosynthesis enzymes. LEC TFs might confer competence to undergo SE by repressing GA levels and, perhaps, by enhancing levels of ABA, creating a cellular environment similar to that of maturation phase zygotic embryos. It is not known whether LEC-mediated competence and induction occur concurrently or in parallel. (From Braybrook, S. A. and J. J. Harada. 2008. *Trends Plant Sci.* 13: 624–630. With permission.)

Baby Boom (BBM)

BBM is a transcription factor expressed in seed and root meristem that was identified as marker for embryo development in *Brassica napus* microspore-derived embryo cultures and can generate somatic embryos on transgenic *Arabidopsis* cotyledons when ectopically overexpressed. The gene also shows preferential expression in the basal region of the *Arabidopsis* embryo and is an auxin-inducible root-expressed gene in *Medicago truncatula.* In *Nicotiana tabacum*, heterologous BBM expression induces spontaneous shoot and callus formation, whereas a cytokinin pulse is required for somatic embryo formation.

The activation of root meristem–expressed genes by BBM is intriguing. BBM is expressed in the root meristem, but BBM overexpression does not appear to induce ectopic root or root meristem formation in *Arabidopsis.* This suggests that many of the root meristem–expressed BBM target genes may play a more general role in maintaining cells in an undifferentiated state. Recently, a number of the BBM target genes have been identified in microarray-based expression studies. Notably, at least six of the BBM target genes have been identified through screening for meristem-expressed genes. Expression of *BBM, TUBBY-LIKE PROTEIN 8* (*TLP8*), an LRR kinase gene (*At5g45780*), XTH9 (*At4g03210*), and a gene coding for a PH domain-containing expressed protein (*At5g47440*) is enriched in the quiescent center, a group of four to seven cells that give rise to the root initial/meristem cells (Nawy et al., 2005).

WUS Homeodomain Protein

The WUS homeodomain protein specifies stem cell fate in the shoot and floral meristem, but also promotes somatic embryo development in seedlings when ectopically expressed. WUS function is not directly linked to embryo identity, but rather to the maintenance of an undifferentiated cell state that responds to different stimuli to change the developmental fate of tissues. Ectopic expression of WUS resulted in enlarged meristems. The *WUS* gene is not expressed in the stem or meristem cells, but rather its expression is restricted to a small group of cells underneath the stem cells during all stages of embryogenesis and postembryogenesis. The unexpected expression pattern led to postulations that WUS promotes and/or maintains stem cell fate by a diffusion mechanism, or acts in a non-cell-autonomous manner.

In some situations the effect of ectopic expression of WUS depends on other proteins or hormones. Expression of WOX5 was induced by auxin in both *Arabidopsis* and *M. truncatula.* Overexpression of WUS causes highly embryogenic callus formation in the presence of auxin, whereas it directly induces somatic embryo formation from different plant organs in the absence of any exogenous auxin. Therefore it appears that WUS can reprogram cell fate, bypassing the auxin requirement or simply taking advantage of the endogenous auxin. Moreover, no callus phase, or at least only a few cell-division cycles are sufficient to induce cells to restart a totally new embryogenic pathway in tissues of plants that overexpress WUS. However, there is at least one part of the plant that WUS cannot reprogram to form embryos: the shoot apical meristem.

WUS can interact with CLAVATA (CLV), and the WUS/CLV self-regulatory loop, in which CLV presumably acts upstream of WUS, appears to be critical for the maintenance of stem cell identity (Figure 23.9). Overexpression of WUS under the control of meristem-specific promoters, such as CLV1, ANT, LFY, AP3, and AG, did not result in any somatic embryogenesis phenotype. The presence or absence of some factors in the shoot apex could favor one (a shoot meristem organizer) or the other WUS function (an embryo organizer).

MADS-Domain Protein AGAMOUS-LIKE15 (AGL15)

MADS-domain proteins are a family of transcriptional regulatory factors found in eukaryotic organisms. In plants, MADS-domain proteins are central players in many developmental processes, including control of flowering time, homeotic regulation of floral organogenesis, fruit development, and seed pigmentation. The MADS-domain protein AGL15 was initially identified using

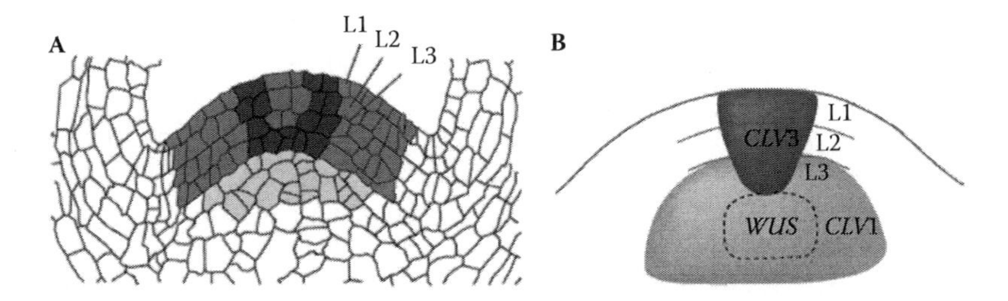

FIGURE 23.9 Organization of the shoot apical meristem. (A) Schematic view of SAM domains. (B) Outline of the central part of (A), showing the approximate mRNA expression domains of *CLV1, CLV3,* and *WUS.* (From Haecker, A., and T. Laux. 2001. *Curr. Opin. Plant Biol.* 4:441–446. With permission.)

differential display of mRNA as an embryo-expressed gene as well as during characterization of MADS-box genes in *Arabidopsis* (Rounsley et al., 1995). In *Glycine max, GmAGL15* mRNA was not detectable at very early stages of seed pod development, but was present in young developing embryos, and the level declined after maturation. Notably, the highest level of *GmAGL15* mRNA accumulation was detected in the somatic embryo culture (Thakare et al., 2008). Likewise, AGL15 promotes somatic embryo development and enhances production of secondary embryonic tissue from cultured zygotic embryos.

Interestingly and perhaps relevant for somatic embryogenesis, AGL15 has been identified as a component of a SERK1 protein complex, and both SERK1 and AGL15 are expressed in response to auxin treatment. Also intriguing are recent results that indicate that LEC2 may directly induce expression of AGL15. Like LEC2 and FUS3, AGL15 impacts upon bioactive GA accumulation, but AGL15 mediates its effect at least in part by directly inducing expression of *AtGA2ox6* (*At1g02400*) that encodes a GA 2-oxidase that catabolizes biologically active GA. Expression of this GA 2-oxidase affects somatic embryo development from the shoot apical meristem (SAM) of liquid culture grown seedlings in the presence of 2,4-D (Wang et al., 2004).

Extracellular Protein as Markers for Somatic Embryogenesis

For induction of somatic embryogenesis in plants, a variety of growth medium systems have been used. Some molecules in these induction systems help to trigger embryogenic potential in plant cells for obtaining reasonable regeneration frequencies and provide information on the molecular mechanisms of plant cell differentiation. Several markers have been reported that are able to distinguish between embryogenic and nonembryogenic cell cultures, such as the embryo-specific genes *SERK, LEC1, FUS3,* and *ABI3.* However, in most cases, the large number of specific proteins differentially expressed in embryogenic and nonembryogenic cells makes it difficult to use them as markers. The characterization of extracellular protein markers for somatic embryogenesis offers the possibility of determining the embryogenic potential of plant cells in culture long before any morphological changes have taken place, and many extracellular protein markers for embryogenic potential have been described such as the following:

1. Arabinogalactan proteins
2. Nonspecific lipid transfer proteins
3. Germins and germin-like proteins

Arabinogalactan Proteins

Secreted glycoproteins play a significant role in somatic embryogenesis by stimulating or inhibiting somatic embryo induction and development. One group of glycoproteins that are involved in plant differentiation is AGPs, representing a heterogeneous group of proteoglycans widely distributed in the plant kingdom. The AGPs are present in cell membranes, in cell walls, and in the intercellular spaces of tissues, and are secreted into the medium in cell cultures. It has been reported that AGPs are involved in processes of plant growth and development, such as cell development and differentiation, fiber development, and somatic embryogenesis. Application of extracellular AGPs to the media of embryogenic suspension cultures of *D. carota*, *Cyclamen persicum*, and *Picea abies* results in increasing the embryogenic potential of less embryogenic lines. Removing AGPs bound to the cell wall decreased the ability of the protoplasts to form somatic embryos, while addition of isolated extracellular AGPs partially reversed the effect of cell wall removal. Immunocytochemistry experiments have demonstrated the developmental regulation of some AGP epitopes in somatic embryogenesis. Cells of embryogenic suspension cultures of *Daucus carota* exhibited a distinct temporal and spatial expression pattern of an AGP epitope detectable with the monoclonal antibody JIM4 (Stacey et al., 1990). There was a remarkable accumulation of JIM4-binding AGPs in the protoderm of globular embryos. The addition of Yariv's reagent, a phenylglycoside that selectively binds to AGPs, blocks somatic embryogenesis in suspension culture (Thompson and Knox, 1998).

Interestingly, experiments with AGPs isolated by using different monoclonal antibodies revealed that they may also inhibit somatic embryogenesis, depending on various culture parameters. Differerent AGP populations were able to increase or decrease embryogenic potential in suspension cultures of *D. carota*. In old cell lines that had lost the ability to develop somatic embryos, embryo formation could be reinitiated by addition of carrot-seed AGPs; certain classes of carrot AGPs might inhibit the formation of embryogenic cells. When added to embryogenic cell populations, AGPs increased the number of cells that morphologically resembled embryogenic cells, while in nonembryogenic suspension cultures, AGPs reinitiated embryogenic potential in *D. carota* (Kreuger and van Holst, 1993). Surprisingly, JIM8-AGPs showed an inhibitory effect on the frequency of embryo development from single cells (Toonen et al., 1997). It seems contradictory that, on one hand, the cell wall epitope is related to embryogenic competence and that, on the other hand, the soluble JIM8-AGP fraction inhibits somatic embryo development. The results indicated that the JIM8 cell wall epitope can be regarded as a marker for embryogenic competence of a cell line in *D. carota* suspension cultures. Precise manipulation of somatic embryogenesis by AGPs in suspension cultures now requires that both promotive and inhibitory AGPs are purified and their precise cellular origin is determined. In this way, manipulation of cell type composition of embryogenic cultures could be carried out in conjunction with addition of exogenous AGPs to obtain more control over the process of somatic embryogenesis.

Nonspecific Lipid Transfer Proteins

Nonspecific lipid transfer proteins (LTPs) represent a protein family that is ubiquitous in plants. These proteins are characterized by their ability to transfer phospholipids between membranes and to bind fatty acids in vitro. Several in vivo functions have been attributed to nonspecific LTPs, including transport of cuticular compounds and inhibition of the growth of bacterial and fungal pathogens. The *ltp* gene has been implicated as a well-known early marker of somatic embryogenesis induction in different systems, being that it is linked to the protoderm layer formation, which exerts a regulatory role in controlling cell expansion during embryo development in developing somatic and zygotic embryos.

EP2 from *D. carota* embryogenic cultures was the first gene encoding an LTP to be isolated and characterized, and it was shown to be secreted extracellularly for the function of transporting lipids or nonpolar molecules from their place of synthesis in the endoplasmic reticulum to various cellular locations (Toonen et al., 1997). The gene is uniformly expressed in PEMs, whereas expression diminishes in nonembryogenic cell lines. Evidence indicated that EP2 is already expressed in precursor cell clusters from which somatic embryos develop. Taken together, a correct expression of *ltp* genes is required for normal embryo development. Five acidic LTP-like proteins have been found in the cell wall and the conditioned medium of microcluster cells from embryogenic suspension cultures of *Draba glomerata* were able to discriminate between embryogenic and nonembryogenic suspension cultures (Tchorbadjieva et al., 2005).

Expression of LTP gene products is restricted to the peripheral layers of young tissues and developing embryos. In *Camellia japonica* leaf cultures, during induction of somatic embryogenesis, LTP genes were found to be necessary for normal somatic embryogenesis to occur. Under- and overexpression of a putative LTP gene affect sequential developmental stages during somatic embryogenesis by changing the morphology and occurrence frequency of somatic embryos (Hjortswang et al., 2002). The expression of a LTP homologous gene in *Gossypium hirsutum* was absent in hypocotyls, nonembryogenic cell tissues, and plantlets, but markedly activated to the highest level of expression in embryogenic cells and preglobular embryos, through transitional PEMs with higher expression, whereas the level of expression was sharply diminished in all postglobular-stage somatic embryos (Zeng et al., 2006). Expression of VvLTP1, a *Vitis vinifera* homologue of AtLTP1, in somatic embryo development demonstrated that this LTP isoform is a marker of protoderm formation and confirmed that this tissue forms sequentially over time. Furthermore, ectopic expression of VvLTP1 under the control of the 35S promoter led to grossly misshapen embryos, which failed to acquire bilateral symmetry and displayed an abnormal epidermal layer (Francois et al., 2008).

Germins and Germin-Like Proteins

Germins and germin-like proteins (GLPs) are members of a superfamily of proteins widely distributed in the plant kingdom. They are functionally diverse but structurally related to members of the cupin superfamily, which were named on the basis of the conserved β-barrel mature cupin domain. Following their initial identification as germination-specific markers in wheat, from which function the name "germin" was given, germins and GLPs were characterized as glycoproteins with oxalate oxidase activity that is frequently retained in the extracellular matrix by ionic bonds. Hence, they are known to play a wide variety of roles as enzymes, structural proteins, or receptors during somatic embryogenesis, salt stress, and pathogen responses. Characterization and cloning of the genes encoding germins and GLPs has facilitated a better understanding of their regulation and raised their potential in biotechnological applications.

The original studies of the role of GLPs in somatic embryogenesis were conducted by Domon et al. (1995). Comparing the profiles of extracellular proteins of nonembryogenic and embryogenic cell lines in *Pinus caribaea* led to the identification of the first GLP protein in somatic embryogenesis, whose cDNA (PcGER1) was later isolated in a library; expression analysis confirmed the embryogenic specificity of this GLP (Neutelings et al., 1998). Further studies showed that PcGER1 expression was related to the cell cycle as shown previously in the case of wheat germin. Later, a similar GLP cDNA, isolated in *P. radiata*, showed high mRNA transcript levels in embryogenic tissue and little or no expression in nonembryogenic (roots, shoots, and needles) or callus tissue (needle and fiber callus culture). Additionally, proteomic methods have been employed to quantitatively assess the expression levels of proteins across four stages of somatic embryo maturation in *Picea glauca* (Lippert et al., 2005), and the GLP protein displayed a significant change in abundance as early as day 7 of embryo development.

In summary, much progress has been made since the prediction of plant cell totipotency by Haberlandt in the early 1900s, but the journey toward unearthing the underlying events of somatic embryogenesis was actually initiated in the late 1960s. Early research on somatic embryogenesis mostly focused on the hormonal regulation of this developmental process, and a repertoire of strategies was developed to regenerate many species via somatic embryogenesis. During these decades, molecular understanding of this developmental program has been greatly based on experiments. To gain better insight into the mechanisms of somatic embryogenesis, a combination of more advanced methods, histological analysis, micromanipulation, and in situ techniques, including polymerase chain-reaction-based subtractive hybridization, will provide critical new information on early embryo gene expression in time and space. Differential display of mRNA, transposon tagging, amplified antisense RNA, and proteome analysis, in some cases coupled with the previously mentioned techniques, have been employed to address the complexity of sequences expressed during somatic embryogenesis. Overexpression and ectopic expression of wild-type and mutant genes represent another approach. Since the recognition of various aspects orchestrating this process, as well as the key genes annotated in somatic embryogenesis, we believe that we are on the way toward manipulating the process of somatic embryogenesis in higher plants and that we will eventually be able to artificially control the plant somatic embryogenesis process.

REFERENCES

Anil, V. S., and K. S. Rao. 2000. Calcium mediated signaling during sandalwood somatic embryogenesis. Role for exogenous calcium as second messenger. *Plant Physiol.* 123:1301–1311.

Aquea, F., and P. Arce-Johnson. 2008. Identification of genes expressed during early somatic embryogenesis in *Pinus radiate. Plant Physiol. Biochem.* 46:559–568.

Braybrook, S. A., and J. J. Harada. 2008. LECs go crazy in embryo development. *Trends Plant Sci.* 13: 624–630.

Casson, S., and K. Lindsey. 2006. The turnip mutant of Arabidopsis reveals that LEAFY COTYLEDON1 expression mediates the effect of auxin and sugars to promote embryonic cell identity. *Plant Physiol.* 142:526–541.

Chugh, A. and P. Khurana. 2002. Gene expression during somatic embryogenesis— recent advances. *Curr. Sci.* 86:715–730.

Domon, J. M., G. D. Neutelings, R. A. David, and H. David. 1995. Three glycosylated polypeptides secreted by several embryogenic cell cultures of show highly specific serological affinity to antibodies directed against the wheat germin apoprotein monomer. *Plant Physiol.* 108:141–148.

Filonova, L. H., P. V. Bozhkov, and S. von Arnold. 2000a. Developmental pathway of somatic embryogenesis in *Picea abies* as revealed by time-lapse tracking. *J. Exp. Bot.* 51:249–264.

Filonova, L. H., P. V. Bozhkov, V. B. Brukhin, G. Daniel, B. Zhivotovsky, and S. von Arnold. 2000b. Two waves of programmed cell death occur during formation and development of somatic embryos in the gymnosperm, Norway spruce. *J. Cell Sci.* 113:4399–4411.

Francois, J., M. Lallemand, P. Fleurat-Lessard, L. Laquitaine, S. Delrot, P. Coutos-Thevenot, and E. Gomes. 2008. Overexpression of the VvLTP1 gene interferes with somatic embryo development in grapevine. *Funct. Plant Biol.* 35:394–402.

Gazzarrini, S., Y. Tsuchiya, S. Lumba, M. Okamoto, and P. McCourt. 2004. The transcription factor FUSCA3 controls developmental timing in Arabidopsis through the hormones gibberellin and abscisic acid. *Developmental Cell* 4:373–385.

Haecker, A., and T. Laux. 2001. Cell–cell signaling in the shoot meristem. *Curr. Opin. Plant Biol.* 4:441–446.

Hagen, G., A. Kleinschmidt, and T. Guilfoyle. 1984. Auxin regulated gene expression in intact soybean hypocotyls and excised hypocotyl sections. *Planta* 162:147–153.

Hirt, H., A. Pay, J. Gyorgyey, L. Bako, K. Nemet, L. Bogre, R. J. Schweyen, E. Heberle-Bors, and D. Dudits. 1991. Complementation of a yeast cell cycle mutant by an alfalfa cDNA encoding a protein kinase homologous to p34cdc2. *Proc. Natl. Acad. Sci. USA* 88:1636–1640.

Hjortswang, H. I., A. Sundas-Larsson, G. Bharathan, P. V. Bozhkov, S. von Arnold, and T. Vahala. 2002. *KNOTTED1*-like homeobox genes of a gymnosperm, Norway spruce, expressed during somatic embryogenesis. *Plant Physiol. Biochem.* 40:837–843.

Kreuger, M., and G. J. van Holst. 1993. Arabinogalactan proteins are essential in somatic embryogenesis of *Daucus carota L. Planta* 189:243–248.

Lippert, D., J. Zhuang, S. Ralph, D. E. Ellis, M. Gilbert, R. Olafson, K. Ritland, B. Ellis, C. J. Douglas, and J. Bohlmann. 2005. Proteome analysis of early somatic embryogenesis in *Picea glauca. Proteomics* 5:461–473.

Lotan, T., M. Ohto, and K. M. Yee. 1998. Arabidopsis LEAFY COTYLEDON1 is sufficient to induce embryo development in vegetative cells. *Cell* 93:1195–1205.

Mayer, U., U. Herzog, F. Berger, D. Inzé, and G. Jürgens. 1999. Mutations in the pilz group genes disrupt the microtubule cytoskeleton and uncouple cell cycle progression from cell division in Arabidopsis embryo and endosperm. *Eur. J. Cell Biol.* 78:100–108.

Murray, A. W. and M. W. Kirschner. 1989. Cyclin synthesis drives the early embryonic cell cycle. *Nature* 339: 275–280.

Nawy, T., J. Y. Lee, J. Colinas, J. Y. Wang, S. C. Thongrod, J. E. Malamy, K. Birnbaum, and P. N. Benfey. 2005. Transcriptional profile of the Arabidopsis root quiescent center. *Plant Cell* 17:1908–1925.

Neutelings, G., J. M. Domon, N. Membre, F. Bernier, Y. Meyer, A. David, and H. David. 1998. Characterization of a germin-like protein gene expressed in somatic and zygotic embryos of pine (*Pinus caribaea Morelet*). *Plant Mol. Biol.* 38:1179–1190.

Quint, M., and W. M. Gray. 2006. Auxin signaling. *Curr Opin Plant Biol.* 9:448–453.

Quiroz-Figueroa, F. R., R. Rojas-Herrera, R. M. Galaz-Avalos, and V. M. Loyola-Vargas. 2006. Embryo production through somatic embryogenesis can be used to study cell differentiation in plants. *Plant Cell Tiss. Org. Cult.* 86:285–301.

Rounsley, S. D., G. S. Ditta, and M. F. Yanofsky. 1995. Diverse roles for MADS box genes in Arabidopsis development. *Plant Cell* 7:1259–1269.

Salaj, J., I. R. von Recklinghausen, V. Hecht, S. C. de Vries, J. H. N. Schel, and A. M. van Lammeren. 2008. AtSERK1 expression precedes and coincides with early somatic embryogenesis in *Arabidopsis thaliana. Plant Physiol. Biochem.* 46:709–714.

Samaj, J., M. Bobák, A. Blehová, and A. Pretová. 2005. Importance of cytoskeleton and cell wall in somatic embryogenesis. In A. Mujib and J., Samaj, (Eds), *Somatic Embryogenesis.* Plant Cell Monogr 2:35–50.

Sato, S., T. Toya, R. Kawahara, F. R. Whittier, H. Fukuda, and A. Komamine. 1995. Isolation of a carrot gene expressed specifically during early-stage somatic embryogenesis. *Plant Mol. Biol.* 28:39–46.

Schmidt, E. D. L., F. Guzzo, M. A. J. Toonen, and S. C. de Vries. 1997. A leucine-rich repeat containing receptor-like kinase marks somatic plant cells competent to form embryos. *Development* 124:2049–2062.

Stacey, N. J., K. Roberts, and J. P. Knox. 1990. Patterns of expression of the JIM4 arabinogalactan-protein epitope in cell cultures and during somatic embryogenesis in *Daucus carota* L. *Planta* 180:285–292.

Steinborn, K., C. Maulbetsch, B. Priester, S. Trautmann, T. Pacher, B. Geiges, F. Kuttner, L. Lepiniec, Y. D. Stierhof, H. Schwarz, G. Jurgens, and U. Mayer. 2002. The Arabidopsis *PILZ* group genes encode tubulin-folding cofactor orthologs required for cell division but not cell growth. *Genes Dev.* 16:959–971.

Suprasanna, P., N. S. Desai, G. Nishanth, S. B. Ghosh, N. Laxmi, and V. A. Bapat. 2004. Differential gene expression in embryogenic, non-embryogenic and desiccation induced cultures of sugarcane. *Sugar Tech.* 6:305–309.

Suzuki, M., H. H. Y. Wang, and D. R. McCarty. 2007. Repression of the LEAFY COTYLEDON 1/B3 regulatory network in plant embryo development by VP1/ABSCISIC ACID INSENSITIVE 3-LIKE B3 genes. *Plant Physiol.* 143:902–911.

Tchorbadjieva, M., R. Kalmukova, I. Pantchev, and S. Kyurkchiev. 2005. Monoclonal antibody against a cell wall marker protein for embryogenic potential of *Dactylis glomerata* L. suspension cultures. *Planta* 222:811–819.

Thakare, D., W. Tang, K. Hill, and S. E. Perry. 2008. The MADS-Domain transcriptional regulator AGAMOUS-LIKE15 promotes somatic embryo development in arabidopsis and soybean. *Plant Physiol.* 146:1663–1672.

Thompson, H. J. M., and J. P. Knox. 1998. Stage-specific responses of embryogenic carrot cell suspension cultures to arabinogalactan protein-binding beta-glucosyl Yariv reagent. *Planta* 205:32–38.

Toonen, M. A. J., T. Hendriks, E. D. L. Schmidt, H. A. Verhoeven, A. van Kammen, and S. C. de Vries. 1994. Description of somatic-embryo-forming single cells in carrot suspension cultures employing video cell tracking. *Planta* 194:565–572.

Toonen, M. A. J., J. A. Verhees, E. D. L. Schmidt, A. van Kammen, and S. C. de Vries. 1997. AtLTP1 luciferase expression during carrot somatic embryogenesis. *Plant J.* 12:1213–1221.

Van Arnold, S., I. Sabala, P. Bozhkov, J. Dyachok, and L. H. Filonova. 2002. Developmental pathways of somatic embryogenesis. *Plant Cell Tiss. Org. Cult.* 69:233–249.

Wang, H., L. V. Caruso, A. B. Downie, and S. E. Perry. 2004. The embryo MADS domain protein AGAMOUS-Like 15 directly regulates expression of a gene encoding an enzyme involved in gibberellin metabolism. *Plant Cell* 16:1206–1219.

Yang, T. B. and B. W. Poovaiah. 2003. Calcium/calmodulin-mediated signal network in plants. *Trends Plant Sci.* 8:505–512.

Yang, X. Y. and X. L. Zhang. 2010. Regulation of somatic embryogenesis in higher plants. *Crit. Rev. Plant Sci.* 29: 36–57.

Zeng F. C., X. L. Zhang, L. F. Zhu, L. L. Tu, X. P. Guo, and Y. C. Nie. 2006. Isolation and characterization of genes associated to cotton somatic embryogenesis by suppression subtractive hybridization and macroarray. *Plant Mol. Biol.* 60:167–183.

Zhu, H. G., L. L. Tu, S. X. Jin, L. Xu, J. F. Tan, F. L. Deng, and X. L. Zhang. 2008. Analysis of genes differentially expressed during initial cellular dedifferentiation in cotton. *Chinese Sci. Bull.* 3:3666–3676.

Zimmerman, J. L. 1993. Somatic embryogenesis: A model for early development in higher plants. *Plant Cell* 5:1411–1423.

24 Embryogenic Callus and Suspension Cultures from Leaves of Orchardgrass

D. J. Gray, R. N. Trigiano, and Bob V. Conger

LABORATORY EXERCISES

Members of the family Poaceae comprise the single most economically important group of plants. This family includes the nutritious cereal grain crops such as corn (*Zea mays* L.), oats (*Avena sativa* L.), rice (*Oryza sativa* L.), and wheat (*Triticum aestivum* L.), as well as the forage grasses such as annual ryegrass (*Lolium multiflorum* Lam.), tall fescue (*Festuca arundinacea* Schreb.), and the wheatgrasses (*Agropyron* spp.). The cereals are important sources of complex carbohydrates, oils, and proteins, whereas the forages are important in dairy and meat production. Sugarcane (*Saccharum officinarum* L.), another member of this family, is an important source of sucrose.

Much research emphasis in the Poaceae has gone toward developing biotechnological approaches to crop improvement. One area of biotechnology that received early attention by researchers was that of in vitro plant regeneration. The grasses and cereals were initially regarded by researchers to be difficult or impossible to regenerate from tissue and cell cultures. However, continuous effort since the early 1980s led to significant advancements, such that plant regeneration has been obtained for all grass and cereal species that have been attempted (Gray, 1990).

Somatic embryogenesis is the most common mode of regeneration for species in the Poaceae. Embryogenic culture systems are useful in the classroom, not only for illustrating plant regeneration but also for studying embryo development and morphology. Since the Poaceae is contained within the monocotyledoneae, a taxonomic division based on embryo morphology, their zygotic embryos are distinctly different from dicotyledonous embryos. In many poaceous plants, the induction of somatic embryogenesis is not simple and reliable enough to be used for classroom exercises. Many species require use of immature zygotic embryos as explants, which can only be obtained by careful cultivation and pollination of source plants. Other species tend to produce somatic embryos that have relatively abnormal morphologies when compared to zygotic embryos. However, a few poaceous species, notably orchardgrass (*Dactylis glomerata* L.), are much easier to manipulate.

Orchardgrass is a perennial cool-season forage species that is grown in temperate regions of the world to produce high-quality hay. It is genetically self-incompatible, which makes breeding of new varieties difficult and time consuming. The potential of using somatic embryos as synthetic seed to efficiently clone outstanding parental lines holds promise in the development of improved varieties (see Gray et al., 1992, for details). The following laboratory exercises cover the methods used to produce somatic embryos from cultures of orchardgrass. Selected clones of this species are highly embryogenic and, as a perennial, can be maintained in the greenhouse as potted plants. These factors account for its convenient use. Leaf tissues are used as explant material for initiating cultures from which morphologically correct somatic embryos are produced. Orchardgrass has "clasping-type" leaves, which are easier to dissect than "whorled-type" leaves. Thus, the orchardgrass system is ideal for demonstrating monocotyledonous somatic embryogenesis.

The following experiments represent a greatly expanded version of exercises that were originally published in *HortTechnology* (Gray et al., 1994) and are arranged to lead students through successful culture establishment, observations of embryogenic callus formation, and somatic embryo development, including the gradient embryogenic response that is typical of this culture system, culture maintenance, plant regeneration, manipulation of liquid suspension cultures, and the production of model "synthetic seeds."

GENERAL CONSIDERATIONS

Initiation of Cultures and Growth of Plants

A special clone of orchardgrass, "Embryogen-P" (Conger and Hanning, 1991), was selected from seedlings for its high embryogenic capacity. It also is possible to generate new embryogenic genotypes from seed of orchardgrass by following the procedure of Hanning and Conger (1983) and recurrently selecting somatic embryos in vitro through multiple subcultures until plants with highly embryogenic leaves are obtained. However, "Embryogen P" is the most convenient source of leaf explant tissue and is available from Dr. Robert N. Trigiano (Department of Entomology and Plant Pathology, 205 Ellington Plant Sciences Building, 2431 Center Drive, University of Tennessee, Knoxville, TN 37901-4500). The clone should be maintained in greenhouse pots in a high-quality potting mix, with regular fertilizer applications, and kept in vigorous growing condition. Alternatively, the clone may be maintained in vitro as described below. Before becoming root-bound, the plant number can be increased by splitting entire plants, including the root ball and transplanting into two or three new pots. Insect and fungal pests should be controlled; however, culture response may be decreased for up to 6 weeks after chemical spraying, especially with systemic fungicides. Therefore, stock plant health and vigor are the most important factors in successful embryogenic culture initiation of orchardgrass.

Culture Media

The culture medium used is Schenk and Hildebrandt (SH) basal salt mixture (see Chapter 3 for composition) and vitamin powder with 6.6 mg/L (30 μM) dicamba (synthetic auxin plant growth regulator [PGR], 30 g/L (88 mM) sucrose, and 0.7% tissue culture-tested agar. This medium is hereafter referred to as SH30, whereas medium used for embryo germination studies and lacking dicamba is designated as SH0. The pH is adjusted to 5.4 before autoclaving. The cooling medium then is poured, 25 mL, into each 100 × 15 mm sterile petri dish. SH30-C is a liquid medium (i.e., without agar) that contains 3 g/L casein acid hydrolysate type 1. The casein hydrolysate is dissolved in 25 mL of distilled water, loaded into a 30 mL syringe, and then dispensed through a sterile 0.2 μm syringe filter into cooled autoclaved SH30 liquid medium. The resulting SH30-C then is sterilely pipetted (25 mL) into each 125-mL erlenmeyer flask. Generally at least four petri dishes or flasks of the appropriate medium are required for each treatment and/or transfer for each student or student team.

Experiment 1. Culture Establishment

Orchardgrass is a tillering-type forage grass in which the plant body consists of dense clumps of tillers. New tillers arise by adventitious budding from the base of previously developing tillers. Each tiller is actually a complete plant in itself, with a root-shoot axis and several flattened, interfolded clasping leaves. Grass leaves grow from a pronounced basal (intercalary) meristem, such that the youngest part of a given leaf is the portion nearest to the shoot apical meristem. Youngest leaves are those nearest to the center. Thus, the youngest, most meristematically active leaf tissues are those nearest the bases of the innermost leaves. Figure 24.1 summarizes the steps of this experiment.

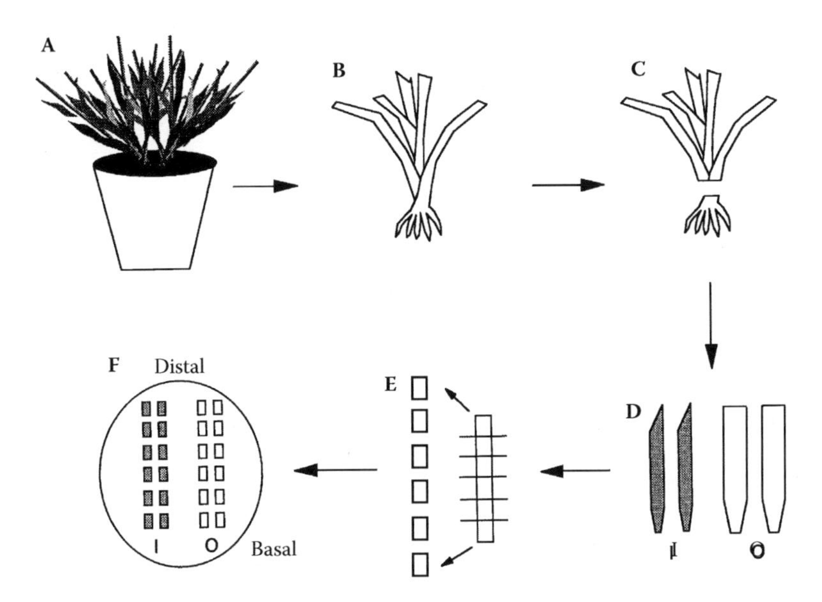

FIGURE 24.1 Diagrammatic representation of orchardgrass leaf explant preparation for embryogenic culture initiation. A. Potted plant of orchardgrass variety Embryogen-P. B. An isolated tiller. C. Remove and discard basal stem and roots, leaving only the clasped leaves. D. Carefully separate leaves and retain the innermost and next leaf out. E. Carefully split each leaf longitudinally into two halves. Cut halves of innermost leaf at an angle to retain its identity, then surface disinfest. F. Cut each leaf half into sections and plate sections of all four halves on a single petri dish containing SH30 medium. Be careful to retain the identity of each leaf and to keep the sections within each leaf in order of most basal to distal section. (From Trigiano, R.N. and D.J. Gray, Eds., *Plant Tissue Culture Concepts and Laboratory Exercises, Second Edition*, CRC Press LLC, Boca Raton, FL, 2000.)

Materials

The following items are needed for each student or team:

- 100 mL of 50% commercial bleach with a drop of surfactant (such as Triton-X100™, Tween 20™ or Ivory Liquid™ soap)
- Twelve (approximately) 100 × 15 mm petri dishes containing SH30 medium
- Stereomicroscope with light source

Follow the protocol provided in Procedure 24.1 to complete this experiment.

Anticipated Results

A measure of bacterial and fungal contamination growing on the medium or from cultured leaf sections is expected. While up to 50% of sections may become contaminated, a more normal contamination rate is 10%. It is important to follow directions for flaming instruments and manipulating leaf sections during transfer in order to eliminate, rather than spread, contamination.

Questions

- Discuss the importance of proper "transfer etiquette" in the establishment of cultures.
- Are the same procedures required in the maintenance of established cultures? Why?
- Did younger leaves become more or less contaminated than older leaves over time? Why?

Experiment 2. The Gradient Embryogenic Response

The embryogenic response in orchardgrass is related to leaf position within a tiller (i.e., the innermost or youngest leaf versus the next leaf out or older leaf) as well as the position of cultured leaf

Procedure 24.1
Embryogenic Culture Establishment

Step	Instructions and Comments
1	Remove individual tillers from a rapidly growing plant. Dissect individual leaves carefully so as to prevent damage and identify the innermost two leaves.
2	Remove the basal-most 3 cm of these leaves and separate the two halves of each longitudinally. Each leaf base should yield two strips of blade tissue approximately 3 cm long x 4 mm wide.
3	Place bleach solution into a sterile beaker and prepare three additional beakers with sterile distilled water for rinsing. Surface disinfest the leaf halves by gently agitating in the bleach solution for 2 min. Sterile forceps may be used to accomplish agitation.
4	Gently rinse the leaf halves three times in sterile water.
5	Remove one leaf half at a time and cut on a hard sterile surface (such as a petri dish) with a fresh, sharp scalpel blade. Cut the basal most part of each leaf half crosswise into six approximately square sections (approximate size = 5 × 5 mm) and discard the remaining portion of each leaf-half.
6	Immediately plate the sections onto SH30 medium. All of the dissection steps should be accomplished as rapidly as possible.
7	Arrange the sections from each leaf-half in separate rows such that basal sections are at one end of a given row and the most distal sections are at the other end. Sections from four leaf-halves should be placed in each petri dish. Seal each petri dish securely with several layers of parafilm.
8	Incubate cultures at 23°C in dark and evaluate daily for bacterial and fungal contamination. Transfer noncontaminated sections to fresh medium as needed. Forceps should be sterilized between moving of each section when transferring noncontaminated sections in order to avoid spreading of inconspicuous contamination to clean sections.
9	Determine the number of sections from each leaf (i.e., innermost versus outermost) that became contaminated weekly. Chart the relative contamination rate as a function of leaf position in the tiller.

sections relative to the shoot apical meristem (i.e., proximal or distal). The following two distinct embryogenic responses are possible (Conger et al., 1983): either (1) an embryogenic callus develops or (2) somatic embryos grow directly from the leaf. Embryogenic callus is white-to-yellowish in color and friable-to-wet in consistency. Often, adventitious roots develop from the leaf sections, but never lead to plant regeneration. Somatic embryos, which are described in more detail below, appear as small (approximately 1 mm), white, organized structures, either embedded in callus or growing directly from the leaf section surface.

Materials

As this experiment utilizes the cultures developed previously, additional materials are not required. Follow the methods outlined in Procedure 24.2 to complete this exercise.

Anticipated Results

The gradient embryogenic response is demonstrated by correlating the relative position in a leaf from which a given section is obtained. A typical gradient response is shown in Figure 24.2. After 4–6 weeks, basal sections produce embryogenic callus. The callusing response progressively diminishes in sections from increasingly more distal positions of the leaf, being replaced by a direct embryogenesis response and, eventually, no embryogenic response at all. The differential embryogenic response is related to meristematic activity in the explant tissue. Leaf sections with abundant undifferentiated meristematic cells tend to produce embryogenic callus. As meristematic activity is decreased, embryogenic response is reduced such that only direct somatic embryos develop on sections with limited meristematic activity. Sections in which meristematic activity is too low or has ceased do not exhibit an embryogenic response.

Procedure 24.2
The Gradient Embryogenic Response

Step	Instructions and Comments
1	Plate leaf sections on SH30 medium as described above such that the sections from each leaf half are kept together in a single row and arranged in sequential order from the most proximal to the most distal section.
2	Place the sections from four leaf halves in each dish (i.e., four rows of sections/dish—see Figure 24.1). Mark each leaf from each tiller so that the identity of the innermost versus the outermost of two leaves per tiller is maintained.
3	Culture and screen for contamination as described in Procedure 24.1.
4	Determine the number of sections from each location within each leaf half that produces embryogenic callus and/or direct embryos at weekly intervals for 8 weeks. Chart the occurrence of embryogenic callus and somatic embryos as a function of time and position of the section in the original tiller.

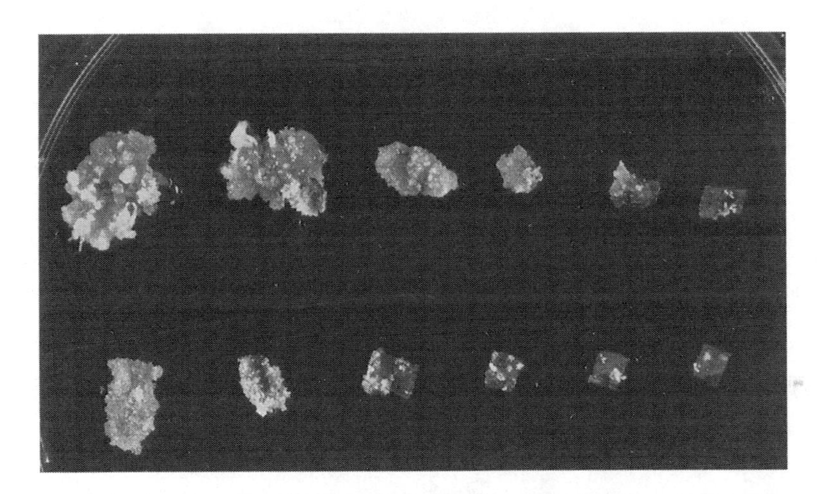

FIGURE 24.2 Gradient embryogenic response exhibited by the younger innermost (upper row of sections) and older next out (lower row of sections) leaves of orchardgrass. Note that the basal most leaf sections (left) produce embryogenic callus, whereas, the more distal leaf sections (right) produce embryos directly. Also, more of the basal most leaf sections produce embryogenic callus than the older leaf sections. (From Trigiano, R.N. and D.J. Gray, Eds., *Plant Tissue Culture Concepts and Laboratory Exercises, Second Edition,* CRC Press LLC, Boca Raton, FL, 2000.)

Questions

- Do sections from younger or older leaves or sections from proximal or distal locations within a leaf produce more callus or direct embryos?
- Which sections respond most rapidly?
- How do observed embryogenic responses of individual sections correlate with the meristematic zones in intact orchardgrass leaves?

Experiment 3. Maintenance of Embryogenic Callus

Embryogenic callus of orchardgrass can be maintained by monthly transfer of callus and/or somatic embryos to fresh SH30 medium. Embryogenic callus is very heterogeneous and continuously sectors into embryogenic and nonembryogenic portions. For maintenance of long-term embryogenic cultures, it is important to recognize and transfer only embryogenic sectors (Gray et al., 1984).

Materials

The following items are needed for each student or team:

- Ten 100 × 15 mm petri dishes containing SH30

Follow the methods in Procedure 24.3 to complete this experiment.

Anticipated Results

A typical embryogenic callus is shown in Figure 24.3. Embryogenic calluses or individual embryos are transferred to fresh SH30 medium in order to maintain and increase the embryogenic line. When transferring callus, it is important to isolate only sectors that are producing somatic embryos.

Procedure 24.3
Embryogenic Callus Maintenance

Step	Instructions and Comments
1	Isolate embryogenic callus and somatic embryos from leaf cultures. Dissect callus utilizing a stereomicroscope to assist in identifying various types of callus and tissue.
2	Separate dry, friable-type callus from wet, mucilaginous-type callus. Identify and separate sectors with numerous root primordia, as well as clumps of large somatic embryos and masses of small somatic embryos embedded in a watery matrix (Gray et al., 1984). Separate the calluses and tissues by type and plate small clumps of each, five clumps per petri dish, onto SH30 medium (produce five cultures). Be sure to also plate single, isolated somatic embryos (5 per petri dish) into 5 petri dishes in order to test their response.
3	Incubate in the dark for 4 weeks as in Procedure 24.1.
4	Determine and describe the different callus-types obtained based on their morphologies. Determine which of these callus types and/or tissues gave rise to embryogenic callus.

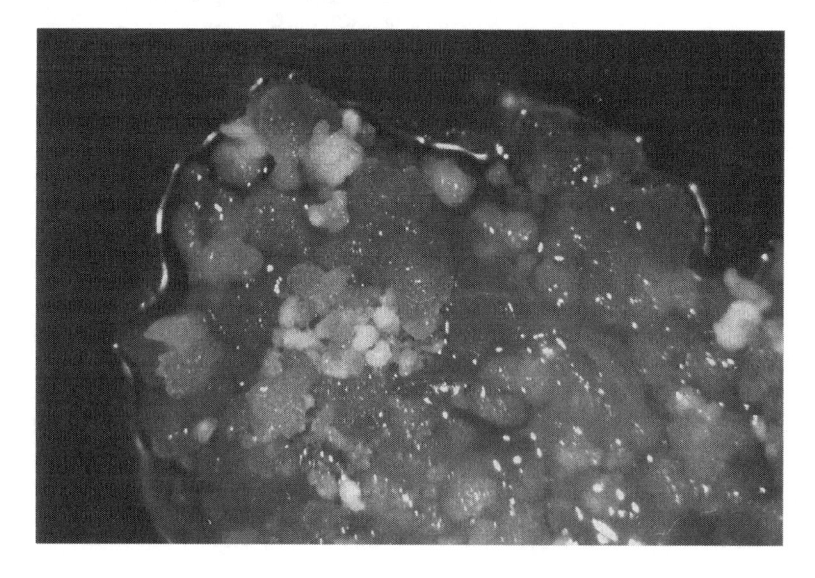

FIGURE 24.3 Embryogenic callus of orchardgrass. (From Trigiano, R.N. and D.J. Gray, Eds., *Plant Tissue Culture Concepts and Laboratory Exercises, Second Edition*, CRC Press LLC, Boca Raton, FL, 2000.)

The watery matrix material that contains masses of small somatic embryos frequently produces the most vigorous and prolific embryogenic callus. Embryogenic callus continues to sector into embryogenic and nonembryogenic callus-types, primarily due to the recallusing of existing somatic embryos. The somatic embryos are anatomically complex and produce a number of different callus types, including embryogenic and rhizogenic (root-forming) calluses, when recultured on SH30.

Questions

- Which of these callus-types produced the most embryogenic callus and somatic embryos?
- Where did embryogenic callus arise from plated somatic embryos?
- What type of tissue isolated from embryogenic callus would be best to use in culture maintenance?
- Why does embryogenic callus continually sector?

Experiment 4. Observations on Monocotyledonous Somatic Embryos

Orchardgrass embryogenic cultures produce monocotyledonous somatic embryos that are of the poaceous type. The typical embryo possesses a scutellum and a coleoptile. While the somatic embryos are small, it is possible to see morphological details using a stereomicroscope. Abnormal embryos are also common. For reference, see Figure 22.6, Conger et al. (1983); Gray et al. (1984); Gray and Conger (1985b); and Trigiano et al. (1989).

Materials

The following items are needed for each student or team:

- Leaf and callus cultures developed through previous experiments

Follow the methods listed in Procedure 24.4 to complete this experiment.

Anticipated Results

When viewed with a stereomicroscope, small white somatic embryos that are morphologically similar to zygotic embryos either appear in embryogenic callus or emerge directly from uncallused leaf sections. Embryos at all developmental stages can be recovered from the cultures (Figure 24.4). The embryos possess a distinct scutellum and a notch, from which a coleoptile develops and enlarges during germination (see Figure 22.6). Direct embryos are frequently attached to the leaf by a distinct suspensor.

Questions

- What types of abnormalities are most common?
- How might such abnormalities have occurred during embryo development?

Procedure 24.4
Observations on Monocotyledonous Somatic Embryos

Step	Instructions and Comments
1	Observe somatic embryos developing from callus and directly from leaf cultures. Correlate specific embryos observed with those illustrated in referenced publications.
2	Identify the scutellum and coleoptile.
3	Observe and describe the most commonly occurring types of abnormal embryos.

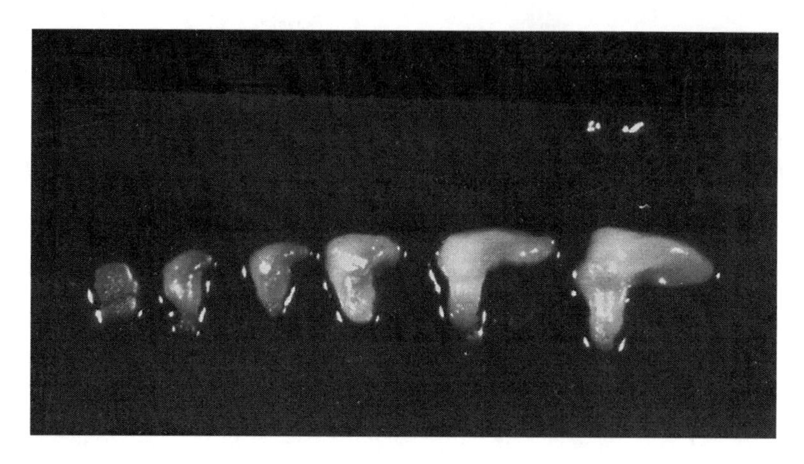

FIGURE 24.4 Somatic embryos isolated from embryogenic callus of orchardgrass. The youngest recognizable embryonic stages are shown on the left. Well developed embryos (right) are white and opaque, and possess a large flattened scutellum and a narrowed embryo axis. (From Trigiano, R.N. and D.J. Gray, Eds., *Plant Tissue Culture Concepts and Laboratory Exercises, Second Edition,* CRC Press LLC, Boca Raton, FL, 2000.)

Experiment 5. Somatic Embryo Germination and Plant Recovery

Orchardgrass somatic embryos germinate readily and in a manner similar to that of zygotic embryos when placed on SH0 medium and incubated in light (cool white fluorescent) at 25°C. Small rooted plants are removed from petri dishes and quickly transferred (to avoid dehydration) to moist potting medium in small pots, which are enclosed in plastic bags and cultured in a lighted incubator as above. When plants begin to grow vigorously, often within a week, they are transferred to greenhouse pots and maintained in the same manner as stock plants.

Materials

The following items are needed for each student or team:

- Source of embryos from previous experiments
- Five 100 × 15 mm petri dishes of SH0

Follow the methods listed in Procedure 24.5 to complete this experiment.

Anticipated Results

Individual embryos are induced to germinate into plants by transfer to medium lacking dicamba. Relative germination responses can easily be determined by counting the number of roots and shoots produced per culture. Resulting plants are easily acclimatized to greenhouse conditions.

Questions

- Why does transfer of embryos to medium lacking dicamba cause germination to occur?
- What is the first sign of germination and when does it occur?

Experiment 6. Initiation and Manipulation of Embryogenic Suspension Cultures

Embryogenic suspension cultures are produced by culturing embryogenic callus in liquid SH30 medium (i.e., lacking agar). The development of somatic embryos can be either stimulated or stopped by adding or removing various organic sources of nitrogen (Gray et al., 1984; Gray and Conger, 1985a; Trigiano and Conger, 1987).

Procedure 24.5
Somatic Embryo Germination and Plant Recovery

Step	Instruction and Comments
1	Transfer well-developed somatic embryos, five per petri dish, to SH0 medium and incubate as above. Alternatively, transfer small clumps of embryogenic callus to the same medium.
2	Observe each daily for signs of germination. Count the number of shoots and roots present after 2, 4, 6, and 8 days to determine whether shoots or roots emerge first. Identify the developing red-pigmented coleoptiles.
3	Count the number of shoots per callus; this provides an indication of the relative number of well-developed somatic embryos per callus. Determine the percentage of somatic embryos that produce acclimatized plants.

Materials

The following items are needed for each student or team:

- Rotary shaker
- Four 125-mL erlenmeyer flasks containing 10 mL liquid SH30
- Twelve 125-mL erlenmeyer flasks containing 20 mL liquid SH30
- Four 125-mL erlenmeyer flasks containing 20 mL liquid SH30-C
- Five or more 25-mL pipettes

Follow the methods given in Procedure 24.6 to complete this experiment.

Anticipated Results

A typical suspension culture is shown in Figure 24.5. Suspension cultures will proliferate and can be maintained in liquid SH30 medium but will not produce recognizable somatic embryos unless an organic nitrogen source, such as casein hydrolysate, which contains a complex array of amino acids, is present. This indicates that the nitrogen source is an important controlling factor in somatic embryogenesis. In fact, somatic embryo development can be repeatedly started or stopped by adding or deleting casein hydrolysate in medium during transfer. When culture material from SH30 and SH30-C medium is placed on solidified SH30 medium, both produce embryogenic callus cultures. But only the tissue from liquid SH30-C medium produce embryos that readily germinate into plants on SH0 medium. This indicates that, although both SH30 and SH30-C liquid culture media support the proliferation of embryogenic cells, only medium with casein hydrolysate promoted development of the cells into somatic embryos.

Questions

- What accounts for the effect of casein hydrolysate on embryo development?
- What do the observed differences between the responses of culture material from the two liquid medium indicate when transferred to the two types of solidified media?

Experiment 7. Encapsulation of Somatic Embryos to Produce Synthetic Seeds

Somatic embryogenesis is an ideal route for vegetative reproduction, since somatic embryos arise from the cells of one "parent." In fact, somatic embryogenesis represents the most efficient vegetative propagation system that can be envisioned, due to the rapid scale-up potential and prolific plant production exhibited by certain systems. In addition, since somatic embryos are nearly identical to zygotic embryos, research had attempted to add desirable seed-like qualities to somatic

Procedure 24.6
Embryogenic Suspension Culture

Step	Instructions and Comments
1	Initiate embryogenic suspension cultures by placing approximately 2 g of rapidly growing embryogenic callus into each 125-mL flask containing 10 mL of SH30 medium. Start with four flasks.
2	Cover the neck of each flask with several layers of sterile aluminum foil and seal the edges of the foil to the flask with parafilm. Rotate the cultures at approximately 75 rpm in the dark.
3	After 2 weeks, add 20 mL of SH30 medium to each flask and rotate at 100 rpm for an additional 2 weeks. Observe the increase of culture mass in rapidly growing cultures.
4	Increase and maintain cultures by pouring half of each rapidly growing culture into new flasks (total of eight flasks) containing 20 mL of SH30 medium at 2-week intervals.
5	Observe the remaining culture material with a stereomicroscope and produce fresh-mounted microscope slides to observe cells with a compound microscope. Inspect for somatic embryos or embryogenic cells, which are characterized as being small and densely cytoplasmic (see chapter nineteen).
6	To induce embryo development, transfer half of each established, proliferating culture into flasks containing fresh SH30 medium and the other half into flasks containing SH30-C medium. Maintain cultures as before and observe twice weekly for the development of small white somatic embryos, which can be readily observed collecting around the rim of the flask.
7	After 4 weeks growth in SH30 and SH30-C media, decant the liquid from flasks and spill the remaining culture mass into an empty, sterile petri dish.
8	Using a spatula and/or tweezers, plate small amounts of material (clumps about 3 mm in diameter) from each medium treatment, five clumps per petri dish, onto either solidified SH30 or SH0 medium and incubate in the light.
9	Observe the response weekly for 4 weeks.

FIGURE 24.5 An embryogenic suspension culture of orchardgrass. (From Trigiano, R.N. and D.J. Gray, Eds., *Plant Tissue Culture Concepts and Laboratory Exercises, Second Edition,* CRC Press LLC, Boca Raton, FL, 2000.)

embryos including a protective coating and ability to become quiescent. Such synthetic seeds would revolutionize certain aspects of agriculture, allowing genetically uniform synthetic seed to be produced indoors at will. The following experiment illustrates an aspect of synthetic seed research, the addition of a protective coating. See Gray and Purohit (1991) for a complete treatment of the subject.

Materials

The following items are needed for each student or team:

- 100 mL of 2% sterile suspension of the sodium salt of alginic acid (Sigma Chemical Co.)
- 100 mL of 25 mM sterile aqueous solution of $CaCl_2$
- 100 mL of liquid SH0 medium
- Ten 100 × 15 mm petri dishes, each containing 25 mL of SH0 medium
- Three 150-mL sterile beakers with magnetic stir bars
- Magnetic stir plate
- Sterile spoonulas
- Sterile 1000-μm nylon screen

Follow the experimental protocol listed in Procedure 24.7. The steps in this procedure are illustrated in Figure 20.6.

Anticipated Results

Immediately after dropping the alginic acid/embryo complex into $CaCl_2$, a cloudy-appearing bead should form around the embryo. This is the salt of alginic acid formed with the divalent cation Ca^{+2}, which is relatively insoluble in water. It may take several minutes for the calcium alginate drop to solidify completely to a soft gel-like consistency. Typically, plants are obtained from encapsulated embryos, but the plant recovery rate often is lower than that of controls.

Encapsulated somatic embryos constitute one type of synthetic seed. The calcium alginate can be regarded as a synthetic seed coat, which, theoretically, offers protection as well as a matrix to hold nutrients, pesticides, etc. See Gray and Purohit (1991) for a detailed discussion.

Questions

- Why did encapsulated embryos differ from controls in plant recovery rate?
- What are some potential advantages of coating somatic embryos?

Procedure 24.7
Encapsulation of Somatic Embryos

Step	Instructions and Comments
1	Harvest somatic embryos from suspension cultures by sieving the contents of a flask through a 1000 μm screen. Individual and small clumps of embryos, cells, and cell aggregates will pass through the screen. Discard the material retained by the screen.
2	Using a stereomicroscope, select individual, morphologically mature embryos and place them into the 2% alginic acid solution, gently stirring. Alternatively, leaf or callus cultures with mature somatic embryos may be used in place of suspension cultures by plucking embryos directly from the cultures and placing them into alginic acid.
3	For a control treatment, place some embryos directly on solidified SH0 medium without alginic acid treatment.
4	Using a wide bore pipette, withdraw a single embryo at a time from the alginic acid solution. Express enough of the embryo/alginic acid suspension to form a small drop at the tip of the pipette, such that the drop contains one embryo.
5	Allow this drop to fall into the gently stirring 25 mM $CaCl_2$ solution. Repeat this step numerous times, trying to perfect the art of forming a uniform drop containing an embryo.
6	Transfer the resulting calcium alginate/embryo beads to a swirling liquid SH0 solution for 15 min and then plate the beads, five per petri dish to solidified SH0 medium and incubate at 25°C in the light.
7	Compare germination responses with control embryos twice weekly for two weeks. Record the number of root and shoot emergences at each time for each treatment.

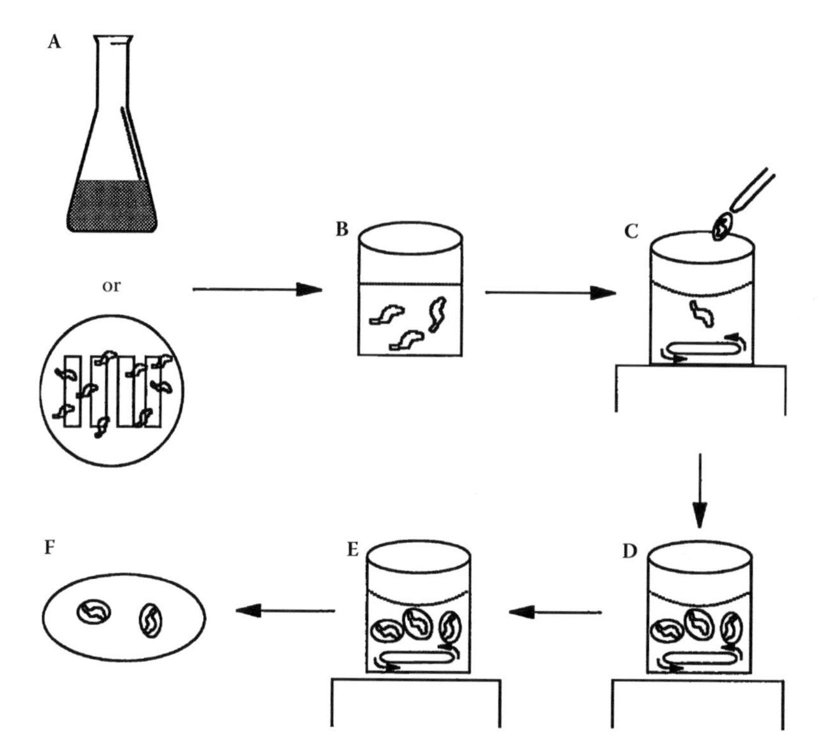

FIGURE 24.6 Encapsulation of orchardgrass somatic embryos in calcium alginate beads. A. Somatic embryos harvested from 5 week-old suspension, leaf, or callus culture. B. Embryos suspended in 2% alginic acid solution. C. Embryos dropped singly into a beaker containing swirling 25 mM $CaCl_2$ solution. D. Alginic acid is converted to calcium alginate, forming bead around embryos. E. Synthetic seeds transferred to liquid SH0 medium to remove excess $CaCl_2$. F. Plants emerging from synthetic seeds on solidified SH0 medium. (From Trigiano, R.N. and D.J. Gray, Eds., *Plant Tissue Culture Concepts and Laboratory Exercises, Second Edition,* CRC Press LLC, Boca Raton, FL, 2000.)

REFERENCES

Conger, B. V. and G. E. Hanning. 1991. Registration of Embryogen-P orchardgrass germplasm with a high capacity for somatic embryogenesis from in vitro cultures. *Crop Sci.* 31:855.

Conger, B. V., G. E. Hanning, D. J. Gray, and J. K. McDaniel. 1983. Direct embryogenesis from mesophyll cells of orchardgrass. *Science* 221:850–851.

Gray, D. J. 1990. Somatic cell culture and embryogenesis in the Poaceae. *Biotechnology in tall fescue improvement.* M. J. Kasperbauer (Ed.), pp. 25–57, CRC Press, Boca Raton, FL.

Gray, D. J. and B. V. Conger. 1985a. Influence of dicamba and casein hydrolysate on somatic embryo number and quality on cell suspensions of *Dactylis glomerata* (Gramineae). *Plant Cell Tiss. Org. Cult.* 4:123–133.

Gray, D. J. and B. V. Conger. 1985b. Time-lapse light photomicrography and scanning electron microscopy of somatic embryo ontogeny from cultured leaves of *Dactylis glomerata* (Gramineae). *Trans. Amer. Microsc. Soc.* 104:395–399.

Gray, D. J., B. V. Conger, and G. E. Hanning. 1984. Somatic embryogenesis in suspension and suspension-derived callus cultures of *Dactylis glomerata. Protoplasma* 122:196–202.

Gray, D. J. and A. Purohit. 1991. Somatic embryogenesis and development of synthetic seed technology. *Crit. Rev. Plant Sci.* 10:33–61.

Gray, D. J., R. N. Trigiano, and B. V. Conger. 1992. Liquid culture of orchardgrass somatic embryos and their use in synthetic cultivar development. SynSeeds: *Applications of Synthetic Seeds to Crop Improvement.* K. Redenbaugh (Ed.), pp. 351–366, CRC Press, Boca Raton, FL.

Gray, D. J., R. N. Trigiano, and B. V. Conger. 1994. Classroom exercises in the study of orchardgrass somatic embryogenesis. *HortTechnology* 4:322–324.

Hanning, G. E. and B. V. Conger. 1983. Embryoid and plantlet formation from leaf segments of *Dactylis glomerata* L. *Theor. Appl. Genet.*, 110: 121–128.

Trigiano, R. N. and B. V. Conger. 1987. Regulation of growth and somatic embryogenesis by proline and serine in suspension cultures of *Dactylis glomerata. J. Plant Physiol.* 130:49–55.

Trigiano, R. N., D. J. Gray, B. V. Conger, and J. K. McDaniel. 1989. Origin of direct embryos from cultured leaf segments of *Dactylis glomerata. Bot. Gaz.* 150:72–77.

25 Direct Nonzygotic Embryogenesis from Leaves and Flower Receptacles of Cineraria

R. N. Trigiano, M. C. Scott, and K. R. Malueg

The florists' cineraria, *Senecio x hybridus* Hyl., is a daisy-like flowering pot plant and a species belonging to the Asteraceae. The most notable horticultural attributes of the plant are its flowers that are produced in a wide range of colors including white, pink, red, purple, maroon, magenta, and several striking hues of blue. Cineraria is a perennial, but usually grown as an annual, and is very well-suited to low light intensities typical of winter production in the Northern hemisphere. Cineraria is usually considered to be a minor crop of little economic importance and is under-utilized as a floricultural crop. However, it is an excellent subject to demonstrate direct somatic (nonzygotic) embryogenesis (Chapters 22 and 23). The following two exercises will illustrate the effects of medium composition on the direct formation of somatic embryos from either leaves or flower receptacles of cineraria.

GENERAL CONSIDERATIONS

GROWTH OF PLANTS

Cineraria seeds are sown sparingly, about 4–5 seeds per inch, in very shallow trenches of prewetted vermiculite or other suitable seedling mix and then lightly covered with medium. We have found that a 11 × 21 inch plastic flat of seedlings will provide enough material for a class of 10–15 students. Cover the trays with plastic wrap and incubate at 21°C with about 50 $\mu mol \cdot m^{-2} \cdot s^{-1}$ of light provided by cool fluorescent tubes. If a growth chamber is unavailable, a benchtop in a cool laboratory will work well. The plastic wrap should be removed after the seedlings emerge in about 10 to 14 days. The first true leaves will be formed after an additional one to two weeks and are suitable for explants. Flower formation requires about 26–28 weeks of growth, including some time in a cooler. Five weeks after sowing, seedlings are transplanted into bedding plant cell packs. After 4 weeks, plants are moved up to 4-inch pots (any soilless medium will do) and, after an additional four weeks of growth, plants are placed in a cooler at 15°C. The plants are moved after six weeks to a cool greenhouse (about 18°C) and should flower between 8–10 weeks later (see Larsen, 1985, for scheduling). For receptacle explants, choose flower buds that are showing slight color but are more or less tightly closed.

EXPLANT PREPARATION AND BASAL CULTURE MEDIA

The first true leaves, including the petiole, may be harvested for explants about 2 weeks after the seedlings emerge. Water plants well the evening before the experiment is to begin. This will ensure that the leaves are fully hydrated and turgid. Surface disinfest the leaves by first immersing them in 70% ethanol for 30 s, followed by gently agitating then in 10% bleach solution for 10 min, and finally rinsing them three times with sterile distilled water. Excise and discard the petiole and bisect

TABLE 25.1
Example of Data Analysis Using a t-test for Paired Variates

Treatments	Mathematical Procedures				
Pair number	"A"[a]	"B"[a]	D = A–B	$D–D_m$[b]	$(D–D_m)^2$
1	1	9	–8	–2	4
2	0	9	–9	–3	9
3	1	12	–11	–5	25
4	1	6	–5	1	1
5	0	0	0	6	36
6	2	5	–3	3	9
			–36	0	84

Note: Calculations: 1. $D_{mean} = -36/6 = -6$; 2. standard deviation $(84/5)^{1/2} = (16.8)^{1/2}$; 3. standard error $(16.8)^{1/2}/(6)^{1/2} = (2.8)^{1/2} = 1.65$; 4. $t = (-6 - 0)/1.65 = -3.64$: Note "0" is used to include the null hypothesis of no difference between treatments; 5. For 5 degrees of freedom and $p = 0.05 = -2.57$ (see Snedecor and Cochran, 1967); 6. Since $t = -2.57$ is greater than -3.64, the result is significant; explants cultured on treatment "B" produced significantly more embryos than were produced on treatment "A."

[a] Number of somatic embryos.

[b] Mean difference between treatments.

the leaves (lamina) longitudinally (lengthwise) into two equal halves. If receptacle tissue is to be used as explants, flower buds with about 2 cm of the subtending pedicle (stem) should be immersed in 70% ethanol for 1 min and then the alcohol ignited by quickly passing the bud through a flame. Soak the buds in 20% Clorox for 10 min followed by three rinses in sterile distilled water. Using aseptic technique, remove the subtending pedicle and strip the calyx (green portion) and all the florets from the receptacle. Cut the receptacle into two equal halves. An alternative disinfecting treatment is to immerse the naked receptacle in 10% Clorox solution for 5 min followed by three rinses with sterile water before bisection.

There are two treatments in both of the experiments included in this chapter, therefore the exercises may be set-up as paired-variate designs. In this experimental design, we make the assumption that an individual leaf or flower receptacle is composed of relatively homogenous tissue in which any portion (e.g., a leaf half) is capable of responding to any treatment in a similar manner compared to the corresponding sister half. Therefore, if the sister halves are placed on different treatments, variations in responses should be attributable to the treatment and not to the differences in the ability of the tissue to respond or to other extraneous factors. This is a very powerful experimental design and eliminates the need for extensive replication. Regardless of the type of tissue used in these experiments, place one-half of the explant onto one medium (treatment) and place the corresponding sister half onto the other (this is the "pair" in the paired-variate design). Label the first petri dish A-1 and the second B-1, where the letter indicates the treatment and the numeral designates the replicate or explant. Data (e.g., the number of somatic embryos) should be analyzed using a t-test for paired-variates (Snedecor and Cochran, 1967, or other comparable statistics textbook). Also consult Table 25.1 for an example of the analysis.

The culture media in the following two experiments are composed of either Murashige and Skoog (MS) (1962) or Schenk and Hildebrandt (SH) (1972) basal salts (see Chapter 3 for composition) supplemented with 30 g (88 mM) sucrose, 1 g (0.55 mM) myo-inositol, 5 mg (2.9 μM) thiamine-HCl, 3 mg (13.5 μM) 2,4- dichlorophenoxyacetic acid (2,4-D), 1 mg (4.5 μM) benzyladenine (BA) and 8 g of phytagar per liter. The pH of the medium is adjusted to 5.8 before sterilization and approximately 10 mL of medium poured into each 60 × 15 mm plastic petri dish.

Experiment 1. The Effect of Basal Medium on Initiation of Somatic Embryos

The composition of the basal medium can have dramatic effects on the response of an explant in culture. This simple experiment is designed to test the ability of cineraria explants to form somatic embryos on two commonly used tissue culture media, SH and MS. The experiment will require about four weeks to complete.

Materials

The following items are needed for each student or team of students.

- Six seedlings or six flower receptacles
- Twelve 60 × 15 mm—6 each containing SH (treatment "A") and MS medium (treatment "B") and both media supplemented with 13.5 μM 2,4-D and 4.5 μM BA
- 10 and 20% Clorox solutions + 0.1% Triton X-100

Follow the experimental outline provided in Procedure 25.1 to complete this exercise.

Anticipated Results

Both leaf explants should begin to curl and become somewhat contorted after 4–5 days in culture. Late in the second week of culture, explants grown on SH medium or treatment "A" (Figure 25.1a) will produce some crystalline-looking callus and occasionally, a somatic embryo, whereas numerous globular-stage yellow somatic embryos will begin to form directly from explants cultured on MS medium or treatment "B" (Figure 25.1b). Four weeks after culture initiation, leaf portions cultured on SH medium will have produced abundant callus, whereas receptacle explants will be curled with little evidence of callus growth (Figure 25.1c). Explants cultured on MS medium will usually be covered with yellow to green globular to heart-shaped somatic embryos that may be beginning callus (Figure 25.1d). Very few, if any, morphologically mature embryos with well-developed cotyledons will be present.

Questions

- Why did MS and not SH medium support the development of somatic embryos? Hint: Consider the nitrogen sources and total salt concentration of each medium.
- Why was the ontogeny of somatic embryos on MS medium arrested in the globular or heart-shaped stage of development? What are some culture strategies that would permit continued morphological development?

Procedure 25.1
The Effect of Basal Medium on Initiation of Somatic Embryos

Step	Instructions and Comments
1	Surface disinfest leaves or flower receptacles with ethanol and bleach as described under explant preparation.
2	Cut the leaves or receptacles into identical "sister halves" and place one half onto each of the experimental media. Label each of the dishes with the treatment "A" or "B" and be *absolutely* certain that the dishes containing the "sister halves" are labeled with the same number.
3	Incubate the cultures in the dark between 22°C and 25°C. Examine weekly for embryo development. If one culture becomes contaminated (e.g., A2), then the corresponding culture (B-2) should also be discarded.
4	After 2 and 4 weeks, count the number of somatic embryos formed on explants cultured on each of the treatments (media). We suggest that the class share and analyze the data using a t-test for paired variates (see Table 25.1) for an example.

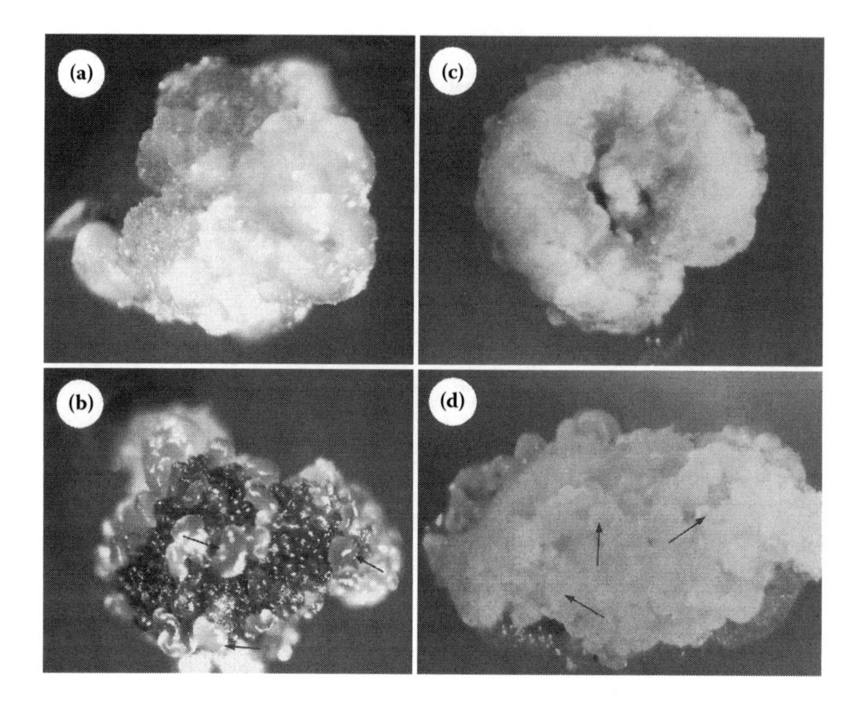

FIGURE 25.1 Somatic embryogenesis from leaf and flower receptacles of cineraria. (a) A leaf explant cultured on SH medium (treatment "A" does not produce somatic embryos). (b) The sister leaf half cultured on MS medium (treatment "B") produced numerous somatic embryos (arrows). (c) One-half of a flower receptacle cultured on SH medium becomes contorted, produces some callus, but does not generate somatic embryos. (d) The other half of the flower receptacle formed many embryos (arrows), which in this case, are beginning to callus. (From Trigiano R.N. and D.J. Gray, 2000, *Plant Tissue Culture Concepts and Laboratory Exercises, Second Edition*, p. 227, CRC Press LLC, Boca Raton, FL.)

Experiment 2. Maturation of Somatic Embryos and Regeneration of Plants

Somatic embryo initiation and development are often affected by nutrition, as demonstrated in the previous exercise, and also by plant growth regulators (PGRs). Sequential media transfer systems are often necessary to promote induction, maturation, and germination of somatic embryos (see Chapter 19). Typically, induction or primary medium for somatic embryogenesis contains relatively high concentrations of PGRs, especially auxin or auxin-like compounds, whereas the levels of PGRs in maturation medium are usually low, and in some, may even be excluded. Germination medium may be the same as maturation medium, may have reduced concentrations of basal salts and sugar, or may have incorporated into it low levels of some PGRs. The following exercise illustrates a three medium scheme to produce somatic embryos that can germinate and form plants and is adapted from a publication by Malueg et al. (1994). The experiment will require about 5 weeks for somatic embryo production or if acclimatized, regenerated plants are desired, about 11 weeks.

Materials

Each student or team of students will need the following materials:

- Eighteen 60 × 15 petri dishes containing MS medium amended with 13.5 µM 2,4-D and 4.5 µM BA
- Six 60 × 15 petri dishes containing MS medium without PGRs, but supplemented with 0.5% (5 g/L) activated charcoal (AC)
- Six 60 × 15 petri dishes containing MS medium without PGRs
- Three GA-7 vessels containing MS medium without PGRs

Procedure 25.2
Maturation of Somatic Embryos and Regeneration of Plants

Step	Instructions and Comments
1	Surface disinfest explants with ethanol and/or bleach as indicated under explant preparation.
2	Bisect explants and place one half on MS medium with PGRs and label "A-1" and the sister half into another dish containing the same medium, but label "B-1." The letters designate different treatments. We suggest that each team prepare six replications and incubate them in the dark at 22–25°C for 2 weeks.
3	Transfer the explants from the petri dishes labeled "A" to medium containing activated charcoal (AC) only and label with "A" and the explant number. Incubate as before for 3 days, then transfer explants to MS medium without PGRs.
4	Transfer the explants from "B" treatment to fresh MS medium with PGRs. Be sure to retain the original lids with labels on all the new petri dishes.
5	Incubate all cultures for an additional 3 weeks in the same conditions described above. Count and categorize somatic embryos by developmental stage (i.e., globular, cotyledonary) for each of the treatments. Analyze data by embryo development stage and total number of embryos using a t-test for paired variates (see Table 25.1).
6	Remove cotyledonary-stage somatic embryos from the explants and transfer to MS medium without PGRs contained in GA-7 vessels. Incubate in 25–75 $\mu mol \cdot m^{-2} \cdot s^{-1}$ at 22°C. When plants are about 2 cm tall, transfer to moist Jiffy 7 peat pellets and cover with plastic bags.
7	After 2 or 3 weeks, plants can be acclimatized to ambient conditions by opening the bags for increasingly longer intervals over a 2-week period. Plants can then be potted in Fafard No. 2 potting medium and grown as discussed under General Considerations, Growth of Plants.

- Jiffy—7 Peat Pellets (Jiffy Products of America, Inc., Chicago, IL)
- Fafard number 2 potting mix (Conrad Fafard, Inc., Agawam, MA)
- Small plastic bags

Follow the experimental protocol outlined in Procedure 25.2 to complete this exercise.

Anticipated Results

Somatic embryos should form directly on all explants capable of responding to the treatments. However, there should be more developmentally advanced (torpedo and cotyledonary stage) embryos on explants that were transferred through the series of three media. Most embryos formed on explants that remained on the initial medium with PGRs should be callused as shown in Figure 25.1d.

About one third of the cotyledonary stage embryos transferred to MS without PGRs should produce a radicle and then a shoot in less than 10 days; the others will only form a radicle or will never germinate. Most of the plants derived from somatic embryos will have multiple stems, should acclimatize easily and produce normal colored and shaped, but smaller flowers.

Questions

- Why do some or most germinating embryos develop multiple shoot apices?
- Why do less than one-half of the embryos placed on germination medium form plants?
- How would you determine if somatic embryos were formed directly or indirectly?

REFERENCES

Larsen, R. 1985. Temperature and flower induction of *Senico x hybridus* Hyl. *Swed. J. Agric. Res.* 15:87–92.

Malueg, K. R., G. L. McDaniel, E. T. Graham, and R. N. Trigiano. 1994. A three media transfer system for direct somatic embryogenesis from leaves of *Senico x hybridus* Hyl. (Asteraceae). *Plant Cell, Tiss. Org. Cult.* 36:249–253.

Murashige, T. and F. Skoog. 1962. A revised medium for rapid growth and bioassays with tobacco tissue cultures. *Physiol. Plant.* 15:473–497.

Schenk, R. U. and A. C. Hildebrandt. 1972. Medium and techniques for induction and growth of monocotyledonous and dicotyledonous plant cell cultures. *Can. J. Bot.* 50:199–204.

Snedecor, G. W. and W. G. Cochran. 1967. *Statistical Methods*, Sixth Edition. The Iowa State University Press, Ames, Iowa, 593 p.

Section IV

Crop Improvement Techniques

26 Protoplasts—An Increasingly Valuable Tool in Plant Research

Jude W. Grosser and Ahmad A. Omar

CONCEPTS

- Protoclonal variation is an excellent source of genetic variation that can lead directly to new cultivars with improved traits such as seedlessness and expanded maturity dates.
- Somatic hybridization has become a key tool for building superior breeding parents needed to maximize genetic variability in progeny for selection and for use in interploid crosses to generate seedless triploids.
- Somatic cybridization has become an effective tool for organelle transfer to improve disease resistance and other important traits.
- Protoplast transformation is an alternative to *Agrobacterium*-mediated transformation, resulting in transgenic plants not containing antibiotic resistance genes.
- Protoplast-based transient assays and studies of protein–protein interactions are making significant contributions in plant physiology.
- Plant pathology is benefiting from protoplast-based disease screening assays that often are much quicker than traditional assays.
- In the future, protoplast-based strategies are expected to enhance interdisciplinary plant science research and facilitate functional genomics studies made possible by genome-sequencing projects.

Plant protoplasts are defined as cells that have had their rigid cellulose cell wall removed without damaging the external cell membrane that surrounds the nucleus and cytoplasm. Under appropriate osmoticum conditions, protoplasts are perfectly spherical in shape. Initial efforts to isolate protoplasts began in the late 1800s using mechanical methods, but these met with limited success. The extraction of cellulase, macerase, and pectinase enzymes from various fungi during the mid-1900s provided new opportunities for protoplast isolation via enzymatic digestion, resulting in the landmark paper on plant protoplast isolation by Edward C. Cocking (Cocking, 1960), who used a crude cellulase preparation to isolate protoplasts from roots of tomato seedlings. Cell-wall digestion is generally carried out in a cocktail solution of enzymes and elevated osmoticum (for plasmolysis, usually sugars or sugar alcohols), empirically determined for any particular species/explant. Protoplasts can now be routinely isolated on a large scale from the most important crops and ornamental species from nearly any plant part or cultured cells. Protoplasts have been used in many applications in plant biotechnology and molecular biology research, including somatic hybridization and cybridization, transformation, transient assays, protein–protein interactions, and screening for disease resistance. This chapter will discuss practical applications of evolving plant protoplast technologies.

PROTOCLONAL VARIATION

Somaclonal variation is defined as variability in plants regenerated from tissue culture that is either induced or uncovered by a tissue-culture process (Larkin and Scowcroft, 1981). Most somaclonal variation is negative, but if enough plants are examined, positive changes can usually be recovered. It was immediately recognized that plants regenerated from protoplasts frequently display phenotypic diversity. Such somaclonal variation has also been termed *protoclonal variation*. The generation of positive variation without the need for a complex breeding strategy can be useful in plant improvement, particularly where it is difficult or impossible to maintain cultivar integrity using conventional breeding approaches. Such is the case with efforts to improve sweet orange and lemon. Initial reports of positive somaclonal variation have been primarily from solanaceous or cereal crops, affecting a wide range of traits, including plant height, overall growth habit, flower, fruit and leaf morphology, juvenility, maturity date, disease resistance, yield, and biochemical characteristics. More recently, long-term studies of somaclonal variation in the woody perennial fruit crops, sweet orange, and lemon (including protoclones). Grosser et al. (2007) have revealed similar variation. Somaclonal variation was compared in populations of "Hamlin" and "Valencia" sweet oranges regenerated from callus cultures and protoplasts via somatic embryogenesis and nucellar stem pieces via organogenesis. Although the majority of the sweet orange somaclones appear to be normal, significant stable variation has been observed in tree and fruit characteristics. Altered tree characteristics included canopy size/shape, leaf size/shape, ploidy level, juvenility/thorniness/vigor, and fruit yield. Altered fruit characteristics included brix (soluble solids)/acid ratio, color (fruit/juice), maturity date, size, rind thickness, rag, juice content, and seediness. The genotypic differences in the level of variation were observed in the regenerated plants, as Valencia populations were much more variable than Hamlin populations. It was thus clear that the protoplast-derived populations exhibited higher levels of useful variation. Several new Valencia sweet orange clones will be released for commercial use as a result of this study, including a seedless protoclone (Figure 26.1), a protoclone that matures 6–8 weeks earlier than standard Valencia, a somaclone with significantly higher yield than standard Valencia, and a protoclone that has significantly higher brix and soluble solids than standard Valencia. Thus, protoclonal variation can be a useful breeding tool, particularly in efforts to improve crops such as sweet orange that are not amenable to conventional breeding. Chemical mutagens such as ethyl methanesulfonate (EMS) have also been tested to augment the genetic variability observed in protoclone populations, but the results from a study with *Brassica napus* did not support this approach (Jain and Newton, 1988).

FIGURE 26.1 Late maturing seedless Valencia sweet orange protoclone.

SOMATIC HYBRIDIZATION AND CYBRIDIZATION

It was quickly realized that the fusion of protoplasts from different parental partners could result in novel additive combinations of genetic material even when the partners were sexually incompatible, potentially overcoming barriers to sexual hybridization often encountered in breeding programs (Figures 26.2 and 26.3). The term *somatic hybrid* was coined to describe novel hybrids obtained from the fusion of somatic cell protoplasts from two different parents. The first somatic hybrid plants were regenerated in tobacco by Carlson et al. (1972). Since then, somatic hybrid plants have been regenerated from hundreds of parental combinations. Thus, plant somatic hybridization via protoplast fusion has become an important tool in plant improvement, allowing researchers to combine somatic cells from different cultivars, species, or genera, resulting in novel genetic combinations including symmetric allotetraploid somatic hybrids, asymmetric somatic hybrids, or somatic cybrids (defined as a plant with the nucleus from one parent and the mitochondrial genome and/or chloroplast genome of the second parent). This technique can facilitate breeding and gene transfer

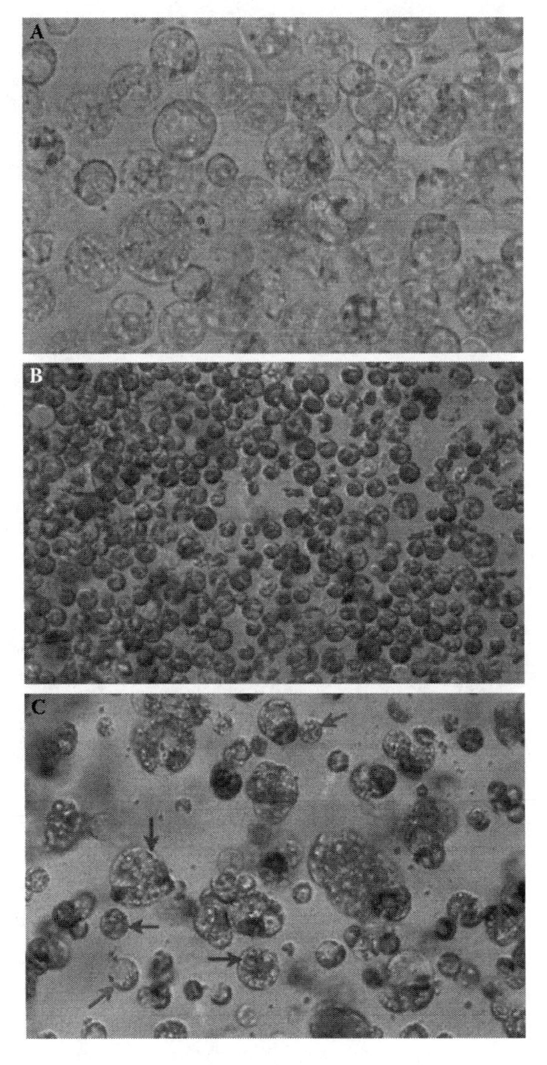

FIGURE 26.2 Panel A: Protoplasts isolated from embryogenic suspension culture. Panel B: Mesophyll protoplasts isolated from citrus leaves. Panel C: Protoplast fusion; red arrows indicate suspension protoplasts, green arrows indicate mesophyll protoplasts, and blue arrows indicated fused cells.

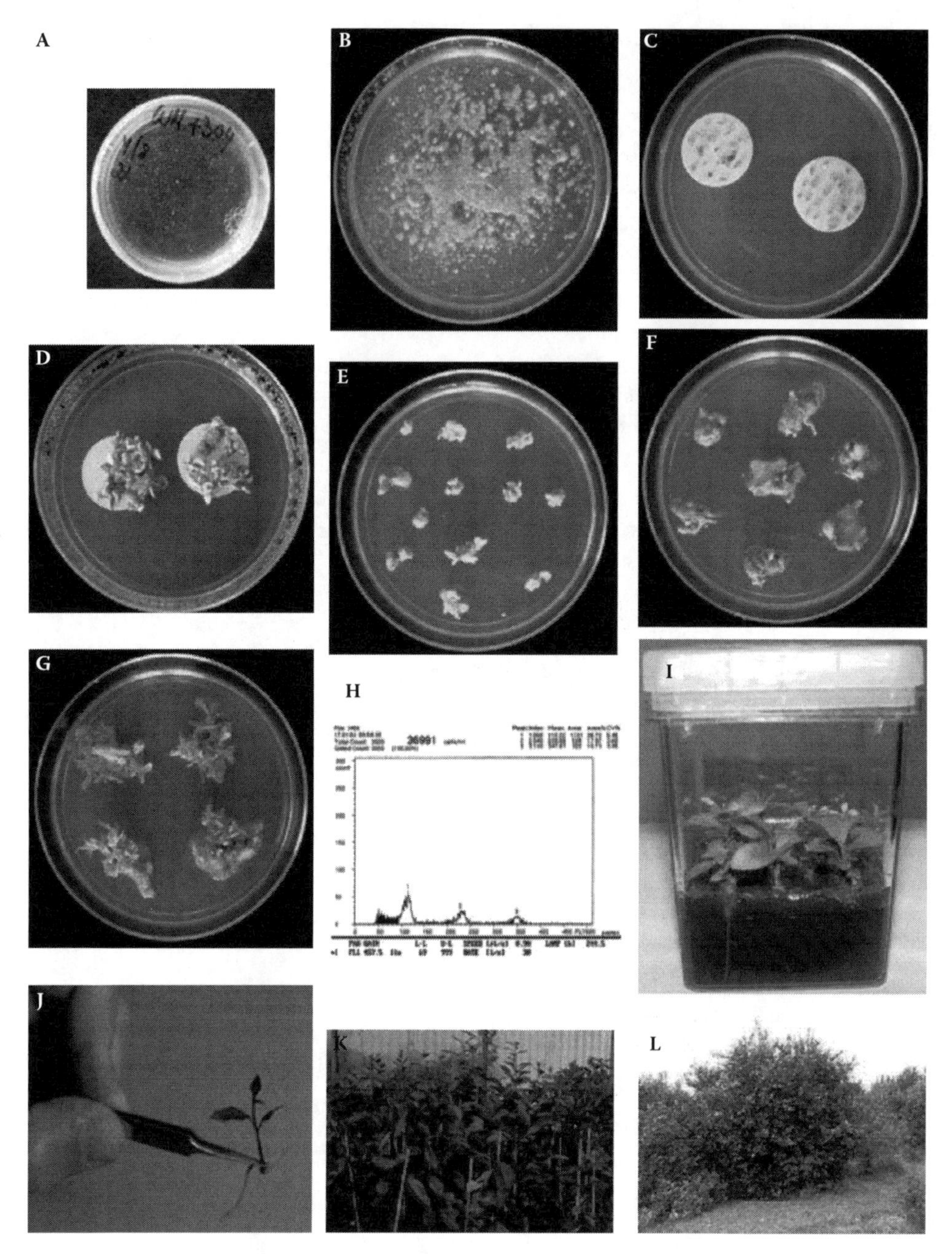

FIGURE 26.3 Regeneration of somatic hybrid plants using protoplast fusion in citrus. Panel A: Protoplasts fused by the PEG method, cultured directly in the fusion Petri dishes in a thin layer of BH3:EME (1:1 v:v) medium. Panel B: Protoplast-derived calli and somatic embryos growing on EME-maltose solid medium 6 to 8 weeks after fusion. Panels C and D: Somatic embryos growing on cellulose acetate membranes laid on EME-maltose medium 8–10 weeks after fusion. Panel E: Somatic embryos enlargement growing on 1500 medium 3–4 months after fusion. Panel F: Somatic embryos germinated on B+ medium 4–5 months after fusion. Panel G: Somatic embryos on DBA3 medium for shoot induction 5–6 months after fusion. Panel H: Ploidy determination by flow cytometry analysis showing diploid, triploid, and tetraploid peaks using Partec (Munster, Germany) tabletop flow cytometer. Panel I: Somatic hybrid plantlets on RMAN medium. Panel J: New somatic hybrid plant ready for transfer into soil. Panel K: Somatic hybrid plants growing in the greenhouse. Panel L: Somatic hybrid tree growing in the field (allotetraploid scion-breeding parent).

by bypassing problems sometimes associated with conventional sexual crossing, including sexual incompatibility, long generation times, nonoverlapping flowering, polyembryony, and male or female sterility. During the past 35 years, many reports have been published that extend the procedures to additional plant genera and tomato, potato, and citrus (Davey and Kumar, 1983; Gleba and Sytnik, 1984; Grosser and Gmitter, 1990, 2005; Grosser et al., 2000; Johnson and Veilleux, 2001, Ollitrault et al., 2007; Orczyk et al., 2003; Waara and Glimelius, 1995). Somatic cybridization and in vitro breeding organelle inheritance has also been recently reviewed (Guo et al., 2004a).

Somatic Hybrid Selection

Many approaches have been developed and tested to facilitate the selection of somatic hybrid and cybrid cells/plants, separating them from cells/plants regenerating directly from nonfused protoplasts. Selection schemes have been reviewed by Davey and Kumar (1983) and include the following: use of selective media for preferred growth of hybrid cells; complementation selection schemes that use albino mutants, auxotrophic mutants, natural or mutant lines resistant to antibiotics, cells resistant to amino acid analogs, cells resistant to toxins; use of irradiated protoplasts or cytoplasts; selection based on identifiable markers; or mechanical isolation of heterokaryons. More recently, it has been shown in citrus that heterokaryon-derived cells exhibit hybrid vigor, giving them a competitive advantage in subsequent plant regeneration (Guo and Grosser, 2005). Combining this factor with the use of routinely available tabletop flow cytometers (Partec tabletop flow cytometer, Model D-48-161, Munster, Germany), which quickly determine ploidy level, eliminates the need for sophisticated selection schemes used previously to facilitate somatic hybrid recovery (Ananthakrishnan et al., 2006). Regenerants can be efficiently screened at the somatic embryo or plantlet stage for the expected somatic hybrid ploidy level.

Applications

Applications of somatic hybridization to crop and ornamental improvement continue to evolve. Initially, the majority of somatic hybridization experiments targeted gene transfer from wild accessions to cultivated selections that were either difficult or impossible to hybridize by conventional methods, including intergeneric combinations (Dudits et al., 1980; Grosser et al., 1996). Transfer of disease-resistant genes has been a primary target (Ananthakrishnan et al., 2006; Fock et al., 2000). Since symmetric somatic hybridization is an additive process, the resulting polyploid products may also receive many undesirable genes from the donor parent. Due to this, it is rare that a somatic hybrid may have direct use as an improved cultivar (Guo et al., 2004a). This is not the case with rootstocks that merely provide the root systems for commercial trees. Tetraploid citrus rootstocks budded with typical diploid commercial scions always show some level of tree size reduction (usually 29%–85% the size of trees budded to comparable diploid rootstocks), by some yet-to-be-determined mechanism (Grosser and Gmitter, 2005). Thus, tetraploid rootstocks have great potential in emerging citrus production and harvesting systems that feature high-density plantings and early fruit production. Our initial approach to rootstock improvement using somatic hybridization was to generate allotetraploid somatic hybrids that combine elite complementary diploid rootstocks (Grosser and Gmitter, 1990). Numerous somatic hybrid rootstocks have been produced and evaluated in our ongoing rootstock improvement program, and two widely adapted, tree-size controlling rootstocks will be released to the industry during the next two years. This technology could be extended to other crops that utilize high-density planting schemes, such as apple and stone fruits.

More recently, it has become clear that the most important application of somatic hybridization in citrus is the building of novel germplasm as a source of elite breeding parents for subsequent use in various types of conventional crosses, for both scion and rootstock improvement. Citrus somatic hybridization is generating key allotetraploid breeding parents for use in interploid crosses to generate seedless triploids (Grosser and Gmitter, 2005). This approach should be extended to other important crops where seedlessness is a priority breeding objective, including table grapes and banana. For example, there are known seedy diploid banana genotypes that have resistance to the major diseases affecting commercial triploid banana culture. These disease-resistant selections could be used to generate allotetraploid somatic hybrid banana breeding parents for subsequent use in interploid crosses to generate disease resistant triploids. Autotetraploids can also be used in interploid crosses for triploid production and they can rarely be obtained as a byproduct of somatic hybridization experiments, and occasionally recovered from direct protoplast culture. However, the use of allotetraploids in interploid crosses offers the advantage of generating much more genetic variability in triploid progeny for selection. Successful somatic hybridization in citrus rootstock improvement has allowed the creation of a rootstock breeding program at the tetraploid level that can generate maximum genetic diversity in zygotic progeny and that has great potential for developing rootstocks to control tree size (Grosser et al., 2003). Thus, the numerous attributes required for development of an improved rootstock can now more easily be packaged together into individual hybrids.

Asymetric Hybridization

As mentioned, somatic cybridization is the process of combining the nuclear genome of one parent with the mitochondrial and/or chloroplast genome of a second parent. Cybrids can be produced by the donor–recipient method (Vardi et al., 1987), cytoplast-protoplast fusion (Lorz et al., 1981), or simply as a by-product of symmetric somatic hybridization (Grosser et al., 1996). A primary target of somatic cybridization experiments has been the transfer of cytoplasmic male sterility (CMS) to facilitate conventional breeding (Sigareva and Earle, 1997), or to reduce the seed content of fruit (Guo et al., 2004b). More recently, preliminary experiments suggest the successful intergeneric transfer of citrus canker resistance encoded in the mitochondrial genome of kumquat to susceptible citrus cultivars via somatic cybridization (Grosser et al., 2008).

Incomplete Asymmetric Somatic Hybridization

Somatic hybridization technology also provides opportunities for transfer of fragments of the nuclear genome, including one or more intact chromosomes from one parent (donor) into the intact genome of a second parent (recipient; Varotto et al., 2001). It is also possible to isolate adequate quantities of microprotoplasts to facilitate microprotoplast-mediated chromosome transfer (MMCT) with subsequent plant regeneration in various crop species (Binsfeld et al., 2000; Louzada et al., 2002; Ramulu et al., 1995). However, a high percentage of regenerated plants from this technique exhibit reduced vitality. Additional research is expected to make this a viable technique for partial genome transfer in crop improvement.

PROTOPLAST TRANSFORMATION

Since protoplasts ("naked cells") are surrounded only by a cell membrane, they can be manipulated in a variety of ways with the advantage of serving as single-cell targets. Protoplasts are frequently obtained from established suspension cell lines initiated from immature embryos, immature inflorescences, mesocotyls, immature leaf bases, or anthers (Maheshwari et al., 1995). Protoplasts can either be transformed by *Agrobacterium* or by direct DNA uptake methods, facilitated by polyethylene glycol (PEG) treatment, electroporation, or liposomes (Shillito, 1999). DNA uptake into protoplasts is now a routine and universally accepted procedure in plant biotechnology for introducing and evaluating both short-term (transient) and long-term (stable) expression of genes in cells and regenerated plants. Moreover, direct DNA uptake into plant cells has been especially important in transforming plants that are not amenable to other methods of gene delivery, particularly,

Agrobacterium-mediated transformation. Protoplasts are ideal candidate cells for direct DNA uptake and the subsequent selection of transgenic events. DNA can be delivered into protoplast cells by either PEG-mediated chemical induction or electroporation. PEG-mediated transformation is simple and efficient, allowing a simultaneous processing of many samples, and yields a transformed cell population with good survival and cell division rates. The method utilizes inexpensive supplies and equipment and helps to overcome an obstacle of host range limitations of *Agrobacterium*-mediated transformation. PEG-mediated DNA transfer can be easily adapted to a wide range of plant species and tissue cultures. The major disadvantage of the PEG-mediated DNA transfer system is the need for large quantities of isolated plasmid DNA.

Electroporation is used to produce stable genetic transformants using protoplasts as target cells. This technique has been used in many plant species, including sugarcane, pea, maize, wheat, tobacco, and citrus to produce stable transgenic plants (Bates, 2008). Although electroporation seems to be an extremely simple and effective method, it has not yet been widely used for plant transformation. Electroporation conditions were optimized for transfection of protoplasts isolated from an embryogenic line of Hamlin sweet orange (*Citrus sinensis*), and stable transgenic plants were obtained (Niedz et al., 2003).

Protoplasts have been transformed with different DNA plasmids such as Ti plasmid from *Agrobacterium tumefaciens*, and genes carried on a simple *E. coli*–based cloning vector that confirmed that Ti-DNA borders were not important for DNA integration into the plant genome (Davey et al., 1989). Protoplast transformation efficiency for recovery of transgenic events is higher because cross-feeding and chimerism between transgenic and wild-type cells are minimized in comparison to transformation systems based on multicellular tissues. However, protoplast transformation frequencies remain low (one in 10^4 protoplasts that develop into stable transformed tissue), and protoplast-to-plant systems with efficient selection need to be improved for many species to recover transformed cells and tissues.

Some treatments could enhance transformation frequency, such as heat shock treatment and irradiation of recipient protoplasts, probably by increasing the recombination of genomic DNA with incoming foreign DNA, or the initiation of repair mechanisms that favor integration. DNA fragment size, carrier DNA, and the nature of the plant genome could also influence transformation frequencies. Large binary vectors used in *Agrobacterium*-mediated transformation for stable transformation often results in a poor transformation frequency when used in protoplast transformation. The smaller the plasmid size used in protoplast transformation, the higher the transformation rate. As nucleases may block DNA uptake into isolated protoplasts, experiments have been undertaken to reduce DNA damage during transformation. Folling et al. (1998) studied PEG-mediated DNA transfer into protoplasts of *Lolium perenne* and reported that plasmids were protected by a combination of high pH (9.0) and reduced temperature (0°C), since such conditions suppressed DNA nicking and improved transformation efficiency. The same authors showed that two nucleases usually associated with isolated protoplasts were involved with DNA degradation, with one being released into the medium and the other localized to the plasma membrane.

Since plants regenerated from protoplast come from a single cell, all cells in the transgenic plant are expected to contain inserted genes of interest. Protoplasts can be cotransformed with more than one gene carried on the same or separate plasmids (Omar et al., 2007). There are many factors that influence protoplast transformation, with the stage of the cell cycle probably being the most important factor. In citrus, newly initiated embryogenic suspension culture lines are preferred because they exhibit rapid proliferation and routinely provide reasonable yields of good-quality protoplasts with high regeneration capacities. Optimal protoplast performance is achieved when isolation is attempted from days 4–10 into a 2-week subculture period. Lines in culture for extended periods of time tend to build up mutations (negative somaclonal variation) that reduce totipotency, and thus, transformation efficiency. We have also observed genotypic effects on culture stability in sweet orange, with protoplast isolation and plant regeneration from "Hamlin" being more consistent over time than with other sweet orange cultivars.

DNA can be delivered into protoplasts isolated from different plant species using several different techniques. There are many studies utilizing protoplast procedures for efficient delivery of plasmids into suspension culture–derived protoplasts and optimization of protoplast-to-plant systems. Many such studies focused on cereals, particularly rice (Davey et al., 2005). PEG has been used to induce DNA uptake into protoplasts isolated from several plant species, which include tobacco (*Nicotiana tabacum*), *Arabidopsis thaliana*, *Datura inoxia*, wheat (*Triticum aestivum*), rice, barley, sugarbeet, apple, sweet potato, and citrus (Davey et al., 2005).

Successful single plasmid transformation and cotransformation of "Hamlin" sweet orange protoplasts with a *Xa21* gene from Rice (*Oryza sativa*) conferring resistance to *Xanthomonas oryzae* and green fluorescent protein (*GFP*) as a reporter gene, using PEG-mediated, direct DNA uptake, has been achieved (Omar et al., 2007) by following the protocol outlined in Figure 26.4. Over 1,000 transgenic plants from 80 independent transformation events have been obtained from cotransformation (using

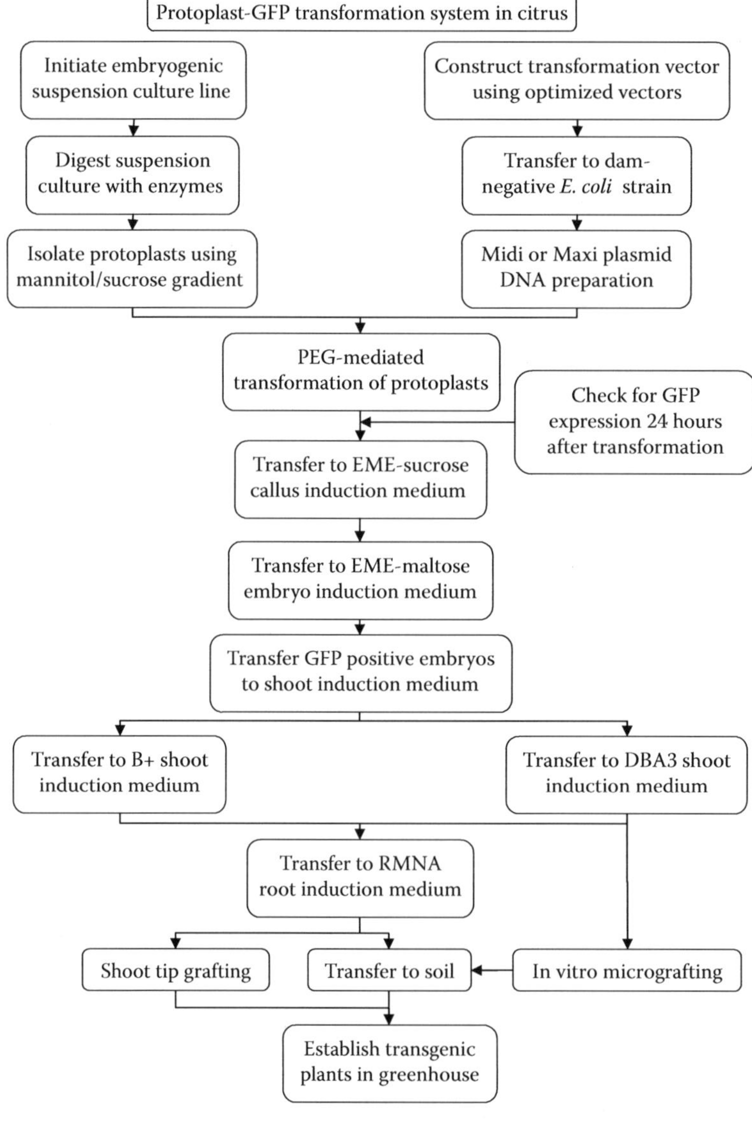

FIGURE 26.4 Schematic representation of the work-flow of the protoplast–GFP transformation system in citrus.

two genes in two separate plasmids) experiments using shoot tip grafting and/or in vitro micrografting. A total of 75 transgenic plants were recovered via grafting from eight independent events from single plasmid transformation experiments. The typical morphology of the regenerated transgenic "Hamlin" sweet orange plants was similar to that of the wild-type plants recovered from control cultures (Figure 26.5). This was the largest population of transgenic plants from a woody perennial fruit tree—"Hamlin" sweet orange, containing two foreign genes (*Xa21* and *GFP*), produced using the protoplast–GFP transformation system. The efficiency was comparable to that of *Agrobacterium*-mediated citrus transformation. Although more plants were regenerated using the cotransformation (dual plasmid) system, the final efficiency for plants containing the gene of interest was low, compared with plants obtained from single plasmid transformation. Western blot analysis showed higher average accumulation levels of the Xa21 protein in transgenic plants from single plasmid transformation events (87.5%) versus plants from cotransformation events (24%).

As with other transformation methods, naked-DNA gene transfer results in only a small proportion of cells being stably transformed. Therefore, it is often necessary to improve the system by including genes that confer a selectable advantage during transformation. Traditionally, antibiotic resistance or herbicide resistance genes have been used for this purpose. These genes can either be introduced on a single plasmid DNA or on separate plasmid DNA (cotransformation). Any antibiotic-resistance or herbicide-resistance selectable genes are not recommended, and consumers prefer transgenic plants to be selectable marker free before they have been released into the environment. Therefore, it is desirable to develop systems that will eventually allow the removal of selectable-marker genes or will not require any antibiotic- or herbicide-resistance genes as selectable markers.

To improve transformation efficiency using the protoplast–GFP system in citrus, we compared two GFP constructs: one endoplasmic reticulum-targeted (ER-GFP) and the other cytoplasmic targeted (Cy-GFP; Omar and Grosser, 2008). We concluded that ER-GFP had brighter fluorescence than Cy-GFP. Although both constructs gave transient expression at the protoplast level, Cy- GFP gave less stable expression than ER-GFP at the microcalli or plant level. Stable expression of GFP after 4 weeks of culture was observed in 1.0% and 0.1% of the transient GFP positive cells in ER-GFP and Cy-GFP experiments, respectively.

Protoplast–GFP transformation has also been successful (although with lower efficiency) in "Valencia" sweet orange (*Citrus sinensis*) to produce transgenic plants that overexpress CsPME4, a sense gene cassette containing a gene-specific sequence from a putative thermostable pectin methylesterase (TSPME) cDNA. The goal of this work was to improve orange juice quality by eliminating or greatly reducing TSPME activity by downregulating this gene (Guo et al., 2005). "W. Murcott" mandarin (a hybrid of "Murcott" tangor and an unknown pollen-parent) is a commercially important fresh citrus cultivar grown in many regions around the world. Mandarin cultivars are generally considered to be the more recalcitrant to transformation using the *Agarobacterium*-mediated system. Successful protoplast–GFP transformation was recently reported for this important cultivar (Omar et al., 2008) and was possible because of the high-performance embryogenic suspension culture line available for this cultivar.

The protoplast transformation system offers several advantages as a tool to improve different plant species. One primary advantage is that antibiotic resistance genes for selection are not required at the cellular level. This provides an advantage over standard citrus transformation methodology using *Agrobacterium*, in which antibiotic-resistance genes are used for selection and to kill *Agrobacterium* following transformation. By using the protoplast–GFP transformation system, we can eliminate the use of *Agrobacterium*, antibiotic-resistance genes, and destructive assay systems associated with using GUS as reporter gene. Transgenic plants containing antibiotic-resistance genes are frowned on by consumers in several markets, especially in the European Union (Schaart et al., 2004). Another advantage of the protoplast transformation system is the ability to transform seedless polyembryonic citrus cultivars because of the explant type required. Efficient juvenile citrus transformation using *Agrobacterium*-mediated transformation requires seed to generate nucellar stem piece explants, and thus, seedless cultivars would not be amenable to this approach. Embryogenic suspension cultures are initiated from ovule-derived callus, so no seed are required.

TRANSIENT ASSAY SYSTEMS USING PROTOPLASTS

The production of transgenic plants has become routine in many laboratories around the globe, providing a powerful tool for studying gene function in plants. Studying gene function and characterization is usually conducted by (1) mutant screens, where an illustrative phenotype is required to elucidate

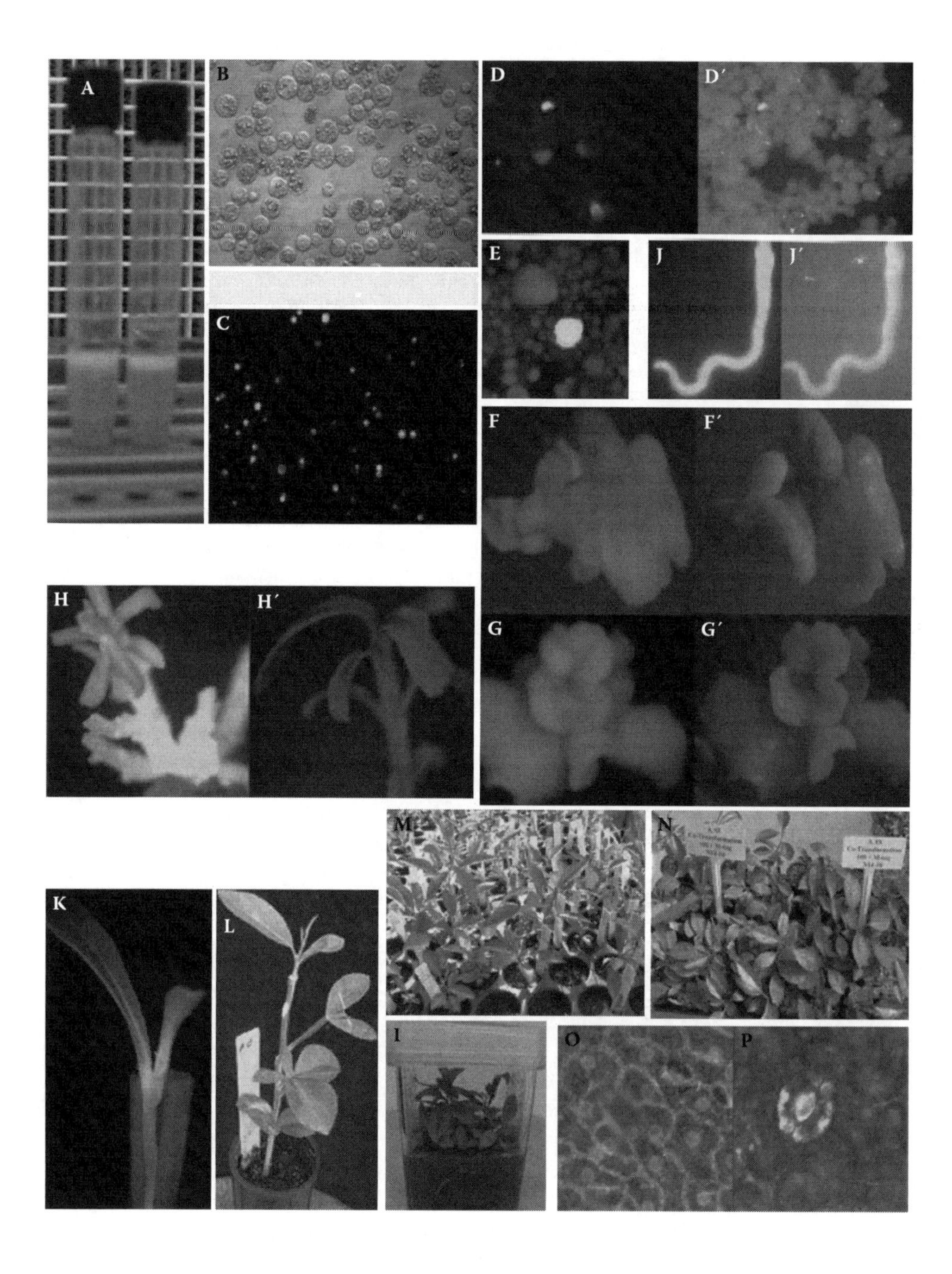

gene function, and (2) the insertion of a transgene into the plant chromosome through plant transformation. These methods are relatively expensive, labor intensive, and time consuming, and thus limit the utilization of this technology for large-scale analyses of plant genes. The use of transient assays offers an opportunity to study large numbers of genes quickly. Transient assays have been optimized for several dicots and monocots to isolate and manipulate protoplasts from different sources to study gene expression.

Transient assays also provide a convenient alternative system to study the functional expression of plant genes. Unlike conventional transgenic approaches, transient assays provide an advantage in that gene activity can be investigated easily and rapidly after DNA delivery. The transient assay based on viral infection or *Agrobacterium* infiltration of leaves is an effective method to study genes of interest and has been proposed for large-scale functional analysis of genes in both dicot (Liu et al., 2002) and monocot plants (Holzberg et al., 2002). Conversely, the use of transient gene expression assays offers an opportunity to study large number of genes quickly, which would be advantageous for evaluating the transcriptional activity of different promoters, and might be especially useful for assaying cell biology and cell-wall traits. Protoplast-based transient assay systems provide powerful tools for many types of assays in plants. They have proven very useful for dissecting a broad range of plant signal transduction pathways, transcriptional regulatory networks, and evaluation of reporter gene expression.

Transient assay using protoplasts has been established in several plant species such as *Arabidopsis*, maize, rice, and tobacco, and are being developed in many other species. In addition to showing a high transformation efficiency, plant protoplasts have also been proven to show similar reactions to that of intact tissues and plants to hormones, metabolites, environmental cues, and pathogen-derived elicitors, providing a powerful and adaptable cell system for high-throughput analyses of plant signal transduction pathways (Sheen, 2001). Since protoplasts can be isolated from different plant parts, transient assays based on protoplasts could be designed to address specific questions. Several studies reported that the transient assay based on mesophyll or cultured cell-derived protoplast has become a powerful tool for rapid gene functional analysis and biochemical manipulation in several plant species. However, in some cases the system has not been adapted due to difficulties in large-scale isolation of protoplasts from leaves or suspension culture cells from certain plant species. Protoplasts have been used as a tool in metabolite transport studies to the vacuole from the apoplast (Etxeberria et al., 2007; Pozueta-Romero et al., 2008). Protoplast transient assays have resulted in mitogen-activated protein (MAP) kinase and transcription factors acting downstream of FLS2, an *Arabidopsis* pathogen-recognition receptor, being characterized using *Arabidopsis* mesophyll protoplasts (Asai et al., 2002). More recently, protoplast transient assays have been automated based on robotic platforms to indentify two transcription factors involved in tobacco jasmonate signaling (De Sutter et al., 2005).

FIGURE 26.5 (Opposite) Regeneration of transgenic citrus plant and monitoring of *GFP* expression from protoplast to plant. Panel A: Protoplast ring in the interface between 25% (w/v) sucrose and 13% (w/v) mannitol. Panel B: Protoplasts after purification. Panel C: Protoplasts cotransformated with two plasmids or single transformed (one plasmid) by direct DNA uptake with PEG after 24 h. Panels D and D': Protoplast-derived calli (transformed or nontransformed) on EME-maltose solid medium 36 d after transformation, under blue light (D) or white light (D'). Panel E: Transgenic (green) and nontransgenic (red) somatic embryos growing on EME medium 6 to 8 weeks after transformation. Panels F and F': Nontransgenic somatic embryos growing on 1500 medium 2–3 months after transformation under blue light (F) or white light (F'). Panels G and G': Transgenic somatic embryos growing on 1500 medium 2–3 months after transformation under blue light (G) or white light (G'). Panels H and H': Embryo-derived transformed (H) and nontransformed (H') shoots on B+ medium 5–6 months after transformation. Panel I: In vitro rooted transgenic citrus plants constitutively expressing the *GFP* gene. Panels J and J': GFP expression in transgic roots under blue light (J) and white light (J'). Panel K: Micrografted transgenic scion (green) on Carrizo seedling (red). Panel L: Transgenic scion growing on Carrizo rootstock in soil 2 weeks after removing the tip. Panels M and N: Transgenic "Hamlin" sweet orange plants in soil. Panel O: ER-targeted *GFP* expression in mature citrus leaf. Panel P: Cytoplasmic targeted *GFP* expression in mature citrus leaf.

Genetic manipulation of the growth and development of grasses for biofuel production is needed for better cellulosic ethanol production, especially to improve cellulose-to-lignin ratios. The latest genomic and biotechnology tools can be used for the biomass production of designer plants for this purpose. There are several genes that can make significant improvements in agronomic and feedstock traits of grasses. This includes any traits that might alter cellulose levels, dwarfism, drought resistance, and pollen alterations that can be introduced via transgenesis. To study all of these genes in a timely manner, protoplast transient assays can be utilized to study gene function and expression in grasses, which could be of crucial importance to bioenergy and biotechnology development (Mazarei et al., 2008).

RNA interference (RNAi) has been shown to occur in many eukaryotes, including mammals and plants. RNAi is now used as a powerful tool for functional genomics in many eukaryotes because of its effectiveness and specificity compared to conventional methods such as antisense technologies. There are many RNAi vectors that can be used to introduce RNAi in plants. However, construction of RNAi vectors and regeneration of transformants can be laborious and time consuming. In addition, the commonly used CaMV 35S RNAi vector cannot be applied to genes whose suppression may result in lethal phenotypes. To avoid some of these problems, a transient gene silencing system using plasmid vectors, virus vectors, and in-vitro-prepared dsRNA has been developed. The introduction of dsRNA into protoplasts isolated from *Arabidopsis thaliana* led to marked silencing in target transgenes (An et al., 2003). Short-interfering RNA (siRNAs) have been investigated in 3-day-old tobacco suspension cell-derived protoplasts, targeting either GFP or red fluorescent protein (DsRed2) from Discosoma using plasmids expressing both reporter genes (GFP and DsRed2). Fluorescence was measured, and siRNA-mediated silencing resulted in a decrease in expression of 58% and 47%, respectively, of GFP and DsRed2 (Vanitharani et al., 2003).

Plant Protein–Protein Interaction Studies Using Protoplasts

As a result of many plant genomics projects initiated in the last decade, several plant genomes have been completely sequenced and global gene expression profiles have been obtained. This progress has provided a wealth of information and resources for functional genomics studies in plants. However, the underlying mechanisms that coordinate all cellular functions supporting the complexity of plant embryogenesis, growth, and response to the environment cannot be understood from only the knowledge of the primary sequences and identification of all proteins but rather from an understanding of gene function. Although stable transformation is currently available for studying gene function in plants, there is a lack of transient assay systems that can be used for large-scale analysis of gene function in plants.

Protein function is often studied via formation of both stable or transient complexes and networks. Protoplast transient assays can be used for in-vivo screening of protein–protein interactions that used to be investigated using the yeast two-hybrid system. Although the yeast two-hybrid system is valuable in detecting protein–protein interactions, it has limitations such as high rate of false positive clones and the lack of plant-specific posttranslational protein modifications (Walter et al., 2004). Moreover, a protoplast two-hybrid (P2H) system can be used to identify weak hetrodimerization events that could not be detected in the yeast two-hybrid system. This information is essential in order to understand protein function at the cellular, tissue, and organism levels. Also, protoplast transient assays associated with fluorescent protein-based reporter systems or the study of defense-related gene expression could be very useful for screening and characterizing genes involved in plant defense signaling pathways. Recently, a new protoplast-based protein–protein interaction detection method coupled with bimolecular fluorescence complementation (BiFC) has been developed and used in plants (Walter et al., 2004). The BiFC method is based on the complementation of the yellow fluorescence protein (YFP) fluorescence activity by the N-terminal (YFPN) and C-terminal (YFPC) halves of the YFP protein when brought together by two interacting proteins that are fused to each half. This method allows visualizing of the protein interaction that takes place in different cellular compartments. It had been used successfully to detect the *Arabidopsis* basic leucine zipper transcription factor bZIP63 and the interaction between α and β units of the *Arabidopsis* protein franesyltransferase (PFT; Bracha-Drori et al., 2004). Using protoplast transient assays to detect protein–protein interactions will facilitate large-scale functional analysis of specific genes in plants.

PLANT DISEASE SCREENING ASSAYS USING PROTOPLASTS

One of the most effective methods of characterizing resistance mechanisms is to determine whether resistance is expressed at the single-cell level or not. If a source of resistance effective at the single-cell level is found, the resistance mechanism can be easily analyzed in protoplasts from callus cells prior to regeneration or from plantlets prior to acclimatization, reducing the time and effort required to screen plants for resistance, especially for woody plants with long generation times. The development of novel strategies against plant viral diseases relies on a better understanding of molecular virus–host interactions at the early stages of plant life. For example, the time required to transform and regenerate citrus trees can be a matter of years, and the time required then to screen for disease resistance can take months. A limiting factor in the development of CTV (Citrus Tristeza Virus)-resistant citrus varieties has been that assays for resistance to CTV-induced diseases are long term, generally requiring one or more years. Protoplast challenge assays have been used to assess CTV replication at the single-cell level to screen citrus relatives for CTV resistance (Albiach-Marti et al., 2004) and to assess CTV resistance in transgenic citrus callus (Olivares-Fuster et al., 2003) and transgenic grapefruit plants that contain virus-derived sequences (Ananthakrishnan et al., 2007). In the latter study, a few plants with altered virus replication profiles were identified, and these plants are now being tested by traditional methods to determine if there is CTV resistance at the whole-plant level. Several researchers have studied isolation and transfection of different protoplasts to study infection and replication and to investigate virus–host interaction such as with *Plum pox virus* (PPV) in *Arabidopsis* (Raghupathy et al., 2006), *Cowpea mosaic virus* (CPMV) in cowpea (Carette et al., 2002), *Cymbidium mosaic virus* in orchid (Steinhart and Renvyle, 1993) and *Brome mosaic virus* in barley (Sivakumaran et al., 2003). If viral resistance occurring at the cellular level can be identified by protoplast assays and validated at the whole-plant level, protoplast assays can be useful in identifying promising materials at an early stage of plant development.

CONCLUDING REMARKS

Protoplast technologies continue to evolve and are now being utilized in many areas of plant research. Protoclonal variation, somatic hybridization and cybridization, and protoplast transformation are facilitating variety improvement and genetic studies in plant-breeding programs. Protoplast-based transient assays and studies of protein–protein interactions are making significant contributions in plant physiology. Plant pathology is benefiting from protoplast-based disease screening assays that often are much quicker than traditional assays. In the future, protoplast-based strategies are expected to enhance interdisciplinary plant science research and facilitate functional genomic studies made possible by genome-sequencing projects.

REFERENCES

Albiach-Marti, M.R., J.W. Grosser, S. Gowda, M. Mawasi, T. Satyanarayana, S. Garnsey, and W.O. Dawson. 2004. Citrus tristeza virus replicates and forms infectious virions in protoplasts of resistant citrus relatives. *Mol. Breed.* 14:117–128.

An, C., A. Sawada, E. Fukusaki, and A. Kobayashi. 2003. A transient RNA interference assay system using *Arabidopsis* protoplast. *Biosci. Biotechnol. Biochem.* 67:1674–2677.

Ananthakrishnan, G., M. Calovic, P. Serrano, and J.W. Grosser. 2006. Production of additional allotetraploid somatic hybrids combining mandarins and sweet orange with pre-selected pummelos as potential candidates to replace sour orange rootstock. *In Vitro Cell. Dev.—Plant*. 42:367–371.

Ananthakrishnan, G., V. Orbovic, G. Pasquali, and J.W. Grosser. 2007. Transfer of CTV-derived resistance candidate sequences to four grapefruit cultivars through Agrobacterium-mediated genetic transformation. *In Vitro Cell. Dev.—Plant*. 43:593–601.

Asai, T., G. Tena, J. Plotnikova, M.R. Willmann, W.L. Chiu, L. Gomez-Gomez, T. Boller, F.M. Ausubel, and J. Sheen. 2002. MAP kinase signalling cascade in Arabidopsis innate immunity. *Nature*. 415:977–983.

Bates, G.W. (2008). Plant transformation via protoplast electroporation. In Hall, R.D. (Ed.), *Plant Cell Culture Protocols*. Humana Press, Totowa, NJ, pp. 359–366.

Binsfeld, P.C., R. Wingender, and H. Schnabl. 2000. Characterization and molecular analysis of transgenic plants obtained by microprotoplast fusion in sunflower. *Theor. Appl. Genet*. 101:1250–1258.

Bracha-Drori, K., K. Shichrur, A. Katz, M. Oliva, R. Angelovici, S. Yalovsky, and N. Ohad. 2004. Detection of protein–protein interactions in plants using bimolecular fluorescence complementation. *The Plant J*. 40:419–427.

Carette, J.E., J. van Lent, S.A. MacFarlane, J. Wellink, and A. van Kammen. 2002. Cowpea mosaic virus 32- and 60-kilodalton replication proteins target and change the morphology of endoplasmic reticulum membranes. *J. Virol*. 76:6293–6301.

Carlson, P.S., H. Smith, and R.D. Dearing. 1972. Parasexual interspecific plant hybridization. *Proc. Natl. Acad. Sci. USA*. 69:2292–2294.

Cocking, E.C. 1960. Method for the isolation of plant protoplasts and vacuoles. *Nature* 187:927–929.

Davey, M.R. and A. Kumar. 1983. Higher plant protoplasts: Retrospect and prospect. *Int. Rev. Cytol*. Suppl. 16:219–299.

Davey, M.R., E.L. Rech, and B.J. Mulligan. 1989. Direct DNA transfer to plant cells. *Plant Mol. Biol*. 13:273–285.

Davey, M.R., P. Anthony, J.B. Power, and K.C. Lowe. 2005. Plant protoplasts: Status and biotechnological perspectives. *Biotechnol. Adv*. 23:131–171.

De Sutter, V., R. Vanderhaeghen, S. Tilleman, F. Lammertyn, I. Vanhoutte, M. Karimi, D. Inze, A. Goossens, and P. Hilson. 2005. Exploration of jasmonate signalling via automated and standardized transient expression assays in tobacco cells. *Plant J*. 44:1065–1076.

Dudits, D., O. Fejer, G. Hadlaczky, C. Koncz, G.B. Lazar, and G. Horvath. 1980. Intergeneric gene transfer mediated by plant protoplast fusion. *Mol. Gen. Genet*. 179:283–288.

Etxeberria, E., P. Gonzalez, and J. Pozueta-Romero. 2007. Mannitol-enhanced, fluid-phase endocytosis in storage parenchyma cells of celery (*Apium graveolens*; apiaceae) petioles. *Am. J. Botany* 94:1043–1047.

Fock, I., C. Collonnier, A. Purwito, J. Luisetti, V. Souvannavong, F. Vedel, A. Servaes, A. Ambroise, H. Kodja, G. Ducreux, and D. Sihachakr. 2000. Resistance to bacterial wilt in somatic hybrids between *Solanum tuberosum* and *Solanum phureja*. *Plant Sci*. 160: 165–176.

Folling, M., C. Pedersen, and A. Olesen. 1998. Reduction of nuclease activity from lolium protoplasts: Effect on transformation frequency. *Plant Sci*. 139:29–40.

Gleba, Y.Y. and K.M. Sytnik. 1984. Protoplast Fusion In R. Shoeman, (Ed.), *Genetic Engineering in Higher Plants*. Springer-Verlag, Berlin.

Grosser, J.W. and F.G. Gmitter. 1990. Protoplast fusion and citrus improvement. In J. Janick, (Ed.), *Plant Breeding Reviews*. Timber Press, Portland, OR, pp. 339–374.

Grosser, J.W., F.A.A. Mourao, F.G. Gmitter, E.S. Louzada, J. Jiang, K. Baergen, A. Quiros, C. Cabasson, J.L. Schell, and J.L. Chandler. 1996. Allotetraploid hybrids between Citrus and seven related genera produced by somatic hybridization. *Theor. Appl. Genet*. 92:577–582.

Grosser, J.W., P. Ollitrault, and O. Olivares-Fuster. 2000. Somatic hybridization in Citrus: An effective tool to facilitate variety improvement. *In Vitro Cell. Dev.—Plant*. 36:434–449.

Grosser, J.W., J.H. Graham, C.W. McCoy, A. Hoyte, H.M. M. Rubio, D.B.B. Bright, and J.L. Chandler. 2003. Development of "Tetrazyg" rootstocks tolerant of the diaprepes/phytophthora complex under greenhouse conditions. *Proc. Fla. State Hort. Soc*. 116:262–267.

Grosser, J.W. and F.G. Gmitter. 2005. "Thinking Outside the Cell"—Applications of somatic hybridization and cybridization in crop improvement, with citrus as a model. *In Vitro Cell. Dev.—Plant*. 41:220–225.

Grosser, J.W., X.X. Deng, and R.M. Goodrich (2007). Somaclonal variation in sweet orange: Practical applications for variety improvement and possible causes. In: I.H. Kahn, (Ed.), *Citrus Genetics, Breeding and Biotechnology*. CAB International, pp. 219–234.

Grosser, J.W., M. Francis, and J.H. Graham. 2008. Transfer of canker resistance from Kumquat to susceptible Citrus via somatic cybridization. *HortScience* 43:1117.

Guo, W.W., X.D. Cai, and J.W. Grosser. 2004a. Somatic cell cybrids and hybrids in plant improvement. In H. Daniell, and C.D. Chase (Eds.), *Molecular Biology and Biotechnology of Plant Organelles.* Kluwer Academic, Dordrecht, The Netherlands, pp. 635–659.

Guo, W.W., Y. Duan, O. Olivares-Fuster, Z. Wu, C.R. Arias, J.K. Burns, and J.W. Grosser. 2005. Protoplast transformation and regeneration of transgenic Valencia sweet orange plants containing a juice quality-related pectin methylesterase gene. *Plant Cell Rep.* 24:482–486.

Guo, W.W. and J.W. Grosser. 2005. Somatic hybrid vigor in Citrus: Direct evidence from protoplast fusion of an embryogenic callus line with a transgenic mesophyll parent expressing the GFP gene. *Plant Sci.* 168:1541–1545.

Guo, W.W., D. Prasad, P. Serrano, F.G. Gmitter, and J.W. Grosser. 2004b. Citrus somatic hybridization with potential for direct tetraploid scion cultivar development. *J. Hort. Sci. Biotechnol.* 79:400–405.

Holzberg, S., P. Brosio, C. Gross, and G.P. Pogue. 2002. Barley stripe mosaic virus-induced gene silencing in a monocot plant. *Plant J.* 30:315–327.

Jain, S.M. and R.J. Newton. 1988. Proto-variation in protoplast derived *Brassica napus* plants. In K.J., Puite, J.J.M. Dons, and H.J. Huizing (Eds.), *Progress in Plant Protoplast Research: Proceedings of the 7th International Protoplast Symposium.* Wageningen, The Netherlands, pp. 403–404.

Johnson, A.A.T. and R.E. Veilleux. 2001. Somatic hybridization and applications in plant breeding. *Plant Breeding Reviews*, John Wiley & Sons, New York, pp. 167–225.

Larkin, P.J. and W.R. Scowcroft. 1981. Somaclonal variation—A novel source of variability from cell culture for plant improvement. *Theor. Appl. Genet.* 60:197–214.

Liu, Y., M. Schiff, and S.P. Dinesh-Kumar. 2002. Virus-induced gene silencing in tomato. *Plant J.* 31:777–786.

Lorz, H., J. Paszkowshi, C. Dierks-Ventling, and I. Potrykus. 1981. Isolation and characterization of cytoplasts and miniprotoplasts derived from protoplasts of cultured cells. *Physiol. Plant.* 53: 385–391.

Louzada, E.S., H.S. del rio, D. Xia, and J.M. Moran-Mirabal. 2002. Preparation and fusion of microprotoplasts from Citrus spp. *J. Amer. Soc. Hort. Sci.* 127:484–488.

Maheshwari, N., K. Rajyalakshmi, K. Baweja, S.K. Dhir, C.N. Chowdhry, and S.C. Maheshwari. 1995. In vitro culture of wheat and genetic transformation: Retrospect and prospect. *Crit. Rev. Plant Sci.* 14:149–178.

Mazarei, M., H. Al-Ahmad, M.R. Rudis, and C.N. Stewart. 2008. Protoplast isolation and transient gene expression in switchgrass, *Panicum virgatum* L. *Biotechnol. J.* 3:354–359.

Niedz, R.P., W.L. Mckendree, and R.G.J. Shatters. 2003. Electroporation of embryogenic protoplasts of sweet orange (*Citrus sinensis* L. Osbeck) and regeneration of transformed plants. *In Vitro Cell. Dev.—Plant.* 39:586–594.

Olivares-Fuster, O., G.H. Fleming, M.R. Albiach-Marti, S. Gowda, W.O. Dawson, and J.W. Grosser. 2003. CTV resistance in transgenic citrus based on virus challenge of protoplasts. *In Vitro Cell. Dev.—Plant.* 39:567–572.

Ollitrault, P., W.W. Guo, and J.W. Grosser. 2007. Recent advances and evolving strategies in Citrus somatic hybridization. In I.H. Kahn, (Ed.), *Citrus Genetics, Breeding and Biotechnology.* CAB International, Egham, U.K., pp. 235–260.

Omar, A.A., M. Calovic, H.A. El-Shamy, H.-J. An, and J.W. Grosser. 2008. Transformation and regeneration of transgenic "W. Murcott" (Nadorcott) mandarin using a protoplast-GFP transformation system. *Hortscience* 43:1155.

Omar, A.A. and J.W. Grosser. 2008. Comparison of endoplasmic reticulum targeted and non-targeted cytoplasmic GFP as a selectable marker in citrus protoplast transformation. *Plant Sci.* 174:131–139.

Omar, A.A., W.Y. Song, and J.W. Grosser. 2007. Introduction of Xa21, a Xanthomonas-resistance gene from rice, into "Hamlin" sweet orange [Citrus sinensis (L.) Osbeck] using protoplast-GFP co-transformation or single plasmid transformation. *J. Hort. Sci. Biotechnol.* 82:914–923.

Orczyk, W., J. Przetakiewicz, and A. Nadolska-Orczyk. 2003. Somatic hybrids of *Solanum tuberosum*—Application to genetics and breeding. *Plant Cell Tiss. Org. Cult.* 73:245–256.

Pozueta-Romero, D., P. Gonzalez, and E. Etxeberria. 2008. The hyperbolic and linear phases of the sucrose accumulation curve in turnip storage cells denote carrier-mediated and fluid phase endocytic transport, respectively. *J. Amer. Soc. Hort. Sci.* 133:612–618.

Raghupathy, M.B., J.S. Griffiths, L.W. Stobbs, D.C.W. Browna, J.E. Brandle, and A. Wang. 2006. Transfection of Arabidopsis protoplasts with a Plum pox virus (PPV) infectious clone for studying early molecular events associated with PPV infection. *J. Virological* Meth. 136:147–153.

Ramulu, K.S., P. Diijkhuis, E. Rutgers, J. Blaas, W.H.J. Verbeek, H.A. Verhoeven, and C.M. Colijn-Hooymans. 1995. Microprotoplast fusion technique: A new tool for gene transfer between sexually-incongruent plant species. *Euphytica* 85:255–268.

Schaart, J.G., F.A. Krens, K.T.B. Pelgrom, O. Mendes, and G.J.A. Rouwendal. 2004. Effective production of marker-free transgenic strawberry plants using inducible site-specific recombination and a bifunctional selectable marker gene. *Plant Biotechnol. J.* 2:233–240.

Sheen, J. 2001. Signal transduction in maize and Arabidopsis mesophyll protoplasts. *Plant Physiol.* 127:1466–1475.

Shillito, R. 1999. Methods of genetic transformations: Electroporation and polyethylene glycol treatment. In Vasil, I. (Ed.), *Molecular Improvement of Cereal Crop.* Kluwer, Dordrecht, The Netherlands, pp. 9–20.

Sigareva, M.A. and E.D. Earle. 1997. Direct transfer of a cold-tolerant Ogura male-sterile cytoplasm into cabbage (*Brassica oleracea* ssp. capitata) via protoplast fusion. *Theor. Appl. Genet.* 94:213–220.

Sivakumaran, K., M. Hema, and C.C. Kao. 2003. Brome mosaic virus RNA syntheses in vitro and in barley protoplasts. *J. Virol.* 77:5703–5711.

Steinhart, W. and T.T. Renvyle. 1993. Cymbidium mosaic virus RNA synthesis in isolated orchid protoplasts. *Virus Res.* 30:205–213.

Vanitharani, R., P. Chellappan, and C.M. Fauquet. 2003. Short interfering RNA mediated interference of gene expression and viral DNA accumulation in cultured plant cells. *Proc. Natl. Acad. Sci. USA.* 100:9632–9636.

Vardi, A., A. Breiman, and E. Galun. 1987. Citrus cybrids: Production by donor–recipient protoplast-fusion and verification by mitochondrial-DNA restriction profiles. *Theor. Appl. Genet.* 75:51–58.

Varotto, S., E. Nenz, M. Lucchin, and P. Parrini. 2001. Production of asymmetric somatic hybrid plants between *Cichorium intybus* L and *Helianthus annuus* L. *Theor. Appl. Genet.* 102:950–956.

Waara, S. and K. Glimelius. 1995. The potential of somatic hybridisation in crop breeding. *Euphytica.* 85:217–233.

Walter, M., C. Chaban, K. Schutze, O. Batistic, K. Weckermann, C. Nake, D. Blazevic, C. Grefen, K. Schumacher, C. Oecking, K. Harter, and J. Kudla. 2004. Visualization of protein interactions in living plant cells using bimolecular fluorescence complementation. *Plant J.* 40:428–438.

27 Demonstration of Principles of Protoplast Isolation Using Chrysanthemum and Orchardgrass Leaves

R. N. Trigiano

The following laboratory exercises use leaves of either chrysanthemum (*Dendrathema grandiflora* Tzvelev) or orchardgrass (*Dactylis glomerata* L.) to illustrate some aspects of protoplast technology; however, any number of plants will be suitable for these exercises. If different plants are selected, choose plants with young leaves that are thin and lack substantial pubescence. We have found that, although protoplasts are relatively easy to isolate from leaf tissue of these two species, it is more difficult to regenerate plants from chrysanthemum (Sauvadet et al., 1990) and nearly impossible from orchardgrass. As such, this laboratory is devised only to demonstrate some of the general principles of protoplast methodologies, that is, donor tissue preparation, preculture, tissue digestion, purification, and assessment of purity and viability. We suggest that the laboratory exercise be completed without regard to aseptic conditions since callus and plants will not be regenerated and that some of the complex protoplast culture media discussed in Chapter 26 are not required. The mechanics of the experiment may be completed in 1 day if some students can help out "off and on" for the entire day or if the instructor prepares some of the initial steps of the procedure in advance of the class meeting. The laboratory may also be conducted as a demonstration.

MATERIALS

The following materials will be required to complete the exercises:

- 30–60 tillers of orchardgrass or about 15 young, expanding, light green leaves of chrysanthemum—just about any cultivar will work
- Pectolyase Y-23 (Seishin Pharm. Co. Ltd., 4-13, Koamicho, Nihonbashi, Tokyo, Japan)
- RS "Onozuka" Cellulase (Yakult Honsha Co., Ltd. 1.1.19 Higashi Shinabashi, Minato-ku, Tokyo, 105, Japan)
- The following four isolation media:

  1. One-half strength SH salts (see Chapter 2) amended with 0.38 M (ca. 7%) mannitol, 3 mM MES, 2 mM $CaCl_2$, pH = 5.7
  2. Same medium as above except use 0.49 M (ca. 9%) mannitol
  3. Same medium as above except use 0.60 M (ca. 11%) mannitol
  4. Same medium as above except use 0.71 M (ca. 13%) mannitol

- The following digestion media:
  - Same as the four isolation media listed earlier except include 1% (w/v) RS cellulase and 0.4% (w/v) Y-23 pectolyase and 0.1% Bovine Serum Albumin (BSA)

    - 1% calcofluor in water
    - 0.05% fluorescein diacetate dissolved in acetone
    - Hemocytometers
    - Compound microscope equipped for epifluorescence and filters
    - Nylon mesh sieves with 40–100 μm cross-sectional openings in beakers
    - Low-speed table-top centrifuge
    - 25- or 50-mL glass or plastic centrifuge tubes
- Any solidified agar medium (0.8% water agar works well) in petri dishes

GENERAL PROCEDURE

The experiment, as listed in the following text, is designed to investigate only the influence of isolation medium, specifically the osmotic pressure, on successful procurement of viable protoplasts. As alternative treatments, the concentration of the hydrolytic enzymes may be varied. This is a little more difficult since the appropriate isolation medium must first be determined. We suggest that each student or student team work with either one or two of the isolation media listed earlier. Students should then share and analyze data, that is, the number of protoplasts, etc. The experiment may be considered as a completely randomized design for statistical analyses. Alternatively, if desired, only one replication of each treatment may be prepared and the class divided into four groups or teams. In this scenario, the experiment can be regarded as a demonstration.

Follow the instructions in Procedure 27.1 to complete this laboratory exercise.

Now that protoplasts are isolated from cells, they must be collected and concentrated by centrifugation. The following two cautionary notes before centrifugation: (1) Be sure that the "centrifuge or rotor is balanced." Insert a tube of equal weight in the rotor station exactly opposite of your tube containing protoplasts. (2) Do not confuse 100× g with 100 rpm (revolutions per minute)—they

Procedure 27.1
Preculture of Plant Material and Enzymatic Digestion of Plant Cell Walls

Step	Instructions and Comments
1	Collect 10 tillers of orchardgrass for each replication of isolation media and dissect the innermost two leaves for culture as outline in Chapter 24. Only use the bottom 2 cm from each of the leaves. For mum, 3–5 leaves of any cultivar for each replication of a treatment will be sufficient.
2	Record the fresh weight of the leaf tissue and place in a 60 × 15 mm petri dish.
3	Add 15 mL of ice cold (4°C) isolation medium and place in the refrigerator for 30 min.
4	Using fine forceps and a scalpel with a #10 blade, cut the leaves into very thin (<1 mm) strips. Orchardgrass leaves are easily "feathered" by cutting parallel to the veins; mum leaves should be cut along the long axis. Care should be taken not to tear the leaves—a sharp cut works best for protoplast formation and thinner strips ensure a higher yield of protoplasts.
5	After all the leaves are cut, use a Pasteur pipette to remove most of the isolation medium and replace it with more of the appropriate cold isolation medium. Repeat this step until the fluid is free of most small debris and green color. For mum, the preparation should be odor free.
6	Prepare digestion media about 1 h before use. Dissolve 200 mg (1%) RS cellulase and 80 mg (0.4%) Y-23 Pectolyase in 20 mL of each of the four isolation solutions—use 100-mL beakers and vigorous stir for at least 30 min. Remove the predigestion media and add the digestion solutions to the cut tissues.
7	Wrap the petri dishes with several layers of Parafilm and incubate at room temperature (22°C–25°C) in the dark. Cultures should remain stationary—shaking the petri dishes at this time is detrimental to protoplast formation.
8	Using a compound microscope, observe the tissue preparation for release of protoplasts. Focus on the scalpel blade marks on the bottom of the petri dish—often the smaller, more cytoplasmic dense protoplasts will sink to the bottom. Also observe the cut edges of the leaves for plasmolyzed cells and the release of protoplasts (Figure 27.1). Repeat this operation and record your observations every 30 min for a total of 3 or 4 h.

are not the same. Check the specifications of the centrifuge (the manufacturer usually provides a conversion chart or a mathematical formula to calculate relative force [g] from rpm) and adjust rpm to achieve 100× g.

Follow the instructions provided in Procedure 27.2 to concentrate the protoplasts.

DETERMINING THE YIELD OF PROTOPLASTS

Protoplasts are now concentrated at the bottom of the tube, and we can estimate the number of protoplasts formed per gram of leaf tissue. This information is necessary to compare treatment effects on protoplast isolation.

Follow the outline in Procedure 27.3 to calculate the yield of protoplasts.

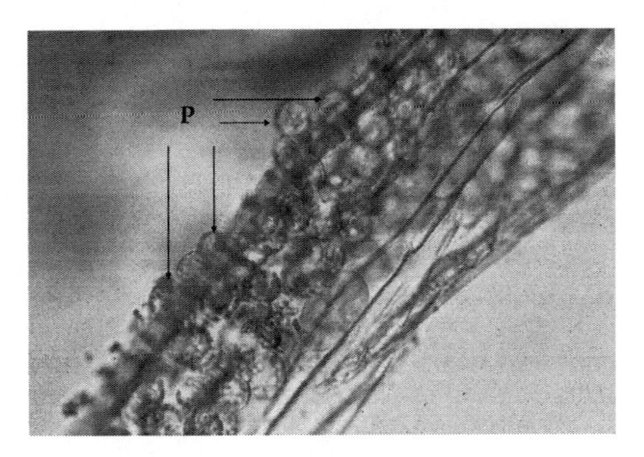

FIGURE 27.1 Protoplasts formed from a strip of a chrysanthemum leaf exposed to release medium with cell wall degrading enzymes for 2 h. P = either protoplasts being released or in the process of forming.

Procedure 27.2
Isolation of Protoplasts

Step	Instructions and Comments
1	After the plant material is digested, place the petri dishes on a rotary shaker revolving at the slowest possible speed for 5 min; the liquid should barely move.
2	Remove as much of the undigested debris as possible using a fine pair of forceps. The debris will be very soft and adhere to the forceps blades, but can easily be removed by running the forcep blades through some agar medium in a petri dish.
3	Gently swirl the contents of the petri dish and using a wide-bore glass pipette equipped with a bulb or pump, carefully remove the liquid from the petri dish.
4	Hold the tip of the pipette against the inside of the nylon mesh sieve (40–100 μm cross-sectional openings) housing and very gently express the liquid medium down the side—protoplasts are very fragile and may be damaged if not handled carefully. If the surface tension between the liquid and sieve prevents passage of the liquid, gently scratch the nylon sieve surface with the forcep blades and the liquid will flow into the collection beaker.
5	Using a wide-bore pipette, gently transfer the sieved medium to either a 25- or 50-mL plastic centrifuge tube. The medium may appear "cloudy"—no problem, this is great! This is a sign that many protoplasts were formed during digestion. Centrifuge the suspension at 100× g for 5 min following the precautions listed earlier.
6	Decant or pipette (long Pasteur pipette) the digestion medium from the tubes without disturbing the green pellets of protoplasts on the bottom. Carefully pipette 20 mL of isolation medium against the inside of the centrifuge tube. Centrifuge as in Step 5 and repeat this washing step twice more. All hydrolytic enzymes must be removed if the protoplasts are to be cultured. Cell regeneration will not occur in the presence of these enzymes.

Procedure 27.3
Protoplast Yield

Step	Instructions and Comments
1	Resuspend the pelleted protoplasts in 5 mL of fresh isolation medium and make a slide to observe the protoplasts (Figure 27.2).
2	Estimate the number of protoplasts in the suspension using a hemacytometer. Add the number of protoplasts in five squares (Figure 27.3). Only count those protoplasts that lie entirely within the boundaries of the counting squares; those touching a line should not be counted. Calculate the mean from at least three samples.
3	Calculate the number of protoplasts per milliliter using the following equation: $X \times 2{,}000 = P$; where X is the mean number of protoplasts and P = number of protoplasts/mL. For example, if $X = 50$; then $P = 10^5$ protoplast ± 10%/ mL. Since the total volume of the suspension is 5 mL, the estimate of total number of protoplasts is 5 mL × 10^5 protoplasts/mL = 5×10^5 or a total of 500,000 ± 10% protoplasts in the tube.
4	Now, estimate the yield of protoplasts per gram of tissue. For example, 0.5 g of tissue was used in the isolation, then 5×10^5 protoplasts divided by 0.5 g = 10^6 or 1,000,000 protoplast per gram of tissue. The yield of protoplasts from the different treatments now can be compared since the data are all expressed as protoplast/gram of tissue.
5	If you complete the assessment of viability, then multiple the percentage of viable protoplasts by protoplast/gram of tissue. This number represents the number of viable protoplasts/gram of tissue.

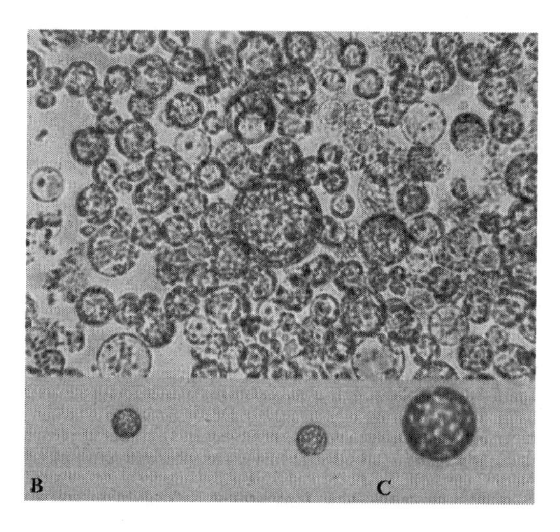

FIGURE 27.2 Freshly isolated protoplasts. (A) Suspension of protoplast in washing medium. Note the different sizes of protoplasts. (B) Small, densely cytoplasmic protoplasts. (C) Enlargement of a single protoplast.

METHODS FOR EVALUATING THE QUALITY OF PROTOPLASTS

When counting protoplasts, you may have noticed that they were almost perfectly spherical; normally, this would indicate that all of the cell wall material was digested by the enzymes. However, "patches" of cell wall may still remain. The best way to detect wall materials is with calcofluor, a chemical that binds with primarily cellulose, a major plant cell wall polymer. The stain is made by mixing 1 mL of 1% aqueous calcofluor with 20 mL of the appropriate isolation medium without the enzymes. Pipette 0.25 mL of protoplast suspension into 1 mL of the stain solution. Swirl to mix, mount a few drops of the mixture on a glass slide, and cover with a #1 or 1½ coverslip. First, observe protoplasts with transmitted visible light using a 10× or 20× objective lens. Next, examine the protoplasts using an ultraviolet light (UV) source with the following filters and mirror: 400–440 nm excitation, 470 nm barrier, and 455 nm dichroic mirror (equivalent to Nikon filter assembly BV-2A). If the wall is not completely digested, the calcofluor bound to cellulose will glow a brilliant blue

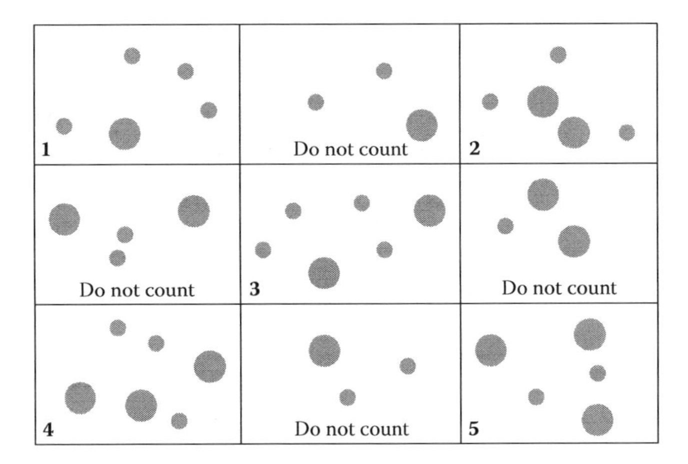

FIGURE 27.3 Representation of the counting areas on a hemocytometer omitting the fine lines. Sum the protoplasts in the five numbered areas—do not count any protoplasts that touch any of the boundary lines of the squares—and multiply by 2,000 to obtain an estimate (±10%) of the protoplasts per milliliter. In this example, there are 27 protoplasts pictured in the counting areas (1–5), therefore an estimated 54,000 protoplasts/mL.

that will not fade with time. Some protoplasts may still retain wall fragments, which are visible as irregularly shaped blue glowing patches. If true protoplasts (i.e., the cell minus the wall), the field will appear dark and the protoplast may not be discernable except for perhaps some red autofloresence of chloroplasts. Observe at least 50 protoplasts and calculate the percentage of protoplasts that are free of cell wall. Incidentally, calcoflour is the same optical brightener that makes white clothing "bright" in natural light or fluoresce with black light.

The protoplast isolation procedure often damages or kills a percentage of cells; therefore, it is advisable to assess the condition of the protoplasts. Mount a small drop of the protoplast suspension on a glass slide, coverslip, and observe using 20 and 40 objective lens. The cytoplasm of some or a few may appear to be streaming or moving in a circle—look for the small "particles" (organelles) in the cytoplasm. This phenomenon is termed *cyclosis* and only occurs in living cells. However, a cell that does not exhibit cyclosis is not necessarily dead. Also, look for "bulges or budding" of the plasmalemma (cell membrane) that may indicate damaged, dying, or fused protoplasts.

There are a number of presumptive tests to determine viability of cells. The tests are presumptive because they do not actually measure which cells are "living" and those cells that are "dead," but usually assess some physiological or physical properties associated with living cells. Most of these tests are based on enzyme activity or permeability of the plasmalemma to certain indicator chemicals, such as dyes or stains.

Fluorescein diacetate has often been used to estimate the percentage of viable protoplasts in a preparation (Widholm, 1972). This technique uses fluorescence and is very efficient—many protoplasts can be examined in a short time. Fluorescein diacetate normally does not fluoresce but, once across the cell membrane, an esterase in the cytoplasm cleaves the acetate groups from the main portion of the molecule. The resulting moiety will fluoresce green when exposed to the proper wavelengths of UV light.

Mix 0.25 mL of 0.05% fluorescein diacetate (dissolved in acetone) with 20 mL of the appropriate isolation medium to make the "stain" solution. Pipette 0.25 mL of the protoplast suspension into 1.0 mL of the stain solution. Swirl gently to mix and mount some of the protoplasts on a glass slide. First, look at the protoplasts using a 10× or 20× objective lens. Now, view the protoplasts using the UV source and the following filters and mirror: 420–490 nm excitation, 520 nm barrier, and a 510 nm dichroic mirror (equivalent to Nikon filter assembly—B-3A). The cytoplasm of "living" protoplasts should fluoresce green (Figure 27.4); vacuoles in the protoplasts will appear dark. Protoplasts

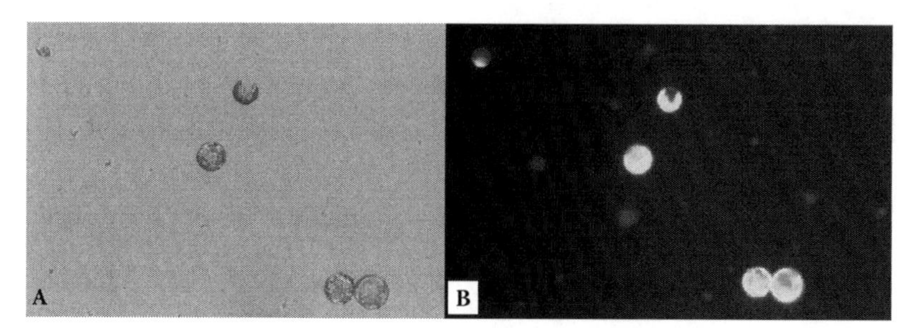

FIGURE 27.4 Determining viability of protoplasts using fluorescein diacetate. (a) Protoplasts viewed with transmitted visible light. (b) Protoplasts stained with fluorescein diacetate and viewed with UV light. The fluorescence light is green.

in which the cytoplasmic esterase is not active are presumably dead and will appear to be dark. Occasionally, the autoflorescence of chlorophyll or other compounds will impart a reddish hue to the protoplasts. Observe at least 50 protoplasts and calculate the percentage of living cells. Multiply the estimate of number of protoplasts by this percentage to obtain an estimate of the number of living protoplasts. From the preceding example, there are 5×10^5 protoplast per milliliter multiplied by 0.95 viable equals 4.75×10^5 living protoplasts per milliliter. A good isolation will yield between 90% and 95% viable protoplasts.

As an alternative to epifluorescence microscopy to determine viability, Evan's blue stain may be used with visible light (Gaff and Okong 'O-Ogola, 1971). Mix 2 mL of 1% aqueous stain with 10 mL of the appropriate isolation medium. Pipette 0.25 mL of the protoplast suspension into 1.0 mL of the stain mixture and allow to stand for about 15 min. The dye will be excluded from the protoplasts with intact and functional plasmalemma (presumed living cells), whereas those protoplasts whose membranes are dysfunctional will turn blue and are considered nonviable.

Anticipated Results

Preculturing and cutting the donor tissue in isolation medium should minimize damage to cells. Rinsing the donor tissue before digestion will remove cellular debris and chemical compounds released from broken cells during cutting operation. Some of these chemicals are thought to be detrimental to the remaining cells.

A very few protoplasts should be released within the first hour and will be difficult to find. However, focusing on a cut edge of the leaf tissue (Figure 27.1) should show cells with various degrees of cell wall degradation. Also, plasmolysis of cells should be easily observed. As the experiment progresses, more and more protoplasts will be released and can easily be seen on the bottom of the petri dish or floating in the medium. The brief shaking at the conclusion of the experiment helps to maximize the release of protoplasts. Because so many factors influence the yield of protoplasts, there is not a way to accurately foretell which release medium will produce the best results. Typically, this protocol can be optimized for both orchardgrass and chrysanthemum leaves to yield between 3×10^5 and 5×10^6 protoplasts per gram of tissue. Although some protoplasts should be produced throughout the range of osmotic pressures developed in the four media, protoplasts in some of the media may be very fragile and may not have survived the centrifugation steps. It is not unusual to observe in the lowest osmotic pressure medium that intact, viable protoplasts will not be produced. Therefore, when approaching the question of how to isolate protoplasts from a "new species," it is often useful to initially try a wide range of media and enzyme concentrations.

Questions

- Why would preculturing the donor tissue in isolation medium at 4°C facilitate production of protoplasts?
- Why are protoplast isolation and release media made with relatively high concentrations (%) of mannitol? Could other osmotically active sugar alcohols be used?
- Why are protoplasts isolated from relatively young, meristematic tissues? Hint: consider cell wall properties also.
- Why is it important to remove all of the enzyme solution before trying to culture protoplasts?
- Why is it important to determine both the yield of protoplasts from the donor tissue and the concentration of protoplasts in the culture medium?
- Are the cell wall degradation enzymes (cellulase and pectinase) used in this experiment pure? If not, what other hydrolytic and oxidative enzymes might be present as contaminants and are these beneficial (aids in liberating protoplasts) or detrimental (may cause harm or death to the cells or protoplasts)?
- What is fluorescence?

REFERENCES

Gaff, D. F., and O. Okong 'O-Ogola. 1971. The use of non-permeating pigments for testing the survival of cells. *J. Exp. Bot.* 22:756–758.

Sauvadet, M.-A., P. Brochard, and J. B. Boccongibad. 1990. A protoplast-to-plant system in chrysanthemum: Differential response among several commercial clones. *Plant Cell Rep.* 8:692–695.

Widholm, J. 1972. The use of fluorescein diacetate and phenosafranine for determining viability of cultured plant cells. *Stain Technol.* 47:189–194.

28 Isolation, Culture, and Fusion of Tobacco and Potato Protoplasts

Richard E. Veilleux and Michael E. Compton

Isolation of plant protoplasts, followed by fusion and subsequent regeneration of somatic hybrids, is a remarkable technique that allows exchange of genetic information between sexually incompatible species. Success in application of these techniques was first achieved with tobacco (*Nicotiana tabacum* L.) for which Nagata and Takebe (1970) developed a protocol for cell wall regrowth and cell division, followed eventually by plant regeneration (Nagata and Takebe, 1970). Carlson et al. (1972) reported the first interspecific somatic hybrids. The list of species amenable to these techniques has grown to include many of the major crops. However, the source of explants for protoplast extraction, the choice of genotype, and the specific conditions of isolation and medium changes must be customized for each species.

The objectives of somatic hybridization may be to develop the following: (1) somatic or parasexual hybrids comprised of the complete genome of two "parents," (2) asymmetric somatic hybrids where there has been only partial genome transfer between species, or (3) cytoplasmic hybrids (cybrids) with the nucleus of one parent and the cytoplasm of another. Somatic hybrids between closely related species are generally polyploid and have the potential of fertility. Potato has been the object of somatic hybridization with a range of related species from sexually compatible tuber-bearing *Solanum* relatives to sexually incompatible nontuberous *Solanum*, including tomato (*Solanum lycopersicum* L.), nightshade (*Solanum nigrum* L.), and tobacco. Both its plasticity in tissue culture and facile vegetative propagation have made the potato attractive to geneticists wishing to examine the potential of somatic hybridization in an important economic crop. Somatic hybrids between potato (*Solanum tuberosum* L.) and a weedy, nontuber-bearing relative, *S. brevidens* Phil., contained the additive number of chromosomes of the two parents (i.e., 72 with 48 from potato and 24 from *S. brevidens*). These somatic hybrids could be backcrossed to potato, resulting in fertile tuber-producing plants with considerable breeding potential as sources of disease resistance (Helgeson et al., 1986).

As the taxonomic divergence between species increases, the possibility of obtaining fertile somatic hybrids decreases and the likelihood of chromosome elimination—that is, partial or complete spontaneous loss of chromosomes of one or the other parent—increases. Such chromosome elimination results in aneuploid regenerants or asymmetric hybrids. An extraordinary example has been the regeneration of plants following protoplast fusion between tobacco and carrot (*Daucus carota* L.) reported by Kisaka and Kameya (1994). The somatic hybrids contained between 23 and 32 chromosomes instead of the additive number of 66 (48 from tobacco and 18 from carrot).

As opposed to somatic hybrids in which nuclear fusion has occurred, cybrids between even distantly related parents are more likely to be fertile (Rose et al., 1990). Nuclei from protoplasts of one fusion partner can be inactivated by irradiation or mutagenic treatment before fusion to prevent transmission of the nuclear genome. Protoplasts of the other fusion partner may be metabolically inhibited by iodoacetate treatment. In this way, unfused protoplasts of either parent will not regenerate, and the preponderance of regenerants will be cybrids. Such a scheme has recently been successful to develop cybrids between potato and either of the two sexually incompatible wild species, *S. bulbocastanum* Dun. and *S. pinnatisectum* Dun. (Sidorov et al., 1994). Complete prevention of transmission of nuclear material of the irradiated protoplasts cannot be guaranteed, such that occasional asymmetric hybrids may result (Matabiri and Mantell, 1994).

The following exercises illustrate the basic procedures of protoplast isolation and fusion. The more elaborate schemes mentioned earlier may be built upon these exercises. The conditions under which donor plants are grown can greatly affect the success with protoplast yield and culture. This is especially true of potato. Therefore, the first exercise has been designed for tobacco protoplast isolation because it is much more tolerant of a range of conditions, making it easier to handle and more likely to succeed as a student lab. The exercise described in Experiment 2 is a small-scale chemical fusion of protoplasts using polyethylene glycol and high Ca^{2+} to facilitate membrane fusion between adjacent protoplasts.

GENERAL CONSIDERATIONS

TOBACCO

Most cultivars of tobacco will respond well to protoplast isolation and culture. However, Nagata and Takebe's (1970) initial protoplast culture of tobacco was hindered by inadvertent choice of a cultivar prone to seasonal fluctuations that prevented repeatability of experiments. We have often used "Samsun," an older cultivar that has been used extensively in tissue culture research. Seeds can be obtained through the USDA-ARS Tobacco Collection, North Carolina State University, 901 Hillboro St., Oxford, North Carolina 17565 ([919] 693-5151).

POTATO

Response of potato to protoplast manipulation is under genetic control (Cheng and Veilleux, 1991), so care must be taken in selection of genotypes; some are completely recalcitrant in protoplast culture, whereas others respond easily. In a comparison of 36 cultivars or breeding lines for ease of manipulation in protoplast culture, Haberlach et al. (1985) listed *S. tuberosum* cv. Russet Burbank, US-W 5328.4, US-W 9546.46, PI 423654, *S. etuberosum* Lindl. PIs 245924 and 245939, and *S. brevidens* Phil. 245763 as highly regenerable. [Of these, Russet Burbank is tetraploid ($2n = 4x = 48$) and the rest are diploid ($2n = 2x = 24$).] Many other genotypes yielded few or no p-calli (protoplast-derived calli) or would not regenerate from p-calli. In vitro copies of regenerable cultivars or species may be obtained from the NRSP-6 Potato Introduction Project, Sturgeon Bay, Wisconsin 54235 ([920] 743-5406). Several genotypes should be used such that each student or student team will extract protoplasts from one genotype, and then fuse his/her protoplasts with those extracted by another student or student team. It may be desirable to use one highly regenerable and one recalcitrant genotype for protoplast fusion. In this way, unfused protoplasts of one parent would not be expected to regenerate in the fusion experiments. Haberlach et al. (1985) list *S. tuberosum* PI 256977 (tetraploid), Kennebec haploid 731.3 (diploid), US-W 9587.24 (diploid), and 77-16 (diploid) as unable to form p-calli (callus formed from protoplasts) after protoplast isolation and culture.

GROWTH OF PLANTS

TOBACCO

The plants to be used for protoplast extraction should be approximately 50–60 days from planting at the time of protoplast isolation. The seedlings are small and slow to start, so many seeds can be planted in a single 10 cm pot for the first month and then transplanted individually to 3.7-L nursery pots. One plant per student should be sufficient. Any of the soilless potting mixes is suitable.

Tobacco is a heavy feeder, so weekly fertilization with a 20:20:20 (N:P:K) liquid fertilizer is recommended. Light and temperature conditions in the greenhouse are not especially critical. Tobacco is susceptible to infestation with whitefly under greenhouse conditions, so treatment with a systemic insecticide (e.g., Marathon, Olympic Horticultural Products, PO Box 230, Mainland, Pennsylvania 19451) shortly after transplanting is recommended. Otherwise, microbial contamination is very likely in protoplast cultures.

POTATO

The best source of plant material for protoplast manipulation of potato is in vitro plantlets. Therefore, an exercise in making approximately 10–15 copies using single nodes of previously established in vitro plantlets of the selected genotypes on propagation medium for each student should be planned approximately 3 weeks before the protoplast fusion lab. Obviously, adequate in vitro multiplication will have been necessary 3–4 weeks prior to this round of multiplication. In vitro potato can harbor bacterial contaminants that are not apparent during routine subculture on basal medium. However, such contaminants often become obvious during protoplast manipulations, resulting in failed experiments. A precaution against such "cryptic" contamination is to use the antibiotic cefotaxime in the propagation medium (Table 28.4) used just prior to the transfer when students will multiply their genotypes.

EXPLANT PREPARATION

TOBACCO

Fully expanded young leaves yield the best protoplasts. Leaves should be harvested and brought to the lab on the day of the exercise.

POTATO

In vitro potato cultures should be placed in the dark in a refrigerator for 48 h prior to protoplast extraction. Examine cultures carefully and remove any obviously contaminated ones.

CULTURE MEDIUM

For tobacco protoplast isolation, five media are needed (Tables 28.1–28.3). Three of these are simple variations of protoplast isolation (PI) medium (Table 28.2) and can be made from the same preparation of PI (preplasmolysis = PI + 13% mannitol; flotation = PI + 20% sucrose; and rinse = PI + 10% mannitol). The other two media (enzyme solution and culture medium) have a different basal composition so must be prepared separately. Only the enzyme solution requires filter sterilization; the other four media can be autoclaved. It is advisable to dispense each medium prior to autoclaving into

TABLE 28.1
Enzyme Medium—Tobacco Laboratory

Component	%
Onozuka R10 cellulase	0.5
Onozuka R10 macerozyme	0.1
mannitol	13.0
pH 5.8	

TABLE 28.2
Protoplast Isolation (PI) Medium—Tobacco Laboratory

Component	mg L^{-1}
$CaCl_2 \bullet H_2O$	1480.0
KH_2PO_4	27.2
KNO_3	101.0
$MgSO_4 \bullet 7H_2O$	246.0
$CuSO_45 \bullet H_2O$	0.025
KI	0.16
pH 5.8	

TABLE 28.3
Protoplast Culture (PC) Medium—Tobacco Laboratory

Component	mg L^{-1}
$Ca(H_2PO_4) \bullet H_2O$	100.0
$CaCl_2 \bullet 2H_2O$	450.0
KNO_3	2500.0
$MgSO_4 \bullet 7H_2O$	250.0
$NaH_2PO_4 \bullet 2H_2O$	170.0
$(NH_3)_2SO_4$	134.0
$CoCl_2 \bullet 6H_2O$	0.025
$CuSO_4 \bullet 5H_2O$	0.025
H_3BO_3	3.0
KI	0.75
$MnSO_4 \bullet 4H_2O$	13.2
$Na_2MoO_4 \bullet 2H_2O$	0.25
$ZnSO_4 \bullet 7H_2O$	2.0
Sequestrene 330	28.0
Sucrose	10,000.0
Glucose	18,000.0
Mannitol	100,000.0
Inositol	100.0
Nicotinic acid	1.0
Pyridoxine-HCl	1.0
Thiamine-HCl	10.0
2,4-D	0.1
NAA	1.0
BA	1.0
pH 5.8	

properly labeled 25 × 150 mL culture tubes with Magenta 2-way caps in aliquots sufficient for each student or student team. If different students or student teams share a common supply of medium, contamination is much more likely.

For potato protoplast isolation and fusion, nine different media (Tables 28.4 through 28.12) are required, including the protoplast propagation medium for the micropropagation step

TABLE 28.4
Propagation Medium for In Vitro Potato Plantlets

Component	Amount (per liter)
MS1 and MS2	20 mL
S3, MS4, MS5	10 mL
Sucrose	20 g
Myo-inositol	100 mg
KH_2PO_4	170 mg
Casein hydrolysate	500 mg
Cefotaxime[a]	250 mg
Agar	7 g
pH 5.7	
Autoclave[b]	

[a] This antibiotic is optional but is especially recommended for propagation of stock plantlets one or more clonal generations before propagation for protoplast extraction to reduce bacterial contamination. It is not autoclavable so must be filter sterilized and added to the autoclaved medium as it cools.

[b] Filter sterilize if using cefotaxime.

TABLE 28.5
Preplasmolysis Medium—Potato Laboratory

Component	Per 100 mL
MS2	6.7 mL
MS5	1.0 mL
KH_2PO_4	3.3 mg
KNO_3	10.1 mg
$MgSO_4$	600.0 mg
Mannitol	9.0 g
MES	58.6 mg
pH 5.8	
Filter sterilize	

conducted 3 weeks before isolation. Four of these (propagation, preplasmolysis, enzyme, and culture) are variations of Murashige and Skoog (1962) basal medium, such that stocks 2, 3, 4, and 5 for MS medium preparation can be used. A modified MS1 stock consisting of 9.5 g KNO_3, 0.85 g KH_2PO_4, and 0.903 g $MgSO_4$ per 100 mL is necessary; this stock differs from routine MS1 by elimination of NH_4NO_3 because of the ammonium sensitivity of protoplasts. Four of these media (preplasmolysis, enzyme, culture, and Ca^{2+}) require filter sterilization. The other five (propagation, flotation, rinse, CPW 13M, and PEG 22.5) can be autoclaved. If the propagation medium contains cefotaxime, this component must be filter sterilized separately and added to the autoclaved propagation medium as it cools. As with the tobacco experiment, it is advisable to dispense these media (prior to autoclaving) into aliquots sufficient for each student or student team. For filter-sterilized media, dispensing into autoclaved culture tubes after filter sterilization is recommended.

TABLE 28.6
Enzyme Medium—Potato Laboratory

Component	Per 100 mL
Mannitol	7.3 g
Glucose	1.8 g
Modified MS1[a]	1.0 mL
MS2	3.0 mL
MS3, MS4, MS5	0.5 mL each
Cellulase Onozuka R-10	1.0 g
Macerozyme R-10	0.1 g
pH 5.8	
Filter sterilize, make fresh on the day of isolation	

[a] MMS1 has 9.5 g KNO_3, 0.85 g $KH2PO_4$, and 0.903 g $MgSO_4$ per 100 mL.

TABLE 28.7
Rinse Medium—Potato Laboratory

Component	Per 200 mL
KCl	4.4 g
MS2	6.0 mL
pH 5.8	
Autoclave	

TABLE 28.8
Flotation Medium—Potato Laboratory

Component	Per 100 mL
Sucrose	7.1 g
MS2	3.0 mL
pH 5.8	
Autoclave	

EXPERIMENT 1. PROTOPLAST ISOLATION AND CULTURE OF TOBACCO

MATERIALS

For each student or student team:

- Five sterile 300 mL fleakers to surface-sterilize leaves (one for EtOH, one for bleach, and three for sterile water)
- 500 mL sterile water
- 100 mL 70% ethanol
- 100 mL 20% commercial bleach
- Scalpel handle with #10 blade

TABLE 28.9
Culture Medium—Potato Laboratory

Component	Per 100 mL
modified MS1	1.0 mL
MS3, MS4, MS5	0.5 mL each
MS2	3.0 mL
Glucose	3.0 g
Sucrose	1.0 g
Sorbitol	5.0 g
Casein hydrolysate	50.0 mg
Glutamine	10.0 mg
Serine	1.0 mg
Myo-inositol	10.0 mg
Thiamine	1.0 mg
NAA	0.125 mg
2,4-D	0.025 mg
Zeatin	0.1 mg
pH 5.8	
Filter sterilize	

TABLE 28.10
CPW 13M Medium—Potato Laboratory

Component	Per liter
KH_2PO_4	27.2 mg
KNO_3	0.101 g
$CaCl_2•2H_2O$	1.48 g
$MgSO_4•7H_2O$	246.0 mg
KI	0.16 mg[a]
$CuSO_4•5H_2O$	0.025 mg[b]
Mannitol	130.0 g
pH 5.8	
Autoclave	

[a] 8 mg/50 mL stock—use 1 mL.
[b] 2.5 mg/50 mL stock—use 0.5 mL.

- Fine (Dumont) forceps
- Ten sterile Pasteur pipettes
- Rubber dispensing bulbs
- Hemacytometer
- Centrifuge
- FDA stain [5 mg FDA (product no. F 7378, Sigma Chemical Co., St. Louis, Missouri) in 1 mL acetone]
- A 250-mL beaker containing a 63-μm filter (such a filter can be constructed by ordering 63 μm wire mesh from the Newark Wire Cloth Co., Newark, New Jersey, cutting the wire mesh into squares just large enough to cover the cap end of a Nalgene 125 mL autoclavable animal watering bottle, and fusing the wire mesh to the plastic on a hotplate covered with

TABLE 28.11
Fusogen PEG 22.5 Medium—Potato Laboratory

Component	mg per 100 mL	%
PEG 8000	22500	22.5
Sucrose	1800	1.8
$CaCl_2 \cdot 2H_2O$	150	0.15
KH_2PO_4	10	0.01
pH 5.8[a]		
Autoclave		

[a]Use 1 N KOH buffer.

TABLE 28.12
Ca^{2+} Washing—Potato Laboratory

Component	per 100 mL	%
$CaCl_2 \cdot 2H_2O$	0.74 g	0.74
Glycine	0.38 g	0.38
Sucrose	11.0 g	11.0
pH 5.8		
Filter sterilize		

aluminum foil. After the mesh has embedded into the melted plastic, the excess mesh can be trimmed with a scissors, and the bottom of the bottle can be cut off with a scalpel. The filter can be autoclaved in the beaker covered with foil and used repeatedly. Alternatively, nylon mesh of approximately the same size can be used, but most nylon cannot be autoclaved so must be sterilized by an ethanol soak.

- Centrifuge tubes (15-mL)
- Disposable plastic petri dishes
- Three 10-mL pipettes
- Two 5-mL pipettes
- 10 mL enzyme solution
- 20 mL preplasmolysis medium
- 10 mL flotation medium
- 30 mL rinse medium
- 25 mL protoplast culture medium
- Calcofluor White stain [0.1% Fluorescent Brightener 28 (product no. F 3543, Sigma Chemical Co., St. Louis, Missouri) in 0.7 M mannitol]

Follow the outline in Procedure 28.1 to complete this experiment.

Anticipated Results

By day 7, the protoplasts should fluoresce blue after staining with Calcofluor White. Protoplasts that have died prior to wall resynthesis do not fluoresce. Some protoplasts should have divided already such that two- or four-celled colonies are visible. If cell colonies can be found, the protoplasts can be diluted with fresh culture medium with lower mannitol to observe the development of p-calli. After approximately 8–10 weeks, p-calli can be transferred to shoot regeneration medium.

Procedure 28.1
Tobacco Protoplast Extraction and Culture

Step	Instructions and Comments
	Day 1
1	Surface sterilize one tobacco leaf by immersion in 70% EtOH for 30 s and transfer to 20% bleach/Tween for 15 min. Rinse three times in sterile, distilled water.
2	Place the tobacco leaf in a sterile petri dish, and peel sections of the lower epidermis from the leaf and/or cut the leaf into 1 mm strips. Even though the peeling is tedious, it is more effective than cutting. The epidermis can be stripped by lifting at a leaf vein with fine forceps and pulling gently across the lamina. This should be repeated until at least one-half of the epidermis has been removed.
3	Add 20 mL preplasmolysis medium and incubate the peeled leaf for 1 to 8 h.
4	Remove preplasmolysis solution using a sterile Pasteur pipette and replace with 10 mL enzyme solution. Wrap the petri dish with parafilm, label, and place in the dark overnight at low speed (40 rpm) on a shaker.
	Day 2
5	Gently agitate the plate of protoplasts in enzyme solution to facilitate the release of protoplasts. Transfer protoplast suspension to the 63 μm filter in a 250 mL beaker using a Pasteur pipette.
6	Add 3 mL rinse medium to the plate with leaf debris, shake vigorously, and combine with the protoplast/enzyme mixture in the 250 mL beaker.
7	Use an additional 2 mL rinse medium to rinse protoplasts off the filter into the beaker by holding the filter with forceps as you rinse. Transfer the protoplast–enzyme mixture to a 15 mL sterile centrifuge tube and spin at 50× g for 10 min.
8	Remove the supernatant and resuspend the *pellet* in 10 mL flotation medium, add 1 mL rinse medium to the top of the protoplasts in flotation medium *dropwise* so as not to mix the two media but rather to have distinct layers of medium (dripping the rinse medium down the side of the tube works better than splashing it on the surface of the flotation medium) and recentrifuge for 10 min at 50 × g. *Viable protoplasts should float to the surface.*
9	Remove the band of green protoplasts with a Pasteur pipette, transfer it to a fresh centrifuge tube, add 10 mL rinse medium, and recentrifuge. Protoplasts should sink to the bottom and form a *pellet.*
10	Rinse the protoplasts again in 10 mL rinse medium. (This second rinse can be omitted if the protoplast pellet is small, i.e., barely visible.) Remove the supernatant and resuspend the protoplasts in approximately 1 mL culture medium.
11	Place a drop of protoplasts on a hemacytometer and estimate protoplast density (number of cells/grid ×10,000). To culture the protoplasts, dilute to 50,000 cells/mL with protoplast culture (PC) medium and transfer to a sterile petri dish.
12	Stain a protoplast sample with FDA stain by diluting one drop of FDA stain with 25 drops of culture medium. Add one drop of protoplasts to one drop of diluted FDA stain on a microscope slide. Observe cell viability under a fluorescence microscope. Viable protoplasts fluoresce green; dead cells fluoresce red due to autofluorescence of chlorophyll. Record the frequency of viable protoplasts in a sample of 100–200 cells. Seal, label, and culture in the dark at 28°C overnight.
	Day 3
13	Move culture to low light (10–20 μmol s^{-1} m^{-2}) at a 16 h photoperiod (e.g., under cool white fluorescent light covered by a layer of cheesecloth) for 2 days.
	Day 5
14	Move culture to higher light intensity (50–75 μmol s^{-1} m^{-2}) by removing the layer of cheesecloth.
	Day 7
15	Examine cultures for cell wall regrowth by staining a sample with Calcofluor White stain and observing under a fluorescence microscope. Is there any evidence of cell wall regrowth or cell division evident by blue fluorescence? If so, record the plating efficiency of the protoplast cultures, that is, the number of dividing cells divided by the total number of cultured cells, in a sample of 100–200 cultured protoplasts. Is there evidence of contamination?

Questions

- Why do the protoplasts float to the interface of the flotation and rinse solutions while the debris sinks to the bottom of the sucrose solution during centrifugation?
- Why do viable protoplasts fluoresce green when stained with FDA stain?
- Report the viability of freshly isolated protoplasts and the plating efficiency after 7 days. Explain why there is a difference between frequency of viable protoplasts and plating efficiency.
- Would you expect the plants regenerated from individual protoplasts derived from a single leaf to be genetically identical? Explain.

EXPERIMENT 2. PROTOPLAST FUSION OF POTATO

Materials

For each student or student team:

- Two scalpel handles with #10 blades
- Fine (Dumont) forceps
- 15 sterile Pasteur pipettes
- Rubber dispensing bulbs
- Hemacytometer
- Centrifuge
- FDA stain
- A 250-mL beaker containing a 63 μm filter, as in Experiment 1
- Centrifuge tubes (15 mL)
- Disposable plastic petri dishes
- Three 10-mL pipettes
- Two 5-mL pipettes
- Calcofluor White stain
- Mineral oil (autoclaved)
- Microscope coverslips (autoclaved in a glass petri dish wrapped in foil)
- 10 mL preplasmolysis medium
- 10 mL enzyme medium containing either RITC or FITC (10 μL of a 3 mg/mL stock in ethanol)
- 30 mL rinse medium
- 10 mL flotation medium
- 10 mL CPW 13M
- 1 mL fusogen (PEG 22.5)
- 5 mL Ca^{2+} wash medium
- 10 mL culture medium

Following the protocols listed in Procedure 28.2 to complete this experiment.

Anticipated Results

After the Ca^{2+} washing medium has been added, protoplast fusion can be observed under an inverted microscope. If the microscope is equipped with fluorescence, FITC-stained protoplasts should fluoresce green, RITC-stained protoplasts should fluoresce red, and fusion products should fluoresce yellow. Without a selection scheme for growing the fusion products exclusively, it would be difficult to recover somatic hybrids from these cultures. Using iodoacetate or irradiation to prevent unfused protoplasts from regenerating complicates the experiment unnecessarily for a student lab. However,

Procedure 28.2
Protoplast Extraction and PEG-Mediated Fusion of Potato

Step	Instructions and Comments
	Day 1—Preparation
1	In the late afternoon, cut in vitro shoots and leaves of potato plantlets into fine pieces using two scalpels in a 10 cm petri dish and add preplasmolysis medium for 1–8 h.
2	Remove preplasmolysis medium with a sterile Pasteur pipette and add the enzyme medium with either FITC or RITC (one for one fusion partner and one for the other) if a fluorescence microscope is available. Incubate in the dark at room temperature on a shaker (60 rpm) for 4–16 h.
	Day 2—Isolation
3	Gently swirl the petri dishes to loosen protoplasts from debris; then transfer protoplasts in enzyme media to a sterile 63 μm filter in a 250 mL beaker.
4	Add 3 mL rinse solution to the petri dish, swirl vigorously, and add this to the enzyme solution. Rinse the filter with an additional 2 mL rinse medium. Remove the filter and pour the protoplast solution into a sterile centrifuge tube. Centrifuge at 50× g for 5 min.
5	Remove supernatant with a Pasteur pipette. Resuspend the pellet in 10 mL flotation medium. Add 1 mL (25 drops) rinse medium dropwise to the top of the sucrose solution by dripping it gently down the side of the centrifuge tube so that the two layers remain intact. Centrifuge at 50× g for 10 min.
6	Collect protoplast band at the interface between rinse and sucrose media with a Pasteur pipette. Transfer protoplasts to a sterile 15 mL centrifuge tube, add rinse medium to 10 mL, and spin at 50× g for 5 min.
7	Remove rinse medium with a Pasteur pipette, being careful not to disturb the pellet. Adjust the protoplast density to 1×10^6 per mL in CPW 13M medium. If insufficient protoplasts have been obtained to reach this density, adjust both fusion partners to the same density as close to this as possible.
	Day 2— Fusion
8	Students should be paired such that each student in a pair has isolated protoplasts from different parents with different stains. The students can then share their protoplasts with each other and retain the remainder for control plates.
9	Place a small drop of sterile mineral oil in the center of a plastic petri dish and gently lower a sterile coverslip onto the oil.
10	Pipet three drops of each of the two protoplast suspensions (one incubated with RITC and isolated by one student in the pair and the other incubated with FITC and isolated by the other student in the pair) onto the coverslip. Leave undisturbed for 10 min to allow the protoplasts to settle and stick to the coverslip.
11	Add six drops fusogen PEG 22.5, drop by drop, surrounding the protoplasts, leaving the last drop for the center. Incubate at room temperature for 20–25 min.
12	Gently add three drops of Ca^{2+} washing medium every 5 min for 20 min to one side of the protoplast culture while removing three drops of fusion medium from the opposite side using a second Pasteur pipette. Agglutination begins during this washing procedure. It is important that the protoplasts are disturbed as little as possible during this process.
13	Replace the Ca^{2+} washing medium with culture medium. Place three drops of culture medium at one side of the protoplast mixture. Then, using a second Pasteur pipette, remove some of the washing medium from the opposite side; repeat two more times. Then flood the coverslip with culture medium and place a few drops of culture medium on the plate surrounding the coverslip to maintain humidity. Seal the plate and incubate in dim light at 25°C.
14	Prepare control plates of the remaining unfused protoplasts by centrifuging for 5 min, removing the supernatant, and replacing it with sufficient culture medium to adjust the density to 2.5×10^5 per mL. Dispense to the appropriate size petri dish, depending on the amount of culture obtained—4-6 mL in a 100 × 20 mm dish; 2–4 mL in a 60 × 15 mm dish; 1–2 mL in a 35 × 10 mm dish. Seal the plate and incubate in dim light at 25°C, as earlier.
	Postfusion handling of protoplasts
15	After 2 and 7 days, the cultures should be checked for cell wall regrowth and cell division.
16	At 7–10 days after isolation, plating efficiency of the fusions and control plates of each parent should be recorded.

17	If the protoplasts are dividing and callus colonies are growing, the cultures should be diluted with fresh culture medium after approximately 2 weeks. This can be done by preparing double strength filter-sterilized culture medium and autoclaving an equal volume of double strength (0.6%) agarose in distilled water. The two can then be mixed to provide a volume of medium equal to what is in the plates of protoplasts. Then, the protoplasts in liquid culture medium are added to the fresh medium to embed the protoplasts in agarose while replenishing the medium simultaneously.
18	After 2 more weeks, the number of visible callus colonies should be recorded for the fusion plates and the controls. Callus morphology often differs among clones for obvious morphological traits such as color, texture, and size. The morphology of callus in each plate should be noted. Fast growing or obviously different callus in the fusion plates may be indicative of putative somatic hybrids.
19	The number of p-calli per plate should be recorded for fusions and controls after 4–6 weeks of culture. If there are too many to count, a sample region of equal size representative of each plate can be scored.

it is possible that the hybrid p-calli will differ morphologically from the control p-calli, such that putative hybrids can be identified.

Questions

- How would you select for fusion products?
- Why is mannitol used in protoplast culture?
- What frequency of fused protoplasts was observed?
- How did the plating efficiency compare between the fusion plates and the control plates?
- Was there a difference in p-calli frequency, morphology, or growth rate between the control plates and the fusion plates?

REFERENCES

Carlson, P.S., H.H. Smith, and R.D. Dearing. 1972. Parasexual interspecific plant hybridisation. *Proc. Natl. Acad. Sci. USA* 69:2292–2294.

Cheng, J., and R.E. Veilleux. 1991. Genetic analysis of protoplast culturability in *Solanum phureja*. *Plant Sci.* 75:257–265.

Haberlach, G.T., B.A. Cohen, N.A. Reichert, M.A. Baer, L.E. Towill, and J.P. Helgeson. 1985. Isolation, culture and regeneration of protoplasts from potato and several related *Solanum* species. *Plant Sci.* 39:67–74.

Helgeson, J. P., G. J. Hunt, G. T. Haberlach, and S. Austin. 1986. Somatic hybrids between *Solanum brevideus* and *Solanum tuberosum*: Expression of a late blight-resistant gene and potato leaf roll resistance. *Plant Cell Rep*. 3: 212–214.

Kisaka, H. and T. Kameya. 1994. Production of somatic hybrids between *Daucus carota* L. and *Nicotiana tabacum*. *Theor. Appl. Genet.* 88:75–80.

Murashige T. and F. Skoog. 1962. A revised medium for rapid growth and bioassays with tobacco tissue cultures. *Physiol. Plant.* 15:473–497.

Nagata, T. and I. Takebe. 1970. Cell wall regeneration and cell division in isolated tobacco mesophyll protoplasts. *Planta* 92:301–308.

Rose, R.J., M.R. Thomas, and J.T. Fitter. 1990. The transfer of cytoplasmic and nuclear genomes by somatic hybridisation. *Aust. J. Plant Physiol.* 17:303–321.

Sidorov, V.A., D.P. Yevtushenko, A.M. Shakhovsky, and Y.Y. Gleba. 1994. Cybrid production based on mutagenic inactivation of protoplasts and rescuing of mutant plastids in fusion products: Potato with a plastome from *S. bulbocastanum* and *S. pinnatisectum*. *Theor. Appl. Genet.* 88:525–529.

29 Haploid Cultures*

Denita Hadziabdic, Phillip A. Wadl, and Sandra M. Reed

CONCEPTS

- Haploids may be produced from male (androgenesis) or female (gynogenesis) gametophytes.
- In vitro androgenesis involves culture of anthers or isolated microspores.
- Gynogenesis may be induced in unfertilized ovule or ovary culture.
- Success of haploid cultures depends on genotype, media conditions, and cultural conditions.
- Stage of microspore development at time of culture is critical to success of androgenesis.
- Chromosome doubled haploids (diploids) are useful in breeding projects.

The ability to produce haploid plants is a tremendous asset in genetic and plant breeding studies. Haploid plants are considered autonomous and saprophytic since they originate from gametic cells, and therefore have the gametophytic chromosome number. The haploid phase dominates the life cycle for some lower plants, such as unicellular green alga (*Chlamydomonas* Ehrenb.). In higher plants, the haploid phase (the embryo sac and pollen grains) is greatly diminished; however, it has an important role in the life cycle. Due to the presence of only one set of chromosomes, heritability studies using haploids are simplified and recessive mutations are easily identified. In addition, doubling the chromosome number to produce doubled haploids results in a completely homozygous plant. Theoretically, the genotypes present among a large group of doubled haploids derived from an F_1 hybrid represent in a fixed form the genotypes expected from an F_2 population. Haploid breeding shortens the breeding process and allows production of homozygous lines from a segregating population in the immediate generation (Jain et al., 1996a).

Haploids have been available for genetic studies for many years. Prior to the 1960s, they were mostly obtained spontaneously following interspecific hybridization or through the use of irradiated pollen, but usually only infrequently and in very small numbers. Haploid methodology took a giant step forward 40 years ago when Guha and Maheshwari (1964) found that haploid plants could be obtained on a regular basis and in relatively large numbers by placing immature anthers of pricklyburr (*Datura innoxia* Mill.) into culture. This work was rapidly expanded using tobacco (*Nicotiana tabacum* L.), which became the "model species" for anther culture experiments. In vitro androgenesis became an indispensable tool for tree breeding because the majority of tree species are outbreeding, highly heterozygous, and have long generation cycles (for review of in vitro androgenesis in tree species see Srivastava and Chaturvedi, 2008). To date, haploids and doubled haploids have been reported in over 200 species (Forster et al., 2007); several good reviews provide lists of these species (Maheshwari et al., 1982; Bajaj, 1983; Heberle-Bors, 1985; Dunwell, 1996). While efforts have been more limited, haploids have also been obtained from in vitro culture of the female gametophyte in over 30 species (Keller and Korzun, 1996; Lakshmi Sita, 1997). Gynogenesis has been successfully applied to several species in which androgenesis is generally ineffective, such as sugar beet (*Beta vulgaris* L.), onion (*Allium cepa* L.), and Gerbera daisy (*Gerbera jamesonii* H. Bolus ex Hook).

Although much of the terminology used in this chapter has been discussed in previous chapters, the in vitro induction of haploids involves a few specialized terms. A haploid is a plant with the

* Mention of trade names of commercial products in this chapter is solely for the purpose of providing specific information and does not imply recommendation or endorsement by the U.S. Department of Agriculture.

gametic or "n" number of chromosomes. Doubled haploids are chromosome doubled haploids or "2n" plants. Androgenesis is the process by which haploid plants develop from the male gametophyte (pollen). When anthers are cultured intact, the procedure is called *anther culture* (AC). Microspore culture involves isolating microspores from anthers before culture and is sometimes referred to as pollen culture. Haploids derived via the female gametophyte (embryo sac) are obtained through a process known as gynogenesis in which haploid cells are stimulated to develop into an embryo in an induced process similar to parthenogenesis.

In vitro gynogenesis involves the culture of unfertilized ovules or ovaries. While both androgenesis and gynogenesis may occur in vivo, the usage of the terms in this chapter will refer to the in vitro induction of haploids via these two mechanisms. Although haploid cultures are produced in vivo from parthenogenesis, pseudogamy or chromosome elimination after wide hybridization, here we will focus on in vitro haploid culture development using androgenesis (anther and microspore culture) and gynogenesis (flower and ovary culture); see Figure 29.1.

This chapter will begin with general discussions of androgenesis and gynogenesis, followed by a review of the factors that affect the successful production of androgenic and gynogenic haploids. Finally, some of the basic procedures used for the in vitro production of haploids will be summarized. Excellent discussions of in vitro haploid production, along with specific protocols for a number of crop species, can be found in Jain et al., 1996a–d, 1997.

ANDROGENESIS

Development of Haploids

Haploid plants develop from anther culture either directly or indirectly through a callus phase. Direct androgenesis mimics zygotic embryogenesis; however, neither a suspensor nor an endosperm is present. At the globular stage of development, most of the embryos are released from the pollen cell wall (exine). They continue to develop, and after 4 to 8 weeks, the cotyledons unfold and plantlets emerge from the anthers. Direct androgenesis is primarily found among members of the Solanaceae and Cruciferae.

During indirect androgenesis, the early cell division pattern is similar to that found in the zygotic embryogenic and direct androgenic pathways. After the globular stage, irregular and asynchronous divisions occur, and callus is formed. This callus must then undergo organogenesis for haploid plants to be recovered. The cereals are among the species that undergo indirect androgenesis.

The early cell divisions that occur in cultured anthers have been studied (for review, see Reynolds, 1990). For species cultured during the uninucleate stage, the microspore either undergoes a normal mitosis and forms a vegetative and a generative nucleus or divides to form two "similar looking" nuclei. In those cases where avegetative and generative nuclei are formed in culture, or where binucleate microspores are placed into culture, it is usually the vegetative nucleus that participates in androgenesis. The only species in which the generative nucleus has been found to be actively involved in androgenesis is black henbane (*Hyoscyamus niger* L.). When similar looking nuclei are formed, one or both nuclei may undergo further divisions. In some cases, the two nuclei will fuse, producing homozygous diploid plants or callus. Since diploid callus may also arise from somatic tissue associated with the anther, diploids produced from anther culture cannot be assumed to be homozygous. To verify that plants produced from anther culture are haploid, chromosome counts should be made from root tips or other meristematic somatic tissues (see Chapter 7). Because haploids derived from diploid species are expected to be sterile or have greatly reduced fertility, pollen staining, which is much quicker and requires less skill than chromosome counting, can also be used to identify and eliminate potential diploids. However, pollen staining may not distinguish between haploids and plants that have reduced fertility because they have a few extra or missing chromosomes (i.e., aneuploids). Haploids and diploids recovered from anther culture may also be distinguished by comparing size of cells, particularly stomatal guard cells, or through the use of flow cytometry.

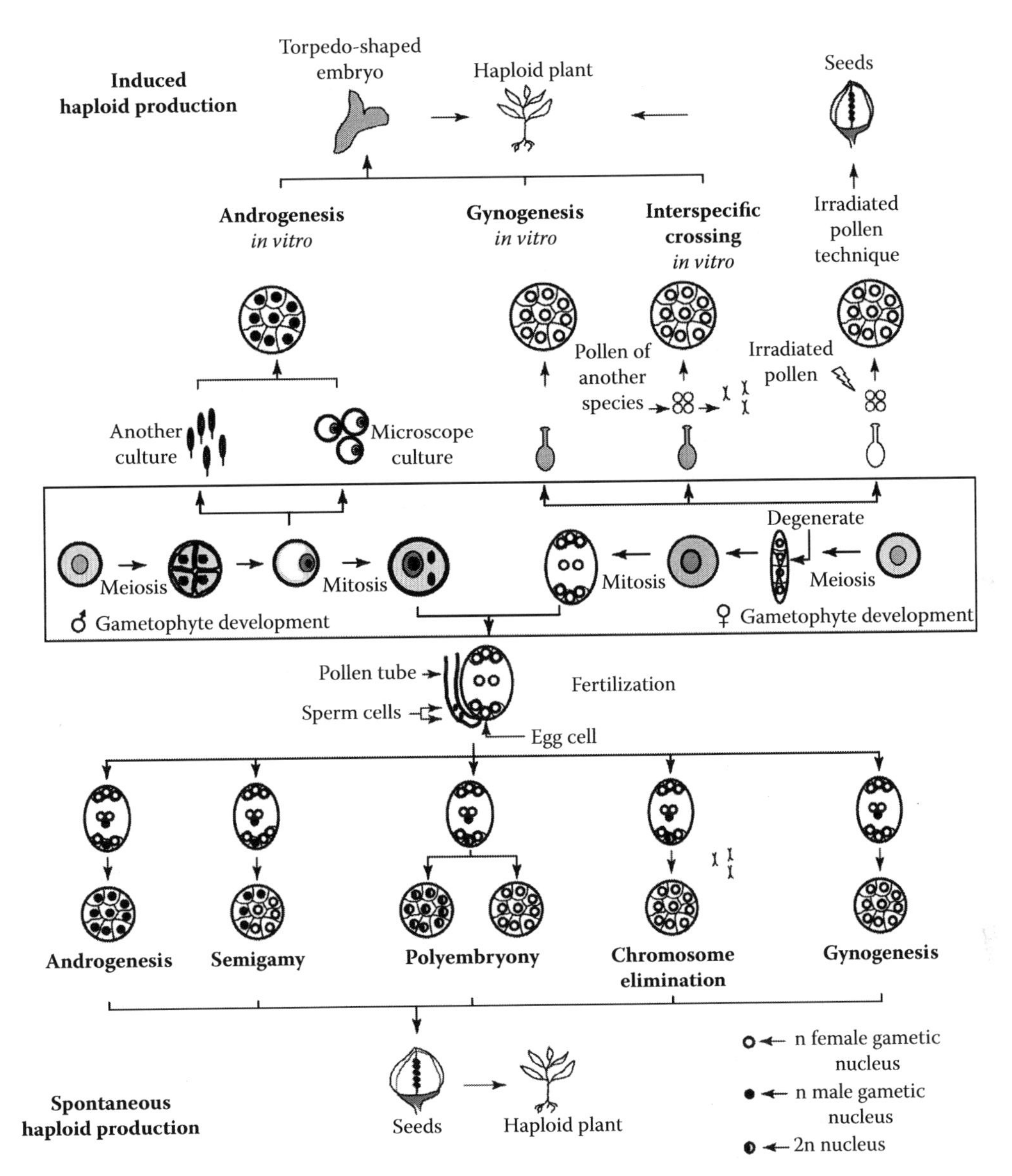

FIGURE 29.1 Different methods of plant haploid production. Haploids can be induced either spontaneously or by various in vitro methods using male and female tissues. In-vitro-cultured anthers or microspores, ovules or ovaries, and irradiated pollen or pollen from another species can be induced via androgenesis, gynogenesis, and distant hybridization, respectively. Androgenesis (male parthenogenesis) is the process in which the embryo contains only paternal chromosomes because the egg has been activated by sperm without fusion of the egg and sperm nuclei. Gynogenesis (spontaneous or induced female parthenogenesis) refers to the embryo that contains only maternal chromosomes because of the failure of sperm cell to fuse with the egg nucleus. Interspecific crossing involves development of haploid embryo by fertilizing an ovule with pollen of another species and the subsequent elimination of the chromosomes of the pollen. Irradiated pollen technique is development of haploid embryo by fertilization of an ovule with irradiated (inactive) pollen. In addition to induced haploid production, spontaneous haploids occur via semigamy, polyembryony, chromosome elimination, gynogenesis, and androgenesis at very low frequencies. Semigamy is an abnormal type of fertilization in which the sperm nucleus penetrates the egg but fails to fuse with the egg nucleus. The resulting embryo consists of both maternal and paternal segments derived independently from the division of sperm and egg nuclei. Polyembryony refers to production of two or more embryos from a single fertilized egg or ovule. (Reprinted from Forster, B.P., E. Heberle-Bors, K.J. Kasha, and A. Touraev. 2007. *Trends in Plant Science* 12:368–375. With permission from Elsevier.)

Problems Associated with Anther Culture

A number of challenges are encountered during or as a result of anther culture, and they range from low yields, production of albino plants, chromosomal abnormalities, and somaclonal variation to overall genetic instability. Although the application of anther culture is widely used, the processes involved are not well understood and number of responsive genotypes can often be limiting factors for breeders. Many of the major horticultural and agronomic crops do not yield sufficient haploids to allow them to be useful in breeding programs. In other species, genetic instability has often been observed from plants recovered from anther and microspore cultures.

The term *gametoclonal variation* has been coined to refer to the variation observed among plants regenerated from cultured gametic cells and has been observed in many species. While often negative in nature, some useful traits have been observed among plants recovered from anther and microspore culture. Gametoclonal variation may arise from changes in chromosome number (i.e., polyploidy or aneuploidy) or chromosome structure (e.g., duplications, deletion, translocations, inversions, etc.). In tobacco, gametoclonal variation has been associated with an increase in amount of nuclear DNA without a concomitant increase in chromosome number (DNA amplification). In many cereals, a high percentage of the plants regenerated from anther culture are albino; changes in cytoplasmic DNA have been associated with this albinism. Good discussions and reviews of gametoclonal variation can be found in several chapters of Jain et al. (1996b).

Gynogenesis

As with androgenesis, gynogenic haploids may develop directly or indirectly via regeneration from callus. (For review, see Keller and Korzun, 1996.) The first cell divisions of gynogenesis are generally similar to those of zygotic embryogenesis. Direct gynogenesis usually involves the egg cell, synergids, or antipodals with organized cell divisions leading first to the formation of proembryos and then to well-differentiated embryos. In indirect gynogenesis, callus may be formed directly from the egg cell, synergids, polar nuclei or antipodals, or may develop from proembryos. Plants regenerated from callus may be haploid, diploid, or mixoploid. As with plants produced from anther cultures, chromosome counts can be used to identify haploids. Distinguishing between homozygous dihaploids, in which chromosome doubling occurred in culture, and diploids that developed from somatic tissue requires the use of molecular markers. Additionally, gynogenesis can be induced by using gamma-irradiated pollen (Winton and Stettler, 1974), which is often used in tree species. Pollen of a related species is irradiated, and later used to pollinate the desired parent to induce gynogenetic haploid production.

The major problems affecting the use of gynogenesis are the lack of established protocols for most species, poor yields, and production of diploid or mixoploid plants. Gametoclonal variation among gynogenic haploids has not been widely studied; however, it has been noted that, unlike androgenesis, gynogenesis of cereal species does not result in the production of albino plants. Even though gynogenesis is the least favored technique due to its low efficiency, production and importance of doubled haploids in species that do not respond to more efficient techniques makes this method desirable, regardless of obstacles associated with this process.

Factors Affecting Androgenesis

Six major factors can be recognized that affect androgenesis:

1. Genotype
2. Physiological condition of donor plants
3. Stage of microspore development at the time of culture
4. Anther pretreatment

5. Composition of culture medium
6. Conditions during culture.

Some factors are more vital and crucial for induction of androgenic structures than others, and genotype is most likely the most important role during this process.

Genotype

The choice of starting material for an anther or microspore culture project is of the utmost importance. In particular, genotype plays a major role in determining the success or failure of an experiment. Haploid plant production via androgenesis has been very limited or nonexistent in many plant species. Furthermore, within a species, differences exist as to the ability to produce haploid plants. Even within an amenable species, such as tobacco, some genotypes produce haploids at a much higher rate than do others. Because of this genotypic effect, it is important to include as much genetic diversity as possible when developing protocols for producing haploid plants via anther or microspore culture.

Physiological Condition of Donor Plants

The age and physiological condition of donor plants often affect the outcome of androgenesis experiments. In most species, the best response usually comes from the anthers obtained from the first flush of flowers produced by a plant early in the season. As a general rule, anthers should be cultured from buds collected as early as possible during the course of flowering. Various environmental factors to which the donor plants are exposed may also affect haploid plant production. In the case of some cereals, for example, vernalization period of the donor plant plays a very important role for regeneration of androgenic plants. Light intensity, photoperiod, temperature, and field versus greenhouse-grown plants have been investigated, and at least for some species, these have been found to influence the number of plants produced from anther cultures. Alternations in the physiology of donor plants by other treatments such as additional salts (tobacco microspore) or 2-chloroethyl phosphonic acid for 48 h at 10°C (rice inflorescence) have been shown to affect androgenesis (Heberle-Bors and Reinert, 1979; Wang et al., 1974). Specific optimum growing conditions differ from species to species and are reviewed by Powell (1990). In general, the best results are obtained from healthy, vigorously growing plants.

Stage of Microspore Development at the Time of Culture

The most critical factor affecting haploid production from anther and microspore culture is the stage of microspore development; for many species, success is achieved only when anthers are collected during the uninucleate stage of pollen development. In contrast, optimum response is obtained in tobacco and *Brassica napus* L. from anthers cultured just before, during, and just after the first pollen mitosis (late uninucleate to early binucleate microspores).

In developing a protocol for anther culture, one anther from each bud is usually set aside and later cytologically observed to determine the stage of microspore development. In many cases, anthers within a bud are sufficiently synchronized to allow this one anther to represent the remaining cultured anthers. Measurements of physical characteristics of the flower, such as calyx and corolla length and anther color, shape, and size are also recorded. Results of the experiments are analyzed to determine which microspore stage was the most responsive. The physical descriptions of the buds and anthers are then examined to determine if this microspore stage correlates to any easily identified inflorescence, flower, or anther characteristic(s). For example, in tobacco, buds in which the calyx and corolla are almost identical in length usually contain anthers having microspores at or near the first pollen mitosis. A researcher wishing to produce a maximum number of haploid plants of tobacco would collect only buds fitting this physical description.

Anther Pretreatment

For some species, a pretreatment following collection of buds, but before surface disinfestation and excision of anthers, has been shown to be beneficial and therefore enhanced the frequency of androgenesis. Application of temperature shock (heat or cold at optimal temperature) may increase the induction of androgenesis if applied as a pretreatment before culturing anthers or directly after culturing. Yields of tobacco haploids are often increased by storing excised buds at 7–8°C for 12 days prior to anther excision and culture (Sunderland and Roberts, 1979). For other species, temperatures from 4°–10°C and durations from 3 days to 3 weeks have been utilized. For Chinese pink (*Dianthus chinensis* L.), buds that were exposed to cold pretreatment (4°C for 3 days), followed by incubation of anthers in darkness, resulted in highest levels of embryogenic callus induction (Fu et al., 2008). For any one species, there may be more than one optimum temperature-length of treatment combination. In general, lower temperatures require shorter durations, whereas a longer pretreatment time is dictated for temperatures at the upper end of the cold pretreatment range mentioned above.

In barley (*Hordeum vulgare* L.), a pretreatment with mannitol was found to be successful in promoting microspore division and regeneration of anther cultures (Roberts-Oehlschlager and Dunwell, 1990).

Composition of Culture Medium

Various basal media with modified components have been used for establishing anther cultures. The list of media suitable for a wide range of species is constantly expanding; however, most media are specific for a few or even only one genotype. In the case of tobacco and a few other species, androgenesis can be induced on a simple medium such as that developed by Nitsch and Nitsch (1969). For most other species, the commonly used media for anther culture include MS (Murashige and Skoog, 1962), N6 (Chu, 1978), or variations on these media. Mordhorst and Lorz (1993) found that the salt composition of media has no effect on the frequency of the initial divisions. Salt composition had moderate to dramatic effects on plating efficiency, embryogenesis and plant regeneration, respectively (Mordhorst and Lorz, 1993). The pH of the medium also has an important role for eliciting successful androgenesis. Usually, pH in acidic range seems to provide optimum growth for anther cultures. In some cases, complex organic compounds, such as potato extract, coconut milk, and casein hydrolysate, have been added to the media. For many species, 58–88 mM (2–3%) sucrose is added to the media whereas other species, particularly the cereals, have responded better to higher (up to 435 mM or about 15%) concentrations of sucrose. The higher levels of sucrose may fulfill an osmotic, rather than a nutritional, requirement. Also, during the induction phase, high sucrose concentrations tend to suppress the divisions of somatic cells and, thereby, promote microspore callusing/embryogenesis. Other sugars, such as ribose, maltose, and glucose, have been found to be superior to sucrose for some species. For rye, higher androgenic capacity was achieved by replacing sucrose with maltose in the induction medium, resulting in both increased induction and regeneration rates (Jain et al., 1996d).

For a few species, such as tobacco, it is not necessary to add plant growth regulators (PGRs) to the anther culture media. Most species, however, require a low concentration of some form of auxin in the media. Cytokinin is sometimes used in combination with auxin, especially in species in which a callus phase is intermediate in the production of haploid plants.

The gelling agent is another important factor that can influence success of anther culture. Therefore, culture media is often solidified using agar. Because agar may contain compounds inhibitory to the androgenic process in some species, the use of alternative gelling agents has been investigated. Ficoll, Gelrite (Merck and Co., Inc., Rahway, New Jersey), agarose, and starch have proven superior to agar for solidifying anther culture media in various species (Calleberg and Johansson, 1996). The use of liquid medium has been advocated by some researchers as a way to avoid the potentially inhibitory substances in gelling agents. Anthers may be placed on the surface of the medium, forming a so-called float culture. Alternatively, microspores may be isolated and cultured directly in liquid medium.

Conditions during Culture

Temperature

Temperature is one of the important aspects of androgenesis that influences the induction of pollen, embryo, and callus development. Anther cultures are usually cultured at 24–25°C, and there are several possible explanations for the success rate of androgenesis related to elevated temperatures (Dunwell et al., 1983). High temperatures may disrupt normal development of somatic anther tissue, synchronize the microspore population, and consequently increase total number of spores at the stage of cell cycle that is susceptible to induction. Additionally, elevated temperatures may result in increased growth rate of haploid embryos when compared to nonhaploids (Jain et al., 1996a). In some species, an initial incubation at a higher or lower temperature has been beneficial. Haploid plant production was increased in *Brassica campestris* L. by culturing the anthers at 35°C for 1 to 3 days prior to culture at 25°C (Keller and Armstrong, 1979). In contrast, androgenesis was promoted in *Cyclamen persicum* Mill. by incubating cultured anthers at 5°C for the first 2 days of culture (Ishizaka and Uematsu, 1993).

Light

Some species respond best when exposed to alternating periods of light and dark, whereas continuous light or dark cultural conditions have proven beneficial in other species. The highest rate of somatic embryogenesis in *Dianthus chinensis* was achieved after maintaining anther culture in darkness for 40 days (Fu et al., 2008). Other physical cultural factors, such as atmospheric conditions in the culture vessel, anther density, and anther orientation, have been studied and found to affect androgenic response in some species; however, species have varied greatly in their response to these physical factors.

Factors Affecting Gynogenesis

Gynogenesis has not been investigated as thoroughly or with as many species as has androgenesis; therefore, less information is available concerning the various factors that contribute to the successful production of haploids from the female as compared to the male gametophyte. However, several factors have been identified that affect the successful production of haploids from the female gametophyte.

Genotype

Several studies have identified genotype as a critical and probably the most important factor in determining the success of a gynogenesis experiment. Not only are there differences between species, but genotypes within individual species have responded differently. In summer squash (*Cucurbita pepo* L.), genotype proved to be a key factor influencing in vitro gynogenesis, indicating that there are differences between ovule and anther culture response within the same genotype (Shalaby, 2007). As with androgenesis, it is important to include a wide range of genotypes in ovule and ovary culture experiments.

Media

Media has also been identified as an important factor in gynogenesis. The most commonly used basal media for recovering gynogenic haploids are MS (Murashige and Skoog, 1962), B-5 (Gamborg et al., 1968), Miller (Miller, 1963), or variations on these media. Sucrose levels have ranged from 58–348 mM (2–12%). In *C. pepo*, as the concentration of sucrose increased, the percentage of ovules forming embryos decreased (Shalaby, 2007). On the other hand, increased sucrose concentrations (8–10%) in the culture medium have been shown to be beneficial for some species like sweet potato (*Ipomoea batatas* L.) (Kobayashi et al., 1993) and onion (*Allium cepa* L.) (Campion et al., 1992).

While gynogenic haploids have been developed in a few species without the use of growth regulators, most species have required auxins and/or cytokinins in the medium. For those species that undergo indirect gynogenesis, both an induction and a regeneration medium may be required. Most ovule and ovary culture experiments have been conducted using solid medium. A list of specific media components used for gynogenesis in several crop species can be found in Keller and Korzun (1996).

Stage of Gametophytic Development

Because the female gametophyte is difficult to handle and observe, determining the optimum stage of gametophytic development for gynogenesis is usually based on other, more easily discerned, characteristics. Performance of ovule and ovary cultures has often been correlated with stage of microspore development. Depending on species, the best results have been obtained when the female gametophyte was cultured from the late uninucleate to trinucleate stage of megaspore development. In other studies, number of days until anthesis has been used as an indicator of stage of gametophytic development. A few gynogenesis studies that involved direct observations of the female gametophyte have been conducted. For several species, gynogenesis was most successful when cultures were initiated when the embryo sac was mature or almost mature (for review, see Keller and Korzun, 1996).

Other Factors

Temperature shock treatments have been found to improve gynogenesis by diverting normal gametophytic development into a saprophytic pathway and therefore promote the formation of haploid embryos. In few species, cold pretreatment of flower buds at 4°C for 4 to 5 days has a strong influence on the gynogenesis process in vitro and has been effective in increasing yields of haploid embryos or callus. However, this has not been widely explored since ideal physical characteristics of a culture are often genotype dependent, requiring specific protocols for most species. Shalaby (2007) investigated the effects of low and high temperature treatments (4° and 32°C) and various duration periods for haploid induction of *C. pepo*. Ovules exposed to 32°C for 4 days produced the greatest number of gynogenic ovules, followed by ovules exposed to 4°C for 4 days. Higher temperatures, when compared to cold treatment, resulted in better embryogenic response (28 and 22%, respectively).

Seasonal effects have been observed in several species. Many of the other factors that affect androgenesis probably also affect gynogenesis; however, in most cases, insufficient data are available to detect trends in response. These variables should be considered when initiating gynogenesis experiments.

GENERAL ANDROGENESIS PROCEDURES

Collection, Disinfestation, Excision, and Culture

Floral buds may be collected from plants grown in the field, greenhouse, or growth chamber. Entire inflorescences or individual buds are harvested and kept moist until ready for culturing. If buds are to be pretreated (i.e., low temperature), they should be wrapped in a moistened paper tissue and placed into a small zipper-type plastic bag.

Flower buds are typically disinfested using a 5% sodium or calcium hypochlorite solution for 5 to 10 min and then rinsed thoroughly in sterile distilled water. Anthers are aseptically excised in a laminar flow hood, taking care not to cause injury. If the anther is still attached to the filament, the filament is carefully removed.

If a solid medium is used, the anthers are gently pressed onto the surface of the medium (just enough to adhere to the medium), but should not be deeply embedded. When using a liquid medium for intact anthers, the anthers are floated on the surface. Care must be taken when moving float cultures so as not to cause the anthers to sink below the surface.

For most species, disposable petri dishes are utilized for anther cultures. For a species with large anthers, such as tobacco, the anthers from 4–5 buds (20–25 anthers) may be cultured together on one 100 × 15 mm diameter petri dish. For species with smaller anthers, or for certain experimental designs, smaller petri dishes or other containers may be more useful. Petri dishes are usually sealed and placed into an incubator; the specific temperature and light requirements of the incubator depend on the species being cultured.

While many of the steps involved in microspore culture are similar to those of anther culture, microspore culture also requires the separation of the microspores from the surrounding anther tissue. Microspores may be squeezed out of anthers using a pestle or similar devise, or a micro-blending procedure may be used. See Dunwell (1996) for a review of literature pertaining to microspore culture.

Determining Stage of Microspore Development

For most species, stage of microspore development can be determined by "squashing" an entire anther in aceto-carmine or propiono-carmine and then observing the preparation under a low-power objective of a light microscope. The early uninucleate microspore is lightly staining with a centrally located nucleus. As the uninucleate microspore develops, its size increases and a large central vacuole is formed. As the microspore nears the first pollen mitosis, the nucleus is pressed up near the periphery of the microspore (Figure 30.1a). Staining will still be fairly light. Pollen mitosis is of short duration, but it may sometimes be observed; it is recognized by the presence of condensed chromosomes (Figure 30.1b). The product of the first pollen mitosis is a binucleate microspore containing a large vegetative and a small generative nucleus. The vegetative nucleus is often difficult to recognize because it is so diffuse and lightly staining. However, this stage may be definitively identified by the presence of the small densely staining generative nucleus (Figure 30.1c). As the binucleate microspore ages, the intensity of the staining increases and starch granules begin to accumulate (Figure 30.1d). Eventually, both nuclei may be hidden by the dark staining starch granules.

Handling of Haploid Plantlets

For species undergoing direct androgenesis, small plantlets can usually be seen emerging from the anthers 4 to 8 weeks after culture (Figure 30.3). When these get large enough to handle, they should be teased apart using fine-pointed forceps and can then either be placed onto a rooting medium (usually low salt, with small concentration of an auxin) or transplanted directly into a small pot filled with soilless potting mixture. The callus produced in species that undergo indirect androgenesis must be removed from the anther and placed onto a regeneration medium containing the appropriate ratio of cytokinin to auxin.

To produce doubled haploid plants, it is necessary to double the chromosome number of the haploids, and for many species, a colchicine treatment is used. Published procedures for producing polyploids from diploids can be modified for use with anther culture derived haploids. For example, it may be possible to use a colchicine treatment designed for small seedlings with haploid plants directly out of anther culture. Alternatively, established procedures using larger plants may be used. In wheat and other cereals, chromosome doubling is induced by initially culturing anthers on a medium containing a low concentration of colchicine. In addition to leading to the direct regeneration of homozygous doubled haploids, the inclusion of colchicine in the medium for the first few days of culture caused a decline in the number of albino regenerants (Barnabás et al., 2001).

General Gynogenesis Procedures

Gynogenesis experiments are usually conducted using unfertilized ovules or ovaries, although entire immature flower buds have been cultured in a few species. It is easier to dissect ovaries than ovules without damaging the female gametophyte. However, in polyovulate ovaries, it may be advantageous to excise the ovules so that they can be in direct contact with the culture medium.

Inflorescences must be collected before pollen shed, unless the species is highly self-incompatible or a male-sterile line is used. In developing a gynogenesis protocol for a species, it may be necessary to collect explants from several days before anthesis to just before anthesis. As discussed earlier, the stage of microspore development is sometimes recorded as an indicator of the developmental stage of the female gametophyte. Procedures used for determining stage of microspore development are described earlier in this chapter.

Disinfestation varies depending on species, growing conditions of explant source, and choice of explant. Woody plant material often requires longer disinfestation times and/or stronger sterilizing agents than herbaceous materials. Tissue from greenhouse-grown plants is usually easier to disinfest than that of field-grown plants. If ovules are to be cultured, a harsh surface sterilization procedure should be applied to ovaries. It should not be necessary to disinfest the ovules since they are presumed to have been removed from a sterile environment inside the ovary. Commonly used sterilizing agents and disinfestation times are presented in Chapter 3 of this book.

Techniques used for the excision of ovules depend on the arrangement of ovules within the ovary. Care must be taken not to let ovules dry out during excision. A solid medium is typically used for gynogenesis experiments; choice of culture vessel depends on size of explant. Disposable petri dishes work well for culturing ovules of small-seeded polyovulate species, whereas test tubes may be preferable for large ovaries.

Handling procedures for gynogenic haploids are similar to those described for androgenic haploids. As plants emerge from cultured ovules or ovaries, they can be transferred to a rooting medium or transplanted directly to a soilless potting mixture. Colchicine or another mitotic inhibitor is typically used for doubling chromosome number to produce doubled haploids.

Doubled Haploids (DH) in Plant Breeding and Genetics

Doubled haploid breeding methods involve making haploid tissues or plants (n) from heterozygous parents and doubling the chromosomes in order to obtain diploid plants (2n), which are referred to as doubled haploids (Baenziger, 1996). After chromosome doubling (formation of identical copy of each haploid chromosome) every gene is homozygous, and therefore each doubled haploid plant is considered to be homozygous.

Doubled haploids can be produced by anther culture or by parthenogenesis by crossing with desired genotypes. In 1946, Chase pioneered the use of haploids in breeding by exploiting a spontaneous parthenogenesis system to produce the first maize doubled haploid inbreds (Chase, 1949). Today, doubled haploid cultivars are the favored choice of many breeders and are often used as parents for F_1 hybrid seed production. Doubled haploid techniques can greatly reduce time needed to obtain stable resistant lines suitable for future breeding efforts on many crops. In maize, for example, by using doubled haploid technology, the time required to produce inbred lines is reduced from six or more generations of selfing to two. Secondly, the effectiveness of selection is improved by availability of higher genetic variance among doubled haploid lines when compared to F_2 or selfed plants.

Doubled haploids in combination with marker-assisted selection can greatly reduce the time needed for improving elite lines that are defective for other desired traits (e.g., disease or drought resistance) as compared to traditional backcrossing techniques that are often used in plant breeding. Reversible male sterility and doubled haploid production are two relatively new but valuable technologies required for F_1 hybrid breeding. Novel techniques such as reverse breeding involve suppression of recombination during microspore formation and recovery of doubled haploids from those spores. Recombinant inbred populations created by reverse breeding can further be screened via molecular markers to identify populations containing complementary combinations of chromosomes and allow an original heterozygous parent of the doubled haploid to be reconstructed by hybridizing the two individuals (Forster et al., 2007).

Doubled haploid lines, due to their homozygosity and uniformity, are often used in genetic studies related to quantitative traits, detection of linkage and gene interactions, production of genetic translocations, substitutions and chromosome addition lines, or as permanent mapping populations. This way, they can be proliferated and reproduced without changing the genetic makeup in the long term (Lapitan et al., 2009; Semagn et al., 2006). The ability to produce homozygous and homogenous lines in shorter time than conventional breeding methods is one of the biggest advantages of using doubled haploids. However, cost of producing those lines in sufficient quantities for breeding program can be a limiting factor for integrating this system.

SUMMARY

Haploids of many plant species can be produced in vitro. Anther culture has been the most widely used in vitro technique for producing haploids, but androgenic haploids have been obtained in a few species through the culture of isolated microspores. While fewer studies have been conducted involving the induction of haploids from the female gametophyte, gynogenesis has proven successful in several species. Yields of androgenic and gynogenic haploids differ greatly depending on species, and are also affected by cultural conditions, such as media formulation, stage of microspore or embryo sac at time of culture, and use of a low- or high-temperature pretreatment. Both androgenic and gynogenic haploids may either arise directly or be produced indirectly through a callus intermediate. In vitro derived doubled haploids of several important crop species are currently produced routinely. Use of doubled haploids in breeding programs of these species has shortened cultivar development time. Expansion of this valuable breeding technique to additional species should occur as continued efforts are made to identify factors critical to in vitro induction of haploidy.

REFERENCES

Baenziger, P.S. 1996. Reflections on double haploids in plant breeding. In S.M., Jain, S. K. Sopory, and R.E. Veilleux (Eds.), *In Vitro Haploid Production in Higher Plants, Vol. 1, Fundamental Aspects and Methods*. Kluwer Academic Publishers, Dordrecht, pp. 35–48.

Bajaj, Y.P.S. 1983. In vitro production of haploids, pp. 228–287. In D.A., Evans, W.R. Sharp, P.V. Ammiarto, and Y. Yamada (Eds.), *Handbook of Plant Cell Culture, Vol. 1: Techniques for Propagation and Breeding*. Macmillan, New York.

Barnabás, B., É. Szakács, I. Karsai, and Z. Bedö. 2001. In vitro androgenesis of wheat: From fundamentals to practical application. *Euphytica* 119:211–216.

Calleberg, E.K. and L.B. Johansson. 1996. Effect of gelling agents on anther culture, pp. 189–203. In S.M., Jain, S. K. Sopory and R.E. Veilleux (Eds.) *In Vitro Haploid Production in Higher Plants, Vol. 1, Fundamental Aspects and Methods*. Kluwer Academic Publishers, Dordrecht.

Campion, B., M.T. Azzimonti, E. Vicini, M. Schiavi, and A. Falavigna. 1992. Advances in haploid plants induction in onion (*Allium cepa* L.) through in vitro gynogenesis. *Plant Sci.* 86:97–104.

Chase, S.S. 1949. The reproductive success of monoploid maize. *Am. J. Bot.* 36:795–796.

Chu, C. 1978. The N6 medium and its applications to anther culture of cereal crops. In *Proceedings of Symposium on Plant Tissue Culture*. Science Press, Peking, pp. 51–56.

Dunwell, J.M. 1996. Microspore culture, pp. 205–216. In S.M., Jain, S. K. Sopory, and R.E. Veilleux (Eds.), *In Vitro Haploid Production in Higher Plants, Vol. 1, Fundamental Aspects and Methods*. Kluwer Academic Publishers, Dordrecht.

Forster, B.P., E. Heberle-Bors, K.J. Kasha, and A. Touraev. 2007. The resurgence of haploids in higher plants. *Trends Plant Sci.* 12:368–375.

Fu, X., S. Yang and M. Bao. 2008. Factors affecting somatic embryogenesis in anther cultures of Chinese pink (*Dianthus chinensis* L.). *In Vitro Cell. Dev. Biol.—Plant* 44:194–202.

Gamborg, O.L, R.A. Miller, and K. Ojima. 1968. Nutrient requirements of suspension cultures of soybean root cells. *Exp. Cell Res.* 50:157–158.

Guha, S. and S.C. Maheshwari. 1964. In vitro production of embryos from anthers of *Datura*. *Nature* 204:497.

Heberle-Bors, E. 1985. In vitro haploid formation from pollen: A critical review. *Theor. Appl. Genet.* 71:361–374.

Heberle-Bors, E. and J. Reinert. 1979. Androgenesis in isolated pollen cultures of *Nicotiana tabacum*: Dependence upon pollen development. *Protoplasma* 99: 237–245.

Ishizaka, H. and J. Uematsu. 1993. Production of plants from pollen in *Cyclamen persicum* Mill. through anther culture. *Jpn. J. Breed.* 43:207–218.

Jain, S.M., S.K. Sopory, and R.E. Veilleux (Eds.). 1996a. *In Vitro Haploid Production in Higher Plants, Vol. 1: Fundamental Aspects and Methods*. Kluwer Academic Publishers, Dordrecht.

Jain, S.M., S.K. Sopory, and R.E. Veilleux (Eds.). 1996b. *In Vitro Haploid Production in Higher Plants, Vol. 2: Applications*. Kluwer Academic Publishers, Dordrecht.

Jain, S.M., S.K. Sopory, and R.E. Veilleux (Eds.). 1996c. *In Vitro Haploid Production in Higher Plants, Vol. 3: Important Selected Plants*. Kluwer Academic Publishers, Dordrecht.

Jain, S.M., S.K. Sopory, and R.E. Veilleux (Eds.). 1996d. *In Vitro Haploid Production in Higher Plants, Vol. 4: Cereals*. Kluwer Academic Publishers, Dordrecht.

Jain, S.M., S.K. Sopory, and R.E. Veilleux (Eds.). 1997. *In Vitro Haploid Production in Higher Plants, Vol. 5: Oil, Ornamental and Miscellaneous Plants*. Kluwer Academic Publishers, Dordrecht.

Keller, E.R.J. and L. Korzun. 1996. Ovary and ovule culture for haploid production. In S.M., Jain, S. K. Sopory, and R.E. Veilleux (Eds), *In Vitro Haploid Production in Higher Plants, Vol. 1, Fundamental Aspects and Methods*. Kluwer Academic Publishers, Dordrecht, pp. 217–235.

Keller, W.A. and K.C. Armstrong. 1979. Stimulation of embryogenesis and haploid production in *Brassica campestris* anther cultures by elevated temperature treatments. *Theor. Appl. Genet*. 55:65–67.

Kobayashi, R.S., S.L. Sinden, and J.C. Bouwkamp. 1993. Ovule cultures of sweet potato (*Ipomoea batatas*) and closely related species. *Plant Cell Tiss. Org. Cult*. 32:77–82.

Lakshmi Sita, G. 1997. Gynogenic haploids in vitro, pp. 175–193. In S.M., Jain, S. K. Sopory, and R.E. Veilleux (Eds.), *In Vitro Haploid Production in Higher Plants, Vol. 5, Oil, Ornamental and Miscellaneous Plants*. Kluwer Academic Publishers, Dordrecht.

Lapitan, V.C., E.D. Redona, T. Abe, and D.S. Brar. 2009. Molecular characterization and agronomic performance of DH lines from the F1 of indica and japonica cultivars of rice (*Oryza sativa* L.). *Field Crop Res*. 112:222–228.

Maheshwari, S.C, A. Rashid, and A.K. Tyagi. 1982. Haploids from pollen grains—retrospect and prospect. *Amer. J. Bot*. 69:865–879.

Miller, C.O. 1963. Kinetin and kinetin-like compounds. In H.F. Liskens and M.V. Tracey (Eds), *Moderne Methoden der Pflanzenanalyse*, Vol. 6. Springer-Verlag, Berlin, pp. 194–202.

Mordhorst, A. and H. Lorz. 1993. Embryogenesis and development of isolated barley (*Hordeurn oulgare* L.) microspores are influenced by the amount and composition of nitrogen sources in culture media. *J. Plant Physiol*. 142:485–492.

Murashige, T., and F. Skoog. 1962. A revised medium for rapid growth and bioassays with tobacco tissue cultures. *Physiol. Plant*. 15:473–497.

Nitsch, J.P. and C. Nitsch. 1969. Haploid plants from pollen grains. *Science*. 163:85–87.

Powell, W. 1990. Environmental and genetical aspects of pollen embryogenesis, pp. 45–65. In Y.P.S. Bajaj, (Ed.), *Biotechnology in Agriculture and Forestry, Vol. 12, Haploids in Crop Improvement*. Springer-Verlag, Berlin.

Reynolds, T.L. 1990. Ultrastructure of pollen embryogenesis. In Y.P.S. Bajaj, (Ed.), *Biotechnology in Agriculture and Forestry*, Vol. 12, Haploids in Crop Improvement. Springer-Verlag, Berlin, pp. 66–82.

Roberts-Oehlschlager, S.L. and J.M. Dunwell. 1990. Barley anther culture: Pretreatment on mannitol stimulates production of microsporederived embryos. *Plant Cell Tiss Org Cult*. 20:235–240.

Semagn, K., A. Bjornstad, and M.N. Ndjiondjop. 2006. Principles, requirements and prospects of genetic mapping in plants. *Afr. J. Biotechnol*. 25:2569–2587.

Shalaby, T.A. 2007. Factors affecting haploid induction through in vitro gynogenesis in summer squash (*Cucurbita pepo* L.). *Sci. Hort*. 115:1–6.

Srivastava, P. and R. Chaturvedi. 2008. In vitro androgenesis in tree species: An update and prospect for further research. *Biotechnol. Adv*. 26:482–491.

Sunderland, N. and M. Roberts. 1979. Cold-treatment of excised flower buds in float culture of tobacco anthers. *Ann. Bot*. 43:405–414.

Wang, J.J., J.S. Sun and Z.Q. Zhu. 1974. On the conditions for the induction of rice pollen plantlets and certain factors affecting the frequency of induction. *Acta Bot Sin*. 16: 43–54.

Winton, L.L., and R.F. Stettler. 1974. Utilization of haploidy in tree breeding. In K.J. Kasha, (Ed.), *Haploids in Higher Plants. Advances and Potential*. Proceedings of the First International Symposium. Guelph University Press, pp. 259–273.

30 Production of Haploid Tobacco and Potato Plants Using Anther Culture*

Phillip A. Wadl, Sandra M. Reed, and Denita Hadziabdic

Haploids are plants that have the gametic, or *n*, number of chromosomes. They are valuable in genetic and breeding studies because recessive genotypes are easily identified. Moreover, completely homozygous plants can be obtained quickly by doubling the chromosome number of the haploids to produce dihaploids, which have the 2n number of chromosomes; these dihaploids are very useful in plant breeding programs (see Chapter 29).

For many years, few haploids were available to geneticists and breeders. This situation began to change in the 1960s with the discovery that haploids of *Datura* species and tobacco (*Nicotiana tabacum* L.) could be produced by culturing anthers. In no other species has anther culture been more successful than tobacco, where large numbers of haploids can be obtained from many different genotypes. Because it is so easy to obtain tobacco and potato haploids, they are the ideal species to demonstrate and learn anther culture methodology.

The following laboratory exercises illustrate the technique of anther culture using tobacco and potato. Students will become familiar with the following techniques and concepts: (1) basic anther culture procedures; (2) determining the optimum stage of microspore development for anther culture; and (3) the use of cold pretreatments for enhancing haploid production from anther culture. Directions are given for student teams. It should take 30 to 60 minutes of laminar flow hood use for each team to complete the culturing stage of each of the experiments. Preparation of media, staging of microspores, and evaluation of cultures will require additional laboratory time, but minimum usage of the flow hood. Depending on class size, facilities, and time available, two to four students per team are recommended.

GENERAL CONSIDERATIONS

GROWTH OF PLANTS

Tobacco

Any cultivar of tobacco may be used for the experiments. After completing the anther culture experiments, allow a few plants to continue to flower. Approximately 1 month after flowering, open-pollinated seed capsules will turn brown and dry. Seeds can be collected and stored for several years under cool, dry conditions. A single capsule will contain approximately 2,000 seeds that can be used as a source of plants for future experiments.

Fill a 10-cm-diameter clay pot with vermiculite and water from bottom by placing the pot in a pan of water. Lightly sow seed on the surface of vermiculite and keep the pot in a saucer of water until the

* Mention of trade names of commercial products in this chapter is solely for the purpose of providing specific information and does not imply recommendation or endorsement by the U.S. Department of Agriculture.

seeds germinate, which should take about 1 week. Continue to water seedlings from the bottom of the pot, but do not allow the vermiculite to be continually soaked nor to dry out completely. Seedlings may be grown in the greenhouse or in a growth chamber at 22 to 26°C with a 16/8-h light/dark cycle.

After about 6 weeks, the seedlings will be large enough to transplant to individual 10- or 15-cm-diameter clay pots. Any commercial soilless potting mixture may be used. A controlled-release all-purpose fertilizer may be incorporated into the medium at this time or pots can be watered twice weekly with an all-purpose liquid fertilizer. Plants should be grown in a greenhouse. Supplemental light is not required during the winter, but may accelerate growth of the plants and hasten flowering. While plants should not be allowed to undergo water stress, take care not to overwater them. The first sign of overwatering in tobacco is wilting, so do not assume that a wilted plant needs more water. Instead, check soil moisture level before watering. Five plants per team of students should be sufficient for the both experiments included in this chapter. Plants should flower 2 to 4 months after transplant to 15-cm-diameter pots, depending on growing conditions, cultivar, and size of pot (plants can be grown in a smaller pot to encourage more rapid flowering).

Potatoes (Veilleux, 1999)

Solanum phureja (potato) grows and flowers best under cool greenhouse conditions, 25°C day/15°C night (77°F/59°F), long photoperiod (16 h) and high light intensity provided by high pressure sodium vapor lamps. In Virginia, there are two greenhouse seasons that induce adequate flowering to be able to conduct anther culture—planting in August for October to November experiments and planting in January for March to April experiments. Supplemental lighting should be provided starting around September 15 for the August planting and as soon as the plants have emerged for the January planting. It can be discontinued after April 1 for the spring planting. Flowering occurs best when the plants are given adequate space for root development, for example, using 7.6 L (2-gal) nursery pots containing 1 part sand to 2 parts soil-less mix (Sunshine, Fisons Horticulture Inc., Vancouver, BC, Canada, or Pro-Mix BX, Premier Brands, Inc., Red Hill, Pennsylvania). Also, weekly fertilization with a water soluble fertilizer such as Peter's Fertilizer Products (W.R. Grace & Co., Fogelsville, Pennsylvania) containing 20 N–8.4 P–14.9 K is required. A systemic insecticide such as Marathon (Olympic Horticultural Products, Mainland, Pennsylvania) should be applied when planting the tubers to control thrips and whiteflies. Some type of support, such as flower support netting (product 761020; Hummert, International, Earth City, Missouri) or stakes, is necessary. Once flowering has started, ~10 to 20 buds per plant per day can be collected three times per week.

EXPLANT PREPARATION AND CULTURE HANDLING

Collect buds soon after the plants begin to flower and transport to the laboratory in plastic bags containing moistened paper tissues. Surface disinfest the buds in groups of 20 to 25. Place a single bud in a sterile 100 × 15 mm plastic petri dish. Holding the bud gently with one set of forceps, remove the calyx (green) and corolla (colored) using another set of sharp fine-tipped forceps. Using the forceps, push the anthers away from the filaments (they should separate easily). Taking care not to squeeze the anthers, lift them gently with the forceps and place on the surface of the agar solidified media (Figure 30.1). Seal each petri dish and place in a 25°C incubator or growth chamber under constant light or 16/8-h light regime.

Cultures can be scored for number of haploids produced approximately 8 weeks after culture initiation. Using fine-tipped forceps, gently separate plants in order to get accurate counts. Unless these plants are to be used for other experiments, cultures can be scored outside of aseptic conditions.

CULTURE MEDIUM

The medium that will be used for the two tobacco experiments is that of Nitsch and Nitsch (1969; see Chapter 2 for composition). Each liter of media will be supplemented with 20 g (58 mM) sucrose and 8 g of agar. Plant growth regulators (PGRs) will not be added. Pour the medium into

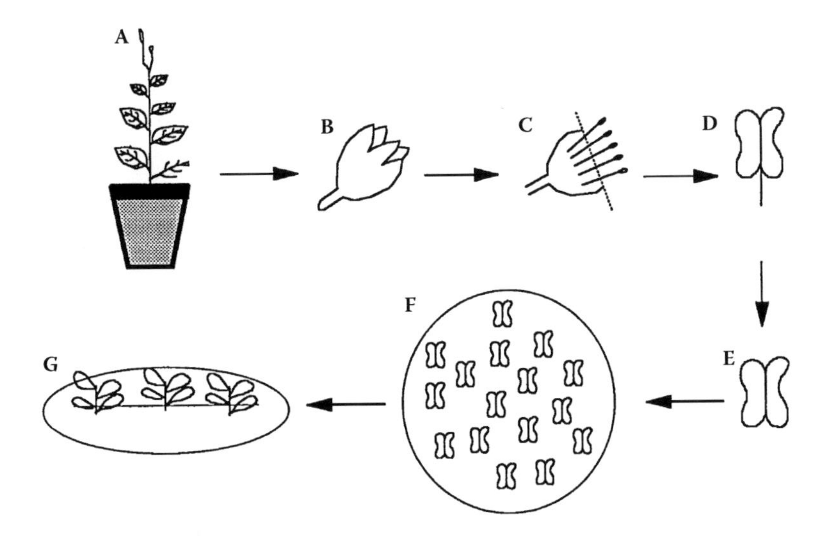

FIGURE 30.1 Diagrammatic representation of the anther culture protocol for tobacco. (A) Flowering diploid plant. (B) Floral bud excised from plant. Calyx and corolla are about equal lengths. (C) Floral buds with calyx and corolla removed. (D) Isolated stamen with anther and filament. (E) Anther removed from filament. (F) Petri dish with cultured anthers. (G) Plantlets emerging from anthers. (From Trigiano, R.N. and D.J. Gray (Eds.) 1996. *Plant Tissue Culture and Laboratory Exercises, First Edition*, CRC Press, Boca Raton, FL.)

TABLE 30.1
Formula Composition of Linsmaier and Skoog (1965) Medium

Chemical	Mg/L
Ammonium nitrate	1650
Boric acid	6.2
Calcium chloride, anhydrous	332.2
Cobalt chloride•$6H_2O$	0.025
Cupric sulfate•$5H_2O$	0.025
Na_2 edta•$2H_2O$	37.26
Ferrous sulfate•$7H_2O$	27.8
Magnesium sulfate	180.7
Manganese sulfate•H_2O	16.9
Molybdic acid (sodium salt)•$2H_2O$	0.25
Potassium iodide	0.83
Potassium nitrate	1900
Potassium phosphate, monobasic	170
Zinc sulfate•$7H_2O$	8.6

100 × 15 mm or 100 × 20 mm sterile plastic petri dishes after autoclaving. Petri dishes should be filled approximately half-way with media and 1 liter of media should yield 20 to 25 dishes of media.

For the two potato experiments, fill nine 125-mL Delong culture flasks with 15 mL autoclaved half-strength Linsmaier and Skoog (LS) (1965; Table 30.1) medium supplemented with 100 mg/L myo-inositol, 0.4 mg/L thiamine, 60 g/L sucrose, 2.5 mg/L N^6-benzyladenine, 0.1 mg/L indole-3-acetic acid, and 2.5 g/L activated charcoal, pH 5.8. *Hint*: make 500 mL of LS medium salts according to Table 30.1 and then add 400 mL of water. Add supplements as described above, adjust pH to 5.8, and bring to 1L with distilled water.

Exercises

Experiment 1. Effect of Stage of Microspore Development on Yield of Haploid Plants of Tobacco

The stage of microspore development at the time of culture is a critical factor affecting androgenesis. Students will be culturing anthers from buds of different developmental stages, identifying the stage of microspore development, and determining whether physical characteristics of the tobacco buds and anthers are correlated to the stage of microspore development and the ability to produce haploid plants.

Materials

The following items are needed for each team of students:

- 100 mL of 20% commercial bleach (i.e., 20 mL bleach plus 80 mL distilled water)
- Nine 100 × 15 mm or 100 × 20 mm petri dishes containing Nitsch and Nitsch medium
- Lighted incubator providing approximately 25°C
- Acetocarmine stain. Combine 45 mL glacial acetic acid, 55 mL of distilled water, and 0.5 g carmine; place in a beaker; cover with aluminum foil and gently boil solution for 5 min under a fume hood. Filter the solution and store in a refrigerator. The solution should be dark red and it may need to be filtered again before use.

Follow the protocols listed in Procedure 30.1 to complete this experiment.

Anticipated Results

Four to six weeks after initiation, the anther cultures, small plantlets, will be seen emerging from some of the anthers (Figure 30.3). Even under the best of conditions not all tobacco anthers will

Procedure 30.1
Effect of Stage of Microspore Development on Yield of Tobacco Haploids

Step	Instructions and Comments
1	Collect 36 tobacco flowers buds. The calyx and corolla on 12 of the buds should be of approximately equal length (medium length buds). Twelve of the buds should have a corolla that extends 5 to 7 mm beyond the calyx (long length buds). The remaining 12 buds should be the largest of the buds in which the corolla is not visible at all beneath the calyx (short length buds).
2	Using an indelible marker, divide the bottom of each of the nine media dishes into four quadrants and label them from 1 to 36, which will correspond to the culture numbers.
3	Surface-disinfect medium length buds in 70% ethanol for 1 min, followed by 20% commercial bleach for 5 min. Rinse twice in sterile distilled water.
4	Follow explant preparation and culture handling instructions given in general directions. Place four of the anthers onto culture medium in one of the numbered quadrants. Place the remaining anther onto a microscope slide, which should be marked with the culture number. Repeat with remaining 11 medium-length buds.
5	Repeat steps 3 and 4 with long-length buds, then repeat with short-length buds.
6	Squash anthers on microscope slides in a small drop of acetocarmine stain, remove debris, and cover with cover slip. Heat slide over steam for a few seconds and observe under the microscope. (For steam, fill a 250 mL flask about half full of water; simmer gently on hot plate.) Using photographs in Figures 30.2a–d and the descriptions presented in Chapter 29, determine stage of microspore development for each anther. If there is not enough time to examine the anthers when cultures are initiated, place them in fixative (3 parts 95% ethanol and 1 part glacial acetic acid; prepare fixative in fume hood) in small vials. After 2 days, replace fixative with 70% ethanol. Leave samples at room temperature for up to a week, then place in refrigerator. Anthers can be removed from fixative or alcohol at any time, squashed and examined following the previous directions.
7	For each culture, record bud length (short, medium, or long) and stage of microspore development.
8	Count and record number of plants in each culture approximately 8 weeks after culture initiation.

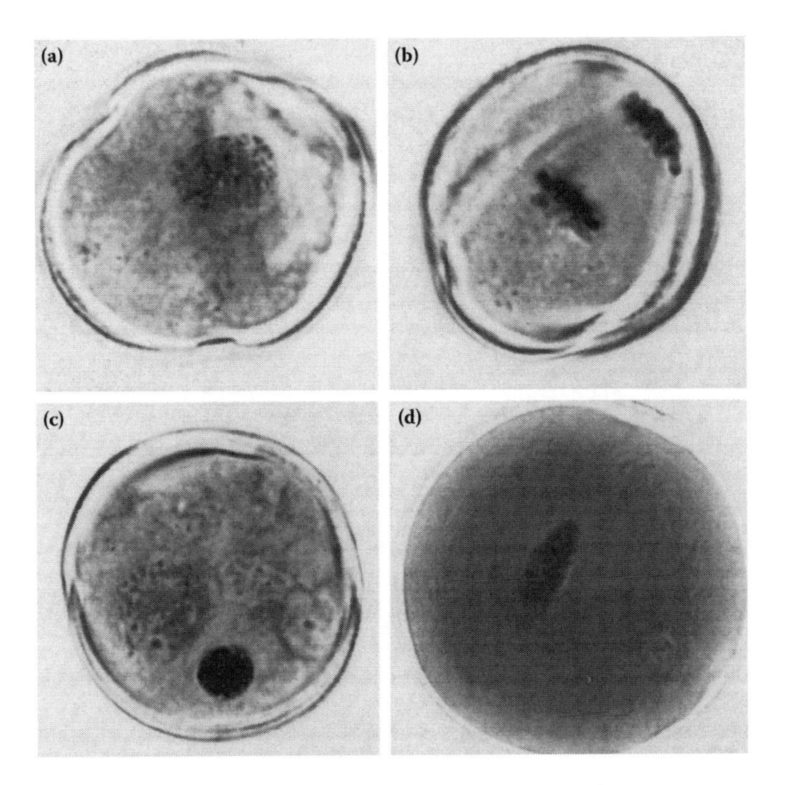

FIGURE 30.2 Microspores of tobacco. (a) Late uninucleate microspores. Vacuole can be seen of right side of the cell. (b) Anaphase of first pollen meiosis. (c) Early binucleate microspore. Large diffuse vegetative nucleus can be distinguished. (d) Mid-binucleate microspore. Staining is darker, obscuring vegetative nucleus, but the generative nucleus is still visible. (From Trigiano, R.N. and D.J. Gray (Eds.) 1996. *Plant Tissue Culture and Laboratory Exercises, First Edition*, CRC Press, Boca Raton, FL.)

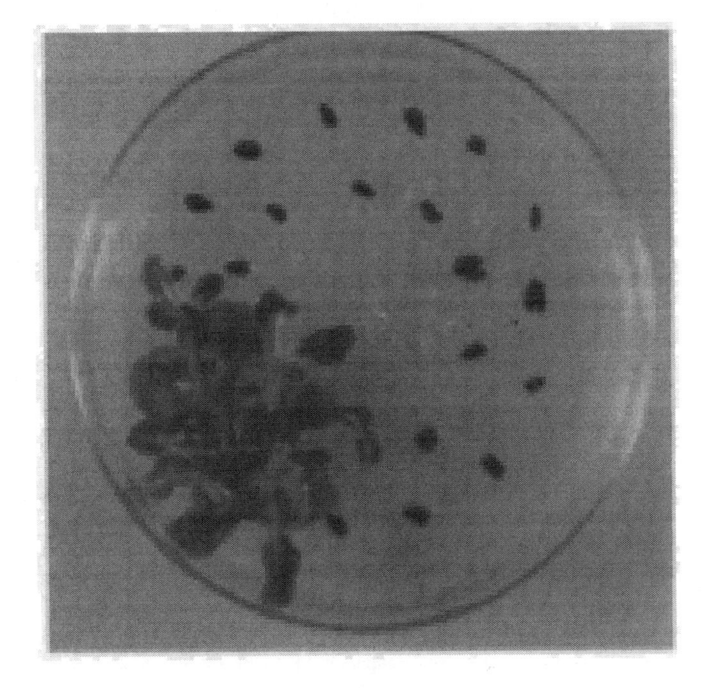

FIGURE 30.3 Putative haploid tobacco plantlets emerging from anthers.

produce haploids. The numbers of plantlets emerging from each anther will vary from none to a few to many—there is no way to predict the results of any given experiment.

Anthers from short length buds will produce few, if any, haploid plantlets. Uninucleate microspores will have been observed from the anthers of this size bud. Anthers from the long-length buds will also produce few, if any, plantlets. This size bud will correspond to a mid-to-late binucleate microspore stage. All or almost all of the plantlets produced in this experiment will come from the anthers in which the calyx and corolla were equal in length. This size bud corresponds to the late-uninucleate to early-binucleate microspore stage, which produces the most haploid plants in tobacco.

Questions

- What is the purpose of trying to correlate a physical characteristic of buds and/or anthers to stage of microspore development?
- How can you be sure that the plants you obtained are haploids?
- If you cultured anthers from a tobacco plant that was heterozygous for the *Su* gene (*SuSu* = albino, *Susu* = light yellow-green leaves, *susu* = green leaves), what would be the expected phenotypes and genotypes of the seedlings obtained from anther culture? What is the expected ratio of the different genotypes?
- How could you produce dihaploids from the haploids obtained through anther culture?

Experiment 2. Effect of Cold Pre-Treatment on Yield of Haploids from Anther Cultures of Tobacco

Subjecting buds to several days of low temperatures after collection, but before culture, increases yields of haploid plants in tobacco and several other species. This experiment will demonstrate the effect that cold pretreatment of flower buds (anthers) has on yields of tobacco haploid plants.

Materials

Each student team will require the materials listed in Experiment 1 and the following:

- Ten 100-mm diameter petri dishes containing Nitsch and Nitsch (1969) medium
- Refrigerator or incubator providing temperature of 7 to 8°C

Follow the directions provided in Procedure 30.2 to complete this experiment.

Procedure 30.2

Effects of Cold Pretreatment on Yield of Haploids from Tobacco Anther Cultures

Step	Instructions and Comments
1	Collect 40 tobacco buds, all of which should have a calyx and corolla of equal length.
2	Place 20 buds in a zipper-type plastic bag with a water-moistened tissue. Seal bag and place in a 7 to 8°C incubator or refrigerator.
3	Sterilize remaining 20 buds as directed in Step 3, Procedure 30.1. Follow explant preparation and culture handling instructions given in general directions. Place the anthers from four buds in the same dish of medium (20 anthers/dish of medium).
4	After 10 to 12 days, repeat step 3 with cold pretreated buds.
5	Eight weeks after initiating cultures, count the number of plantlets from both the untreated and cold pretreated buds. Calculate the mean number and standard deviation of plants/anther for both groups of buds.

Anticipated Results

The mean number of plants/anther should be significantly greater for the buds that received a cold pretreatment; however, differences between treatments will not be as great as observed in Experiment 1. It may be advisable to pool results from all the teams in the class prior to the statistical analysis in order to see the effect of the cold pretreatment.

Questions

- Why were buds collected when the calyx and corolla were approximately equal in length?
- What temperatures and lengths of treatment would you test if you were trying to work out an anther culture procedure for a new species?

Experiment 3. Effect of Stage of Microspore Development on Yield of Haploid Plants of Potato

Materials

The following items are needed for each team of students:

- *Solanum phureja* Juz. & Buk. PI 225669 (Owen et al., 1988)—Seeds available from the USDA-ARS National Plant Germplasm System (http://www.ars-grin.gov/npgs/searchgrin.html)
- 100 mL of full-strength commercial bleach
- Nine 125 mL Delong culture flasks containing 15 mL autoclaved half strength Linsmaier and Skoog (1965) medium as described previously
- Acetocarmine stain (see instructions in Experiment 1)

Follow the protocols listed in Procedure 30.3 to complete this experiment.

Anticipated Results

Five to six weeks after initiating anther cultures, small embryos will be seen emerging from some of the anthers. The number of embryos produced will vary. Anthers from short length buds will produce few, if any, embryos. Tetrad or uninucleate microspores will have been observed from the anthers of this size bud. The long length buds will also produce few, if any, embryos. This stage

Procedure 30.3

Effect of Stage of Microspore Development on Yield of Potato Haploids

Step	Instructions and Comments
1	Collect three potato flower buds for each of the following lengths: short length (3 to 4 mm long), medium length (4 to 7 mm long), and long length (7 to 8 mm long).
2	Using a permanent marker, label each culture flask corresponding to the culture number.
3	Surface disinfect small buds in 70% ethanol for 30 s, followed by 5 min with full strength bleach containing Tween 20. Rinse twice in sterile distilled water.
4	Follow explant preparation and culture handling instructions given in general directions. Place four of the anthers into one of the labeled culture flasks. Place the remaining anther onto a microscope slide, which should be marked with the culture number. Repeat with the remaining two short-length buds.
5	Repeat steps 3 and 4 with medium-length buds, and then repeat with long-length buds.
6	Follow the instructions and comments outlined in step 6, Procedure 30.1.
7	For each culture, record bud length (short, medium, or long) and stage of microspore development.
8	Grow cultures at room temperature (about 25°C) in the dark on a rotary shaker at 125 to 150 rpm for 5 to 6 weeks. Harvest the embryos by pouring the contents of the flask through a mesh sieve. Count and record the number of embryos in each culture approximately 6 weeks after culture initiation.

Procedure 30.4
Effect of Cold Treatment on Yield of Haploids from Potato Anther Cultures

Step	Instructions and Comments
1	Collect 8 potato buds, all of which should be 4 to 7 mm long (medium length).
2	Place four buds into a zipper type plastic bag with water-moistened tissue. Seal bag and place in a 4°C refrigerator.
3	Sterilize remaining four buds as directed in step 3, Procedure 30.2. Follow explant preparation and culture handling instructions given in general directions. Place the anthers into one of the labeled culture flasks. Repeat with the remaining three medium length buds.
4	After 3 days, repeat step 3 with cold-treated buds.
5	Grow cultures at room temperature in the dark on a rotary shaker at 125 to 150 rpm for 5 to 6 weeks. Count and record the number of embryos in each culture approximately 5 to 6 weeks after culture initiation. Calculate the mean number and standard deviation of embryos/anther for both groups of buds.

of microspore development will correspond to the binucleate stage. Most, if not all of the embryos produced from this experiment will come from anthers derived from the medium-length buds. This size bud corresponds to the uninucleate to late uninucleate microspore stage, which produces the most embryos in potato (Sopory et al., 1978). If regeneration of the embryos is required, refer to the procedures outlined by Veilleux (1999).

Questions

See questions under "Experiment 1. Effect of Stage of Microspore Development on Yield of Haploid Plants."

Experiment 4. Effect of Cold Pretreatment on Yield of Haploids from Anther Cultures of Potato

The following items are needed for each team of students:

Materials

- *Solanum* phureja Juz. & Buk. PI 225669 (Owen et al., 1988)—Seeds available from the USDA-ARS National Plant Germplasm System (http://www.ars-grin.gov/npgs/searchgrin.html)
- 100 mL of full strength commercial bleach
- Refrigerator providing temperature of 4°C

Follow the protocols listed in Procedure 30.4 to complete this experiment.

Anticipated Results and Question

See results and question under "Experiment 2. Effect of Cold Pre-Treatment on Yield of Haploids from Anther Cultures of Tobacco."

REFERENCES

Linsmaier, E.M. and E. Skoog. 1965. Organic growth factor requirements of tobacco tissue culture. *Physiol. Plant* 18:100–127.

Nitsch, J. P. and C. Nitsch. 1969. Haploid plants from pollen grains. *Science* 163:85–87.

Owen, H.R., R.E. Veilleux, D. Levy, and D.L. Ochs. 1988. Environmental, genotypic, and ploidy effects on endopolyploidization within a genotype of *Solanum phureja* and its derivatives. *Genome* 30:506–510.

Sopory, S.K., E. Jacobsen, and G. Wenzel. 1978. Production of monohaploid embryoids and plantlets in cultured anthers of *Solanum tuberosum*. *Plant Sci. Letters* 12:47–54.

Veilleux, R.E. 1999. Anther culture of potato and molecular analysis of anther-derived plants as laboratory exercises for plant breeding courses. *HortTechnology* 9:585–588.

31 Embryo Rescue

Tom Eeckhaut, Katrijn Van Laere,
and Johan Van Huylenbroeck

CONCEPTS

- Combining genes from different species is a major tool for innovating plants.
- It is difficult to sexually cross different plant species.
- Immature fruits can and will spontaneously abort.
- In vitro culture replaces nutrients normally provided by the seed parent.
- A medium for embryo culture should be kept as simple as possible.
- Embryo rescue can yield many hybrids but is no guarantee that you can obtain viable seedlings.

SPECIES: WHAT'S IN A WORD?

Generally, individuals or populations are considered as members of the same species when they can sexually recombine and generate offspring which, in their turn, are fully fertile. For instance, horses and donkeys can mate and produce offspring (mules), but as mules themselves are sterile, we can not consider horses and donkeys as belonging to the same species. The fertility problems of mules, or interspecific hybrids in general, are caused by inadequate chromosome pairing during meiosis, as maternal and paternal chromosomes are too different for this so-called synapsis.

Throughout the plant and animal kingdom, this rough definition of "species" is used to classify individual genotypes in the same or different group. However, as traditionally observed in biology, exceptions to this rule are common and different plant species are often sexually compatible, provided they are very narrowly related. Often, they also need to share the same chromosome number. When hybrids are not sterile, interspecific hybrids could very well gain an advantage over both ancestral species, whenever they combine beneficial genes from both sides. As they indeed offer "the best of two worlds" they can even suppress both of their progenitor species. A new species has arisen—in fact, nearly all current plant species are what we call "paleopolyploids," polyploids that have spontaneously formed in nature through the combination of two species and the subsequent formation of unreduced gametes. But that's another story.

WHY RESCUE EMBRYOS?

At any rate, nature tells us that variation to the largest possible extent is desired and that through rather uncontrolled gene and species mingling new successful genotypes can be created. Humanity, in its search for higher yields, disease resistance, prettier flowers, or even more enjoyable hobbies, has mapped the variation present in different plant species groups and tried to make interesting new combinations for its own benefit. The main differences with natural experiments is that we want it done faster and more controlled.

However, in doing so, plant breeders are confronted with lots of barriers that either inhibit (1) a normal pollination of the seed parent, (2) the growth of the pollen tube in the style, (3) the subsequent fusion of the male and the female gamete, (4) the development of the fertilized ovule to a mature seed, (5) the growth of this mature seed to a normal plant, and (6) the fertility of the obtained interspecific hybrids. The former three problems are referred to as "prezygotic incongruity," whereas the latter three problems are grouped as "postzygotic incongruity" problems.

The purpose of this chapter is not to provide a straight-forward protocol of how to deal with any of those problems. As is probably clear, their nature is very diverse, and different obstacles do require different approaches. Problems (1), (2), and (3) can generally be overcome by applying a range of pollination techniques and (6) by the application of unreduced gametes or chromosome doubling by mitosis inhibitors. Problem (5) is harder to circumvent, but sometimes carrying out the reciprocal cross offers a way out. This chapter deals with the common method to deal with the impeded development of a zygote to a viable seed. Indeed, this development is frequently inhibited by lack of endosperm (nurse tissue) formation or spontaneous abortion. Both problems can be resolved by dissecting either the young ovule or the young embryo out of the immature fruit on the seed parent, and putting it on an artificial medium in optimally controlled circumstances. This is what we call "embryo rescue." The term "embryo rescue" is restricted to only those cases where the embryos, if not rescued, are endangered and would not form seedlings.

Next to rescuing hybrid embryos the main applications of embryo culture are solving the problems of low seed set, seed dormancy, slow germination, germination of obligatory parasites, haploid breeding, shortening the breeding cycle, and vegetative propagation (Zenkteler, 1990; Pierik, 1999). The main factors affecting its success are the genotype and growth conditions of the mother plant, the developmental stage of the embryo at isolation, the composition of the nutrient media and the culture environmental conditions (oxygen, light, and temperature; Pierik, 1999).

Embryo Rescue Practically: How to Proceed in Greenhouse and Lab

In the greenhouse, interspecific crosses are usually made. Proper fertilization techniques include, of course, emasculation of the seed parent and attaching a label with all necessary info. Sometimes flowering periods are different, which can be overcome by pollen storage (usually at 4°C) or flowering time management (e.g., by storage in a fridge or altered application of growth inhibitors). Evidently, good plant practices also include to keep the infection pressure as low as possible. Droplet irrigation can be a good tool to accomplish that, compared to normal watering with a garden hose, which may spread microorganisms over the flowers.

The optimum harvest period of the immature fruits is just before spontaneous abortion. In other words, when the first fruits fall, harvest the remainder as soon as possible for in vitro initiation. This way, they have exploited nutrients provided by the seed parent as much as possible; the less immature seeds are, the higher their chances for "rescuing" in vitro. If you have no idea when to begin, a traditional approach is to harvest the fruits at regular time intervals, independent of their developmental state, and initiate them until all remaining fruits abort. This would allow you to determine the minimal age of the ovules for embryo rescue.

As mentioned before, also the exact genotype combination matters for the potential success of the whole procedure. In other words, never limit yourself to the use of only 1 seed and/or pollen parent. Use at least 4 different genotypes of both seed and pollen parents; moreover, pollinations should also be performed reciprocally. Not only does the direction of the cross influence possible prezygotic incongruity, also the occurrence of albinism is affected by it. Of course, sometimes the number of genotypes is limited to less than 4 and making reciprocal crosses is impossible due to unilateral sterility of a parent.

Growth conditions in the greenhouse are obviously depending on the species you work with. In vitro, usually standard culture situations can be maintained: 21-25°C, 16h light regime or 24 h dark depending on the species requirements for germination, light intensity 40-100 μmol m^{-2} s^{-1} photosynthetic active radiation.

In the lab, usually standard sterilization procedures (15 min in a solution of 1% NaOCl) are sufficient to eradicate microorganisms, especially when appropriate culture practices in the greenhouse have been followed (see above). Moreover, seed capsules tend to be significantly more sterile than vegetative plant parts, provided they have not burst open. Mostly, the use of a binocular is indispensable. Provide suitable care to the establishment of a sterile work surface (e.g., sterile paper). The most suitable recipients for ovule or embryo culture are small petri dishes. In case of bacterial or fungal contamination, it allows you to keep the losses restricted.

More in detail, embryo rescue can be performed in one step (embryo culture) or two steps (ovule culture + subsequent embryo culture). The most applied culture methods are ovary slice culture, ovule with placenta culture, single ovule culture, and embryo sac culture. Once the embryo can be dissected, it is transferred to new medium, independent of its developmental stage (globular, heart-shaped, torpedo-shaped, or cotyledonary; Figure 31.1).

Finally, medium composition significantly influences the outcome of the embryo culture. Especially sugar content is important, as immature embryos are known to require higher osmotic strength (lower potential) of the medium compared to the relatively mature ones. High osmotic strength of the medium prevents precocious germination of young embryos and supports normal embryonic growth. Therefore sugar concentration is at least as important for the osmolarity of the medium as for providing nutrition, at least when culturing immature embryos.

Plant hormones are seldom applied. Cytokinins and auxins can induce unwanted formation of callus. Gibberellic acid is sometimes used to lift seed dormancy. Extra vitamins are rarely included in the medium. Occasionally, casein hydrolysate (around 1 g/l) is present in the medium to provide organic N-sources. Others add coconut milk, which is however difficult toward replication of the experiment due to its varying composition.

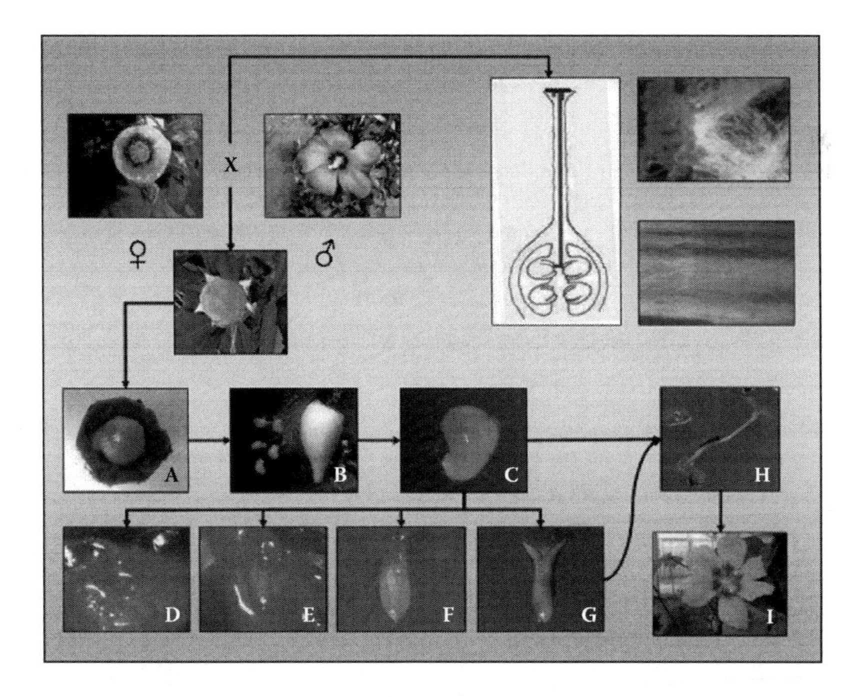

FIGURE 31.1 Different stages in interspecific hybridization. After verification of prezygotic incongruity (above, right) crosses are made and fruits appear on the seed parent (above, left). Complete fruits (A) or dissected ovules (B, C) can be used for in vitro initiation; whenever large enough or after temporary ovule culture embryos can be cultured in the globular (D), heart-shaped (E), torpedo-shaped (F), or cotyledonary (G) stage. Successful embryo rescue yields seedlings (H) but the interspecific cross itself can only be considered a success when the seedlings flower (I) and are fertile.

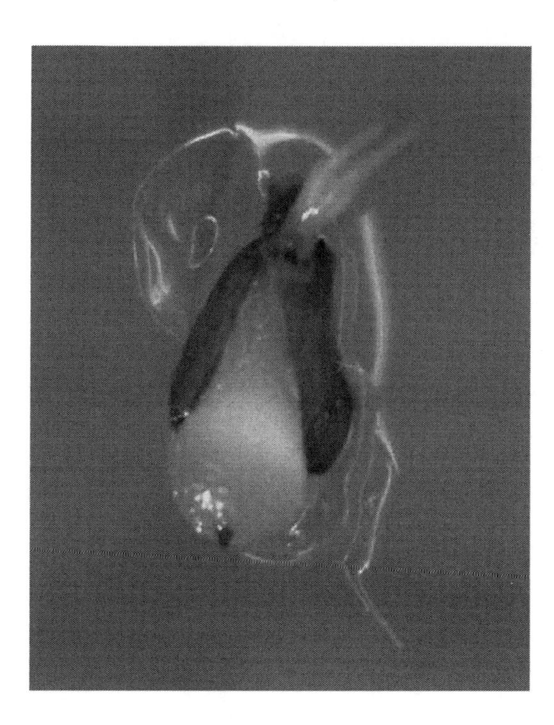

FIGURE 31.2 Germinating immature rhododendron ovule.

Once seeds have germinated (Figure 31.2), seedlings can be multiplied and acclimatized. This requires no special adjustments to normal multiplication and rooting media. It should, however, be taken into account that one works with interspecific hybrids, which usually tend to be vulnerable and/or less vigorous. In other words, throughout the whole procedure usually significant amounts of plants can get lost. This is normal, and it is even recommendable that no extra efforts are spent trying to save and/or multiply plantlets that stand no chance ex vitro. Remember, plants that do not form functional gametes are useless unless they are, by themselves, suitable cultivars that can be vegetatively propagated. The chance to obtain such a genotype without making any further crossing is extremely low. Stated otherwise: It is better to focus on the genotypes that grow well. Whether the embryo rescue protocol is in the end useful is qualitatively defined "Are there any fertile hybrids yielded from the ovules I put in vitro?" rather than quantitatively "How many fertile hybrids do I obtain from the ovules I put in vitro?"

EXERCISE

EXPERIMENT 1. RESCUING IMMATURE *RHODODENDRON* EMBRYOS

Introduction

Interspecific *Rhododendron* embryos can be rescued in vitro on a standard medium enriched with gibberellic acid (GA_3). However, the optimum dose of gibberellic acid needs to be determined experimentally for each hybrid group or subgenus within *Rhododendron*. The purpose of this exercise is to determine the optimum content of GA3, so we can maximize the number of immature seeds that germinate. To optimize the "embryo rescue" system we use seeds generated after a normal compatible pollination. Indeed, it is best to design an embryo rescue protocol by means of crosses upon which embryo rescue would normally not be required, in order to have a reference.

Materials

- Woody Plant Medium (WPM) with 20 g/l sucrose, 7 g agar (agarose/MC29/...), pH 5.4
- Petri dishes (diameter 5 cm)
- 5 glass bottles of 0.5 l
- Autoclave, a laminar flow hood, a binocular, scalpels, pincets, a glass bead sterilizer, sterile water, sterile paper, and sterile 0.2 µm filter
- Gibberellic acid GA_3, a few drops of KOH 1N
- Commercial bleach solution, detergent (e.g., Tween), distilled water
- pH meter, denatured ethanol (70%)
- 2 cross compatible *Rhododendron* plants

Follow the protocols listed in Procedure 31.1 to complete this experiment.

Procedure 31.1
Rescuing Immature *Rhododendron* Embryos

Step	Instructions and Comments
1	Cross the compatible rhododendron plants (for instance, two garden rhododendrons, two dwarf rhododendrons, or two deciduous azaleas) with one another by hand pollination (you don't need to do this reciprocally; choose a seed and a pollen parent. Hint: choose two cultivars that produce lots of pollen, they are usually the most fertile). Normally, pollination of 25–30 flowers should yield plenty of material for the experiment. Leave the fruits on the seed parent for min 3 to max 4 months (the normal maturation period is 9 months).
2	Prepare the medium. First, make 2 liters of basal medium containing WPM + 20g/l sucrose + 7g/l agar (pH 5.4)—see Chapter 3. Divide it into 5 bottles (5 x 0.4 l) and autoclave those (121°C, 30 min, 500 hPa).
3	Enrich the media with gibberellins under the laminar flow. Gibberellins are not heat tolerant and should be sterilized by filters. Dissolve 100 mg into a few drops of KOH (1N) which you dilute with distilled water up to 10 mL (final concentration 10 mg/mL). Sterilize this solution through a 0.2 µm filter and store it (if necessary) in the freezer.
4	Supplement the five bottles, after the media are autoclaved but not yet solidified (the temperature should be low enough for you to keep the bottle in your hand without too much discomfort) with the gibberellin solution. Make solutions of 0, 10, 20, 50, and 100 mg/L. Pour the media into the petri dishes (around 10 mL/ petri dish, you don't need to use a pipette) and label the petri dishes. You will have about 40 petri dishes per treatment.
5	Surface sterilize the rhododendron fruit by the following procedure: (1) removing any hairs on the fruit with a razor blade, (2) rinsing the fruit in ethanol 70%, and finally (3) soaking them in a solution with 1% active NaOCl (usually 10% of a commercially available bleach solution) and a drop of detergent (e.g., Tween or a commercial dishwashing detergent), in a closed recipient (e.g., Meli-jar, falcon tube) for 15 minutes. (4) Rinse the fruits three times with sterile water.
6	Prepare the working surface (flow): clean it with 70% ethanol, install the binocular microscope, and use sterile paper on it as a sterile cutting surface. Sterilize your equipment (scalpels, pincets) by dry heating in a glass bead sterilizer (250°C), but have it cooled down before using (let it rest on a supporting device without getting into contact with the flow surface for a few minutes).
7	The rhododendron fruit is typically five-lobed (see Figure 31.3). Cut it into five parts by making longitudinal incisions (the figure shows a cross section) in a way that you are left with 5 separate lobes. If you put those fruit parts on their backs and get rid of the partitions on the left and right, you can see the placenta with two rows of ovules attached. Just one fruit may contain up to 100–200 immature seeds/ovules.
8	Now dissect the ovules and put them on the medium (20 ovules/petri dish, so you need around 800 ovules). Close the petri dishes with low-density polyethylene foil or Urgopore tape.
9	Put all petri dishes in a climate room (23 ± 2°C) under a 16 h photoperiod at 30–100 µMol $m^{-2} s^{-1}$ photosynthetic active radiation, supplied by cool white fluorescent lamps.

FIGURE 31.3 Cross section (xs) of a rhododendron fruit.

Anticipated Results

Germination will start after 2–3 weeks. It is advisable to count the seedlings after 4 weeks. You can expect to see a gradual increase of germination along with the gibberellin concentration in the medium. However, at a certain moment the concentration will be supra-optimal, meaning the efficiency will decrease after a certain concentration. Usually the optimal dosage is around 50 mg/l.

IS IT ALWAYS USEFUL?

One should of course not bother to design an embryo rescue medium when prezygotic barriers can not be overcome, meaning that there is no formation of a zygote. In that case it is recommendable to attempt somatic hybridization (protoplast fusion and regeneration). For the transfer of a single gene, without any polygenic traits being involved in its expression, genetic modification is scientifically a better alternative.

Questions

- Which plant species are more suitable for embryo rescue: plants with many ovules or plants with few ovules?
- If cross A × B is successful through embryo rescue, does this imply that (a) you will need embryo rescue for B × A, (b) the A × B interspecific hybrid will be sterile, (c) endosperm is formed, (d) none or the above, or (e) all of the above?
- Mannitol is often used as ingredient of embryo rescue media. Why would that be done, as mannitol is in fact metabolically inactive?

REFERENCES

Eeckhaut, T., K. Van Laere, J. De Riek, and J. Van Huylenbroeck. 2007. Overcoming interspecific barriers in ornamental plant breeding. In Teixeira da Silva J (Ed.), *Floriculture, Ornamental and Plant Biotechnology: Advances and Topical Issues,* First Edition, Global Science Books, London, UK, pp. 540–551.

Pierik, R. 1999. *In Vitro Culture of Higher Plants*, First Edition, Kluwer Academic Publishers, Dordrecht, The Netherlands, 360 pp.

Sharma, D., R. Kaur, and K. Kumar. 1996. Embryo rescue in plants—a review. *Euphytica* 89: 325–337.

Sharma, H. 1995. How wide can a wide cross be? *Euphytica* 82: 43–64.
Van Tuyl, J. and M. De Jeu. 1997. Methods for overcoming interspecific crossing barriers. In Shivanna, K. and Sawhney, V. (Eds), *Pollen Biotechnology for Crop Production and Improvement*, First Edition, University Press, Cambridge, U.K., pp. 273–292.
Zenkteler, M. 1990. In vitro fertilization and wide hybridization in higher plants. *Crit. Rev. Plant Sci.* 9: 267–279.

32 Promoters and Gene Expression Regulation

Zhijian T. Li and D.J. Gray

CONCEPTS

- Gene expression is a highly regulated cellular process requiring the involvement of many structural components including promoters. Promoters are of paramount importance for their structural complexity and extraordinary capability to regulate gene expression.
- A promoter comprises two discrete functional subunits, namely the core promoter and the enhancer. The core promoter contains sequence elements for interaction with RNA polymerase, while the enhancer comprises many unique regulatory elements essential for modulation of transcriptional activity.
- Promoters can be classified into three major groups and include regulated, constitutive, and inducible promoters. The mode of action employed by these promoters is mainly dictated by what types of regulatory elements are present and how these elements are arranged within the promoter region.
- Gene expression can be regulated at the promoter level. This can be accomplished by employing DNA sequence geometric alternation, regulatory element partition, core promoter structure, and interaction between associated regulatory factors.

Gene expression refers to a cellular process by which inheritable genetic information is converted into functional and structural proteins. Many complex processes including chromatin remodeling, transcription, RNA processing, translation, and posttranslational modification are involved in the regulation of gene expression (Fickett and Hatzigeorgiou, 1997). Among these processes, transcription controls the precise copying of the DNA-encoded message onto a complementary RNA transcript that is eventually used as template for protein synthesis. This transcriptional control is accomplished by highly sophisticated and multifaceted regulatory mechanisms. Over the years, transcription has long been recognized as the most important process in gene expression due to its commanding regulatory capability to achieve both precise and timely transcription initiation and to ensure rate-limited RNA transcript production (Lee and Young, 2000).

Transcription of protein-coding genes entails the active participation of several structural and functional components, including RNA polymerase II, other associated interactive protein factors, and a regulatory sequence located upstream of the transcribed gene, which is commonly called the *promoter*. Although these components are all required for the formation of transcription initiation complex, the promoter contains the primary triggering signals essential for the recruitment of transcription factors and RNA polymerase components and serves as a platform to accommodate the physical assembly of transcription machinery. Therefore, promoters possess the pivotal mechanistic capacity to modulate transcription processes and influence gene expression both qualitatively and quantitatively.

Over the years, a large number of promoters from a wide variety of plant species and associated organisms have been isolated and analyzed in detail. Information on these promoters can be readily accessed through several publicly available databases (Yamamoto and Obokata, 2008). These investigative efforts greatly advanced our understanding of the central role of promoters to regulate gene expression during growth and development and in response to environmental cues in plants. In addition, a number of well-characterized promoters have been widely used in genetic engineering endeavors to direct expression of transgenes conferring novel desirable phenotypes and value-added traits.

In this chapter the basics of structural organization and major categories of promoters found in plants are described. Mechanisms of gene expression regulation at the promoter level will also be discussed.

BASIC STRUCTURE OF PROMOTERS

A promoter is a stretch of DNA comprising many unique sequence motifs (Figure 32.1). These motifs can provide biological functions on the same molecule by serving as signaling or recognition sites for regulatory proteins and RNA polymerase, and are therefore normally called *cis*-acting elements (CAEs). Regulatory proteins that recognize CAEs and interact with basal RNA polymerase to form productive transcription initiation complexes are usually named as *trans*-acting factors (TAFs) for their ability to act on a different molecule. Although all promoters can invariably activate transcription, it is well established that promoter activity and mode of action primarily depend on the structural organization of promoter sequences (i.e., the combination of different types of CAEs) and their distribution pattern within the promoter region.

CORE PROMOTERS

From a functional standpoint, a typical promoter is composed of two subunits: a core promoter and an enhancer. A core promoter refers to a DNA sequence spanning approximately 100 base pairs upstream of the transcription start site (TSS) and is commonly defined by its minimal transcriptional activity. The majority of core promoters characterized thus far contain a TATA-box consensus element, a minimal number of CAEs such as CAAT-box element and a transcriptional initiation site termed initiator with a core sequence TCATC or TCACTC (INR). However, some core promoters lack the TATA-box element (TATA-less promoter), but normally contain an INR (Smale, 1997). Both TATA-box element and INR serve as specific recognition sites for preinitiation complex (PIC), which comprises RNA polymerase II (pol II) and a number of general transcription factors (GTFs) such as TFIIA, -B, -D, -E, -F and -H (Fickett and Hatzigeorgiou, 1997). These sequence elements within a core promoter are determinants of the precise location of TSS and, to a certain degree, capable of influencing promoter strength and mode of action (Guilfoyle, 1997; Smale, 2001).

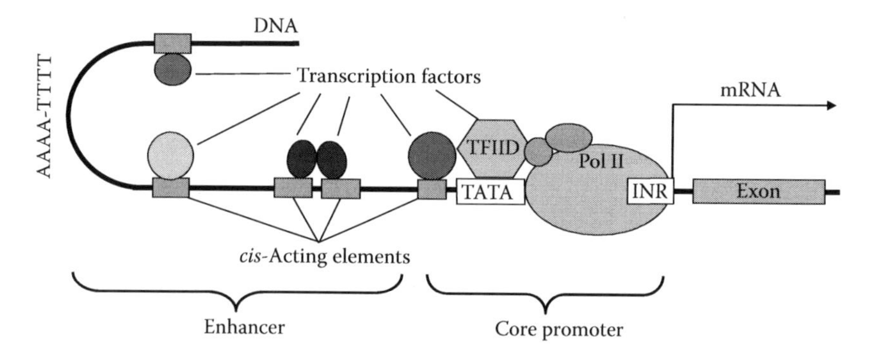

FIGURE 32.1 Schematic representation of promoter region and transcription initiation machinery. *Cis*-acting elements (CAEs) and transcription factors (TAFs) are indicated by boxes on DNA strand and various shaped objects. Sequence-dependent TAFs recognize corresponding CAEs and bind to DNA within enhancer region. Poly A/T tracts allow for distal enhancer CAEs to interact with proximal CAEs via curved DNA. These DNA-protein interactions trigger assembly of functional RNA polymerase II complex on core promoter region containing a TATA-box and an initiator site (INR) and subsequent initiation of transcription.

Enhancers

An enhancer is often referred to as the controlling region of a promoter. The location of an enhancer can be either upstream or downstream of an associated core promoter. Enhancers are commonly defined by their functional influence, either enhancing or suppressing, upon a corresponding core promoter. Many techniques including sequence deletion/substitution analyses, protein-DNA binding assays, and computational sequence prediction have been used to identify and characterize enhancers (Fickett and Hatzigeorgiou, 1997; Valouev et al., 2008).

An enhancer usually contains a unique sequence context intercalated with many regulatory CAEs. The number and types of CAEs present dictate the biological functionality of an enhancer. Thus far, a large number of CAEs from many unique plant promoters have been identified with their biological functions well characterized. These CAEs have been deposited in various databases for public access. For instance, up to 469 discrete CAEs derived from plant promoters have been compiled in a database called PLACE version 30.0 (http://www.dna.affrc.go.jp/PLACE/; K. Higo, NIAS, Ibaraki, Japan). The majority of CAEs are recognition/binding sites or anchoring points for TAFs, whereas other CAEs act as architectural components to alter particular DNA geometry conducive to the assembly of productive transcription machinery (Schwechheimer et al., 1998).

As a side note, plants are programmed to produce a large number of highly diversified TAFs to interact with all sorts of CAEs. For instance, up to 5% of the *Arabidopsis* genome is devoted to encode more than 1,500 TAFs. Almost half of these TAFs are unique in plants with the rest showing no significant sequence similarity among TAFs commonly found in other eukaryotes (Riechmann et al., 2000). This indicates the prodigious complexity of interactive mechanisms between CAEs and TAFs developed by plants to regulate gene expression.

Functionally, most enhancers can interact with TAFs to exert programmed functions in a position- and orientation-independent fashion and over a long distance (Blackwood and Kadonaga, 1998). In addition, enhancers can regulate promoter activity either positively by interacting with activating TAFs (activators) or negatively by working with repressing TAFs (repressors) (Lemon and Tjian, 2000). Structural enhancer elements with unique nucleotide sequence patterns such as poly-A and poly-T tracks can induce intrinsic DNA curvature. Such bended DNA can confer architectural properties important for transcription activation (Perez-Martin and de Lorenzo, 1997). On the other hand, different enhancers within the genome can be delineated by blockers known as insulators (West et al., 2002). Obviously, the enhancer is of paramount importance for coordinating interactions between a promoter and associated regulatory protein factors to provide sophisticated controlling mechanisms to modulate transcription activation.

MAJOR TYPES OF PROMOTERS

Promoters in plants can be classified into three major groups according to their mode of action. Tissue-specific and developmentally regulated promoters become functionally active only in particular types of cell, tissue, or organ, and at a certain stage of plant development. Constitutive promoters, on the other hand, are active in all types of cell and at all stages of plant development, providing a relatively constant level of gene expression. Lastly, inducible promoters increase activity greatly after coming into contact with eliciting chemicals or physical signals. In addition, some promoters may have multiple modes of action due to the presence of multifunctional regulatory elements.

Seed Storage Protein Promoter for Spatial and Temporal Expression

In plants, the production and accumulation of seed storage proteins is under strict spatial and temporal regulation. In other words, storage proteins have to be produced within the right growth stage and by specific types of cell in the developing seeds. It becomes an intriguing question as to how and by what means plants achieve such specific expression control at the promoter level. Thus, over the years, a number of promoters associated with seed storage protein genes such as β-*phaseolin* (*phas*) of common bean (*Phaseolus vulgaris* L.) and *glutelin* (*GT-1*) of rice (*Oryza sativa* L.) have been used as a model for extensive studies to elucidate both structure organization and mode of action for developmental and tissue-specific gene expression regulation.

Seed storage protein promoters contain many different types of CAEs. Some belong to the highly conserved general transcription activating CAEs commonly found in most eukaryotic promoters such as G-box element with a core sequence CACGTG (Williams et al., 1992) and CAAT-box element (Maity and de Crombrugghe, 1998); others are seed-specific CAEs including RY element, E-box, CACA element, ACGT motif, A/T-rich motif, vicilin-box (Guilfoyle, 1997). The majority of these CAEs are located within a span of 300 base pair region upstream of TSS. However, critical CAEs were also found in 800 base pair region of the *phas* promoter (Burrow et al., 1992) and even in 5 kb region of the *Gt-1* promoter (Zheng et al., 1993) upstream of TSS. Deletion and/or alternation of these regulatory sequence motifs often abolish the normal control manner of the promoter. Apparently, seed-storage protein promoters from different plant species have idiosyncratic structural characteristics tailored to their genetic milieus.

Seed specific transcription activation is regulated by the presence of and dynamic interplay between positive and negative CAEs within the promoter region, while some other CAEs manipulate the level of expression quantitatively (Guilfoyle, 1997). However, specific sequence context made up by interspersed spacer sequences also influence promoter activity. When portions of upstream controlling region from a seed-specific promoter were fused with a nonseed core promoter, seed-specificity of gene expression was altered. In other cases, alternating a few nucleotides surrounding the CAEs present in seed-specific promoters also resulted in dramatic change in binding affinity of corresponding TAFs (Williams et al., 1992; Kawagoe et al., 1994). Such structure-function relationship highlights the importance of sequence environments employed by seed-specific promoters for stringent gene expression control.

CaMV 35S Promoter for Constitutive Expression

All plant promoters isolated thus far control gene expression differentially in a developmental-regulated, tissue-specific, and often species-limited manner. Therefore, much of the search for strong constitutive promoters focused on expression control systems derived from plant viruses. The cauliflower mosaic virus (CaMV) 35S promoter is perhaps the most powerful promoter identified so far for use in plants. Its transcription activity is virtually unaffected by any environment factors, tissue types, or host species. This promoter has been dissected and analyzed in great details to unravel the structural organization, functionality of critical regulatory CAEs, and molecular mechanisms by which the promoter impacts gene expression in plants.

The 35S promoter was originally isolated as a ~900 base pair fragment from the DNA virus. However, subsequent expression analysis delimited the full-level activity to an approximately 340 base pair upstream region relative to TSS of the 35S transcript. This particular region was subsequently divided into two domains: domain A (–90 to +1) capable of conferring minimal expression and thus referred to as a core promoter; and domain B (–343 to –90) functioning as an enhancer responsible for the majority (99%) of promoter activity (Benfey and Chua, 1990).

The 35S core promoter contains a typical TATA-box element and an INR. In addition, there are two CAAT-box elements and two activation sequence-1 (*as-1*) motifs marked by a core sequence TGACG (Benfey et al., 1989). In spite of the existence of these critical elements, however, this core promoter is only capable of supporting about 1% of total activity attainable from the full promoter. Apparently, the presence of these controlling elements and recognition sites in the core promoter is far from sufficient for effective recruitment of TAFs for the establishment of productive transcription machinery. Additional CAEs must be required to significantly potentiate promoter activity. Nevertheless, the 35S core promoter has been widely used as a fusion partner to construct hybrid promoters in which enhancer fragments from other promoters including plant promoters are placed upstream of the 35S core promoter. Such hybrid or fusion promoters can provide an invaluable tool for studying the functional properties of enhancers under investigation.

The 35S enhancer is very unique in that it has an unusually high number of CAE repeats clustered within a relatively short nucleotide sequence. For instance, up to 10 copies of the GTGG-containing SV-40 enhancer-like motif were found within a 270 base pair region (Serfling et al., 1985). The presence of these clustered CAEs within the 35S enhancer drastically improves DNA-protein interaction and recruitment of TAFs leading to an exceptionally high efficiency of transcription activation (Benfey et al., 1989). The synergistic and additive effect of clustered CAEs on transcription activation was further exemplified by the exponential increase of promoter activity with the use of 35S promoter-derived enhancer duplications (Kay et al., 1987; Mitsuhara et al., 1996). More recently, the application of duplicated 35S enhancers for transcription enhancement was further culminated by novel bidirectional duplex promoters designed to facilitate simultaneous expression of multiple transgenes at high levels (Li et al., 2004).

Thus far, the 35S promoter remains the most widely utilized and consumer accepted promoter in transgenic plants to support high-level, constitutive transgene expression. It also has been used as an effective transcription-activating component in many fusion promoter systems. The majority of transgenic plants commercially grown today utilized this promoter due to its powerful promoter activity unaffected by environmental factors. This promoter, being a marvel of nature, will also serve as an excellent model or blueprint for the design and development of synthetic promoters to offer customized, application-tailored transcription control.

Promoters Responding to Biotic and Abiotic Inducers

Among the inducible promoters studied thus far, light-regulated promoters received greater attention due to their involvement in photosynthesis and growth and development. Examples of well-characterized light-regulated promoters include those associated with genes encoding the small subunit of ribulose-1,5-bisphosphate carboxylase/oxygenase (*rbcS*), the chlorophyll *a*/*b* binding proteins (*Lhc*, formerly called *Cab*), Chalconesynthase (*chs*), and phytochrome A (*Phy-A*) (Arguello-Astorga and Herrera-Estrella, 1998). These promoters can respond to different wavelengths of light and regulate gene expression positively or negatively. In most cases, a promoter fragment of about 300 base pairs in length upstream from TSS is sufficient to direct light inducible expression. Sequence organization of these promoters is similar to that of seed-specific promoters (i.e., they all contain both positive and negative regulatory CAEs). These CAEs include major light responsive CAEs such as GT-1 (GGTTAA), I-box, or GATA-box (GATAA) motifs, as well as a number of other general CAEs such as G-box and A/T-rich motifs that are commonly found in other promoters incapable of responding to light. However, no single CAE found in light-regulated promoters alone can activate light-dependent gene expression (Guilfoyle, 1997). Thus, transcription activation in response to light signal requires specific combination and coordinated action of multiple CAEs.

Over the years, a large number of promoters have been identified that can be induced by a variety of environmental signals such as temperature and water stresses or exposure to various macromolecule inducers. Sequence and functional analysis of these promoters unraveled many unique CAEs and their combination patterns responsible for gene activation in response to these environmental cues. For instance, the heat shock response element (HSE) composed of multiple GAA repeats in alternating palindromic orientation is present in almost all heat-responsive gene promoters including heat shock protein (*hsp*) promoters in plants. Unlike other highly regulated promoters, these HSEs appear to be the major determinant of heat-dependent transcription activation, whereas other concomitant CAEs may contribute to the modulation of expression levels.

A large number of hormone-inducible promoters have also been characterized and they, too, contain many CAEs with unique core sequences and specific responsiveness to growth hormones. A particular auxin responsive element (AuxRE), as demonstrated in the octopine synthase (*ocs*) promoter of *Agrobacterium*, was found to comprise direct repeats of a core motif (TGACGTAA) separated by a 4-base pair spacer. Alternation of these sequence elements abolished auxin inducibility. Noticeably, this *ocs* element well resembles the *as-1* motif found in 35S promoter and therefore is often referring to as *ocs/as-1* element. On the other hand, many plant-derived auxin-regulated promoters harbor different types of AuxREs. These elements normally comprise a unique core sequence (TGTCTC) and are arranged in modular structure with variegated patterns. These AuxREs tend to work in conjunction with general CAEs such as G-box to achieve auxin-dependent activation with finely tuned expression levels. Similar modular structure and organization of responsive elements have been found in promoters that can be induced by abscisic acid, gibberellic acid, ethylene, salicylic acid, jasmonic acid, and pathogen elicitors. Unlike developmentally regulated promoters, in which both positive and negative elements mediate promoter activity, the majority of responsive elements found in hormone-inducible promoters tend to regulate transcription activity positively (Guilfoyle, 1997). A better understanding of the relationship between structure and function for inducible promoters will provide important insights into the underlying mechanisms of hormone regulation of gene expression in plants. Such knowledge will undoubtedly facilitate efforts to manipulate gene expression for crop improvement via transgenic approaches.

Chemically inducible promoters have gained a lot of attention in recent years due to their unparalleled advantages in gene expression control. These promoters normally remain inactive but become functional quickly upon induction, like a controllable switch. Therefore robust, high levels of overexpression can be achieved on demand, while plants can save energy and cellular resources if the expression machinery is not turned on. Tissue-specificity of induced gene expression can also be manipulated by localized application of inducers. Chemically inducible promoters are commonly constructed based on the repressor-activator combination principle. For instance, tetracycline and lactose repressor-operator systems were borrowed from prokaryotes, such as *E. coli*, to construct inducible promoter fusions using the highly active 35S promoter as a fusion partner. While in other cases, copper or ethanol-activating promoter elements from yeast and fungus were utilized as controlling switch in 35S-derived fusion promoters. In addition, activating regions of hormone-inducible promoters isolated from nonplant hosts have also been tested to develop chemically inducible promoters in plants (Jepson et al., 1998). Over the years, a number of chemically inducible promoters have been successfully utilized in transgenic plants to control expression of various transgenes conferring resistance to insect pests, disease pathogens, herbicides, and production of value-added compounds.

Gene Expression Regulation at the Promoter Level

Promoters utilize multiple mechanisms to modulate gene expression both qualitatively and quantitatively. One important mechanism through which promoters regulate transcription activation is by changing DNA conformation or geometry to facilitate the assembly of transcription initiation complexes. These DNA topographic changes within promoter or enhancer region can alter accessibility of CAEs and modulate molecular interactions between DNA elements and regulatory proteins. Some well-known DNA conformational changes include DNA curvature, kinking, twisting, looping, and stretching that can be induced by different mechanisms. For instance, the formation of DNA-protein complexes as a result of TAF binding at corresponding CAE sites can cause DNA convolution or twisting, thus altering adjacent molecular geometry that becomes more conducive to binding by other co-factors. Also, A/T-rich sequences such as A- and T-tracts (at least A_4 or T_4) are capable of inducing strong, intrinsic DNA curvature. In fact, the majority of promoters identified in plants contain more significantly curved sequences than non-promoter DNA (Pandey and Krishnamachari, 2006). There are specific functions associated with these unique structural properties. Using the 2S albumin gene promoter isolated from grape (*Vitis vinifera* L.) as an example, deletion of an A/T-rich region abolished both high levels of expression activity and seed specificity suggesting the important regulatory functions of DNA bending for the overall promoter activity (Li et al., 2007). In recent years, a lot has been accomplished to determine the influence of DNA geometry on promoter activity using prokaryotic and animal model systems. However, many challenges still remain in efforts to define precisely the DNA architectural requirements to achieve optimized transcription activation by plant promoters.

CAEs are literally like hands and fingers of a promoter providing an effective tool to accurately and discriminatingly attract corresponding TAFs to regulate transcription. Promoters contain many different combinations and distribution patterns of CAEs. The types, numbers, sequential order, distribution density, and overlapping patterns of CAEs, as well as interspersion of spacer sequences, are all unique to each individual promoter. These are critical factors affecting the type and consequence of DNA-protein interactions and thus determining the mode of transcription activation. The unique CAE distribution pattern allows promoters to selectively detect the availability of particular TAFs due to compartmentalization and/or translocation of regulatory proteins triggered by environmental conditions or developmental cues, and then either represses or activates transcription at desired levels. The recently revealed defense gene activation mediated by plant immune receptors provided an excellent example for such dynamic and highly elaborated TAF availability-mediated transcription activation (Caplan et al., 2008).

Core promoter structure and interaction with enhancers can also impact transcription. As mentioned previously, core promoters contain TATA-box and/or INR elements as recognition sites for basal transcription initiation complex. However, the sequence composition and duplication of TATA-box can influence transcription efficiency. For instance, there are three repeated, clustered TATA-box elements in *phas* core promoter. Sequence substitution analysis revealed that all these elements were functionally active contributing in part to high levels of gene transcription (Grace et al., 2004). The 35S core promoter (domain A) is capable of directing root-specific gene expression at a relatively high level. Expression over the entire plant can, however, only be achieved when the same core promoter is sequentially linked to the 35S enhancer (domain B). This combinatory expression pattern suggests that core promoters even with minimum regulatory elements are capable of synergistically interacting with the upstream CAEs to substantiate programmed expression pattern (Benfey et al., 1989).

Promoters are discrete structural components. Each promoter contains unique signature sequence features and is capable of directing expression of a specific gene via structure-bound mechanisms. However, it should be pointed out that different promoters can also work together cohesively with a high degree of synergy, to be a part of a larger mechanism within the organismic system. According to genome-wide promoter studies in organisms such as yeast and humans, global control of gene expression critical for the survival of the organism relies on a so-called combinational transcriptional network by which many transcription factors work coordinately to modulate the accessibility of CAEs, thus allowing various promoters to be turned on or off in a spatially and temporarily regulated fashion. Such well-orchestrated regulatory networks ensure stringent and concordant regulation of gene expression throughout the course of growth and development (Pilpel et al., 2001; Werner, 2001). And CAEs of promoters are one of the important components of this global regulatory machinery.

In the last two decades, the majority of gene expression studies to investigate aforementioned structure-function relationships of plant promoters mainly relied on the use of macromolecule-level techniques such as DNA-protein binding assays and transgene expression from promoter fusion constructs. These studies provided a tremendous amount of information and fostered great advances in our understanding of transcription regulation in general at cellular and organismal levels. In recent years, many single-molecule tracing techniques have been developed to allow the visualization and real-time monitoring of the action of transcriptional macromolecules during transcription process (Bai et al., 2006). The use of such advanced techniques will shed more light onto the molecular bases of enormously complex CAE-TAF interaction, transcription activation, and gene expression regulation at atomic level.

REFERENCES

Arguello-Astorga, G., Herrera-Estrella, L. 1998. Evolution of light-regulated plant promoters. *Annu. Rev. Plant Physiol. Plant Mol. Biol.* 49:525–555.

Bai, L., Santangelo, T. J., Wang, M. D. 2006. Single-molecule analysis of RNA polymerase transcription. *Annu. Rev. Biophys. Biomol, Struct.* 35:343–360.

Benfey, P. N., Chua, N. H. 1990. The cauliflower mosaic virus 35S promoter: combinational regulation of transcription in plants. *Science* 250:959–966.

Benfey, P. N., Ren, L., Chua, N. H. 1989. The CaMV 35S enhancer contains at least two domains which can confer different developmental and tissue-specific expression patterns. *EMBO* 8:2195–2202.

Blackwood, E. M., Kadonaga, J. T. 1998. Going with the distance: A current view of enhancer action. *Science* 281:60–63.

Burrow, M. D., Sen P., Chlan, C. A., Murai, N. 1992. Developmental control of the β-phaseolin gene requires positive, negative, and temporal seed-specific transcriptional regulatory elements and a negative element for stem and root expression. *Plant J.* 2:537–548.

Caplan, J., Padmanabhan, M., Dinesh-Kumar, S. P. 2008. Plant NB-LRR immune receptors: From recognition to transcriptional reprogramming. *Cell* 3:126–135.

Fickett, J. W., Hatzigeorgiou, A. G. 1997. Eukaryotic promoter recognition. *Genome Res.* 7:861–878.

Grace, M. L., Chandrasekharan, M. B., Hall, T. C., Crowe, A. J. 2004. Sequence and spacing of TATA box elements are critical for accurate initiation from the β-phaseolin promoter. *J. Biol. Chem.* 279:8102–8110.

Guilfoyle, T.J. 1997. The structure of plant promoters. *Genetic Eng. Princ. Meth.* 19:15–48.

Jepson, I., Martinez, A., Sweetman, J. P. 1998. Chemical-inducible gene expression systems for plants—a review. *Pestic. Sci.* 54:360–367.

Kawagoe, Y., B.R. Campell, and N. Murai, 1994. Synergism between CACGTG (G-box) and CACCTG *cis*-elements is required for activation of the bean seed storage protein β-*phaseolin* gene. *Plant J.* 5:885–890.

Kay, R., Chan, A., Daly, M., McPherson, J. 1987. Duplication of CaMV 35S promoter sequences creates a strong enhancer for plant genes. *Science* 236:1299–1302.

Lee, T. I., Young, R. A. 2000. Transcription of eukaryotic protein-coding genes. *Annu. Rev. Genet.* 34:77–137.

Lemon, B., Tjian, R. 2000. Orchestrated response: a symphony of transcription factors for gene control. *Genes Dev.* 14:2551–2569.

Li, Z. T., Dhekney, S., Dutt, M., VanAman, M., Tattersall, J., Kelley, K. T., Gray, D. J. 2007. Isolation and characterization of the 2S albumin gene and promoter from grapevine. 2005 Proc. Int. Symp. on Biotechnol. of Temperate Fruit Crops and Tropical Species. *ACTA Hort* 738:759–765.

Li, Z. T., Jayasankar, S., Gray, D. J. 2004. Bi-directional duplex promoter with duplicated enhancers significantly increase transgene expression in grape and tobacco. *Transg. Res.* 13:143–154.

Maity, S. N., de Crombrugghe, B. 1998. Role of the CCAAT-binding protein CBF/NF-Y in transcription. *TIBS* 23:174–178.

Mitsuhara, I., Ugaki, M., Hirochika, H., Ohshima, M., Murakami, T., Gotoh, Y., et al. 1996. Efficient promoter cassettes for enhanced expression of foreign genes in dicotyledonous and monocotyledonous plants. *Plant Cell Physiol.* 37:49–59.

Pandey, S. P., Krishnamachari, A. 2006. Computational analysis of plant RNA pol-II promoters. *BioScience* 83:38–50.

Perez-Martin, J., de Lorenzo, V. 1997. Clues and consequences of DNA bending in transcription. *Annu. Rev. Microbiol.* 51:593–628.

Riechmann, J.L., J. Heard, G. Martin, L. Reuber, C. Jiang, J. Keddie, L. Adam, O. Pineda, O.J. Ratcliffe, R.R. Samaha, et al. 2000. *Arabidopsis* transcription factors: Genome-wide comparative analysis among eukaryotes. *Science* 290:2105–2110.

Robert, F., Douziech, M., Forget, D., Egly, J. M., Greenblatt, J., Burton, Z. F., Coulombe, B. 1998. Wrapping of promoter DNA around the RNA polymerase II initiation complex induced by TFIIF. *Mol. Cell* 2:341–351.

Schwechheimer, C., Zourelidou, M., Bevan, M. W. 1998. Plant transcription factor studies. *Annu. Rev. Plant Physiol. Plant Mol. Biol.* 49:127–150.

Serfling, E., Jasin, M., Schaffner, W. 1985. Enhancers and eukaryotic gene transcription. *Trends Genet.* 1:224–230.

Smale, S. T. 1997. Transcription initiation from TATA-less promoters within eukaryotic protein coding genes. *Biochem. Biophys. Acta* 1351:73–88.

Smale, S. T. 2001. Core promoters: Active contributors to combinational gene regulation. *Genes Dev.* 15:2503–2508.

Valouev, A., Johnson, D. S., Sundquist, A., Medina, C., Anton, E., Batzoglou, S., Myers, R. M., Sidow, A. 2008. Genome-wide analysis of transcription factor binding sites based on ChIP-Seq data. *Nature Meth.* 5:829–834.

Werner, T. 2001. The promoter connection. *Nat. Genet.* 29:153–159.

West, A. G., Gaszner, M., Felsenfeld, G. 2002. Insulators: Many functions many mechanisms. *Genes Dev.* 16:271–288.

Williams, M. E., Foster, R., Chua, N. H. 1992. Sequences flanking the hexameric G-box core CACGTG affect the specificity of protein binding. *Plant Cell* 4:485–496.

Yamamoto, Y. Y., Obokata, J. 2008. ppdb: A plant promoter database. *Nucl. Acids Res.* 36:D977–D981.

Zheng, Z., Kawagoe, Y., Xiao, S., Li, Z., Okita, T., Hau, T. L., Lin, A., Murai, N. 1993. 5' Distal and proximal cis-acting regulator elements are required for developmental control of a rice seed storage protein glutelin gene. *Plant J.* 4:357–366.

33 Genetic Engineering Technologies

Zhijian T. Li, Sadanand A. Dhekney, and D.J. Gray

CONCEPTS

- Over the last three decades, great advancements have been made to develop transformation methodologies. This has resulted in the identification of several reliable transformation techniques that can be used routinely to transform plant cells and to produce transgenic plants of many crop species.
- Genetic engineering technologies complement conventional breeding efforts by providing unique tools to stably incorporate foreign genetic materials and novel characteristics into target plants without the hindrance of biological barriers.
- Transfer of foreign DNA into plant cells and plastids can be accomplished by a variety of biological, physical, and chemical means, among which the most commonly used are *Agrobacterium*-mediated transformation, protoplast-mediated transformation, and microprojectile bombardment.
- The incorporation of transformation technologies into contemporary plant improvement programs has yielded new cultivars with improved agronomic traits.

Genetic engineering of crop plants represents a major milestone in modern agricultural science. The advent of recombinant DNA technology in the early 1970s and the subsequent development of DNA transfer techniques provided exciting opportunities for plant scientists to isolate and utilize useful genes from both prokaryotic and eukaryotic organisms to confer novel traits on plants. Technological advancements in plant tissue culture techniques facilitated introduction of both native and foreign genes into the plant genome and the production of transgenic plants. Transgenic plants expressing novel traits now are being widely cultivated for their improved yield, quality, and other value-added characteristics. It should be noted, however, that in most instances genetic engineering techniques provide only an alternative approach to conventional breeding programs.

In crop improvement, conventional breeding and hybrid seed production are the mainstays in ongoing efforts directed toward varietal development (Morandini and Salamini, 2003). Nonetheless, modern genetic engineering technologies offer several unique advantages over conventional hybridization approaches. For example, in vitro DNA transfer techniques permit the introduction of genes and other genetic elements among sexually unrelated organisms, thereby bypassing biological barriers. Such genetic manipulation can be accomplished using a large quantity of plant materials in a relatively small space with a year-round artificially controlled growth environment. Hence, use of genetic engineering techniques can complement and expedite conventional breeding programs by increasing the diversity of genetic resources, enhancing efficiency, and reducing length of time needed to introgress desirable traits into existing elite crop varieties. Genetic engineering also allows utilization of exotic genes for the development of transgenic plants to produce proteins with novel nutritive, pharmaceutical, agrichemical, and industrial characteristics (Fischer and Emans, 2000).

Although several studies reported transfer and expression of prokaryotic genes, the first expression of eukaryotic genes into genetically modified plant cells was demonstrated in transgenic cells of sunflower containing a gene encoding the seed storage protein phaseolin (Murai et al., 1983). In these experiments, transformation was accomplished using the soil-borne phytopathogenic bacterium *Agrobacterium tumefaciens* to transfer the desired genes and cause them to be incorporated into the plant genome. This DNA transfer technique was developed as a result of years of extensive studies on crown gall disease and molecular mechanisms that control tumor formation (Ream and Gordon, 1982). *Agrobacterium*-mediated transformation has become the most commonly used method of creating transgenic plants. Desired gene expression was derived from so-called hybrid genes in which target gene coding sequences were operably linked to functional promoter and terminator sequences. In the case of transgenic sunflower, gene transcription and translation were confirmed by accumulation of phaseolin gene-specific mRNA and production of phaseolin protein in transformed tissues. However, the first actual transgenic plants were subsequently obtained using tobacco, which was easier to regenerate from cells.

Stimulated by these and more recent studies, plant scientists worldwide have continued to develop and refine transformation technologies in order to enable the generation of transgenic plants from numerous species. Today, three major DNA transfer methods are widely utilized. These include *Agrobacterium*-mediated transformation, direct protoplast-mediated DNA transfer, and microprojectile bombardment-mediated transformation. In addition, several alternative DNA delivery systems have also been developed. Successful utilization of these DNA transfer techniques has resulted in examples of transgenic plants among almost all major crop species. In this chapter, emphasis is given to discussion of major transformation methods currently being used to transfer foreign DNA into plant cells and plastids.

AGROBACTERIUM-MEDIATED TRANSFORMATION

Agrobacterium tumefaciens is a Gram-negative, soilborne, phytopathogenic bacterium responsible for inciting crown gall disease in a large number of gymnosperms and angiosperms. The development of crown gall disease occurs through an intricate interplay between bacterial genetic elements and plant host responses. Early molecular studies revealed that this bacterium was capable of transferring a short piece of DNA (T-DNA) from its tumor-inducing (Ti) plasmid into the genome of susceptible host plant cells. The T-DNA contains genes encoding proteins that are involved in the biosynthesis of phytohormones (the oncogenes) and novel conjugates of organic acids and amino acids or sugars called opines (the opine synthesis genes). These phytohormones and bacterial metabolites are necessary for survival and proliferation of bacterial cells in the modified host cell environment. They also stimulate tumor formation (i.e., "crown gall"). In addition, genes located in the virulence (*vir*) region of the Ti plasmid encode a number of virulence proteins (Vir proteins) that mediate the T-DNA transfer process. Activation of these genes is controlled by plant regulatory factors that are produced by infected host plant cells.

For producing transgenic crop plants, there are several advantages associated with (*Agrobacterium*-mediated) transformation. T-DNA transfer is a highly sophisticated biological process that requires the participation and coordination of not only many bacterial proteins but also several important proteins produced by the plant cells. Thus, the efficacy of *Agrobacterium*-mediated transformation can be readily enhanced by optimizing conditions conducive to the production of these participating proteins. *Agrobacterium*-mediated transformation results in transfer of DNA with defined ends and with minimal rearrangement. Relatively large segments of DNA can also be accommodated within the T-DNA region and subsequently transferred. In addition, compared to other gene transfer techniques, only one or a few copies of the transferred genes are generally integrated into plant chromosomes with *Agrobacterium*-mediated transformation. With a minimum degree of gene disruption resulting from genomic integration of transgenes, transgenic plants with normal agronomic performance and fecundity can be obtained.

THE T-DNA TRANSFER PROCESS

The T-DNA transfer process is composed of the following three major steps:

1. Activation of virulence genes
2. T-strand processing and transfer
3. T-DNA integration into the plant cell genome (Figure 33.1)

For successful DNA transfer, *Agrobacterium* detects the presence of low molecular mass phenolic and sugar compounds (e.g., acetosyringone) produced by wounded plant cells. This molecular sensing process is mediated by an inner membrane protein, VirA, which subsequently transduces information to a transcriptional activator, VirG, by a mechanism involving protein phosphorylation. Activated VirG then triggers transcription and expression of the *vir* region contained in the Ti plasmid leading to production of up to 20 Vir proteins from six operons; *virA*, *-B*, *-C*, *-D*, *-E*, and *-G*. These proteins perform a variety of functions to ensure the success of T-DNA transfer. Of particular interest is the site-specific endonuclease VirD2. This protein, assisted by VirD1 protein, recognizes the T-DNA border sequences (25-bp direct repeats at its ends), creates a nick site, and initiates the formation of a single-strand (ss) copy of T-DNA (T-strand). VirE2 is a sequence nonspecific ssDNA binding protein (Grange et al., 2008). It functions to bind ssDNA regardless of sequence, forming a fully coated VirE2:T-strand complex (T-complex) that effectively protects T-DNA from nucleolytic degradation during the transfer process. The T-complex is then exported from the bacterial cell to plant cells via a VirB/D4-mediated type IV secretion system (Christie, 2004). Once inside the plant cell, a number of host proteins, including VirE2-interacting proteins (VIP1 and VIP2), are involved in the active transport of T-complex into the nucleus (Li et al., 2005; Anand et al., 2007). Nuclear localization signals (NLSs) present in VirE2 and VirD2 are also recognized by plant-produced importins (Importin α and β) to form a stable nuclear pore-targeting complex for nuclear import (Bhattacharjee et al., 2008). In addition, several other Vir proteins are also translocated into the

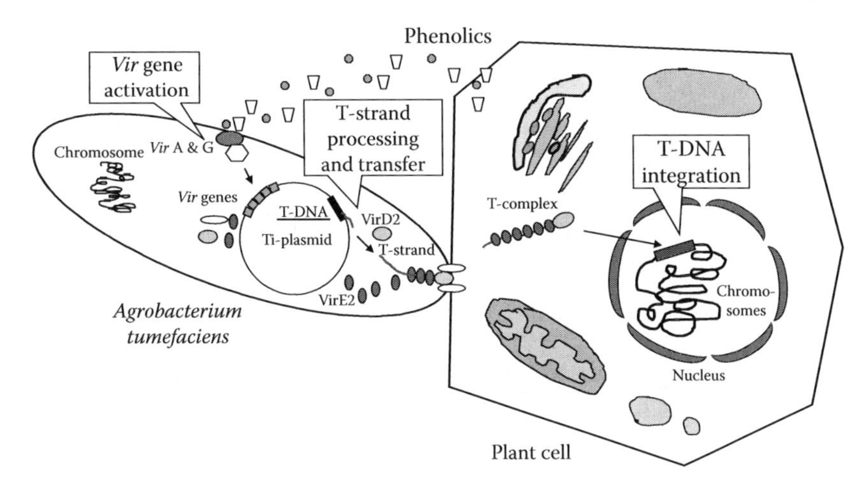

FIGURE 33.1 Schematic representation of the *Agrobacterium* T-DNA transfer process. *Agrobacterium*-mediated T-DNA transfer can be viewed as a three-step process. During the first step, bacterial attachment to plant cells facilitates interactions between plant inducers, including low molecular mass phenolic and sugar compounds and the signal sensing *vir*A product. The latter then activates the transcription activator *vir*G product. The activated *vir*G product induces transcription of the rest of *vir* loci located on the Ti plasmid. In the second step, a T-strand containing a single strand copy of T-DNA is cut out from the Ti plasmid by the *vir*D2 product and transferred from bacterial cell to plant cell through the coordinated action of a variety of other *vir* products, including the single-strand DNA binding protein VirE2. In the final step, T-DNA is introduced into the nucleus and subsequently integrated into the plant cell chromosome.

plant cells to participate in T-DNA integration process. For instance, the VirF protein is capable of entering plant nucleus and functions to disassemble the T-complex by degrading VirE2 and VIPs via a target-specific proteolysis (Tzfira et al., 2004). Throughout DNA transfer process, a single VirD2 protein covalently binds to the right border of T-strand and facilitates its transfer into the nucleus and subsequent integration in the plant genome (Figure 33.1). Additional details regarding *Agrobacterium*-mediated T-DNA transfer and related cellular processes can be found in an excellent review by Citovsky et al. (2007).

TRANSFORMATION STRATEGIES

Different strategies have been developed to make use of *Agrobacterium*-mediated T-DNA transfer. It was first essential to modify the native T-DNA to remove tumor-inducing genes responsible for phytohormone biosynthesis so that the crown gall disease did not occur. These unwanted genes then could be replaced with target genes encoding products of interest. Ti plasmids from which tumor-inducing genes have been removed are called *disarmed* Ti plasmids.

One early strategy of *Agrobacterium*-mediated DNA transfer utilized the so-called cointegrating Ti plasmid vectors (Rogers et al., 1988). In this method, an intermediate plasmid carrying the right border sequence, a portion of Ti plasmid DNA, and sequences for the origin of replication (replicon) from *Escherichia coli* were used to provide a vehicle to manipulate genes of interest in *E. coli* cells. A disarmed Ti helper plasmid carrying the left border was also prepared in an appropriate *Agrobacterium* strain. In addition, a transfer helper plasmid providing plasmid DNA transfer functions between bacterial cells was maintained in *E. coli* cells. When all three types of bacteria containing these specialized plasmids were mixed together, the intermediate plasmid was transferred from *E. coli* cells into the *Agrobacterium* cells where it was cointegrated in the T-DNA region of the disarmed Ti plasmid through a process called homologous recombination. *Agrobacterium* cells harboring cointegrated Ti plasmids that contain a fully functional T-DNA region with both the right and left borders are identified by selection for antibiotic resistances encoded by the different integration partners and are propagated for subsequent use in transformation experiments.

A more commonly used strategy involves binary plasmid vectors (Bevan, 1984). This type of plasmid combines both right and left borders of the T-DNA and compatible replicons from both *E. coli* and *Agrobacterium* in a single plasmid molecule, such that it can be propagated and maintained in both hosts for DNA manipulation and subsequent transfer. However, use of the binary vector system requires an *Agrobacterium* host strain that carries a wild-type Ti plasmid or a disarmed Ti plasmid in order to provide functions of the *vir* genes essential for the T-DNA transfer process.

FACTORS AFFECTING *AGROBACTERIUM*-MEDIATED TRANSFORMATION

T-DNA transfer is a complex biological process necessitating activation of a large number of Ti plasmid and related bacterial-encoded genes that involve active molecular interactions between *Agrobacterium* and host plant cells. Accordingly, development of efficient *Agrobacterium*-mediated transformation systems for different crop species can only be accomplished if the essential requirements for improving *vir* gene expression and plant cell response to *Agrobacterium* infection are met. Among factors that affect the efficacy of DNA transfer into plant cells and the obtainment of transgenic plants are virulence of the *Agrobacterium* strain, plant genotypic response, tissue culture/infection procedures, selection, and regeneration of transformants.

In general, there are differences in levels of virulence among *Agrobacterium* strains. The virulence of a particular *Agrobacterium* strain represents its capability to transfer T-DNA through the bacterial cell wall and introduce it into affected plant cells. Hence, screening different *Agrobacterium* strains for high levels of virulence and infectivity is frequently utilized as a crucial first step in development of an efficient transformation system for a particular plant species. The degree of virulence is primarily attributed to level of Vir protein expression from the Ti plasmid. For example, the supervirulence of a wild-type strain A281 was found to be associated with expression of *virG* and *virB* loci of its Ti plasmid pTiBo542. Consequently, use of a disarmed version of this supervirulent Ti plasmid (pEHA101) resulted in virulent strain EHA101, which has been used to transform numerous plant species (Hood et al., 1986). In addition, several techniques have also been developed to enhance the virulence of a chosen *Agrobacterium* strain. For instance, culturing the bacterium in the presence of plant phenolic compound inducers, such as acetosyringone, resulted in dramatically enhanced *vir* gene expression and subsequently led to a higher transformation frequency (Engstrom et al., 1987). In another approach, key virulence genes such as *virG* encoding a transcriptional activator were modified for constitutive expression in bacterial cells to improve bacterial virulence and host range (Hansen et al., 1994). The use of these modified supervirulent binary vectors ultimately resulted in the successful transformation and recovery of trangenic plants from recalcitrant monocotyledonous species including rice (Hiei et al., 1997).

Judicious selection of the receptor genotype and explant type and age, along with physiological state and pretreatment of receptor tissues, are also a critical factor affecting successful transformation using *Agrobacterium*. While mechanisms still remain unknown, different genotypes of a receptor species often display different levels of susceptibility to *Agrobacterium* infection. Many economically important crop species and elite genotypes of particular species are highly recalcitrant to *Agrobacterium*-mediated transformation. Hence, a few susceptible genotypes (model genotypes) were typically first utilized in order to identify factors that limit the efficacy of T-DNA transfer and to develop optimized procedures for transformation of other genotypes. Cells from various tissues or tissues subjected to various in vitro culture conditions or physical or chemical pretreatments prior to transformation may become differentially susceptible to transformation and plant regeneration. In addition, it is important to define tissue culture regeneration systems to maximize the population size of transformable and regenerable cells and to identify effective pretreatments that enhance infection and DNA uptake. Recent efforts to identify plant genes and proteins responsible for resistance to *Agrobacterium* infection may facilitate the eventual transformation of recalcitrant species and increase transformation frequencies in general.

Protoplast-Mediated Transformation

Protoplasts are cells without cell walls (Chapter 26). In a protoplast system, the only barrier between the living protoplasm and the external environment is the plasma membrane. Thus, the semifluid nature of the plasma membrane allows direct movement of macromolecules, such as DNA, into protoplasts by using relatively simple physical or chemical treatments. When cultured under suitable conditions, protoplasts are capable of cell wall regeneration and subsequent growth into whole plants. These unique characteristics of protoplasts permit development of efficient protoplast-mediated DNA transfer and transformation systems. Protoplast-mediated transformation has been successfully utilized in numerous molecular and genetic studies of genes and genetic elements to elucidate gene functionality and mechanisms regulating gene expression. Efficient protoplast-mediated transformation systems have also been utilized for transformation of a number of major crop species, including rice (Li et al., 1990).

Protoplast Isolation and Regeneration

Protoplasts can be isolated from a variety of plant materials, including leaves, callus tissues, suspension cultured cells, somatic embryos, pollen, cotyledons, and other storage organs. Leaf tissue is an excellent source of protoplasts due to the relative abundance of more or less uniform mesophyll cells and a relatively simple anatomical structure that facilitates protoplast release. When leaves of young plants are used, surface sterilization is normally needed, followed by removal of the epidermis, enzymatic digestion, and ultimately the purification of protoplasts. The use of in vitro shoot cultures or plantlets can reduce problems of contamination during and subsequent to protoplast isolation. However, suspension cultures are considered to be a better protoplast source material for transformation studies because they provide high quality, actively growing cells, often with synchronized cell division cycles. In addition, protoplasts from suspension cultures may have a higher potential for rapid cell division and a higher competency for DNA uptake and integration. These characteristics result in a better efficiency of transformation and transgenic plant regeneration. (Details of protoplast isolation and culture are provided in Chapter 27.) After their isolation, it is important to clean protoplasts thoroughly to remove cell debris and residual enzymes that may otherwise interfere with DNA transfer or hamper cell wall regeneration during subsequent protoplast culture. The combination of filtration through nylon mesh (40 μm) and simple low speed centrifugation in an isotonic washing solution proved sufficient to ensure efficient recovery of viable protoplasts for use in transformation studies (Li et al., 1990, 1995).

Regeneration of plants from isolated protoplasts is a prerequisite for the successful utilization of protoplast-mediated transformation for crop improvement. Over the years, a wide range of techniques has been developed for regeneration of plants from protoplasts. In particular, immobilization of protoplasts in a low melting point agarose gel to facilitate selection and restrict movement of transformed protoplasts during the culture process, and the utilization of nurse cells to enhance protoplast regeneration have been successfully applied to several important crop species (Li et al., 2000).

Protoplast-Mediated DNA Transfer

Electroporation and polyethylene glycol (PEG) treatments are the two most widely used methods to deliver DNA into protoplasts. Cationic liposome-mediated transformation of protoplasts is less frequently used. Electroporation treatment provides a technically simple way to introduce DNA into protoplasts through electrically induced membrane pores. Electroporated protoplasts can be immediately subjected to culture treatments facilitating the recovery of transformed cells. Transformation frequency following electroporation is often affected by several factors including electroporation voltage, current and pulse duration, DNA concentration, and buffer type. The development of an efficient electroporation procedure necessitates the optimization of these parameters.

PEG-mediated DNA uptake offers an inexpensive, simple, and efficient approach for introducing DNA into protoplasts (Negrutiu et al., 1987). PEG molecules chemically induce the formation of membrane pores and thus allow the movement of macromolecules into protoplasts. PEG-mediated DNA uptake was up to 100 times higher than that obtained using electroporation (Hayashimoto et al., 1990). PEG concentration is the most important parameter affecting transformation frequency. For example, it was found that 30% (w/v) PEG dramatically reduced protoplast viability, whereas less than 10% (w/v) PEG was ineffective in improving DNA uptake by rice protoplasts (Hayashimoto et al., 1990). High concentrations of PEG can also induce undesirable protoplast fusion during the transformation process. Thus, one has to utilize an appropriate PEG concentration during the transformation process in order to obtain a high transformation efficiency while maximizing protoplast viability after treatment.

Microprojectile Bombardment

Microprojectile bombardment is a technique by which solid particles such as tungsten or gold microspheres or microparticles (carriers) are coated with DNA molecules and accelerated by high-pressure helium gas or propulsion of an explosive charge at high velocity into plant cells. These particles have sufficient momentum to penetrate cell walls and move themselves into target cells without causing lethal damage. Once inside the cells, DNA molecules are released from their carriers and integrated into the chromosome by cellular components. Transformed cells are then identified and transgenic plants subsequently recovered through appropriate tissue culture selection and regeneration procedures. As with other direct DNA transfer systems, microprojectile bombardment permits introduction of naked DNA into plant cells, making construction of gene expression units relatively simple because the need for the *Agrobacterium* binary vector system is eliminated. It also allows simultaneous introduction of multiple transgenes or large DNA fragments into plant cells in a single operation. This technique enables the use of a wide range of regenerable plant tissues as transformation target materials for studies of gene expression and obtainment of transgenic plants (Taylor and Fauquet, 2002). Microprojectile bombardment is especially useful for the transformation of plastids (Maliga, 2004).

PARTICLE BOMBARDMENT TECHNOLOGY

A particle bombardment device was first demonstrated by Sanford and coworkers (Sanford et al., 1987). This device utilized tungsten particles (microcarriers) that were coated with DNA and placed onto the surface of small plastic bullets (macrocarriers). A gunpowder charge was then employed to propel the plastic bullets at high velocity into a stopping plate that effectively stopped the plastic bullets but allowed the tungsten particles to be released with sufficient speed and force on impact to penetrate the target tissues placed below. Based on a similar mechanism, several modified devices were subsequently developed in efforts to provide a better control for the bombardment process, improve transformation efficiency, and increase operational safety while reducing the cost.

BioRad (Hercules, California) markets the Biolistic PDS-1000/He particle delivery system. This device utilizes helium gas to propel microcarriers in combination with plastic rupture discs that break at specified pressures ranging from 450 to 2200 psi. A detailed description of the device and operation procedures can be found on the BioRad Web site (http://bio-rad.com/LifeScience/pdf/Bulletin_9075.pdf). Parameters that affect DNA delivery, damage to target tissues due to microcarrier impact, and transformation efficiency can be experimentally manipulated and optimized. This device continues to be popular in plant biotechnology laboratories and is capable of efficiently and reproducibly transforming numerous crop species. BioRad has also developed a hand-held, helium-powered device called the Helios Gene Gun for transforming target tissues and organs that are not suitable for placement in a vacuum chamber. This device has been used primarily to study transient gene expression and for the inoculation of plant viral pathogens.

Several variations of the original particle gun have been developed. The Particle Inflow Gun (PIG) device developed by Vain et al. (1993) differs from the BioRad device in that it utilizes a burst of helium to carry microcarriers into a vacuum chamber, thus eliminating the need for rupture discs or macrocarriers and associated parts. The PIG device is relatively inexpensive, easy to operate, and has been used successfully to produce transgenic plants in several major crop species. Construction and operation of a simplified PIG device was described by Gray et al. (1994).

Another variant of the particle gun is named ACCELL Technology and utilizes shock waves generated by high voltage electrical discharge to propel microcarriers into the target tissues (McCabe and Christou, 1993). Because this device does not rely on gaseous propellants to transfer power to microcarriers, damage to the target tissues can be greatly minimized. In addition, the electrical discharge mechanism provides a high degree of controllability making it possible to deliver microcarriers into desirable layers of plant cells within target tissues. Transgenic plants containing transgenes with low-copy number integration were produced in several crop species using the ACCELL system.

FACTORS AFFECTING MICROPROJECTILE BOMBARDMENT-MEDIATED TRANSFORMATION

The successful application of microprojectile bombardment technology to plant genetic engineering largely depends on development of a set of optimized parameters associated with the bombardment process. These parameters include choosing appropriate type and size of microspheres, coating procedures, DNA concentrations, microcarrier dispersing and loading procedure, vacuum levels, helium gas pressures, electrical discharge settings, and the number of repeated bombardments. It is important to make certain that the DNA of interest can be delivered into target cells of selected tissue without causing excessive cellular damage that could otherwise impair the subsequent regeneration process.

The ability to deliver foreign DNA into cells by microprojectile bombardment greatly increases the range of cell and tissue types and the number of elite genotypes that can be used in transformation experiments. However, selection of appropriate target cells and tissues is crucial for achieving high transformation efficiencies and ultimately transgenic plants. There are several criteria to be evaluated in choosing an appropriate target tissue. Target tissues should have the following capabilities: (1) integrating introduced foreign DNA into its genome and germline cells; (2) expressing selectable marker genes that have been engineered into the transferred DNA so that transformants can be successfully selected; and (3) proliferating and regenerating into normal transgenic plants at a high frequency. Embryogenic and meristematic tissues are most frequently selected for use as target materials for the production of transgenic plants using microprojectile bombardment technology.

Alternative Direct DNA Transfer

Over the years, the previously discussed transformation methods have been widely utilized in efforts to produce transformed tissues and transgenic plants of numerous crop species. However, there are certain limitations associated with these methods. For instance, the use of *Agrobacterium*-mediated transformation is limited by target tissue availability and low transformation efficiencies associated with poor susceptibility to infection by recalcitrant species and genotypes. Protoplast-mediated transformation often requires elaborate and time-consuming transformant selection regime and plant regeneration systems that can increase the occurrence of somaclonal variations. Microprojectile bombardment is frequently associated with high costs for acquiring the operating apparatus and associated accessories. To circumvent these problems, several alternative transformation methods based on the use of direct DNA transfer system were developed.

Silicon-carbide whisker-mediated transformation is perhaps the best documented alternative method of direct DNA transfer. It involves a brief vortex of transformation mixture containing needle-like micro-bodies, DNA, and target tissues. DNA is transported into target cells through holes generated in cell walls by silicon-carbide whiskers (Kaeppler et al., 1990). This efficient and inexpensive method has been used successfully to produce fertile transgenic maize plants (Frame et al., 1994). Transformation via DNA infiltration and electroporation of cells and intact tissues, electrophoresis and imbibition of embryos, microprojectile bombardment of pollen tube, and liposome-mediated transformation have also been attempted. However, the application of these techniques in practical transformation of major crop plants requires further technical development.

PLASTID TRANSFORMATION

Plastids are unique organelles in plant cells specialized to provide important functions in photosynthesis and biosynthesis of a variety of cellular macromolecules including amino acids, starch, lipid, and pigments. A plastid contains its own genome commonly referred to as plastome or ptDNA. A typical plastid genome is made up of a circular double-stranded DNA with a size of 120 to 180 kb encoding roughly 120 genes. It should be pointed out that in addition to these plastid genes, several thousands of nuclear genes are also involved in the differentiation and development of plastids. Unlike the nuclear genome, a plastid genome is marked by a dynamic change of polyploidy, meaning the number of plastome copies can be altered drastically depending on the cell type and development stage of plastids. For instance, a plastid in a meristematic cell contains 1–2 plastome copies, whereas each chloroplast in a mesophyll cell harbors up to 10–65 plastome copies depending on the cell age. The number of plastome copy in chloroplasts can also decrease significantly during chloroplast division. In plants, the number of chloroplasts per mesophyll cell ranges from 10 to 70, thus the chloroplast genome can reach up to several thousand plastome copies.

Chloroplast transformation offers unique benefits for many important applications in both basic and applied research. For instance, the functional attributes of any chloroplast genes can be determined using chloroplast genome manipulations involving gene replacement, sequence mutation, and overexpression. In addition, transgenes coding for useful value-added proteins and/or important agronomic traits can be integrated into the chloroplast genome to achieve a desired expression profile that is normally unobtainable via nuclear transgene expression.

Regarding the terminology used in plastid transformation, a cell containing transformed plastid(s) is termed a transplastomic cell. If all the plastids within a cell are uniformly transgenic for a foreign gene, that cell is called a homoplastomic cell, whereas a heteroplastomic cell refers to one containing both non-transformed and transformed plastomes. The unusually high level of polyploidy in the semi-autonomous, self-reproducing plastids posts extra challenges for genetic transformation, particularly with respect to areas including DNA delivery, selection and propagation of transformed plastids, and obtainment of homoplastomic plants.

Plastid transformation in plants mainly focuses on transgene insertion and expression in chloroplasts due to the abundance of organelles and the large number of chloroplast DNA copies (cpDNA) in each cell. Currently, chloroplast transformation mainly relies on the use of physical methods including particle bombardment using leaves as target tissues and PEG-mediated protoplast transformation. An extended period of antibiotic selection is often employed for chloroplasts in treated tissues to reach the homoplastomic state, i.e., all the chloroplast genomes in a cell are transformed. Transplastomic shoots or plants are subsequently regenerated from the resulting transplastomic explants (Maliga, 2004).

Factors Affecting Plastid Transformation

Appropriate design of transformation vectors is critical for integration of transgenes into the plastid genome and subsequent recovery of transplastomic plants. Insertion of transgenes into the chloroplast genome is often accomplished through a process of sequence-dependent DNA recombination, i.e., displacement of chloroplast genomic sequences by exogenous DNA with sequence homology. In practice, chloroplast targeting sequences corresponding to certain regions of cpDNA are used to flank both sides of a transfer DNA containing both marker and target genes. These cpDNA fragments are obtained from intergenic regions in the chloroplast genome. In doing so transfer DNA will only be inserted into these intergenic regions to avoid disruption of gene coding regions after transformation.

Cre-*loxP* recombination system can also be incorporated into the vector design to create (marker-free) transplastomic plants. In a Cre-*loxP* system, a DNA fragment bordered by a *loxP* site at each end will be cleaved and removed by Cre protein, a P1 phage site-specific recombinase (Nagy, 2000). Accordingly, *loxP* sites can be placed at each end of the marker gene expression unit in the chloroplast transformation vector, which will be used to generate transplastomic plants. Nuclear transgene corresponding to the Cre protein engineered to carry a chloroplast signal peptide will be introduced into the nuclear genome of the transplastomic plants. Expression and targeting of the Cre protein to transplastomic chloroplasts will eventually facilitate the removal of the marker gene from the transformed chloroplast genome.

The use of selectable marker genes and antibiotics is also an important consideration in chloroplast transformation. Selectable marker genes commonly used in chloroplast transformation include those conferring resistance to stryptomycin, spectinomycin, and lincomycin. These antibiotics are preferred because of their ability to inhibit callus formation, greening, and subsequent shoot regeneration from leaf explants. It is estimated that, under normal selection conditions with these antibiotics, up to 20 cell division cycles are needed to achieve the homoplastomic state. In addition, reporter markers including the visible marker green fluorescent protein (GFP) of *Aequorea victoria* and the assayable marker β-glucuronidase (GUS) can also be used to facilitate identification of transformants and analysis of gene expression in transplastomic plants. The use of these reporter markers provides additional tools for verification of transplastomic clones under the circumstances where spontaneous mutation to antibiotic resistance occurs in non-transformed target cells. Spontaneous mutation to spectinomycin resistance is very common in many plant species.

Thus far, routine transformation of chloroplasts can be achieved mainly in tobacco. Although sporadic success in obtaining transplastomic plants from other species including potato, tomato, and rice has been reported, more efforts are still needed to tackle many technical problems, especially the lack of effective and refined systems for selection, regeneration, and maintenance of stable transplastomic clones in important plant species. In addition, the fidelity and stability of gene expression in transplastomic plants remains problematic due to relatively higher rates of spontaneous DNA recombination in plastid genome and cpDNA variations associated with cellular growth and development and paternal plastid transfer. Therefore, more efforts are still needed to resolve outstanding issues in plastid transformation in order to enhance the reliability and applicability of this technology for crop improvement.

CONCLUDING REMARKS

Over the past three decades, great advancements have been made by plant scientists through concerted efforts to develop transformation methodologies. This has resulted in the identification of several reliable techniques that can be used routinely in many crop species to transform plant cells and to produce transgenic plants. Continuous refinement of these techniques and discovery of new approaches will further enhance our ability to genetically modify existing elite genotypes and expedite efforts to improve agronomic performance and value of crop plants. Despite these advances, genetic engineering of crop plants continues to face a variety of challenges. For example, factors affecting the level and stability of transgene expression remain to be fully elucidated. Innumerable in depth studies of transgene expression in transgenic plants have revealed that plants employ sophisticated internal mechanisms to cope with the intrusion of exogenous DNA and associated gene expression components. Molecular strategies that facilitate sustainable transgene expression in transgenic crop plants are needed. In addition, further studies are needed in areas such as the removal of objectionable marker genes from transgenic plants (Zhuo et al., 2002), multigene engineering for development of novel biological and metabolic pathways (Daniell and Dhingra, 2002), and the introduction of artificial chromosomes into crop plants. With technological advancements in these and other areas, genetic engineering of plants will continue to play an increasingly important role in global crop improvement efforts.

REFERENCES

Anand, A., et al. 2007. *Arabidopsis* Vir2 interacting protein2 is required for *Agroabcterium* T-DNA integration in plants. *Plant Cell* 19:1695–1708.

Bevan, M.W. 1984. Binary *Agrobacterium* vectors for plant transformation. *Nucl. Acid Res.* 12:8711–8721.

Bhattacharjee, S. et al. 2008. IMPa-4, an Arabidopsis importin α isoform, is preferentially involved in Agrobacterium-mediated plant transformation. Plant Cell online version: www.plantcell.org/cgi/doi/10.1105/tpc.108.060467

Christie, P.J. 2004. Type IV secretion: The Agrobacterium VirB/D4 and related conjugation systems. *Biochim. Biophys. Acta* 1694: 219–234.

Citovsky, V., S.V. Kozlovsky, B. Lacroix, A. Zaltsman, M. Dafny-Yelin, S. Vyas, A. Tovkach, and T. Tzfira, 2007. Biological systems of the host cell involved in *Agrobacterium* infection. *Cell. Microbiol.* 9:9–20.

Daniell, H., and A. Dhingra. 2002. Multigene engineering: Dawn of an exciting new era in biotechnology. *Curr. Opin. Biotechnol.* 13:136–141.

Engstrom, P., et al. 1987. Characterization of *Agrobacterium tumefaciens* virulence proteins induced by the plant factor acetosyringone. *J. Mol. Biol.* 197:635–45.

Fischer, R. and N. Emans. 2000. Molecular farming of pharmaceutical proteins. *Transgenic Res.* 9:279–299.

Frame, B.R., et al. 1994. Production of fertile transgenic maize plants by silicon carbide whisker-mediated transformation. *Plant J.* 6:941–948.

Grange, W., et al. 2008. VirE2: A unique ssDNA-compacting molecular machine. *PLoS Biol.* 6:e44.

Gray, D. J., E. Hiebert, C. M. Lin, M. E. Compton, and D. W. McColley. 1994. Simplified construction and performance of a device for particle bombardment, *Plant Cell Tiss. Org. Cult.* 37:179–184.

Hansen, G., A. Das, and M.D. Chilton. 1994. Constitutive expression of the virulence genes improves the efficiency of plant transformation by *Agrobacterium. Proc. Natl. Acad. Sci. (U S A).* 91:7603–7607.

Hayashimoto, A., Z. Li, and N. Murai. 1990. A polyethylene glycol-mediated protoplast transformation system for production of fertile transgenic rice plants. *Plant Physiol.* 93:857–863.

Hiei, Y., T. Komari and T. Kubo. 1997. Transformation of rice mediated by *Agrobacterium tumefaciens*. *Plant Mol. Biol.* 35:205–218.

Hood, E.E. et al. 1986. The hypervirulence of *Agrobacterium tumefaciens* A281 is encoded in a region of pTiBo542 outside of T-DNA. *J. Bacteriol.* 168:1291–1301

Kaeppler, H.F. et al. 1990. Silicon carbide fiber-mediated DNA delivery into plant cells. *Plant Cell Rep.* 9:415–418.

Li, J. et al. 2005. Uncoupling of the functions of the *Arabidopsis* VIP1 protein in transient and stable plant genetic transformation by *Agrobacterium*. *Proc. Natl. Acad. Sci. (U S A).* 102:5733–5738.

Li, Z., M.D. Burow, and N. Murai. 1990. High frequency generation of fertile transgenic rice plants after PEG-mediated protoplast transformation. *Plant Mol. Biol. Rep.* 8:276–291.

Li, Z. et al. 1995. Improved electroporation buffer enhances transient gene expression in *Arachis hypogaea* protoplasts. *Genome* 38:858–863.

Li, Z. et al. 2000. Transgenic peanut (Arachis hypogaea). In Y.P.S. Bajaj, Ed. *Biotechnology in Agriculture and Forestry*, Vol. 46, Transgenic Crops I. Springer-Verlag, Berlin. pp. 209–224.

Maliga, P. 2004. Plastid transformation in higher plants. *Annu. Rev. Plant Biol.* 55:289–313.

McCabe, D. and P. Christou. 1993. Direct DNA transfer using electrical discharge particle acceleration (ACCELLTM technology). *Plant Cell. Tiss. Org. Cult.* 33:227–236.

Morandini, P. and F. Salamini. 2003. Plant biotechnology and breeding: Allied for years to come. *Trends Plant Sci.* 8:70–75

Murai, N. et al. 1983. Phaseolin gene from bean is expressed after transfer to sunflower via tumor-inducing plasmid vectors. *Science* 222:476–482.

Nagy, A. 2000. Cre recombinase: The universal reagent for genome tailoring. *Genesis* 26:99–109.

Negrutiu, I. et al. 1987. Hybrid genes in the analysis of transformation conditions. I. Setting up a simple method for direct gene transfer in plant protoplasts. *Plant Mol. Biol.* 8:363–373.

Ream, L.W. and M.P. Gordon. 1982. Crown gall disease and prospects for genetic manipulation of plants. *Science* 218:854–859.

Rogers, S.G. et al. 1988. Use of cointegrating Ti plasmid vectors. In *Plant Molecular Biology Manual*. Eds. S.B. Gelvin, and R.A. Schilperoort, Kluwer Acad. Publ. Dordrecht. pp. A2/1–12.

Sanford, J. et al. 1987. Delivery of substances into cells and tissues using particle bombardment process. *J. Part. Sci. Tech.* 5:27–37.

Taylor, N.J. and C.M. Fauquet, 2002. Microprojectile bombardment as a tool in plant science and agricultural biotechnology. *DNA Cell Biol.* 21:963–977.

Tzfira, T. et al. 2004. Involvement of targeted proteolysis in plant genetic transformation by *Agrobacterium*. *Nature* 431:87–92.

Vain, P. et al. 1993. Development of the particle inflow gun. *Plant Cell. Tiss. Org. Cult.* 33:237–246.

34 Transformation of Plant Meristems

Jean H. Gould

MERISTEM TRANSFORMATION USING *AGROBACTERIUM*

The evolved gene transfer mechanism of *Agrobacterium tumefaciens* has become a transformation method of choice. Its use is simple; gene transfers are low copy and result in permanent heritable genetic changes. *Agrobacterium*-mediated transformation was once thought to be limited to species within the pathogen's native host-range. This was later expanded to include gymnosperm, most dicotyledon, and a few monocotyledon species. The important cereal crops were thought to be refractive to transformation until the early 1990s. In addition, it was thought that plant meristems excluded the *Agrobacterium* tumor-inducing principle in the same way virus was excluded from plant meristems. Because of these understandings, *Agrobacterium* transformations were thought to be limited to plant varieties known to be within *Agrobacterium*'s host-range, and limited to varieties that could be regenerated from callus and through somatic embryogenesis.

Aside from the genotype limitation to plant regeneration, plants regenerated from callus and somatic embryogenesis are prone to culture-induced mutations or "somaclonal variation." In the late 1980s, these limitations to host-range species, and to varieties that could be regenerated from callus, characterized the *Agrobacterium* problem, and this problem became a bottleneck in the transformation of important crop species and varieties.

In hindsight, use of plant shoots and meristems for *Agrobacterium*-mediated transformation would appear to have been an obvious approach due active cell division and to the ease of plant regeneration from shoots. Plant regeneration in culture from plant shoot apices, shoots produced through organogenesis, and other shoot meristematic tissue is well established. The method of meristem and shoot tip culture became popular following the discovery by Morel (Morel and Martin, 1952) that virus-free germplasm could be obtained from heavily infected plants through isolation and culture of shoot meristem tissue. The use of isolated shoots eventually became a cornerstone of the plant micropropagation industry (Murashige, 1974). Direct regeneration of shoots into plants in culture is known to be genotype-independent. Regenerated plants have the lowest mutation rates, and often low-grade germplasm contaminants can be eliminated.

The method of shoot meristem transformation using *Agrobacterium* was originally developed to circumvent the genotype limitations to plant regeneration from callus and somatic embryogenesis. Meristem, or shoot apex transformation, was first used with a known *Agrobacterium* host-range species (*Petunia hybrida* L.) to determine if inoculation of this tissue would produce transgenic plants (Ulian et al., 1988). Meristem, or shoot apex inoculation, was also used to determine if a nonhost species, a cereal (*Zea mays* L.), would be transformed using *Agrobacterium* (Gould et al., 1991). In both cases, normal plants were quickly generated in culture. Regenerated plants set seed and produced transgenic progeny that exhibited a Mendelian inheritance pattern of transferred genes. Both *Petunia* and corn transformation procedures are described in the following exercises.

Meristem and shoot apex culture was originally developed to remove the viral load from infected germplasm (Morel and Martin, 1952) and is the method of choice in the nursery industry for production of clonal lines that are true to type (Murashige, 1974). In cereals, as in other plant species, dividing cells in the apical meristem contain a competent cell type that accepts and sustains transformation by *Agrobacterium* (Smith and Hood, 1995).

Genotype fidelity can be lost in tissue culture, especially during long culture passages through callus and somatic embryogenesis. In vivo, genetic fidelity is most closely maintained in the meristems of plants. Regeneration of plants from differentiated meristems and shoots is direct and rapid. Time in culture is minimized. Plants regenerated from shoots have a lower incidence of mutation (Murashige, 1974). Culture-induced mutations are often permanent and can range from sterility to many phenotypic changes, and changes that impact yield (Bregitzer et al., 2002). Chromosome rearrangements and loss of chromosome material have been observed in plants regenerated in culture. Activation of latent retrotransposons in plant genomes during stress, infection, and dedifferentiation in tissue culture was found to produce permanent mutations similar to those seen in plants regenerated in culture (Hirochicka, 1993). This mechanism may account for some of the mutations that occur in tissue culture.

Terminology related to plant meristems can be confusing. The terms *meristem, shoot apex, shoot tip, shoot*, and other related terms are often used interchangeably and in reading the literature, it is important to understand what is actually described. A shoot apex contains the true meristem, leaf primordia, and elongating leaves (Figure 34.1). The shoot apex can be of almost any size, from a millimeter to 1 cm, while the true meristem is less than a millimeter. The true meristem of a plant consists of the meristematic dome containing the L1, L2 layers of the tunica and the cells of the corpus. The L1 layer will become the epidermal cell layer of the plant. The cells of the L2 layer become incorporated into the germline of the plant, while the corpus produces the body of the plant (i.e., parenchyma and the vascular tissue). In the case of some virus eradication work, the explant material used is a small shoot apex, but can often be referred to as a meristem.

At this time, *Agrobacterium* and its evolved transformation mechanisms are the only known example in nature where an organism can transfer genes to cells of another kingdom (in this case from prokaryotic cells to eukaryotic cells) and where the transferred genes become permanently incorporated into the host cell's DNA. In the case of viruses, genes are transferred but the genes do not become permanently incorporated into the host's genome. Of the many key elements, the right and left border (RB, LB) regions define the genetic region of a plasmid that will be transferred to the plant. Induction and maintenance of bacterial virulence and transformation competence in the plant

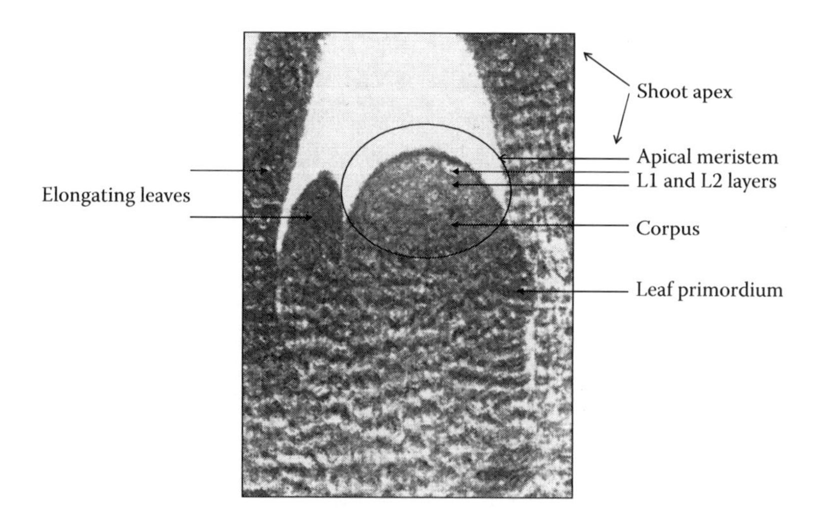

FIGURE 34.1 Image of a *Zea mays* shoot apex and showing the true meristem, L1 and L2 layers of the tunica, corpus region, leaf primordium, and elongating leaves. (From Gould, J. et al. 1991. *Plant Physiol.* 95:424–426. With permission.)

shoot are critical considerations. There are many steps in the transformation process that cannot be described here. In addition, some steps are still areas of active research. For more information, consult one of the many excellent review articles that are available (for example, McCullen and Binns, 2006; Tzfira and Citovski, 2003).

This chapter will provide an opportunity to use shoot meristem transformation with *Petunia hybrida* L., or *Zea mays* L. There are a number of different approaches to shoot isolation and inoculation and these options are described so that the best method can be chosen. The chapter includes transient expression assays so that the initial stages of a transformation can be observed. Continued plant selection, culture, and rooting procedures are provided to enable recovery of intact plants.

Lab Equipment Needed

- Autoclave or other device (i.e., pressure cooker) for equipment and media sterilization
- Dissection microscope or illuminated 3X magnifying lenses (Bausch & Lomb) inexpensive and often easier to use
- Distilled or reverse osmosis water
- Incubator for *Agrobacterium* set at 19 to 22°C, or an area in lab maintained at 25–26°C with no light, low light, or in a drawer
- Laminar air flow hood, clean room, or clean box
- Lighted shelving for cultures, light intensity of 70 to 100 μmol/ms, 16 h day
- pH meter
- Spectrophotometer for O.D. (optical density) readings

GENERAL CONSIDERATIONS

Seed Germination

It is important to obtain high quality seed that is in good condition. Seed grown under cool and dry conditions, such as seed sold commercially in seed packets, is best. In the South, corn seed collected from the field in summer will be internally contaminated and difficult to disinfect.

Shoot Isolation, Inoculation, and the L2 Layer

The shoot isolation procedure is intended to open and expose a facet of the meristem to *Agrobacterium*. A shoot meristem consists of the apical dome and tissues known as *tunica* and *corpus* (Figure 34.1). The cell layers in the tunica, L1 and L2, divide in two dimensions in such a way as to form single cell sheets that cover the plant. The cells of the corpus divide in three dimensions to produce the body of the plant. In general, if a 1 mm shoot apex is prepared, the cut base is usually the tissue that *Agrobacterium* accesses because the shoot is covered with elongating leaves (Figure 34.4a), the corpus cell layers will be in contact with *Agrobacterium*. The L1 and L2 layers of the tunica may or may not become transformed. On the other hand, if the top or the side of the meristem is pierced or opened and then inoculated (Figures 34.4b and 34.5d), the tunica containing the L1 and L2 layers, as well as the corpus cell layers, can come into contact with *Agrobacterium*. It is important to transform the L2 cell layer, because these cells ultimately become incorporated into the plant's gem-line. Transformation of these cells is critical for inheritance of transferred traits by the plant's progeny. In the procedures described below, several inoculation procedures are described. A simple shoot explant of approx 1 mm in height (Figure 34.4a.) can be used with *Petunia* and corn. A variation of this method is wounding the apex and meristem of the 1 mm apex with a hypodermic needle that pierces the meristematic dome of a shoot from the top of the explant (Figure 34.4b). In another procedure, which is best used with corn (Figure 34.5), tissue near the meristem will be sliced away to provide a broader region of access to *Agrobacterium* infection.

Transformation Competence

As with many procedures, there are a few steps that make a difference in the transformation competence of both *Agrobacterium* and plant cells. In this procedure, the following two elements are important: (1) induction of the virulence functions in *A. tumefaciens*; and (2) wounding and induction of cell division in the plant shoot meristem.

Acetosyringone (As; Sigma) is a phenolic compound that helps induce the *Agrobacterium* virulence cascade. Along with As, pH and a sugar are also important. Virulence induction media (VIM) typically contain As, a buffer MES at pH 5.4, and glucose. Other VIM contain Murashige and Skoog medium (MS; Murashige and Skoog, 1962), sucrose, and MES buffer at pH 5.4. VIM should be made up fresh and filter sterilized. It can also be stored frozen in 1 to 15 mL aliquots. When As is used alone in culture media, it should be added after autoclaved culture media and allowed to cool. In the laminar air-flow hood, attach a syringe filter disk to a syringe body and push the solution through the filter into the cooled medium, mix well, and dispense the media into culture dishes, 25 to 30 mL per dish.

Cocultivation temperature is important. *Agrobacterium* tumefaciens is a soil organism and adapted to cool soil temperatures. Depending on the strain of *Agrobacterium* used, virulence proteins function well at 20°C; however, the function of different classes of vir genes are inhibited at temperatures over 20°C, while other classes are inhibited by temperatures over 26°C (Fullner and Nestor, 1996; Baron et al., 2001). These temperatures can easily be met and exceeded on a lighted culture shelf, within culture dishes, and in some laboratories in the southern United States during summer months.

The optimum temperature for T-DNA transfer is 19–20°C (Fullner and Nestor, 1996). If possible, incubate the newly inoculated cultures (the cocultivation step) at these temperatures in darkness for the first three to four days. If incubation at 19–20°C is not feasible, incubate the cultures at room temperature in darkness or reduced light, such as in a drawer in the lab. During this cocultivation stage, do not put cultures under lights because temperatures will build up inside the culture.

Wounding and cell division activity is important in the transformation competence of plant cells. Wounding and inoculation approaches are described below. Cell division can be promoted by adding a cytokinin, such as kinetin or benzyladenine, to VIM and the cocultivation medium, described below.

About Selection

In *Agrobacterium* transformations, selection genes can sometimes be confusing. There can be from two to four or more selection genes to consider, depending on the bacterial strain and other plasmids used. One antibiotic can be used to select for *A. tumefaciens,* rifampicin in the case of EHA105. One or two antibiotics are used to select for various plasmids, while a third agent, an antibiotic, herbicide, or other selection gene, is used to select for transformed cells and transgenic plants. Refer to diagrams of EHA105 (Figure 34.2) and the binary plasmid pCNL56 (Figure 34.3). In the case of pCNL56, the plasmid resistance gene is to kanamycin (neomycin phosphotransferase I; nptI). Another selection gene, an antibiotic, herbicide, or other agent, is used for selection of transgenic plant tissues. In the case of pCNL56, this gene is also a kanamycin resistance gene (neomycin phosphotransferase II; nptII).

So, if the antibiotic kanamycin is being used to select for the plasmid, and also for transformed plant tissues, why are two different kanamycin genes needed? The reason is that each resistance gene will be active in a different place and time. The kanamycin gene (nptI) used for selection of the pCNL56 plasmid is driven by a bacterial promoter and protects bacteria from kanamycin toxicity when the plasmid is in the bacteria. Also, this gene and other elements of the pCNL56 plasmid are not within the right and left borders (RB and LB) of the plasmid and are usually not transferred to

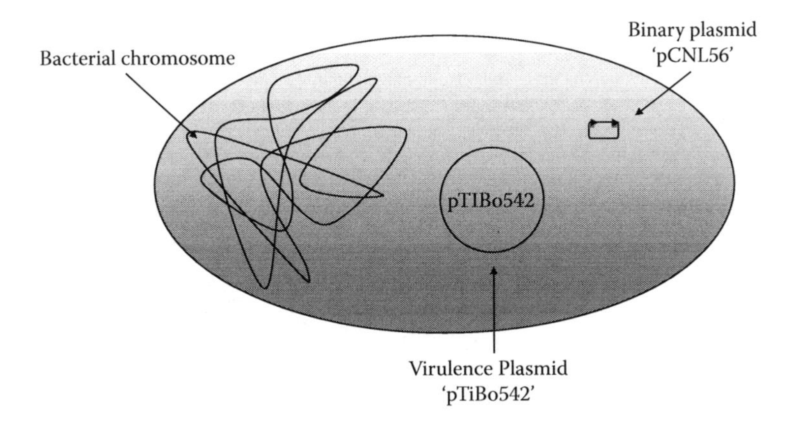

FIGURE 34.2 Diagram of *Agrobacterium tumefaciens* EHA105 showing three different genetic domains: the bacterium's chromosome, the plasmid pTiBo543, which contains the bacterium's virulence genes, and the small pCNL56 binary plasmid. The binary plasmid system consists of the pTiBo542 virulence plasmid and pCNL56 carrying the genes for transfer to the plant.

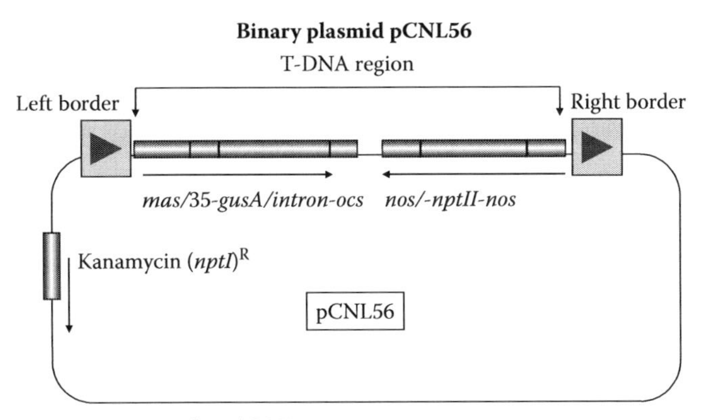

T-DNA = Transferred DNA

Kanamycin^R = a kanamycin resistance gene (*nptI*; Neomycin phosphotransferase I). This gene is driven by a bacterial promoter

nptII = a modified kanamycin resistance gene (neomycin phosphotransferase II). This gene is driven by a plant-style promoter, the nopaline synthase promoter (*nos*).

FIGURE 34.3 Diagram of the pCNL56 plasmid indicating the right (RB) and left (LB) borders, and the T-DNA region containing the promoters and genes that will be transferred to the plant and expressed in the plant: the MAS 35S promoter, the GUS gene plus intron, the *ocs* terminator; the *nos* promoter, the *nptII* gene (kanamycin resistance), *nos* terminator. The *nptI* gene (kanamycin resistance) outside of RB and LB provides resistance to the pCNL56 plasmid, and is not transferred to the plant.

the plant. The nptII resistance gene that is used to protect transgenic plant cells expressing this gene is driven by a promoter that functions in plant cells. These genes are placed between the RB and LB of pCNL56 and are transferred to plant cells. Although a plasmid such as pCNL56 contains many genetic elements, in general, only the DNA placed between the RB and LB will be transferred to the plant. The RB has many functions in transformation and incorporation of the transferred DNA into the plant's genome. This description is an over simplification but is used here to outline some of the processes involved in *Agrobacterium*-mediated plant transformations.

Antibiotics and Filter Sterilization

In general, antibiotics in solution are unstable and should be made up just prior to use. Also, most antibiotics are not heat stable and should be added to cooled autoclaved media. Solutions of the antibiotics should be filter sterilized using a sterile 0.22-μm syringe filter to remove bacteria and spores, and added to cooled (~55°C) autoclaved medium. Dissolve the antibiotics and other heat liable compounds in 2–3 mL deionized water. In the laminar air-flow hood, attach the filter disk to a syringe body and push the solution through the filter into the cooled medium, mix media well, and dispense into culture dishes, 25 to 30 mL per dish.

The penicillin derivatives carbenicillin and cefotaxime are frequently used to kill *Agrobacterium* after cocultivation with plant tissue. The penicillin-based antibiotics target bacterial cell wall synthesis and do not harm plant cells. Carbenicillin and cefotaxime are frequently cited in the literature but can be expensive. Clavamox, Augmentin, and Timentin are penicillin-based antibiotics and are also effective. Clavamox and Timentin can be obtained through a local veterinarian.

If Clavamox is used instead of carbenicillin, use the following procedure. Clavamox cannot be filter sterilized because it is in tablet form and contains particulate fillers; however, the tablet is wrapped and sterile. The outside of the wrapping is not sterile. Spray the outside of the sealed tablet with 70% ethanol. When dry, open the wrapping in the laminar flow hood using aseptic technique, sterile scalpel, and forceps. Remove the tablet and place it into 10 mL sterile water in a sterile 50 mL conical screw cap tube. Shake occasionally. The tablet will disintegrate in 15–20 minuets. Only after the tablet has disintegrated should you add the suspension to cooled autoclaved medium. Don't add the tablet directly to media unless the tablet has completely disintegrated, because once in the media, it will not dissolve. Mix well. (Derived from Gould and Magallanes-Cedeno, 1999.)

Genes and Promoters

Kanamycin is effective in *Petunia* and *Zea* meristem transformations; geneticin (G418) (Sigma) can be used with kanamycin resistance genes and may be more effective with *Zea* and other cereals. The popular visual marker gene (GUS, β-glucuronidase, *uidA* or *gusA*) has been in use since 1987 (Jefferson et al., 1987), and many reliable methods for this reporter gene are available. The standard gene constructs developed for use in tobacco and other dicot species contain the kanamycin resistance gene (neomycin phosphotransferase II, *npt II*) driven by the nopaline synthase (*nos*) promoter. The intron-GUS construct driven by the CaMV 35S promoter can be used with both *Petunia* and corn; however, expression of GUS using this promoter can be low in cereal shoots and leaves. The binary plasmid pCNL56 contains the MAS35S promoter to drive GUS expression and is effective in both *Petunia* and *Zea* shoots (Lee et al., 2007).

Chimeras

As with callus-based transformations (Schmulling and Schell, 1993), the plants transformed and regenerated using the shoot or meristem inoculation procedures can be chimeric for the transferred genes. Chimeras are plants composed of different genotypes. A visual example of a chimera is a variegated plant, which contains green sectors with normal chlorophyll content and white sectors with low or no chlorophyll content. In this case a sector or layer can be mutated. In a transgenic plant, a sector, or a layer, of the meristem can be transformed by one or many different genomic insertion events, while some sectors may not be transgenic. Regenerating transgenic plants (T_0) can contain many different sectors of cells. If the different chimeric sectors produce flowers, the genetic composition of the L2 layer of each sector producing a flower will be represented in the flower's ovules and pollen. In this way, the chimeric sectors of a T_0 plant will segregate in the T_1 and the T_2 generations, if the flowers are self-pollinated. The T_1, T_2, etc., progeny will not be chimeric. The original transgenic plant (T_0) can be composed of many different transformation events, while the progeny will contain one or more of these events, but not all. Therefore, the T_0 parent will have a different genotype than its progeny. Progeny will either be transgenic or not-transgenic.

Exercises

The following exercises are presented in this chapter:

- Preparation of *Agrobacterium tumefaciens*, Procedure 1,
- Transformation of *Petunia hybrida* shoots, Experiment 1,
- Transformation of *Zea mays* shoots Experiment 3, and
- Transient expression of the transferred GUS gene, Experiment 4.

Materials for Preparation of *Agrobacterium* Tumefaciens Cultures, Procedure 1

- *Agrobacterium tumefaciens* EHA105 (Hood et al., 1993) (Figure 34.2), contains the "super-virulent pTiBo542 virulence plasmid, that harbors a binary plasmid, such as pCNL56 (Li et al., 1992; Figure 34.3), which carries intron-GUS and kanamycin resistance genes (Jefferson et al., 1987).
- Seeds, *Petunia hybrida* L., and *Zea mays* L., described as follows.
- Agar solidified and liquid LB medium (Table 1): Bacto Agar (Gibco, BRL Life Sciences) for bacteria, Antibiotics: kanamycin, Bacto-tryptone (Gibco BRL), Bacto-yeast extract (Gibco BRL), Sodium Chloride (NaCl).
- 100 × 15 mm petri dishes containing agar solidified LB medium and 50 mg/L kanamycin (Table 34.1).
- 1 mL of Virulence Induction Media plus kinetin (VIM+Kin) in 1.5 mL micro-centrifuge tubes (Table 34.2): Acetosyringone (3', 5'-Dimethoxy-4'-hydroxy-acetophenone, Sigma-Aldrich); MES buffer; glucose; kinetin; Silwet-77 (Lehle Seeds). VIM+Kin can be prepared ahead of time and stored frozen in 1.5 mL micro-centrifuge tubes.
- Scalpels, forceps, inoculation loops, and dish spreaders
- Ethanol burner or bunsen burner
- Incubator for *Agrobacterium* co-cultivation set to 19 to 21°C. If this is not available, use a lab drawer (25–27°C).
- Sterile, 0.22-µm syringe filter (Acrodisk, VWR) and syringes
- Autoclaveable biohazard-type bags
- Alcohol or bunsen burners
- Sterile distilled or deionized water, household bleach, dishwashing detergent, Parafilm

TABLE 34.1
Composition of LB Medium for *A. Tumefaciens*

10 g Bacto-tryptone
5 g Bacto-yeast extract
10 g NaCl
15 g/L Bacto Agar

Note: Start with 800 mL of water and add ingredients. Adjust pH to 7.5 with NaOH; add water to bring volume to 1L. Autoclave 20 min under standard conditions. Agar will not dissolve until after medium has been autoclaved. Also, 15 g/L is a high agar concentration. Mix well until the agar has dissolved. Allow medium to cool (~55°C, cool enough to be held in hand). Prepare 50 mg kanamycin into 1 to 5 mL water and add to cooled medium through the 0.22 µm syringe filter to sterilize the solution. Mix medium well to distribute the antibiotic and pour into sterile petri dishes. (Derived from Maniatis et al. [1982].)

TABLE 34.2
Composition of Virulence Induction Medium (VIM + Kin) for *A. Tumefaciens*

75 mL of 75 mM MES buffer
2 g glucose
10 mg Acetosyringone (3′, 5′-Dimethoxy-4′-hydroxy-acetophenone)
0.5 mL kinetin (kinetin stock: 1 mg kinetin/5 mL)
0.01% mL Silwet-77

Note: Start with 75 mL of 75 mM MES Buffer and add glucose. Check and adjust pH = 5.4 and add more MES buffer to make 100 mL. Autoclave, allow to cool. Prepare 2–3 mL of a solution of 10 mg acetosyringone and 0.5 mL of the kinetin stock (1 mg/5 mL) in 5 mL distilled of deionized water and add to cooled medium through a 0.22 μm filer, to sterilize the solution. Dispense into 1 mL aliquots in sterile 1.5 mL micro-centrifuge tubes and store frozen. (Derived from Gao and Lynn, 2005.)

- Permanent marking pen (Sharpies)
- Gloves

Complete the protocols in Procedure 34.1.

Experiment 34.1 Transformation of *Petunia hybrida* Shoots

Petunia hybrida. Seed of any *Petunia* variety will work in this procedure. We used “Rose Flash” (Ulian et al., 1989; F1 hybrid: Single Grandiflora), Deep Rose, Ball Seed Co., West Chicago, Illinois). At the time, this variety was not known to be regenerable, and its ability to be transformed by *A. tumefaciens* was not known; however, it is a member of the tobacco family, which is known to be in the *Agrobacterium* native host range.

The *Petunia* procedure requires 5 media: seed germination (SG), preculture, and cocultivation (MS + BA), selection #1 (MS + BA + Kan200 + C), selection #2 (MS + BA + Kan100 + C), and rooting (MS + Kan100 + C), described in the following text.

Procedure 34.1
Agrobacterium tumefaciens: Culture and Induction of Virulence

Step	Instructions and Comments
1	For each student or team, prepare two dishes of Agrobacterium EHA105 harboring a binary plasmid (i.e., pCNL56) that contains an intron-GUS (the E. coli Beta-glucuronidase; uidA) gene (Vancanneyt et al., 1990; Jefferson et al., 1987) and an antibiotic gene (i.e., the kanamycin resistance gene [neomycin phosphotransferase II, NPTII, nptII]) genes, onto medium containing kanamycin (LB + Kan). (1) The antibiotic gene for selection for the binary plasmid will be outside of the RB and LB and will have a bacterial promoter. In the case of pCNL56, this is a kanamycin resistance gene nptI. (2) The gene for antibiotic or herbicide selection that is to be transferred into the plant will be located between the RB and LB, and will have a promoter that functions in plants. In the case of pCNL56, this is another kanamycin resistance gene derived from nptII. Inoculate LB medium with Agrobacterium to produce confluent cultures. This can be done with a sterile inoculation loop. Cover as much of the medium with bacteria as possible (the loop will need to be dipped in the bacterial culture repeatedly). Alternatively, Agrobacterium can be suspended in liquid LB and spread over the medium.
2	Allow bacteria to grow for 3 days at room temperature (24 to 26°C) in a lab drawer. After 3 days, bacteria should cover the dish. Avoid incubating *Agrobacterium* under lights or at temperatures of 27°C or higher, because this will inactivate one of the virulence functions of *Agrobacterium* that is necessary for gene transfer.
3	When cultures are covered with bacteria, scrape the new growth of bacteria from each culture dish with a spatula, and mix the contents into 1 mL of virulence induction medium (VIM + Kin) in a 1.5 mL micro-centrifuge tube. Vortex tubes to suspend the bacteria. There should be at least 1 to 2 tubes of *Agrobacterium* suspended in VIM + Kin for each student or team. You can use this suspension immediately, or you can leave the bacteria in the medium up to 2 h for virulence induction.

Materials

Antibiotics: carbenicillin* and kanamycin (PhyoTechnology Laboratory, Shawnee Mission, Kansas, or Sigma Aldrich, St. Louis, Missouri). Alternatively, 250 mg/L Clavamox (Smith Kline Beecham Veterinary) can be used instead of 500 mg/L carbenicillin. Calvamox is available through a local veterinarian or a veterinary clinic pharmacy and may be less expensive. Timentin can also be used.

- One package of *Petunia hybrida* seeds (Rose Flash, Ball Seed Co.), depending on the size of the class
- Murashige Skoog (1962) basal medium pre mixed (PhytoTechnology Laboratory, or Sigma Aldrich)
- Benzyladenine (N6 Benzyladenine, BA) (PhytoTechnology Laboratory, or Sigma Aldrich)
- Commercial bleach (5.25% Sodium hypochlorite)
- Ethanol 95%
- Plastic wrap, parafilm, or Micropore tape (3M)
- Petri dishes, sterile polystyrene, 100 x 20 mm for plants
- Pipettors, 100 µL
- Plastic bags, quart size
- Scalpels and forceps
- Sucrose
- Gloves
- Syringes and 0.22 µm syringe filters (Acrodisc)
- Autoclaveable biohazard-type bags
- Alcohol or bunsen burners
- Sterile distilled or deionized water, household bleach, dishwashing detergent, Parafilm or plastic wrap, petri dishes 100 × 15 mm and 100 × 20 mm
- Sterile insulin hypodermic needles (optional)

Media

- *Petunia* seed germination medium (SG): 7g/L agar (Sigma Agar) suspended in distilled water, autoclave 20 minutes under standard conditions. Prepare 4 dishes (100 x 15 mm or 100 x 20 mm sterile plastic petri dishes) for each student or team.
- *Petunia* preculture and cocultivation medium (MS+BA): MS (see note at end of Media), 0.1 mg/L N6 benzyladenine (BA) 4 dishes (100 x 20 mm petri dishes) for each student or team.
- *Petunia* selection medium #1 (MS + BA + Kan200 + C): MS plus 0.1 mg/L N6 Benzyladenine, 200 mg/L kanamycin, 500 mg/L carbenicillin*. Four to eight 100 × 20 mm petri dishes for each student or team.
- *Petunia* selection medium #2 (MS + BA + Kan100 + C): MS plus 0.1 mg/L N6 Benzyladenine, 100 mg/L kanamycin, 500 mg/L carbenicillin*.
- *Petunia* rooting medium (MS+Kan100+C); MS, 100 mg/L kanamycin, 500 mg/L carbenicillin*.
- Four 100 x 20 mm petri dishes per student or team

Note: All MS media (Murashige and Skoog, 1962; MS) described above contain MS salts, MS vitamins and m-inositol (available as Basal MS from suppliers, PhytoTechnology Laboratory, or Sigma Aldrich), and 30 g/L sucrose, pH = 5.7 ± 0.1, and 8 g/L Sigma Agar (Sigma Aldrich).

Petunia Shoot Isolation and Inoculation Procedure

The shoot isolation procedure exposes the meristem area of the shoot to wounds, which opens a facet of the meristem to *Agrobacterium* transformation. Under magnification, remove 1 mm of the shoot apex from *Petunia* seedlings. Remove the largest leaves leaving the smaller elongating and primordial leaves (as in Figure 34.4a). Place a drop (10–20 μL) of *Agrobacterium* suspended in VIM + BA, on the shoot. This procedure places *Agrobacterium* at the base of the shoot, which allows access to the corpus cell layers in the meristem. To access the L2 layer, pierce the meristem once with a sterile insulin hypodermic needle (Optional; Figure 34.4b). *Petunia* shoots are small, tender, and wound easily.

- *Petunia* Seed Germination Medium (SG): 7g/L agar (Sigma Agar) suspended in distilled water, autoclave 20 minuets under standard conditions. Prepare 4 dishes (100 x 15 mm or 100 x 20 mm sterile plastic petri dishes) for each student or team.
- *Petunia* preculture and cocultivation medium (MS + BA + As): MS (note: below), 0.1 mg/L N6 benzyladenine (BA), 20 mg/L acetosyringone (add after autoclaving, described below), 8 g/L Sigma agar in 4 petri dishes (100 × 20 mm petri dishes) for each student or team.
- *Petunia* selection medium #1 (MS + BA + Kan200 + C): MS plus 0.1 mg/L N6 Benzyladenine, 200 mg/L kanamycin, 500 mg/L carbenicillin, 8 g/L Sigma agar in four to eight 100 × 20 mm petri dishes for each student or team.
- *Petunia* selection medium #2 (MS + BA + Kan100 + C): MS plus 0.1 mg/L N6 Benzyladenine, 100 mg/L kanamycin, 500 mg/L carbenicillin, 8 g/L Sigma agar.
- *Petunia* rooting medium (MS + Kan100 + C); MS, 100 mg/L kanamycin, 500 mg/L carbenicillin*, 8 g/L Sigma agar.

Follow the protocols listed below in Procedure 34.2 to complete this experiment.

Experiment 34.2 Transformation and Regeneration of *Zea mays* L. Shoots

Start with fresh commercially grown and packaged seed available in garden stores or through a seed catalogue: Funk's G90 (Gould et al., 1991), Pioneer Hi-Bred lines, B73, and Shrunken 1 have been used. Seed collected from plants in the field will be heavily contaminated. We used Funk's "G90" because this variety was available at the time, and it was not known to be a transformable or regenerable genotype. All cereals were thought to be refractory to *Agrobacterium*-mediated transformation.

The procedure requires four media: Seed Germination (SG), Preculture and Cocultivation (MS + Kin), Selection (MS + Kan + C), Rooting ½ MS + C.

Materials

- One to three packages of *Zea mays* L. seeds, depending on the size of the class
- Antibiotics: carbenicillin* and kanamycin (PhytoTechnology Laboratory or Sigma) Alternatively (see *Note* following), 250 to 500 mg/L Clavamox (Smith Kline Beecham Veterinary) can be used instead of 500 mg/L carbenicillin. Timentin can also be used.
- Sterile water, household bleach, dishwashing detergent, plastic wrap, or parafilm, 50 mL conical sterile plastic centrifuge tubes, 100 × 20 mm sterile petri dishes
- Seed germination medium (SG): 7g/L agar (Sigma agar) in distilled water, autoclaved and dispense into 100 x 200 mm petri dishes. Four dishes for each student or team.
- *Zea* Preculture and Inoculation medium (MS+Kin+As): MS (see note below), 0.1 mg/L kinetin, 20 mg/L acetosyringone (added after autoclaving, described below), 8 g/L Sigma agar. Four dishes for each student or team in 100 × 20 mm petri dishes.
- *Zea* Selection medium (MS + Kin + Kan10 + C): MS plus 0.1 mg/L kinetin, 10 mg/L kanamycin, 500 mg/L carbenicillin, or 250 mg/L Clavamox, 8 g/L Sigma agar. Four to eight dishes for each student or team, 100 × 20 mm petri dishes.

Procedure 34.2
Transformation of *Petunia hybrida* Shoots and Regeneration of Plants

Step	Instructions and Comments
1	Place approx 0.2 mL of *Petunia* seeds in 1.5 mL microcentrifuge tube.
2	Add soapy water and shake or vortex to mix contents.
3	Rinse seeds in water, and allow seeds to soak for 15–30 min.
4	Remove water and surface sterilize seeds in 20% household bleach for 10–15 min.
5	Rinse three times with autoclaved sterile water.
6	Suspend seeds in a cool sterile 0.1% agar solution, mix seeds, and allow seeds to become suspended in the agar.
7	Using an automatic pipette, take 1 mL of the seed/agar suspension and distribute onto culture medium at the rate of approximately 25 seeds/dish to germinate (*Petunia* seed germination medium [SG]).
8	Germinate for 7 days, then excise shoots of 0.3 × 0.6 mm that consist of the apical dome and primordial leaves, as in Figure 34.4a. Shoot can be pierced, as in Figure 34.4b. Place shoots onto preculture and cocultivation medium (MS + BA).
9	After 2 days, inoculate with 10 µL *Agrobacterium* suspended in VIM + BA for 15 min and transfer shoots to fresh MS + BA medium.
10	After 2 days, transfer shoots onto *Petunia* selection medium #1 (MS + BA + Kan200 + C) containing 200 mg/L kanamycin, and 500 mg/L carbenicillin, or 250 mg/L Clavamox.
	In 3 weeks the primary leaves of the remaining shoots will be enlarged and bleached. Remove bleached leaves, and culture tissues that are green. Transformed shoots will first appear as green regions at the base of bleached leaves. These regions can be seen by turning the dish over to see the bases of the shoots. Transfer tissue to selection medium #2 (MS + BA + Kan100 + C).
11	After 7 days, single or multiple green shoots will develop from the base of some of the explants. Transfer green shoots to rooting medium that contains kanamycin (MS + Kan100 + C). Shoots that root in the presence of kanamycin should be transgenic.

- *Zea* Rooting medium (½ MS + C): ½ MS basal medium, 15 g/L sucrose, and 500 mg/L carbenicillin, or 250 mg/L Clavamox, 8 g/L Sigma agar
- Gloves and 0.22-µm syringe filters
- 100 × 200 mm petri dishes and small glass (jelly and ½ pint) Mason or Kerr canning jars

Note: All MS media (MS) described contains MS salts, MS vitamins, and m-inositol. This basal MS is available from many suppliers. Add to this 30 g/L sucrose, adjust pH = 5.7 ± 0.1, and 8 g Sigma Agar. If Clavamox is used instead of carbenicillin, do not filter sterilize. This antibiotic is in tablet form, and there is insoluble material that will clog the filter. Tablets are individually wrapped. The inside of the wrapping and tablet is sterile. Cut open the wrapping in the laminar flow hood using sterile scalpel and forceps. Remove the tablet and drop it into 10–25 mL sterile water in a sterile 50 mL conical screw cap tube. The tablet will slowly disintegrate in 15–20 minutes. Once the tablet has disintegrated, the suspension should be added directly to cooled autoclaved medium and mixed well. Don't add the tablet directly to media because it will not dissolve.

Zea mays Shoot Isolation and Inoculation Procedures

The *Zea mays* shoot is larger than the *Petunia* shoot, which makes isolation and inoculation easier. The basic approaches that can be used with *Zea* shoot meristems are illustrated in Figures 34.4a,b and 34.5 to expose a facet of the shoot meristem to *Agrobacterium* infection. The first procedure (Figure 34.4a) will place *Agrobacterium* over the external surface of the shoot apex; most of the access to the meristem will be through the shoot base. This position allows access to the corpus cell layers in the meristem. Procedure in Figure 34.4b imposes a longitudinal wound in the shoot, using a small hypodermic needle, which will allow *Agrobacterium* access to the L1 and L2 cell layers as

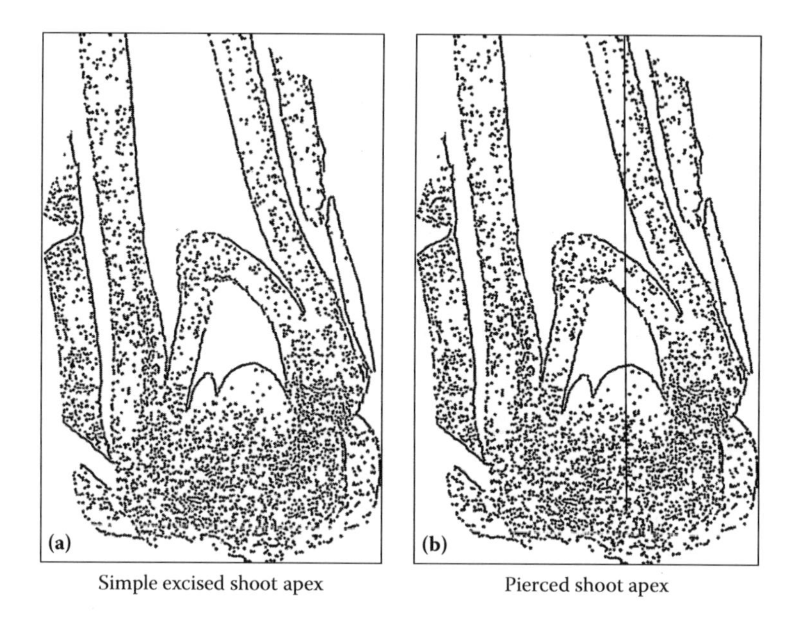

FIGURE 34.4 Diagrams: (a) A simple isolated shoot apex (approx. 0.3 × 0.6 mm) that can be directly inoculated with *Agrobacterium*. This procedure is recommended for use with petunia; (b) The same shoot apex, pierced using an insulin hypodermic needle through of the meristem region is recommended with *Zea mays*. (From Gould, J. et al. 1991. *Plant Physiol.* 95:424–426. With permission.)

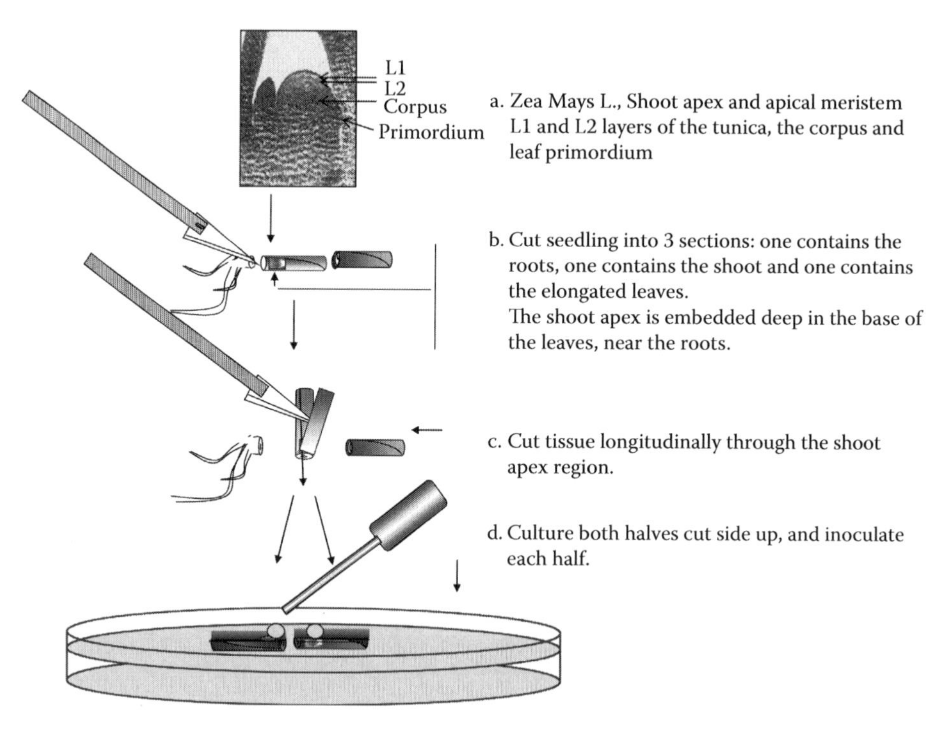

FIGURE 34.5 Outline of the preparation and inoculation of a split shoot of *Zea mays*. (a) Shoot apex, the L1, L2, corpus, and leaf primordium indicated, and the location of the shoot apex at the base of the germinating seedling. (b) Remove the roots and elongated leaves of the seedling. (c) Cut the tissue of the middle section longitudinally through the shoot apex and attempt to cut the meristem in half. (d) Culture both halves, cut side up, and inoculate the apex region of both with *Agrobacterium* in VIM + Kin. (From Gould, J. et al. 1991. *Plant Physiol.* 95:424–426. With permission.)

well as the corpus cells. The L2 cell layer ultimately becomes incorporated into the plant's gem-line and transformation of these cells is critical for transfer traits into the plant's progeny.

Follow the protocols listed in Procedure 34.3 to complete this experiment.

Shoot isolation (Figure 34.4a) was the first procedure developed (Gould et al., 1991), requires practice, and is not as efficient as other methods. Because the principal access for *Agrobacterium* is through the explant base, more transformations occur in the corpus region, while transformation is less likely in the L1 and L2 layers. The procedures in Figures 34.4b and 34.5 were developed later and may be easier.

Follow the experimental process in Procedures 34.3a and 34.b to complete this experiment.

Experiment 34.3 Histochemical Staining for GUS Activity in *Petunia* or *Zea* Shoots

GUS histochemial chemical staining can be used to show the transient expression of transferred genes, as well as the expression of genes stably incorporated into a plant genome (Jefferson et al., 1987). Transient expression is typically assayed 3–4 days following inoculation and cocultivation and is usually more intense than expression seen weeks later. In transient expression, the transferred genes are present and expressed in nuclei but have not yet been incorporated into the cell's genome. In this experiment, transient expression will provide a snapshot showing the transformation point(s) and progression of nuclear access by the T-DNA.

Procedure 34.3

Transformation of *Zea mays* Shoots and Regeneration of Plants

Step	Instructions and Comments
1	Empty one 15 g seed packet and divide seeds into 5 to 10, 50 mL plastic centrifuge tubes depending on the number of students or teams.
2	Add soapy water, cap, shake contents, and let sit for 5 min.
3	Drain soapy water from seeds.
4	Add 20% v/v household bleach and let sit for 15–20 min.
5	Rinse three times with autoclaved sterile water in laminar air flow hood.
6	Sterilization can be repeated in 24 hr to kill germinating fungal spores.
7	Place seeds on seed germination medium (SG). Incubate at room temperature, under a 16 hour day for 3–7 days. Germination will take longer if room temperature is low.
8	After 3-7 days, isolate shoots using shoot isolation outlined in Figure 34.4a or 34.5. Inoculate each shoot with 10 μL of *Agrobacterium* suspended in VIM+Kin, described above.
9	Culture inoculated shoots onto the preculture and inoculation medium containing MS, 0.1 mg/L kinetin (MS + Kin). Place shoots on the medium lengthwise, cut side up, so that cut area is not in contact with the medium. Shoots remain in contact with *Agrobacterium* for 3 days at 19–22°C, or room temperature in dark.
10	Transfer shoots to the same medium as above, to which has been added 7.5 to 10 mg/L, kanamycin; 500 mg/L, carbenicillin, or 250 mg/L Clavamox (MS + Kin + Kan + C).
11	Reculture shoots to fresh medium every 7 to 10 days, to renew antibiotics, the culture medium and to exchange the culture atmosphere. If regenerating shoots push against the petri dish lid, larger culture vessels, such as glass canning jars, can be used.
12	Shoots will root spontaneously within 3–4 weeks from inoculation.
13	Transfer rooted shoots to hydrated Jiffy peat pellets or small pots containing soil. Place the peat pellets within a larger vessel (½ pint canning jar or Magenta box). Cover canning jar with sterile petri dish lids and wrap with plastic wrap or parafilm.
14	Transfer regenerated plants to soil in 1 gallon pots when roots begin to grow out of the peat pellets. Cover potted plants with plastic bags to allow plants to adapt to conditions outside culture.

Note: Maize roots are sensitive and transferring the plants from culture to soil and from small pots to larger pots should be avoided or done carefully. Use of Jiffy peat pellets minimizes damage in the transfer and seedling can be established directly in a 1 gal pot. Keep plants in a well-lighted area at room temperature.

Procedure 34.3a
Zea mays: Shoot Apex, Isolation and Inoculation

Step	Instructions and Comments
1	Cut a shoot of approximately 1 mm from the seedling near the base. Depending on the variety you have and if the seedling has been grown in light, it may have a red or pink section between the white root and the green leaves. Cereals have leaves that enclose the shoot apex, and it can be difficult to find. There will be a slight bulge at the intersection of white and green tissues, and this area may be red. This region covers the shoot apex.
2	Remove the large elongating leaves to create a >1 mm explant seen in 34.4a.
3	Pierce the shoot base with a fine sterile hypodermic needle, Figure 34.4b.
4	Inoculate the entire shoot with 10 to 15 μL *Agrobacterium* suspended in VIM + Kin.
5	Culture shoots on MS medium containing 1 mg/L kinetin (MS + Kin).
6	After 3 to 7 days, transfer shoots to the same medium as above, that includes 7.5 mg/L, kanamycin; 500 mg/L carbenicillin, or 250 mg/L Clavamox (MS + Kin + Kan + C).

Procedure 34.3b
Zea mays: Split Shoot Isolation, Inoculation

Step	Instructions and Comments
1	The shoot meristem of the seedling will be buried in the base of the leaves not far from the roots. There will be a slight bulge at the intersection of white and green tissues. This is tissue covers the shoot apex. Remove the roots and most of the upper green cylinder of leaf material from 1 cm above the roots leaving a 1 cm cylinder explant (Figure 34.5a,b).
2	Slice the 1 cm cylinder in half lengthwise. Although you will have attempted to cut the shoot in half, it won't be exact. One side will contain the meristem or most of the meristem. Look at the explant and determine which half contains the shoot and stab the shoot with a hypodermic needle (Figure 34.5c).
3	Culture both halves of the tissue with the cut side up on MS medium containing 1 mg/L kinetin (MS + Kin) and 8 g/L agar, pH 5.7 ± 0.1 (Figure 34.5d).
4	Inoculate shoot and meristem area with 10 or 15 μL *Agrobacterium* suspended in VIM + Kin.
5	After 3 to 7 days, transfer shoots to the same medium as above, that includes 7.5 mg/L, kanamycin, 500 mg/L carbenicillin, or 350 mg/L Clavamox* (MS + Kin + Kan + C).

TABLE 34.3
Preparation of the X-Gluc Histochemical Reagent for GUS Assays

X-Gluc histochemical reagent: 10 mL 50 mM phosphate buffer, pH 7.0. Prepare just before use and filter sterilize.

1. Combine:
   39 mL 0.2 M NaH_2PO_4 (31.2 g/L)
   61 mL 0.2 M Na2HPO4 (28.39 g/l)
   (10 mM EDTA, 0.5 mM ferricyanide, 0.5 mM ferrocyanide; these compounds are toxic and optional.)
   This makes 100 mL of a 200 mM phosphate buffer.
   Dissolve 5 mg 5-bromo-4-chloro-3-indolyl glucuronide (X-Gluc).
2. Dilute the 200 mM phosphate buffer 1:4 with distilled H_2O.
3. Filter sterilize using a 0.22 um syringe filter.
4. Dispense into sterile 15 or 50 mL centrifuge tubes.

The following solution is needed for each student, or team of students.

- 40 mL of 0.1 M sodium phosphate buffer at pH=7.0.
- 40 mg of X-Gluc (5-bromo-4-chloro-3-indoyl β-d-glucuronic acid)

Source: Derived from Jefferson et al. (1987).

Stable expression, characteristic of genomic incorporation of transferred genes, can be assayed in regenerating plant tissues, regenerated plants (T_0), or in their seedling progeny. In genomic incorporation, expression of the GUS gene will be lower than that observed during transient expression. One of the reasons behind this difference is that transcription of the GUS gene is under the control of the transferred promoter, the MAS35S promoter in the case of pCNL56, and will show developmental and tissue specific expression characteristic of that promoter.

Materials

- The experiment requires a solution of X-Gluc (5-bromo-4-chloro-3-indolyl-beta-D-glucuronic acid) to visualize expression of the β-glucuronidase (GUS) gene. This solution should be prepared just before use and filter sterilized (Table 34.3). The GUS expression assays require the following materials for each student or group of students. Solution should be prepared just before use and filter sterilized. The X-Gluc solution can be stored frozen for up to 1 week.
- NaH2PO4 (Sigma #S9638)
- Na2HPO4 (Sigma #S0876)
- EDTA
- X-Gluc (5-bromo-4-chloro-3-indolyl b-D-glucuronide) (Sigma #B6650)
- 0.1% Triton X-100 or other detergent
- 1 mL Pipettors or transfer pipettes (sterile)
- 24-well sterile dishes (Sigma #M9655); 2–3 groups can share one dish
- Access to at least one stereo-type dissection microscope. Illuminated magnifying lenses (3X Bausch & Lomb) are inexpensive and often easier to use.

Place noninoculated tissues (controls) and inoculated tissues to be assayed into sterile microcentrifuge tubes, petri dishes, or the wells of a 96-well dish. Allow one control shoot for every 10 inoculated shoots. Cover tissue with the X-Gluc solution (approximately 100 to 200 μL). Incubate at 37°C overnight. The origin of the GUS gene is *E. coli*; therefore, the enzyme has the temperature optimum at human body temperature (37°C). After overnight incubation, visualize the blue stained cells using a dissecting type microscope or an illuminated 3× lens. If the blue color is faint, incubate overnight again and observe the next day. If blue color is developed, soak stained tissues in 75% followed by 95% (v/v) ethanol to remove the masking effects of chlorophyll.

Note on sterile technique and sterilization: The substrates used in the GUS assays require an enzyme that breaks the glucuronide bond to allow the production of a colored or fluorescent product. The GUS and similar enzymes are native to many bacteria and fungi that are present in the lab. If tissues become contaminated, intense and abundant spotty staining can be observed due to infection and can be mistaken for transformation. Therefore, plant tissues, utensils, solutions, and procedures must be sterile. X-Gluc is not heat stable and cannot be autoclaved. Solutions should be filter sterilized (0.22 um filter, Acrodisc syringe filter), and dispensed into sterile containers (1.5 mL microcentrifuge tubes). (Derived from Jefferson et al. [1987].)

Follow the protocols listed in Procedure 34.4 to complete this experiment.

Results and Subsequent Procedures

After 3 weeks of selection on kanamycin, green shoot tissue should be regenerating. Survival on kanamycin is the first step in the selection process. Surviving and regenerating shoots can be seen by turning the culture dish upside down. Transgenic shoots will regenerate from the small green centers that can be seen at the base of tissues. Once regenerated, surviving green shoots can be transferred to rooting medium that contains kanamycin. Root formation on kanamycin is the second selection step. Shoots are more likely to be transgenic if they can produce roots in the presence of kanamycin. Shoots surviving antibiotic selection and having produced roots (4–6 weeks in culture)

Procedure 34.4
Transient and Stable Expression Assays for GUS Using X-Gluc

Step (a)	Instructions and Comments
	Transient Expression Assay
1	After 3 days of cocultivation with *Agrobacterium*, take 3 to 5 shoots from each student, or group, to assay. Aseptically, remove tissue from culture, place 3–5 shoots into 0.5 mL of sterile X-Gluc solution. Also, place a number of noninoculated control shoots. Use more solution if more tissue is used.
2	Incubate tissues at 37°C, or room temperature, for 24 h or more for color to develop.
3	After 24 h, observe shoots under magnification (dissection microscope or illuminated lens). If the blue color is faint, return the tissues, in X-Gluc, to the incubator overnight or another 24 h.
4	If a blue color is observed, transfer stained tissue to 70%–75% ethanol to remove the green chlorophyll background. An ethanol dehydration series can be used, which helps to preserve tissue shape and structure: 30% ethanol (1 hour), 50% ethanol (1 h), followed by 70%–75% ethanol and can be stored in 95% ethanol (optional).
Step (b)	Stable Expression Assay
1	After 4 to 5 weeks, shoots may be forming roots in culture. Tissues such as root tips and leaves can be excised aseptically and assayed to preserve the life of the plant. Alternatively, the entire plant can be stained. Plant tissues do not survive the GUS assay.
2	Place tissue in a sterile petri dish, 94-well dish, or 1.5 mL micro-centrifuge tube. Add freshly made, sterile X-Gluc solution to cover tissue. The volume of X-Gluc needed will depend on the size of the tissues to be stained.
3	Incubate at 37ºC, or room temperature, 24 h and observe using a dissection microscope or illuminated magnifying lens, as described in Steps 4 and 5 above.

are ready to transfer to soil. Flowering can begin 8 weeks after transfer to soil and senescence begins after 12 weeks.

Progeny Germination and Assays

Germinate mature seeds on an MS based, hormone-free medium containing thiamine-HCl, 0.1 mg/L, and sucrose, 20 g/L. Alternatively, germinate seeds on MS media containing kanamycin.

After 1–2 weeks, seedlings will be ready for transfer to soil. Prior to transfer, remove and assay leaf tissue and root tips for assays of reporter gene activity (i.e., GUS using the X-Gluc colorimetric assay [Jefferson et al., 1987]).

SUGGESTED READING

Baron, C., N. Domke, M. Michael Beinhofer, and S. Siegfried Hapfelmeier. 2001. Elevated temperature differentially affects virulence, VirB protein accumulation, and T-pilus formation in different *Agrobacterium tumefaciens* and *Agrobacterium vitis Strains. J. Bact.*, 183:6852–6861.

Bregitzer, P., S. Zhang, M-J. Cho, P. Lemaux. 2002. Reduced somaclonal variation in barley is associated with culturing highly differentiated, meristematic tissues. *Crop Sci.* 4:1303–1308.

Fullner, K. J. and E. W. Nester. 1996. Temperature affects the T-DNA transfer machinery of *Agrobacterium* tumefaciens, *J Bacteriol.* 178:1498–1504

Gao, R. and D. G. Lynn. 2005. Environmental pH sensing: Resolving the VirA/VirG two-component system inputs for *Agrobacterium* pathogenesis. *J. Bacteriol.* 187:2182–2189

Gould, J., M. Devey, O. Hasegawa, E. Ulian, G. Peterson, and R. Smith. 1991. Transformation of *Zea mays* L., using *Agrobacterium tumefaciens* and the shoot apex. *Plant Physiol.* 95:424–426.

Gould, J. and M. Magallanes-Cedeno. 1999. Protocol: Adaptation of cotton shoot apex culture to *Agrobacterium*-mediated transformation. *Plant Mol. Biol. Rep.* 16:283.

Hewezi, T., A. Perrault, G. Alibert, and J. Kallerhoff. 2002. Dehydrating immature embryo split apices and rehydrating with *Agrobacterium tumefaciens*: A new method for genetically transforming recalcitrant sunflower. *Plant Mol. Bio. Rep.* 20:335–345.

Hewezi, T., F. Jardinaud, G. Alibert, and J. Kallerhoff. 2003. A new approach for efficient regeneration of a recalcitrant genotype of sunflower (*Helianthus annuus*) by organogenesis induction on split embryonic axes. *Plant Cell Tiss. Org. Cult.* 73: 81–86.

Hirochika, H. 1993. Activation of tobacco retrotransposons during tissue culture. *EMBO J.* 12:2521–2528.

Hood, E.E., S. Gelvin, L. Melchers, and A. Hoekema. 1993. New *Agrobacterium* helper plasmids for gene transfer to plants. *Transgenic Res.* 2:208–218.

Jefferson, R., T. Kavanagh, and M. Bevan. 1987. GUS Fusions: β-glucuronidase as a sensitive and versatile gene fusion marker in plants. *EMBO J* 6:3901–3907.

Lee, L.-Y., M.E. Kononov, B. Bassuner, B.R. Frame, K. Wang, and S.B. Gelvin. 2007. Novel plant transformation vectors containing the super promoter. *Plant Physiol.* 145:1294–1300.

Li, X.-Q., Liu, C.-N., Ritchie, S. W., Peng, J.-Y., Gelvin, S. B., Hodges, T. K. 1992. Factors influencing *Agrobacterium*-mediated transient expression of *gus*A in rice. *Plant Mol. Biol.* 20: 1037–1048.

Maniatis, T., E. F. Fritsch, and J. Sambrook. 1982. Bacteriology. Molecular Cloning. IA.1.

McCullen, C.A. and A.N. Binns. 2006. *Agrobacterium tumefaciens* and plant cell interactions and activities required for interkingdom macromolecular transfer. *Annu. Rev. Cell Dev. Biol.* 22:101–127.

Morel, G. and C. Martin. 1952. Guerison de dalhias atteints d'une maladie a virus. *C. R. Acad. Sci.* 238:1324–1325.

Murashige, T. 1974. Plant propagation through tissue culture. *Annu. Rev. Plant Physiol.* 25: 135–166.

Murashige, T. and F. Skoog. 1962. A revised medium for rapid growth of and bioassays with tobacco tissue cultures. *Physiol Plant* 15:473–497.

Schmulling, T. and J. Schell. 1993. Transgenic tobacco plants regenerated from leaf disks can be periclinal chimeras. *Plant Mol. Biol.* 21:705–708

Smith, R., J. Gould, and E. Ulian. 1992. Transformation of Plants Via the Shoot Apex, TAMUS. U.S. Patent No. 5,164,310. Patent Application filed, June 1988.

Smith, R.H. and E.E. Hood. 1995. *Agrobacterium tumefaciens* transformation of monocotyledons, *Crop Sci.* 35: 301–309.

Tzfira, T. and V. Citovsky. 2003. The *Agrobacterium*-plant cell interaction. Taking biology lessons from a bug. *Plant Physiol.* 133:943–947.

Ulian, E., R. Smith, J. Gould, and T. McKnight. 1988. Transformation of plants via the shoot apex. *In Vitro, Cell Dev. Biol.* 24:951–954.

Vancanneyt, G., R. Schmidt, A. O'Connor-Sanchez, L. Willmitzer, and M. Rocha-Sosa. 1990. Construction of an intron-containing marker gene: Splicing of the intron in transgenic plants and its use in monitoring early events in *Agrobacterium*-mediated plant transformation. *Mol. Gen. Genet.* 220:245–250.

35 Genetic Transformation of Chysanthemum and Tobacco Using *Agrobacterium tumefaciens*

Margaret M. Young, Linas Padegimas, Nancy A. Reichert, and R. N. Trigiano

In the previous decades, many methodologies were developed to introduce or move a wide variety of genes from one species into another. Many of these techniques have been discussed in other chapters and will not be considered further in this chapter. However in the following experiments, we will use *Agrobacterium tumefaciens*, a soil-inhabiting, pathogenic (disease-causing) bacterium occurring naturally in many areas of the world, as the agent/vector or shuttle that transfers gene(s) between the two species. This bacterium can be aptly described as nature's own and the original genetic engineer since during pathogenesis (disease development) several genes located on a plasmid (a small circular piece of DNA) within the bacterium are delivered and incorporated into the genome of host plant cells. These genes cause an unusually high number of host cell divisions (hyperplasia) to occur and, as a result, a tumor or gall is formed. Hence, the common name of the disease is crown gall. We can take advantage of this natural gene delivery system. Via molecular biological techniques, genes that incite disease can be removed from the t-DNA region of the plasmid and replaced with gene(s) and promoters of our choice.

In recent years there has been increased interest in the transformation of ornamental plants, and the chysanthemum (*Dendranthema grandiflora* Tzvelev.) has not been the exception (e.g., Aida, et al., 2008a,b; de Jong et al., 1993; Lowe et al., 1993; Seo et al., 2007; Sun et al., 2009; Urban et al., 1994). Most of the interest in genetically transforming chysanthemum thus far has been in the areas of creating new flower colors and conferring resistance to tomato spotted wilt virus. Before chysanthemum could be genetically engineered, a reliable regeneration system had to be developed. The plant regeneration system described in Chapter 21 is ideally suited in this regard for the following reasons: chysanthemums are very susceptible to infection by certain strains of *A. tumefaciens*; shoots are formed directly (without an intervening callus phase); shoots develop from the cut edges of the explant, which are easily infiltrated with a bacterial vector; and the leaf tissues of some cultivars have high regenerative capacity. For a more complete discussion of plant transformation with *A. tumefaciens*, refer to Chapters 33, 34, and 36.

A word of caution before beginning the experiments outlined in this chapter. When working with any plant pathogen, in this case *A. tumefaciens*, precautions and care must be taken to limit contamination of the work area and "bystander plants," both in the laboratory and in the environment. Inoculated plants and contaminated instruments and cultures must never leave the laboratory. All cultures of bacteria (except stock cultures), plant tissues, and plants exposed to the bacterium must be destroyed, by autoclaving at the conclusion of the experiments. All instruments and glassware also should be decontaminated by autoclaving.

The experiments outlined in this chapter are designed for smaller, more advanced classes (fewer than 10 students) or students participating in special topics or advanced individual study problems. Generally, they require more developed skills, careful attention to detail, and closer supervision by the instructor. These experiments are adapted primary from those published by Trigiano and May (1994).

EXPERIMENT 1. TRANSFORMATION OF CELLS USING "WILD-TYPE" *AGROBACTERIUM TUMEFACIENS*

This experiment is designed as a paired-variate (Chapters 5 and 25) and uses the chysanthemum cultivars "Regal Jamestown" or "Spirit Lake" and "Vail" or "Little Rock." We suggest that half of the students or team of students work with one cultivar, while the other half works with the remaining cultivar.

MATERIALS

The following items are required for each student or team of students:

- Liquid and solidified AB medium (Table 35.1)
- *A. tumefaciens* strain 281 (Hood et al., 1986) or any other virulent strain may be secured from any number of sources, including a plant pathology or plant sciences department, but first ascertain whether or not a USDA permit is required for this pathogen.
- 125- mL erlenmeyer flasks
- 0.22- or 0.45-μm filters and 10 cc (mL) syringes
- Autoclavable biological hazard bags
- Several pieces of sterile 9-cm filter paper
- Four sterile 50- mL beakers
- Six 60 x 15 mm petri dishes containing MS medium amended with 1.0 μM (0.22 mg/L) BA and 11.5 μM (2 mg/L) IAA (shoot initiation medium)
- Six 60 x 15 mm petri dishes containing shoot initiation medium and 100 mg/L cefotaxime
- Six 60 x 15 mm petri dishes containing MS medium without plant growth regulators (PGRs), but amended with 100 μg/ mL cefotaxime
- Liquid MS medium with 100 μg/ mL cefotaxime

TABLE 35.1
Composition of AB Medium for Growth of *Agrobacterium Tumefaciens*

20× salts (A)	20× buffer (B)	Vitamin
(500 mL)	(500 mL)	(100 mL)
10 g NH_4Cl	30 g K_2HPO_4	20 mg biotin
12.5 g $MgSO_4 \cdot 7H_2O$	11.5 g NaH_2PO_4	(filter sterilize)
1.5 g KCl		
0.1 g $CaCl_2$		
0.025 g $FeSO_4$		

Note: Start with 400 mL of water and add 25 mL of each the 20× salts (A) and buffer (B). Dissolve 2.5 g of D-glucose, bring volume to 500 mL and add 7.5 g of agar. Autoclave, allow to cool, but not harden and aseptically pipette 0.1 mL of biotin stock into the medium. Swirl to mix biotin and disperse agar and pour into vessels. (Derived from White, F.F. and E.W. Nester. 1980. *J. Bacteriol.* 141:1134–1141.)

Note: Gloves and a particle mask should be worn when working with antibiotics and PGRs. These antibiotic solutions should be sterilized using a syringe and either a 0.22- or 0.45-µm filter and added after the basal medium has been autoclaved and cooled, but not hardened (see DVD—Medium Preparation for the proper technique for using filter sterilizing liquids).

Follow the protocols outlined in Procedure 35.1 to complete this experiment.

Anticipated Results

Leaf sections of "Regal Jamestown" or "Spirit Lake" not treated with bacteria ("B" petri dishes) should not have formed shoots, but may have a few adventitious roots and, perhaps, some scanty callus along the cut edges. Leaf sections of "Vail" or "Little Rock" not treated with bacteria ("B" dishes) should have produced plentiful shoots, a few roots and little or no callus at the cut edges (Figure 35.1a). Successful transformation of cells within the leaf explants of the corresponding pairs of both "Regal Jamestown" or "Spirit Lake" and "Vail" or "Little Rock" ("A"

Procedure 35.1
Transformation of Chysanthemum Cells Using "Wild Type" *Agrobacterium Tumefaciens*

Step	Instructions and Comments
1	Surface disinfest mum leaves with 10% commercial bleach and a few drops of liquid detergent for 5 min and then rinse three times with sterile distilled water.
2	Excise four midrib sections from each leaf and place two sections per 60-mm petri dish containing MS medium with 1.0 µM BA and 11.5 µM IAA. Label the dishes "1A" and "1B," etc. The number indicates the leaf number and the "A" and "B" constitute the "pair" in the design of the experiment. Using this design, specific results may be attributed to a treatment and not to other variables, such as physiological status of the explant. Maintain the identity of the pairs and incubate the cultures at 25°C with 25 $\mu mol \cdot m^{-2} \cdot s^{-1}$ for 3 days.
3	Twenty-four h before leaf sections are to be inoculated with bacteria, start liquid cultures of *A. tumefaciens* A281— supervirulent (Hood et al., 1986) or other. Select a dish with a "well-grown lawn" of bacterial colonies and pipette about 1.0 mL of sterile AB medium onto the surface of the agar. Rub gently with a sterile glass rod or an inoculation loop to dislodge bacteria. Aseptically transfer about 0.5 mL of the bacterial suspension to 50 mL of AB medium in each of two 125-mL erlenmeyer flasks. Incubate on a rotary shaker at 75–100 rpm at 25°C for 24–36 h when the suspension appears cloudy or almost opaque. Do not use suspension older than 48 h. *Note*: Many procedures will require that the bacterial suspension be adjusted to a specified colony forming unit (cfu) concentration or optical density at a specified wavelength, usually between 500 and 600 nm (see Urban, L.A., J.M. Sherman, J.W. Moyer, and M.E. Daub. 1994. *Plant Sci.* 98:69–79). Although in most experiments cfu should be standardized, in our experience this step is not necessary for the successful completion of this laboratory exercise.
4	Three days after the initiation of leaf section cultures, aseptically transfer about 10 mL of the bacterial suspension to sterile 50-mL beakers. Immerse all leaf sections from the petri dishes labeled "A," one at time, in the suspension of bacteria for 10 s. Remove excess fluid and bacteria by blotting the leaf sections on sterile, dry filter paper in petri dishes. Return the sections to their original petri dishes. Using a different sterile forceps, submerge leaf sections from petri dishes labeled "B" in sterile AB medium contained in sterile 50-mL beakers for 10 s; blot dry and return to their original dishes. Incubate all petri dishes as before for a maximum of 2 days. The suspension of bacteria and filter papers should be autoclaved before discarding.
5	Filter sterilize 25 mL of MS liquid medium containing 100 µg/ mL cefotaxime (Sigma Chemical Co, St. Louis, MO) into each of two sterile 50-mL beakers. Vigorously agitate the leaf sections that have been cocultivated with bacteria in the solution and blot dry with sterile filter paper. Use different forceps and solutions for washing bacteria-treated and bacteria-free leaf sections. Incubate all sections on MS medium amended with PGRs and 100 µg/ mL cefotaxime for 9 days in the conditions described in Step 2.
6	Transfer all leaf sections to MS medium without PGRs, but containing 100 µg/ mL. Incubate all petri dishes for an additional 21 days or for a total of 35 days from the initiation of cultures.

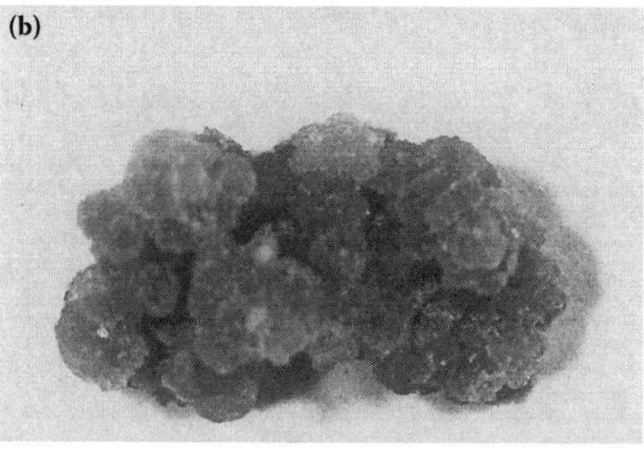

FIGURE 35.1 Transformation of mum leaf explants of "Vail" or "Little Rock" using *A. tumefaciens* wild type A281. (a) Explant treated only with AB medium (without bacteria) produced shoots whereas (b) the explant treated with bacteria formed copious amounts of callus. (From Trigiano, R.N. and D.J. Gray. 1996. *Plant Tissue Culture Concepts and Laboratory Exercises.* CRC Press, Boca Raton, FL.)

petri dishes) is indicated by the production of copious amounts of yellow-green callus that is typically not produced during shoot formation (Figure 35.1b). Callus growth may be patchy on some explants. Shoots and roots should not be present on explants of any of the cultivars used in this treatment.

Callus and/or the explants treated with bacteria may be transferred to MS medium without PGRs supplemented with 100 µg/ mL cefotaxime every 3–4 weeks. Callus should continue to grow (autotrophic for PGRs) and can be maintained for indefinite periods. At the conclusion of the experiment, autoclave all cultures including the petri dishes containing used medium.

Questions

- Why is the treatment of dipping leaf sections into sterile AB medium without bacteria included in the experiment?
- Why is it necessary to maintain cefotaxime in the medium after cocultivation with the bacteria?
- Why is the callus produced on the leaf sections autotrophic with regard to PGRs?

EXPERIMENT 2. REGENERATION OF TRANSGENIC CHYSANTHEMUM PLANTS

The protocol used in this experiment is similar to the protocol used in Experiment 1 except that *A. tumefaciens* "wild-type" A281 is replaced with EHA105 (Urban et al., 1994), a disarmed A281 lacking genes encoding PGRs but harboring a gene (and CMV-35s promoter) for neomycin phosphotransferase (NPT II). This gene when incorporated in the plant genome will confer resistance to kanamycin, an antibiotic. The experiment is designed as a paired variate (see Experiment 1); two leaf sections cocultivated with bacteria and two leaf sections not treated with bacteria from the same leaf will serve as the pairs. Only the cultivars "Vail" or "Little Rock" are used in this exercise.

Materials

In addition to the items listed under Experiment 1, each student or team of students will require the following:

- Six 60 x 15 mm petri dishes containing MS medium without PGRs and supplemented with 100 μg/mL cefotaxime and 50 μg/ mL kanamycin (Sigma)
- Disarmed strain of A281 harboring the NPTII gene (EHA105) (see suggested procurement procedures listed in Experiment 1)

Note: Gloves and a particle mask should be worn when working with antibiotics. These solutions should be filter-sterilized and added to medium after autoclaving and cooling.

Follow the protocols listed in Procedure 35.2 to complete this experiment.

Anticipated Results

Shoots should have formed after 35 days from both the bacteria and AB-medium-only-treated leaf sections. Shoots originating from AB-medium-only-treated sections should appear bleached or white. Experiments in our laboratory have shown that shoots formed from untransformed cells of "Vail" or "Little Rock" are sensitive to as little as 5–25 μg/ mL kanamycin. If any green shoots are present in this treatment, they are probably "escapes" and should be retested on medium without PGRs, but containing kanamycin and cefotaxime. If sensitive, most of the newly produced leaves will be bleached white, and the shoots will not produce roots.

Most shoots formed from explants treated with bacteria should be bleached, but putatively transformed shoots will remain dark green (Figure 35.2). Some of these will be escapes and all green shoots should be excised and recultured to medium containing cefotaxime and kanamycin for an additional 3 weeks. If these shoots are escapes, then the new growth will be white and roots will not be formed. If the NPTII gene has been successfully incorporated into the genome, then the shoots will remain dark green and the shoots will produce roots. From our experience, about 2–5% of the shoots formed will be transgenic. Remember to autoclave all plants and materials from this experiment when finished.

Procedure 35.2
Regeneration of Transgenic Chysanthemum Plants

Step	Instructions and Comments
1	Complete steps 1–4 as outlined in Procedure 35.1 except inoculate leaf sections with the disarmed A281 stain harboring the NPTII gene (EHA 105).
2	Complete step 5 as outlined in Procedure 35.1.
3	Transfer all leaf sections to MS medium without PGRs, but containing 100 μg/mL cefotaxime and 50 μg/mL kanamycin. Incubate sections as before for an additional 21 days. Count the number of white and green shoots for each pair (Figure 35.2).

FIGURE 35.2 Transformation of leaf explants of "Vail" or "Little Rock" using a disarmed A281 and harboring the NPTII gene. Green shoots (t) may have been transformed (putative) to contain the NPTII gene. White shoots (n) are not transformed. (From Trigiano, R.N. and D.J. Gray. 1996. *Plant Tissue Culture Concepts and Laboratory Exercises*. CRC Press, Boca Raton, FL.)

Questions

- Why are there "escape" shoots in the kanamycin treated explants?
- Why is it necessary to provide physiological/molecular evidence in addition to data from kanamycin selection to conclusively demonstrate transformation?

ALTERNATIVE EXPERIMENT 2. REGENERATION OF TRANSGENIC CHYSANTHEMUM PLANTS

A more complete version of the above experiment is to use the following four treatments of leaf sections that are incorporated into a randomized complete block design or an incomplete block design (Chapter 5): (1) cocultivation with bacteria, cultured on MS plus antibiotics; (2) without bacteria (immersed in AB medium), incubated on medium lacking antibiotics; (3) without bacteria, cultured on MS amended with cefotaxime; and (4) without bacteria, incubated on MS supplemented with kanamycin. Place one leaf section per 60 × 15 mm petri dish.

White and green shoots are obtained after 35 days with treatment 1. Green shoots should be excised and cultured on medium containing both antibiotics. Only green shoots are produced in treatments 2 and 3 and generally more shoots are produced in treatment 3 (cefotaxime) than treatment 2 (without the antibiotic). Green and white shoots are produced in treatment 4, but when the green shoots are transferred to fresh medium containing kanamycin, all new growth will be white and rooting will not occur.

EXPERIMENT 3: TISSUE CULTURE AND TRANSFORMATION OF TOBACCO (*NICOTIANA TABACUM*) WITH *AGROBACTERIUM TUMEFACIENS*

Tobacco is a good model system for tissue culture and transformation as transgenic plants can be observed within 8 weeks. In this lab, we will be looking at the effect of inoculation time (5 or 30 min) on GUS histochemical expression and shoot production in leaf explants of tobacco (*Nicotiana tabacum*). We will be using an *A. tumefaciens* culture containing the plasmid vector pCAMBIA2301 www.cambia.org. This vector contains the *gus* coding sequence (with catalase intron) between the 35S promoter and *nos* 3' end. Expression of the *gus* gene can be observed as early as 24 h after inoculation/transformation. The plasmid construct also contains the *nptII* gene driven by the CaMV

35S promoter and 3' untranslated sequences. This allows the leaf explants to survive on kanamycin (antibiotic). The vector was mated into *A. tumefaciens* strain EHA105. Your control will be uninoculated explants. We suggest using the tobacco cultivars Xanthi or Hicks.

Materials

The following items are required for each student or team of students:

- Sterile tobacco plants (see Procedure 35.3 for putting tobacco plants into culture)
- *Agrobacterium* strain EHA105 (see suggested procurement procedures listed in Experiment 1) containing the pCAMBIA2301 plasmid. This plasmid can be bought for educational purposes from CAMBIA (www.cambia.org) and transferred into EHA105 using triparental mating (Goldberg and Ohman, 1984) or electroporation (Mersereau et al., 1990).
- Thirteen 100 x 15 mm petri dishes containing MS medium amended with 36.91 μM (7.5 mg/L) 2iP and 0.54 μM (0.1 mg/L) 4-CPA (TSM: shoot initiation medium)
- Liquid TSM with 1000 mg/L cefotaxime (50 mL: for washing explants)
- Several pieces of 9-cm sterile filter paper
- Eleven 100 x 15 mm petri dishes containing TSM with 100 mg/l kanamycin and 500 mg/L cefotaxime (TSCK: selection medium). Extra dishes need to be poured 3–6 weeks after start of experiment.
- Ten magenta boxes (Magenta Corp. GA-7 boxes, 77 x 77 x 97 mm) containing MS with 50 mg/L kanamycin and 250 mg/L cefotaxime (MS½CK: rooting medium). Needed 4–6 weeks after start of experiment
- Five magenta boxes containing MS without PGR (for rooting of control explants). Needed 4–6 weeks after start of experiment
- Sterile GUS assay solution (pH 7.0: 0.1 M NaPO4 buffer, 10 mM EDTA, 0.5 mM potassium ferrocyanide, 0.5 mM potassium ferricyanide, 1.0 mM X-Glucuronide and 0.1% Triton X-100 [Stomp, 1992]). Dissolve X-Glucuronide in N,N dimethyl formamide (DMF) before addition to other components. Make immediately before use.
- Five sterile 6-well dishes

Follow Procedure 35.4 to complete this experiment.

Anticipated Results

Uninoculated leaf explants of tobacco on TSM (controls) should start expanding in 1 week. Meristemoids (green bumps) and a small amount of greenish white callus should be produced all over the explants in 2 weeks. In 6 weeks, the meristemoids would have differentiated into numerous

Procedure 35.3
Aseptic Germination of Tobacco Seeds

Step	Instructions and Comments
1	Wrap tobacco seeds (20–30) in four layers of cheese cloth. Place in sterile beaker.
2	Surface sterilize seeds with 20% commercial bleach and 0.5% SDS solution for 10 min. Wash with sterile, distilled water three times.
3	Place wrapped seeds on a sterile 100 x 15 mm petri dish and remove layers of cheese cloth using sterile forceps. Plate 5 seeds in each magenta box containing MS medium with no PGRs.
4	Place magenta boxes under 16 h photoperiod (cool white fluorescent light; minimum 25 $\mu mol \cdot m^{-2} \cdot s^{-1}$) for 2 months at 24°C.
5	Plants can be maintained indefinitely if nodal sections are recultured every 3–4 months.

Procedure 35.4

Tissue Culture and Transformation of Tobacco with *Agrobacterium tumefaciens*

Step	Instructions and Comments
1	For each person/group: remove young leaves from a sterile plant and place on a sterile 100 x 15 mm petri dish. (See below if using nonsterile plants.) Trim all edges of the leaves and cut into 8–10 mm squares. Place 90 leaf sections top side down (greener side touching the medium) onto nine dishes of TSM (10 per dish; inoculated explants). Push the explants gently into the medium. Place 40 leaf pieces onto four dishes of TSM (uninoculated control). Label dishes and randomly incubate under 16 h photoperiod (cool white fluorescent light; minimum of 25 $\mu mol \cdot m^{-2} \cdot s^{-1}$) at 24°C for 1 day.
	Alternatively, if sterile plants in culture are not available, germinate tobacco seeds in soil and grow for 2–3 months. Remove young leaves from the plants and wash in 70% ethanol for 30 s. Pour off ethanol and surface sterilize with 10% commercial bleach and 0.5% SDS solution for 10 min. Rinse 5 times with sterile water.
2	Start liquid cultures of EHA105 containing pCAMBIA2301 24 h before leaf inoculations. Using a freshly streaked dish (overnight growth), take a colony and inoculate 50 mL of liquid LB plus 60 mg/L kanamycin and 10 mg/L rifampicin. Grow overnight at 26°C with vigorous shaking. Grow cultures to OD_{600nm} = 0.6–0.8, centrifuge at 3500 rpm for 10 min, and resuspend in liquid TSM. *Note*: it may not be necessary to check the OD if the cultures have been grown for at least 18 h and are cloudy. However, this only happens if freshly streaked dishes are used for inoculation. Do not use cultures that are older than 48 h.
3	One day after cutting leaf sections, pour approximately 30 mL of the resuspended *A. tumefaciens* culture into a separate sterile 100 x 15 mm petri dish. Immerse all 90 leaf sections of tobacco of each group into the culture (not the controls). Make sure no air bubbles are trapped on the surface of the sections and that they are not sticking to one another. With gentle stirring, incubate for 5 or 30 min (inoculation time).
4	Transfer all sections for each time parameter onto sterile filter paper contained in a 100 x 15 mm dish. Blot the explants on sterile filter paper. Place the explants (90) from each time parameter back onto TSM media (greener side down; can use same original dishes). Label dishes, wrap and incubate as before. Autoclave all materials that came in contact with the *A. tumefaciens* solution, i.e., culture, petri dishes, filter paper, etc.
5	Two days after inoculation, pour 50 mL of liquid TSM plus 1000 mg/L cefotaxime into a sterile beaker. Starting with the least contaminated ones, rub as much *A. tumefaciens* off inoculated explants by dragging them on top of the media. Rinse inoculated explants in the liquid medium (about 2 min) then blot on sterile filter paper. Transfer inoculated explants to TSCK (selection medium), maintaining proper orientation, 10 explants per dish. Culture under the conditions described previously. Transfer 20 control explants onto 2 new TSCK dishes, maintaining proper orientation and culture as described above.
6	Three days after placement on selection medium, randomly select 15 explants from each set of inoculations and 5 from the controls (those on TSM and TSCK dishes). Place the explants separately in a sterile 6-well dish and add sterile GUS assay solution to cover the explants (GUS transient expression). Nescofilm the dish and place at 37°C overnight in the dark (dish can be wrapped in aluminum foil). The next day, incubate tissues in absolute ethanol:acetic acid (3:1, v/v) solution if blue staining is obscured. Under the microscope or using magnifying lens, count number of blue spots or estimate percentage area covered by blue each explant / inoculation time. Check both sides of the explants.
7	Every three weeks for 12 weeks, take notes on all cultures. Compare control leaf explant responses on TSM with those on TSCK. Compare controls to inoculated sections. Note the appearance of meristemoids (green bumps) and eventually green and white shoots.
	Any shoots arising from the controls on TSM, transfer to MS for root formation. Any shoots arising from the inoculated treatments on TSCK, transfer to MS½CK (rooting medium). All remaining explants that are alive, cut off dead areas and transfer to respective medium (TSCK or TSM). Remove one arising leaf from every shoot from the inoculated explants and place in GUS assay solution as before. Place one leaf also from shoots arising from the controls. Take notes on GUS spots as before. *Note:* the entire leaf arising from inoculated explants should be blue (GUS stable expression).

green shoots. These shoots should be transferred to rooting medium (MS). Uninoculated leaf explants on TSCK should start dying in 2 weeks (turn brown or white, with no callus or meristemoids) and be completely dead in 6 weeks. Inoculated leaf explants on TSCK should start producing meristemoids in 2–3 weeks, usually in one to several areas over the explants. The rest of the explants will start to die. Remove the dead areas before transferring to fresh selection medium after 3 weeks. This will allow the rest of the explants to expand and produce several green shoots. These shoots should be transferred to selection medium (MS½CK) for rooting. These plants can be transplanted to a soil-less mix, acclimatized, and allowed to flower to produce transgenic tobacco seeds. Explants inoculated for 5 min in *A. tumefaciens* culture may produce fewer green shoots and more explants will die compared to those inoculated for 30 min. If the explants are not washed thoroughly with TSM and cefotaxime after inoculations, *A. tumefaciens* contamination maybe more apparent in explants inoculated for 30 min (slimy whitish bacteria). Control explants should show no blue spots after GUS analyses 5 days after inoculation. Explants inoculated for 30 min usually show more blue spots than those inoculated for 5 min (GUS transient expression). Once shoots are formed, leaves can be removed and tested for the *gus* gene. Again, no blue GUS spots should be seen on leaves from control explants, but those explants inoculated should produce leaves that become completely blue after GUS analyses (stable expression). The entire plant can also be tested for GUS analyses at the end of the experiment (Figure 35.3).

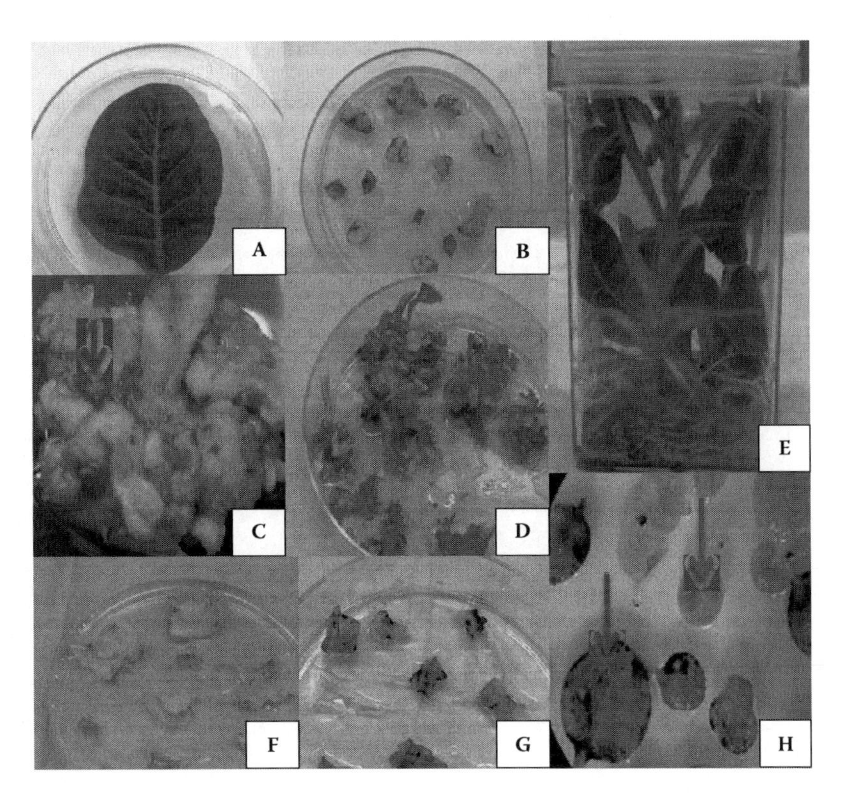

FIGURE 35.3 Tissue culture and transformation of tobacco with *Agrobacterium tumefaciens*. (A) Sterile leaf before all edges are removed. (B) Inoculated leaf explants after 2 weeks on selection medium. Note some plants (parts of explants) are still green. (C) Meristemoids (green bumps; red arrow) on inoculated explants after 3 weeks on selection medium. (D) Inoculated explants producing shoots after 6 weeks on selection medium. (E) Rooted transgenic plant. (F) No GUS transient expression on control explants. (G) Inoculated explants showing GUS transient expression (blue spots). (H) Stable GUS expression in leaves of shoots recovered after 8 weeks. Note that the leaves from the inoculated explants (red arrow) are blue; whereas those of control (green arrow) are not expressing the *gus* gene.

Questions

- What was/were the role(s) of the uninoculated controls in this experiment?
- Were there differences in the production of meristemoids and shoots on inoculated tissues and uninoculated tissues? If yes, why? If no, why not?
- Comparing overall regeneration of normal and transgenic plants, did one set take longer to appear and develop? If so, which one and why? Do the regenerated plants look similar or different?
- What are estimated transformation efficiencies (% explants that have been transformed) for each inoculation time? Can this be determined? Why or why not?

EXPERIMENT 4. USING PCR TO DECIDE WHICH TOBACCO PLANT IS TRANSGENIC FOR THE *GUS* GENE

Agrobacterium tumefaciens was used to transform leaf explants of tobacco. The construct contained the *gus* and *nptII* genes. Each student or team of students will be given two tobacco plants (A and B) and will have to decide which one is transgenic for the *gus* gene by extracting genomic DNA and conducting PCR analyses.

Materials

The following items are required for each student or team of students:

- One transgenic tobacco plant with the *gus* gene and one control plant. We suggest using transgenic plants recovered from Experiment 2.
- Master Plant Leaf DNA Purification kit from Epicentre Biotechnologies (Madison, Wisconsin).
- 1.5 mL tubes, 0.2 mL PCR tubes and micro-pestles (Fisher)
- Water bath at 70C, ice
- Isopropanol, 70% ethanol and sterile, deionized water (Sigma)
- 10X PCR buffer, 25 mM MgC12, 10 mM dNTP, Taq polymerase, 6X loading dye, GeneRuler 100 bp Plus DNA Ladder (MBI Fermentas, Glen Burnie, Maryland)
- 4 μM GUS DIR (5'-GCGTCCGATCACCTGCGTCAATGTAATGTTCT-3') primer and;
- 4 μM GUS REV (5'-ATTGTTTGCCTCCCTGCTGCGGTTTTTCAC-3') primer. Primers can be bought from many companies including IDT (Coralville, Iowa).
- Thermocycler
- Agarose, 10X TAE buffer and ethidium bromide (Fisher)
- Gel electrophoresis unit and UV light (Fisher)

Follow Procedure 35.5 to complete this experiment.

Anticipated Results

After addition of the Elution Buffer to the precipitate, the plant genomic DNA should dissolve in 5–20 min (if not, leave overnight at 4C).

After gel electrophoresis and visualization of DNA bands, you should see a band at 510 bp (use the marker lane to verify) in the positive sample (pCAMBIA2301), and either A or B sample lanes (depending on which is transgenic: Figure 35.4). You should not see a band in the negative control lane and either A or B sample lanes (depending on which plant is not transgenic). If the control lane and/or the nontransgenic plant have bands, contamination occurred in one or several steps. It is recommended that you begin again from the plant extraction step.

Procedure 35.5
Using PCR to Decide which Tobacco Plant Is Transgenic for the *Gus* Gene

Step	Instructions and Comments
1	Give each student/group a set of the tobacco plants (one is transgenic for the *gus* gene; the other is not and will be the control). Label the plants A or B (Instructor: keep track as to which one is transgenic). You can use plants regenerated from Experiment 3. Ensure that the work area is clean. Work on a new lab diaper and wipe down all supplies with DNA Away solution (Fisher). It is always best that one person make aliquots from the solutions in the kits (DNA extraction and PCR) at the beginning to prevent contamination. Discard any excess solutions from the used aliquots once you have completed the experiment.
2	Extract plant genomic DNA using the Master Plant Leaf DNA Purification kit from Epicentre Biotechnologies (Madison, WI). This kit does not require liquid nitrogen. It does require a water bath set at 70C, ice, micro-pestles, isopropanol and 70% ethanol. Store the extracted DNA at 4C until use. Ensure that you maintain labeling A or B of the tubes.
3	Set-up a PCR master mix. Make a master PCR mix according to the table below. N = number of samples; N + 2 = number of samples + controls. Your controls will be a tube containing plasmid pCAMBIA2301 that has a known *gus* gene; and a tube with the reaction mix plus water instead of plasmid/DNA sample. ***Change tips after addition of each solution. µl*** Deionized sterile water Deionized sterile water $10.3 \times (N + 2) =$ 10X PCR Buffer $2.5 \times (N + 2) =$ 25 mM $MgCl_2$ $1.5 \times (N + 2) =$ 10 mM dNTP mix $0.5 \times (N + 2) =$ 4 µM GUS DIR $2.5 \times (N + 2) =$ 4 µM GUS REV $2.5 \times (N + 2) =$ Mix by vortexing at speed 3 and add 5 U/µl Taq DNA polymerase $0.2 \times (N + 2) =$ **Final volume** (Each tube should have 20 µL of master mix)
4	Aliquot 20 µL of master mix to each labeled PCR tube. Add 5 µL of your extracted plant genomic DNA to the labeled tubes. Add 5 µL of the positive/negative controls to their labeled tubes. The positive control is plasmid pCAMBIA2301; negative control is the reaction mix plus water. Follow instructions for using the thermocycler. Amplification of the 510 bp *gus* fragment: 36 cycles of 10 s at 95C, 30 s at 60C, 2 min at 72C and a final extension at 72C for 4 min. Store PCR tubes at –20°C until use.
5	Prepare a 1.4% agarose gel. Wear gloves as ethidium bromide is a known carcinogen. Add 0.7 g of agarose in a 125 mL flask. Add 5 mL of 10X TAE buffer and 45 mL of distilled water to the flask. Bring solution to a boil in microwave (about 1 min; in 10–20 s increments) to melt the agarose. Ensure that the agarose is melted completely. Let the agarose solution cool for 5–10 min. Add 1 µL (4 mg/mL) of ethidium bromide solution and swirl to mix. Slowly pour the solution into the electrophoresis chamber containing gel comb. Remove any air bubbles. Let solidify for 40 min then cover with 1X TAE + 0.1 µg/ mL ethidium bromide (gel surface submerged 1–2 mm). Use within 4 days.
6	Remove PCR tubes from -20C. Add 4 µl of 6X loading dye to each tube. Mix with the pipette tip. Mark an additional tube M* (marker). M*—GeneRuler 100 bp Plus DNA Ladder [3000, 2000, 1500, 1200, 1000, 900, 800, 700, 600, 500, 400, 300, 200, and 100 bp]). To it, add 9 µl water, 1 µl 100 bp Plus DNA Ladder and 2 µl of 6X loading dye solution. Mix well and briefly centrifuge. Carefully load 12 µl of each PCR sample and the marker in separate wells. Electrophorese gel at 120 V for approx. 60 min (until blue dye has migrated near the end of gel). Observe gel and take picture under *UV* light. You can also do GUS histochemical analyses of both plants (Procedure 35.4) to verify the presence of the *gus* gene.

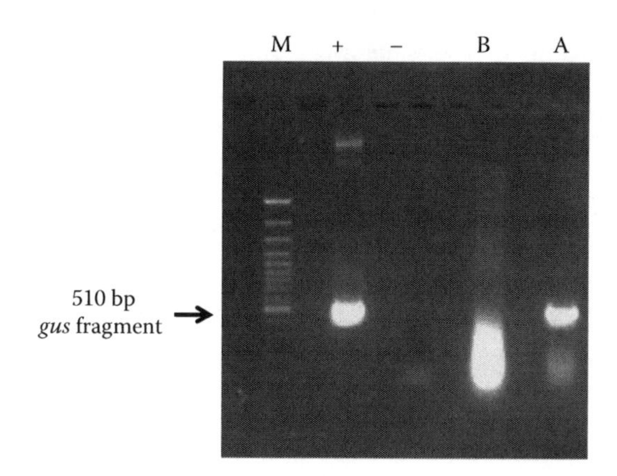

FIGURE 35.4 Using PCR to determine which transgenic tobacco plant (A or B) has the *gus* gene. Gel picture after PCR and electrophoresis: (M) is 100 bp Plus DNA Ladder; (+) is DNA from plasmid pCAMBIA2301; (–) is reaction mix plus water; (B) is genomic DNA from nontransgenic plant B; (A) is genomic DNA from transgenic plant A.

Questions

- Did you observe any contamination of your PCR samples? Why or why not?
- Which plant A or B was transgenic? Give reasons.
- Which plant A or B was not transgenic? Why was DNA extracted from this plant?
- Discuss what other molecular techniques (apart from PCR analyses) that could be used to confirm tobacco transformation?

REFERENCES

Aida, R, M. Komano, M. Saito, K. Nakase, and K. Murai. 2008a. Chysanthemum flower shape modification by suppression of chysanthemum—AGAMOUS gene. *Plant Biotechnol.* 25: 55–59.

Aida, R., T. Narumi, H. Ohsubo, H. Yamaguchi, K. Kato, A. Shinmyo, and M. Shibata. 2008b. Improved translation efficiency in chysanthemum and torenia with a translational enhancer derived from the tobacco alcohol dehydrogenases gene. *Plant Biotechnol.* 25:69–75.

de Jong, J., W. Rademaker, and M. F. Van Wordragen. 1993. Restoring adventitious shoot formation on chysanthemum leaf explants following co-cultivation with *Agrobacterium tumefaciens*. *Plant Cell, Tiss. Org. Cult.* 32:263–270.

Goldberg, J.B. and D.E. Ohman. 1984. Cloning and expression in *Pseudomonas aeruginosa* of a gene involved in the production of alginate. *J. Bacteriol.* 158:115–1121.

Hood, E.E., G. L. Helmer, R.T. Fraley, and M. D. Chilton. 1986. The hypervirulence of *Agrobacterium tumefaciens* A281 is encoded in a region of pTiBo542 outside of t-DNA. *J. Bacteriol.* 168:1291–1301.

Lowe, J.M., M. R. Davey, J. B. Power, and K.S. Blundy. 1993. A study of some factors affecting *Agrobacterium* transformation and plant regeneration of *Dendrathema grandiflora* Tzvelev (syn. *Chysanthemum morifolium* Ramat.). *Plant Cell, Tiss. Org. Cult.* 33:171–180.

Mersereau, M., G.J. Pazour, and A. Das. 1990. Efficient transformation of *Agrobacterium tumefaciens* by electroporation. *Gene*. 90:149–151.

Seo, J., S. W. Kim, J. Kim, H. W. Cha, and J.R. Liu. 2007. Co-expression of flavonoid 3', 5'-hydroxylase and flavonoid 3'-hydroxylase accelerates decolorization in transgenic chysanthemum petals. *J. Plant Biol.* 50:626–631.

Stomp, A. 1992. Histochemical localization of β-glucuronide. Pp. 103–113 in S.R. Gallagher (ed.), GUS protocols: using the GUS gene as a reporter of gene expression. Academic Press. 221 p.

Sun, L., L. Zhou, M. Lu, M. Cai. X.W. Jiang, and Q.X. Zhang. 2009. Marker-free transgenic chysanthemum obtained by *Agrobacterium*-mediated transformation with twin T-DNA binary vectors. *Plant Molec. Biol. Rep.* 27:102–108.

Trigiano, R.N. and R.A. May. 1994. Laboratory exercises illustrating organogenesis and transformation using chysanthemum cultivars. *HortTechnology* 4: 325–327.

Urban, L.A., J.M. Sherman, J.W. Moyer, and M.E. Daub. 1994. High frequency shoot regeneration and *Agrobacterium*-mediated transformation of chysanthemum (*Dendranthema grandiflora*). *Plant Sci.* 98:69–79.

White, F.F., and E.W. Nester. 1980. Hairy root: Plasmid encodes virulence traits in *Agrobacterium rhizogenes* *J. Bacteriol.* 141:1134–1141.

36 Genetic Transformation of Tobacco and Production of Transgenic Plants

Sadanand A. Dhekney, Zhijian T. Li, and D. J. Gray

Genetic transformation involves transfer of a DNA sequence into a plant cell and its subsequent integration into the host genome. Transformation serves as an important tool in plant biology for studying gene function and expression. It also allows for the addition of a single or few desirable traits in the existing elite varieties of plants without altering their genetic constitution. The essential prerequisites for a gene transfer system include the availability of a target tissue, a method to introduce foreign DNA, and a procedure to select transformed cells and regenerate transgenic plants (Birch, 1997). Although several methods have been used for delivery of transgenes into plant cells, *Agrobacterium*-mediated transformation has been the preferred method for a number of species. In nature, *Agrobacterium tumefaciens* is a pathogenic bacterium of many dicotyledonous and a few monocotyledenous plants. During the process of infection, *Agrobacterium* transfers a part of its DNA (T-DNA) present on a tumor-inducing (Ti) plasmid to the nuclear genome of the host plant. The components needed for transfer of T-DNA, known as the *vir* (virulence) genes, are present on the Ti plasmid in *Agrobacterium*. Transfer of the T-DNA region is achieved by the activation and action of the *vir* genes. Proliferating plant cells at the site of infection result in formation of tumors or crown galls. Infected cells produce amino acids known as opines that are utilized by the bacterium. The genes for opine synthesis are present in the T-DNA region, which is transferred to the host genome (Gelvin, 2003). The T-DNA transfer mechanism is utilized for the genetic transformation of crop plants, with the T-DNA region responsible for tumor formation being replaced by a T-DNA containing the desired genes of interest to be studied. Such *Agrobacterium* strains can no longer cause tumors upon infection and are termed *disarmed* strains. Genes of interest can either be inserted into the T-DNA region of a Ti plasmid or the T-DNA region of a smaller plasmid that is then introduced into disarmed *Agrobacterium*. The latter, known as a binary vector system, is more commonly adopted for transformation (Hellens et al., 2000).

The concept of transgenic plant production was first demonstrated successfully in tobacco (Horsch et al., 1984). Since then, tobacco has been a model system for studying genetic transformation and gene integration (Horsch et al., 1985; Burrow et al., 1990).

Tobacco is readily amenable to transformation and can be easily regenerated in tissue culture within 6–8 weeks. Each independent plant line can produce a large number of seeds that can be used to create a population of transgenic plants within a relatively short period of time.

The success of a transformation system depends on the ability to successfully recover transformants after cocultivation. Certain genes are incorporated, along with the gene of interest, into the T-DNA region to confirm the presence of the T-DNA region in plant cells and/or confer a competitive advantage to the growth of transgenic cells over nontransformed cells. These include reporter genes, which indicate the presence of the T-DNA and selectable marker genes that most often confer antibiotic resistance to transformed cells. Two reporter genes, such as the β-glucuronidase (*GUS*) gene (Jefferson et al., 1987) and the Green Fluorescent Protein (*GFP*) gene (Stewart, 2001) are most commonly used in transformation studies. The *GUS* gene, isolated from *E. coli*, produces the β-glucuronidase enzyme that can cleave a colorless substrate 5-Bromo 4-Chloro- 3-indolyl- β-glucuronic acid (X-Gluc), resulting in the production of intense blue color. Transgenic cells expressing *GUS* will produce a blue coloration in the presence of the substrate, thereby facilitating transgene detection. The *GFP* gene, isolated from the Pacific jellyfish, *Aequoria victoria,* is commonly used as a reporter gene in transformation studies. Plant cells expressing *GFP* produce a bright green fluorescence that can be observed under a stereo microscope equipped for epifluorescence illumination. Alternatively, *GFP* can be visualized in plant cells using a relatively inexpensive detection system (Gray et al., 2005). Selectable marker genes such as neomycin phosphotransferase II (npt II) and hygromycin phosphotransferase (hpt) genes are frequently used along with reporter genes in genetic transformation (Benveniste and Davies, 1973; van den Elzen et al., 1984). Transgenic cells carrying these marker genes can selectively grow on the culture medium containing kanamycin or hygromycin antibiotics, while inhibiting the growth of nontransformed cells.

This laboratory experiment will introduce students to the following concepts:

1. *Agrobacterium*-mediated transformation and production of transgenic plants
2. Screening transgenic plants from nontransformed plants by utilizing reporter-selectable marker gene systems

GENERAL CONSIDERATIONS

PREPARATION OF TOBACCO PLANT MATERIAL

The following steps are used to obtain aseptic tobacco seedlings. Tobacco cultivar "Samson" was used in experiments; however, other cultivars may be suitable. Collect approximately 100 seeds in a 1.5 mL microcentrifuge tube and add 1 mL 2.5% sodium hypochlorite (50% commercial bleach) solution. Surface disinfest seeds by constant agitation on a shaker for 10 min. Remove the bleach solution by washing three times (10 min each) with 1 mL sterile distilled water. Transfer the seeds by evenly spreading them on petri dishes containing agar-solidified MS (Murashige and Skoog, 1962) medium. Seal the petri dishes with Parafilm and place them in a culture room at 25°C with an 18-h cool white fluorescent light (60 μmol m^{-2} s^{-1})/6-h dark cycle. Seeds will germinate after 7–8 d. Leaves from 3–4 week-old seedlings are used as explants for the transformation procedure. Tobacco plants can be maintained in vitro by transferring shoot cuttings to GA 7 Magenta vessels containing 50 mL MS medium for future use; however, the best results of transformation and regeneration are observed when explants derived from young seedlings are utilized.

PREPARATION OF *AGROBACTERIUM* CELLS FOR TRANSFORMATION

Agrobacterium strains LBA 4404 or EHA105 are most commonly used for transformation. LBA 4404, along with the binary plasmid pBI121 (containing the *GUS* and *npt II* genes), can be purchased from Invitrogen Corporation (www.invitrogen.com, Catalog No 18313-015). Alternatively, plasmids containing the *GFP* gene, along with selectable marker genes, can be purchased from CAMBIA (www.cambia.org). Initiate an *Agrobacterium* culture in 50 mL YEB medium (5 g/L beef extract, 1 g/L yeast extract, 5 g/L peptone, 5 g/L sucrose, 0.3 g/L$MgSO_4 \cdot 7H_2 0$) containing the appropriate antibiotics (100 mg/L streptomycin for LBA 4404, or 20 mg/L rifampicin for EHA 105). Antibiotics are dissolved in appropriate solvent (water or DMSO), filter sterilized using a 0.2 μm nylon sterile filter (Fisher Scientific Company Catalog No. 09-719C), and added to the medium after cooling the medium to 50–55°C. Place the culture on a rotary shaker at 180 rpm and 28°C. After 24 h of growth, centrifuge the culture and transfer the pellet to a 1.5 mL microcentrifuge tube. Wash the cells by resuspending the pellet in 1 mL chilled TE buffer. Centrifuge the cells to form a pellet. Discard the supernatant and resuspend cells in 50 μL LB medium (10.0 g/L tryptone, 5.0 g/L yeast extract, 5.0 g/L NaCl, 1.0 g/L sucrose). Add plasmid DNA (300 ng–1 μg) to cells and mix well. Incubate the mixture on ice for 1 h. Transfer cells to a freezer (–50 to –70°C) for 10 min. Thaw cells by placing the mixture at 37°C for 5 min. Add 300 μL LB medium to the mixture and incubate cells on a shaker at 180 rpm (28°C) for 4 h. After 4 h, spread 100 μL of transformed cells onto 100 X 15 mm petri dishes containing 15 mL LB agar medium (as earlier with the addition of 15.0 g/L agar) supplemented with filter-sterilized antibiotics (100 mg/L streptomycin and 50 mg/L kanamycin for LBA 4404, or 20 mg/L rifampicin and 50 mg/L kanamycin for EHA 105). Incubate petri dishes in the dark at 28°C for 60–72 h. Bacterial colonies will be observed growing on the medium. Transfer a single bacterial colony to a 125 mL conical flask containing 30 mL liquid MG/L medium (composition described in the following section) with appropriate antibiotics. Incubate on a rotary shaker at 180 rpm at 28°C for 16–20 h. The bacterial culture should appear cloudy at the end of the culture period. Transfer 800 μL bacterial culture and 200 μL sterile glycerol to 1.5 mL microcentrifuge tubes. Vortex to mix, and label, and store tubes at –70°C for use in plant transformation experiments.

Materials

The following materials are used for multiple experiments. Recipes for media are per liter final volume. Media are autoclaved at 121°C and 15 psi pressure for 20 min.

- Liquid MG/L medium
  Mannitol—5.0 g, l-Glutamate—1.0 g, Tryptone—5.0 g, Yeast extract—2.5 g, NaCl—5.0 g, KH_2PO_4—150.0 mg, $MgSO_4.7H_2O$—100.0 mg, Fe-EDTA—2.5 mL (To make a stock solution of Fe-EDTA, dissolve 7.44 g of $Na_2EDTA.2H_2O$ and 1.86 g $FeSO_4.7H_2O$ in sterile distilled water and make final volume to 1L.)
- MS medium
  MS basal medium with vitamins (Phytotechnology Laboratories Catalog No. M519), 0.1 g myo-inositol, 30 g sucrose, 7.0 g T.C agar (Phytotechnology Laboratories Catalog No. A175), pH adjusted to 5.8
- MSTcck medium
  MS medium, 1.0 mg BAP, 0.1 mg NAA, 200 mg each of carbenicillin and cefotaxime, 100 mg kanamycin
- GUS assay buffer
  a. Sodium phosphate buffer (50 mM, pH 7.0)
     $NaH_2PO_4{\cdot}H_2O$—0.269 g, $NaH_2PO_4{\cdot}7H_2O$—0.818 g. Dissolve in sterile distilled water and make final volume to 100 mL.
  b. Substrate staining solution
     X- Gluc (5-Bromo 4-Chloro- 3-indolyl- β glucuronic acid sodium salt (Phytotechnology Laboratories Catalog No. 129541-41-9). Dissolve 5 mg X-Gluc powder in 50 μL dimethlyformamide or dimethyl sulfoxide. Mix the substrate staining solution with 10 mL sodium phosphate buffer and vortex. (*Note*: The GUS assay buffer can retain activity for one month if stored in the dark at 4°C.)

The following materials will be needed for a group of 2–4 students:

- Five young (2–3-week-old) tobacco seedlings
- Five 100 × 15 mm plastic petri dishes (Fisher Scientific Catalog No. 08-757-12)
- One 50 mL centrifuge tube
- Three 125 mL conical flasks
- Five 1.5 mL microcentrifuge tubes
- Five petri dishes containing LB agar medium supplemented with antibiotics

- Ten petri dishes containing MSTcck medium
- Ten GA 7 Magenta vessels containing 50 mL MS medium
- 30 mL liquid MG/L medium containing 100 mg L^{-1} streptomycin and kanamycin
- 30 mL liquid MS medium

Follow the protocol described in Procedure 36.1 to complete the experiment.

Anticipated Results

Bacterial growth may be observed on explants after 2–3 d of culture on MSTcck medium. Any bacterial growth can be minimized by transferring explants to fresh medium. After 2 weeks of culture, callus formation will be observed from edges of leaf disks. Calli will give rise to shoot primordia, which will subsequently produce shoots. If explants have been transformed with a plasmid containing the *GFP* gene, transgenic cultures will produce bright green fluorescence, which can be observed with a stereo microscope equipped for eplifluorescence illumination. Transgenic as well as nontransformed shoots will be observed growing on the culture medium. Growth of nontransformed shoots occurs due to their cross-protection by adjacently growing transgenic cells and shoots. Transgenic shoots will produce a prolific and extensive root system when transferred to MScck medium, whereas nontransformed shoots will fail to produce any roots and appear bleached over a period of time. Cultures from the positive control should produce callus and shoots similar to

Procedure 36.1

Agrobacterium-Mediated Transformation of Tobacco

Step	Instructions and Comments
1	Start *Agrobacterium* culture by adding 50 µL frozen stock culture to a conical flask containing 30 mL liquid MG/L medium and appropriate antibiotics. Incubate the flask on a shaker at 180 rpm and 28°C for 24 h. The bacterial culture should appear cloudy at the end of the culture period.
2	Transfer the culture to a 50 mL centrifuge tube and spin at 6000 rpm for 8 min at room temperature. Discard the supernatant and resuspend the pellet in 30 liquid MS medium. Transfer the contents of the tube to a 125 mL conical flask and incubate for an additional 4 h on a rotary shaker under the same conditions as above.
3	Collect leaves from 2–3-week-old tobacco seedlings. Wound leaves using a sterile paper punch to obtain 1 cm leaf disks as explants. Transfer explants to a petri dish and add 5 mL *Agrobacterium* solution. Mix thoroughly by swirling and incubate for 7 min.
4	Remove excessive bacterial solution by blotting explants on a sterile filter paper and transfer explants upside down to petri dishes containing MST medium. Incubate dishes in the dark for cocultivation at 28°C for 48–72 h.
5	Following cocultivation, blot explants on sterile filter paper to remove any bacterial overgrowth and transfer five explants upside down to each petri dish containing MSTcck medium. As positive and negative controls for the experiment, transfer leaf disks that were not cocultivated onto petri dishes containing MST and MSTcck medium. Maintain cultures under conditions described earlier (for tobacco seedlings) for 3–5 weeks.
6	Wounded edges of the explants will produce calli, which will give rise to shoots. Excise 3–5 cm-long shoots using a sharp scalpel and transfer individual shoots to Magenta boxes containing MScck medium.
7	Transgenic shoots will produce roots in the medium. Transfer plants with a well-developed root system to plug trays containing autoclaved potting mix. Maintain plants in a growth room at 25°C with an 18 h cool white fluorescent light (60 µmol m^{-2} s^{-1})/6 h dark cycle under 100% relative humidity conditions for 1 week.
8	Transfer plants to a greenhouse after 1 week. Place plants away from direct sunlight for 1 week for hardening and then transfer under regular conditions of light and relative humidity.
9	To screen transgenic plants from nontransformed plants using the GUS assay, collect leaf samples from the first young leaf of each transgenic plant using a clean paper punch and immerse them in centrifuge tubes containing 100 µL GUS assay buffer. Incubate tubes at 37°C for 3 h to overnight.
10	Score samples by checking blue color at the wounded edges of leaf disks. Any blue coloration observed would indicate expression of transgenic GUS protein by the integrated transgene.
11	Transgenic plants will flower in 8 weeks. Collect any seeds obtained from transgenic plants separately and dry them in an oven overnight at 50°C. Store seeds in brown paper bags at room temperature for additional analysis.

cocultivated explants, whereas no growth should be observed from the negative control. When leaf samples from independent transgenic plant lines are analyzed for *GUS* expression, variations will be observed in the intensity of blue color produced among transgenic plants. This is attributed to the random integration of transgenes at various locations in the plant genome. Transgenes integrated at certain regions on the chromosomes can result in greater expression levels compared to transgenes integrated at other regions.

Questions

- How many shoots are produced from each cocultivated leaf disk?
- How many of these shoots produce a prolific root system in MScck medium?
- What is the percentage of transgenic shoots produced from each leaf disk?
- How are transgenic shoots differentiated from nontransformed shoots in the rooting medium?
- What is the difference between a reporter gene and a selectable marker gene?

REFERENCES

Benveniste, R. and Davies, J. 1973. Mechanisms of antibiotic resistance in bacteria. *Annu. Rev. Biochem.* 42:471–506.

Birch, R.G. 1997. Plant transformation: Problems and strategies for practical applications. *Annu. Rev. Plant Physiol. Plant Mol. Biol.* 48:297–326.

Burrow, M.D., Chlan, C.A., Sen, P., Lisca, A., and Murai, N. 1990. High frequency generation of transgenic tobacco plants after modified leaf disk cocultivation with *Agrobacterium tumefaciens*. *Plant Mol. Biol. Rep.* 2:124–139.

Gelvin, S.B. 2003. *Agrobacterium*-mediated plant transformation: The biology behind the "Gene-Jockeying" tool. *Micro. Mol. Biol. Rev.* 1:16–37.

Gray. D.J., Jayasankar, S., and Li, Z.T. 2005. A simple illumination system for visualizing green fluorescent protein. Chapter 22. In R.N. Trigiano, and D.J. Gray, (Eds.), *Plant Development and Biotechnology*, CRC Press, Boca Raton, FL.

Hellens, R., Mullinex, P., and Klee, H. 2000. A guide to *Agrobacterium* binary Ti vectors. *Trends in Plant Sci.* 5:446:451.

Horsch, R.B., Fraley, R.T., Rogers, S.G., Sanders, P.R., Lyod, A., and Hoffman, N.L. 1984. Inheritence of functional foreign genes in plants. *Science*. 223:496–498.

Horsch, R.B., Fry, J.E., Hoffman, N.L., Eichholtz, D., Rogers, S.G., and Fraley, R.T. 1985. A simple and general method for transferring genes into plants. *Science* 227:1229–1231.

Jefferson, R.A., Kavanaugh, T.A., and Bevan, M.W. 1987. GUS fusions: Glucuronidase as a sensitive and versatile gene fusion marker in higher plants. *EMBO J.* 6:3609–3907.

Murashige, T. and Skoog, F. 1962. A revised medium for rapid growth and bio assays with tobacco tissue cultures. *Physiol. Plant.* 15:473–497.

Stewart, C.N. 2001. The utility of green fluorescent protein in transgenic plants. *Plant Cell Rep.* 20:376–382.

van den Elzen, P.J.M., Townsend, J., Lee, K.Y., and Bedbrook, J.R. 1984. A chimeric hygromycin resistant gene as a selectable marker in plant cells. *Plant Mol. Biol.* 5:299–302.

37 Genetically Modified Controversies

Sensational Headlines versus Pragmatic Research

H.A. Richards, L.C. Hudson, M.D. Halfhill, and C.N. Stewart, Jr.

CONCEPTS

- Genetically modified (GM) plants have been widely grown throughout the world and extensively in the United States.
- No measureable harm has been detected from the cultivation of GM plants.
- Popular press and public perceptions often focus on the sensational side.
- Established methods to test food safety of GM plants and products are available.
- Testing ecological interactions between GM plants and wild plants is complicated.
- While today's GM plant products are safe, biotechnology risk assessment research should be continued.

Controversies over scientific issues have often been initiated by sensational media reports of research that capture the imagination of the public. Before the application of biotechnology, the biology of crop plants had never been the focus of controversy and debate. But with recent attention drawn to plausible harm to foodstuffs and the environment, the previously unheralded subject of agriculture has been transformed into media fodder. We wish to illuminate the facts and discuss the research behind the issues by focusing on a few timely plant biotechnology controversies. Fortunately, it seems that, in the past few years, media coverage of plant biotechnology has been moving toward more objectivity and less sensationalism, and the expectation of it as a technology and industry has become more mature (Stewart and Littmann, 2008).

BIOTECHNOLOGY IN THE SPOTLIGHT

A remarkable development occurred in the early 1980s—the first genetically engineered plants were produced. For the potential impact that this technology would have on agriculture as we know it, the event was rather unheralded. As biotechnology grew in application, it remained nearly a non-topic for news media and the general populace. There were rumblings from some concerned parties, environmental groups for one, but in general, genetically engineered foods were accepted into use with little fanfare or concern. Opponents tended to be thought of as technophobes or extremists. However, the situation exploded in 1999, and awareness of genetically modified (GM) food skyrocketed as news media outlets reported that GM crops devastated monarch butterfly populations and could be toxic to men, women, children, and babies.

The question remains to be answered if all this attention is warranted. Clearly, if genetically engineered crops are driving butterflies to extinction, or if eating corn chips made from GM corn made people sick, then there is cause for alarm. But, as is often the case, these reports are not reflective of the scientific experimentation behind them, nor do they indicate the levels of safeguards in place to protect the population from potential risks. Since 2001, the debate has ebbed as other world events and issues have taken priority in the media spotlight. However, the issue has not been settled. The purpose of this article is to outline the risks posed by food and agricultural biotechnology and discuss examples of the research that has been conducted to evaluate these risks. The more popular aspects of the GMO controversy will be addressed by outlining the experiments that are the basis for the hysteria and by reporting the research conducted as follow-up to these preliminary studies. Finally, we will provide our perspective on these issues and offer opinions as to the lessons learned and the future direction of plant biotechnology.

THE INCENTIVE

The transfer of genetic materials to create a new plant variety is not a new concept in agriculture. Intensive breeding programs have long been in place to develop crops that perform better in the field. This technology is limited by sexual compatibility and the difficulties in obtaining the desired result from the transfer of thousands of genes from parents to offspring. Irradiation and mutagenization have been used to create heritable changes in the genetic structure of crops to generate new traits. These techniques are limited by their lack of predictability and precision. The incentive for the development and use of biotechnology is to overcome these limitations. Because this type of gene transfer does not rely on compatible crosses, it is possible to introduce traits from any species into plants, and the transfer involves a single gene or handful of genes, which provides more predictable and precise outcomes for the next generation of plants. This technology does not solve all of the problems associated with modern agriculture, but it does offer a number of potential benefits that make it worth pursuing. However, because it is conceptually different from the technologies employed before, the question becomes—What new complications and risks are posed by biotechnology, and how can they be effectively managed to allow its safe use now and in the future?

THE ENVIRONMENT

Risk can be defined as a function of the probability of a potential hazard occurring and the damage incurred if the hazard occurs—the exposure model. One advantage in evaluating environmental risks of biotechnology is that the parameters of the risk equation are similar to those for crops produced via traditional breeding or mutation techniques. When a new variety is created, then it is evaluated to determine how its new characteristics will affect its performance in the field. This includes risk analysis such as increased invasiveness or sexual compatibility with wild relatives. The difference with varieties created through biotechnology is that the potential list of transferred characteristics is greater and more diverse. Also, the traits are more discreet because they are linked to a single gene or a handful of genes. Therefore, genetically engineered crops must be evaluated on a case-by-case basis, as the potential hazards will differ for different transgenes. The risk associated with an identified hazard will be defined by the probability of occurrence. So the key to risk assessment of biotechnological products is to first identify potential hazards and evaluate the likelihood of that situation occurring. If the probabilities are very low, then risk would be low, too.

Volunteerism and invasiveness of new varieties could lead to economic complications or the persistence of the plants in the environment. These potential hazards are of concern when transgenes are added that increase the fitness of the plants to either agricultural treatments or environmental stresses. Traits such as herbicide resistance could make volunteers (uncollected seeds from the previous crop) difficult to manage or remove. Traits such as drought tolerance may allow crops to be grown in new climates, but could also lead to those plants escaping into new environments and adversely affecting the native species. Research into these risks involves characterizing the performance of fitness-enhancing transgenes and evaluating management strategies to reduce their occurrence.

HYBRIDIZATION

Potential complications of hybridization with transgenic crops can be divided into two categories—crosses with nontransgenic plants of the same species and crosses with related wild species. The first category could cause problems for farmers who wish to keep crops free of genetically engineered plants, particularly organic farmers that must abide by a low GM plant threshold. The second category encompasses the risk of transgene escape from crops to weedy relatives, the concern being the generation of "superweeds." These hazards are addressed through containment strategies to limit the movement of transgenic material. To define the risk of the creation of a superweed, the effects of the transgenic traits in agricultural or ecological environments must be considered. Plants expressing nonfitness-enhancing traits, such as increased carotenoid expression, do not pose this type of risk, but insect-resistance or salt-tolerance genes may need consideration. In addition, hybridization requires sexual compatibility. In the United States, there is no concern that soybeans will hybridize with anything other than other soybeans; therefore, the superweed risk is so low as to be negligible. A significant amount of research has been conducted on these issues for crop species that have wild relatives near cultivation (Stewart et al., 2003; Warwick et al., 2003).

TRANSGENE FLOW FROM GENETICALLY MODIFIED CANOLA

Although maize and soybean have no wild relatives that occur near cultivation in the Unites States, this is not to say that all GM crops are free of potential consequences from hybridization. Canola (*Brassica napus* L.) is an emerging crop that could pose potential risks if agricultural acreages expand in geographic regions with naturally occurring wild relatives. Canola is of particular interest as a potential source for transgene escape because of several factors. Approximately 11% of the 25 million hectares of canola produced globally is transgenic, and it is even higher in certain countries, for example, 60% of the Canadian production (Warwick et al., 2003). The crop is predominantly selfing with outcrossing averaging 30% (Beckie et al., 2003), forms a persistent seed bank, and produces large weedy volunteer or feral populations particularly in the first year after canola production. Transgenic volunteers are a potential problem for several reasons. The GM volunteers represent a management concern because they can be a "weedy" problem in the next year's crop and can be difficult to control due to transgenic traits such as herbicide tolerance. More importantly, the volunteers also represent a recurring source of transgenes via hybridization to other canola varieties and wild relatives.

Hybridization from genetically modified canola to other canola varieties has been documented in several scientific studies. In one famous case, Hall et al. (2000) found volunteer individuals that were resistant to three different herbicides as the result of hybridization of three herbicide-tolerant varieties over several seasons. Therefore, in areas where more than one type of herbicide-tolerant canola variety has been grown, multiple herbicide-tolerant volunteer individuals have arisen as the result of intraspecific hybridization (Hall et al., 2000; Beckie et al., 2003). Also, the unintended presence of transgenes in certified seed stocks has the potential to allow the flow of transgenes in unexpected ways. Beckie et al. (2003) found an example where certified seed of a Roundup Ready (glyphosate-tolerant) canola variety had a small percentage of Liberty Link (glufosinate-tolerant) seeds. As a result of the unexpected transgene, a higher quantity of double-resistant volunteers was found in the following field season than would be expected due to hybridization with surrounding canola fields. The agronomic characteristics of canola production, combined with the use of multiple GM varieties and the occurrence of unexpected transgenes in certified seed lots, demonstrate the necessity to monitor canola as a potential source of transgene flow, at least until the consequences can be determined.

Wild relatives also have the potential to be the recipient of transgenes from GM canola varieties. In many areas, wild relatives such as *Brassica rapa* L. (bird rape, field mustard), *Raphanus raphanistrum* L. (wild radish), and *Sinapis arvensis* L. (wild mustard) occur in or near canola cultivation. Several studies have shown that hybridization between canola and bird rape (*B. rapa*) occurs under field conditions and results in the production of hybrid populations (Warwick et al., 2003). Hybridization occurs at a range of frequencies based on the ratio of parental crop species to the wild relative, and is as high as 93%. Transgenic Bt canola varieties have been shown to hybridize with *B. rapa* under field conditions, and the frequency of hybridization was reported to range from 1–17% based on the transgenic variety (Warwick et al., 2003). In a recent report, Warwick et al. (2003) have demonstrated the transfer of an herbicide (glyphosate)-tolerance gene from commercial fields of canola to a naturally occurring population of *B. rapa* in Quebec, Canada. With relatively high hybridization frequencies observed in these experiments, transgenic hybrid populations should be expected when canola and naturally occurring wild relatives occur in close proximity.

Warwick et al. (2008) confirmed this hypothesis by demonstrating the persistence and stable incorporation of an herbicide-resistance transgene in the canola wild relative *B. rapa*. This report indicated that the trait persisted over a 6-year period in the absence of herbicidal selection pressure. It is clear that if GM traits are to be used with any effectiveness in crops such as canola, biocontainment practices should be considered to minimize transgene escape to wild relatives. Even though the incidence of introgression was low (one event), this might be the harbinger of things to come.

Transgenes Discovered in Mexican Corn

Transgene flow from transgenic crop varieties to wild relatives is of particular interest because of the potential ramifications both inside and outside agriculture. However, there have been some cases where suspected transgene flow within a crop species has generated both scientific and media attention. Quist and Chapela (2001) reported the introgression of the cauliflower mosaic virus (CaMV) 35S promoter into the genome of native Mexican landraces of corn. The corn sampled was from the state of Oaxaca, which was remotely located from areas where transgenic corn had been grown legally. In fact, the cultivation of GM crops had been illegal in Mexico since 1998. The paper initiated a wave of public and scientific debate, and after a half year of scrutiny, ended with the de facto retraction of the work by the editor of the journal.

The central conclusion of the paper was that transgenic DNA sequences were present in traditional Mexican corn landraces. In addition, the authors concluded that transgenes had been introgressed into the genome of individuals in this population, and that the transgenic material was moving within the genome. These conclusions resulted in a firestorm of public scientific criticism. The predominant argument was that the methodology used for this research was not appropriate for the conclusions drawn (Metz and Fütterer, 2002; Kaplinsky et al., 2002). First, the authors only reported evidence of the CaMV 35S promoter and not actual transgenes in Mexican corn. In addition, PCR was the method used for detection, which creates the distinct possibility that contamination of trace quantities of the cauliflower mosaic virus itself (ubiquitous in the environment) may have resulted in false positive results.

Critics argued that, before conclusions of the presence and introgression of transgenes can be drawn, more stringent and direct tests would be required. For instance, a direct assay for the presence of a transgene, such as Bt or Roundup Ready genes, would have been more indicative of transgenic material. Instead of PCR, a genomic hybridization technique such as Southern blot analysis would have greatly reduced the likelihood of false positives and would have been more convincing. The authors published a follow-up article in which they conducted a DNA dot blot assay; however, the result was still based on the CaMV 35S promoter, and the data could not indicate insert size, copy number, or integration into the genome, which resulted in continued debate and criticism.

Two follow-up communications in *Nature* re-evaluated the data presented by Quist and Chapela (2001). In the original paper, nested PCR suggested that a few kernels on any given ear of corn contained the CaMV 35S promoter. Metz and Fütterer (2002) argued that, if this result was accurate, then introgression into the genome could not have occurred because the expected result would have been the presence of the transgenic sequences in most or all of the kernels. The original publication also suggested that the CaMV 35S promoter was changing locations within the genome of Mexican corn. These data were collected using inverse PCR, which uses known sequences (in this case the CaMV 35S promoter) to determine the identity of adjacent sequences of DNA. However, Quist and Chapela (2001) did not find transgenes located near these putative transgenic promoters, as critics argued would be expected (Metz and Fütterer, 2002; Kaplinsky et al., 2002). Additionally, Kaplinsky et al. (2002) indicated that the article mistakenly reported a gene sequence as one commonly used in plant genetic engineering (*adh1* intron) when sequence similarity actually suggested that the sequences were from naturally present corn genes (*adh1* and *bronze1* genes). The intense criticism of the methodology and conclusions in the paper resulted in its de facto retraction, though the authors remained steadfast by their assertion that transgenic material *was* present in Mexican corn landraces.

More recently, another group investigated this claim. A systematic survey of the frequency of transgenes in currently grown landraces of maize was conducted (Ortiz-García et al., 2005). Maize seeds from 870 plants in 125 fields and 18 localities in the state of Oaxaca during 2003 and 2004 were examined. This group screened 153,746 sampled seeds for the presence of two transgene elements from the 35S promoter of the cauliflower mosaic virus and the nopaline synthase gene (nopaline synthase terminator) from *Agrobacterium tumefaciens*, one or both of which is present in all transgenic commercial varieties of maize. No transgenic sequences were detected with highly sensitive PCR-based markers, appropriate positive and negative controls, and duplicate samples for DNA extraction. It was concluded that transgenic maize seeds were absent or extremely rare in the sampled fields.

In contrast, Serratos-Hernandez et al. (2007) tested Mexican corn for the presence of recombinant proteins. They sampled maize fields in the Mexican Federal District and used the ELISA technique to test for three transgenic proteins. Of the 42 fields sampled, they detected Cry1Ab/Ac in two fields and CP4 EPSPS in one, and concluded that the recombinant proteins were present at a rate of 1% of maize. Positive protein tests are not direct evidence of the transgene, as false positives can occur due to cross reactivity; therefore, genetic studies are required as a follow-up. However, this research highlights the need for regular and vigilant research into transgene persistence in agriculture. It also highlights the paradox that Quist and Chapela (2001) might have been correct in their finding of escaped transgenes in Mexican landrace maize, but their incorrect data nullified the impact of the correct data.

Nontarget Organisms

When transgenic crops are utilized, especially those designed for pest control, the potential exists for them to have adverse effects on species that were not intended to be harmed (Losey et al., 1999). This hazard is similar to the risks associated with the application of chemical pesticides in modern agriculture. In fact, GM crops have been developed to reduce the application of these chemicals so as to reduce costs to farmers and lessen the impact of pest control agents on the environment. Nevertheless, pesticidal transgenic plants receive the lion's share of the attention and criticism. Research has been conducted on the toxicity of these materials to agriculturally benign species and on their persistence in the environment. Until recently there has been little public debate until the time the monarch butterfly became the subject of significant scientific and media interest regarding nontarget effects from biotech crops. This issue focused on the pollen of Bt corn, which could blow onto milkweed leaves, the exclusive diet of monarch caterpillars.

The monarch butterfly became the focus of attention in 1999 when a published article captured media attention. *Bacillus thuringiensis* (Bt) is a naturally occurring soil-borne bacterium, which produces a crystal-like protein (Cry proteins) that selectively kills a specific group of insects

(Lepidopteran caterpillars), and is not harmful to agriculturally beneficial insects, animals, or humans, and does not persist in the environment. Liquid and granular formulations of the Bt proteins have been used successfully for 50 years on a variety of crops. Bt corn refers to corn that has been enhanced through biotechnology to produce its own Bt insecticidal proteins.

Insect resistance, conferred via expression of a variety of *Bacillus thuringiensis* (Bt) delta-endotoxins, is the second most commonly used trait, after herbicide resistance, in commercial GM crops. Four Bt delta-endotoxin genes (*cry1Ab*, *cry1Ac*, *cry2Ab*, and *cry9C*) are currently used commercially in maize and cotton to protect against lepidopteran pest attack (Shelton et al., 2000). Many more Bt GM plants expressing delta-endotoxins are being developed, particularly annual crop species that are grown on a large scale around the world. In recent years, vegetative insecticidal proteins (VIP) have been found to have potent broad spectrum activity against insects (Estruch et al., 1996). VIP genes are not homologous to *Cry* and *Cyt* genes and bind to cell membrane proteins in a different way from the other toxins (Lee et al., 2003).

Regulation for Bt corn and other transgenic crop species falls under the jurisdiction of the Environmental Protection Agency (EPA) in the United States, who is responsible for determining the environmental effect. Before the approval of Bt corn in 1995, 1996, and 1998, the EPA concluded that the product does not cause any "unreasonable adverse effects to nontarget organisms based on evaluations of toxicity and exposure." These conclusions were based on the fact that corn pollen is heavy and can move only a short distance, cornfields typically contain a low concentration of weeds, and monarch larvae exposure would be limited to milkweeds within or in close proximity to cornfields during pollen shed.

In May 1999, a communication in the journal *Nature* (Losey et al., 1999) raised concerns regarding nontarget effects of transgenic crops containing genes from *Bacillus thuringiensis.* A laboratory feeding experiment performed by researchers at Cornell University found that monarch butterfly larvae that were given no choice but to feed on milkweed leaves dusted with high levels of pollen from Bt corn had slower growth rates and a higher morality rate than those larvae consuming leaves with no pollen or non-Bt pollen. This group of scientists suggested that there was a potential risk to monarch butterfly populations, if demonstrated in the natural environment, since migratory patterns include the central United States, where the majority of the corn is grown.

In order to address public concerns about the monarch, researchers from nine universities, the U.S. Department of Agriculture (USDA), and Agriculture Extension conducted numerous field test in 1999 and reported their findings at the Monarch Butterfly Research Symposia and later in the *Proceedings of the National Academy of Sciences, USA*. This collaborative work found that monarch larvae have very little exposure to corn pollen under field conditions (Sears et al., 2001) since the majority of corn pollen settles in the immediate vicinity of the field. Pollen movement is limited; however, wind can disperse pollen away from the field, depositing it over a broad area, creating low pollen concentrations on surfaces. Further field studies demonstrated that milkweed leaves captured only 30% of corn pollen available, and that wind and rain can reduce this amount another 90%. Bt endotoxin degrades in pollen upon exposure to sunlight (Pleasants et al., 2001).

Monarch butterflies may have exposure to small quantities of Bt corn pollen on milkweed, but at amounts below the threshold to harm the larvae. Monarch larvae can safely consume milkweed leaves with up to 1,100 Bt pollen grains per square centimeter (Hellmich et al., 2001), which is high for field levels. Surveys in Maryland were conducted in 81 fields, showing that the mean pollen level inside the cornfields or around the edge was 56 pollen grains, and only 10 leaves out of 127 leaves examined had higher levels (Pleasants et al., 2001). Hellmich et al. (2001) found that monarch larvae would avoid eating milkweed with Bt or non-Bt corn pollen on the surface if milkweeds free from pollen were available. Field studies conducted in Canada and the United States sponsored by the USDA and the Agricultural Biotechnology Stewardship Technical Committee concluded that monarch survival, weight gain, and milkweed consumption were similar to monarch larvae feeding for five consecutive days on milkweed plants in Bt and non-Bt cornfield during pollen shed (Hellmich et al., 2001; Sears et al., 2001). These experiments confirmed that milkweed leaves in close proximity to Bt cornfields contained pollen levels too low to impact the normal development of monarch butterfly larva.

To address the concern that chronic low-level exposure could have adverse affects, a consortium of scientists conducted a formal risk assessment of the impact of Bt corn pollen on monarch butterfly populations. These studies suggested that the impact of acute toxicity and exposure to Bt pollen from current commercial levels of maize is negligible and does not pose an unreasonable risk to monarch populations (Dively et al., 2004). These studies analyzed the hazard associated with the most extreme exposure scenario that might occur under field conditions and allow for a more precise upper bound for the overall risk to monarch populations to be derived. The results of five studies conducted over two years at three locations indicated that 23.7% fewer larvae exposed to Bt pollen during the first week of anthesis reached the adult stage compared to unexposed controls. Exposure to Bt pollen also prolonged the development time of larvae by 1.8 days and reduced the weights of both pupae and adults by 5.5%. The sex ratio and wing length of adults were unaffected. Chronic exposure of monarch larvae throughout their development to Bt corn pollen is detrimental to only a small fraction of the breeding population because the risk of exposure is low. When this impact is considered over the entire range of the American corn belt, the ecological outcome is very small and not likely to affect monarch populations in North America.

The potential negative impact of other Bt crops and GM toxins have been evaluated. Cry1Ac oilseed rape ingested by the nontarget aphid *Myzus persicae* had no effect on the ability of the parasitoid *Diaeretiella rapae* to control this pest (Schuler et al., 2001). Cry3A potato fed to the Colorado potato beetle had no effect on its carabid predator, *Lebia grandis* (Riddick and Barbosa, 2000). Field studies with Cry3A potatoes showed no significant impacts on beneficial predator species compared with the reductions noted with the regular permethrin applications required with a conventional crop (Reed et al., 2001). Cry1Ab rice plants did not affect the development or survival of the predator *Cytorhiunus lividipennis* feeding on brown planthoppers, *Nilaparvata lugens*, which were not susceptible to Cry1Ab but excreted it with their honeydew (Bernal et al., 2002). Field tests with rice expressing Cry1Ab and Cry1Ac showed no effects on the population dynamics of five spider species (Liu et al., 2002). Honey bees (*Apis mellifera* L.), key pollinators for many important agricultural crops, were reported to not be appreciably affected by Bt proteins found in transgenic pollen (Duan et al., 2008).

A large meta-analysis of Bt transgene environmental impact data from field experiments all over the world, conducted by the National Center for Ecological Analysis and Synthesis (NCEAS), found that GMO crops could have a net environmental benefit (Marvier et al., 2007). An analysis of 42 field experiments indicated that plants modified to produce Bt toxin reduced the need for large-scale insecticide application, and organisms such as ladybird beetles, earthworms, and bees in locales with Bt crops fared better in field trials than those within locales treated with chemical insecticides. Similarly, a large multiyear analysis of Bt cotton growing areas in China confirmed that the cultivation of this transgenic crop decreased economic losses of caterpillars on other crops; the caterpillar load was decreased over the region (Wu et al., 2008). As more data of this nature are collected, better cost/benefit analyses of transgenic agriculture can be conducted.

Resistance

As with any chemical pesticide treatment, the reality exists that eventually pests will evolve resistance to the treatment, a fact that keeps the chemical industry searching for new insecticides and herbicides to meet the demand of modern agriculture. The same hazard exists for transgenic crops. To address this risk, the same types of management strategies used for chemical resistance management are used for genetically engineered crops. Because transgenic crops have been in widespread use for nearly a decade, data are now being collected on the occurrence of resistance. These data will be valuable in determining the long-term effectiveness of specific GM crops, as well as being helpful in developing new management strategies.

Bt Resistant Insects

Concerns exist that the utility of transgenic Bt crops will be short lived due to pest evolution coinciding with the widespread implementation of transgenic Bt cotton, maize, and potato (Gould, 1998). Although insecticidal crop varieties (mainly Bt) have been widely implemented in the United States, no pest control failures have occurred due to pest resistance to recombinant proteins produced in transgenic plants. This is not to say Bt resistance alleles are not present in insect pest populations. Resistance to Bt has already evolved in the diamondback moth (*Plutella xylostella*, DBM) due to the use not of transgenic plants but of Bt foliar sprays under field conditions. Also, Bt-resistant DBM populations have the potential to survive and cause damage to transgenic plants. Resistance alleles have been detected in bollworms (*Helicoverpa zea*) collected near areas of Bt cotton production in North Carolina (Burd et al., 2003). The management of pest resistance to Bt transgenic plants is the essential next step in allowing the continued use of Bt in pest control.

Understanding the genetic mechanism for Bt resistance in insect populations is vitally important to develop strategies that control the resistance. Dominant resistance alleles would be considerably more problematic than recessive alleles because the insect only needs one copy of a dominant allele to be resistant to the Bt protein. Also, at least half of the offspring of a resistant parent who has a dominant allele will be resistant to the Bt protein, which could allow a rapid population shift to the resistant phenotype. Recessive alleles still must be managed, but the effects of a recessive gene might be slower in a population because individuals must have two copies of the recessive alleles to be resistant to the protein. In both cases, management strategies for the control of resistance alleles in pest populations involve nontransgenic refuges to allow the survival of susceptible individuals.

In the case of DBM, resistance alleles have been shown to be recessive (Shelton et al., 2000). They and others have suggested a Bt management strategy in which transgenic plants must produce a high lethal dose combined with a nontransgenic refuge in order to delay the onset of Bt resistance within insect populations. This strategy effectively dilutes recessive resistance genes by providing an abundance of susceptible individuals as mates for rare homozygous resistant individuals. The offspring of this mating would be heterozygous for the resistance alleles, and would be susceptible to the Bt toxin. In a recent report, Burd et al. (2003) have found evidence that dominant resistance alleles have been found in bollworm populations. Dominant resistance alleles could be an eminent problem to the utility of Bt crops because heterozygous individuals that arise from the mating of resistant and susceptible individuals will be resistant. The control of dominant resistance alleles will require stringent management strategies that have larger refuges to allow the survival of larger populations of susceptible individuals. If not properly managed, the utility of Bt transgenic plants in pest control could be lost, and agricultural systems will have to return to chemical-based sprays to control pest populations.

ROUNDUP RESISTANT WEEDS

A parallel situation of Bt resistance exists for herbicide resistance management. Herbicide resistance refers to the ability of a plant biotype to survive and reproduce under a normally lethal dose of herbicide. Over the past two decades, there had been a dramatic increase in the number of weeds showing herbicide resistance. Among these, resistance to the herbicide glyphosate is currently of greatest concern. The widespread adoption of herbicide-resistant crops such as Roundup-Ready soybean, corn, cotton, and oilseed rape has greatly improved the effectiveness of weed management. However, greater glyphosate usage has played a role in the evolution of glyphosate resistance in weedy species (Dill, 2005).

It is interesting that, while so many scientists were studying intra- and interspecific transgene flow, herbicide overuse was selecting glyphosate-resistant weeds. One example of a glyphosate-resistant weed is the case of horseweed (*Conyza canadensis*, Asteraceae). Even though this has absolutely nothing to do with transgene flow from GM plants to non-GM plants, the evolution of herbicide-resistant biotypes might be accelerated and exacerbated by the habitual applications of a single herbicide and the practice of conservation tillage. Roundup Ready crops enable the use of a single herbicide: Roundup. Can a plant species that is already a problematic weed become herbicide tolerant so that it is no longer killed by spraying herbicide on it?

Millions of acres in the United States are planted every year with herbicide-tolerant soybean, corn, and cotton. There is a tremendous benefit to farmers for growing these plants. They can treat for weeds by applying a single herbicide sprayed over the top of the crop. Weeds are killed, but the crop thrives because of the one or two added genes that render them tolerant to an herbicide such as Roundup (glyphosate). In addition, there is a twofold environmental benefit to this practice as well. The first is that glyphosate is a nontoxic chemical (except to plants) that degrades rapidly in the environment, which is in contrast with less environmentally friendly herbicides. The second, and even larger, benefit to the environment is that farmers can better practice no-till farming. Farmers will commonly remove weeds from fields by tilling the ground. In the absence of GM herbicide-tolerant crops, tillage reduces weed load so that chemical applications are kept at a minimum. The downside of tillage is increase in topsoil erosion. In the past 10 years, the amount of no-till and reduced-tillage agriculture has increased more than threefold to 52 million acres in the United States. In 2002, Roundup Ready soybean varieties were grown on about the same acreage. Erosion has historically been a problem in places like the Mississippi River watershed, which includes western Tennessee. In western Tennessee, reduced-tillage agriculture has been practiced for many years to keep fertile and fragile soils from washing down the Mississippi River to the Delta. Also, in no-till agriculture, there are generally greater weed problems. Conservation tillage increases the importance of certain weeds such as horseweed, also known as mares-tail.

Horseweed has, until recently, been controllable by glyphosate. And, in fact, it was doubted that a glyphosate-resistant broadleaf weed such as horseweed would spontaneously emerge because of the herbicide's mode of action. However, in Delaware (Van Gessel, 2001), and later that same year (2000) in western Tennessee, a few glyphosate-resistant horseweed plants were discovered. By the following year, glyphosate-resistant horseweed was found on tens of thousands of acres. In 2002, upwards of a half a million acres of crops in Tennessee alone contained Roundup-tolerant horseweed. It appears as if a single genetic locus in horseweed allows resistant biotypes to survive Roundup applications, but prior to Roundup Ready crop cultivation, the resistance gene was very rare in the weed populations. However, the use of Roundup year after year was selected for the rare resistance gene, which spread rapidly across populations of horseweed, just as population genetics would predict. Little is known about the molecular biology of glyphosate resistance in horseweed, except that it is a nontarget mechanism (see Yuan et al., 2007). However, unlike the predominant paradigm in Bt resistance management in which the resistance in recessive, herbicide resistance seems to be dominant or semidominant. In our opinion, the risk of herbicide-resistance management, while not unlike the risk conferred by conventional agriculture, is the greatest risk of biotechnology, albeit indirectly. Environmental risks seem to be borne out of using an agricultural biotechnology as a silver bullet, the final solution. Most of the risks that find a high profile in the popular media and promulgated by environmental activists are not great, while Bt and herbicide resistance management do seem to be very important risks that need to be managed hand in glove with the deployment of biotechnology.

THE FOOD SUPPLY

Whenever a new product or agricultural commodity enters the food supply, there is concern as to whether the food is safe for consumers. Products of biotechnology should be no different; however, they are much more heavily scrutinized than their traditional counterparts because of public reservations and confusion over recombinant DNA technology. While these precautions may have added costs to consumers and delayed the availability of some products, ultimately, the public needs to know that regulatory agencies are working to protect them and the food supply.

As such, genetically engineered foods are evaluated for toxicity and allergenicity before approval for consumption.

Toxicity research follows the same design as for testing other food additives or agricultural chemicals. Allergenicity testing poses unique difficulties in that testing traditional foods for new allergens has not been previously feasible and, therefore, a well-establish system does not yet exist. Furthermore, transgenic crops may contain proteins that have not been consumed by humans before, which limit the available allergenicity research tools. The concept of substantial equivalence may be useful in evaluating GM foods. Essentially, it holds that a product that is similar in composition to its processor may be considered to have the same level of benefit and risk. Basically, if a new soybean variety is created, it is still considered a soybean. Transgenic foods will be nearly identical to their processor except for the changes introduced by biotechnology.

ASSESSING TOXICITY

There is, unfortunately, a lack of peer-reviewed published data on GM food safety. There may be several reasons for this, the foremost likely being that the research is expensive and often results in negative data that are difficult to use in obtaining external funding or for publishing. Therefore, much of this research is conducted by private companies that are typically reticent to publish on proprietary products. As such, opponents make the argument that insufficient research has been conducted in this area and that the research conducted lacks accountability. To dispute this claim, more research on the matter needs to be published in peer-reviewed journals, which requires more private company publications, more public research dollars into the effort, and more journals willing to publish this type of data. Due to the current climate toward biotechnology, an environment of openness and accountability must be established to address public concerns.

The research that has been published to date has suggested that GM foods are safe for consumption. Safety evaluations on transgenic plants have focused on several aspects of the crop. For instance, studies have been conducted on the safety of consuming genetically engineered DNA. In one study, mice were fed recombinant GFP DNA for eight generations (Hohlweg and Doerfler, 2001) without deleterious effects on the animals. In general, because transgenic DNA accounts for less than 1/10,000th of the total DNA humans consume on a daily basis, it is difficult to imagine it posing a health risk.

Most of the food safety research has been conducted on the engineered product or on the specific recombinant compound that was added. Assessing the safety of a gene product has followed established toxicological practices. Compounds such as enolpyruylshikimate-3-phosphate synthase (an herbicidal agent based on existing plant enzymes) and Bt (an insecticidal agent used in agriculture for 40 years) have an established safe history of consumption and use, which would not expect to be altered in transgenic plants (Cockburn, 2002). If the transgene product is novel to the food supply or changes the metabolic profile of the plant, then classical toxicity assessment would be conducted, including acute oral and repeat dose studies (Cockburn, 2002).

The research that has generated more interest and controversy is that on novel crops or the derived GM foods themselves. Most of these studies have not indicated any evidence of toxicity upon consumption. Examples include the Flavr Savr tested on tube-feeding mice, Bt corn feed given to broiler chickens, insecticidal peas (expressing α-amylase) fed to rats, and GM potatoes expressing soybean glycinin given to rats (Besler et al., 2001). Probably the most tested GM crop is herbicide tolerant soybeans (Roundup Ready), which have been fed to broiler chickens, catfish, dairy cows, rats, and mice (Besler et al., 2001). In all of these studies, there were no indications of toxicity or compromised health. The most significant criticism has been that detailed histology of the stomach and intestinal tracts had not been conducted and that the conclusions of safety are overstated. However, these studies were an acceptable quality for effective safety assessment research.

The research that has generated the most attention among the studies that have been published is the one that produced a positive result (Ewen and Pusztai, 1999). This article is still a cornerstone of opponents in the GM foods debate. The authors found that feeding mice GM potatoes expressing an insecticidal protein (lectin) compromised the intestinal tract. This effect was not observed in mice fed control potatoes or control potatoes and purified lectin protein. The conclusion was that the genetic modification itself was responsible for the observed toxic effect. The publication became a lightning rod for both praise and criticism. The consensus in the scientific community was that the experiments were not conclusive and were insufficiently designed. Whether or not the conclusions were warranted, condemnation of the paper did not dilute its effect on the public or the debate. The research published by Ewen and Pusztai raises concerns, and more complete experiments should be conducted to either support or dispute their findings.

To date there is little follow-up research on the GM potatoes expressing lectin; however, several experiments on genetically modified foods and ingredients have been conducted and published (Chassy et al., 2004; Flachowsky et al., 2007; Konig et al., 2004; van Eenennaam, 2006). These data indicate no significant, unintended differences between genetically modified and conventional varieties in composition, digestibility, or animal health and performance. These findings support the substantial equivalence hypothesis of genetically modified foods.

Ultimately, there are systems in place that allow for the screening of products for toxicity before they are approved for food supply. The likelihood that a harmful GM product would enter the food supply is remote, and it is certainly less than the chance of other novel food products to cause harm. The question becomes—to what degree of scrutiny should GM foods be subjected to? Currently, testing of GM foods is more rigorous than that of their conventional, irradiated, and mutagenized counterparts. To the authors of this article, that seems to be a sufficient standard.

POTENTIAL ALLERGENICITY

The potential introduction of food allergens into the food supply by biotechnology is perhaps the most significant concern that has developed in public consciousness. This issue is complicated because understanding of the biology of IgE-mediated allergenicity is limited (Ladics et al., 2003). Models to test for and identify food allergens are relatively new and are only now in the process of validation. As such, potential allergenicity has been a focal issue of biotechnology opponents, despite the fact that all novel food products (regardless of origin) are a potential source of new allergens. However, because genetically engineered crops will possess only a handful of modified traits, it is possible to propose models to systematically test those novel proteins. One standard for allergenicity assessment is the decision tree format. The tree relies upon evaluating several parameters of the transgene and its product, which includes the source organism of the gene (whether or not it is a common allergen), sequence homology of the gene to known allergens, resistance to peptic digestion, and studies involving sensitized human serum and skin tests. These data are used to classify a protein's relative risk of allergenicity, and this classification is influenced by the level of expression of the novel trait and the results of animal model testing.

Sequence comparison as an assessment tool relies upon the existence of a database of sequenced food allergens and alignment/comparison programs such as FASTA. Proteins that have a high overall identity to a food allergen (about a 70% match in amino acids) may have similar three-dimensional structures and have potential cross-reactivity to immunoglobulins. While a lack of amino acid similarity does not indicate that a protein is not a potential allergen, it is data that can contribute to an overall assessment strategy and provide direction for future evaluation. This methodology is difficult to validate because it has not been used to identify a novel food allergen as of yet. While the principle seems sound, without a positive control to demonstrate that it can be effective, questions will remain as to its effectiveness. However, sequence comparison is still a valuable assessment tool to characterize novel proteins, as it provides potential targets to direct future research toward.

Stability to peptic digestion is an important characteristic in food allergen identification. Food allergens, relative to other food proteins, are stable to pepsin (Besler et al., 2001). This characteristic is thought to be important to increase the level of allergen exposure to the intestinal immune tissues (Ladics et al., 2003). Therefore, it stands to reason that a novel protein sharing this characteristic would be at an elevated level of allergenicity risk. This parameter should be used in conjunction with other criteria to determine potential exposure, as the assay is not intended to mimic the range of the human digestive process. Another factor contributing to intestinal exposure would be the level of protein expression. For example, if a transgenic protein consisted of 40% of the total protein of a food product, then it would be assumed that a significant amount of that protein would survive digestion regardless of its digestive stability. Conversely, a protein consisting of 0.01% of the total protein would constitute much less potential exposure. The parameters used during simulated gastric digestion, such as pH, can dramatically affect the results. Therefore, it is important that a standard protocol be adopted that maximizes the opportunity to identify stable proteins and allows for data comparison between studies.

Bolstering these analyses would be reliable animal models for allergenicity assessment. Unfortunately, though several are in development, none is as of yet validated as effective for human allergen identification. One strategy is to expose strongly IgE-expressing mice or rats to intrapenitoneal injections of the target protein to characterize its potential to elicit IgE responses (Kimber et al., 2000). Preliminarily, these studies have demonstrated that these models can detect differences between major food allergens and nonallergens in IgE response. The concern about their appropriateness stems from the uncertainty as to how reflective rodent physiology is to humans (Ladics et al., 2003). Nonrodent models have focused on high-IgE-producing dogs and swine. These models would be more reflective of human physiology but are not as well characterized and are considerably more expensive (Ladics et al., 2003; Asero et al., 2007). While animal models are likely to be critical for future analysis of allergenicity, currently they are in the developmental stage, which means that regulatory agencies should remain cautious when approving novel proteins for consumption. Proteins raising red flags in existing assessment protocols should be withheld from the food supply until these new models are ready for application. The research in the field of allergenicity assessment has progressed significantly. As research continues, more will be learned about food allergens and the biology of their effects, which should result in improved detection methodologies.

Starlink Corn

Starlink corn, commercialized by Aventis CropScience in the United States, has been modified through well-recognized genetic techniques to produce the protein Cry9C from the bacteria *Bacillus thuringiensis* due to the proteins known insecticidal properties. Cry9C is a variant of a number of Bt toxins, including the commercially used Cry1A mentioned previously. Starlink corn was registered in 1998 for industrial uses and animal feed with the EPA under the Federal Insecticide, Fungicide, and Rodenticide Act. The EPA concluded, after granting the registration of Starlink, that the Cry9C protein met the safety standard for use in field corn for animal feed based on the toxicology data and limited exposure expected with animal feed use, and that there was reasonable certainty that no harm would result in exposure to the human population. The EPA did not extend the exemption to human food because there was concern that the Cry9C protein is resistant to simulated gastric digestion.

On September 18, 2000, a press conference was held announcing that taco shells purchased from a local grocery store contained trace amounts of GM corn DNA associated with Starlink. This event was convened by Genetically Modified Food Alert, a consortium of seven consumer organizations based out of Washington, D.C., who notified Kraft foods of their findings. A few days later, Kraft Foods announced its voluntary recall of Taco Bell Home Originals taco shells and taco dinners sold nationwide, which resulted in the return of 2.5 million boxes of taco shell products. Following

Kraft's action, a number of other food manufactures issued recalls for products made from corn, resulting in the eventual recall of nearly 300 food products.

An investigation done by the Centers for Disease Control (CDC) established that 28 people who filed adverse event reports after eating a product containing the Cry9C protein had experienced an allergic reaction. However, their study could not confirm a link between Cry9C and the production of detectable amounts of the Cry9C-specific antibody in blood serum from the patients. The CDC stated that, although their results did not provide evidence that the allergic reactions experienced by these people were associated with the Cry9C protein, the possibility could not be completely ruled out, leaving the EPA with the responsibility to decide how to regulate plants containing the Cry9C protein (CDC, 2001). Following this string of events, Aventis stopped the sales of Starlink seed and agreed to purchase Starlink corn in order to isolate it form the U.S. food supply. A few months later, Aventis announced that is was canceling the registration of Starlink corn, resulting in the corn being banned for any agricultural purpose.

The Consumers Union of Japan reported that Starlink protein had been detected in cornmeal, and exports to Japan, the largest foreign market for U.S. corn, dropped by about two-thirds, while South Korea, the second largest consumer of U.S. corn, banned it all together. Many proposals for reform surfaced in light of the Starlink recalls, introducing legislation making it mandatory for the FDA to review GM foods (S3184: The Genetically Engineered Foods Act of 2000) and a promise to reintroduce a bill to establish one agency with primary responsibility for food safety. The issue of food labeling was also raised after this controversy; however, the FDA reaffirmed its decision to not require special labeling of all bioengineered food but proposing guidelines on voluntary labeling.

CONCLUSIONS

While media attention over GM foods has declined over the last few years, unresolved issues remain pertaining to the debate over the application of genetically engineered crops. While it is generally believed that biotechnology is, or will be, a valuable addition to modern agriculture, there is also agreement that products of biotechnology must be evaluated for safety of the environment, animals, and people. This technology has elicited more concern than previous agricultural technologies in the public arena and, as such, they are more heavily scrutinized than other products. For this scrutiny to serve its purpose, that is, increase safety, it must be founded on scientific principles.

Research on the environmental impact of GM crops is steadily increasing. Risk assessment models are being developed based on both previous experience with similar products and on new insights specific to GM plants. High-profile issues, such as the monarch butterfly controversy, are receiving significant scientific attention to provide a reasonable and measured understanding of the actual impact. However, potentially more problematic issues such as invasiveness, hybridization, and resistance are not being ignored. In addition, public concern has spurred the development of a rigorous food safety evaluation paradigm that is applied to genetically engineered plants and the resulting food products. This system is more stringent than previous safety evaluation requirements, and involves not only the study of individual components and proteins but also research on the whole plant or food product.

While these developments are encouraging and reassuring, it is important that this work continue. To do so, more data must be published in peer-reviewed journals, and more funding must be made available to public scientists to conduct this type of research. Sensational headlines may grab the public's attention and, in the short term, affect policy, but it is continued diligence in research that will ultimately persuade the populace and set the policy. Therefore, follow-up research is critical. The scientific community was very effective in evaluating the monarch butterfly issue but not nearly as successful in food safety. Dedicated research and quality data will allow for biotechnology to be safely put into practice and, as such, allow for its acceptance by the public.

REFERENCES

Asero, R., B.K. Ballmer-Weber, K. Beyer, A. Conti, R. Dubakiene, M. Fernandez-Rivas, K. Hoffmann-Sommergruber, J. Lidholm, T. Mustakov, J.N.G. Oude Elberink, R.S.H. Pumphrey, P.S. Stahl, R. van Ree, B.J. Vlieg-Boerstra, R. Hiller, J.O. Hourihane, M. Kowalski, N.G. Papadopoulos, J.M. Wal, C.E.N Mills, and S. Vieths. 2007. IgE-mediated food allergy diagnosis: Current status and new perspectives. *Mol. Nutr. Food Res.* 51: 135–147.

Beckie, H.J., S.I. Warwick, H. Nair, and G. Sequin-Swartz. 2003. Gene flow in commercial fields of herbicide-resistant canola (*Brassica napus*). *Ecol. Appl.*13: (5) 1276–1294.

Bernal C.C., R.M. Aguda, and M.B. Cohen. 2002. Effect of rice lines transformed with *Bacillus thuringiensis* toxin genes on the brown planthopper and its predator *Cyrtorhinus lividipennis*. *Entomol. Exp. Appl.* 102:21–28.

Besler, M., H. Steinhart, and A. Paschke. 2001. Stability of food allergens and allergenicity of processed foods. *J Chromatog. Biomed. Sci. Appl.* 756:207.

Burd, A.D., F. Gould, J.R. Bradley, J.W. Van Duyn, and W.J. Moar. 2003. Estimated frequency of nonrecessive *Bt* resistance genes in bollworm, *Helicoverpa zea* (Boddie) (Lepidoptera: Noctuidae) in eastern North Carolina. *J Econ. Entomol.* 96:137–142.

Centers for Disease Control (CDC). 2001. Investigation of human health effects associated with potential exposure to genetically modified corn: A report to the US Food and Drug Administration form the Centers for Disease Control and Prevention.

Chassy, B., J.J. Hlywka, G.A. Kleter, E.J. Kok, H.A. Kuiper, M. McGloughlin, I. C. Munro, R.H. Phipps, and J.E. Reid. 2004. Nutritional and safety assessments of foods and feeds nutritionally improved through biotechnology: An executive summary. *Compr. Rev. Food Sci. Food Saf.* 3:25–104.

Cockburn, A. 2002. Assuring the safety of genetically modified (GM) foods: The importance of an holistic, integrative approach. *J Biotechnol.* 98:79–106.

Dill, G.M. 2005. Glyphosate-resistant crops: History, status and future. *Pest Manag. Sci.* 61: 219–224.

Dively, G.P., R. Rose, M.K. Sears, R.L. Hellmich II, D.E Stanley-Horn, D.D. Calvin, J.M. Russo, and P.L. Anderson, 2004. Effects on monarch butterfly larvae (Lepidoptera: Danaidae) after continuous exposure to Cry1Ab-expressing corn during anthesis. *Environ. Entomol.* 33:1116–1125.

Duan, J. J., M. Marvier, J. Huesing, G. Dively, and Z. Y. Huang. 2008. A meta-analysis of effects of Bt crops on honey bees (*Hymenoptera: Apidae*). *PLoS One* 3(1): e1415.

Estruch, J.J, G.W Warren, M.A. Mullins, G.J. Nye, J.A. Craig, and M.G. Koziel. 1996. VIP3A, a novel *Bacillus thuringiensis* vegetative insecticidal protein with a wide spectrum of activities against lepidopteran insects. *Proc. Natl. Acad. Sci. USA.* 93:5389–5394.

Ewen, S.W.B., and A. Putztai. 1999. Effects of diets containing genetically modified potatoes expressing *Galanthus nivalis* lectin on rat small intestines. *Lancet* 354:1353–1355.

FAO/WHO. 2001. Evaluation of Allergenicity of genetically modified foods. Report of the joint FAO/WHO expert consultation on allergenicity of foods derived from biotechnology. Food and Agric. Org. of the UN/WHO, Rome.

Flachowsky, G., K. Aulrich, H. Bohme, and I. Halle. 2007. Studies on feeds from genetically modified plants (GMP)—contributions to nutritional and safety assessment. *Anim. Feed Sci. Technol.*133:2–30.

Fu, T.J., U.R. Abbott, and C. Hatzos. 2002. Digestibility of food allergens and non-allergenic proteins in simulated gastric fluid and simulated intestinal fluid—a comparative study. *J. Agric. Food Chem.* 50:7154.

Gould, F. 1998. Sustainability of transgenic insecticidal cultivars: Integrating pest genetics and ecology. *Annu. Rev. Entomol.* 43:701–726.

Hall, L., K. Topinka, J. Huffman, L. Davis, and A. Good. 2000. Pollen flow between herbicide-resistant *Brassica napus* is the cause of multiple-resistant *B. napus* volunteers. *Weed Sci.* 48: 688–694.

Hellmich R.L., B.D. Siegfried, M.K. Sears, D.E. Stanley-Horn, M.J. Daniels, H.R. Mattila, T. Spencer, K.G. Bidne, and L.C. Lewis. 2001. Monarch larvae sensitivity to *Bacillus thuringiensis* purified proteins and pollen. *Proc. Natl. Acad. Sci. USA* 98: 11925–11930.

Helm, R.M., G. Furuta, J.S. Stanley, J. Yui, G. Cockrell, C. Connaughton, G.A. Bannon, and A.W. Burks. 2002. A neonatal swine model for peanut allergy. *J Allergy Clin. Immunol.* 109:135–142.

Hohlweg, U. and W. Doerfler. 2001. On the fate of plant or other foreign genes upon the uptake in food or after intramuscular injection in mice. *Mol. Gen. Genet.* 265:225–233.

Hoyle, M., K. Hayter, and J.E. Cresswell. 2007. Effect of pollinator abundance on self-fertilization and gene flow: Application to GM canola. *Ecol. Appl.* 17: 2123–2135.

Kaplinsky, N., D. Braun, D. Lisch, A. Hay, S. Hake, and M. Freeling. 2002. Maize transgene results in Mexico are artifacts. *Nature* 416:601.

Kimber, I., K.T. Atherton, J.G. Kenna, and R.J. Dearman. 2000. Predictive methods for food allergenicity: Perspectives and current status. *Toxicology* 147:147–150.

Konig, A., A. Cockburn, R.W.R. Crevel, E. Debruyne, R. Grafstroem, U. Hammerling, I. Kimber, I. Knudsen, H.A. Kuiper, A.A.C.M Peijnenburg, A.H. Penninks, M. Poulsen, M. Schauzu, and J.M. Wal. 2004. Assessment of the safety of foods derived from genetically modified (GM) crops. *Food Chem. Toxicol.* 42:1047–88.

Ladics, G.S., M.P. Holsapple, J.D. Astwood, I. Kimber, L.M.J. Knippels, R.M. Helm, and W. Dong. 2003. Approaches to the assessment of the allergenic potential of food from genetically modified crops. In *Proceedings of the 41st Annual Meeting of the Society of Toxicology*, Nashville, TN.

Langhof, M., B. Hommel, A. Husken, J. Schiemann, P. Wehling, R. Wilhem, and G. Ruhl. 2008. Coexistence in maize: Do non-maize buffer zones reduce gene flow between maize fields? *Crop Sci.* 48:305–316.

Lee, M.K., F.S. Walters, H. Hart, N. Palekar, and J.S. Chen. 2003. The mode of action of the *Bacillus thuringiensis* vegetative insecticidal protein Vip3A differs from that of Cry1Ab δ-endotoxin. *Appl. Environ. Microbiol.* 69(8):4648–4657.

Liu, Z.C., G.Y. Ye, and C. Hu. 2002. Effects of Bt transgenic rice on population dynamics of main non-target insect pests and dominant spider species in rice paddies. *Acta Phytophylacica Sin.* 29:138–44.

Losey, J.O., L. Rainer, and M. Carter. 1999. Transgenic pollen harms monarch larvae. *Nature* 399:214.

Marvier, M., C. McCreedy, J. Regetz, and P. Kareiva. 2007. A meta-analysis of effects of Bt cotton and maize on non-target invertebrates. *Science* 316:1475–1477.

Mercer, K.L. and J.D. Wainwright. 2008. Gene flow from transgenic maize to landraces in Mexico: An analysis. *Agric. Ecosyst. Environ.* 123:109–115.

Metz, M., and J. Fütterer. 2002. Suspect evidence of transgenic contamination. *Nature* 416: 600–601.

Ortiz-García, S., E. Ezcurra, B. Schoel, F. Acevedo, J. Soberon, and A.A Snow. 2005. Absence of detectable transgenes in local landraces of maize in Oaxaca, Mexico. 2003–2004. *Proc. Natl. Acad. Sci. USA* 102, 12338–12343.

Pleasants, J.M., R.L. Hellmich, G.P. Dively, M.K. Sears, D.E. Stanley-Horn, H.R. Mattila, J.E. Foster, P. Clark, and G.D. Jones. 2001. Corn pollen deposition on milkweeds in and near cornfields. *Proc. Natl. Acad. Sci. USA* 98: 11919–11924.

Quist, D. and I.H. Chapela. 2001. Transgenic DNA introgressed into traditional maize landraces in Oaxaca, Mexico. *Nature* 414:541–543.

Reed, G.L., A.S. Jensen, J. Riebe , G. Head, and J.J. Duan. 2001. Transgenic Bt potato and conventional insecticides for Colorado potato beetle management: Comparative efficacy and non-target impacts. *Entomol. Exp. Appl.* 100:89–100.

Riddick, E.W. and P. Barbosa. 2000. Cry3Aintoxicated *Leptinotarsa decemlineata* (Say) are palatable prey for *Lebia grandis* Hentz. *J. Entomol. Sci.* 35:342–46.

Schuler T.H., I. Denholm, L. Jouanin, S.J. Clark, A.J. Clark, and G.M. Poppy. 2001. Population-scale laboratory studies of the effect of transgenic plants on nontarget insects. *Mol. Ecol.* 10:1845–53.

Sears, M.K., R.L. Hellmich, D.E. Stanley-Horn, K.S. Oberhauser, J.M. Pleasants, H.R. Pleasants, H.R. Mattila, B.D. Siegfried, and G.P. Dively. 2001. Impact of *Bt* corn pollen on monarch butterfly populations: A risk assessment. *Proc. Natl. Acad. Sci. USA* 98: 11937–11942.

Serratos-Hernandez, J.A., J.L Gomez-Olivares, N. Salinas-Arreortua, E. Buendía-Rodríguez, F. Islas-Gutierrez, and A. de-Ita. 2007. Transgenic proteins in maize in the soil conservation area of Federal District, Mexico. *Front. Ecol. Environ.* 5: 247–252.

Shelton, A.M., J.D. Tang, R.T. Roush, T.D. Metz, and E.D. Earle. 2000. Field tests on managing resistance to *Bt*-engineered plants. *Nat. Biotechnol.* 18:339–342.

Stewart, C.N., Jr., M.D. Halfhill, and S.I. Warwick. 2003. Transgene introgression from genetically modified crops to their wild relatives. *Nat. Rev. Genet.* 4: 806–817.

Stewart, C.N., Jr. and M. Littmann. 2008. Media treatment of plant biotechnology. http://agribiotech.info/moreIssues.htm (invited and peer-reviewed agricultural biotechnology white paper) (March 8, 2008).

Van Eenennaam, A. 2006. Genetic engineering and animal agriculture. Univ. Calif. Agric. *Nat. Resour. Agric. Biotechnol. Calif. Ser.*, Publ. 8184.

Van Gessel, M.J. 2001. Glyphosate-resistant horseweed from Delaware. *Weed Sci.* 49:703–705.

Warwick, S.I., A. Legere, M.J. Simard, and T. James. 2008. Do escaped transgenes persist in nature? The case of an herbicide resistance transgene in a weedy *Brassica rapa* population. *Mol. Ecol.*, 17: 1387–1395.

Warwick, S.I., M.J. Simard, A. Légère, H.J. Beckie, L. Braun, B. Zhu, P. Mason, G. Seguin-Swartz, and C.N. Stewart Jr. 2003. Hybridization between transgenic *Brassica napus* L. and its wild relatives: *B. rapa* L., *Raphanus raphanistrum* L., *Sinapis arvensis* L., and *Erucastrum gallicum* (Willd.) O.E. Schulz. *Theor. Appl. Genet.* 107(3): 528–39.

Wu, K.M., Y.H. Lu, H.Q. Feng, Y.Y. Jiang, and J.Z. Zhao. 2008. Suppression of cotton bollworm in multiple crops in China in areas with Bt toxin-containing cotton. *Science* 321(5896):1676–1678.

Yuan, J.S., P.J. Tranel, and C.N. Stewart Jr. 2007. Non-target site herbicide resistance: A family business. *Trends Plant Sci.* 12:6–13.

38 Cryopreservation of Plant Cells, Tissues, and Organs

Barbara M. Reed, M.N. Normah, and Svetlana V. Kushnarenko

CONCEPTS

- Cold tolerance and dehydration tolerance can be induced in many types of plants, and this improves their recovery from cryopreservation.
- Meristematic cells and cells with few or small vacuoles are most suitable for cryopreservation.
- Osmotic dehydration, air drying, and cryoprotection decrease the water content of cells and allow vitrification of the cytoplasm.
- Cryoprotectants may remain outside the cell and cause osmotic dehydration.
- Cryoprotectants may enter the cell and add to the cell solution, thus reducing the chance of crystallization.
- Culture techniques are very important in successful cryopreservation protocols.
- Seeds can be desiccation and/or cold tolerant, or desiccation and/or cold sensitive.
- Embryos/embryonic axes of desiccation and cold sensitive seeds can be cryopreserved using the desiccation technique.

Cryopreservation, storing biological materials at liquid nitrogen temperatures (–196°C), and once a dream for the future, is now a viable option for most types of plant materials (Reed, 2008). The progression of methods development required about 30 years to make cryopreservation a viable storage technique for a wide range of plant species and for organized tissues. The earliest studies of plant cryopreservation were in the 1960s. Sakai (1960) showed that cold-acclimated dormant buds could be slowly cooled to low temperatures, immersed in liquid nitrogen, rewarmed, and retain viability. Quatrano (1968) published the first account of suspension-cell culture cryopreservation, and this led to the cryopreservation of many additional cell types over the next 15 years. Cryopreservation of actively growing shoot tips developed during the late 1970s with the use of controlled rate cooling (Kartha et al., 1979; Sakai et al., 1978). Several methods were developed in the early 1990s that allowed for rapid cooling of shoot tips, cells, or callus cultures (Dereuddre et al., 1990; Langis and Steponkus, 1990; Sakai et al., 1990). These techniques expanded the number of species and the types of plants that could be cryopreserved. Cryopreservation is now used for storing laboratory specimens, germplasm collections, stock culture collections, and rare or endangered species (Reed, 2008).

TISSUE CULTURE ASPECTS

Many of the plant materials stored by cryopreservation come from tissue culture systems. It is vitally important that the culture system produce healthy plants for the process. Cultures in less than optimum condition have lower regrowth than those in good condition. Small plant cells with few vacuoles are the easiest to cryopreserve. Most cryopreservation protocols involve preconditioning stages in culture. These may include growth on highly osmotic media, cold acclimation, or extended culture without transfer. Sequential transfer to medium with increasing sucrose concentrations, cold acclimation, growth on abscisic acid, or glycerol are used to increase osmotic tolerance when plants are desiccation sensitive. Most of these techniques improve cryopreservation tolerance by adding solutes to the cytosol or developing desiccation tolerance in the cells. Recovery following cryopreservation is also dependent on in vitro culture systems in most cases. The optimal growth room temperature and lighting as well as the proper recovery medium can greatly affect regrowth.

FREEZING IN NATURE

The success of a cryopreservation protocol is based mainly on cellular water content. In order to successfully cryopreserve a plant, the water in the cell must not form ice. Ice formation and the resulting crystals cause physical damage to the cell membranes and organelles. In nature, cooling of plants does not result in ice formation until several degrees below zero, and many temperate plants can avoid freezing down to −15°C. At some point, water molecules are nucleated or seeded, and coalesce to form ice crystals. Nucleation may be caused by a physical movement or by the presence of dust or bacteria. Many temperate woody plants can cool to about −40°C (supercooling) before homogeneous (spontaneous) ice nucleation occurs. This temperature of homogeneous ice nucleation determines the tree line on mountains.

If we consider a plant cell as a watery solution in a bag (cell membrane), in a box (cell wall), then we can determine ways to remove water from the cell without damaging the cell structures. Two ways to accomplish this are osmosis and cryoprotection. These factors are both at work during natural cold acclimation and freezing avoidance in plants. Natural cryoprotectants in plants are sugars, proteins, and amino acids. As plants cold acclimate with cooler temperatures and shorter day lengths, the cells naturally increase the amount of these cryoprotectant chemicals in the cell solution. As the temperature decreases below zero, ice forms in the intercellular spaces, producing an osmotic gradient, and water moves out of the cells and into the growing ice mass. As this occurs, the cytoplasm becomes more concentrated and resistant to freezing. For temperate zone trees, this resistance can remain until −40°C, when homogeneous nucleation occurs. For many plant types, cold temperatures are damaging even without ice, and for others the dehydration of the cytoplasm can cause lethal damage. Cryopreservation techniques aim to overcome these damaging effects.

Orthodox seeds (desiccation and freeze tolerant) can be stored under conventional gene bank conditions for centuries, at 3%–7% seed water content and −20°C; however, cryopreservation is the only available option for long-term storage of nonorthodox or recalcitrant seeds (intolerant to desiccation and freezing). Desiccation of excised embryos and embryonic axes is considered one of the simplest techniques. This procedure has been used for successful cryopreservation of zygotic embryos and embryonic axes excised from seeds of many nonorthodox plant taxa such as rubber, coffee, olive, almond, tea, and *Citrus* (Normah and Makeen, 2008).

GENERAL STEPS IN CRYOPRESERVATION

All cryopreservation techniques are based on a similar premise: condition the plant cells and dehydrate them to the point that they will remain alive, but the cytoplasm will vitrify (become a glass) on contact with liquid nitrogen. The physiological condition of the plant is the first concern. Ideally, the plant is naturally adapted to cold or dry conditions. In reality, that is not the case for most in vitro grown plants or cell cultures. The first step is pretreatment. This involves growing the plant under conditions that improve resistance to cold or dry conditions. Cultures can be cold-acclimated with long cold nights and short warm days, grown for a long time without transfer, or cultured on osmotic media. The pretreatment method chosen should be suitable for the type of plant in question. A temperate plant will usually respond to cold acclimation, while a tropical plant might die from such treatment. A plant conditioned to grow in moist environments would require a more gradual increase in osmotic medium than a naturally desiccation tolerant plant.

Once the plant is pretreated, the shoot tips can be further conditioned by preculture for a few days on specialized media. Standard preculture for many controlled-rate cooling protocols includes medium with 5% dimethyl sulfoxide (DMSO) for 2 days. Vitrification protocols may include high molarity sucrose preculture.

Exposure to cryoprotectant solutions varies with the technique. Controlled rate cooling cryoprotectants are relatively mild, and exposure is often 30 min to as long as 2 h. Vitrification solutions are highly osmotic, and exposure is usually carefully timed to avoid overexposure that results in damage or death of explants. Once the optimum cryoprotectant exposure is achieved, slow controlled cooling is initiated, or vials are plunged in liquid nitrogen in the case of vitrification protocols.

Rewarming is a critical procedure in all of the techniques. Rapid rewarming of the vials without overheating the plant materials requires exposure to hot water (~40°C) for 1–2 min followed by cool water (~25°C) to protect from overheating. Movement of vials from the liquid nitrogen to the hot water should be immediate, or the vitrified solutions may crystallize rather than liquefying and death will result.

Removal of the cryoprotectant and reculture of the plant material is the next important step. For dedifferentiated cultures, decanting and not rinsing may produce better regrowth than rinsing. For shoot cultures complete rinsing is important. Culture explants under low or no light for the first week, followed by standard growth conditions favors regrowth. The recovery medium is often the standard multiplication medium; however, in some cases, auxin in the medium can cause callus formation rather than shoot tip growth and should be avoided.

CRYOPRESERVATION TECHNIQUES

There are three main types of cryopreservation techniques, controlled cooling, vitrification, and desiccation, and some involve combinations of these (Table 38.1). The first type of cryopreservation developed for organized tissues involved controlled rate cooling. This technique is applied mostly to cell cultures and to organized tissues of temperate plants, both dormant buds and shoot tips of in vitro grown plantlets.

In controlled rate cooling, the samples are preconditioned, cryoprotected, and then slowly cooled to the freezing point of the cryoprotectant where ice is initiated (about –9°C). The ice nucleation initiates ice in the intercellular spaces and begins the osmotic dehydration of the cytoplasm. Then they are slowly cooled to –35°C or –40°C before plunging in liquid nitrogen (Figure 38.1). The speed of

TABLE 38.1
Commonly Used Cryopreservation Techniques

Technique	Procedure
Controlled rate cooling (slow cooling, two-step cooling)	Dormant wood collected from trees during the coldest winter months are given additional chilling, dehydrated gradually by slow cooling, and stored in liquid nitrogen vapor. Thawed buds are grafted or budded onto rootstocks for recovery. Shoot tips, callus, or cell cultures are pretreated in cryoprotectants and frozen at <1°C/ min to –40°C, and then plunged into liquid nitrogen.
Solution-based vitrification (rapid cooling)	Several protocols and cryoprotectants are available. After a brief loading phase and a brief cryoprotectant pretreatment, the vials or foils are directly plunged into liquid nitrogen.
Encapsulation-dehydration (vitrification)	Explants are encapsulated in alginate and dehydrated osmotically for about 20 h and then dried in airflow or over silica gel before direct immersion in liquid nitrogen.
Encapsulation vitrification	Encapsulated explants are immersed in vitrification solutions and directly plunged in liquid nitrogen.
Desiccation	Orthodox seeds are dried over silica gel until they retain 3%–7% seed water content. Recalcitrant seed embryonic axes are dried to 7%–20% water (fresh weight basis) before direct immersion in liquid nitrogen.

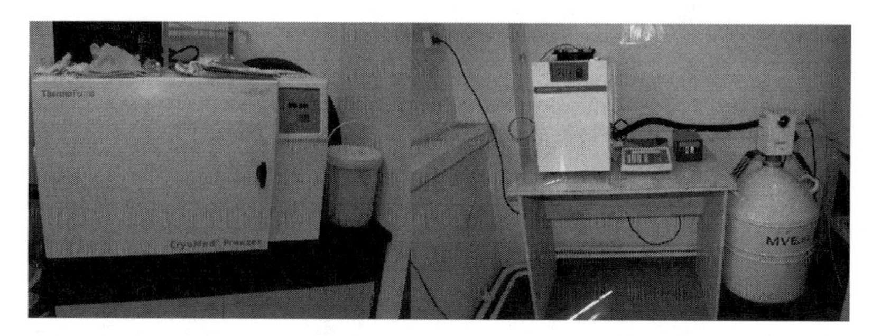

FIGURE 38.1 Cryogenic freezers used for controlled rate of cooling of cell cultures and shoot tips.

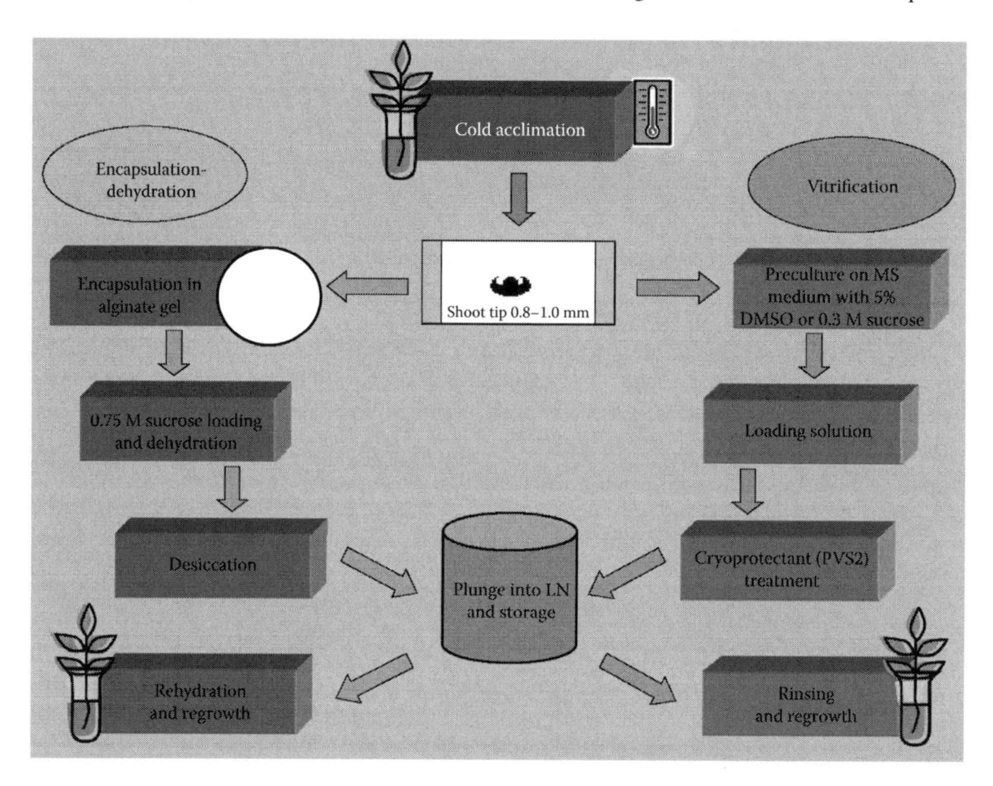

FIGURE 38.2 Diagram of procedure flow for the encapsulation-dehydration and vitrification protocols.

cooling determines the amount of time that the dehydration can continue. If the rate is too fast, then freezable water will remain in the cell and crystallize at low temperatures. If the rate is too slow, then the cell might be lethally dehydrated. At the terminal transfer temperature (–35°C to −40°C), the plunge into liquid nitrogen causes the cytoplasm to vitrify (turn to a glass) and avoid ice crystal damage. Rewarming must be fast so that the cytoplasm liquefies rather than crystallizing. Critical steps of the procedure include preculture or pretreatment, cryoprotectant exposure, and cooling rate.

Vitrification techniques involve the transition of cellular liquids to amorphous glass without crystal formation. Water can transition from liquid to solid (ice) or liquid to an amorphous glass (vitrified). A vitrified solution is in an amorphous state that has the physical properties of a liquid. This amorphous state can easily revert to a liquid or to ice crystals if the conditions change (during slow warming, for example). Glass is an excellent freeze avoidance system since a glass does not increase in volume like ice. Vitrified samples must be quickly rewarmed, or ice formation will occur. There are two main techniques that use this principle, solution-based and desiccation-based vitrification (Figure 38.2).

TABLE 38.2
Cryoprotectants Commonly Used in Plant Cryopreservation in Combination with the Normal Growth Medium

Cryoprotectant	Composition	Sample Type	Technique
DGS	1 M DMSO 1 M Glycerol 2 M Sucrose	Dedifferentiated cultures	Controlled rate cooling
DGP	1 M DMSO 1 M Glycerol 2 M Proline	Dedifferentiated cultures	Controlled rate cooling
DGSP	1 M DMSO 1 M Glycerol 1 M Sucrose 1 M Proline	Dedifferentiated cultures	Controlled rate cooling
DGlu	2.5 M DMSO 1.1 M D-Glucose	Dedifferentiated cultures	Controlled rate cooling
DMSO	5 or 10 % DMSO	Embryogenic cultures of conifers, algal cultures, mosses	Controlled rate cooling
PGD	10% PEG 10% Glucose 10% DMSO	Cell cultures, somatic embryos, shoot tips	Controlled rate cooling
DMSO-sucrose	10% DMSO 0.6 M sucrose	Embryogenic cultures	Controlled rate cooling
Sucrose	0.75M sucrose	Cell cultures, somatic embryos, shoot tips, algal cultures, ferns, and mosses	Encapsulation-dehydration
PVS2	30% glycerol (w/v) 15% ethylene glycol (w/v) 15% DMSO (w/v) in medium containing 0.4 M sucrose (pH 5.8)	Cell cultures, somatic embryos, shoot tips	Vitrification and encapsulation-vitrification
PVS3	40% glycerol (w/v) 40% sucrose (w/v)	Shoot tips	Vitrification

Solution-based vitrification protocols use loading solutions of sucrose and glycerol to add solutes to the cell and to osmotically dehydrate the cell, followed by highly viscous cryoprotective solutions that will vitrify as the samples are plunged into liquid nitrogen. The main solution used is Plant Vitrification Solution #2 (PVS2) (Sakai et al., 2008; Sakai et al., 1990). In any vitrification technique, the preculture of plants, loading solution, time and temperature of the cryoprotectant solution, and rewarming procedure are all critical to success.

Encapsulation-dehydration is a vitrification technique that employs sucrose loading and dehydration to remove water from the cells, and the cytoplasm forms a glassy state (vitrifies) on exposure to liquid nitrogen. In this technique, in vitro plantlets are pretreated by cold acclimation or sucrose preculture, and then shoot tips are removed and encapsulated in alginate beads. The beads are cultured overnight in a 0.75M sucrose solution to add sucrose to the cells, and then they are blotted dry and allowed to desiccate under laminar flow or with silica gel until they have about 20% moisture content. At this point the cells will vitrify on contact with liquid nitrogen (Dereuddre et al., 1990; Engelmann et al., 2008).

Cryopreservation of seeds is accomplished using desiccation methods. This normally involves direct drying of orthodox seeds over silica gel, or in the case of recalcitrant seeds, the embryonic axis is removed and dried under laminar flow, over silica gel, or flash dried (Normah and Makeen, 2008; Walters et al., 2008).

CRYOPROTECTION

All forms of cryopreservation require some type of cryoprotection. In the case of dormant buds from field-grown trees, the natural cold acclimation and climatic conditions provide the cryoprotection. Orthodox seeds are naturally dry and require only some additional desiccation for optimum cryopreservation. In nondormant plants, acclimation can still play a role; however, supplementary cryoprotection is also necessary. A wide range of cryoprotectant solutions are available for use (Table 38.2). Cryoprotectants can be penetrating and add to the osmolality of the cell or nonpenetrating and aid in osmotic dehydration of the cell. Penetrating cryoprotectants such as glycerol, sugars, and DMSO are useful for increasing solutes in the cell and in the case of DMSO for conditioning membranes. Nonpenetrating cryoprotectants serve to osmotically dehydrate the cells. Vitrification solutions are highly viscous and cause rapid dehydration of cells, so their use can be toxic if timing and temperature of application are not carefully controlled.

Cryopreservation techniques are now well developed and can be used for both temperate and tropical plant materials with only modest modifications. The choice of technique is dependent on the resources of the laboratory, the type of plant material, and the characteristics of the plant.

REFERENCES

Dereuddre, J., C. Scottez, Y. Arnaud, and M. Duron. 1990. Resistance of alginate-coated axillary shoot tips of pear tree (*Pyrus communis* L. cv. Beurre Hardy) in vitro plantlets to dehydration and subsequent freezing in liquid nitrogen. *C. R. Acad. Sci. Paris* 310:317–323.

Engelmann, F., M. T. Gonzalez Arnao, Y. Wu, and R. H. Escobar. 2008. Development of encapsulation dehydration. In Reed, B. (Ed.), *Plant Cryopreservation: A Practical Guide*, Springer, New York.

Kartha, K. K., N. L. Leung, and O. L. Gamborg. 1979. Freeze-preservation of pea meristems in liquid nitrogen and subsequent plant regeneration. *Plant Sci. Lett.* 15:7–15.

Langis, R. and P. Steponkus. 1990. Cryopreservation of rye protoplasts by vitrification. *Plant Physiol.* 92:666–671.

Normah, M. N. and A. M. Makeen. 2008. Cryopreservation of Excised embryos and embryonic axes. In Reed, B. (Ed.), *Plant Cryopreservation: A Practical Guide*, Springer Science + Business Media LLC, New York, pp. 211–240.

Quatrano, R. S. 1968. Freeze preservation of cultured flax cells utilizing dimethylsulfoxide. *Plant Physiol.* 43:2057–2061.

Reed, B. M. 2008. *Plant Cryopreservation: A Practical Guide*. Reed, B. M. (Ed.). Springer Science + Business Media LLC, New York, p. 513.

Sakai, A. 1960. Survival of the twig of woody plants. *Nature* 185:393–394.

Sakai, A., D. Hirai, and T. Niino. 2008. Development of PVS-based vitrification and encapsulation-vitrification protocols. In Reed, B. (Ed.), *Plant Cryopreservation: A Practical Guide*, Springer, New York, pp. 33–58.

Sakai, A., S. Kobayashi, and I. Oiyama. 1990. Cryopreservation of nucellar cells of navel orange (*Citrus sinensis* Osb. var. *brasiliensis* Tanaka) by vitrification. *Plant Cell Rep*. 9:30–33.

Sakai, A., M. Yamakawa, D. Sakata, T. Harada, and T. Yakuwa. 1978. Development of a whole plant from an excised strawberry runner apex frozen to -196 C. *Low Temp. Sci. Ser*. 36:31–38.

Walters, C., J. Wesley-Smith, J. Crane, L. Hill, P. Chmielarz, N. W. Pammenter, and P. Berjak. 2008. Cryopreservation of recalcitrant (i.e. desiccation-sensitive) seeds. In Reed, B. (Ed.), *Plant Cryopreservation: A Practical Guide*, Springer Science + Business Media LLC, New York, pp. 465–484.

39 Cryopreservation of In Vitro Grown Shoot Tips

Svetlana V. Kushnarenko, Barbara M. Reed, and M.N. Normah

Shoot tip cryopreservation methods are well developed for many crops (Reed, 2008). These techniques especially work well for vegetatively propagated plants that cannot be stored as seeds (fruits, berries, and some vegetables). Techniques for storing shoot apices are more labor intensive than protocols for seeds. In vitro plantlets contain large amounts of water, and for successful cryopreservation that content must be reduced. A first step for reducing water in plant cells of in vitro plantlets is cold acclimation (CA). This is useful for temperate plants. Plantlets are usually cultured at low temperature in darkness or under a short photoperiod for 2 to 4 weeks. Constant low temperatures of 5°C with an 8 h photoperiod (Wu et al., 1999) or 4°C in darkness (Paul et al., 2000) were reported for acclimating apple shoots. A comparison of various CA temperature regimes for in vitro pear shoots showed that alternating temperatures (22°/–1°C) significantly improved the cold hardiness and regrowth of cryopreserved shoot tips compared to CA at constant low temperatures (Chang and Reed, 2000). Preculture of shoot tips on highly osmotic media can often substitute for the cold acclimation step. In addition, pretreatments may include media with cryoprotectants added.

Cryopreservation of shoot tips can be accomplished by any of the main methods mentioned in Chapter 38. The vitrification-based methods require only tissue culture laboratory facilities and little special equipment, so they are easy to perform in most laboratories. Both techniques involve a direct plunge into liquid nitrogen and a rapid rewarming of the samples. The glass formed by the plunge changes back into the liquid form without the formation of damaging ice crystals.

The following experiments illustrate two popular techniques that are usually successful with a wide range of plant materials. In vitro grown shoot tips of a wide range of plants can be cryopreserved with these methods. Temperate plants can be cold-acclimated and pretreated on osmotic medium, while tropical or cold sensitive plants are only exposed to the osmotic medium. Desiccation-based protocols are highly successful with drought-tolerant plants such as grasses, but are also successful for most other plants as well.

The following experiments require aseptic conditions and sterile tools. Pay special attention to following the steps as written. Small changes in the protocols can reduce or eliminate survival of the shoot tips. Practice in dissection of shoot tips is important. The experiments should be performed only after students can dissect shoots and have 100% survival of the dissected and in vitro cultured shoot tips.

EXPERIMENT 1: CRYOPRESERVATION OF SHOOT TIPS BY VITRIFICATION

This procedure demonstrates one of the methods of shoot tip cryopreservation—vitrification with preculture on Murashige and Skoog (1962) medium with 0.3 M sucrose. We suggest that all students work with the same cultivar, but the experiment includes two durations of cryoprotectant pretreatment (for 2 teams of students). Use temperate plantlets with easily accessed shoot tips (apple, pear, and currant work well).

Materials

The following items are needed for each student or team of students (*to be prepared in advance by a laboratory assistant*):

- In vitro plantlets grown for 2 weeks under cold acclimation conditions (22°C 8 h days and −1°C 16 h nights) prior to cryopreservation
- Stereomicroscope
- Sterile pieces of moist paper for shoot tip dissection (approximately 6 × 6 cm pieces of paper towels or filter papers autoclaved in a glass petri dish with water)
- One petri dish (7 cm diameter) of MS medium with 0.3 M sucrose for preculture
- Liquid MS medium with 0.8 M sucrose at pH 5.8 for preparing the cryoprotectant
- Liquid MS medium with 1.2 M sucrose at pH 5.8 for rinsing
- One sterile petri dish with sterile dry filter paper for drying shoot tips
- 2 M glycerol in MS medium with 0.4 M sucrose
- Sterile Millipore filter for cryoprotectant sterilization
- Sterile 100-mL beaker
- Sterile petri dish
- Sterile 1-2 mL pipette
- Sterile 5 mL syringe, forceps, and scalpel
- Sterilizer for forceps and scalpel
- 50 mL graduated cylinder with base
- MS medium with plant growth regulators (PGRs) for recovery in petri dish or 24-cell plates
- Cryovial holder frozen in a block of ice
- Block of ice for cryoprotectant cooling
- Wide-neck Dewar vessel (1-2 L capacity) with liquid nitrogen (LN)
- Four sterile cryovials for two durations of cryoprotectant treatment
- Two aluminum canes for cryovials
- Two plastic 1-L beakers for water bath and a thermometer
- Cryovial holder
- Parafilm
- Marker
- Protective equipment (gloves, apron, and goggles)

Note: The experiment will take two days. All procedures will be carried out in the laminar flow hood. Follow the instructions outlined in Procedure 39.1 to complete this exercise.

Anticipated Results

Regrowth of shoots following LN will be apparent 1–2 weeks after rewarming (Figure 39.1). You can use a magnifying glass or dissecting scope to distinguish between living and dead shoot tips. The surviving shoot tips usually retain green color but damaged shoot tips became brown. After 3–4 weeks you can observe the formation of new green shoots and can count the percentage of regrowth. The control shoot tips usually grow faster than LN treated shoot tips. The percentage of recovered shoot tips depends on cryoprotectant pretreatment duration. In our laboratory, 80 min of PVS2 pretreatment is better than 20 min.

Questions

- What types of damage might occur with too little or too much PVS2 exposure?
- Which cryoprotectant treatment time was more effective for post-cryopreservation regrowth? For the control regrowth?
- Would you expect to see ice formation in tissues or solutions with the vitrification method?

Procedure 39.1
Cryopreservation by Solution-Based Vitrification

Step	Instruction and Comments
1	1st day: Dissection and preculture of shoot tips. Dissect 50 shoot tips (0.8–1 mm long) from in vitro grown cold-hardened plantlets and place on MS medium with 0.3 M sucrose in a petri dish. For dissection you could use sterile syringe or sterile scalpel with narrow edge blade, and sterile forceps. Dissect on a moist sterile filter paper with a stereomicroscope.
2	Place petri dish with shoot tips in cold-hardening conditions for two days.
3	3rd day: Pretreatment, cooling, and rewarming procedure: Make the cryoprotectant (PVS2) for 50 mL in graduated cylinder. (a) Weigh empty 50 mL graduated cylinder. (b) In this cylinder, weigh 15 g glycerol (30%, density 1.2613) (this solution is very viscose; therefore, it is better to weigh it). (c) Add 6. 8 mL ethylene glycol (15%). (d) Add 6.8 mL DMSO (15%). (e) Bring to volume 50 mL with liquid MS medium containing 0.8 M sucrose. (f) Cover the neck of cylinder with parafilm tightly, and carefully mix the solution. *Important safety tip*: Wear gloves for this procedure, and wash them before touching other material or telephone, etc.
4	Filter-sterilize cryoprotectant through sterile filter into sterile 100 mL beaker, and place in refrigerator at 4°C for 30 min. Wash out the cylinder immediately yourself; do not leave it for others. In each experiment, fresh cryoprotectant should be used; do not store cryoprotectant.
5	Label four cryovials (1 and 3 for control; 2 and 4 for cryopreserved shoot tips), and place in a cryovial holder.
6	Unscrew lids, and to each cryovial add 1 mL of 2 M glycerol in MS medium with 0.4 M sucrose using a sterile pipette. To keep the pipette sterile, place sterile pipette tip first in a sterile petri dish with the lid covering much of the pipette. Then add shoot tips (5 shoot tips in cryovial 1 for the 20 min PVS2 control, 15 shoot tips in cryovial 2 for PVS2 plus liquid nitrogen, 5 shoot tips in cryovial 3 for the 80 min PVS2 control; 15 shoot tips in cryovial 4 for PVS2 plus liquid nitrogen). Retain 10 untreated shoot tips for the dissection control. Screw on the lids and hold cryovials for 20 min at 25°C. Shoot tips should be picked up very carefully with a sterile syringe or knife blade tip.
7	Cool the beaker of cryoprotectant for 30 min, and then place in hood on the block of ice.
8	Remove 2 M glycerol from the cryovials using sterile pipette. A small amount of the solution can remain in the cryovials.
9	Transfer the cryovials to the holder frozen in a block of ice.
10	Add chilled cryoprotectant PVS2 in each cryovial. *Wear gloves.* The vial should be filled to more than 1 mL. Close lids. *Start timer.* Time for cryoprotectant pretreatment: for cryovials 1 and 2, 20 min at 0°C; for cryovials 3 and 4, 80 min at 0°C.
11	After 10 min, mix the cryoprotectant in cryovials 1 and 2, and bring the fluid level to the 1 mL mark using a sterile pipette.
12	Fill small Dewar with LN.
13	After 20 min in PVS2, put experimental cryovial 2 (with 15 shoot tips) on a cane and submerge in LN. Leave it there for 5–10 min or more as schedules permit. Wear protective gloves, apron, and goggles.
14	Immediately add 1.2 M sucrose liquid medium up to the top of cryovial 1 (with 5 PVS2 control shoot tips).
15	Immediately pipette out solution from the cryovial, and replace with 1.2 M sucrose liquid medium two times.
16	Transfer rinsed shoot tips onto sterile dry filter paper in petri dish using pipette, and then place shoot tips on recovery medium using a sterile syringe or tip of scalpel blade. Place five of the untreated shoot tips onto the recovery medium at this time as an untreated control.
17	Transfer the cane with cryovial 2 from LN to plastic beaker with 45°C water, and stir for 1 min.
18	Move the cane to 25°C water for 1 min with stirring.
19	Dry the outside of the cryovial, and move into the laminar hood.
20	Repeat steps 14–16 for cryovial 2.
21	After 40 min, mix cryoprotectant in 3 and 4 cryovials, and bring volume to the 1 mL mark on the tubes using sterile pipette.

(Continued)

Procedure 39.1 (Continued)
Cryopreservation by Solution-Based Vitrification

Step	Instruction and Comments
22	After 80 min in PVS2, put experimental cryovial 4 (with 15 shoot tips) on a cane, and submerge in LN for 5–10 min or more as schedules permit. Wear protecting gloves, apron, and goggles.
23	Repeat steps 14–20 for cryovials 3 and 4.
24	Label petri dishes with date of freezing procedure, plant genotype, and name of student. Cover the petri dish with parafilm to prevent contamination or drying of the medium.
25	Record the data on the data sheet (Table 39.1) weekly for 6 weeks.

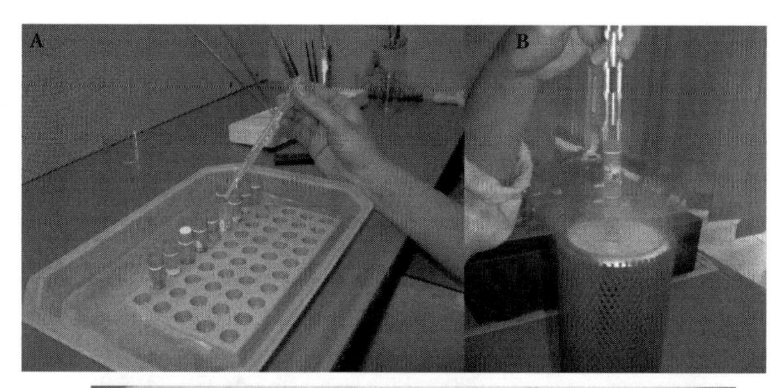

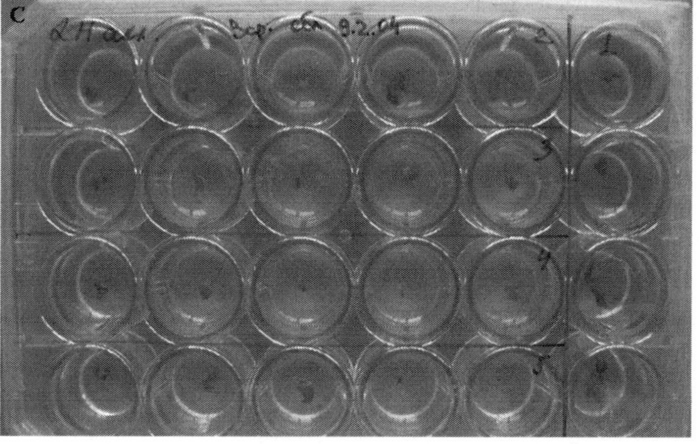

FIGURE 39.1 A. Cryovials held in ice block. B. Plunging cane into the Dewar. C. Regrowth of shoots in a 24-well plate following each of five steps of the vitrification protocol. 1—Unfrozen control. 2—After preculture on MS medium with 0.3 M sucrose. 3—After 2 M glycerol. 4—After PVS2 treatment. 5—After LN.

EXPERIMENT 39.2: CRYOPRESERVATION OF SHOOT TIPS BY ENCAPSULATION-DEHYDRATION

This procedure demonstrates the desiccation-based method of shoot tip cryopreservation: encapsulation-dehydration (Figure 39.2). We suggest that one-half of the students or team of students work with one cultivar, while the other half work with another cultivar. Use temperate plantlets with easily accessed shoot tips (apple, pear, and currant work well). This exercise will require 2 consecutive days.

Materials

The following items are needed for each student or team of students (*to be prepared in advance by a laboratory assistant*):

- In vitro plantlets grown for 2 weeks under cold-hardening conditions (22°C 8 h days and −1°C 16 h nights) prior to cryopreservation
- Stereomicroscope
- Sterile pieces of moist paper for shoot tip dissection (approximately 6 × 6 cm pieces of paper towel or filter paper autoclaved in petri dish with water)
- One petri dish of regular MS medium without PGRs for holding shoot tips temporarily
- Liquid MS medium with 0.75 M sucrose for pretreatment (75 mL in 125 mL flask)
- Liquid MS medium for rehydrating the beads after rewarming (10–25 mL)
- Liquid MS medium without calcium and with 3% low-viscosity alginate and 0.75 M sucrose (in a flask). This is very difficult to dissolve, so heat medium and add alginate slowly, boil to dissolve
- Liquid MS with 100 mM calcium chloride medium for forming beads (in a flask)
- Two sterile 250-mL beakers for forming beads in calcium chloride and for draining beads
- One sterile 100-mL beaker or small petri dish for alginate solution
- Two sterile transfer pipettes for forming beads and rehydration of beads
- Two sterile sieves or tea strainers for removing beads from solution
- One sterile petri dish for holding beads during dehydration
- One sterile petri dish with sterile filter paper for draining beads
- One small sterile petri dish for bead rehydration
- Sterile dry filter paper in sterile petri dish for draining beads
- Standard MS medium with PGRs for recovery in petri dish or 24-cell plates
- One sterile 5-mL syringe, forceps, and scalpel

TABLE 39.1
Sample Data Sheet for Vitrification[a]

Sample data sheet

Plant (Accession)____________ Cryo date ____________

Meristem date ____________ Technique____________

Pretreatment____________ Student (or Student team)____________

Vial No.	Treatment	No. Shoot Tips	Number of Surviving Shoot Tips/Week 1	2	3	4	5	6	%Regrowth at 6 wks	Notes
	Control: no PVS2 no liquid nitrogen	10								
1	20 min PVS2 no liquid nitrogen	5								
2	20 min PVS2 liquid nitrogen	15								
3	80 min PVS2 no liquid nitrogen	5								
4	80 min PVS2 liquid nitrogen	15								

[a] Note the number of new shoots growing, any callus production, contaminated shoot tips, and calculate the percentage of regrowth.

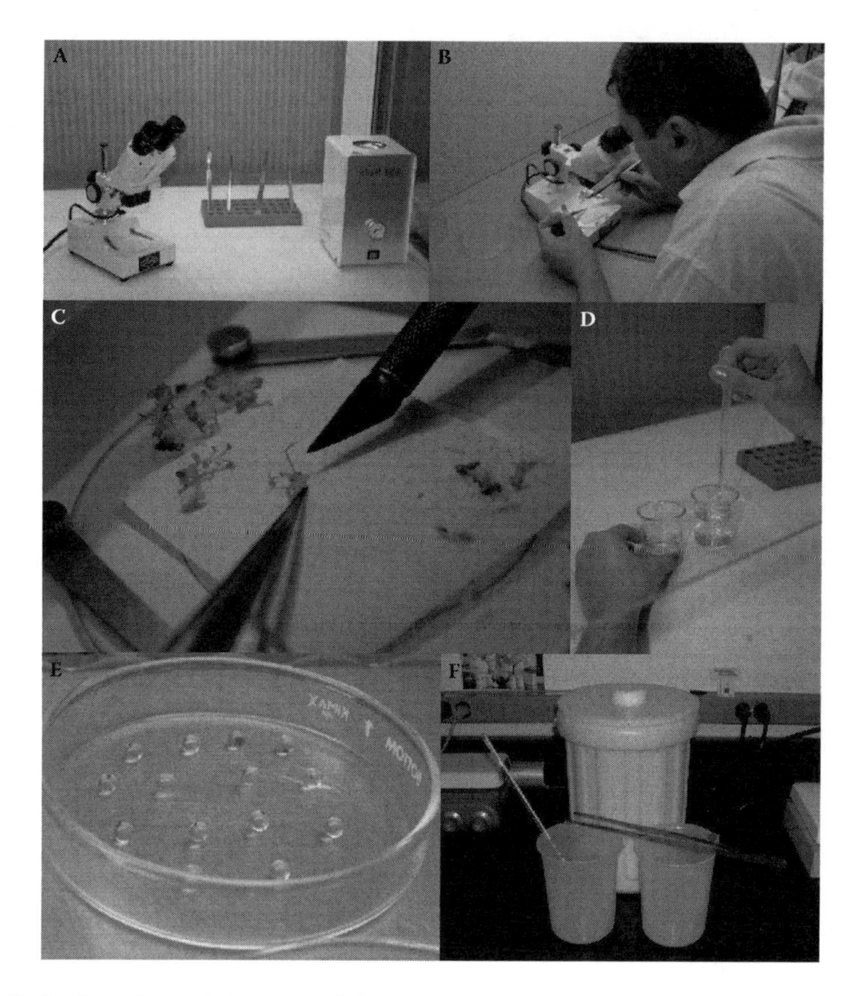

FIGURE 39.2 Procedures of the encapsulation-dehydration protocol. A. Laminar flow hood set up for dissection. B. Dissecting shoot tips. C. Shoot tip ready for dissection. D. Making beads by dripping alginate with shoot tips into calcium chloride solution. E. Beads drying on open petri dish. F. Dewar and equipment for rewarming vials.

- Sterilizer for forceps and scalpel
- Wide-neck Dewar (1–2-L capacity) with liquid nitrogen (LN)
- Two sterile cryovials
- One aluminum can for cryovials
- Marker
- Protective clothes (gloves, apron, and goggles)

(*Note:* The experiment will take 2 days. All procedures will be carried out in laminar flow hood.)

Follow the instructions outlined in Procedure 39.2 to complete this exercise.

Anticipated Results

Regrowth of shoots following rewarming takes 1–2 weeks. You can use a dissecting scope to distinguish between living and dead shoot tips. The surviving shoot tips usually retain green color, but damaged shoot tips became brown. After 3–4 weeks, you can observe formation of green shoots and can count regrowth percentage. The control shoot tips usually grow faster than shoot

Procedure 39.2
Cryopreservation by Encapsulation Dehydration

Step	Instruction and Comments
1	1st day: Dissection of shoot tips and alginate bead formation. Dissect 40 shoot tips (0.8–1.0 mm long) from in vitro grown cold-hardened plantlets, and place onto petri dish of MS medium without PGRs until enough are collected. For dissection, use a sterile syringe or sterile scalpel with narrow edge blade, and sterile forceps.
2	Place the solutions in the laminar flow hood: alginate, calcium chloride and MS medium with 0.75 M sucrose. All these solutions should be at room temperature.
3	Add 25 mL alginate solution to a small sterile beaker or petri dish. Place 100 mL calcium chloride solution in a sterile 250 mL beaker.
4	Transfer dissected shoot tips into the alginate solution using a sterile syringe or scalpel. Carefully add 35 shoot tips to the alginate solution without introducing any air bubbles. The remaining 5 shoot tips will be dissection controls and should be plated directly on the recovery medium.
5	Using a sterile transfer pipette, collect single shoot tips (try to avoid bubbles), and drop them into the 250 mL beaker of liquid MS with 100 mM calcium chloride to make beads. Hold the pipette vertically to form uniform alginate beads.
6	Leave the beads in the solution for 20 min to firm up.
7	Drain off the solution using a sterile sieve or tea strainer, and place the beads onto a sterile petri dish. Include 30 empty beads for moisture determination (step 15).
8	Using a sterile forceps transfer beads to a 125 mL flask of liquid MS medium with 0.75 M sucrose, and place this flask on a shaker for 18 h (for example, in at 3 PM, out at 9 AM).
9	2d day: Dehydration, cooling, and rewarming procedure. Using a sterile sieve, collect beads, drain them, and briefly place on sterile dry filter paper in petri dish to absorb excess moisture.
10	Set aside five beads as a sucrose-treated control and plate on recovery medium. Place remaining beads in a sterile petri dish so they do not touch each other. Put the open petri dish near the laminar filter. There should be no extra moisture in the dish (absorb with sterile filter paper if necessary). Dry them in the open petri dish under the laminar air flow for 3 h (room temperature 23°C–25°C), then remove half of the beads and follow steps 11–13. Dry the remaining beads for 4 h, and follow steps 11–13.
11	Place five dry beads (dried control) in a small sterile petri dish, and add liquid MS medium to cover beads. Rehydrate for 5 min, and then place beads on recovery medium to grow (shoot tip facing down).
12	Place 10 dry beads in a cryovial, put on cane, and submerge in liquid nitrogen for 5–10 min or more as schedule permits. Wear protective gloves, apron, and goggles.
13	Take cane with cryovials from LN, warm cryovials in 40°C water for 2 min, and then add liquid MS medium to the tubes to rehydrate for 5 min and place beads on recovery medium.
14	Label petri dish using marker: date of freezing procedure, plant genotype, and name of student. Cover the petri dish with parafilm to prevent contamination and drying of the medium.
15	Moisture content determination Weigh three sets of five empty beads for each drying time in aluminum weigh boats. Place the boats in a 103°C oven for 16 or 24 h. Cool the beads in a desiccator for a minimum of 5 min and reweigh. Moisture content (%) can be calculated on dry-weight basis.
16	Record data using the format shown in Table 39.2.

tips following LN. The control shoot tips plated before encapsulation, after sucrose pretreatment, and after drying but without liquid nitrogen will provide an indication of any injury caused by dissection or dehydration.

Questions

- What is the function of the pretreatment in 0.75 M sucrose?
- Why are the alginate beads dried in the laminar air flow? What could be used as an alternative drying method?
- Which drying time was more effective for cryopreservation regrowth? For the control regrowth without liquid nitrogen?

TABLE 39.2
Sample Data Sheet for Encapsulation Dehydration[a]

Plant (Accession) ____________ Experiment date ____________
Meristem date ____________ Pretreatment ____________
Technique ____________ Student (or Student team)____________

Treatment	No. Dissected Shoot Tips	Number of Growing Shoot Tips Weeks						% Regrowth	Notes
		1	2	3	4	5	6		
Control: Encapsulated, no sucrose, no drying, no liquid nitrogen	5								
Control: Encapsulated, sucrose treated, no drying, no liquid nitrogen	5								
Control: Encapsulated, sucrose treated, dried 3 h, no liquid nitrogen	5								
Dried 3 h and liquid nitrogen	10								
Control: Encapsulated, sucrose treated, dried 4 h, no liquid nitrogen	5								
Dried 4 h and liquid nitrogen	10								

[a] Note the number of new shoots growing, any callus production, contaminated shoot tips, and calculate the percentage of regrowth. The dissection control shoots should be monitored for regrowth. Eighty to 100% regrowth is required for the experiment to be valid.

- Why is it not necessary to use chemical cryoprotectants (such as PVS2) with this procedure?
- Would you expect to see ice formation in tissues when using the encapsulation-dehydration method?

REFERENCES

Chang, Y. and B. M. Reed. 2000. Extended alternating-temperature cold acclimation and culture duration improve pear shoot cryopreservation. *Cryobiology* 40:311–322.

Murashige, T. and F. Skoog. 1962. A revised medium for rapid growth and bio assays with tobacco tissue cultures. *Physiol Plant.* 15:473–497.

Paul, H., G. Daigny, and B. Sangwan-Norreel. 2000. Cryopreservation of apple (*Malus x domestica* Borkh.) shoot tips following encapsulation-dehydration or encapsulation-vitrification. *Plant Cell. Rep.* 19:768–774.

Reed, B. M. 2008. *Plant Cryopreservation: A Practical Guide*. Reed, B. M. (Ed.). Springer Science + Business Media LLC, New York, p. 513.

Wu, Y., F. Engelmann, Y. Zhao, M. Zhou, and S. Chen. 1999. Cryopreservation of apple shoot tips: Importance of cryopreservation technique and of conditioning of donor plants. *CryoLetters* 20:121–130.

40 Cryopreservation of Orthodox and Recalcitrant Seed

M.N. Normah, W.K. Choo, Svetlana V. Kushnarenko, and Barbara M. Reed

Conservation of plant germplasm is very important in maintaining biodiversity, especially of endangered species or valuable crop cultivars. Seed is the most preferred plant propagule for germplasm conservation due to low storage cost, ease of handling, and regeneration of whole plants from genetically diverse materials (Chin, 1994; Pritchard, 1995). While orthodox seeds (desiccation and freeze tolerant) are tolerant to storage under conventional gene bank conditions for as long as centuries, with 3%–7% seed water content at −20°C (FAO/IPGRI, 1994), cryopreservation is the only available option for long-term storage of nonorthodox or recalcitrant seeds (intolerant to desiccation and freezing).

Cryopreservation uses an ultra low temperature (mostly in liquid nitrogen at −196°C or above liquid nitrogen at −160°C) to minimize cellular metabolism of the preserved seeds or seed parts so that they will resume normal life processes when warmed to room temperature.

The most damaging event during cryopreservation is irreversible injury caused by the formation of intracellular ice crystals. For this reason, modern techniques in cryopreservation emphasize tolerance to dehydration and a high cooling rate. The purpose of these procedures is to reduce or prevent intercellular ice formation when the explants are exposed to liquid nitrogen. This process is straightforward for orthodox seeds with naturally low water contents and cold tolerance because they only require minimal drying and a direct plunge in liquid nitrogen. Whole seed storage of a nonorthodox species must take into account desiccation and freezing sensitivity as well as relatively large seed sizes. Excised embryos and embryonic axes of recalcitrant species can be stored and retrieved in vitro as an alternative to storing whole seeds.

Desiccation of excised embryos and embryonic axes is considered one of the simplest techniques, with dehydration either in the airflow of a laminar flow cabinet, over silica gel or with a flow of sterile compressed air prior to cryoexposure (Normah and Makeen, 2008; Walters et al., 2008). This procedure is used for cryopreservation of zygotic embryos and embryonic axes excised from seeds of many nonorthodox plant taxa. Optimum whole plant recovery following cryopreservation is generally attained at water contents between 7% and 20% on a fresh weight basis (equivalent to 0.08–0.25 g $H_2O.g^{-1}$dw). This usually requires desiccation in laminar airflow for 2 to 4 h. However, embryos and embryonic axes excised from seeds of several recalcitrant species such as tea, chayote, and almond can survive cryogenic exposure after dehydration down to far lower water contents (7%–15%; 0.08–0.18 g $H_2O.g^{-1}$dw).

Moisture content of the excised embryos and embryonic axes is the most critical factor for a successful cryopreservation protocol. The desiccation conditions, such as the rate of desiccation (Pammenter and Berjak, 1999), should also be optimized to recover vigorous plantlets after cryopreservation. The influence of seed tissue dehydration rate under slow, fast, and ultra-rapid drying conditions on desiccation tolerance has been well documented (Makeen et al., 2005). Fast and ultra-rapid desiccation conditions are also successful because they minimize exposure of the tissues to deleterious degradative processes (Pritchard and Manger, 1998).

In the following exercises, we will demonstrate orthodox and recalcitrant seed cryopreservation. We will also observe the effect of moisture content and rate of desiccation (under laminar air flow and over silica gel) on viability of axes from recalcitrant seeds. Direct plunging into liquid nitrogen will be the cryopreservation method in both exercises.

EXPERIMENT 40.1 CRYOPRESERVATION OF ORTHODOX SEEDS

Grass, wheat, barley, or vegetable seeds (radish or cabbage) without dormancy requirements are best for this exercise. Some examples of seeds and germination are shown in Figures 40.1–40.3. The aim is to compare germination and vigor of seeds before and after exposure to LN.

MATERIALS

The following materials will be needed for each student or team of students:

- Protective clothes (gloves, apron, and goggles)
- 120 air-dried seeds
- Wide-neck Dewar (1-2 l capacity) with LN
- Three cryovials or aluminum foil packets
- Racks for cryotubes or long-handle forceps
- Two 100 mL beakers with two pieces (15 × 15 cm) of gauze
- Detergent liquid (soap) for surface sterilizing seeds
- 150 mL 15% H_2O_2 for surface-sterilizing seeds
- Six sterile petri dishes (9 mm diameter) with filter paper for germination of seeds
- Forceps
- Marker
- 200 mL sterile water for petri dishes

Follow Procedure 40.1 outlined above to complete this exercise.

ANTICIPATED RESULTS

The germination percentage of cryopreserved orthodox seeds does not decrease as compared with unfrozen control. Sometimes, germination of seeds declines with storage and cryopreservation can actually improve germination (Figure 40.2B).

QUESTIONS

- Why do we use dry seeds, not imbibed seeds, for cryopreservation?
- Why can we warm the cryovials slowly at room temperature rather than quickly in warm water?

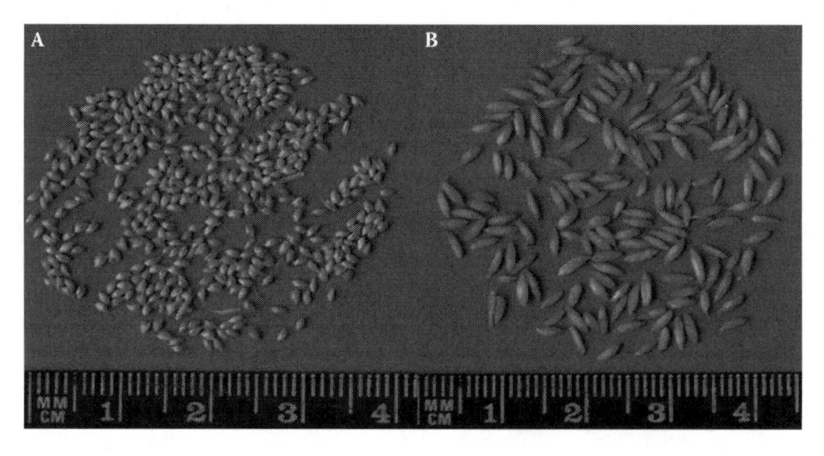

FIGURE 40.1 Air-dried seeds of (A) *Phleum pretensem* and (B) *Arrhenatherum elatius*.

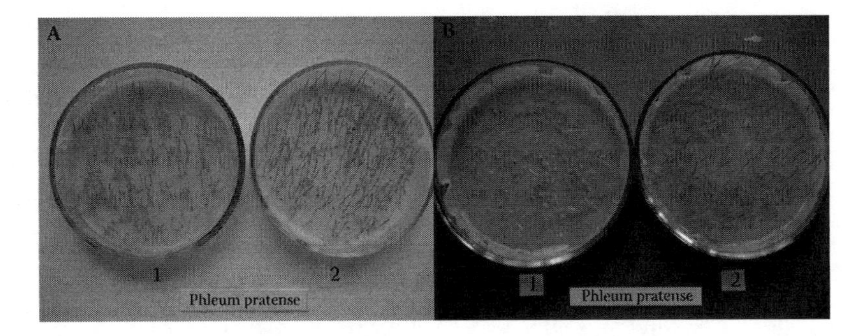

FIGURE 40.2 Germination of *Phleum pratense* seeds after 7 days. A—Seeds collected in 2004; B—seeds collected in 2002. 1—Control seeds; 2—following 1 month in LN nitrogen.

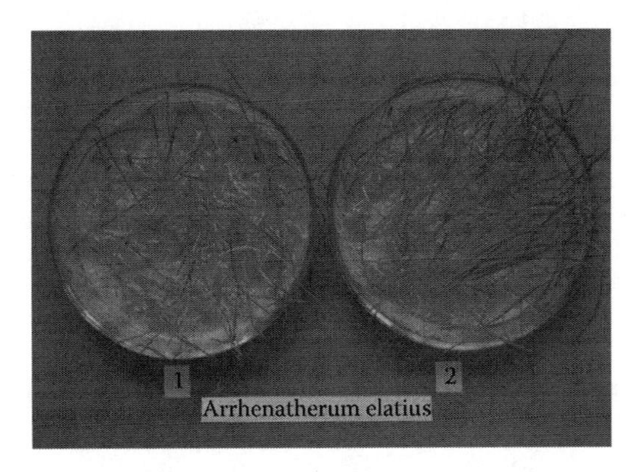

FIGURE 40.3 Germination of *Arrhenatherum elatius* seeds for 10 days. 1—Control seeds; 2—following 1 month in LN nitrogen.

Procedure 40.1
Cryopreservation of Orthodox Seeds

Step	Instructions and Comments
1	Place 60 air-dried seeds in three cryovials or in foil packets (20 seeds per cryovial).
2	Place cryovials in cryocane and submerge in LN for 5 min.
3	Rewarm cryovials with seeds at room temperature for 1 h.
4	Place 60 control seeds in 100 mL beaker, and fix gauze on the top with a rubber band.
5	Rinse seeds in tap water with 1–2 drops of detergent, and rinse with tap water for 20 min.
6	Soak in 15% H_2O_2 for 15 min, and then rinse with sterile water.
7	Decant water, and place seeds in three sterile petri dishes on sterile filter paper (20 seeds per dish).
8	Add some sterile water to each petri dish. Note: Seeds must always be moist, but water should not cover seeds.
9	Take notes on lids of petri dishes: Indicate control seeds, date of the experiment, and name of student.
10	Place petri dishes with control seeds in germinator or incubator at 30°C with 8 h day length.
11	After rewarming the 60 frozen seeds, repeat steps 4–8.
12	Take notes on lids of petri dishes: Indicate frozen seeds, date of the experiment, and name of student.
13	Place petri dishes with frozen seeds in the incubator.
14	Record germination after a 7-day incubation period.

- What would you do differently if you were working with seeds that have dormancy requirements?
- Did the germination percentage change after cryopreservation? What might cause an increase or decrease in germination?
- What might be the reason for the differences in germination of the seeds in Figure 40.2?

EXPERIMENT 40.2 CRYOPRESERVATION OF EMBRYONIC AXES OF *CITRUS AURANTIFOLIA** BY AIR AND SILICA GEL DESICCATION

The following objectives of the experiment are to provide

1. Practical aspects of cryopreservation of embryonic axes by desiccation-freezing technique using two methods of desiccation
2. Information for satisfactory assessment of the recovery of axes from cryoexposure.

The experiments outlined in this chapter are designed for second or third year undergraduate students as a practical class of 3–4 h. We suggest that the students work in groups of four; each pair will work on different desiccation methods.

MATERIALS

The following items are required for each group of students:

- Seeds freshly separated from fresh fruits (remove the seeds from the freshly harvested fruits, and then wash them in running tap water for 1 h; *to be prepared in advance by a laboratory assistant*)
- Sterile glass petri dishes (diameter 7 cm and 9 cm) with filter papers and sterile tissue papers (autoclaved)
- Sterile 250 mL beakers and 500 mL bottle (for seeds' surface sterilization)
- Sterile propylene cryovials, cryocanes, and cryovials rack
- Sterile distilled water
- Tools (scalpels, sterile blades, and forceps)
- 95% (v/v) ethanol for flaming the tools
- 200 mL 80% (v/v) ethanol and Tween 20 (for seeds sterilization)
- 200 mL 20% (v/v) commercial bleach (Clorox), freshly prepared before use
- Silica gel: wrap batches of 15 g silica gel in aluminum foil bags, already autoclaved. To prepare the silica gel, place the foil bags inside an autoclavable plastic container and autoclave at 121°C for 20 min. After that, place silica gel packages inside the oven set at 103° C for 24 h or until use. *This should be done ahead of time by a laboratory assistant.*
- Aluminum foil boats (about 4 × 4 cm) for water content determination
- Murashige and Skoog (MS: 1962) medium with 0.5 mgL^{-1} (2.2 μM) BA in petri dishes
- Stereomicroscope
- Liquid nitrogen (located adjacent to the water bath)
- Water bath preset at 40°C
- An oven preset at 103°C
- A desiccator to cool down the dried embryonic axes (for moisture content determination)

* Other Citrus species can be used for this experiment. The survival percentage and the moisture content for best survival will be different from the anticipated result of *C. aurantifolia* below. A preliminary test by the instructors would help in determining the hours of desiccation necessary for significant results to be obtained by the students.

Each group will be divided into two pairs of students: one pair will complete the laminar flow desiccation, and the other silica gel desiccation.

Each team of students should get 135 seeds. There are 45 seeds for moisture test, 45 for liquid nitrogen (LN) treatment, and 45 for the control (Figure 40.4).

Follow the protocol outline in Procedure 40.2 to complete this experiment.

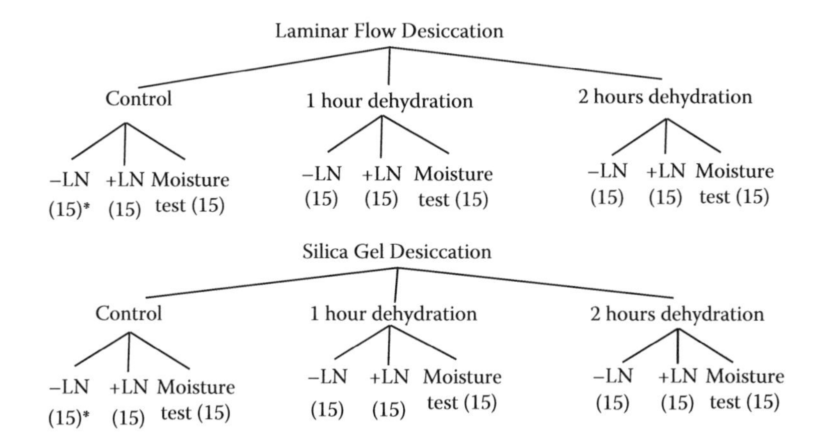

FIGURE 40.4 Number of seeds/axes required for each step of the experiment.

Procedure 40.2
Cryopreservation of Recalcitrant Seed Axes

Step	Instructions and Comments
	Sample preparation
	Surface sterilization of whole seeds, excision, desiccation, and freezing of embryonic axes
1	In the laminar flow hood, place the whole seeds inside the bottle, pour in 200 mL 80% ethanol, and shake gently for 2 min. Decant and add 200 mL of 20% commercial bleach, with 2 drops of Tween 20 for 20 min with shaking every 5 min.
2	Remove the disinfectant, and rinse the seeds with sterile distilled water 3 times until the foam made by Tween 20 disappears.
3	Keep the sterilized seeds in petri dishes.
4	Under the stereomicroscope, place a few seeds in a glass petri dish with a filter paper, and, with the aid of a pair of forceps and a scalpel, excise the biggest embryonic axes, leaving a small block of the cotyledon attached to the axes (Figure 40.4). Place the excised embryonic axes (in a row of 10 axes) inside a petri dish with a filter paper moistened with few drops of sterile distilled water. Close the petri dish. Hold in the petri dish until enough axes are excised for desiccation.
	Desiccation under laminar airflow
1	Transfer the axes to glass petri dishes containing dry filter papers and desiccate 0, 1, and 2 h (can go to 3 h with more seeds and time).
2	When enough for one treatment (45 axes needed), desiccate the longest period of desiccation first (2 h). Out of 45 axes, after desiccation, 15 axes are needed for moisture test (see moisture content determination below), 15 axes for LN treatment, and 15 axes for control (refer to Figure 40.5). Always label petri dishes and cryovials with desiccation period.
3	After the desiccation period, place embryonic axes inside cryovials, mount the cryovials in the cryocanes, and directly plunge them into the LN tank for 1 h (minimum). Rapidly thaw the cryovials in a 37°C water bath for 1–2 min, and then culture the axes on the prepared medium.
4	Culture the controls without LN treatment.

(Continued)

Procedure 40.2 (Continued)
Cryopreservation of Recalcitrant Seed Axes

Step	Instructions and Comments
5	Label each petri dish with name, desiccation period, LN or without LN, and date.
6	Seal all petri dishes with parafilm, and keep the cultures in the culture room (culture conditions: 25 ± 1°C under 16 h light/8 h dark photoperiod with light intensity of 25 µmol $m^{-2}s^{-1}$).
	Desiccation over silica gel
1	Place the batches of axes inside the glass petri dish (7 cm diameter) on filter paper placed on the silica gel. (Figure 40.6)
2	Seal each petri dish with parafilm, and leave for the prescribed desiccation period of 0, 1, and 2 h. Thereafter, follow the procedure as previously described. Determine the water content after each desiccation period.
	Moisture content determination
1	Weigh 5 embryonic axes replicated 3 times for each drying time in aluminum weigh boats.
2	Place the boats in a 103°C oven for 16 or 24 h.
3	Cool the axes in a desiccator for a minimum of 5 min, and reweigh.
4	Moisture content (%) can be calculated on fresh weight or dry weight basis.

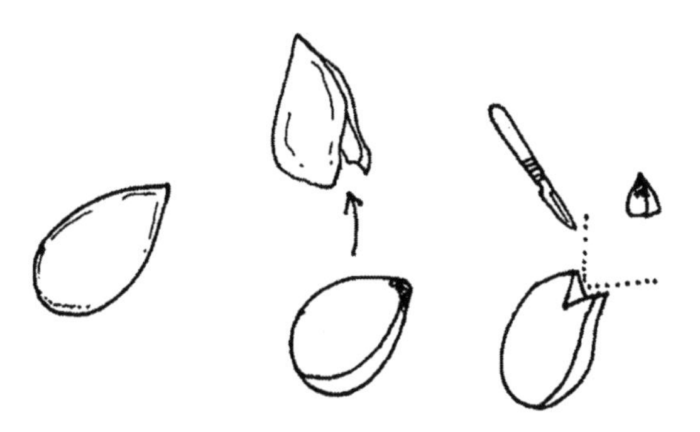

FIGURE 40.5 Excision of embryonic axes of *C. aurantifolia*.

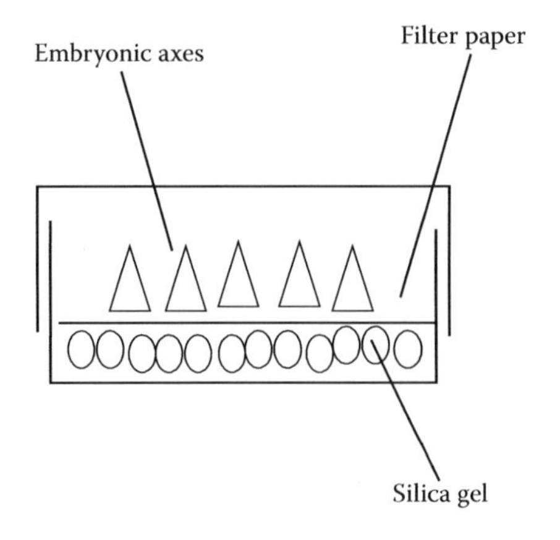

FIGURE 40.6 Desiccation of embryonic axes over silica gel in a petri dish.

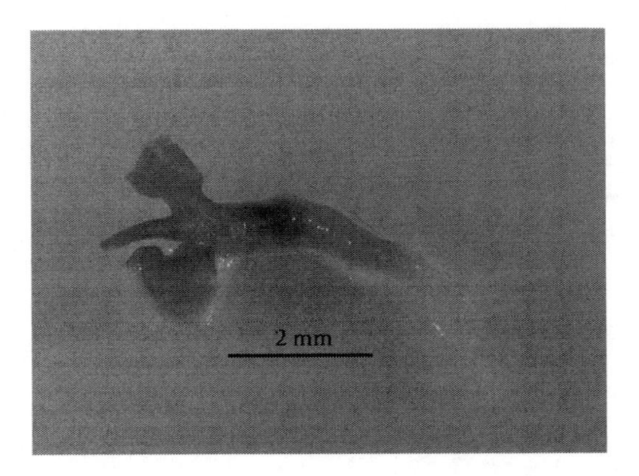

FIGURE 40.7 Growth of embryonic axes after cryopreservation (1 week in culture after rewarming).

Observations

Observe the embryonic axes at 1, 3, 7, 10, and 14 days. Look for any change of color of axes and also for any growth and differentiation. Data at the end of the observation period will be expressed as a percentage of explants that are viable (axes turn green) and regrowth percentage (axes germinated into normal seedlings; Figure 40.7). Data from each pair of students (different methods of desiccation) in a group should be combined and compared. Discuss results of the two methods with respect to moisture content, rate of desiccation, and survival/viability.

Anticipated Results

The percentage of normal seedlings recovered from the frozen axes is 70%–90% obtained at moisture content of 7.8% and 11.5% (= 0.095 and 0.139 g $H_2O.g^{-1}dw$) for laminar flow desiccation and a relatively lower survival (50%–65%) at a similar moisture content for silica gel desiccation.

Questions

- How does rate of desiccation affect survival of embryonic axes after cryopreservation?
- Why is it important to desiccate the embryonic axes before cryopreservation?
- Explain why excised embryos/axes are considered a good choice of explants for cryopreservation of recalcitrant seeds.

REFERENCES

Chin, H. F. 1994. Seed banks: Conservation of the past for the future. *Seed Sci. Tech.* 22:385–400.

FAO/IPGRI. 1994. Genebank standards Food and Agriculture Organization/International Plant Genetic Resources Institute, Rome.

Makeen, A. M., M. N. Normah, S. Dussert, and M. M. Clyde. 2005. Cryopreservation of whole seeds and excised embryonic axes of *Citrus suhuiensis* cv. limau langkat in accordance to their desiccation sensitivity. *CryoLetters* 26:259–268.

Murashige, T., and F. Skoog. 1962. A revised medium for rapid growth and bio assays with tobacco tissue cultures. *Physiol Plant.* 15:473–497.

Normah, M. N., and A. M. Makeen. 2008. Cryopreservation of excised embryos and embryonic axes. In Reed, B. (Ed.), *Plant Cryopreservation: A Practical Guide*, Springer Science + Business Media LLC, New York, pp. 211–240.

Pammenter, N. W., and P. Berjak. 1999. A review of recalcitrant seed physiology in relation to desiccation-tolerance mechanisms. *Seed Sci. Res.* 9:13–37.

Pritchard, H. 1995. Cryopreservation of seeds. In J.G., D., and M. M.R. (Eds.), *Methods in Molecular Biology: Cryopreservation and Freeze-Drying Protocols*, Humana Press, New Jersey, pp. 133–144.
Pritchard, H., and K. R. Manger. 1998. A calorimetric perspective on desiccation stress during preservation procedures with recalcitrant seeds of *Quercus robur* L. *CryoLetters* 19**:** 23–30.
Walters, C., J. Wesley-Smith, J. Crane, L. Hill, P. Chmielarz, N. W. Pammenter, and P. Berjak. 2008. Cryopreservation of recalcitrant (i.e. desiccation-sensitive) seeds. In Reed, B. (Ed.), *Plant Cryopreservation: A Practical Guide*, Springer Science + Business Media LLC, New York, pp. 465–484.

41 Plant Biotechnology for the Production of Natural Products*

Ara Kirakosyan, E. Mitchell Seymour, and Peter Kaufman

CONCEPTS

- Basic knowledge of plant cell and tissue cultures is essential for understanding production of secondary products.
- Plant cell culture is an excellent model for profiling of secondary metabolites.
- Many factors determine biosynthesis and accumulation of natural products in plant cells.
- Plant cells can be micromanipulated for efficient production systems for secondary products
- Bioreactors are used to study growth and production kinetics of cell suspension cultures.
- Plant secondary metabolism can be engineered.

Achievements today in plant biotechnology have already surpassed all previous expectations. Plant biotechnology has thus emerged as an exciting area of research by creating unprecedented opportunities for the manipulation of biological systems of plants. It is a forward-looking research area based on promising accomplishments in the last several decades. Plant biotechnology is changing plant science in the following three major areas: (1) control of plant growth and development (vegetative, generative and propagation); (2) protection of plants against the environmental threats of abiotic or biotic stresses; and (3) expansion of ways by which specialty foods, biochemicals, and pharmaceuticals are produced. In order to determine the current status of plant biotechnology, it must emphasize the difference between the traditional concept of biotechnology and its current status. Early directions of plant biotechnology, which mostly focused on in vitro cell and tissue cultures and their production of important products, now has a new direction. The current state of plant biotechnology research using a number of different approaches includes high throughput methodologies for functional analysis at the levels of transcripts, proteins and metabolites, and methods for genome modification by both homologous and site-specific recombination. Plant biotechnology allows for the transfer of a greater variety of genetic information in a more precise, controlled manner. The potential for improving plant and animal productivity and their proper use in agriculture relies largely on newly developed DNA biotechnology and molecular markers. These techniques enable the selection of successful genotypes, better isolation and cloning of favorable traits, and the creation of transgenic organisms of importance to agriculture and industry.

Many scientists have now combined extensive research experience using plant tissue and cell culture with a deep knowledge of natural products in order to develop the current strategies cited above. This is enabling us to follow up in greater detail points of interest, both theoretical and practical. A number of methods were developed and validated in association with the use of genetically

* This chapter is condensed from Kirakosyan, A. 2006, Plant Biotechnology for the Production of Natural Products. In *Natural Products from Plants*, Second Edition. pp. 221–262,CRC Press, Boca Raton, FL (with permission and adapted for use in this book).

transferred cultures in order to understand the genetics of specific plant traits. Such relevant methods can be used to determine the markers that are retained in genetically manipulated natural products and to determine the elimination of marker genes and procedures for characterization of chromosomal aberrations in genetically manipulated plants. A number of transgenic plants were developed with beneficial characteristics and significant long-term potential to contribute both to biotechnology and to fundamental studies. Therefore, the presentation of all the major achievements in plant biotechnology together will be beneficial for natural products research.

In this chapter, we discuss the most up-to-date information on basic and applied research in plant biotechnology. This will reveal strategies for development of this field, traditional and high-throughput approaches, and future trends.

PLANT BIOTECHNOLOGY: FROM BASIC SCIENCE TO INDUSTRIAL APPLICATION

Because the different fields of plant science have become well-developed in the last several decades, many opportunities are now available to make significant progress in plant biotechnology. Plant biotechnology aims to impart an understanding of the basic principles of plant and molecular biology and to apply these principles to the production of healthy plants in a safe environment for food and nonfood applications (Figure 41.1).

Important aspects of this are the design of transgenic plants and related technology. Different strategies, using in vitro systems, have been extensively studied with the objective of improving the production of natural products. Thus, specific processes have been designed to meet the requirements of plant cell and organ cultures in bioreactors. Moreover, the recent emergence of recombinant DNA technology has opened a new field, whereby it is now possible to directly modify the expression of genes related to natural product biosynthesis. The focus here is on metabolic and genetic engineering of biosynthetic pathways, so as to improve the production of high-value secondary metabolites in plant cells.

There are, however, some limitations concerning the use of genetically modified (GM) plants (Chapter 37). While some scientists find this to be the most forward-looking route in plant

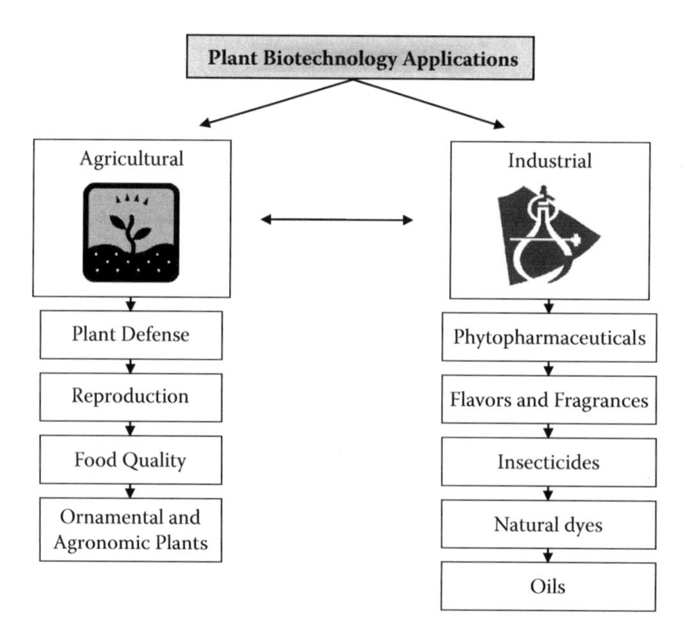

FIGURE 41.1 A schematic representation of plant biotechnology applications.

biotechnology, others are opposed to this exploration. In any case, the inclusion of foreign gene(s) in plant genomes gives us unique opportunities to up-regulate metabolite biosynthesis, and thus, to reveal the nature, functional consequences, and physiological importance of secondary metabolites in plants. Genetic modification technology aims to add or enhance beneficial characteristics in current plant varieties so as to obtain high metabolite producer varieties, which would otherwise be slow, costly, or impossible to achieve through conventional plant breeding (Sonnewald, 2003).

Basic Knowledge of Plant Cell Culture as a Tool for Biosynthetic Studies and Plant Cell Biotechnology

Plant tissues excised from plants can be cultured in vitro and regenerated to whole plants if the culture medium contains suitable nutrients and plant growth hormones. This is due to plants having a unique property called *totipotency*, that is, the ability of plant cells to develop into whole plants or plant organs. Regeneration of plants from callus tissue is usually achieved either by organogenesis (Chapters 12 and 19) or by somatic embryogenesis, which is a special case of organogenesis (Chapters 22 and 23). In the case of organogenesis, plant organs and tissues, such as shoots, roots, and vascular tissue, connecting shoot and root, are formed independently of each other. On the other hand, plant organs regenerated via somatic embryogenesis are thought to originate from a single cell in a callus or from suspension-cultured cells. Due to this unique property of plant cells, in vitro cultures are now used to manipulate the biosynthetic potential of plants.

In recent years, more plant cell culture studies have also utilized plant protoplasts (Chapters 26–28). Isolated protoplasts from plant cells can now be micromanipulated for somatic cell hybridization. However, the development of methods for protoplast isolation, culture, and subsequent plant regeneration are still critical for the successful somatic hybridization of higher plants.

Therefore, in plant cell biotechnology, most emphasis is given to plant cells that are cultured in suspension. Cell suspension cultivation offers a unique possibility for the production of natural products economically on a large scale. However, plant cell suspension cultures, in general, are less productive due to the fact that undifferentiated cells are not able to produce a wide variety of secondary metabolites. Besides, other problems arise in this kind of cultivation system. These problems are concerned with aseptic cultivation, specific design of bioreactors, stability of cell lines, and finally, differences in the cell cycle that are most efficient for optimizing cell suspension cultivation conditions. Thus, plant cell suspension cultivation protocols still need to be very well designed using genetically stable cell lines with highest yields of desired secondary metabolites in order to elaborate conditions for long-term cultivation.

MORPHOLOGICAL DIFFERENTIATION

One of the main problems facing practical applications in the use of plant cell cultures for the production of phytochemicals is that undifferentiated cell cultures do not often form such compounds. However, after the development of shoots or roots from callus tissue, or from cell suspension cultures, the regenerants are able to accumulate secondary metabolites in special types of cells, tissues, or organs. This is due to the fact that the major secondary metabolites in plants are accumulated in special morphological structures within intact plant tissues. This obstacle was the main stumbling block for large-scale cultivation of cell suspension cultures and their possible industrial applications. However, nowadays, much research is concentrated on deriving fully or partly morphologically differentiated cell aggregates or organ cultures. These kinds of cultures have turned out to be high producers of particular metabolites.

VARIATIONS IN CALLUS CULTURES

Callus cultures are derived from intact plant organs, sterile germinated seedlings, or from individual cells cultivated in suspension. Callus culture tissue actually consists of two different groups of cells: parenchyma, which forms soft callus, and deep-green structures consisting of tightly packed meristematic cells located within the soft callus tissue. These tightly packed structures in callus cultures probably represent somatic embryoids (embryos) at an early stage of development. It is known that for indirect embryogenesis to occur, first the formation of clusters from embryogen-determinated cells, which had been redifferentiated, is universal (Vasil et al., 1990). Usually, these cells proliferate to form larger clusters. The tight callus is able to undergo morphogenesis; however, this occurs only when the right composition of phytohormones is selected. The cells proliferate and form larger clusters, which continue to increase in size until they reach a certain critical biomass. Depending on the culture conditions (light, composition of medium, pH, aeration), plant cell lines can differ in aggregate size and in uniformity of cell type.

CELL TYPES IN SUSPENSION CULTURE

Generally, cell suspension cultures are classified as *homologous* and *heterologous* cell cultures. The differences between these two types of cultures in terms of the morphology and uniformity of cell types have been well characterized. Homologous cultures consist of a fine cell suspension culture of mostly homogenous populations of cells. Heterologous cultures, on the other hand, consist of different types of cells made up of clusters as well as cell aggregates. These cell cultures may produce some of the desired secondary metabolites in various amounts; others, in contrast, do not produce them. An explanation for the different biosynthetic abilities of these kinds of cell suspension cultures is that cells do not produce some compounds until full or partial differentiation has occurred, as described above. On the other hand, it is possible that the production of some compounds could be triggered in a critical situation when the biosynthetic ability of the cells must be turned on under the influence of biotic or abiotic factors. Critical to cell suspension culture biosynthetic potential is how and from what part of the intact plant the cell cultures are derived. If the cells are derived from a reproductive part of the plant that synthesizes a particular metabolite(s), this kind of cell culture could be considered to be a "producer" cell suspension line. If, however, the cell suspension culture is derived from callus culture that is, in fact, a nonproducer of a particular metabolite(s), this culture could be considered as a "nonproducer" cell suspension line. However, this classification has not proven to be valid for all the cases discussed. In other words, it is *not* a general rule. This is because the type of culture media formulation, or even elicitation, could trigger biosynthesis and production of some metabolite(s).

There is one interesting example of cell suspension type of cultivation involving globular structures cultivated in liquid media that may have more practical applications in boitechnology. This is a different type of cultivation of plant cells than has been introduced heretofore, but is now being extensively studied. The enhancement of secondary metabolite production in liquid-cultivated cell aggregates differs from that in shoots or callus (Vardapetyan et al., 2000). Recently, we have reported that suspension cultures of *Hypericum perforatum* (St. John's wort) with compact globular structures had a higher total content of some important secondary metabolites than unorganized cell suspension cultures (Vardapetyan et al., 2000). This finding parallels observations those made for two other plant systems, namely, *Catharanthus roseus* (Madagascar pink) and *Rhodiola sachalinensis*, in which compact globular structures constitute a very good system for secondary metabolite synthesis (Verpoorte, 1996; Xu et al., 1999). Actually, the globular structures reach the highest possible critical size during the cell culture process, which does not change after further subculturing. Long-term cultivation of these cultures has shown that further accumulation of biomass is due to an increase in the number of globules (Vardapetyan et al., 2000). It is noteworthy that these globular structures are fully differentiated structures in which their shape appears to be like that of a raspberry (*Rubus* spp.) aggregate fruit.

IN VITRO CULTURES AND THEIR APPLICATIONS

The main application of cell cultures can be attributed to their biosynthetic abilities and large-scale cultivation. The question then arises, do cells cultivated in vitro produce desired natural products? The evidence that plant cell cultures are able to produce secondary metabolites was experimentally proven by Zenk and coworkers (Zenk, 1991). Their work disproved the controversial theory that says only differentiated cells or specialized organs are able to produce secondary compounds. Currently, many kinds of secondary compounds are successfully isolated from suspension-cultured cells. Moreover, there are also de novo synthesized compounds isolated from and characterized in several kinds of plant cell cultures (Dias et al., 1998; Petersen and Alfermann, 2001).

The importance and potential of plant cell and tissue culture for the production of natural products are now proven for some plants and their metabolites, despite some limitations and drawbacks in this technique (Table 41.1). Until now, many phytopharmaceutical compounds have traditionally been obtained from plants growing in the wild or in field-cultivated plants. However, in light of diminishing plant resources in natural and wilderness areas due to clear-cutting of temperate and tropical forests worldwide, and increasingly higher costs of obtaining secondary metabolites from plants growing in the wild, plant biotechnologists have opted to grow these plants in cell cultures. While this method is great for micropropagation of endangered plant species, it is very labor-intensive, costly, and gives notoriously low yields of secondary metabolites as compared with intact plants (Kaufman et al., 1999). It is well-known that processes using large-scale plant cell cultures could be economically feasible, provided the cells have a high growth rate coupled with significant metabolite production rates. Therefore, the first question to be answered before developing plant cell biotechnology further for industrial applications is whether or not large-scale culture in bioreactors is economically feasible (Verpoorte et al., 1994; Verpoorte, 1996).

TABLE 41.1
Comparison of Advantages and Disadvantages of Three Different Culture Regimes for Secondary Metabolite Production from Plants

Mode of Cultivation	Advantages	Disadvantages
Field cultivation in a nursery	Can help to improve conservation of plant biodiversity. Can use low winter temperatures to break seed dormancy. Low cost of production of one-year-old plants.	Short growing season. Problems of disease, insect attack, and herbivory. No control of weather conditions.
Greenhouse cultivation	Can use illumination and environmental parameters to control and regulate both growth of plants and secondary metabolite production. Growing of rare, endemic, or threatened plant species is possible.	Higher energy and labor costs than field cultivation, including automated greenhouse installation and operation. Regrowing plants after leaf harvest is not possible.
Cell/tissue/organ culture	Can genetically modify culture to enhance metabolite biosynthesis. Plant micropropagation is possible resulting in genetic reproducibility. Elicitation is more convenient. Cultivation in bioreactors and enzyme-catalyzed modification of precursors into desired products are possible.	The process is labor-intensive and is expensive due to the high cost of culture media constituents and the requirement for sterile culture conditions. Stability problems of cell lines. Low yield of end-products in bioreactors.

In addition, cell suspension cultures can be used for biotransformation of added substrates, to search for new compounds not present in the intact plant. And finally, one can use plant cells for the isolation of enzymes that are responsible for important metabolic pathways and to use them in chemical synthesis of natural products (reviewed by Alfermann and Petersen, 1995).

Plant Cell Biotechnology for the Production of Secondary Metabolites

The presence of valuable chemicals in plants stimulates interest on the part of industries in the fields of pharmaceuticals (as drug sources), agrochemicals (for the supply of natural fungicides and insecticides), nutrition (for the acquisition of natural substances used for flavoring and coloring foods), and cosmetics (natural fragrances).

The world market for biotechnological products has increased greatly in recent decades. For example, in 2000, biopharmaceuticals represented a global market valued at over $12 billion (U.S. currency). Since then, the industry has expanded considerably despite being severely limited by the manufacturing capacity and cost of the production systems currently in place. Therefore, an alternative source for desired secondary metabolites is of great interest. Cell and tissue cultured plant materials can be an attractive alternative as a production system, as well as a model system to study the regulation of natural product biosynthesis in plants so as to ultimately increase yields. Thus, plant biotechnology can supply information to optimize phytochemical production in plant cell and tissue culture through sustainable, economically viable cultivation. However, trials with different plant cell cultures initially failed to produce high levels of the desired products.

Several medicinal plants are employed in our studies concerning plant cell biotechnology. These include Hawthorn (*Crataegus*) that produces proanthocyanidins and several kinds of flavonoids used for treatment of heart disease; St. John's wort (*Hypericum perforatum)*, which produces antidepressant and anticancer compounds like hyperforin and hypericins; Flax (*Linum* spp.) for the production of cytotoxic lignans such as podophylotoxin and 5′ methoxy-podhophylotoxin; kudzu (*Pueraria montana*) as a source of the isoflavones, daidzein, genistein, and their respective glucoside conjugates, daidzin (daidzein-7-O-glucoside) and genistin (glucosyl-7-genistein) plus puerarin (daidzein-8-C-glucoside). Each of these plants has made important contributions to the pharmaceutical industry. In all plant cell studies, the up-regulation of biosynthesis of several compounds using genetic and epigenetic approaches are now being considered as viable approaches.

Factors Determining the Accumulation of Secondary Metabolites by Plant Cells

The failure to produce high levels of the desired products by cell cultures is mainly due to our insufficient knowledge regarding how plants regulate natural product biosynthesis. The important thing is to elucidate the factors that control the accumulation of secondary metabolites in particular plant species.

Biotic factors are among the environmental factors that affect to a greater extent the production of phytochemicals. Therefore, it is highly probable that there is a relationship with defensive responses that is manifested either in phytoalexin production or compounds produced along the signal transduction pathway. An approach to characterize the biotic parameters that may elicit the plant's defensive mechanisms may be revealed by an analysis of the expression of certain genes involved in the process and by correlation of gene induction with particular metabolite levels.

Applied environmental stress factors along with biotic factors can affect the up-regulation of biosynthesis of secondary metabolites both in intact plants and in cell cultures. For example, in our investigations, two species of *Crataegus* (hawthorn) were chosen for applied environmental stress treatment experiments in order to enhance the levels of polyphenolics in leaves of hawthorn intact plants. One-year-old plants of hawthorn (*C. laevigata* and *C. monogyna*) were subjected to water deficit (continuous water deprivation), cold (4°C), flooding (immersion of roots of plants in water), or herbivory (leaf removal) stress treatments (each of 10 days' duration) in order to assess their effects on levels of polyphenolics in the leaves. Cold and water stresses caused the accumulation of several secondary compounds in both

Crataegus species. Flooding and herbivory caused no net increases, and in some cases, decreases in levels of polyphenolics. Such environmental stress factors could also positively influence secondary metabolite biosynthesis in in vitro cultures derived from young leaves of hawthorn.

The effects of both environmental and genetic factors were also found to influence the levels of the dianthrones, hypericin and pseudohypericin, in field-cultivated *H. perforatum* (Buter et al., 1998). Kirakosyan et al. (2003) examined *H. perforatum* populations based on phytochemical analysis and identified parameters that may enhance production. For example, having more leaf surface in comparison to stem tissue in the sample shifts the proportion of hypericin recovered. This is because of the fact that the glands containing hypericin are generally located along the leaf margins. Branching and gland number per leaf are not affected, which has also been corroborated in another study (Cellarova et al., 1992).

Hypericin and other related compounds vary to a greater extent in somaclones (Chapter 43) originating from the same genotype rather than from different genotypes (Cellarova et al., 1994). The biosynthesis of dianthrones and phloroglucinols has also been studied with in-vitro-grown *H. perforatum* seedlings at early stages of development (Kosuth et al., 2003). Here, it was estimated that peculiarities in reproductive development pathways do not primarily affect the formation of secondary metabolites, but could significantly contribute to genetic variation (Kosuth et al., 2003). Several other factors can also influence production of hypericins and hyperforin. These include light intensity, light quality, and temperature (reviewed by Kirakosyan et al., 2004). The effect of light intensity on the levels of leaf hypericins was examined for *H. perforatum* grown in a sand culture system with artificial lighting by Briskin and Gawienowski (2001). This study clearly demonstrated that increasing the light intensity results in a continuous increase in the levels of leaf hypericins. In shoot cultures, hypericin and pseudohypericin levels were not significantly different from each other when plants were grown either in direct light (185 $\mu E \cdot m^{-2} \cdot s^{-1}$) or under partial light (88 $\mu E \cdot m^{-2} \cdot s^{-1}$), although a general trend indicates that lower light levels may cause an increase in hypericin and pseudohypericin biosynthesis, especially in the case of pseudohypericin in plantlets grown at 25°C (Sirvent, 2001). The effect of light is closely connected with the effect of temperature, but the differences in these reports may also be due to the use of different cell lines, types of cultures, or differences in extraction and harvest methodologies.

Plant Cell Culture as a Method for Studying Biosynthesis and the Production of Secondary Metabolites

Plant cell culture provides an opportunity for extensive manipulation to enhance production of natural products over levels found in intact plants, and to identify parameters for enhancing productivity. It is possible to gain knowledge of the mechanisms that regulate the metabolic flux by careful manipulation of the plant cell culture conditions and by genetic engineering of the plant cells. These approaches offer the possibility to elucidate biosynthetic pathways and to quantify the flux of biosynthetic intermediates through a pathway. They also allow one to become acquainted with the techniques and equipment used to monitor metabolic flux in plant cell cultures.

Consequently, the development of a bioinformatics program that is based on a cellular and molecular level can be a good strategy. Using established cell cultures, it is possible to define the rate-limiting step in biosynthesis by determining accumulation of presumed (labeled) intermediates, characterize the rate-limiting enzyme activity, and probably relate it to the corresponding gene for eventual manipulation. Generally, this approach works for known pathways. Therefore, step-by-step identification of all enzymatic activities that are specifically involved in the pathway is more appropriate and has been carried out successfully. It is also quite common that blockage of one pathway leads to diversion of the substrate to alternative pathways. This would make it very difficult to identify the rate-limiting step in synthesis of a particular metabolite. It may also be that the pathway is subject to developmentally controlled flux at entry, as, for example, through the activity of transcription factors. This kind of research must, therefore, focus on metabolic regulation by first establishing the pathways at the level of intermediates and enzymes that catalyze their formation.

The subsequent step is the selection of targets for further studies at the level of the genes. The studies on regulation of metabolite biosynthesis might eventually lead to transgenic plants or plant cell cultures with an improved productivity of the desired compounds. This knowledge is also of interest in connection with studies on the role of secondary metabolism for plants, and may contribute to a better understanding of resistance of plants to diseases and various herbivores.

Strategies to Improve Metabolite Production

Zenk and coworkers (1977) suggested a strategy to improve the production of secondary metabolites in cell cultures that is being used by many researchers today. This strategy generally includes screening and selection of high metabolite-producing cell lines and analyses of culture conditions that enhance production levels of the metabolites of interest.

What has been changed to date? More important tools, such as new approaches based on genetic and metabolic engineering, have been introduced. Noteworthy, Zenk's strategy involves several approaches that are connected to plant cell culture. These include the following:

1. Plant screening for natural products accumulation
2. Use of high producer plants for initiation of callus cultures
3. Analysis of derived cultures
4. Establishment of cell suspension cultures
5. Analysis of metabolite levels in cell suspension cultures
6. Selection of cell lines based on single cells
7. Analysis of culture stability
8. Further improvement of product yields

However, follow-up research employing this strategy was marginally successful. This is based on the fact that not all desired natural products can be produced by plant cell biotechnology. For instance, only a few compounds, such as shikonin, berberin, vinblastine, vincristine, and taxol, have had successful industrial applications to date. We still do not know why the production of some compounds, like those just cited, can be feasibly employed in large-scale processes, whereas the production of other compounds, using this same approach, is either impossible or not economically feasible for industrial trials.

This strategy also focuses on developing cultures from elite germplasm and optimizing production strategies for important plant constituents through genetic and culture manipulations. Therefore, the aim of comparative studies of different medicinal plants is to obtain a detailed biochemical analysis on genetically distinct populations in order to identify superior plant germplasm sources. In this connection, investigators have been able to take advantage of the wide range of biosynthetic capacities within cultures, either by selection or by screening germplasm for highly productive cell lines, as for example, in production of taxol from *Taxus* cell cultures (Kim et al., 2005).

In light of the above observations, our recommendation to improve secondary metabolite production in in vitro plant cultures is based on optimization of production strategies being introduced for particular plant constituents. Studies are directed toward delineating the optimal conditions for production of economically viable amounts of given phytochemicals. Such optimization of culture conditions includes a variety of media formulations and environmental conditions. These variables to be tested include light intensity and quality; temperature; length of culture period, including kinetics of production; concentration and source of major limiting nutrients such as phosphate, carbon, and nitrogen; concentration and source of micronutrients, vitamins, and plant growth regulators; and presence or absence of fungal or bacterial elicitors. Once in vitro plant cultures have been established for production of economically viable amounts of the phytochemicals, the development of a transformation system will expedite genetic enhancement of such cultures to further increase yields for greater profitability.

SELECTION OF ELITE GERMPLASM FOR AN EFFICIENT PRODUCTION SYSTEM

The vast majority of medicinal plants are collected in the wild in a process known as *wildcrafting*. Likewise, the genetic potential for many medicinal plants has yet to be tapped to identify superior germplasms, whether for traditional cultivation or for derivation of superior plant cell culture lines. However, as the market expands and as new clinical trials expand market size, there is a vast opportunity to introduce more economically viable and environmentally sustainable production strategies using genetically superior material for production of phytochemicals. Development of elite germplasm that can be manipulated for more efficient production of phytochemicals would aid in the development of plant cell culture as an alternative technology to wildcrafting and low economic return practices.

MICROMANIPULATION OF HIGHER PLANT CELLS FOR PRODUCTION SYSTEMS

Plant cell culture may be a reasonable candidate for commercial realization if the natural resources are limited, de novo synthesis is complex, and the product has a high commercial value. Over the past 20 years, advances have been made toward enhancing productivity in plant cell culture. Initial attempts included applying selection pressure for faster growing and better producing cell lines, optimization of media, and multistage production where cells are initially cultured to optimize growth, and then switched to media optimized for productivity.

ELICITATION

More recently, emphasis has been placed on the use of elicitors that can trigger the defense response to induce overproduction of valuable secondary metabolites in plants or plant cell cultures. Elicitation was originally accomplished by adding crude fungal cell wall extracts to cell cultures (Radman et al., 2003). Induction can be mediated within plant tissues by salicylic acid (SA), jasmonic acid (JA), and possibly, nitric oxide (reviewed by Kirakosyan et al., 2004) as well as other compounds. The impact of different elicitors on the quantity and distribution of secondary metabolites can provide valuable information regarding biosynthetic pathways, in addition to elevating the production of a desired compound (Moreno et al., 1996). In a recent research review, Zhao et al. (2005) summarizes all up-to-date progress made on several aspects of elicitor-mediated signal transduction mechanisms leading to production of plant secondary metabolites.

Screening and Selection of Cultured Plant Cells in order to Increase Yields of Phytochemicals

Optimization of culture conditions has been carried out in a variety of media formulations and environmental conditions. The Plackett & Burman technique is particularly useful in that it allows for testing of multiple variables within a single experiment (Plackett and Burman, 1946). This method relies on the following characteristics: each variable is present at a high level in half of the test cultures or at low level or none in the other half. Any two variables will be present in 25% of the test cultures; both will be absent in 25%, and only one variable is present in the remaining 50% of the test cultures. Since the production of secondary metabolites can be followed by HPLC, a medium can be selected which supports good growth and production of secondary metabolites. Those variables to be tested can include light, temperature, length of culture period including kinetics of production, concentration and source of major limiting nutrients such as phosphate, carbon, and nitrogen, and concentration and source of micronutrients, vitamins, and plant growth regulators. Screening for high productivity can be accomplished on several levels. In some cases, high-producing plants, calli, or cell clones can be obtained from single cells and subsequently used for screening of high-producing strains (Figure 41.2).

FIGURE 41.2 High metabolite producing callus (a) and shoot cultures (b) of *Hypericum perforatum*. Shoot cultures shown in (b) were cultivated on agar (top) and in liquid (bottom).

Selection of high-producing cell lines by culturing cells/tissues on media containing certain additives, such as biosynthetic precursors or toxic analogues, has interesting possibilities (Verpoorte, 1996). However, in this case, the successes are rather limited because of the instability of many precursors or toxic effects of some constituents to the cells. For rapid selection of high-producing cells, flow cytometry could also be used. This technique is based on the fact that cells contain fluorescent products (e.g., thiophenes), and therefore, it is possible to separate these cells from others. In this case, however, the cell line can become unstable and even exhibit atrophy. Within addition, difficulties connected to cell differentiation or morphogenesis can occur in these cultures. Therefore, stability problems of cell lines have probably made researchers reluctant to develop extensive screening programs, leaving this as the last step prior to an industrial application (Verpoorte, 1996).

Growth and Production Kinetics of Plant Cell Cultures in Bioreactors

A bioreactor is a controlled apparatus, such as a large fermentation chamber or closed container, for growing various organisms, including plant cell cultures that are used in the biotechnological production of value-added substances. A number of factors should define the choice of a bioreactor that is to be used for particular plant cell culture lines. These include cell type, nature of desired products, and scale of operation. Generally plant cell culture bioreactors are of two types: (1) those that are used for cultivation of primary cultures that originate from callus tissue and (2) those that are used for the cultivation of cell suspensions (e.g., cell lines already established and maintained in an erlenmeyer flask). In some cases, the bioreactor may be modified to grow both primary cultures and suspended cells. The main objectives for bioreactors and their application are to maintain sterile cultures and to set up the conditions that could affect cell growth and secondary metabolite productivity.

A number of physical, chemical, and biochemical interactions exist in a bioreactor. This affects the design and setup of bioreactors for particular plant cell types and desired product production. For example, mixing rates and aeration (O_2 supply) have to be considered very carefully for given cell types, which often have different oxygen demands. There are waste and toxic product accumulations that must also be considered. However, it is most important to determine the cell growth rates and product formation kinetics that depend on nutritional requirements of the plant cells.

Plant cells are sensitive to hydrodynamic stresses that depend on the type and diameter of the bioreactor, and possible occurrence of shear stresses where cells are stirred by propellers, as well as the origin and physiological status of the cultivated cells. These stress conditions may inhibit both cell growth and product formation. Bioreactor-based plant bioprocesses are not only an excellent model for quantitative interpretation of growth kinetics and product biosynthesis but also can be a vehicle for scaled-up production of valuable bioactive natural compounds, mass propagule production, and applications using genetically engineered cells and tissues.

Researchers now focus on current, compelling challenges for bioreactor-based processing of plant cell and organ cultures. Plant cell cultivation in bioreactors is far less routine than microbial cultures, despite the fact that many complex, high-value plant metabolites cannot be synthesized by simpler microbial cell culture systems. Most suspension-cultured plant cells fail to detach from one another completely after cell division. Instead, they form multicellular aggregates, with rigid cell walls and surface-to-volume ratios that make them more brittle than bacterial cells. The high speed agitation required for oxygenation and mixing generates a shear stress that is detrimental to most plant cells. Furthermore, high rates of aeration tend to strip gaseous metabolites like CO_2 and ethylene (C_2H_4) out of the culture media, which can reduce the capacity for plant cells to produce metabolites (Schlatmann et al., 1995). Innovative problem-solving strategies and specific needs for sensors and instrumentation to streamline the bioprocess are now being explored. There are many slightly modified or novel designs of bioreactors for plant cells. A few of some successful bioreactor designs are described below.

BATCH SYSTEMS

In batch cultivation, an inoculum of known density is "seeded" into a specified volume of preconditioned medium in the bioreactor. Ideally, nothing is added or removed from the bioreactor during the course of cultivation. However, in practice, additions of air and acids or bases for pH control are made. Batch cultivation of suspension cells can be carried out in two types of bioreactors: stirred tank and air lift reactors (Vogel and Tadaro, 1997).

Air lift bioreactors. In air lift bioreactors, sterile air is introduced at the bottom of the vessel within the draught tube. A reduction in density of the aerated contents in the draught tube results in a circulation of the culture through the draught tube and down in the outer zone of the vessel. Advantages are that there are no moving parts or mechanical seals, there is adequate oxygen transfer, there are low hydrodynamic shear forces, and there is low power input per unit volume.

Stirred tank characteristics. The size of the stirred tank for a given suspension cell culture process depends upon previous planning estimates and calculations. These determine the quantity of raw material necessary to meet the demands of product formation. These estimates take into consideration the following: (1) product titer, (2) yields through the purification process, and (3) losses that may occur during formulation, final fill operations, or sampling required for quality control and in process monitoring.

This information provides the total number of liters of cell culture medium required. Based on the mode of operation (batch, semicontinuous, batch, or perfusion), the batch size and batches per year can be established. The largest cell suspension cultures in operation today range in volume from 1,000 to 10,000 L. Reactors of similar design on a laboratory scale and pilot plant scale are necessary to ensure scalability of critical parameters, to provide inoculum for the larger reactors, and to supply preclinical and clinical trial material prior to the commercial production phase conducted on a larger scale. Stirred bioreactors for plant cell culture can be designed in virtually any size required to meet the particular need. Stirred cell culture tanks are almost always cylindrical vessels with a ratio of height to diameter between 3:1 (Vogel and Tadaro, 1997).

In addition, a number of substrates are available for cell attachment, spreading, and growth. The simplest batch systems are stationary flasks or rotating bottles made of plastic. The plastic has to be wettable and the surface treated to carry negative charges. Batch cultivation in packed bed reactors can also be carried out using surface area packing materials such as sponges, steel springs, porous ceramic particles, resins of nanoparticle size, and calcium alginate gels. Alternatively, cells may be grown on micro carriers, which are kept in suspension in stirred tank reactors (Vogel and Tadaro, 1997).

LONG-TERM CONTINUOUS CULTIVATION

An alternative approach to batch cultivation is to continuously add fresh medium to the cells and to remove either medium mixed with cells or cell-free medium from the bioreactor. As in batch cultivations, the pH, temperature, and dissolved oxygen need to be monitored and controlled. In addition, automated pumps are required to control the addition of fresh nutrients and for the removal of waste and end-products (Vogel and Tadaro, 1997).

Continuous Flow Stirred Tank Reactors (Chemostat)

In the continuous flow, stirred tank reactor (CSTR or chemostat), fresh medium is fed into the bioreactor at a constant rate, and medium mixed with cells leaves the bioreactor at the same rate. A fixed bioreactor volume is maintained, and ideally, the effluent stream should have the same composition as the bioreactor contents. The culture is fed with fresh medium containing one, and sometimes two, growth-limiting nutrients such as sucrose. The concentration of the cells in the bioreactor is controlled by the concentration of the growth-limiting nutrient. A steady-state cell concentration is reached where the cell density and substrate concentration are constant.

In vitro plant cell culture is currently carried out for a diverse range of bioreactor designs, ranging from batch, airlift, and stirred tank to perfusion and continuous flow systems. For a small-scale operation, both conventional and novel bioreactor designs are relatively easy to operate. For a larger scale operation, problems of maintaining bioreactor sterility and providing an adequate oxygen supply to the cells have yet to be resolved (Vogel and Tadaro, 1997).

METABOLIC ENGINEERING OF PLANT SECONDARY METABOLISM

A new direction of research in plant cell biotechnology, namely, plant metabolic engineering, is currently progressing rapidly. Rational engineering of secondary metabolic pathways requires a thorough understanding of the whole biosynthetic pathway and unraveling of the regulatory mechanisms. Recent achievements have been made in the altering of various pathways by use of specific genes encoding biosynthetic enzymes, or genes that encode regulatory proteins (Maliga and Graham, 2004; Verpoorte and Memelink, 2002). In addition, new anti-sense genes are used to block competitive pathways. Hence, this could increase the total flux toward the desired secondary metabolites (Verpoorte et al., 2000). Shifting attention from recombinant proteins to metabolic engineering introduces new challenges. A better understanding of the basic metabolic process could be key information needed to produce high value natural products. There is another important factor concerning the accumulation and storage of desired secondary metabolites in plants. Secondary metabolites in cell and tissue cultures are usually stored intracellularly, as for example, in vacuoles or multicellular cavities, and transporters probably play an important role in the sequestration of secondary metabolites (Kunze et al., 2002). Moreover, many biosynthetic pathways in plants are long and complicated, requiring multiple enzymatic steps to produce the desired end product. The major aims for engineering secondary metabolism in plant cells are to increase the content of desired secondary compounds, to lower the levels of undesirable compounds, or to introduce novel compound production into specific plants.

Plant metabolism, however, concerns thousands of interacting pathways and processes. Therefore, engineering even known metabolic pathways will not provide the expected results. Extensive metabolic profiling must be more systematic and involve considerable analysis in this case. "Productive metabolic engineering," therefore, must be based on a systems biology approach involving integrated metabolomics, proteomics, and transcriptomics approaches (Carrari et al., 2003; Dixon, 2005). Despite major advances in metabolic engineering, still only a few secondary metabolic pathways have been enzymatically characterized and the corresponding genes cloned. In this context, the biosynthetic pathways for alkaloids, flavonoids, and terpenoids are presently the best-characterized at the enzyme and gene levels. Metabolic engineering is a potentially very powerful tool for the regulation of secondary metabolism in transgenic plants, and it will certainly have many applications in the future (Verpoorte and Alfermann, 2000).

MOLECULAR FARMING

Plant cell biotechnology research has as one of its important goals to develop also a new molecular farming industry. The use of plants as bioreactors offers the global health care industry the most promising system for mass-producing many of the phytopharmaceuticals and is increasingly being used as a safe and inexpensive alternative for the production of valuable proteins. This unique system can offer strict control over the timing and pattern of gene expression, which can be restricted to particular plant organs, such as seeds, leaves, and roots.

With the advent of gene transfer technology, plant-based manufacture of recombinant proteins has become an attractive alternative. Molecular farming is a relatively new area of science and a new industry. It can be defined as the production of novel products in plants and, in the majority of instances, these products are of nonplant origin. This field provides a fundamental understanding of how plants can be exploited for the production of foreign proteins. Plants are induced to produce these products through the insertion and expression of new genes that are responsible for final product formation. Thus, molecular farming involves the genetic modification of the "host" plant with subsequent stable product formation. Nevertheless, a number of factors may contribute to the efficiency of recombinant protein expression in plant organisms. Parameters such as transient versus stable expression, codon optimization, organelle targeting, tissue-diverse versus tissue-specific expression, and activatable/inducible versus constitutive expression should all be taken into consideration when designing a plant-based expression system.

The categories of products that are currently being produced either commercially or experimentally in plants include human and animal therapeutics (including vaccines), diagnostics, industrial proteins, and other industrial products (such as bioplastics). It is important to mention here that such foreign gene products now being produced in transgenic plants should be limited to nonfood plants.

CONCLUSIONS

While plant natural products are useful for treatment of many human ailments, they are often made in only trace amounts within the specific species that produce them. Therefore, plant biotechnology is expected to play a major role in the production of natural products through bioengineering.

Although we still have limited knowledge about plant secondary metabolism, its regulation, functional consequences, and physiological importance, recent achievements in plant genomics, proteomics, metabolomics, and systems biology are now filling these voids. New technologies are emerging that enhance our understanding of the factors that control the accumulation of secondary metabolites, the molecular mechanisms concerning gene expression, signal transduction regulation, and rate-limiting enzymes found within a diverse network of biosynthetic pathways in plants. Molecular biology strategies are being used to produce beneficial products, such as phytopharmaceuticals, vaccines in nonfood plants, natural pesticides, or food additives. Therefore, our challenge will be to integrate different disciplines in plant science in order to unravel metabolic networks and to elucidate the biosynthesis and molecular regulation of several important secondary metabolites in plants. This will lead to successful applications in metabolic engineering and molecular farming. It will further reduce our dependence on accruing large amounts of plant biomass, and hence, reduce loss of important biological resources. Biotechnology is now playing an ever increasing role in the search for biologically active natural products from plants and other sources. Because plant natural products have great chemical diversity, their sustainable supply from original biological sources is increasingly becoming a problem.

REFERENCES

Alfermann, A.W., and M. Petersen. 1995. Natural product formation by plant cell biotechnology. *Plant Cell. Tiss. Org. Cult.* 43: 199–205.

Braz-Filho, R. 1999. Brazilian phytochemical diversity: Bioorganic compounds produced by secondary metabolism as a source of new scientific development, varied industrial applications and to enhance human health and the quality of life. *Pure Appl. Chem.* 71:1663–1672.

Briskin, D. and M. Gawienowski. 2001. Differential effects of light and nitrogen on production of hypericins and leaf glands in *Hypericum perforatum. Plant Physiol. Biochem.* 39:1075–1081.

Buter, B., C. Orlacchio, A. Soldati, and K. Berger. 1998. Significance of genetic and environmental aspects in the field cultivation of *Hypericum perforatum. Planta Med.* 64:431–437.

Carrari, F., E. Urbanczyk-Wochniak, L., Willmitzer and A.R. Fernie. 2003. Engineering central metabolism in crop species: Learning the system. *Metab. Eng.* 5: 191–200.

Dias, A.C.P., F.A. Tomas-Barberan, M. Fernandes-Ferreira, and F. Ferreres. 1998. Unusual flavonoids produced by callus of *Hypericum perforatum. Phytochemistry* 48:1165–1168.

Dixon, R.A. 2005. Engineering of plant natural product pathways. *Curr. Opin. Plant Biol.* 8:329–336.

Jenett-Siems, K, R. Weigl, A. Bohm, P. Mann, B. Tofern-Reblin, S.C. Ott, A. Ghomian, M. Kaloga, K. Siems, L. Witte, M. Hilker, F. Muller, and E. Eich. 2005. Chemotaxonomy of the pantropical genus *Merremia* (*Convolvulaceae*) based on the distribution of tropane alkaloids. *Phytochemistry* 66:1448–1464.

Kaufman, P.B., L.J. Cseke, S. Warber, J.A. Duke, and H.L. Brielmann. 1999. *Natural Products from Plants.* CRC Press, Boca Raton, FL.

Kim, B., D.M. Gibson, and M.L. Shuler. 2005. Relationship of viability and apoptosis to taxol production in *Taxus* sp. suspension cultures elicited with methyl jasmonate. *Biotechnol. Prog.* 21:700–707.

Kirakosyan, A. 2005. Plant biotechnology for the production on natural products appearing in natural products from plants. In Cseke, L.J., A. Kirakosyan, P.B. Daufman, S.L. Warber, J.A. Duke, and H.L. Brielmann (Eds.), *Natural Products form Plants*, Second edition, CRC Press, Boca Raton, FL, pp. 221–262.

Kirakosyan, A., D. M. Gibson, and T. Sirvent. 2003. A comparative study of Hypericum perforatum plants as sources of hypericins and hyperforins. *J. Herbs Spices Med. Plants* 10:73–88.

Kirakosyan, A., T.M. Sirvent, D.M. Gibson, and P.B. Kaufman. 2004. The production of hypericins and hyperforin by in vitro cultures of St. John's wort (*Hypericum perforatum*). *Biotechnol. Appl. Biochem.* 39:71–81.

Kosuth, J., J. Koperdakova, A. Tolonen, A. Hohtola, and E. Cellarova. 2003. The content of hypericins and phloroglucinols in *Hypericum perforatum* L. seedlings at early stage of development. *Plant Sci.* 165:515–521.

Kunze, R., W.B. Frommer, and U.I. Flügge. 2002. Metabolic engineering of plants: The role of membrane transport. *Metab. Eng.* 4: 57–66.

Maliga, P., and I. Graham. 2004. Molecular farming and metabolic engineering promise a new generation of high-tech crops. *Curr. Opin. Plant. Biol.* 7:149–151.

Moreno, P.R.H., C. Poulsen, R. van der Heijden, and R. Verpoorte. 1996. Effects of elicitation on different metabolic pathways in *Catharanthus roseus* (L.) G. Donn cell suspension cultures. *Enz. Microb. Technol.* 18:99–107.

Petersen, M. and A.W. Alfermann. 2001. The production of cytotoxic lignans by plant cell cultures. *Appl Microbiol. Biotechnol.* 55:135–142.

Plackett, R.L. and J.P. Burman. 1946. The design of optimum multifactorial experiments. *Biometrica* 33:305–325.

Radman, R., T. Saez, C. Bucke, and T. Keshavarz. 2003. Elicitation of plants and microbial cell systems. *Biotechnol. Appl. Biochem.* 37:91–102.

Schlatmann, E., J.L. Vinke, H.J.G. Ten Hoopen, and J.J. Heijnen. 1995. Relation between dissolved oxygen concentration and ajmalicine production rate in high-density cultures of *Catharanthus roseus*. *Biotechnol. Bioeng.* 45:435–439.

Sirvent, T. 2001. Hypericins: A family of light-activated anthraquinones in St. John's wort (*Hypericum perforatum* L.) and their importance in plant/pathogen/herbivore interactions, p. 213, Cornell University, Ithaca, New York.

Sonnewald, U. 2003. Plant biotechnology: From basic science to industrial application. *J. Plant Physiol.* 160:723–725

Vardapetyan, H. R., A.B. Kirakosyan, and A.G. Charchoglyan. 2000. The kinetics regularities of the globular structures growth in cell cultures of *Hypericum perforatum* L. *Biotechnologia* 4:53–58.

Vasil, V., F. Redway, and I.K. Vasil. 1990. Regeneration of plants from embryogenic suspension culture protoplasts of wheat (*Triticum aestivum* L.). *Bio/Technology* 8:429–434.

Verpoorte R. 1996. Plant cell biotechnological research in the Netherlands. In F. DiCosmo, and M. Misawa (Eds.), *Plant Cell Culture Secondary Metabolism*. CRC Press, Boca Raton, FL, p. 660.

Verpoorte, R. 2000. Pharmacognosy in the New Millennium: Leadfinding and biotechnology. *J. Pharm. Pharmacol.* 52: 253–262.

Verpoorte, R. and A.W. Alfermann. 2000. *Metabolic Engineering of Plant Secondary Metabolism*. Kluwer Academic Publishers Group. The Netherlands, p. 296.

Verpoorte, R. and J. Memelink. 2002. Engineering secondary metabolite production in plants. *Curr. Opin. Biotechnol.* 13:181–187.

Verpoorte, R., R. van der Heijden, J.H.C. Hoge, and H.J.G. Ten Hoopen. 1994. Plant cell biotechnology for the production of secondary metabolites. *Pure Appl. Chem.* 66: 2307–2310.

Verpoorte, R., R. van der Heijden, and J. Memelink. 2000. Engineering the plant cell factory for secondary metabolite production. *Transgenic Res.* 9:323–343.

Verpoorte R., R. van der Heijden, H.J.G. Ten Hoopen, and J. Memelink. 1999. Metabolic engineering of plant secondary metabolite pathways for the production of fine chemicals. *Biotechnol. Lett.* 21: 467–479.

Vogel, H.C., Tadaro, C.L. 1997. *Fermentation and Biochemical Engineering Handbook—Principles, Process Design, and Equipment*, Second edition. William Andrew Publishing/Noyes, p. 801.

Von Eiff, M., H. Brunner, A. Haegeli, U. Kreuter, B. Martina, B. Meier, and W. Schaffner. 1994. Hawthorn/Passionflower extract and improvement in physical capacity of patients with dyspnoea Class II of the NYHM functional classification. *Acta Therap.* 20:47– 66.

Xu, J. F., P.Q. Ying, A.M. Han, and Z.G. Su. 1999. Enhanced salidroside production in liquid-cultivated compact callus aggregates of *Rhodiola sachalinensis*: Manipulation of plant growth regulators and sucrose. *Plant Cell. Tiss. Org. Cult.* 55, 53–58.

Zenk, M.H., H. El-Shagi, H.Arens, J. Stockigt, E.W. Weiler, and B. Deus. 1977. Formation of indole alkaloids serpentine and ajmalicine in cell suspension cultures of *Catharanthus roseus*. In Barz, W., E. Reinhard, and M.H. Zenk (Eds.), *Plant Tissue Culture and its Biotechnological Application*. Springer Verlag, Berlin, Germany, pp. 27–43.

Zhao, J., L.C. Davis, and R. Verpoorte. 2005. Elicitor signal transduction leading to production of plant secondary metabolites. *Biotechnol. Adv.* 23:283–333.

42 Pigment Production in *Ajuga* Cell Cultures

Mary Ann Lila and Randy B. Rogers

Within the genus *Ajuga,* a number of selections have been cultivated in vitro as sources of valuable secondary metabolites (phytochemicals). Phytochemicals such as ferulic acid, and flavonoids, including acylated anthocyanin pigments with a higher degree of stability, are harvested from cell cultures of *A. reptans,* and from purple foliage-cultivars of the related species *A. pyramidalis* (Callebaut et al., 1997; Madhavi et al., 1996, 1997). Cell and root cultures from *A. turkestanica* and *A. reptans* are also sources of adaptogenic phytoecdysteroids, which alleviate fatigue and enhance the nonspecific resistance of a consumer (Cheng et al., 2008; Gorelick-Feldman et al., 2008). In the following laboratory exercises, a few selections from the genus *Ajuga* are used to demonstrate pigment accumulation in microshoots as well as unorganized (callus and suspension) cultures, and to demonstrate the influence of experimental factors on the quality of product.

Experiment 1 (the main exercise) explicitly illustrates the influence of chemical microenvironmental factors—specifically, carbon source—on the levels of pigmentation produced by *Ajuga* callus (originating from leaf explants). Pigment accumulation in nature and in vitro is often a response to environmental stress and, consequently, pigment expression in cell cultures can conspicuously mark the outcome of experiments that impose a chemical or physical pressure on cultured cells. In this case, to illustrate the effect of the chemical microenvironment on secondary metabolite accumulation, medium composition is deliberately manipulated to enhance anthocyanin accumulation in vitro.

As is the case for some other in vitro pigment production systems, a two-phase culture protocol is used to elicit pigment accumulation. A "growth phase" is followed by a "product accumulation phase" on a production medium that favors accumulation of pigment. For some cell cultures, little or no new growth occurs during the product accumulation phase. Elevated levels of sucrose or other sugar (as a carbon source and/or osmotic agent), a change in the nitrate-to-ammonium ratio, and changes in the growth regulator composition are frequently needed in the production phase of a two-phase culture system to stimulate pigments in vitro.

Experiment 2 demonstrates the influence of explant size and source on productivity and pigment expression in amorphous masses of callus. Experiment 3 features a variegated chimeral *Ajuga* genotype and illustrates the variation in productivity and performance of callus isolated from different sectors of this chimeral leaf. A final exercise, Experiment 4, examines the influence of physical microenvironmental treatments (light wavelengths) on anthocyanin production from cell cultures.

GENERAL CONSIDERATIONS

PLANT MATERIALS

Ajuga pyramidalis "Metallica Crispa" and the variegated chimeral selection *Ajuga reptans* "Burgundy Glow" are both readily available at most garden centers as flats of stoloniferous, spreading plants ready to plant in the landscape (Figure 42.1). To avoid confusion, identified stock plants should be obtained from a reputable source because many retail shops sell flats labeled as "Purple Ajuga," without clear distinction between *A. pyramidalis* "Metallica Crispa" and the phenotypically

FIGURE 42.1 A: The pigmented foliage of Ajuga makes it a popular landscape selection. B: Adventitious microshoots of variegated chimera "Burgundy Glow" Ajuga allow propagation of color variants, and this same pigmentation can be stimulated in cell cultures derived from leaf explants.

similar bronze/purple selections of *A. reptans*. *Ajuga reptans* and *A. pyramidalis* freely hybridize, resulting in much confusion in the trade. Again, although any of these selections will yield a good experimental response in the exercises following, it is important to make sure that all stock plants are of a single genotype to avoid a source of uncontrolled variation unrelated to treatment effects.

Stock plants are planted on 5 inch centers in a 1:1:1 mix (peat:perlite:soil) in a shallow flat and fertilized once per week with a 20:20:20 fertilizer. Both greenhouse plants and microcultured stock plants can be used as explant sources in these exercises. Approximately 2–3 healthy plants per student should be maintained to ensure sufficient new foliage for explanting trials. Greenhouse plants require low maintenance and may be divided and replanted approximately once per year to maintain stock. However, maximal success in explanting from the greenhouse is achieved when young, vigorously growing new leaves from new transplants are used as donor material. While natural light is adequate in a greenhouse maintained at 25°–28°C, supplemental lighting encourages more vigorous growth during the low-light levels of midwinter.

Microcultured stock plants are initiated by explanting meristem tips (devoid of all leaves), or nodes from runners, each approximately 1–2 cm long. Any explant material taken from the greenhouse stock plants should not be in contact with the soil; runners or plants overhanging the sides of the compartments are ideal for this purpose. Alternatively, *Ajuga* plants can be potted and placed in a growth chamber to avoid any overhead watering in the greenhouse until new growth is initiated, and then explants can be taken from the new growth. Surface disinfest the explants by treatment in 15% commercial bleach with 0.1% Tween 20 (a few drops per liter). Agitate explants in the disinfestant for 15 min followed by three rinses in sterile double-distilled water. After removing injured tissue, the explants can be placed in 25 × 150 mm culture tubes containing 10 mL of basal media composed of Woody Plant Medium (WPM) basal salts (Lloyd and McCown, 1980; see Chapter 2 for composition) amended with 88 mM (30 g) sucrose, 0.55 mM (0.1g) myo-inositol, 0.28 mM (50 mg) l-ascorbic acid, 1.48 μM (0.5 mg) thiamine-HCl, 2.44 μM (0.5 mg) pyridoxine-HCl, 4.06 μM (0.5 mg) nicotinic acid, 26.7 μM (2 mg) glycine, 1 μM (0.203 mg) 2iP, 150 mg/L PVP-T (polyvinylpyrrolidone-10), and 7 g/L of agar; pH 5.7 prior to autoclaving. After cultures begin to produce new growth, 4–5 explants are subcultured into each GA7 vessel (Magenta Corp, Chicago) with 50 mL of basal medium. The microcultured plants are maintained indefinitely at 25°C under 80 μmol m^{-2} s^{-1} irradiance as shoot cultures by subculture of meristems every 4–6 weeks to fresh medium. As an added advantage for use in yearly plant tissue-culture classes, microcultured stock plants can be maintained in long-term cold storage to minimize the need for subculture upkeep. After subculture, microplants are placed in the normal culture room (25°C) as described for approximately 3 weeks until new growth and reestablishment is visible. Subsequently, these cultures are parafilmed around the cap and can be stored at 4°C in the dark for 4–6 months. When removed from cold storage and reintroduced in the culture room, the *Ajuga* plants resume growth readily.

EXERCISES

EXPERIMENT 1: EFFECT OF THE CARBON SOURCE ON PIGMENT EXPRESSION IN *AJUGA* CELL CULTURES

This experiment was designed to illustrate the effects of carbon source level on anthocyanin production in callus of *A. pyramidalis* "Metallica Crispa."

The experiment begins at the explanting stage (initiation of callus from vegetative leaf disks) and continues through a final phase of pigment accumulation in callus colonies. Greenhouse leaves may be surface disinfested and used as a source of explants, or presterile microcultured leaves may be explanted (Figure 42.2). Alternatively, students can begin with precultured callus already growing in the dark (label E on Figure 42.2), to shorten the exercise. When starting with dark-grown callus, students can proceed immediately to the subculture of callus to different treatments of the production medium. The same experiment can be performed by subculturing the white friable callus to liquid production media without agar and producing pigment in suspension culture. In either case, at the end of the subculture cycle, the tissues are ground with a mortar and pestle, extracted in methanol, and the absorbance is measured spectrophotometrically to determine anthocyanin content (Figure 42.3).

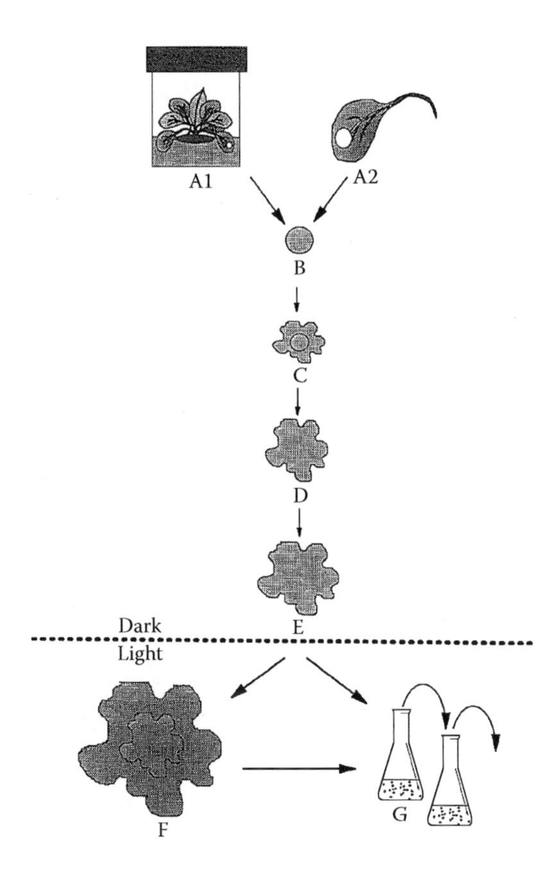

FIGURE 42.2 Schematic representation of the pigment production protocol for *Ajuga*. A: Explants obtained from either presterile microcultured plants (A1), or from greenhouse or growth chamber specimens (A2). B–D: Callus is excised from mother tissue and cultured independently for additional rounds of culture growth in the dark. E: Friable, ivory-colored callus is gently lifted from the callus induction medium, subcultured to pigment production medium, and placed on illuminated culture room shelves. F: Friable, colored callus may be maintained on solid pigment production medium for subsequent subcultures, G: or introduced into liquid suspension cultures.

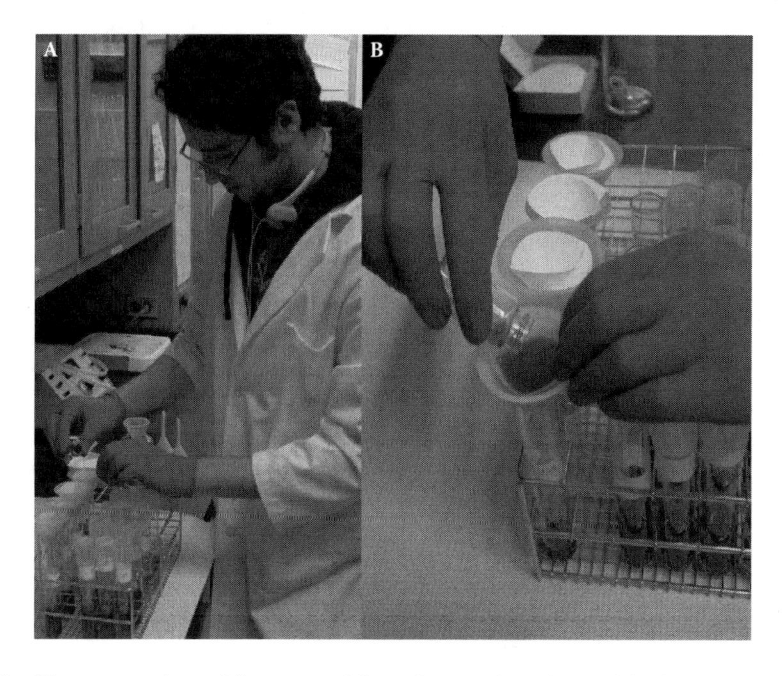

FIGURE 42.3 Pigments can be rapidly extracted from tissue-cultured material. Tissue from either callus or suspension-cultured cells is extracted in acidified methanol (A). The extract is filtered to compare treatments by spectrophotometric estimation of pigment concentration (B).

Materials

The following items are needed for each team:

- One or two shoot cultures of *A. pyramidalis* "Metallica Crispa" in GA7 vessels
- 1 cm diameter cork borer
- Fifteen GA7 vessels containing a callus-induction medium, which substitutes 2.3 μM 2,4-D and 3.5 μM kinetin as the growth regulators
- Eight GA7 vessels each containing experimental pigment production media, all with 2.26 μM IAA and 3.49 μM zeatin, but modified so that two vessels each contain either 10, 30, or 70 g/L sucrose or 30 g/L galactose

Follow the protocols listed in Procedure 42.1 to complete this experiment.

Anticipated Results

Cell cultures generated from the pigment-producing plant material will produce pigments readily in both cell and suspension cultures under light (Figure 42.4). After as little as 4 days under lights, however, callus from each of the four carbon-source treatments should exhibit marked differences in anthocyanin production. The galactose treatment will support cell growth equally well as sucrose treatment at the same concentration, as illustrated by similar increases in FW and DW, but will stimulate significantly greater anthocyanin expression. Typically, 15%–20% greater anthocyanin concentrations will be produced in the galactose treatment as compared to sucrose treatment at 30 g/L. It is important that anthocyanin concentrations be compared on the basis of yield per treatment rather than on a milligram anthocyanin/gram FW or DW basis because different carbon sources and levels will support different callus growth rates as well as different levels of anthocyanin accumulation. The treatment containing low sucrose concentrations (10 g/L) will exhibit significantly less growth during the 2 weeks on pigment production medium. Callus in this treatment will be visibly duller, with less pigment concentration.

Procedure 42.1

Pigment Expression in *Ajuga* Callus and Chemical Microenvironmental Effects

Step	Instructions and Comments
1	Collect five identical leaf disks using a 1 cm diameter corkborer from presterile *Ajuga pyramidalis* "Metallica Crispa" shoot cultures. (Alternatively, surface-disinfested leaves from greenhouse specimens can be used.) Place each disk on the surface of callus induction medium [WPM amended with 88 mM (30 g/L) sucrose, 0.55 mM (0.1 g/L) myo-inositol, 0.28 mM (50 mg/L) l-ascorbic acid, 1.48 μM (0.5 mg/L) thiamine-HCl, 2.44 μM (0.5 mg/Ll) pyridoxine-HCl, 4.06 μM (0.5 mg/L) nicotinic acid, 26.7 μM (2 mg/L) glycine, 2.3 μM (0.5 mg/L) 2,4-D, 3.5 μM (0.75 mg/L) kinetin, 150 mg/L PVP-T (polyvinylpyrrolidone-10), and 7 g/L of agar; pH 5.7 prior to autoclaving].
2	Incubate cultures in darkness at 25°C for 3 weeks.
3	Separate ivory-white friable callus from mother tissue and transfer to fresh induction media. Continue incubation in darkness, subculturing the callus to fresh media at 2 week intervals for 4 weeks.
4	After a total of 7 weeks from initiation, transfer callus masses to pigment production media treatments [callus induction media substituting 2.26 μM (0.4 mg/L) IAA and 3.49 μM (0.76 mg/L) zeatin as PGRs]. The four treatments are sucrose levels of 10, 30, or 70 g/L, or a galactose level of 30 g/L. For each treatment, five 2.5 cm diameter callus masses are transferred to each of two GA7 vessels containing treatment media (two replicate cultures/treatment, with five subreplicates [callus masses]/vessel).
5	Incubate the cultures under cool white fluorescent lamps providing a PPF of 100 μmol m^{-2} s^{-1} at 25°C for 2 weeks.
6	Evaluate growth and pigment production with an initial visual assessment of pigment content in the intact culture. Follow with fresh and dry weight analysis of two subreplicates from each culture. Grind the remaining three subreplicates with a mortar and pestle, store overnight in 10 mL ice-cold acidified methanol (1% HCl), filter, then measure the absorbance of the solution at 535 nm with a spectrophotometer to assess differences in the relative anthocyanin content between treatments.

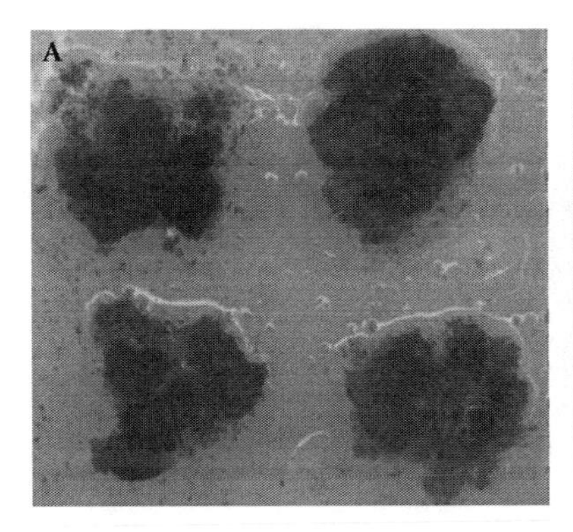

FIGURE 42.4 A: Callus generated from microplant foliage can readily express pigment upon exposure to light. B: Suspension cultures derived from this callus provide a continuous source of readily extractable pigmented tissue.

Questions

- What function does the carbon source (sugar) play in the pigment production stage of culture for *Ajuga* callus? Is the sugar acting mostly as a metabolic agent or as a stress-evoking (osmotic) agent at the 30 g/L concentration? What about at the 70 g/L concentration?
- Why was callus initiated in darkness? How might initiation of callus from explants in the light change the outcome of these experiments?

Experiment 2: Explant Size and Source Effects on In Vitro Pigment Production

The feasibility of stimulating in vitro secondary metabolites from unorganized cells can hinge on the ability to produce uniform, prolific, rapidly growing callus cells, of good quality, from tissues of the target species. Explant size and donor plant physiology can influence the initiation and growth rate of parenchymous callus cells, friability of callus masses, and propensity for the tissue to revert to regeneration (organogenesis) instead of callus production. The characteristics of the callus depend on a complex relationship between the explant tissue used to induce callus, the composition of the medium, and microenvironmental conditions. In this experiment, a range of explant sizes from two sources (greenhouse plants and microcultured plants) are compared in terms of callus induction rate, overall callus production, and the ability of the callus to express anthocyanin pigments. The experiment results in similar trends when using *A. pyramidalis* "Metallica Crispa" or when using purple or bronze-leafed cultivars of *A. reptans*.

Materials

The following items arc nccdcd for each team:

- One or two shoot cultures of *Ajuga* in a GA7 vessel
- One or two greenhouse plants of *Ajuga*
- 10% commercial bleach solution and volumes of sterile double-distilled water
- Screw top glass jars (Bellco) of 250-mL
- Cork borers of diameter 2, 5, 7, and 10 mm
- Sixteen GA7 vessels containing callus initiation medium (2.3 μM 2,4-D and 3.5 μM kinetin)
- Eight GA7 vessels containing pigment production medium (2.26 μM IAA and 3.49 μM zeatin with 30 g/L sucrose)

Follow the procedures outlined in Procedure 42.2 to complete this experiment.

Procedure 42.2
Explant Size and Source Effects on in Vitro Pigment Production by *Ajuga* Cultures

Step	Instructions and Comments
1	Surface disinfest leaves of greenhouse plant material with a 15 min immersion in 10% commercial bleach with a few drops per liter of Tween 20, followed by three 5 min rinses with sterile distilled water.
2	Using cork borers, collect 2, 5, 7, and 10 mm diameter leaf disks (five replicates of each size) surface disinfested greenhouse plant material. Repeat leaf disk collection from presterile shoot cultures. Place disks on the surface of callus induction medium (see Procedure 42.1, Step 1).
3	Incubate cultures in darkness at 25°C for 4 weeks. Inspect frequently for the appearance of callus along the cut edges of leaf sections to determine the effects of explant size and type on the rate of callus initiation. Record onset of callus initiation.
4	For one-half of the subreplicates, separate callus from leaf tissue and collect fresh and dry weights. Transfer remaining subreplicates to pigment production medium [callus induction medium substituting of 2.26 μM (0.4 mg/L) IAA and 3.49 μM (0.76 mg/L) zeatin as PGRs].
5	Incubate cultures under cool white fluorescent lamps at a PPF of 100 μmol $m^{-2}sec^{-1}$ for 2 weeks at 25°C. Inspect daily for pigment production.
6	Evaluate growth and pigment production. Evaluations may include visual, fresh weight, dry weight, and/or relative anthocyanin content via extraction and measurement of absorbance by a spectrophotometer as described previously (Procedure 42.1, Step 6).

Anticipated Results

Callus initiation rate, overall productivity, and tissue quality are all affected by the size and source of the leaf disk explant. Callus growth should be directly proportional to explant size for both greenhouse and micropropagated explants. Small explants produce substantially less biomass within a typical culture period. Although more callus is eventually produced from greenhouse explants, the rate of initiation is slightly faster from microcultured explants. Uniform callus tissue should be apparent after only 3 days on 7–10 mm microcultured leaf disks, but callus appears only sporadically around the perimeter of same-sized greenhouse disks. This result may be because any tissues damaged during surface disinfestation (of a greenhouse leaf prior to explanting) may grow less actively.

Pigment production is usually markedly better from callus generated on medium-sized greenhouse explants (5 or 7 mm disks) as compared to the same-diameter disks from microcultured explants. Residual influence of the elevated levels of stored carbohydrates in greenhouse explants may be responsible for this response. Pigment intensity is not significantly different for any of the microcultured explant sizes, although significantly more color-producing biomass is generated from larger explant sizes. Callus initiates rapidly from the larger 10 mm explants from both the microcultured and greenhouse explants; however, pigment expression is apparent earlier from the smaller explant sizes.

Questions

- Why does explant size and source (in vivo stock plant vs. in vitro microplant) affect the productivity in a pigment production system?
- Pigment produced in a callus often has a simpler profile (a less complex anthocyanin structure) than pigment in tissues of the plant in vivo. How do you think pigments from *Ajuga* callus would compare to pigments found in leaves? Flowers?

Experiment 3: Productivity of Variant *Ajuga* Callus Lines

To amplify levels of production for pigments and other secondary metabolites, highly productive callus sectors may be preferentially selected at the time of subculture, while the remaining callus is excluded. An example of variation of selected callus lines is illustrated in Figure 42.5. The selection process and the repeated subcultures needed to stabilize a callus line tend to require long periods of time and do not fit within the time constraints of a typical semester. To illustrate the potential variability of callus lines generated from a single plant, the variegated chimeral *A. reptans* “Burgundy Glow” may be used

FIGURE 42.5 Callus selected for medium, dark, and low pigmentation. Colored sectors in the pale callus are typical of the type of callus that is isolated during the long-term selection process.

to quickly generate two or three callus lines with variable growth rates and pigment production levels. This cultivar of *A. reptans* and its variants perform quite well in the shoot culture and cold storage regimes outlined for *A. pyramidalis* "Metallica Crispa" with the exception that growth regulators are omitted from the shoot proliferation media to minimize production of adventitious shoots.

Ajuga reptans "Burgundy Glow" cultures can also provide excellent illustrations of other tissue-culture topics. Adventitious shoot production from leaves, roots, and shoots occurs rapidly within 3 weeks when 0.5 mg/L BAP is included in the media. The separation of the chimera from leaf segments results in as many as three distinct variants (bronze, pink, and, occasionally, a pale green, nearly albino) plants (see Figure 42.1 B). The nonphotoautotrophic variants (pink and pale green) have low levels of chlorophyll and do not survive outside the culture medium (Lineberger and Wanstreet, 1983), so they provide excellent visual examples during discussions of carbon sources and photosynthesis in culture. Additionally, shoots arising from axillary buds typically exhibit chimeral variegation, and adventitious buds produce variants, providing visual markers for an easily evaluated demonstration of the potential somaclonal variability that may occur in chimeral plants in culture. While time constraints would likely prohibit the production of callus from these variants for use in subsequent experiments within a single semester, incorporating this adventitious/axillary bud exercise would reinforce precisely how the variants were produced.

In this experiment, leaf disks from individual colored sectors in the chimera are used as explants for either callus production directly or for generation of adventitious shoots that in turn can be used to generate variant callus lines. The visibly different callus lines are compared in terms of growth rate and pigment production on solid media. It should be emphasized that this is an illustrative example of variation that might be achieved through selection.

Materials

The following items are needed for each team of students:

- Three or four shoot cultures of *A. reptans* "Burgundy Glow" in GA-7 vessels
- Twelve GA-7 vessels of 50 mL of callus induction media per variant
- Four GA-7 vessels of 50 mL of pigment production media per variant

Follow the instruction in Procedure 42.3 to complete this experiment.

Anticipated Results

After 3 weeks of growth, callus of the bronze variant will have accumulated approximately twice the fresh weight mass of the slower-growing pink and pale green variants. By plotting the growth curves over a slightly longer (4 week) period, not only the differences in growth rate but differences in shapes of the growth curves become evident. After 3 weeks, the growth rate of the bronze variant callus begins to decelerate, whereas production in the callus of the pink and pale green variants continues to climb.

Each variant also exhibits a different level of pigment production. The bronze variant exhibits the highest anthocyanin concentration, while the pink and pale green variants have concentrations approximately 80% and 40% that of the bronze variant, respectively.

Since both the growth rate and anthocyanin concentration of these callus lines vary, an even more striking example can be presented by estimating overall yield (mass × anthocyanin concentration). Such a comparison should indicate that the potential yield from the bronze variant callus to be 2.5 and 5 times that of the pink and pale green lines, respectively.

Questions

- When these variant callus lines are placed in solution culture and subcultured at the same frequency, the pink and pale green lines tend to die off after the first cycle. What would cause this? How could the regime be altered to maintain these variants in solution culture?
- Could the productivity of the variant callus lines be improved?

Procedure 42.3
Productivity of Variant *Ajuga* Callus Lines

Step	Instructions and Comments
1	Explants that are exclusively pink, pale green, or bronze can be preferentially excised from individual leaves of *A. reptans* "Burgundy Glow" to initiate callus, or, alternatively, shoots of uniform pink, pale green, or bronze phenotype may be regenerated from selected disks and subsequently used as explants for callus initiation. Initiate four cultures of each variant callus line by collecting leaf disks from pink, pale green, or bronze sectors of the foliage from presterile *Ajuga* shoot cultures. Place each disk on the surface of the callus induction medium (see Procedure 42.1, Step 1).
2	Incubate cultures in darkness at 25°C for 3 weeks.
3	Separate ivory white friable callus from mother tissue and transfer to fresh induction media. Continue incubation in darkness, subculturing the callus to fresh media at 2 week intervals for 4 weeks.
4	After a total of 7 weeks from initiation, transfer five 2.5 cm diameter callus masses to pigment production media [callus induction medium substituting 2.26 μM (0.4 mg/L) IAA and 3.49 μM (0.76 mg/L) zeatin as PGRs].
5	Incubate the cultures under cool white fluorescent lamps providing a PPF of 100 μmol $m^{-2}sec^{-1}$ at 25°C.
6	Harvest one culture for each variant on a weekly basis for 4 weeks. Evaluate growth and pigment production with an initial visual assessment of pigment content of the intact culture followed by fresh and/or dry weight analysis of two subreplicates in each culture to compare growth. Grind the remaining three subreplicates with a mortar and pestle, store overnight in 10 mL ice-cold acidified methanol (1% HCL), filter, then measure the absorbance of the solution at 535 nm with a spectrophotometer to assess differences in the relative anthocyanin content between treatments.

Experiment 4: Influence of Light Wavelengths on Growth and Pigment Production of *Ajuga* Cell Cultures

Secondary products, including anthocyanins and other pigments, are often preferentially produced at specific ranges of light wavelengths. Optimum lighting conditions and requirements are often underestimated factors of in vitro pigment production regimes and present particular problems when scaling an existing protocol up to larger volumes. In this experiment, narrow wavelength lamps are used as light sources. Growth and pigment accumulation are evaluated for *A. pyramidalis* "Metallica Crispa" callus cultures on solid media.

Materials

The following are needed for each team or student:

- Standard broadband cool white fluorescent lamps
- Narrow-wavelength lamps in red (660 nm), blue (480 nm), and ultraviolet (313 nm) corresponding to Sylvania fluorescent lamps 2364, 2440, and 2096
- Shielding for each light source to prevent contamination from other light sources
- One callus culture of *Ajuga* in a GA7 vessel per light source to be tested
- One GA7 vessel containing pigment production media per light source to be tested
- UV-B transparent plastic film (Reynolds 910, Richmond, Virginia) to wrap tops of GA7 vessels (replacing standard lids)

Follow the protocols outlined in Procedure 42.4 to complete this experiment.

Anticipated Results

In as little as 3 days, callus under ultraviolet light will be the first treatment to exhibit pigment production, followed by the blue, white, and red light treatments, in that order. Indicative of a slower growth

Procedure 42.4
Influence of Light Wavelengths on Growth and Pigment Production of *Ajuga* Cell Cultures

Step	Instructions and Comments
1	Initiate one culture per light source to be tested by transferring five 2.5 cm diameter clumps of callus from existing cultures to fresh pigment production media [WPM amended with 88 mM (30 g/L) sucrose, 0.55 mM (0.1 g/L) myo-inositol, 0.28 mM (50 mg/L) l-ascorbic acid, 1.48 μM (0.5 mg/L) thiamine-HCl, 2.44 μM (0.5 mg/L) pyridoxine-HCl, 4.06 μM (0.5 mg/L) nicotinic acid, 26.7 μM (2 mg/L) glycine, 2.26 μM (0.4 mg/L) IAA and 3.49 μM (0.76 mg/L) zeatin, 150 mg/L PVP-T (polyvinylpyrrolidone-10), and 7 g/L of agar; pH 5.7 prior to autoclaving].
2	Incubate cultures 10–12 cm below the selected light sources at 25°C for 2 weeks providing a PPF of 50–100 μmol m^{-2} s^{-1}.
3	Inspect cultures frequently (daily, if possible) for pigment production and browning.
4	Harvest cultures from each light source for data collection. First, visual data regarding the pigment content and fresh weight data are collected to compare growth. Then grind the callus with a mortar and pestle, store overnight in 10 mL ice-cold acidified methanol (1% HCl), filter, then measure the absorbance of the solution at 535 nm with a spectrophotometer to assess differences in the relative anthocyanin content between treatments.

rate, the fresh weight of callus from the ultraviolet light treatments will be approximately 75% that of all the other treatments after 14 days. Other treatments will not differ significantly in fresh weight. Pigment production will be significantly different for each light treatment with cultures in the blue light producing the most pigment, and the white, red, and ultraviolet producing approximately 65%, 55%, and 25%, respectively, of the amount of pigment measured in the blue light treatment. Brown necrotic areas should become visible in the ultraviolet light treatment by day 10.

Questions

- Which light source is the best promoter of pigment production? Why would this type of light have a negative impact on growth? Which would be the best choice for a production regime?
- Adequate light is relatively easy to provide in this exercise with solid media or in liquid culture involving small volumes. What problems would you envision in the scale-up of larger volumes of liquid media such as in a bioreactor for commercial production? How easy would it be to enhance certain wavelengths of light?

ACKNOWLEDGMENTS

Many thanks to L.A. Spomer for constructing the schematic diagram illustrated in Figure 42.2.

REFERENCES

Callebaut, A., N. Terahara, M. de Haan, and M. Decleire. 1997. Stability of anthocyanin composition in *Ajuga reptans* callus and cell suspension cultures. *Plant Cell Tiss. Org. Cult.* 50:195–201.

Cheng, D.M., G.G. Yousef, M.H. Grace, R.B. Rogers, J. Gorelick-Feldman, I. Raskin, and M.A. Lila. 2008. In vitro production of metabolism-enhancing phytoecdysteroids from *Ajuga turkestanica. Plant Cell Tiss. Org. Cult.* 93:73–83.

Gorelick-Feldman, J., D. MacLean, N. Illic, A. Poulev, M.A. Lila, D. Cheng, and I. Raskin. 2008. Phytoecdysteroids increase protein synthesis in skeletal muscle cells. *J. Agric. Food Chem.* 56:3532–3537.

Lineberger, R.D. and A. Wanstreet. 1983. Micropropagation of *Ajuga reptans* "Burgundy Glow." Research Circular # 27419-22. Ohio Agricultural Research and Development Center.

Lloyd, G. and B. McCown. 1980. Woody plant medium. *Comb. Proc. Int. Plant Propagator's Society* 30:421–427.

Madhavi, D.L., S. Juthangkoon, K. Lewen, M.D. Berber-Jiménez, and M.A.L. Smith. 1996. Characterization of anthocyanins from *Ajuga pyramidalis* "Metallica Crispa" cell cultures. *J. Agric. Food Chem.* 44:1170–1176.

Madhavi, D.L., M.A.L. Smith, A.C. Linas, and G. Mitiku. 1997. Accumulation of ferulic acid in cell cultures of *Ajuga pyramidalis* "Metallica Crispa." *J. Agric. Food Chem.* 45:1506–1508.

43 Variation in Tissue Culture

Margaret A. Norton and Robert M. Skirvin

CONCEPTS

Somaclonal variation is:

- A variation in plants caused by tissue culture
- Most often associated with adventitious regeneration

Somaclonal variation can:

- Arise de novo or from preexisting variation
- Be caused by genetic mutations or by epigenetic change
- Be identified by changes in phenotype or changes in genotype
- Be beneficial or detrimental

Tissue culture is an excellent way to micropropagate thousands of copies of a single plant. However, as with all good things, sometimes micropropagation goes awry. Sometimes cloned plants don't come out of tissue culture exactly like their parent. This problem has been termed *somaclonal variation.*

WHAT IS SOMACLONAL VARIATION?

Briefly defined, somaclonal variation is variability caused by tissue culture. It is most frequently associated with cultures that rely on adventitious regeneration. Adventitious regeneration means that shoots, roots, or somatic embryos arise from tissues that would not normally form them. A leaf, for example, doesn't usually give rise to a new shoot, nor do nonreproductive parts of a plant generally form embryos. The first review article on somaclonal generation appeared in 1978 (Skirvin, 1978); since then the literature on the subject has increased exponentially (maybe not literally, but it seems like it).

HOW FREQUENT IS SOMACLONAL VARIATION?

Somaclonal variation occurs frequently in most tissue cultures: Estimations range from 0.05% to 3% or greater for obvious variation. It is probably higher for less noticeable biochemical changes. Some researchers speculate that every callus culture differs in at least a small way from the parent plant. The amount of variation depends strongly on genotype. Other factors are also implicated, such as length of time in culture, growth regulator concentration in the medium, and whether multiplying shoots originate from axillary buds or are adventitious in origin.

TYPES OF SOMACLONAL VARIATION

One thing to keep in mind is that somaclonal variation can affect virtually any portion of the genome, but unless it affects a visible or measurable trait, we will not know it is there without doing extensive molecular analysis.

Many somaclonal variants are obvious to the eye. These include changes in morphology, such as growth habit, coloration, and leaf shape. For example, when plants are regenerated in vitro from leaf pieces of the blackberry, *Rubus laciniatus* "Thornless Evergreen," the resulting plants have either a

FIGURE 43.1 Gross morphological changes in *Rubus laciniatus* Willd. "Thornless Evergreen" blackberry regenerants. The stem section on the right is from the parent plant, which has a trailing growth habit, whereas the stem section on the left is from a dwarf regenerant. Note the shortened internodes on the latter.

dwarf, compact growth habit or a normal, vigorous, trailing growth habit (Figure 43.1). Red bananas (*Musa* spp.) often produce variant green plants during micropropagation. Commercial micropropagators of tropical houseplants such as the arrowhead vine (*Syngonium*) and *Philodendron* must scrupulously rogue plants to remove any that are not true-to-type, such as plants with variations in leaf size, shape, or color. (They often save these rogued plants, if they are attractive, to use as new cultivars for their market.)

Detecting other forms of somaclonal variation requires closer observation and measurement. Some somaclonal variation may result in changes in physiology, such as alterations in endogenous hormone content or secondary products. Other kinds of somaclonal variation include changes in yield and changes in environmental tolerances, such as salt tolerance, cold tolerance, herbicide tolerance, insect resistance, and pathogen resistance. There are many traits that can show up in a plant as a result of somaclonal variation. These are just a few.

WHAT CAUSES SOMACLONAL VARIATION?

Somaclonal variation is caused either by a change in gene expression (epigenetic variation) or by a change in the genes themselves (genetic variation) induced during tissue culture.

Genetic Variation

Genetic variation involves actual physical changes in DNA sequences that are lasting and heritable. The first kind of genetic change that probably comes to mind is mutation. Mutations may occur when one base is substituted for another (point mutation; Figure 43.2), when a section of DNA is lost (deletion mutation) or gained (insertion mutation), or when an insertion or deletion of extra base pairs causes the three base-pair code to shift, resulting in a frameshift mutation (Figure 43.3).

Another kind of genetic change occurs in tissue culture when previously inactive transposable elements reactivate. A transposable element is a section of DNA that can clip itself out of its place on a chromosome and move to another location. When it moves, it leaves behind a few base pairs that can interfere with the coding region of a gene, resulting in partial or full loss of the gene's function. In addition, when it inserts itself into a new site on a chromosome, it can land within the coding region of a gene, resulting in loss of function of that gene.

Gene amplification is yet another type of genetic change that can happen when plants are cultured in vitro. Gene amplification is the term used to describe the production of multiple copies of a gene in response to an environmental challenge. For example, when the herbicide glyphosate is added to carrot suspension cultures stepwise in gradually increasing amounts, the number of copies of a gene that confers tolerance to glyphosate increases in response (Shyr et al., 1992).

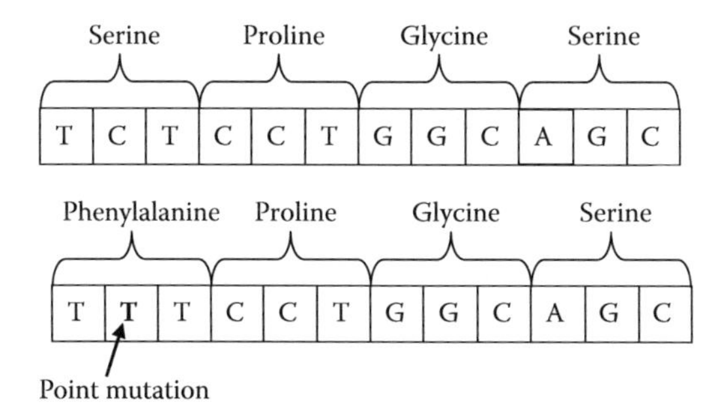

FIGURE 43.2 Point mutation. A thymine (T) has been substituted for a cytosine (C) in the first codon, resulting in a change from serine (TCT) to phenylalanine (TTT).

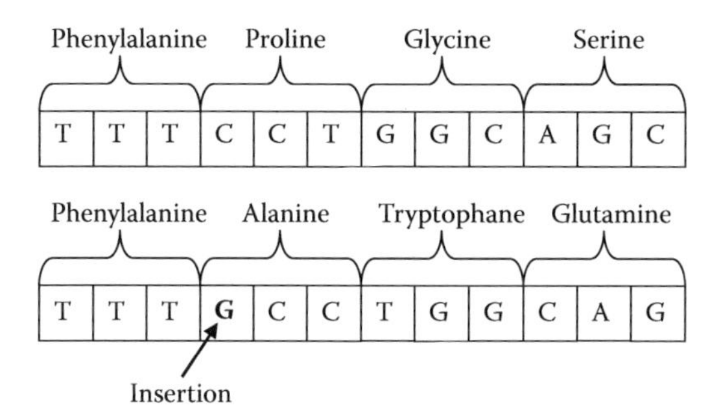

FIGURE 43.3 Frameshift mutation. A guanidine (G) has been inserted after the first codon, resulting in a shift of the following base pairs and completely changing the coding for the remainder of the coding sequence.

Karyotypic changes are major genetic changes in the genome that can occur in tissue culture when cell division results in changes in the number or structure of entire chromosomes (Lee and Phillips, 1987). Changes in ploidy are one type of karyotypic change. A cell that loses or gains one or more chromosomes is called an *aneuploid.* Common types of aneuploidy include loss of one chromosome (monosomy) or gain of one chromosome (trisomy). A monosomic plant can show variation through unmasking of recessive traits; that is, if the one remaining chromosome has recessive alleles that were previously hidden, these recessive traits will be expressed.

Polyploidy is another form of karyotypic change. Polyploidy occurs when an entire set of chromosomes in a cell are gained. Ploidy is referred to by the number of sets of chromosomes in a cell: haploid (half the normal complement of cells—i.e., an entire set has been lost), diploid (the normal complement of chromosomes—i.e. two sets), triploid, tetraploid, etc. Changes in ploidy can lead to loss of fertility and gross morphological changes. This can be useful or not—triploid watermelons, for example, are seedless, and many tetraploid flowering plants have larger flowers and fruits.

Epigenetic Variation

Epigenetic variation involves changes in gene expression such as gene activation or gene silencing. Because epigenetic variation does not permanently alter the sequence of base pairs in a plant's DNA, it is potentially reversible. Epigenetic variation is usually not considered to be heritable, though methylation-induced changes in some plants have remained stable through several cycles of self-pollination.

Most investigation of epigenetic variation has focused on methylation of genes. During methylation, a methyl group attaches to a cytosine base after the DNA is replicated. The methyl group protrudes from the DNA helix and binds to proteins that then act to wind the DNA into heterochromatin. Recall that heterochromatin is DNA that is tightly packed together and euchromatin is DNA that is unwound for replication. Because heterochromatin doesn't present its DNA strands for transcription, methylated genes are silenced.

Methylation can persist through DNA replication and be passed on through both mitosis and meiosis. However, many researchers have reported that tissue culture affects the frequency of methylation. Sometimes methyl groups are added to DNA in tissue culture, but more frequently they are lost. Auxins in the tissue culture medium increase methylation, whereas cytokinins seem to have no effect.

Preexisting versus Induced Mutation

While much of the genetic and epigenetic variation in tissue culture seems to be induced by the tissue culture environment itself, some somaclonal variation arises from preexisting differences among cells in the explant. One example is the separation of chimeras during micropropagation or regeneration. Chimeras are plants that have two genetically dissimilar tissues growing side by side. Chimeras occur when a cell in the apical meristem mutates and grows to form a continuous patch of mutated cells extending from the apical meristem down the stem. A chimera in which the mutated patch occupies part of one layer of the meristem is termed a *mericlinal* chimera. If the patch of mutated cells occupies part of all layers of the meristem, it is a *sectorial* chimera. If the mutated cells occupy one or two entire layers of the meristem, it is called a *periclinal* chimera.

During normal axillary bud formation, all layers of the plant's meristem are incorporated into the developing bud, ensuring that the axillary bud will reproduce the layered structure of the plant's apical meristem. However, when plants are micropropagated at a fast rate, the layered structure in the axillary buds can break down, resulting in shoots that are either all mutated tissue, all wild-type (i.e., nonmutated) tissue, or a rearranged chimera (Figure 43.4). In addition, when plants regenerate adventitiously from a chimera by organogenesis or somatic embryogenesis, they will express one or the other genotype because adventitious regeneration usually originates from a single cell.

During its life, a plant can accumulate mutations in locations other than the apical meristem. By chance, if an explant contains one of these somatic mutations and regenerates adventitiously from that cell, the resulting plant will also have a genotype different from the mother plant.

FIGURE 43.4 Chimeral separation. *Pyrus communis* L. "Louise Bonne" is a periclinal chimera that has an albino LIII. In tissue culture, the chimeral structure may break down, leading to the appearance of all albino tissue.

HOW IS SOMACLONAL VARIATION IDENTIFIED?

Somaclonal variation can be difficult to detect. Recall that not all somaclonal variation is obvious to the eye. Gross physical changes, such as leaf color, may be detectable in vitro, but most somaclonal variation won't be visible until the explants are removed from tissue culture and grown on in soil.

For example, a chimeral variegated hosta with subtle color differences, such as a light green leaf with a dark green border, may appear to be all light green when it is grown in tissue culture, yet when it is transferred to soil and allowed to grow for a year, the original color differences re-emerge. In this case, what may appear to be somaclonal variation when the plant was in tissue culture turns out not to be somaclonal variation at all. The chimeral nature of the plant wasn't lost, just masked by the tissue culture conditions.

The simplest way to detect somaclonal variation is to transfer the plants to soil and monitor them for phenotypic differences. Such an approach is most useful for nurseries that micropropagate herbaceous plants and plants with a short juvenile period. Off-type plants can be easily rogued within a few months, saving valuable nursery space.

However, plants with a longer juvenile period, such as trees, don't lend themselves to this approach. Particularly with plants such as fruit trees or timber trees, the grower may have to wait 5 to 10 years or longer to detect changes such as flavor, yield, or vigor. This is a considerable expenditure of time and money, and it can be devastating to a grower to invest this much in a crop only to be disappointed when it doesn't produce as expected.

Researchers are now investigating molecular techniques to detect somaclonal variation early. Random Amplified Polymorphic DNA (RAPD) analysis is one such technique that has been used in the past, although it has been largely superseded by newer techniques. Polymerase Chain Reaction (PCR) is used with random primers—that is, random sequences of base pairs, usually ten. The primers hybridize to DNA extracted from young plant tissues, and during a DNA amplification step, these primers produce short fragments of DNA. The fragments are separated by electrophoresis on an agarose gel. If the resulting banding pattern shows differences (polymorphisms), this indicates somaclonal variation.

Another technique used to detect somaclonal variation is Restriction Fragment Length Polymorphism (RFLP). DNA is extracted from young leaves of regenerated or micropropagated plants and subjected to restriction enzyme digestion, which cleaves DNA at specific base pair sequences. The resulting fragments are separated by electrophoresis. All identical plants will have the same banding pattern on the gel. Any polymorphisms among the bands represents somaclonal variation.

Two similar techniques, Amplified Fragment Length Polymorphism (AFLP) and Methylation Sensitive Amplified Polymorphism (MSAP), are also being used to detect somaclonal variation. With AFLPs, restriction enzymes are used to cleave DNA, after which adaptors are ligated to the cut ends of the DNA. The primers used with AFLPs have the same adaptors. When the primers anneal to the DNA fragments, any primers that correspond to the adaptor and the attached portion of the plant's DNA sequence will amplify. AFLPs are considered to be very sensitive, more so than RFLPs.

MSAPs are similar to RFLPs except that in place of one of the restriction enzymes, a pair of isoenzymes is used. One enzyme of this pair will cut at a specific site on DNA even if the DNA is methylated, and the other enzyme will not. This produces a polymorphism that allows a researcher to detect whether or not a site is methylated.

SSRs (Simple Sequence Repeats, sometimes called microsatellites) are yet another tool being used to detect somaclonal variation. SSRs are short sequences of no more than six base pairs that repeat themselves, sometimes up to 100 times or more, at various sites in a genome. The repeats mutate frequently, so they exhibit many polymorphisms and work well as molecular markers.

Molecular techniques can detect polymorphisms with great sensitivity. However, it is important to realize that molecular markers are not genes. While markers may indicate differences among segments of DNA, they may not actually be within a gene or even very close to one. If the molecular markers are not closely linked to the coding region of a gene, they may indicate polymorphisms in plants in which no phenotypical difference appears. Unless molecular markers are linked to phenotypically important traits, their usefulness may be limited.

HOW IS SOMACLONAL VARIATION PREVENTED?

Micropropagation

Somaclonal variation occurs most often in plants exposed to high concentrations of growth regulators, particularly auxins, and in cultures in which plants regenerate from callus or from single cells. Therefore, a good rule for minimizing somaclonal variation during micropropagation is to use the lowest concentration of plant growth regulators that will give an adequate multiplication rate. It may be possible to achieve a very high multiplication rate, but it is usually not wise to do so. High multiplication rates may disrupt the layered structure of the meristem or cause shoots to form adventitiously from single cells, increasing the possibility that somaclonal variation will appear.

In addition, during micropropagation, the propagator should be selective about which shoots to choose for the next round of multiplication. If the shoots don't obviously originate from axillary buds, it may be best to discard them. In strawberries, for example, shoots can arise from axillary buds, and at the same growth regulator concentration, other shoots may arise adventitiously from stipules. Those shoots from stipules should be discarded because of their adventitious origin and potential for off types. Because callus has been associated with somaclonal variation, those shoots that originate from a callus intermediary should also be discarded.

Professional micropropagators also reinitiate their cultures frequently. The longer a culture remains in vitro, the greater the chance of somaclonal variation. If the somaclonal variation appears early during micropropagation, it may be proliferated accidentally, resulting in a large number of variant plants. It is best to reinitiate cultures at least once per year.

Somatic Embryogenesis and Organogenesis

Because somatic embryos and regenerated shoots and roots originate adventitiously, it is impossible to prevent somaclonal variation in these regeneration systems. Instead, stringent rogueing must be used to remove any obvious variants. For those traits not easily visualized, the best solution is to test regenerants at an early stage for clonal fidelity using molecular techniques. These tests are costly and not foolproof, but they are the best solution for detecting somaclonal variation at this time.

Callus Cultures

It is probably impossible even to minimize somaclonal variation from callus culture because callus is initiated and maintained using strong auxins such as 2,4-dichlorophenoxyacetic acid (2,4-D). High levels of auxin are especially implicated in somaclonal variation, particularly epigenetic variation caused by methylation. In addition, the nature of callus cultures—especially liquid cultures—is such that certain variant cells are inadvertently selected simply because they survive best in the tissue culture environment. If you choose to use callus culture, expect to get variants.

SOMACLONAL VARIATION—NUISANCE OR ASSET?

Most somaclonal variation is detrimental. The purpose of micropropagation, for example, is to produce mass quantities of identical plants for market. Significant variation among tissue cultured plants can cost a propagator money as well as his good reputation. For example, daylilies are often micropropagated, but the problem of somaclonal variation has led many propagators to abandon in vitro methods and instead advertise their plants as not being from tissue culture. Researchers developing transgenic plants may successfully transform a species with a desirable gene only to discover that unwanted somaclonal variation has turned up in their plants.

However, somaclonal variation as a cause of variability in plants holds promise as a source of novel traits if it is approached carefully. In vitro selection is one way to isolate desirable variation. Callus, suspension cultures, or regenerated plants are challenged with a selection agent, and surviving plants are selected and tested for the desired trait. For example, a pathogen culture filtrate may be added to a suspension culture and coincubated for several days to several weeks. Theoretically, cells that are highly susceptible to the pathogen will die. Any surviving cells are then regenerated into plants, which are grown on in soil and tested with the pathogen itself. Some regenerated plants may show increased tolerance to the pathogen. Using this technique, plants have been developed that express increased salt tolerance, cold tolerance, insect tolerance, pathogen tolerance, and other traits.

Somaclonal variation may also be used to increase levels of secondary products in callus cultures. For example, callus may be selected for pigment expression, such as anthocyanins (red) or carotenoids (yellow to dark orange). One caveat with in vitro selection, however, is that variation expressed in tissue culture may not survive the transition to whole plant.

A classic demonstration of this is an experiment performed by Mok et al. (1976), in which carrot cells were selected for pigment production. A red carrot gave rise to callus lines that ranged in color from red through orange, pink, and yellow; orange carrots produced callus lines colored light orange to dark orange; and yellow and white carrots produced callus lines of varying shades of yellow. However, when cells from these different colored calli were regenerated into whole plants, all resulting carrot plants were the same color as the original plants from which they were taken.

The authors speculated that this phenomenon might be due to aneuploidy. They suggested that the cells that produced the most pigment were aneuploid and were incapable of regenerating. Only normal diploid cells produced whole plants. Another possible explanation presented was that the particular stress of the tissue culture environment caused cells to develop more pigment. The authors noted that strong auxin enhanced carotenoid expression in the carrot callus, whereas kinetin, a cytokinin, inhibited it.

Other explanations come to mind for the difference in pigment production in the various carrot calli. It is possible that the cells producing carotenoids were differentiated cells containing chromoplasts. It is important to remember that while callus is a mass of unorganized cells, the individual cells are not necessarily undifferentiated. There is a difference between unorganized and undifferentiated cells. Undifferentiated cells have not become determined in their fate: they remain as simple cells without a specific function. Unorganized cells may be a mixture of differentiated or undifferentiated cells or both; the individual cells have simply not come together as a tissue to perform a specific function in an organism. A callus may contain differentiated xylem cells, for example, but these xylem cells are not organized into a functional xylem system.

Those cells that overproduce secondary products may be differentiated cells. However, before organogenesis or somatic embryogenesis take place, the cell or cells involved must undergo dedifferentiation and redifferentiation. In other words, a differentiated callus cell will begin to divide, and in the process it returns to an undifferentiated state. As the regeneration process continues, cells redifferentiate into specialized cells other than pigment-producing cells. Thus, the pigment-producing cells resume their proper place in the whole plant.

SUMMARY

Somaclonal variation is a fact of life in plant tissue culture. Because the ultimate cause of somaclonal environment is the chemical and physical tissue culture environment itself, any manipulation of plants in vitro carries with it the possibility of genetic or epigenetic change in plant DNA. Growers and researchers who need identical copies of a plant must understand somaclonal variation in order to minimize it. Researchers who are searching for novel traits to enrich the genotype of a plant species must understand somaclonal variation in order to maximize it.

Much remains to be learned about somaclonal variation. As molecular techniques continue to be refined, variation will be detected earlier and with greater precision. However, true control of somaclonal variation awaits the day when it can be detected and prevented or utilized at will.

REFERENCES AND SUGGESTED READING

Bender, J. 2004. DNA methylation and epigenetics. *Annu. Rev. Plant Biol.* 55:41–68.

Brettel, R.I.S., and E.S. Dennis. 1991. Reactivation of a silent Ac following tissue culture is associated with heritable alterations in its methylation pattern. *Mol. Gen. Genet.* 229:365–372.

Jain, S.M., D.S. Brar, and B.S. Ahloowalia (Eds.). 1998. *Somaclonal Variation and Induced Mutations in Crop Improvement.* Dordrecht & Boston: Kluwer Academic Publishers.

Kaeppler, S.M., H.F. Kaeppler, and Y. Rhee. 2000. Epigenetic aspects of somaclonal variation in plants. *Plant Mol. Biol.* 43:179–188.

Kaeppler, S.M., and R.L. Phillips. 1993. Tissue culture-induced DNA Methylation variation in maize. *Proc. Natl. Acad. Sci.* 90:8773.

Karp, A. 1989. Can genetic instability be controlled in plant tissue cultures? *Int. Plant Tiss. Cult. Assoc. Newsletter* 58:2–11.

Lee, M., and R.L. Phillips. 1987. Genomic rearrangements in maize induced by tissue culture. *Genome* 29:122–128.

Mok, M.C., W.H. Gabelman, and F. Skoog. 1976. Carotenoid synthesis in tissue cultures of *Daucus carota* L. *J. Amer. Soc. Hort. Sci.* 101:442–449.

Rani, V. and S.N. Raina. 2000. Genetic fidelity of organized meristem-derived micropropagated plants: A critical reappraisal. *In Vitro Cell. Dev. Biol. Plant* 36:319–330.

Shyr, Y.Y.J., A.G. Hepburn, and J.M. Widholm. 1992. Glyphosate selected amplification of the 5–enolpyruvyl-shikimate-3-phosphate synthase gene in cultured carrot cells. *Mol. Gen. Genet.* 232:377–382.

Skirvin, R.M. 1978. Natural and induced variation in tissue culture. *Euphytica* 27:241–266.

Section V

The Business of Biotechnology

44 Biotechnology Entrepreneurship in the 21st Century

From Bench to Bag

David W. Altman

CONCEPTS

- Secure intellectual property rights through a license or an asset purchase.
- Establish standard protocols for data storage and retention, conflict of interest, recordkeeping, notification of inventions, disclosure and financial reporting.
- Develop a business plan, preferably with comprehensive strategic planning.
- Complete business formation documents and intellectual property filings.
- Identify and gain commitments for a management team with diverse fields of expertise.
- Raise adequate capital from an appropriate source after considering all options in addition to venture capital.
- Maintain focus during start-up by solidifying intellectual property, demonstrating proof-of-concept with the science and operating within fiscally prudent guidelines.

Now that you've come up with the greatest innovation the plant tissue culture and biotechnology universe has ever seen, what can you do to make sure that it gets into the hands of real people? Like many other imponderables, the answer depends on your perspective. Of course, the scientist thinks that business components are relatively mundane and should be utilitarian and subordinate to the technology. The business men and women often want to box up scientists and to keep them in the laboratory and out of trouble. Finally, investors who come up with capital to drive development are a mixed bag, but generally assume the science is great and are looking for a star team that works well together with both business and science elements.

This chapter will provide an outline for use by any stakeholder in a how-to format. The intention is to describe entrepreneurship conditions and processes for a typical U.S.-based plant tissue culture program in a stepwise fashion, but many of the principles and components could be adapted for other life sciences or broader international situations. The first section will deal with protocols and systems that must be put in place in any modern plant science laboratory even before the thought occurs about a commercialization route. These requirements should be taught to all students so that they in turn will not face major problems due to ignorance of standard protocols. Such standards should be considered in the same light as laboratory safety procedures and other typical operational items. The sections that follow will then be specific to the formation and initiation of a business and include the steps of (1) planning, (2) personnel, (3) legal filings, (4) capital, and (5) start-up. These five sections will assume that rights have been secured for any inventions in the new venture from an employer or other owners of discoveries to be used in the business.

Before proceeding, the predominance of intellectual property considerations that are unique to plant biotechnology, and most life sciences research programs, must be emphasized. From a business perspective, the focus in these types of research companies would be described as dependent on intangible assets. Major entrepreneurship differences arise as a consequence of this distinction, which contrast with companies dependent primarily on tangible assets such as a manufacturing operation making widgets or a retail shop selling knickknacks. This central element often is ignored and limits the potential of many good concepts. The other salient point is that the plant biotechnology practitioner doesn't really need a bread-and-butter type of business expertise for key aspects of putting the enterprise together because plant biotechnology, or life sciences generally, require certain unique business skills, such as intellectual property management and financial guidance for extensive periods without revenues. Chances of success are improved with special attention to enlisting business expertise that includes experience in a life science field; sometimes no other viable option exists.

STANDARD PROTOCOLS AND SYSTEMS FOR RESEARCH GROUPS

Except for the lone scientist working in his garage or in similar circumstances, nearly everyone working in plant biotechnology belongs to either profit or nonprofit organizations. Each type of organization has sets of rules that govern aspects of research touching on commercial endeavors, even if the researcher has no apparent interest at a particular point in time. At the beginning of employment or a short-term engagement, each individual should determine procedures and policies in place for commercialization activities in their organization. In particular, every scientist should familiarize himself/herself with requirements for conflict of interest, recordkeeping, data storage and retention, notification of inventions, and financial reporting.

While most corporate organizations pay attention to these activities, public institutions often either don't take a proactive role in disseminating these policies or have outdated or inaccurate information. For any institution with government funding, the United States has certain requirements as a result of the Bayh-Dole Act of 1980 and guidelines from agencies such as the National Institutes of Health (NIH), National Science Foundation (NSF), U.S. Department of Agriculture (USDA), and others. Any competent group or research leader should have a reasonable understanding of these federal guidelines and be able to provide satisfactory explanation of requirements as well as where to go for more in-depth information when the need arises. These policies prevent individuals from engaging in activities that could get them into trouble while venturing into entrepreneurship.

Most business organizations would insist on compliance with established guidelines in order to put any agreement in place, such as a research agreement, a licensing agreement, or other types of support (see Harrison, 1999, for review of industry requirements for compliance from public institutions). Sometimes, public institutions allow financial support from businesses as a gift, which can greatly simplify some of the minimal elements for commercialization, but this option usually provides a stopgap that shouldn't obviate the need for best practices.

The next important concern is to protect any discovery or invention that might occur. This principle applies even if you want to give the technology away. For this information, companies have legal departments that provide the necessary guidance, and public institutions usually have a technology transfer office. If these sources aren't helpful, then access internet information from the U.S. Patent and Trademark Office (see www.uspto.gov) or refer to references about patents and other forms of intellectual property (IP). Although patents are the most commonly recognized commercial IP assets, other forms of IP include trade secrets, copyrights, trademarks, plant variety protection (PVP) certification and a special type of U.S. patent protection for plants called "plant patents." Plant biotechnology can derive particular IP protection from PVPs or plant patents that generally depend on whether the plant species are vegetatively propagated or sexually propagated, and this protection can be a very valuable asset that would increase the long-term viability of any entrepreneurial venture.

With more utilization of digital information exchange and other Internet-based methods for marketing, social and professional networking, and similar activities, copyright protection takes on more significance. In the U.S., copyrights are registered with the Library of Congress (see www.copyright .gov). A common misconception is that registration is really not necessary. However, if an individual wants to avoid postings on the Internet, Web site materials, instruction manual text, and other items from being used indiscriminately, then registration creates a relatively simple deterrent. In addition, unless the registration is completed within 3 months of publication, the owner of the copyright cannot collect statutory damages or attorney's fees in the event of litigation, which means prompt filings, or before the 3-month deadline has elapsed, are highly recommended. Simply using the symbol © with the year of publication provides basic protection, but does not confer full protection.

In order to properly protect any future IP asset, the laboratory needs to have standardized procedures for collection of data. In most instances, you might not be able to use the actual date of discovery from the research unless these procedures are appropriately followed. The recording of data is particularly critical. Each laboratory notebook should be bound because loose-leaf notebooks or unbound files are more difficult to verify and are not considered adequate by patent examiners and other authorities. With the same rationale, pencils are not acceptable for recording purposes, so pens or other indelible markers are necessary. Corrections in bound laboratory notebooks should not obliterate the original entry and should be accompanied by a brief notation. Pages in the bound notebooks should be initialed, dated, and countersigned by another individual. The notebooks are best numbered sequentially and maintained at a separate location when completed and after being copied; only the copies should remain in the laboratory for routine reference. Other considerations exist for use of data books, including procedures for inserting computer printouts, photographs, and other materials, so researchers need to have training in how to record appropriately all data for purposes of protecting the IP.

Procedures also need to be established for data exchange, public dissemination, disclosure to business groups and similar aspects of modern biotechnology research. Without providing a comprehensive description, several highlights will be emphasized. Assuming that you have taken steps to protect discoveries, that you are following guidelines in place for your employer, and that you are adequately collecting and recording results, the next criteria is to establish rules of engagement outside of the laboratory group. Most employers, either public or private, require assignment of rights for discoveries to the employer, and all research managers need to establish documentation for the employees they supervise to verify such assignment.

The next item is to decide who can be contacted externally and what procedures are necessary for making contact with other commercial groups, sending off publications, or conducting other similar activities. In the private sector, business development personnel take care of managing these details, but public researchers usually must take the initiative to comply with procedures established within their institutions. The first consideration for third party contacts would be conflict of interest. In brief, you can't fairly split the IP and commercialization rights so that you do the same venture, or in some cases overlapping ventures, with the same discoveries without specialized and carefully crafted agreements, and frequently such agreements are ambiguous at best. You also can't give unfair advantages that are outside of the legitimate assigned rights to special friends or to other groups with whom you wish to curry favor.

After passing the conflict-of-interest test, documentation procedures are needed to go forward with contacts that require disclosure of discoveries. Many employers require a discovery disclosure form or notification. In this manner, the appropriate manager can ascertain if any IP protection is necessary in advance of disclosure, or if the researcher can proceed without IP protection. At this point, in the case of a publication, an abstract submission, or other types of similar disclosures that might impact future commercialization, the correct procedure is probably straightforward—namely send it off. However, if you exchange certain materials (DNA, probes, proteins, new chemical entities, cell culture lines, etc.), then a material transfer agreement (MTA) could be required in advance. If the researcher is going to enter into discussions, negotiations, or similar activities for possible entrepreneurship, then a confidential disclosure agreement (CDA), which might also be called a nondisclosure agreement (NDA), would be required. Format for the MTA, CDA, or NDA document is usually standard for most employers and can be obtained from the offices previously indicated.

PLANNING

Hopefully, all of the preceding hasn't dulled the enthusiasm for sailing off to the world of entrepreneurial ventures. The prerequisites are daunting, but by following a prudent set of procedures ahead of time, the intrepid novice or the seasoned veteran already would have a good beginning. Before getting too overconfident, don't forget the obvious need to do sufficient planning. Although planning might seem simple, the business world has particular parameters for typical new venture development, usually condensed down into the process of strategic planning and creation of a business plan. While not always necessary, the discipline of strategic planning and writing a formal business plan improves chances of success, and many investors will insist on such materials before consideration of any proposed funding.

Planning for an entrepreneurial effort depends considerably on an assessment of your goal with the technology. If the goal is to find a partner with business experience, then planning might be simply to research who are the principle players in the area of technology. The Internet is a good starting point to gather information, and for those demanding more detail, usually industry reports and other assessments by specialized consulting groups can be found. Obviously, many veterans of entrepreneurial ventures tend to wax favorably on the virtues of launching a start-up over the more conservative track of finding an experienced partner. Timmons and Spinelli (2008) point out that the United States is unique in creating a remarkable environment that fosters entrepreneurship and label this activity "the great equalizer and mobilizer of opportunity." With that being understood, the exact formula for success has been elusive, and the option to get a partner rather than take the plunge should not be dismissed flippantly.

If you have resources and stamina to complete a comprehensive strategic plan, then several authorities on the process and the anticipated components of resulting documentation could get you started (see Thompson and Strickland, 2007). Usually, a strategic plan would provide an assessment of the current status, an evaluation of internal and external factors, consideration of viable options for the business opportunity, and recommended strategy, with benefits and costs, to execute the most advantageous option. The first item would be to develop the mission and vision for a new venture, as well as objectives of the business. External factors include an analysis of the market, customers, competition, and the unique unmet need that your innovation will address. The internal factors can address management, marketing and sales, operations, product development, distribution and other services. Finally, based on these other components, careful financial projections and an ownership plan would be essential to delineate.

PERSONNEL

Most scientists consider the really essential element of a new venture to be the technology. However, the investment community is going to assume the technology is viable, at least prior to initiating full diligence, and will focus on the management team. The necessity for cohesiveness and experience as an attribute of the management team cannot be over-emphasized.

This aspect of entrepreneurship means that commitments are necessary from individuals with widely different backgrounds. Typically, science, business, regulatory affairs, law, and other disciplines are not used to cross-fertilization, but this convergence is the desired outcome for a new life science venture. The founder or founders for a new venture can come from any of these backgrounds and the task for the entrepreneur is to gather other individuals to round out the management group. Usually, attempting any delay in putting the team together will compromise success of a new venture.

New ventures also have growth stages, similar in many respects to those of other new things. Just as a baby might have different needs compared to the toddler, the teenager, the young adult, the middle-aged individual, etc., a new venture does not have the same management requirements as a mature company. Individuals who desire stability and security would not typically be well suited for new ventures. Preferably, experience is an important attribute, and because many new ventures fail for a variety of reasons, the litmus test should not necessarily be experience with a successful new venture.

Another option would be to seek help from specialists to assemble a management team. Certain consulting firms can provide temporary management for start-up ventures in the formative period. This method of finding appropriate personnel has an added advantage of making a transition to new management easier as the company grows and has a need for management appropriate to another development stage. Venture capitalists who specialize in very early stage projects can often be another source of personnel, although this option can have other risks by potentially creating a conflict of interest with the new recruits. Specifically, the founder might be questioning if team members provided by venture capitalists serve company interests over those of the investment group. Another possibility would be to utilize an executive search firm or other "headhunters." This option can be very viable for those new ventures that have resources to hire such agencies. However, some search firms might defer fees, accept some equity, or make other flexible arrangements that render their services more feasible.

Several pitfalls confront inventors/founders and should be avoided in putting together a management group. Even if some of these cautions seem obvious, they must be mentioned, hopefully to provide guidance for readers without extensive human resources experience. Friends and relatives should not be the first option to complete a management team, and such selections will be viewed skeptically by investors and other stakeholders. Particularly in making critical decisions about personnel, the prudent entrepreneur should not cut corners in recruitment. Some pointers include the following: (1) avoid over-reliance on the interview and be sure to standardize your interview questions as typically recommended by human resource specialists; (2) check references carefully before making a decision; (3) involve all of the stakeholders, founders, and other members of the extended management group; (4) make sure that a penchant for entrepreneurship and risk-taking is an important trait for selection; (5) choose individuals who mesh with other team members; and (6) seek diversity of skills within the team.

LEGAL FILINGS

While getting caught up in aspects of a new venture, certain legal filings are necessary to protect the innovations in a reasonable fashion. All new ventures must be registered to conduct business, so certain filings are perfunctory for any type of new venture. With biotechnology and other life sciences, the IP component has been mentioned and will be readdressed as well.

Business registration usually begins by conceiving a legally-acceptable name for the business. Typically, each state has a department that registers corporations and determines if a selected name is acceptable. Although every state can feasibly provide this service, several states are considered the preferred jurisdiction for registration among investors and others. Since a new venture is not required to have its principal place of business in the state of registration, Delaware is a frequent choice despite its size; there are several others, such as Nevada, that serve well for life science ventures.

After securing a name, the next decision is to decide the best type of incorporation for the company to select. This function doesn't always require a lawyer, and specialized Internet businesses can help the entrepreneur complete registration for minimal expense. However, usually the entrepreneur can benefit from sound legal counsel and tax advice or accounting expertise when making this decision. One important factor is to limit liabilities by incorporating a business venture. Many new ventures become established either as a limited liability partnership, a C-corporation or an S-corporation, rather than a single proprietorship or a partnership. Because most new ventures usually must seek outside funding, the entrepreneur should carefully consider a business incorporation that will be conducive to receiving funding in exchange for equity in the venture. Upon completing the appropriate incorporation, be sure to register in all states where you will have business activities. This registration is sometimes referred to as DBA or "doing business as."

An absolutely essential item for most life sciences ventures would be filings for IP rights. Most individuals must seek outside legal advice for this component of the business. However, for the entrepreneur on an insufficient budget and assuming that an employer or institution hasn't already started the process, several stopgap measures can suffice. Contrary to popular belief, a lawyer is not required to file a patent. The U.S. Patent and Trademark Office allows registered patent agents or inventors to file patents as well. Patent agents typically have lower fees than patent lawyers, but the adage about getting what you pay for might provide a moment to pause. A full patent application also is not necessary because some protection for at least 1 year in the United States can be obtained by filing a provisional patent that has a lower submission fee. In any case, doing nothing is the worst option and will be punished in the marketplace. Thus, minimal filings, although not the preferred method, should be provisional patents submitted by the inventor or inventors.

CAPITAL

Capital is an essential item to launch an entrepreneurial venture. This requirement for any start-up business can be one of the most baffling and difficult for the plant biotechnology practitioner who normally doesn't think about funding requirements of commercialization. Funding procedures for entrepreneurship are usually not similar at all to grant writing or other noncommercial fundraising activities. Many references about new ventures give an introduction to methods to raise capital, so various options will only be briefly outlined.

The first option would be to look at internal sources for getting past the initial steps to launch a new business. Most individuals have access to some funds by leveraging their own personal credit capabilities. Such an option would include home equity lines of credit, credit card debt, personal loans, and other similar sources. Many start-up businesses have only needed to tap these types of reserves in order to get beyond the beginning expenses to launch operations. The advantage of using these resources is that the entrepreneur doesn't need to convince other parties of the viability of the business model.

The other option could be a secured loan from a financial institution such as a bank. Unfortunately, most banks are very conservative and unfamiliar with intangible assets, and the financial crisis in this decade has exasperated this risk-avoidance mentality. Therefore, the usual mode of operation is that these institutions typically offer loans only when the loans really aren't necessary as the primary source of capital. However, sometimes the entrepreneur might be able to offer reasonable security to collateralize a commercial loan or line of credit, so this option shouldn't be dismissed as completely impossible.

A likely scenario to obtain capital would be to approach venture capital firms that specialize in high-risk new ventures. However, these firms only deal with high risks because of anticipated high returns. Therefore, such firms will demand a substantial return on their investment, and the plant biotechnology practitioner must understand that this requirement is normal for this source of capital. In this situation, an experienced business person has the best chance to obtain more advantageous terms for funding.

In addition, specifics of the deal are extremely important to evaluate the opportunity to obtain required funds. The most important element is determination of the pre-money valuation of the business opportunity, which is the value that both parties agree to place on the entire enterprise prior to investment of venture capital. Following many of the suggestions in this "cookbook" set of requirements should result in a higher pre-money value because the savvy entrepreneur would have protected the IP, assembled a credible management team, and put together a professional business plan after thoughtful strategic planning, besides having a great innovation from the beginning.

Well-structured deals have other important details, but everything depends on obtaining a satisfactory pre-money valuation. These specifics include the form of investment vehicle, favorable board representation, antidilution protection, reasonable commitment for assistance with future funding rounds, and other aspects. The completion of this task should be the responsibility of the business people in the management team, but every entrepreneur should try to keep abreast of the issues and to stay informed with the status of negotiations with the venture capital groups. Without a term sheet or other written commitment, all covenants are negotiable and should be carefully considered before reaching a final decision about acceptance.

Finally, even when you have agreed upon a term sheet, this acceptance does not necessarily mean that the fundraising task has been accomplished. Nothing is final until closing papers are signed by all parties to a funding transaction. In particular, most term sheet agreements are contingent upon due diligence by investors. This process is when every aspect of the business is examined in sometimes excruciating detail. If the entrepreneur has completed extensive planning and accumulated the necessary documentation, due diligence will proceed relatively smoothly.

An alternative approach could be to complete a private placement memorandum (PPM), with the intention of raising funds in specified units at a predetermined, fixed price per unit. The only caveats would be to comply with all legal requirements for such a document. Many attorneys could assure the required compliance, and set up a process to screen each potential investor as an "accredited investor" according to the Securities Act of 1933 and subsequently amended. Key requirements for individuals are either to have a net worth in excess of $1 million dollars or annual income in excess of $200,000.

There are occasionally some other options to obtain capital. A large, established company in this industry might be willing to buy your new venture or other aspects of the business that has been developed. Usually, this option is better to pursue once the venture has been able to operate for a period of time, so that milestones have been achieved and the innovation has been validated. However, many companies have divisions that act like regular venture capital firms, and the advantages might include an ability of larger companies to bring experienced business development groups into the venture with a proven capability to take discoveries to the marketplace. The founders might still maintain involvement through a consulting contract or other advisory roles.

START-UP

After all anticipation and work, the novice entrepreneur might think that actual start-up of operations represents achievement of the primary goal to launch the new venture. Actually, in reality, this start-up phase represents only the beginning of the next stage for business development. Start-up companies should never forget the sense of urgency that resulted in their successful launch. Frequently, start-ups don't have an effective implementation plan, and this deficiency can lead to pitfalls as well.

The most common error would be to forget the need to position the new venture for the next round of financing. Usually, a single capital raise is insufficient to bring the new company to the point of becoming a viable enterprise. The risk is that the science might prevent the young company from staying focused on the main aspects to ensure long-term success. For life sciences companies, the start-up phase usually requires progress for enhancing intellectual property and for demonstrating proof-of-concept with basic innovations that provided the reason to commercialize something.

Furthermore, no prerequisite keeps your planning and basic strategy from undergoing a midcourse correction. Often, ability to evaluate progress and to adapt to changing dynamics is the most important attribute of a start-up venture. In addition, all entrepreneurs should carefully monitor cash expenditures, usually represented by an assessment of burn rate.

With all due respect to the science, entrepreneurship is governed by the dictum that "cash is king." With life science and biotechnology ventures, cash can be king, queen, the prime minister, grand poobah, and anything else. A frequent pitfall is to slide into the trap of perennially burning cash, avoiding marketing, and forgoing the milestone of your first customers. The glory days when venture money fed high burn rates of companies with vague revenue models are probably a thing of the past, if such a time ever existed in the first place! While scientific progress is important, be sure someone is monitoring cash, aggressively seeking customers and revenues, and holding down excessive expenditures. Sometimes this requirement can come into conflict with long-term goals, but the company needs to survive in order to reach the future.

Another pitfall comes from the nature of funding that usually provides capital to fuel new biotechnology ventures. Venture funding usually means that equity is diluted for founders and start-up employees who have put their heart and soul into creating the business. Repeated financing rounds can excessively dilute founder and employee equity holdings, so these infusions should be precious and limited to avoid excessive dilution. One frequently neglected aspect is the importance of debt financing as an alternative to equity financing. Although the business might not have been able to raise debt financing for the initial funding, once the start-up phase has begun the venture should have an aggressive, sustained effort to open lines of credit, to utilize chattel loans, to be granted favorable payment terms with vendors, and to exploit every opportunity to establish a sound credit rating. This requires good financial leadership from the company's CFO or other management with this responsibility for the enterprise.

In an era of stimulus funding on top of other government grants, another possibility might be to seek grant funding for commercialization, such as a grant under the Small Business Innovation Research Program (see http://www.sba.gov/aboutsba/sbaprograms/sbir/sbirstir/), which are usually administered in collaboration with a federal department or other research institution including the Department of Agriculture, National Science Foundation, or others. One precautionary tip for those tempted by the public trough would be to consider utilization of an adequately trained accountant or bookkeeper who specifically is competent with government accounting standards. These standards are in addition to the more well-known standards of Generally Accepted Accounting Principles or GAAP (which eventually will be replaced by International Financial Reporting Standards or IFRS) and would typically be found in the relevant Federal Acquisition Requirements (FAR) and Cost Accounting Standards (CAS). Many accountants or bookkeepers might really think they are able to apply these standards, but often mistakes occur which can come back to annoy ventures for many years to come.

Operations in general require specialized attention because mistakes can be costly to correct at a later stage, and in some cases, might even jeopardize the future of the venture. Biotechnology ventures will probably rely on IP rights of some sort, and these rights are intangible assets that must be appropriately entered on the company books. Many of the plain-vanilla accountants and bookkeepers are oblivious to specialized requirements for a life science business, which can include atypical accounting standards for revenue recognition, amortization/expense for research costs, stock options, start-up equity covenants, and other items. Similarly, the management of companies that often have few initial sources of revenues is a skill infrequently acquired by the average MBA. Be sure to insist on these features being addressed early and appropriately.

In the current era after financial scandals typified by Enron, WorldCom, Parmalat, and many others, start-up ventures have been tarred by the same brush. The practical result for operations has been the imposition of new rules for internal controls. The Sarbanes-Oxley or SOX regulations are principally directed to larger, public companies, but even small companies should do an assessment to address applications to the enterprise. The key idea would be to show some good-faith effort in documenting internal controls and basic governance policies. Even if your venture can technically be exempt for the current point of development of the business, banks, investors, government funding agencies, and others will look favorably on minimal compliance. In any case, following these policies and procedures represents best business practices in addition to being a regulatory consideration.

CONCLUDING COMMENTS

Entrepreneurship can be a stimulating and rewarding culmination of scientific achievements for plant biotechnology. Although scientific publications and other such end products have great value and contribute to our advancement of knowledge, without finding a commercial outlet the discoveries from the laboratory might not be easily accessible to the general public. Business is a democratic method to disseminate inventions, at least in the United States and most of the other nations around the world.

The plant biotechnology practitioner brings a vital element to the commercialization effort and is indispensable to the development of a product. However, the preceding material might lead to the obvious conclusion that some other team members and components are necessary to obtain a successful commercialization effort. Each person should do soul searching at the start to decide if they have the temperament and stamina to try the new venture approach, and nothing is wrong with choosing the path to find an established business to buy the discovery and to do the heavy lifting to complete the development part of research and development.

Finally, the ultimate key to success in a life science venture will always be the quality of the people involved in the enterprise. A science-based endeavor can ill afford to forget about this overriding consideration. People in an entrepreneurial business need to be motivated and to work together well as a team. Cohesiveness of the team will be critical when the new business faces inevitable periodic crises that characterize entrepreneurship. Find the best people you can, reward them, and provide incentives as much as you can. Cover various business aspects as well as needed scientific expertise when you assemble the team. Also, because the entrepreneur cannot be an expert in all fields required for any business, don't be reticent in seeking outside expertise or in finding other resources within reasonable budget parameters. In the end, these guidelines can improve your chances of success, and hopefully the element of serendipity will then be more likely to find its place in your new commercial enterprise.

REFERENCES

Harrison, C.H. 1999. Industry-sponsored academic research in the health sciences: Regulatory, policy and practical issues in contract negotiations. *J. Biolaw Business* 2:9–25.

Thompson, Jr., A.A. and A.J. Strickland. 2007. *Crafting and Executing Strategy: The Quest for Competitive Advantage*. McGraw-Hill Irwin, New York, p. 992.

Timmons, J.A. and S. Spinelli. 2008. *New Venture Creation: Entrepreneurship for the 21st Century*. Irwin McGraw-Hill, New York, p. 704.

45 Intellectual Property Protection for Plants

Chris Eisenschenk

CONCEPTS

- Plant patents, utility patents, Plant Variety Protection Act (PVPA) certificates, and trademarks are forms of federal intellectual property protection that are available for plants and plant materials.
- Unauthorized sexual reproduction of plants can be controlled by utility patents and/or PVPA certificates.
- Unauthorized asexual reproduction of plants can be controlled by utility patents and/or plant patents.
- Of all the forms of federal intellectual property protection available for plants, a utility patent provides the strongest intellectual property protection and is the most difficult to obtain.

This chapter will seek to address the various types of federal intellectual property protection that are available for plants and plant materials. While some forms of state intellectual property rights exist, a discussion of these rights is beyond the scope of this chapter. Additionally, this chapter is not intended to be an exhaustive discussion of the various forms of intellectual property protection available for plants and is not to be construed as legal advice regarding any form of intellectual property discussed herein.

The types of federal intellectual property protection available for plants and plant materials include patents (utility patents and/or plant patents), Plant Variety Protection Act (PVPA) certificates, and trademark protection. A discussion of copyrights is also included in this chapter to familiarize the reader with this form of intellectual property protection (primarily with respect to representations of plants produced for marketing or other purposes). Most commonly, utility and plant patents are used to protect intellectual property rights; however, the PVPA provides a form of intellectual property protection that is analogous to patent rights, albeit with a few exceptions that may impact one's decision to pursue PVPA protection for plants.

COPYRIGHT PROTECTION AND PLANTS

Federal copyright protection exists for "original works of authorship fixed in any tangible medium of expression, now known or later developed, from which they can be perceived, reproduced or otherwise communicated, either directly or with the aid of a machine or device" (see 17 U.S.C. §102(a) (1988)). Thus, where an author fixes an image of a plant or depicts the image of a plant in some form of tangible medium (an "original work"), a copyright in that original work is obtained at the moment of its fixation within the tangible medium. However, it is important to note that the scope of protection for the copyrightable material (e.g., an "original work" depicting a plant) only extends to the "original work" and not the plant itself.

17 U.S.C. §102(a) (1988) provides examples of "original works." This nonlimiting list includes the following: literary works, musical works (including lyrics), dramatic works (including accompanying music), pantomimes and choreographic works, pictorial, graphic and sculptural works, motion pictures and other audiovisual works, sound recordings, and architectural works (see 17 U.S.C. §102(a) (1988)). However, specifically excluded from copyright protection is any idea, procedure, process, system, method of operation, concept, principle, or discovery, regardless of the form in which it is described, explained, illustrated, or embodied in such work (see 17 U.S.C. §102(b) (1988)). Additionally, the term "tangible medium" refers to any physical medium, nonlimiting examples of which include paper, video/audio tapes, or electronic storage medium (e.g., computer disks).

The owner of the copyrighted "original work" has the exclusive right, subject to certain limitations, to do and/or to authorize any of the following: (a) to reproduce the copyrighted work in copies; (b) to prepare derivative works based upon the copyrighted work; (c) to distribute copies of the copyrighted work to the public by sale or other transfer of ownership, or by rental, lease, or lending; (d) in the case of literary, musical, dramatic, and choreographic works, pantomimes, and motion pictures and other audiovisual works, to perform the copyrighted work publicly; (e) in the case of literary, musical, dramatic, and choreographic works, pantomimes, and pictorial, graphic, or sculptural works, including the individual images of a motion picture or other audiovisual work, to display the copyrighted work publicly; and (f) in the case of sound recordings, to perform the copyrighted work publicly by means of a digital audio transmission (see 17 U.S.C. §106 (1988)). However, the author of an "original work" depicting a plant should also be aware of the "fair use doctrine" that allows for the limited reproduction of the copyrighted work without the consent of the author. The "fair use doctrine" allows for the reproduction of a copyrighted work for purposes including, but not necessarily limited to, criticism, comment, news reporting, teaching, scholarship, or research. Whether the reproduction of a copyrighted work is a "fair use" is determined on a case-by-case basis and includes an analysis of the following factors: (a) the purpose and character of the use, including whether such use is of a commercial nature or is for nonprofit educational purposes; (b) the nature of the copyrighted work; (c) the amount and substantiality of the portion used in relation to the copyrighted work as a whole; and (d) the effect of the use upon the potential market for or value of the copyrighted work (see 17 U.S.C. §107 (1988)).

Finally, federal copyright protection exists as of the moment that the "original work" is fixed in a tangible medium and there is no longer any requirement for marking the first publication of a work, published as of March 1, 1998, with a copyright symbol "©," the term "Copyright" (or the abbreviation "Copr."), the author's name and the date of publication. Additionally, one should note that there is no requirement for copyright registration with the Copyright Office in the Library of Congress in order to establish one's copyright for an "original work." However, registration of one's copyright with the Copyright Office and proper marking of the "original work" with a copyright notice provides a number of advantages in case of infringement or unauthorized reproduction of the work. For example, no lawsuit for infringement of the copyright in any United States work shall be instituted until registration of the copyright claim has been made (see 17 U.S.C. § 411). Additionally, statutory damages and attorneys' fees are not available to a copyright owner unless registration of the copyright has preceded the infringement, although a three-month grace period exists in this regard (see 17 U.S.C. §§ 412 and 504-505). Finally, the copyright statutes provide that no weight shall be given to such a defendant's assertion of a defense based on innocent infringement if a copyright notice as specified in 17 U.S.C. § 401(b)–(c) has been placed on the published copy or copies to which a defendant in a copyright infringement suit had access.

Thus, one should recognize that Web sites, photographs, marketing materials, sketches, artwork, or other forms of expression that depict a plant produced by a grower (and are fixed in a tangible medium) are a protectable form of intellectual property associated with the plant. One should also be aware of the rights that are associated with this aspect of intellectual property protection and consider enforcing these rights if infringers are identified. Additional information regarding copyrights can be accessed on the Internet at www.copyright.gov.

TRADEMARKS AND THEIR USE WITH PLANTS

The term "trademark" includes any word, name, symbol, or device, or any combination thereof (a) used by a person, or (b) which a person has a *bona fide* intention to use in commerce and who applies to register the trademark on the principal trademark register to identify and distinguish his or her goods, including a unique product, from those manufactured or sold by others and to indicate the source of the goods, even if that source is unknown (see 15 U.S.C. § 1127). Thus, a trademark serves to identify the source or point of origin of a particular product sold in the stream of commerce.

As would perhaps be expected, the U.S. Patent and Trademark Office (USPTO) has promulgated guidelines addressing the use of varietal or cultivar names as trademarks in the Trademark Manual of Examination Procedures (TMEP)—4th Edition (2005). As indicated in Section 1202.12—Varietal and Cultivar Names (Examination of Applications for Seeds and Plants):

> Varietal or cultivar names are designations given to cultivated varieties or subspecies of live plants or agricultural seeds. They amount to the generic name of the plant or seed by which such variety is known to the public. These names can consist of a numeric or alphanumeric code or can be a "fancy" (arbitrary) name. The terms "varietal" and "cultivar" may have slight semantic differences but pose indistinguishable issues and are treated identically for trademark purposes.
>
> Subspecies are types of a particular species of plant or seed that are members of a particular genus. For example, all maple trees are in the genus *Acer.* The sugar maple species is known as *A. saccharum*, while the red maple species is called *A. rubrum*. In turn, these species have been subdivided into various cultivated varieties that are developed commercially and given varietal or cultivar names that are known to the public.
>
> If the examining attorney determines that wording sought to be registered as a mark for live plants, agricultural seeds, fresh fruits, or fresh vegetables comprises a varietal or cultivar name, then the examining attorney must refuse registration, or require a disclaimer, on the ground that the matter is the varietal name of the goods and does not function as a trademark under §§1, 2 and 45 of the Trademark Act, 15 U.S.C. §§1051, 1052 and 1127. *See Dixie Rose Nursery v. Coe* , 131 F.2d 446, 55 USPQ 315 (D.C. Cir. 1942), *cert. denied* 318 U.S. 782, 57 USPQ 568 (1943); *In re Hilltop Orchards & Nurseries, Inc.* , 206 USPQ 1034 (TTAB 1979); *In re Farmer Seed & Nursery Co.* , 137 USPQ 231 (TTAB 1963); *In re Cohn Bodger & Sons Co.* , 122 USPQ 345 (TTAB 1959). Likewise, if the mark identifies the prominent portion of a varietal name, it must be refused. *In re Delta and Pine Land Co.,* 26 USPQ2d 1157 (TTAB 1993) (Board affirmed refusal to register DELTAPINE, which was a portion of the varietal names Deltapine 50, Deltapine 20, Deltapine 105 and Deltapine 506).
>
> A varietal or cultivar name is used in a plant patent to identify the variety. Thus, even if the name was originally arbitrary, it "describe[s] to the public a [plant] of a particular sort, not a [plant] from a particular [source]." *Dixie Rose,* 131 F.2d at 447, 55 USPQ at 316. It is against public policy for any one supplier to retain exclusivity in a patented variety of plant, or the name of a variety, once its patent expires. *Id.*

In trademark law, a descriptive trademark conveys information regarding an ingredient, quality, characteristic, function, feature, purpose, or use of the product or service. Descriptive trademarks are, generally, not eligible for registration because the mark must become associated with the product or service among consumers (e.g., acquire secondary meaning). With respect to a view that the cultivar name is generic, trademark law holds that generic terms or common words for the products or services cannot function as a trademark because it would prevent others from rightfully using the common name for the product or service that they make. Thus, in view of the guidance provided to trademark examining attorneys by the MTEP, it is clear that cultivar names will be considered generic or, at best, descriptive and ineligible for trademark protection.

Thus, while one may not be able to trademark a particular cultivar name, it should be understood that one can still identify a particular cultivar as originating with a particular source or point of origin. In this context, a nursery or grower of a particular cultivar may trademark a different name and use it in conjunction with the cultivar name to identify the source or point of origin of the cultivar produced by the nursery or grower. This combination of a trademark and the cultivar name would then allow the consumers of the plant to identify the source or point of origin of the cultivar. With respect to marking requirements under trademark law, that portion of the combined trademark-cultivar name that constitutes the trademarked portion should be identified with a ® or ™ symbol on the tag or label to designate a registered or common law trademark, respectively. The portion of the combined trademark-cultivar name that identifies the actual cultivar name on the tag or label should identify the cultivar using single quotation marks according to the industry norm.

THE PLANT VARIETY PROTECTION ACT

The Plant Variety Protection Act (PVPA) is a form of federal intellectual property that is available to growers of new sexually reproduced plants. The PVPA is administered through the Plant Variety Protection Office (PVPO) of the U.S. Department of Agriculture and provides a breeder somewhat patent-like protection for any sexually reproduced or tuber-propagated plant variety (other than fungi or bacteria).

Thus, if a plant variety is new, distinct, uniform, and stable as defined in 7 U.S.C. § 2402 of the PVPA*, it will be eligible for a certificate under the Act. Growers should also be aware of the various statutory bars that would prevent one from obtaining a PVPA certificate. These include making the plant variety available to persons, for purposes of exploitation, within the United States more than 1 year prior to the filing date of the PVPA application; making the plant variety available to persons outside the United States more than 4 years prior to the filing date of the PVPA application (although certain exceptions may exist for tubers); or making a tree or vine available to persons outside the United States more than 76 years prior to the filing date of the PVPA application.

Turning now to the scope of protection for plants for which PVPA certificates have been issued, courts have held that the PVPA protects plants and seed that differ "from either of the parent plants as well as from other plants produced from other seeds resulting from the cross-pollination" and that "[p]lants true-to-type, although different in a strict genetic sense, are protectable under the PVPA" [see *Imazio Nursery, Inc. v. Dania Greenhouses*, 69 F3d. 1560 (Fed. Cir. 1996)]. Thus, the

---

*§ 2402. Right to plant variety protection; plant varieties protectable

(a) In general

The breeder of any sexually reproduced or tuber propagated plant variety (other than fungi or bacteria) who has so reproduced the variety, or the successor in interest of the breeder, shall be entitled to plant variety protection for the variety, subject to the conditions and requirements of this chapter, if the variety is—

(1) new, in the sense that, on the date of filing of the application for plant variety protection, propagating or harvested material of the variety has not been sold or otherwise disposed of to other persons, by or with the consent of the breeder, or the successor in interest of the breeder, for purposes of exploitation of the variety—

(A) in the United States, more than 1 year prior to the date of filing; or

(B) in any area outside of the United States—

(i) more than 4 years prior to the date of filing, except that in the case of a tuber propagated plant variety the Secretary may waive the 4-year limitation for a period ending 1 year after April 4, 1996; or

(ii) in the case of a tree or vine, more than 6 years prior to the date of filing;

(2) distinct, in the sense that the variety is clearly distinguishable from any other variety the existence of which is publicly known or a matter of common knowledge at the time of the filing of the application;

(3) uniform, in the sense that any variations are describable, predictable, and commercially acceptable; and

(4) stable, in the sense that the variety, when reproduced, will remain unchanged with regard to the essential and distinctive characteristics of the variety with a reasonable degree of reliability commensurate with that of varieties of the same category in which the same breeding method is employed.

PVPA allows for some genetic variation within the plants and/or seeds of a variety protected by a PVPA certificate. As discussed below, such genetic variation is not permitted for plants protected under the Plant Patent Act (PPA).

With respect to the rights conferred by the owner of a protected variety, the 7 U.S.C. § 2541 indicates that:

> ... it shall be an infringement of the rights of the owner of a protected variety to perform without authority, any of the following acts in the United States, or in commerce which can be regulated by Congress or affecting such commerce, prior to expiration of the right to plant variety protection but after either the issue of the certificate or the distribution of a protected plant variety with the notice under section 2567 of this title:

1. sell or market the protected variety, or offer it or expose it for sale, deliver it, ship it, consign it, exchange it, or solicit an offer to buy it, or any other transfer of title or possession of it;
2. import the variety into, or export it from, the United States;
3. sexually multiply, or propagate by a tuber or a part of a tuber, the variety as a step in marketing (for growing purposes) the variety;
4. use the variety in producing (as distinguished from developing) a hybrid or different variety therefrom;
5. use seed which had been marked "Unauthorized Propagation Prohibited" or "Unauthorized Seed Multiplication Prohibited" or progeny thereof to propagate the variety;
6. dispense the variety to another, in a form which can be propagated, without notice as to being a protected variety under which it was received;
7. condition the variety for the purpose of propagation, except to the extent that the conditioning is related to the activities permitted under section 2543 of this title;
8. stock the variety for any of the purposes referred to in paragraphs (1) through (7);
9. perform any of the foregoing acts even in instances in which the variety is multiplied other than sexually, except in pursuance of a valid United States plant patent; or
10. instigate or actively induce performance of any of the foregoing acts.

Damages that one can obtain for infringement of a PVPA certificate shall be adequate to compensate for the infringement but in no event less than a reasonable royalty for the use made of the variety by the infringer, together with interest and costs as fixed by the court (see 7 U.S.C. § 2564). It is also possible to obtain an injunction prohibiting an infringer from continuing to make and use the seeds or plants protected under the PVPA.

It is also important to recognize that the PVPA provides several exceptions/exemptions for acts of infringement. For example, it is not considered an act of infringement for a farmer to save and replant seed protected by a PVPA certificate*. Additionally, a *bona fide* sale of seed produced on a farm (either from seed lawfully obtained for seeding purposes or from seed produced by descent from the lawfully obtained seed and produced on the farm) for other than reproductive purposes, made in channels

---

*7 U.S.C. 2543 Right to save seed; crop exemption

Except to the extent that such action may constitute an infringement under subsections (3) and (4) of section 111 [7 U.S.C. 2541(3), (4)], it shall not infringe any right hereunder for a person to save seed produced by the person from seed obtained, or descended from seed obtained, by authority of the owner of the variety for seeding purposes and use such saved seed in the production of a crop for use on the farm of the person, or for sale as provided in this section. A bona fide sale for other than reproductive purposes, made in channels usual for such other purposes, of seed produced on a farm either from seed obtained by authority of the owner for seeding purposes or from seed produced by descent on such farm from seed obtained by authority of the owner for seeding purposes shall not constitute an infringement. A purchaser who diverts seed from such channels to seeding purposes shall be deemed to have notice under section 127 [7 U.S.C. 2567] that the actions of the purchaser constitute an infringement.

usual for such other purposes, shall not constitute an infringement*. A third, and potentially important, exception/exemption from infringement is a research use exception. 7 U.S.C. § 2544 states that the use and reproduction of a protected variety for plant breeding or other *bona fide* research shall not constitute an infringement. Thus, for both growers and academics, the use of PVPA protected plants and seeds for plant breeding or other true research purposes do not constitute an act of infringement. As we shall see, this is not the case of plants protected by either a utility or plant patent.

PLANT PATENTS

Plant patents are granted to those who invent or discover, and asexually reproduce, any new and distinct variety of plant (see 35 U.S.C. § 161†). Whereas the PVPA can be used to protect a sexually reproduced plant, a plant patent provides the right to exclude others from asexually reproducing the new and distinct plant variety and parts thereof for a 20-year period (see 35 U.S.C. § 163‡). However, asexually reproduced plants can also be protected by utility patents as 35 U.S.C. § 161 is not an exclusive form of protection which conflicts with the granting of utility patents to plants (*see Ex parte Hibberd*, 227 U.S.P.Q. 443 (Bd. Pat. App. & Int. 1985)). Additionally, plants capable of sexual reproduction (i.e., from seed) can be protected under the Plant Patent Act provided that the plant has been asexually reproduced.

As set forth in 35 U.S.C. § 161, potentially patentable plants include cultivated sports, mutants, hybrids, and newly found seedlings. Fungi and bacteria are not considered "plants" under this aspect of the patents statutes; however, utility patents can be used to protect isolated fungi and bacteria. Additionally, the single claim of a plant patent only protects the entirety of the plant, not subparts of the plant (e.g., seeds, flowers, fruit, etc.)§. It should also be noted that plants identified in the wild are not eligible for plant patent protection. Rather, only plants identified in cultivated areas are eligible for protection under the Plant Patent Act. Interestingly, neither the cultivated area in which the plant is discovered nor ownership of the plant claimed within the plant patent application need be owned by the person who identified the plant (i.e., the Applicant for the plant patent; see *Ex parte Moore*, 115 U.S.P.Q. 145 (Pat. Off. Bd. App. 1957)). Specifically excluded from protection under the Plant Patent Act are tuber propagated plants. Examples of tuber propagated plants include Irish potato and the Jerusalem artichoke. Because these plants are asexually propagated by the same part as is sold for food, the plants are considered ineligible for protection under a plant patent (see Manual of Patent Examination Procedure (M.P.E.P.), section 1601, August 2006, Revision 5).

Among the requirements that must be fulfilled to obtain a plant patent are that: (1) the plant is capable of stable asexual reproduction; (2) the plant was invented or discovered and, if discovered, that the discovery was made in a cultivated area; (3) the plant is not a plant which is excluded by statute, where the part of the plant used for asexual reproduction is not a tuber food part; (4) the person or persons filing the application are those who actually invented the claimed plant, that is, discovered or developed and identified or isolated the plant, and asexually reproduced the plant; (5) the plant

* Ibid.

† 35 U.S.C. § 161 – Patents for Plants

Whoever invents or discovers and asexually reproduces any distinct and new variety of plant, including cultivated sports, mutants, hybrids, and newly found seedlings, other than a tuber propagated plant or a plant found in an uncultivated state, may obtain a patent therefore, subject to the conditions and requirements of this title.

‡ 35 U.S.C. § 163. Grant.

In the case of a plant patent, the grant shall include the right to exclude others from asexually reproducing the plant, and from using, offering for sale, or selling the plant so reproduced, or any of its parts, throughout the United States, or from importing the plant so reproduced, or any parts thereof, into the United States.

§ 37 C.F.R. § 1.164. Claim.

The claim shall be in formal terms to the new and distinct variety of the specified plant as described and illustrated, and may also recite the principal distinguishing characteristics. More than one claim is not permitted.

has not been sold or released in the United States more than one year prior to the effective filing date of the application; (6) the plant has not been enabled to the public, that is, by description in a printed publication in this country more than one year before the application for patent with an offer to sale; or by release or sale of the plant more than one year prior to application for patent; (7) the plant be shown to differ from known, related plants by at least one distinguishing characteristic, which is more than a difference caused by growing conditions or fertility levels, etc.; or (8) the invention would not have been obvious to one skilled in the art at the time of invention by applicant*.

Infringement of a plant patent results whenever someone asexually reproduces, by any means, a plant that is protected by a plant patent. It should also be noted that "experimental use" defenses are exceptionally narrow and would, likely, not be available should one be involved in litigation involving an allegedly infringed plant patent†. Should one be successful in proving infringement, the patent statutes provide for a variety of remedies. These remedies include enjoining the infringer from producing any additional plants and requiring the payment of a monetary award to the patent holder. Monetary damages can be calculated in a variety of ways including actual damages (e.g., lost sales) or a determination of what a reasonable royalty would have been for the plants sold by the infringer. In case of willful infringement, the damages can be trebled, and it may also be possible to recover attorney's fees in instances where a court finds exceptional circumstances.

UTILITY PATENTS

As noted above, utility patents are also a form of federal intellectual property protection that is available for plants and plant materials. For plants, utility patents are most often directed to transgenic plants (e.g., plants containing herbicide-resistance genes or genes that confer resistance of insect pests). It is far less common for utility patents to be sought for nontransgenic plants, although a utility and plant patent can be obtained for a given plant if desired. To be eligible for patent protection, an invention must be one of the statutory classes of invention‡ and the specification that describes the inventions must be written such that it teaches those skilled in the art how to make and use the invention that is claimed (this is commonly referred to as the "enablement requirement")§. In some cases, the deposit of seed will also be required to "enable" a patent application. The specification must also provide what is referred to as the best mode of carrying out the invention that is known to the inventors at the time the application is filed, and it must also contain an adequate written description of that which the inventor considers to be the invention¶. Additional requirements for patentability include a requirement for novelty** and

---

* See U.S. Patent and Trademark Office publication "General Information About 35 U.S.C. 161 Plant Patents," accessible at www.uspto.gov/web/offices/pac/plant/.

† *Madey v. Duke University*, 307 F.3d 1351, 1362 (Fed. Cir. 2002) (stating that regardless of whether a particular institution or entity is engaged in an endeavor for commercial gain, so long as the act is in furtherance of the alleged infringer's legitimate business and is not solely for amusement, to satisfy idle curiosity, or for strictly philosophical inquiry, the act does not qualify for the very narrow and strictly limited experimental use defense.

‡ 35 U.S.C. 101 Inventions patentable
Whoever invents or discovers any new and useful process, machine, manufacture, or composition of matter, or any new and useful improvement thereof, may obtain a patent therefor, subject to the conditions and requirements of this title.

§ 35 U.S.C. 112 Specification
The specification shall contain a written description of the invention, and of the manner and process of making and using it, in such full, clear, concise, and exact terms as to enable any person skilled in the art to which it pertains, or with which it is most nearly connected, to make and use the same, and shall set forth the best mode contemplated by the inventor of carrying out his invention . . .

¶ Ibid.

** 35 U.S.C. 102 Conditions for patentability; novelty and loss of right to patent
A person shall be entitled to a patent unless —
(a) the invention was known or used by others in this country, or patented or described in a printed publication in this or a foreign country, before the invention thereof by the applicant for patent, or
(b) the invention was patented or described in a printed publication in this or a foreign country or in public use or on sale in this country, more than one year prior to the date of the application for patent in the United States . . .

nonobviousness*. As with plant patents and the PVPA, it is possible to obtain a utility patent for a plant that has been disclosed or made available to the public provided that a patent application is filed within one year of the date of disclosure (see 35 U.S.C. § 102(b)). Additional information regarding utility patents can be accessed at the United States Patent Office Web site at www.uspto.gov/go/pac/doc/general/.

Upon the grant of a patent, the patent owner has the right to exclude others from making, using, selling, importing, or offering to sell the patented invention. Thus, it would be an act of infringement to make, use, import, offer to sell, or sell a patented plant in the United States (see 35 U.S.C. § 271). Should such an act occur, the patent owner has a right to recover damages as discussed above with respect to the infringement of plant patents. Namely, one can obtain injunctive relief that precludes the infringer from making, using, selling, offering to sell, or importing the plant. One can also obtain a monetary award of actual damages (e.g., lost sales) or a reasonable royalty for the plants sold by the infringer. As discussed above, treble damages can be obtained in some cases as can attorney's fees (in instances where a court finds exceptional circumstances).

CONCLUSION

The Supreme Court has compared and contrasted rights conferred by the PVPA, plant patents and utility patents in a decision entitled *J.E.M. Ag Supply, Inc. v. Pioneer Hi-Bred Int'l, Inc.*, 534 U.S. 124, 140, 142 (2001). In this decision, the Supreme Court has written:

> The PVPA also contains exemptions for saving seed and for research. A farmer who legally purchases and plants a protected variety can save the seed from these plants for replanting on his own farm. See § 2543 ("[I]t shall not infringe any right hereunder for a person to save seed produced by the person from seed obtained, or descended from seed obtained, by authority of the owner of the variety for seeding purposes and use such saved seed in the production of a crop for use on the farm of the person ..."); see also Asgrow Seed Co. v. Winterboer, 513 U.S. 179, 115 S.Ct. 788, 130 L.Ed.2d 682 (1995). In addition, a protected variety may be used for research. See 7 U.S.C. § 2544 ("The use and reproduction of a protected variety for plant breeding or other bona fide research shall not constitute an infringement of the protection provided under this chapter"). The utility patent statute does not contain similar exemptions.

The Supreme Court continues, at pages 142–143,

> To be sure, there are differences in the requirements for, and coverage of, utility patents and PVP certificates issued pursuant to the PVPA. These differences, however, do not present irreconcilable conflicts because the requirements for obtaining a utility patent under § 101 are more stringent than those for obtaining a PVP certificate, and the protections afforded by a utility patent are greater than those afforded by a PVP certificate. Thus, there is a parallel relationship between the obligations and the level of protection under each statute.
>
> It is much more difficult to obtain a utility patent for a plant than to obtain a PVP certificate because a utility patentable plant must be new, useful, and nonobvious, 35 U.S.C. §§ 101-103. In addition, to obtain a utility patent, a breeder must describe the plant with sufficient specificity to enable others to "make and use" the invention after the patent term expires. § 112. The disclosure required by the Patent Act is "the quid pro quo of the right to exclude."

* *35 U.S.C. 103 Conditions for patentability; non-obvious subject matter*

(a) A patent may not be obtained though the invention is not identically disclosed or described as set forth in section 102 of this title, if the differences between the subject matter sought to be patented and the prior art are such that the subject matter as a whole would have been obvious at the time the invention was made to a person having ordinary skill in the art to which said subject matter pertains. Patentability shall not be negatived by the manner in which the invention was made . . .

Kewanee Oil Co. v. Bicron Corp., 416 U.S. 470, 484, 94 S.Ct. 1879, 40 L.Ed.2d 315 (1974). The description requirement for plants includes a deposit of biological material, for example, seeds, and mandates that such material be accessible to the public. See 37 CFR §§ 1.801-1.809 (2001); see also App. 39 (seed deposits for U.S. Patent No. 5,491,295).

By contrast, a plant variety may receive a PVP certificate without a showing of usefulness or nonobviousness. See 7 U.S.C. § 2402(a) (requiring that the variety be only new, distinct, uniform, and stable). Nor does the PVPA require a description and disclosure as extensive as those required under § 101. The PVPA requires a "description of the variety setting forth its distinctiveness, uniformity and stability and a description of the genealogy and breeding procedure, when known." 7 U.S.C. § 2422(2). It also requires a deposit of seed in a public depository, § 2422(4), but neither the statute nor the applicable regulation mandates that such material be accessible to the general public during the term of the PVP certificate. See 7 CFR § 97.6 (2001).

Because of the more stringent requirements, utility patent holders receive greater rights of exclusion than holders of a PVP certificate. Most notably, there are no exemptions for research or saving seed under a utility patent. Additionally, although Congress increased the level of protection under the PVPA in 1994, a PVP certificate still does not grant the full range of protections afforded by a utility patent. For instance, a utility patent on an inbred plant line protects that line as well as all hybrids produced by crossing that inbred with another plant line. Similarly, the PVPA now protects "any variety whose production requires the repeated use of a protected variety." 7 U.S.C. § 2541(c)(3). Thus, one cannot use a protected plant variety to produce a hybrid for commercial sale. PVPA protection still falls short of a utility patent, however, because a breeder can use a plant that is protected by a PVP certificate to "develop" a new inbred line while he cannot use a plant patented under § 101 for such a purpose. See 7 U.S.C. § 2541(a)(4) (infringement includes "use [of] the variety in producing (as distinguished from developing) a hybrid or different variety therefrom"). See also H. R. Rep. No. 91-1605, p. 11 (1970), U. S. Code Cong. & Admin. News 1970, pp. 5082, 5093; 1 D. Chisum, Patents § 1.05[2][d][i], p. 549 (2001).

Thus, it is apparent that the exclusionary rights conferred by various forms of intellectual property protection for a plant can have significant ramifications for both a grower and those who seek to infringe that grower's rights to the plant that has been developed. For example, a grower may wish to rely on a plant patent and a PVPA certificate to protect the plant and its progeny. However, as noted above, certain limitations exist with respect to the intellectual property rights conferred by plant patents and the PVPA. To retain complete control of the intellectual property rights associated with a plant, a utility patent may provide the most appropriate vehicle to accomplish this goal because a utility patent provides one with the right to exclude others from making, using, offering for sale, or selling the invention in the United States or importing the invention into the United States (regardless of whether the plant was asexually or sexually reproduced).

Index

E

F

G

H

I

N

O

T

U

V

W

X

Y

Z

9781420083262